T. S. Bowers · K. J. Jackson · H. C. Helgeson

Equilibrium Activity Diagrams

for Coexisting Minerals
and Aqueous Solutions at Pressures
and Temperatures to 5 kb and 600 °C

Springer-Verlag
Berlin Heidelberg New York Tokyo 1984

Teresa S. Bowers
Division of Geological
and Planetary Sciences
California Institute of Technology
Pasadena, CA 91125, USA

Kenneth J. Jackson
Department of Earth Sciences
Lawrence Livermore National Laboratory
University of California
Livermore, CA 94550, USA

Prof. Harold C. Helgeson
Department of Geology and Geophysics
University of California
Berkeley, CA 94720, USA

ISBN 3-540-13796-3 Springer-Verlag Berlin Heidelberg New York Tokyo
ISBN 0-387-13796-3 Springer-Verlag New York Heidelberg Berlin Tokyo

Library of Congress Cataloging in Publication Data. Bowers, T. S. (Teresa S.). Equilibrium activity diagrams. Bibliography: p. Includes index. 1. Phase diagrams. 2. Mineralogical chemistry. I. Jackson, K. J. (Kenneth J.). II. Helgeson, Harold C. III. Title. QD503.B68. 1984. 549'.133. 84-20209.

This work is subject to copyright. All rights are reserved, whether the whole or part of the material is concerned, specifically those of translation, reprinting, re-use of illustrations, broadcasting, reproduction by photocopying machine or similar means, and storage in data banks. Under § 54 of the German Copyright Law where copies are made for other than private use a fee is payable to 'Verwertungsgesellschaft Wort', Munich.

© by Springer-Verlag Berlin Heidelberg 1984.
Printed in Germany.

The use of registered names, trademarks etc. in this publication does not imply, even in the absence of a specific statement, that such names are exempt from the relevant protective laws and regulations and therefore free for general use.

Offsetprinting: Weihert-Druck GmbH, Darmstadt.
Bookbinding: Konrad Triltsch, Graphischer Betrieb, Würzburg.
2132/3130-543210

PREFACE

This book represents a revision and expansion of an earlier set of diagrams for temperatures from 25° to 300° C along the equilibrium vapor-liquid curve for H_2O (Helgeson, Brown, and Leeper, 1969). The activity diagrams summarized in the following pages were generated over a six year period from 1977 to 1983 in the Laboratory of Theoretical Geochemistry (otherwise known as Prediction Central) at the University of California, Berkeley. They represent the culmination of research efforts to generate a comprehensive and internally consistent set of thermodynamic data and equations for minerals, gases, and aqueous solutions at high pressures and temperatures. Among the many who contributed to the successful completion of this book, we are especially indebted to David Kirkham, John Walther, and George Flowers, who wrote program SUPCRT, Tom Brown, who created program DIAGRAM, and Eli Messinger, who generated the Tektronix plot routine to construct the diagrams. Ken Jackson and Terri Bowers both devoted an enormous amount of time and effort over the past six years to produce the diagrams in the following pages; some of which went through many stages of revision. Consequently, they appear as senior authors of this volume. It should be mentioned in this regard that their equal dedication to the project made it necessary to determine their order of authorship by flipping a coin. My own role in generating the diagrams was largely supervisory, although I am responsible for the Introduction to the book. We would like to express our appreciation to Carol Bruton, Dara Gilbert, Mike O'Leary, Tom Welsh, Laurel Goodwin, Miki Moore, Ken Herkenhoff, Robert Woodward, Mary Gilzean, Cathy Wilson, Eric Oelkers, Barry Draper, Sydney Dent, and Barbara Ransom, all of whom assisted to one degree or another in generating the diagrams. Thanks are also due Mike Ryan, Greg Williams, Mike Salas, Robert Wong, Joachim Hampel, and Joan Bossart for drafting, photographic reproduction, and typing the manuscript. We are particularly grateful to Robert Woodward and Joan Bossart for their exceptional patience, talent, and dedicated effort toward completion of this project. The research responsible for the diagrams was supported by the National Science Foundation (NSF grants GA 35888, GA 36023, DES 74-14280, EAR 77-14492, and EAR 81-15859), the donors of the Petroleum Research Fund administered by the American Chemical Society (PRF grants 5356 AC2 and 8927 AC2C), the Miller Institute and Committee on Research at the University of California, Berkeley, and the Division of Physical Research of the U.S. Energy Research and Development Administration in conjunction with the Lawrence Berkeley and Livermore Laboratories of the University of California (ERDA contracts UCB-ENG-4288 and W-7405-ENG-48). Computing funds contributed by the University of California, Berkeley, are also acknowledged with thanks. Finally, we would like to express our appreciation to Konrad Springer for his interest and support in undertaking publication of this book.

June, 1984, Berkeley   Harold C. Helgeson

TABLE OF CONTENTS

Introduction	IX
Figures	XVII
Table 1. Fugacity coefficients of H_2O and CO_2	XXXI
Table 2. List of minerals and their abbreviations and formulas	XXXII
Table 3. Summary of hydrolysis reactions	XXXVI
References	XLI
Activity diagrams	1
Fugacity diagrams	291
Appendix: Equilibrium constants for hydrolysis reactions	307
Index of diagrams	369

INTRODUCTION

Activity diagrams representing equilibrium among minerals and aqueous solutions facilitate considerably description and interpretation of phase relations in geologic systems. In addition, they afford a convenient frame of reference for predicting the consequences of mineral reactions, both in laboratory experiments and geochemical processes. Recent advances in theoretical geochemistry make it possible to generate activity diagrams for pressures and temperatures to 5 kb and 600°C. Diagrams of this kind are presented below for a wide variety of multicomponent systems at the pressures and temperatures shown in figure 1. The schematic arrangement of the diagrams depicted in this figure corresponds to that on two facing pages of the book. The three schematic diagrams at the bottom of the left side of figure 1 pertain to the liquid phase region along the vapor-liquid equilibrium curve for H_2O. Except in cases where the presence of amorphous silica is specified, or systems containing CO_2 in which the activity of H_2O (a_{H_2O}) is less than unity, the diagrams in each set on two facing pages represent phase relations in a single system subject to the constraints specified in the caption at the bottom of each page. If the presence of amorphous silica is specified, diagrams are depicted only on one page for a given system at temperatures $\leqslant 300°C$. For systems containing CO_2 in which $a_{H_2O} < 1.0$, diagrams are shown only on one page for a given system at temperatures $\geqslant 400°C$. In the latter cases, a value of the mole fraction of CO_2 (X_{CO_2}) in the aqueous phase of 0.01, 0.05, 0.1, 0.3, 0.5, 0.7, 0.9, 0.95, or 0.99 is specified in the caption at the bottom of each page of diagrams. The activities of H_2O and CO_2 for the stipulated values of X_{CO_2} were computed for $X_{CO_2} + X_{H_2O} = 1$ using equations and parameters taken from Flowers (1979) and Flowers and Helgeson (1983), which provide for nonideal mixing of CO_2 and H_2O. The values of the fugacity coefficients used in the calculations are given in table 1. The axes of the activity diagrams for specified values of X_{CO_2} are labeled in terms of σ_i, which is defined by (Walther and Helgeson, 1980)

$$\sigma_i \equiv a_{H_2O}^{(n_i - Z_i n_{H^+})} \tag{1}$$

where n_{H^+} and n_i stand for the solvation numbers of the hydrogen ion and the ith aqueous species other than the hydrogen ion, respectively, and Z_i refers to the charge on the subscripted species. By specifying the descriptive variables as $\log(a_i/(\sigma_i a_{H^+}^{Z_i}))$ in the diagrams for stipulated values of X_{CO_2}, all phase relations can be portrayed accurately, regardless of the actual state of solvation of the species in solution (Walther and Helgeson, 1980).

The compositions of the bulk of the rock-forming minerals in the Earth's crust can be described in terms of the system $MgO-CaO-FeO-Na_2O-K_2O-Al_2O_3-SiO_2-CO_2-H_2O$. If the component HCl is added to the system, the major compositional characteristics of natural aqueous solutions can also be described in terms of these components.[1] Provision for ore deposits and oxidation-reduction equilibria can be incorporated by adding the components

---

[1] The term component is used in the present communication in its strict thermodynamic sense. A thermodynamic component of a mineral corresponds to a chemical formula unit representing one of the minimum number of independent variables required to describe the composition of the mineral. All stoichiometric minerals are thus composed of a single component corresponding to the formula of the mineral. However, because the term component has no physical connotation, the minerals themselves are *not* components.

..., ..$_2$SO$_4$, Cu$_2$S, PbS, ZnS, Ag$_2$S, etc. This set of components is one of several alternate sets that can be chosen to describe the same geologic system (Helgeson, 1970a).

The activities of individual ions, complexes, or both may be used as descriptive variables in equilibrium activity diagrams. Four of the components listed above do not dissociate to an appreciable extent in acid aqueous solutions. These are the components SiO$_2$, H$_2$S, CO$_2$, and H$_2$O, which occur primarily as neutral molecules. At low temperatures, the remainder of the components are essentially completely dissociated. CO$_2$, H$_2$S, and SiO$_{2(aq)}$ dissociate at low temperatures only in alkaline solutions, where CO$_2$ reacts with OH$^-$ to form HCO$_3^-$, which then dissociates to H$^+$ and CO$_3^{--}$ with increasing pH. Similarly, H$_2$S dissociates to H$^+$ and HS$^-$ as pH increases in alkaline solutions. At low temperatures and high pH, SiO$_{2(aq)}$ is apparently present primarily as H$_3$SiO$_4^-$. At high temperatures, molecular species predominate over their ionic counterparts in hydrothermal solutions.

Homogeneous equilibrium in an aqueous solution can be represented by writing reversible reactions among the thermodynamic components of the solution and their dissociated counterparts. For example, we can write for the component Na$_2$O

$$Na_2O + 2 H^+ = 2 Na^+ + H_2O \tag{2}$$

for which

$$\log (a_{Na^+}/a_{H^+}) = (\log K_{(2)} + \log (a_{Na_2O}/a_{H_2O}))/2 \tag{3}$$

where K$_{(2)}$ refers to the equilibrium constant for reaction (2). Note also that we can write

$$Na_2O + H_2O = 2 Na^+ + 2 OH^- \tag{4}$$

for which

$$\log (a_{Na^+} a_{OH^-}) = (\log K_{(4)} + \log (a_{Na_2O} a_{H_2O}))/2 \tag{5}$$

where K$_{(4)}$ denotes the equilibrium constant for reaction (4). The ionic species in reactions (2) and (4) predominate at low temperatures in acid and alkaline solutions, respectively. However, in the supercritical region the molecules NaCl and HCl predominate over Na$^+$ and H$^+$, respectively, in hydrothermal solutions (Quist and Marshall, 1968; Helgeson, 1969; Helgeson, Kirkham, and Flowers, 1981; Franck, 1982). Equilibrium among these molecular species can be expressed as

$$Na_2O + 2 HCl = 2 NaCl + H_2O \tag{6}$$

for which

$$\log (a_{NaCl}/a_{HCl}) = (\log K_{(6)} + \log (a_{Na_2O}/a_{H_2O}))/2 \tag{7}$$

where K$_{(6)}$ refers to the equilibrium constant for reaction (6). Similarly, equilibrium between the component Na$_2$O and the NaOH molecule can be represented by

$$Na_2O + H_2O = 2 NaOH \tag{8}$$

for which

$$\log a_{NaOH} = (\log K_{(8)} + \log (a_{Na_2O} a_{H_2O}))/2 \tag{9}$$

where K$_{(8)}$ designates the equilibrium constant for reaction (8). Combining equations (3), (5), (7), and (9) leads to

$$\log (a_{Na^+}/a_{H^+}) - ((\log K_{(2)})/2) = \log (a_{NaCl}/a_{HCl}) - ((\log K_{(6)})/2) \quad (10)$$

and

$$\log (a_{Na^+}a_{OH^-}) - ((\log K_{(4)})/2) = \log a_{NaOH} - ((\log K_{(8)})/2) \quad (11)$$

which relate the various descriptive variables to one another. If we now take account of

$$\mu_{Na_2O} = \mu°_{Na_2O} + 2.303\,RT \log a_{Na_2O} \quad (12)$$

and

$$\mu_{H_2O} = \mu°_{H_2O} + 2.303\,RT \log a_{H_2O}, \quad (13)$$

we can write for constant pressure and temperature,

$$d \log (a_{Na^+}/a_{H^+}) = (d \log (a_{Na_2O}/a_{H_2O}))/2 = d \log (a_{NaCl}/a_{HCl})$$
$$= (d\mu_{Na_2O} - d\mu_{H_2O})/2(2.303)\,RT \quad (14)$$

and

$$d \log (a_{Na^+}a_{OH^-}) = (d \log (a_{Na_2O}a_{H_2O}))/2 = d \log (a_{NaOH})$$
$$= (d\mu_{Na_2O} + d\mu_{H_2O})/2(2.303)\,RT \quad (15)$$

where μ_{Na_2O}, $\mu°_{Na_2O}$, μ_{H_2O}, and $\mu°_{H_2O}$ stand for the chemical potential and standard chemical potential of the subscripted components, respectively, R refers to the gas constant, and T stands for temperature in Kelvins.

Equations (14) and (15) relate the change in the ratio or product of the activities of the aqueous species corresponding to descriptive variables in various equilibrium activity diagrams to the change in the chemical potentials of Na$_2$O and H$_2$O in the system. Reactions and equations analogous to reactions (2), (4), (6), and (8), and equations (3), (5), (7), and (9) through (15) can be generated for any thermodynamic component. For most purposes in geochemistry, the ratios of the activities of individual cations commonly found in minerals to that of H$^+$ (raised to the power of the charge on the cation) are probably the most suitable choice of descriptive variables for activity diagrams at temperatures < 600°C. These ratios can be converted to other ratios with the aid (for example) of equation (10) and equilibrium constants taken from sources cited in Helgeson and Kirkham (1976).

Except for systems in which CO$_2$ is a component and values of X_{CO_2} in the fluid are specified, the activity and fugacity diagrams presented in the following pages were generated for $a_{H_2O} = 1$, relative to the liquid standard state (see below). It can be shown that the activity of H$_2$O$_{liquid}$ departs only slightly from unity in most natural aqueous solutions at the temperatures and pressures considered in the present study (Helgeson, 1969, 1982a; Helgeson, Kirkham, and Flowers, 1981).

The diagrams that follow are based on a standard state for minerals and liquid H$_2$O of unit activity of the pure solid or liquid at any pressure and temperature, but that for gases is one of unit fugacity of the hypothetical perfect gas at 1 bar and any temperature. It follows that the activities of components corresponding to stoichiometric minerals and pure liquid H$_2$O are unity, and the fugacities of gases are equal to their activities at all pressures and temperatures. The standard state for aqueous species corresponds to unit activity of the species in a hypothetical one molal solution referenced to infinite dilution at any pressure and temperature. The activity coefficients of aqueous species thus approach one at both high and low temperatures and pressures as the activity of the solvent (relative to the liquid standard state) approaches unity. In contrast, the fugacity coefficients of the components of gas mixtures approach the fugacity coefficients of the pure gases (which are not necessarily unity) as the mole fractions of the gases approach unity at any pressure and temperature.

The thermodynamic data and equations for minerals, gases, and aqueous species used to locate the stability field boundaries in the activity and fugacity diagrams presented below were taken from Helgeson (1982b, 1984), Helgeson and Kirkham (1974a), Walther and Helgeson (1977), Helgeson and others (1978), Bird and Helgeson (1980), and Helgeson, Kirkham, and Flowers (1981). Equations and data taken from these sources were also used to generate the curves shown in figures 2 through 14, as well as the equilibrium constants given in the Appendix. The calculations were carried out on the CDC 6400 computer (since replaced) at the University of California, Berkeley, with the aid of programs SUPCRT (written by D.H. Kirkham, J.V. Walther, and G.C. Flowers) and DIAGRAM (written by T.H. Brown). The diagrams were generated on a Tektronix digital plotter using a software package written by Eli Messinger. Because the diagrams were generated by a computer, the labels on the abscissa and ordinate of each diagram are printed in computer notation; e.g., $\log (a_{Mg^{++}}/a_{H^+}^2)$ appears as LOG (A(MG++)/(A(H+))**2). The stoichiometric minerals considered in the present study are listed in table 2, together with their abbreviations and formulas. The hydrolysis reactions for the minerals are shown in table 3, and the equilibrium constants for the reactions are given in the Appendix.

The equilibrium constant for a given stability field boundary on a logarithmic activity diagram can be generated by adding and/or subtracting values of log K taken from the Appendix. For example, the equilibrium constant for the stability field boundary corresponding to

$$\underset{(tremolite)}{Ca_2Mg_5Si_8O_{22}(OH)_2} + Mg^{++} + 2H^+ = \underset{(talc)}{2Mg_3Si_4O_{10}(OH)_2} + 2Ca^{++} \tag{16}$$

can be calculated by subtracting 2 log $K_{talc\ hydrolysis}$ from log $K_{tremolite\ hydrolysis}$ in the Appendix. All of the equilibrium constants listed in the Appendix are consistent with the liquid standard state for H_2O. However, a number of the activity diagrams in the following pages were generated by adopting the gas standard state for H_2O and CO_2, which permits calculation of the fugacities of H_2O and CO_2 from specified mole fractions of the gases (for $X_{H_2O} + X_{CO_2} = 1$) using the fugacity coefficients shown in table 1 and the relation

$$f_i = \chi_i X_i P \tag{17}$$

where P stands for pressure and f_i, χ_i, and X_i refer to the fugacity, fugacity coefficient, and mole fraction of the ith gas component. The difference in the standard molal Gibbs free energy of $H_2O_{(liquid)}$ and $H_2O_{(gas)}$ divided by RT is depicted as a function of temperature and pressure in figure 2. If the values shown in this figure are multiplied by the number of moles of H_2O in a given hydrolysis reaction in table 3, addition of the resulting values to those of log K for the reaction given in the Appendix leads to corresponding values of log K consistent with the gas standard state for H_2O. Similarly, log K values given in the Appendix for hydrolysis reactions written in terms of HCO_3^-, HS^-, and $SO_4^=$ can be converted to log K values for corresponding reactions involving $CO_{2(gas)}$, $S_{2(gas)}$, and $O_{2(gas)}$ by taking account of the equilibrium constants plotted in figures 3 through 5. In certain cases, diagrams are presented below for specified values of $\log(a_{H^+}a_{HS^-})$. Corresponding values of $\log a_{H_2S_{(aq)}}$ and $\log f_{H_2S_{(g)}}$ can be generated from the curves shown in figures 6 and 7.

Thermodynamic relations and methods used in the construction and interpretation of equilibrium activity diagrams have been summarized in a number of publications (c.f., Garrels and Christ, 1965; Helgeson, 1968, 1970a; Helgeson, Brown, and Leeper, 1969; Walther and Helgeson, 1980; Aagaard and Helgeson, 1983). The position and slope of a given stability field boundary on such diagrams is determined by the stoichiometry of the minerals and the equilibrium constant for the reversible reaction chosen to represent the equilibrium state. For example, equilibrium among clinochlore, muscovite, quartz, and an aqueous phase can be expressed as

$$3 \; \underset{(clinochlore)}{Mg_5Al(AlSi_3O_{10})(OH)_8} + 2K^+ + 28 \; H^+$$

$$= 2\underset{(muscovite)}{KAl_2(AlSi_3O_{10})(OH)_2} + 15 \; Mg^{++} + 3 \; \underset{(quartz)}{SiO_2} + 24 \; H_2O \qquad (18)$$

for which the law of mass action can be written for unit activity of SiO_2, $Mg_5Al(AlSi_3O_{10})(OH)_8$, and $KAl_2(AlSi_3O_{10})(OH)_2$ as (Walther and Helgeson, 1980)

$$\log(a_{Mg^{++}}/(\sigma_{Mg^{++}}a_{H^+}^2)) = (\log K_{(18)} + 2\log(a_{K^+}/(\sigma_{K^+}a_{H^+})) - 24 \log a_{H_2O})/15 \qquad (19)$$

where $\sigma_{Mg^{++}}$ and σ_{K^+} are defined by appropriate statements of equation (1) and $K_{(18)}$ refers to the equilibrium constant for reaction (18) at the pressure and temperature of interest. If the activity of H_2O (relative to the liquid standard state) in the aqueous phase is unity, or nearly so (which is commonly the case in geologic systems--see above) equation (19) reduces to

$$\log(a_{Mg^{++}}/(a_{H^+}^2)) = (\log K_{(18)} + 2\log(a_{K^+}/a_{H^+}))/15 \qquad (20)$$

However, if CO_2 or some other component is present in mole fractions greater than ~ 0.01, the activity of H_2O may differ appreciably from unity. Under these circumstances, equation (19) must be used to describe accurately the equilibrium state. This is particularly true for reactions in which the reaction coefficient of H_2O is large. Under these circumstances, the effect of small departures of a_{H_2O} from unity on the positions of stability field boundaries may be magnified considerably.

The value of $(\log K_{(18)})/15$ in equations (19) and (20) determines the intercept of the linear curve separating the stability fields of clinochlore and muscovite on activity diagrams with $\log(a_{Mg^{++}}/(\sigma_{Mg^{++}}a_{H^+}^2))$ or (for unit a_{H_2O}) $\log(a_{Mg^{++}}/a_{H^+}^2)$ on the ordinate and $\log(a_{K^+}/(\sigma_{K^+}a_{H^+}))$ or (again for unit a_{H_2O}) $\log(a_{K^+}/a_{H^+})$ on the abscissa. The slope of the curve is equal to 2/15, which is related to the slopes of corresponding curves on other activity diagrams representing the same equilibrium state by the quotient of appropriate differential equations like those shown in equation (14). In all cases, a value is specified for a_{H_2O}, which is a constant for each diagram. In contrast, the activities of (in this case) aluminum species in the aqueous phase vary from place to place in the diagram. Alternate diagrams with different descriptive variables such as $\log(a_{Al^{+++}}/a_{H^+}^3)$ on the axes are depicted below to facilitate determination of the extent of this variation. These include diagrams generated by choosing alternately Al_2O_3, MgO, CaO, and SiO_2 as the balancing component in the reversible reactions representing the stability field boundaries on the diagrams. For example, equilibrium among K-feldspar, muscovite, and an aqueous phase can be expressed as

$$3 \; \underset{(K-feldspar)}{KAlSi_3O_8} + 2 \; H^+ = \underset{(muscovite)}{KAl_2(AlSi_3O_{10})(OH)_2} + 2 \; K^+ + 6 \; SiO_{2(aq)} \qquad (21)$$

or

$$KAlSi_3O_8 + 2\,Al^{+++} + 4H_2O = KAl_2(AlSi_3O_{10})(OH)_2 + 6\,H^+ \quad , \qquad (22)$$
$$(K-feldspar) \hspace{4cm} (muscovite)$$

which correspond to only two of several reversible reactions that could be written to describe the equilibrium state. If we now set $a_{KAlSi_3O_8} = a_{KAl_2(AlSi_3O_{10})(OH)_2} = a_{H_2O} = 1$ and designate the equilibrium constants for reactions (21) and (22) as $K_{(21)}$ and $K_{(22)}$, respectively, it follows that we can write

$$\log(a_{K^+}/a_{H^+}) = (\log K_{(21)} - 6\log a_{SiO_{2(aq)}})/2 \qquad (23)$$

and

$$\log(a_{Al^{+++}}/a_{H^+}^3) = -\log K_{(22)}/2 \qquad (24)$$

which describe the stability field boundary separating K-feldspar and muscovite on $\log(a_{K^+}/a_{H^+})$ vs $\log a_{SiO_{2(aq)}}$ and $\log(a_{K^+}/a_{H^+})$ vs $\log(a_{Al^{+++}}/a_{H^+}^3)$ diagrams, respectively.

With the exception of analcime and cordierite (see below), only stoichiometric minerals were considered in constructing the activity diagrams presented in the following pages. Provision for solid solution can be incorporated in the diagrams by taking account of activity coefficients in a statement of the law of mass action for a given stability field boundary. In some cases it may suffice simply to contour stability field boundaries for various activities of a given thermodynamic component of a solid solution corresponding to 0.9, 0.5, etc. Although many of the minerals considered in the calculations exhibit variable composition in nature, if the extent of solid solution is of the order of five or ten percent (or less), solid solubility has a slight effect on most of the phase relations depicted in the diagrams. Possible exceptions to this generalization are iron-magnesian phyllosilicates such as chlorite and biotite, which can be taken into account with the aid of equations summarized by Aagaard and Helgeson (1983), Helgeson and Aagaard (1984), and others. Several examples of activity diagrams in which provision for solid solution has been incorporated can be found in Bird and Helgeson (1981), Aagaard and Helgeson (1983), and Frisch and Helgeson (1983).

Although an aqueous phase does not appear explicitly in the activity diagrams that follow, it is a coexisting phase in all stability fields; i.e., the aqueous phase is saturated with the minerals appearing in the stability fields of the diagrams. The dashed lines labeled in lower case letters represent saturation limits in the aqueous phase for various minerals (identified in the caption at the bottom of the page) that do not occupy stability fields in the diagrams. Fluids with $\log(a_i/(\sigma_i a_{H^+}^{Z_i}))$ less than those limits are undersaturated with respect to the minerals represented by the labels on the saturation limits, and those at greater values of $\log(a_i/(\sigma_i a_{H^+}^{Z_i}))$ are supersaturated. Both stable and metastable equilibria among minerals and the aqueous phase are thus represented in the diagrams.

A number of the activity diagrams depicted below include provision for equilibrium among the aqueous phase and ubiquitous minerals that coexist with those shown in the stability fields of the diagram. For example, the presence of quartz may be specified (as in reaction 18). In others, amorphous silica and albite, quartz and albite, quartz and K-feldspar, or corundum, diaspore,[2] or gibbsite (depending on which of the latter phases is stable at the pressures and temperatures shown in figure 8) are stipulated to be in equilibrium with the

---

[2] Boehmite is apparently metastable in the presence of H_2O at all pressures and temperatures (Delany and Helgeson, 1978).

aqueous phase, which is also in equilibrium with the various minerals shown in the stability fields of the diagrams. Such specifications increase considerably the variety of phase relations that can be represented in two dimensions. For example, equilibrium among amphibole, plagioclase, garnet, and an aqueous phase can be represented on a $\log(a_{Ca^{++}}/a_{H^+}^2)$ vs $\log(a_{Mg^{++}}/a_{H^+}^2)$ diagram by taking account of

$$\underset{(pargasite)}{NaCa_2Mg_4Al(Al_2Si_6O_{22})(OH)_2} + Ca^{++} + 6\,H^+$$

$$= \underset{(grossular)}{Ca_3Al_2Si_3O_{12}} + \underset{(albite)}{NaAlSi_3O_8} + 4\,Mg^{++} + 4\,H_2O \ . \tag{25}$$

For unit activities of $NaCa_2Mg_4Al(Al_2Si_6O_{22})(OH)_2$, $NaAlSi_3O_8$, $Ca_3Al_2Si_3O_{12}$, and H_2O, the law of mass action for reaction (25) can be written as

$$\log(a_{Ca^{++}}/a_{H^+}^2) = 4\log(a_{Mg^{++}}/a_{H^+}^2) - \log K_{(25)} \tag{26}$$

where $K_{(25)}$ represents the equilibrium constant for reaction (25). Despite the fact that $\log(a_{Na^+}/a_{H^+})$ is not a descriptive variable in the diagram, pargasite may occupy a stability field because reaction (25) is balanced simultaneously on Al_2O_3 and Na_2O.

Except where noted otherwise in the page captions and Index citations, all of the minerals listed in table 2 were considered in the calculations. In certain cases diagrams are presented below in which metastable chrysotile and/or metastable 7A-clinochlore were considered in lieu of their stable counterparts (antigorite and 14A-clinochlore, respectively). Comparison of the phase relations shown in these diagrams with those in which antigorite and/or 14A-clinochlore were considered permits evaluation of the consequences of metastability in these systems. Except where cited in conjunction with their monoclinic analogs, enstatite and ferrosilite are used in the following pages to refer to the stable polymorphs of $MgSiO_3$ and $FeSiO_3$ at any pressure and temperature. In contrast, zoisite is used to denote only the orthorhombic form of $Ca_2Al_3Si_3O_{12}(OH)$. The terms K-feldspar, albite, dolomite, and epidote refer to $KAlSi_3O_8$, $NaAlSi_3O_8$, $CaMg(CO_3)_2$, and $Ca_2FeAl_2Si_3O_{12}(OH)$ in their stable states of order/disorder at any pressure and temperature. Stability field boundaries for the completely disordered and/or ordered analogs of these minerals can be generated from those shown in the diagrams for the minerals in their stable states of order/disorder by taking account of the equilibrium constants plotted in figures 9 through 12.

The terms analcime and cordierite are used below to refer to the equilibrium composition of $NaAlSi_2O_6\cdot(H_2O)_n$ and $Mg_2Al_3(AlSi_5O_{18})\cdot(H_2O)_n$, respectively, where n corresponds to the mole fraction of $NaAlSi_2O_6\cdot H_2O$ or $Mg_2Al_3(AlSi_5O18)\cdot H_2O$ in the minerals. The compositions were computed for $a_{H_2O} = 1$ assuming ideal mixing of the components which is consistent with (Helgeson and others, 1978),

$$\frac{X_{NaAlSi_2O_6\cdot H_2O}}{1 - X_{NaAlSi_2O_6\cdot H_2O}} = K_{(29)} \tag{27}$$

and

$$\frac{X_{Mg_2Al_3(AlSi_5O_{18})\cdot H_2O}}{1 - X_{Mg_2Al_3(AlSi_5O_{18})\cdot H_2O}} = K_{(30)} \tag{28}$$

where $K_{(29)}$ and $K_{(30)}$ correspond to the equilibrium constants for

$$NaAlSi_2O_6 + H_2O = NaAlSi_2O_6 \cdot H_2O \tag{29}$$

and

$$Mg_2Al_3(AlSi_5O_{18}) + H_2O = Mg_2Al_3(AlSi_5O_{18}) \cdot H_2O \ , \tag{30}$$

respectively. The equilibrium mole fractions of $NaAlSi_2O_6$ and $Mg_2Al_3(AlSi_5O_{18})$ in the solid solutions are shown as a function of temperature and pressure in figures 13 and 14.

The fugacities of CO_2, O_2, and S_2 (f_{CO_2}, f_{O_2}, and f_{S_2}, respectively) are used as descriptive variables in a number of the diagrams presented below. The fugacities of these gas species are related to their activities by (for example)

$$f_{O_2} = a_{O_2} f°_{O_2} \tag{31}$$

where $f°_{O_2}$ stands for the fugacity of O_2 in the standard state. For the gas standard state adopted in the present study (see above), $f°_{O_2} = 1$, so $f_{O_2} = a_{O_2}$. The ordinate and abscissa in each of the fugacity diagrams are labeled in computer notation; e.g., $\log f_{O_2}$ appears as LOG F(O2).

The systems represented by the diagrams summarized below are designated in the page captions and Index by the thermodynamic components considered in the calculations. Of these, the balancing component is shown in parentheses. The components represented by the descriptive variables on the ordinate and abscissa are shown in italics in the Index, as (for example) in the system $HCl-H_2O-Na_2O-(Al_2O_3)-K_2O-SiO_2$ at quartz saturation. Accordingly, all reactions among the minerals in this system that contain Al_2O_3 were balanced on this component to generate the stability field boundaries in the activity diagrams. All other minerals in the system are represented by saturation limits as a function of $\log(a_{Na^+}/a_{H^+})$ and/or $\log(a_{K^+}/a_{H^+})$, which correspond to the descriptive variables in the diagrams. It follows from equation (3) and its counterpart for $\log(a_{K^+}/a_{H^+})$ that with $a_{H_2O} = 1$, those variables represent the activities of Na_2O and K_2O, respectively. HCl and H_2O are components of all the systems considered in the following pages.

Part of the purpose of generating the theoretical activity diagrams summarized in the present communication is to stimulate critical comparison of the predicted phase relations with those observed in laboratory experiments and geologic systems. It should perhaps be emphasized in this regard that analytical data for natural aqueous solutions can be used to calculate the activities of species in these solutions. The requisite dissociation constants and activity coefficients for aqueous species can be generated from experimental data and theoretical equations (Helgeson, 1969; 1982a; Helgeson and Kirkham, 1974b, 1976; Helgeson, Kirkham, and Flowers, 1981). Such calculations make it possible to compare equilibrium activity ratios taken from activity diagrams with those in natural solutions, and to interpret directly phase relations observed in geologic systems in terms of their idealized counterparts in activity diagrams. This approach affords considerable insight into the chemical environment in which geochemical processes occur. It also facilitates prediction of mass transfer in geologic systems (c.f. Helgeson, 1968, 1970b, 1979; Helgeson, Garrels, and Mackenzie, 1969; Helgeson, Brown, Nigrini, and Jones, 1970; Helgeson and Murphy, 1983).

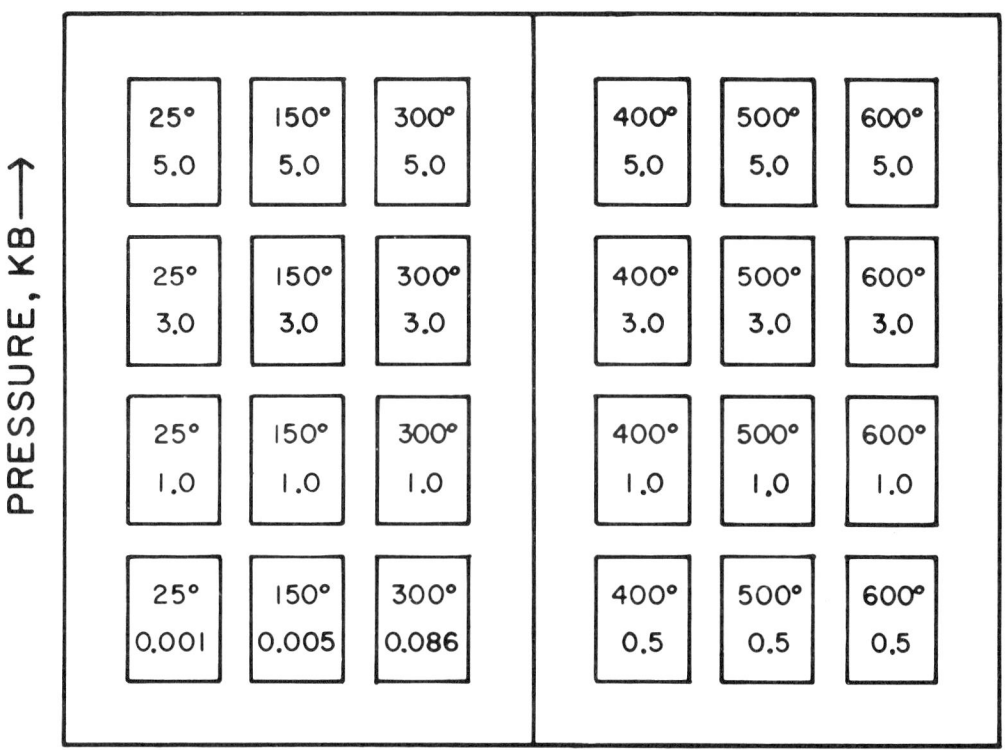

Fig. 1. Schematic arrangement of the activity diagrams for a given system on two facing pages of the present communication. Each small rectangle represents an activity diagram for the indicated temperature (in °C) and pressure (in kb).

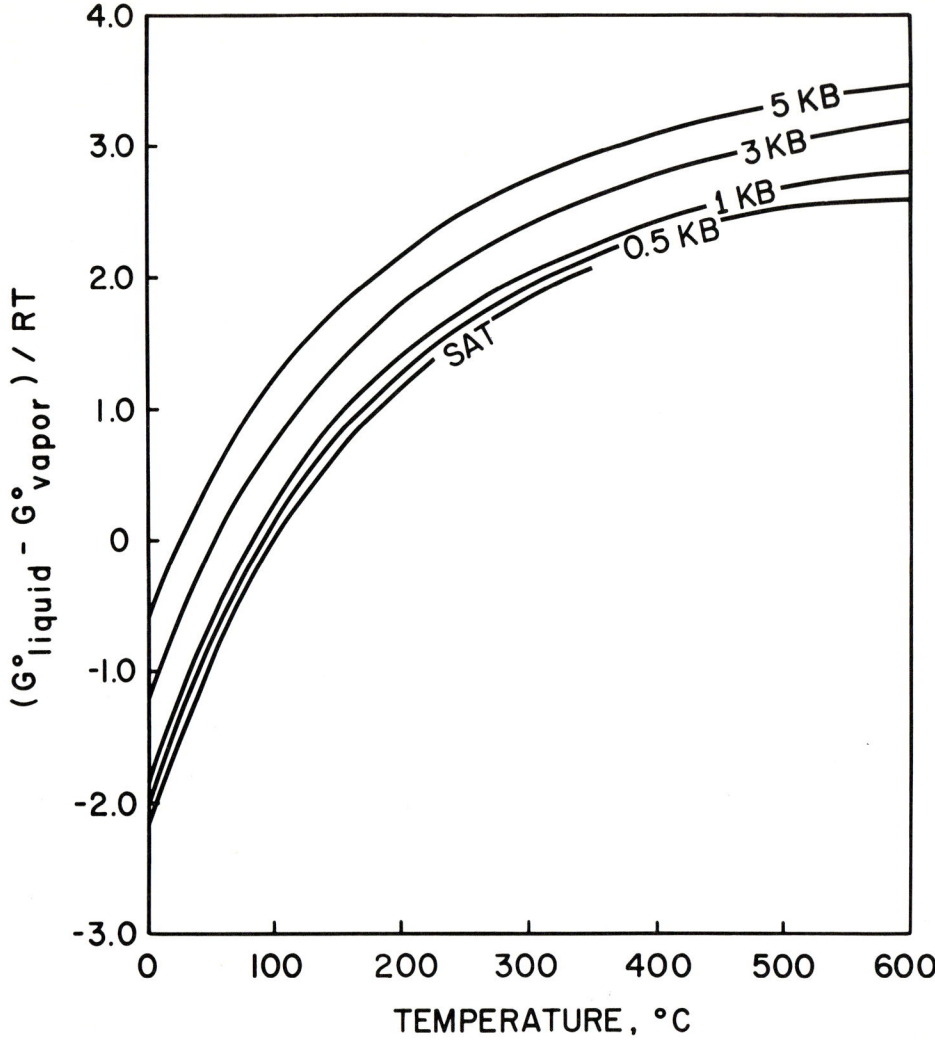

Fig. 2. Standard molal Gibbs free energy difference for $H_2O_{(liquid)}$ and $H_2O_{(gas)}$ divided by RT as a function of temperature at constant pressure. SAT refers to the vapor-liquid equilibrium curve for H_2O.

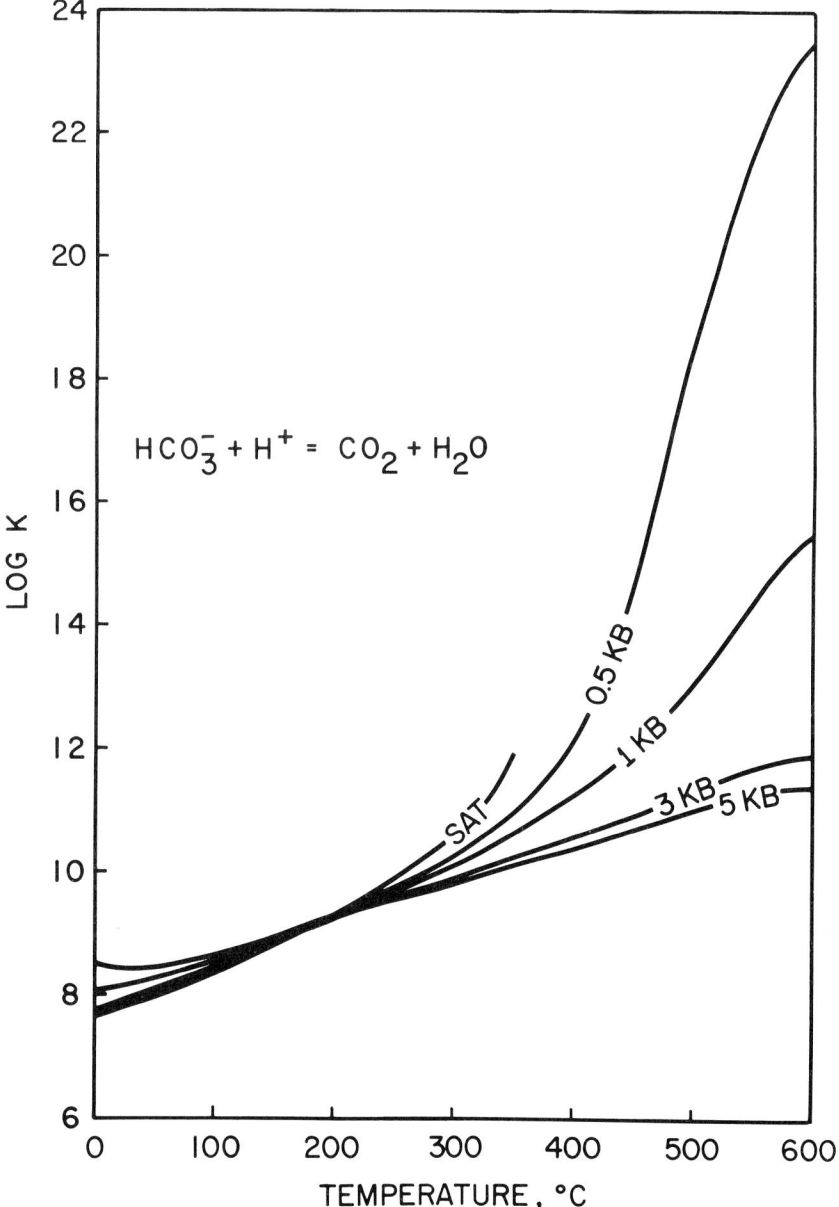

Fig. 3. Equilibrium constant for the reaction $HCO_3^- + H^+ = CO_{2(gas)} + H_2O_{(liquid)}$ as a function of temperature at constant pressure (see caption of figure 2).

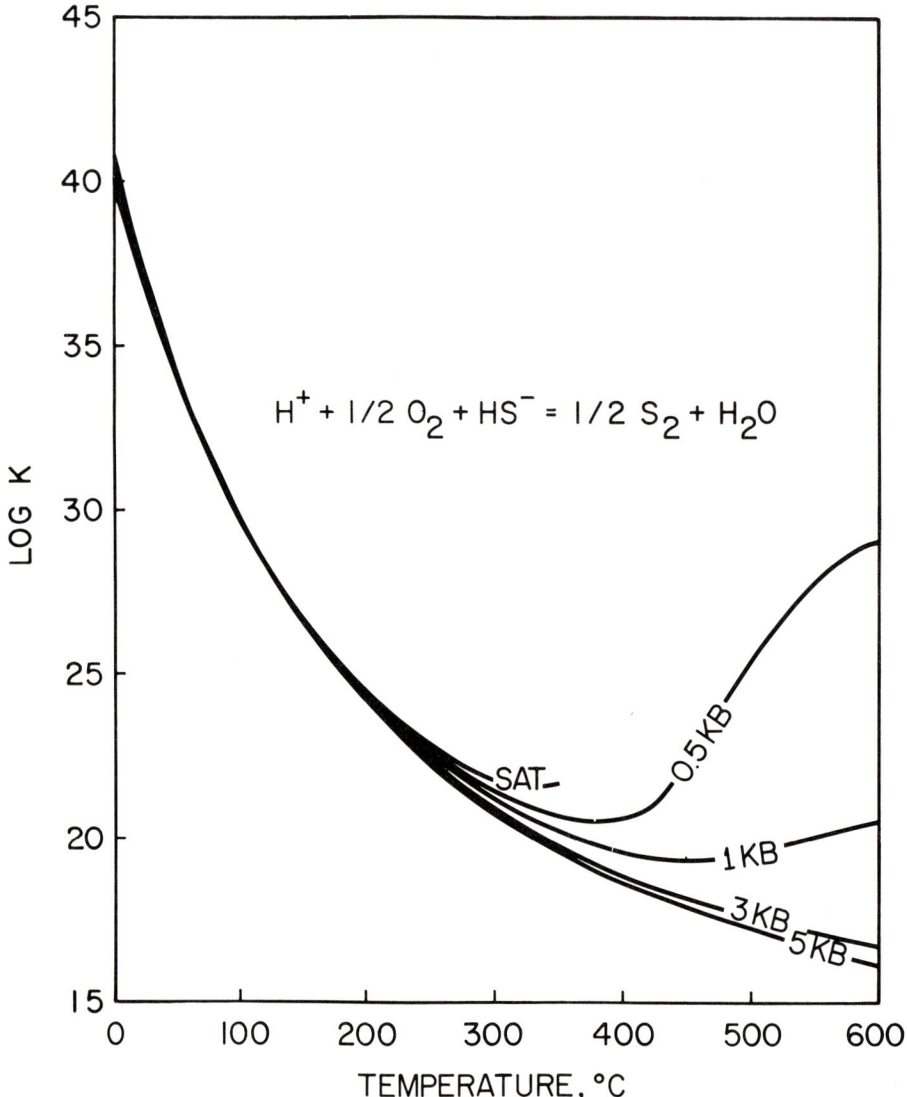

Fig. 4. Equilibrium constant for the reaction $H^+ + 1/2\ O_2 + HS^- = 1/2\ S_{2(gas)} + H_2O_{(liquid)}$ as a function of temperature at constant pressure (see caption of figure 2).

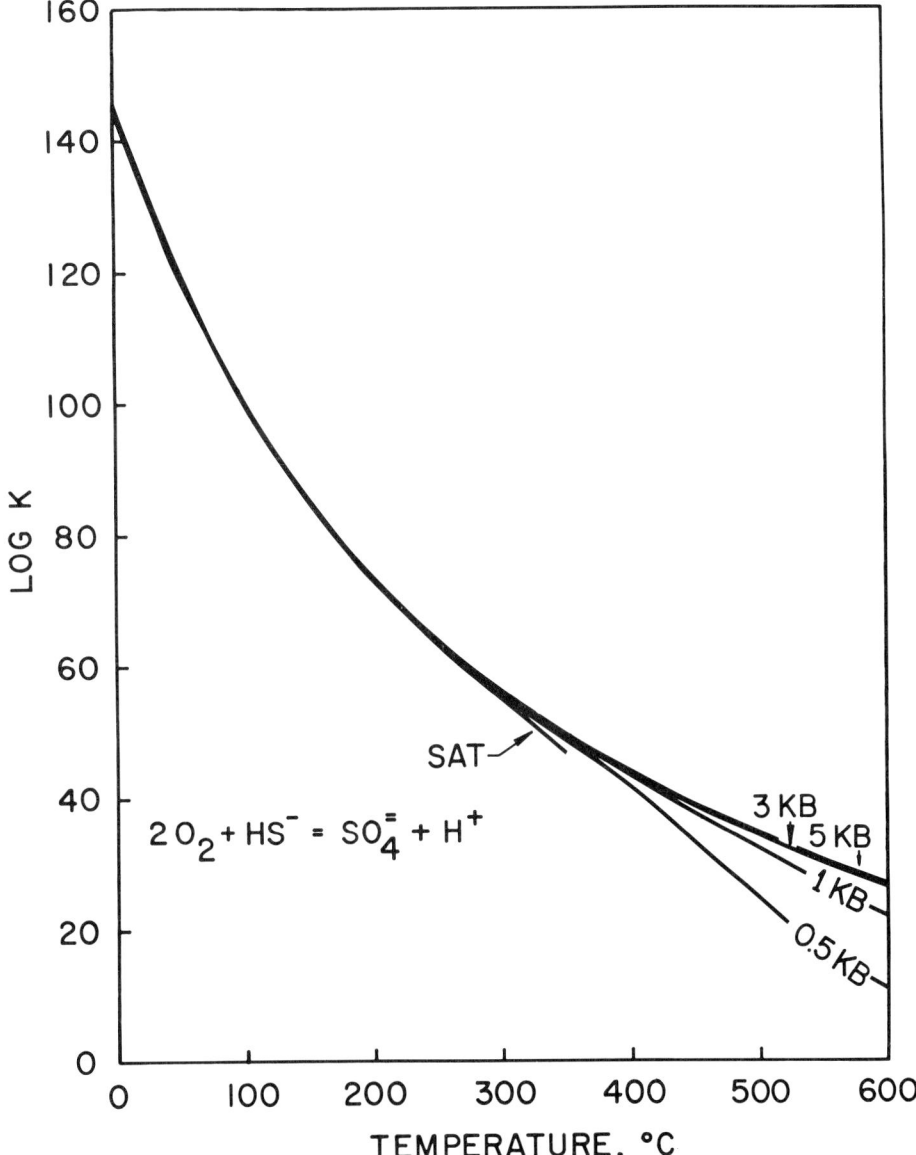

Fig. 5. Equilibrium constant for the reaction $2\,O_{2(gas)} + HS^- = SO_4^= + H^+$ as a function of temperature at constant pressure (see caption of figure 2).

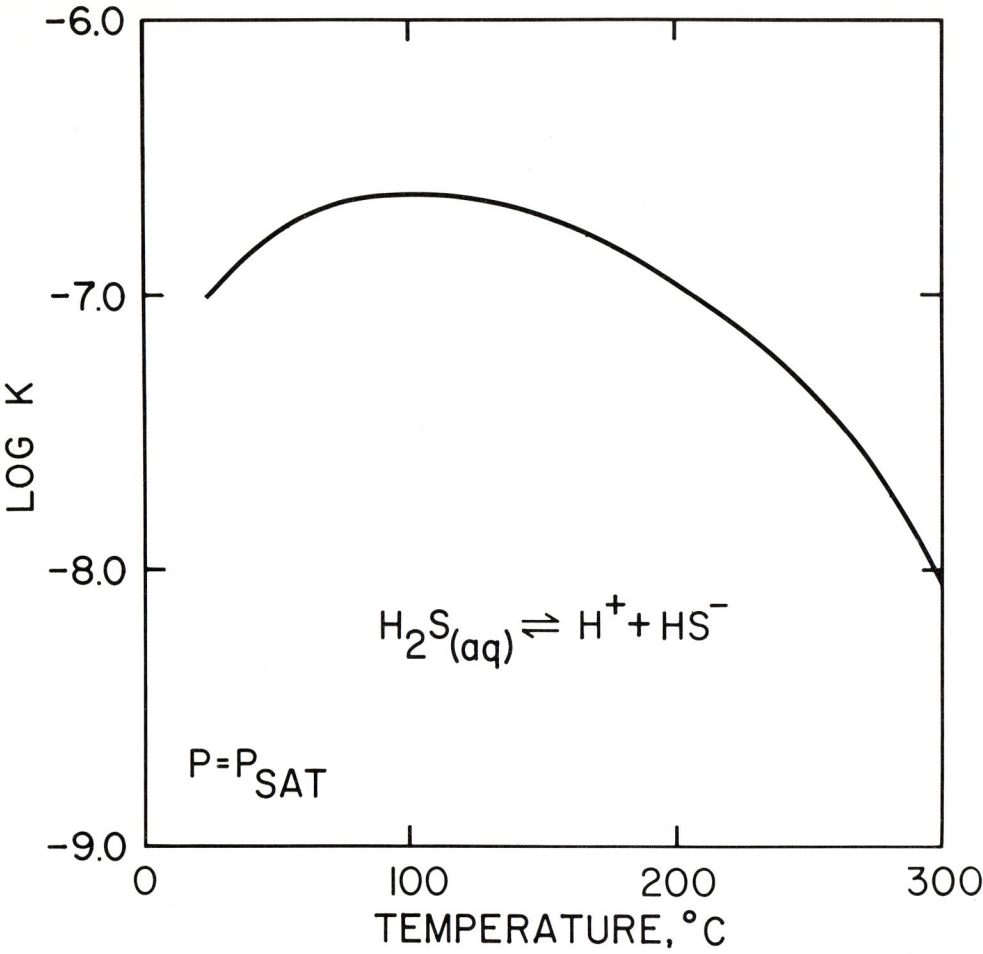

Fig. 6. Equilibrium constant for the reaction $H_2S_{(aq)} = H^+ + HS^-$ as a function of temperature at pressures corresponding to the vapor-liquid equilibrium curve for H_2O (Helgeson, 1969).

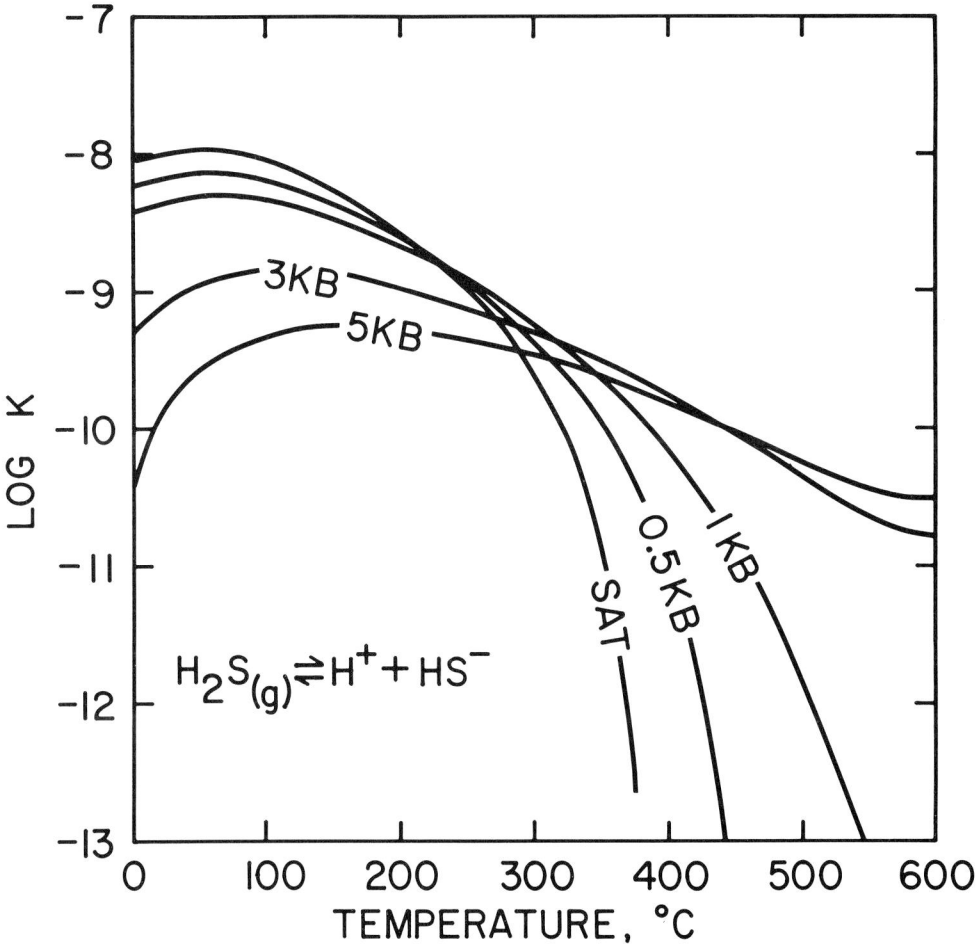

Fig. 7. Equilibrium constant for the reaction $H_2S_{(gas)} = H^+ + HS^-$ as a function of temperature at constant pressure (see caption of figure 2).

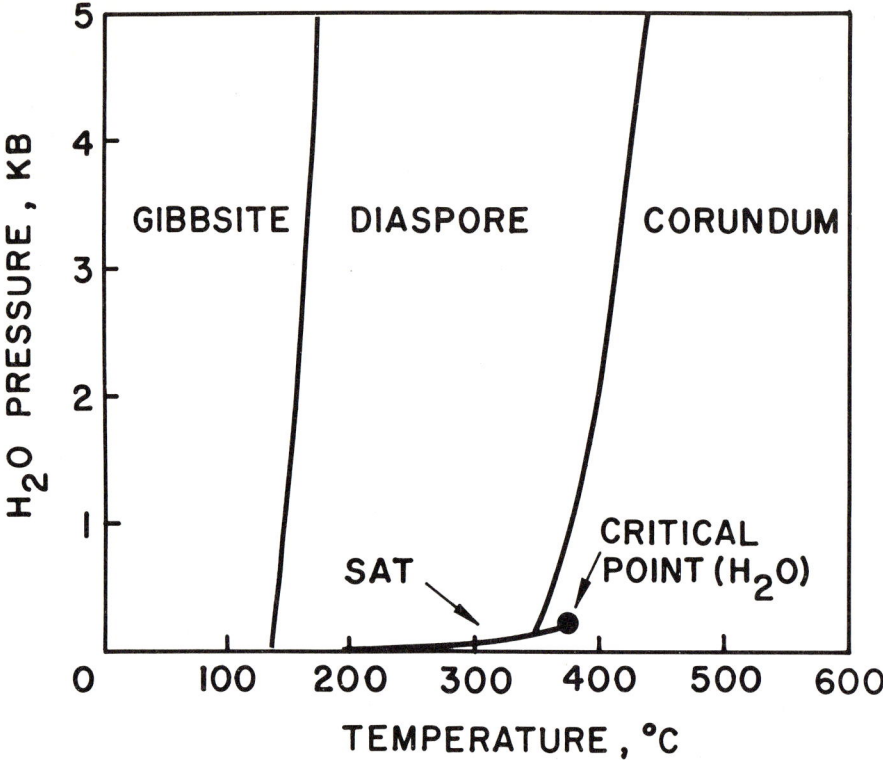

Fig. 8. Stability fields of gibbsite, diaspore, and corundum in the system $Al_2O_3 - H_2O$ as a function of temperature and H_2O pressure (Helgeson et al., 1978)--see caption of figure 2.

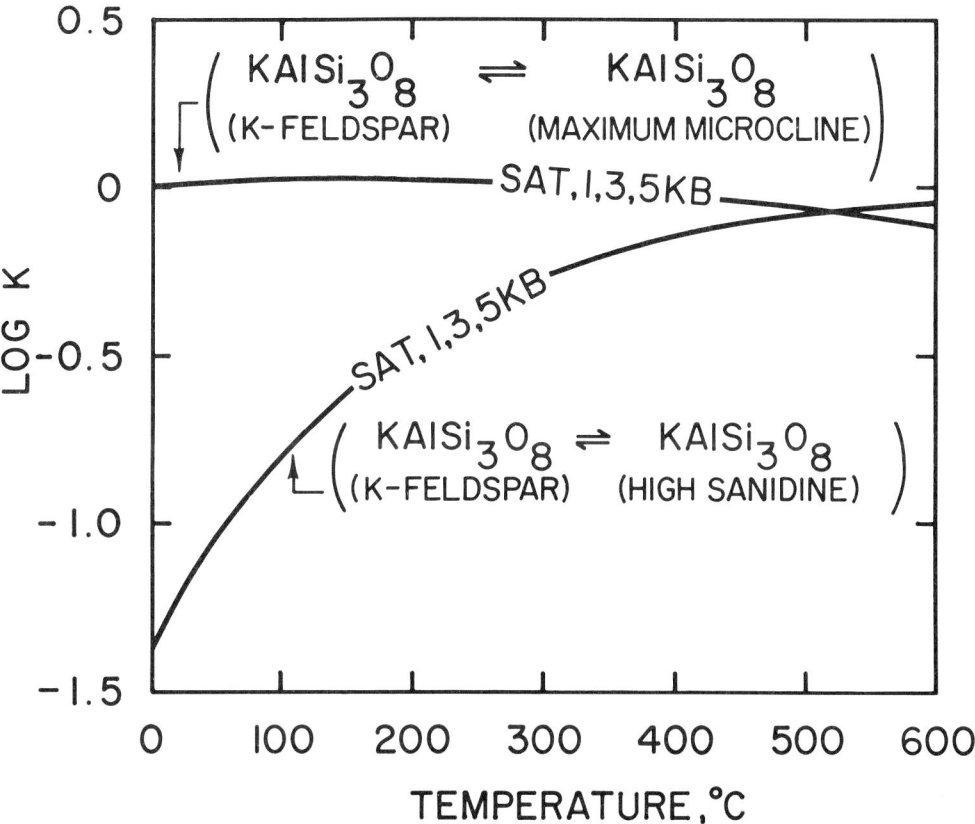

Fig. 9. Equilibrium constants for the metastable coexistence of completely ordered (microcline) and disordered (high sanidine) $KAlSi_3O_8$ with K-feldspar in its stable state of substitutional order/disorder as a function of temperature at constant pressure (see caption of figure 2).

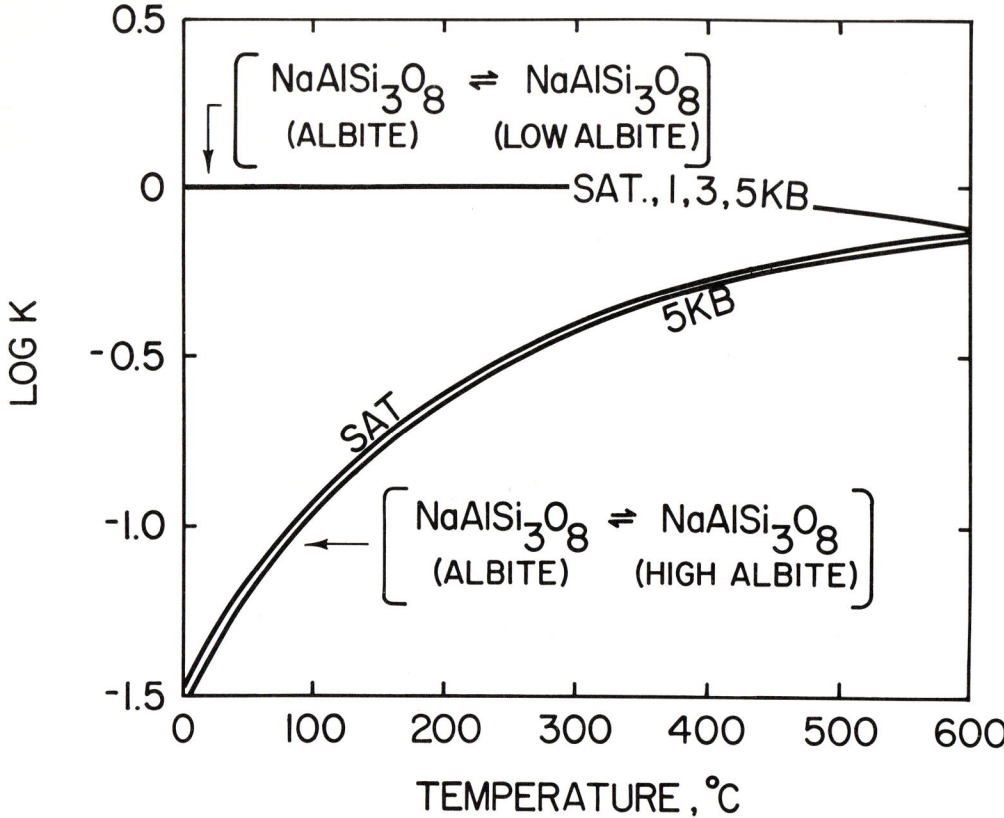

Fig. 10. Equilibrium constants for the metastable coexistence of completely ordered (low albite) and disordered (high albite) $NaAlSi_3O_8$ with albite in its stable state of substitutional order/disorder as a function of temperature at constant pressure (see caption of figure 2).

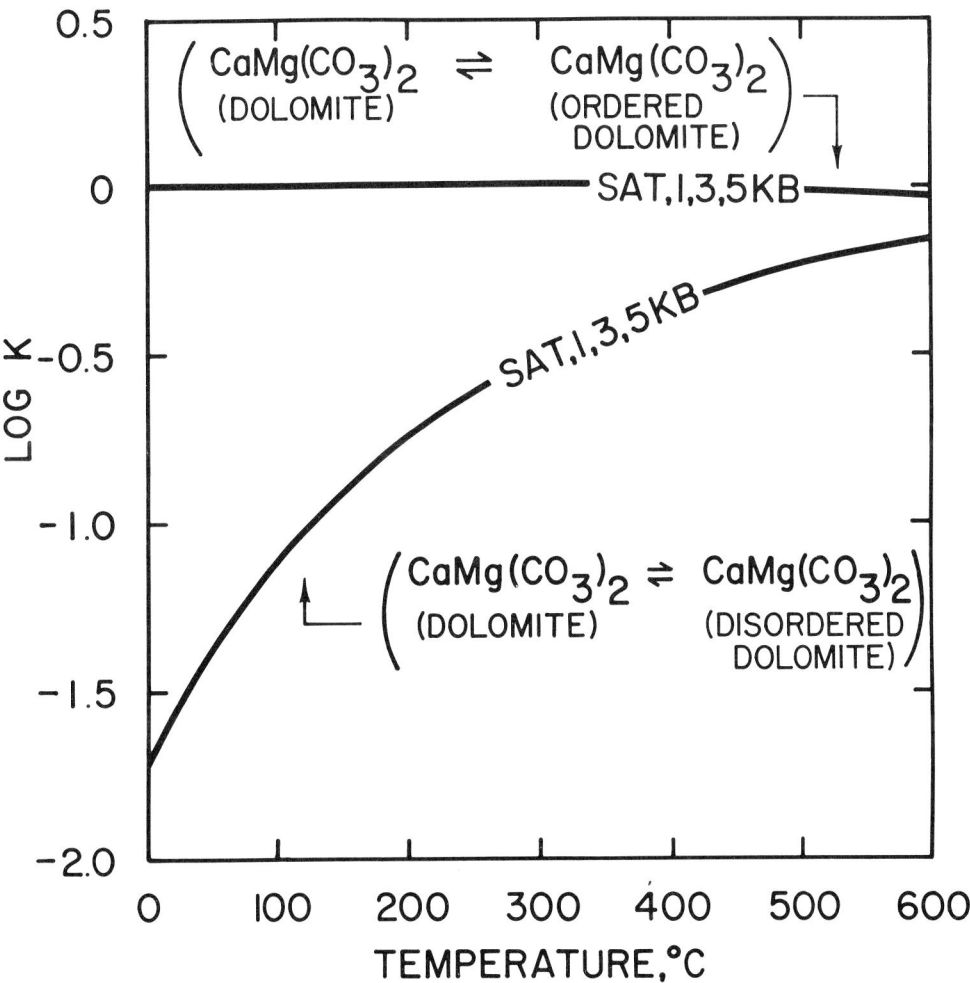

Fig. 11. Equilibrium constants for the metastable coexistence of completely ordered and disordered $CaMg(CO_3)_2$ with dolomite in its stable state of substitutional order/disorder as a function of temperature at constant pressure (see caption of figure 2).

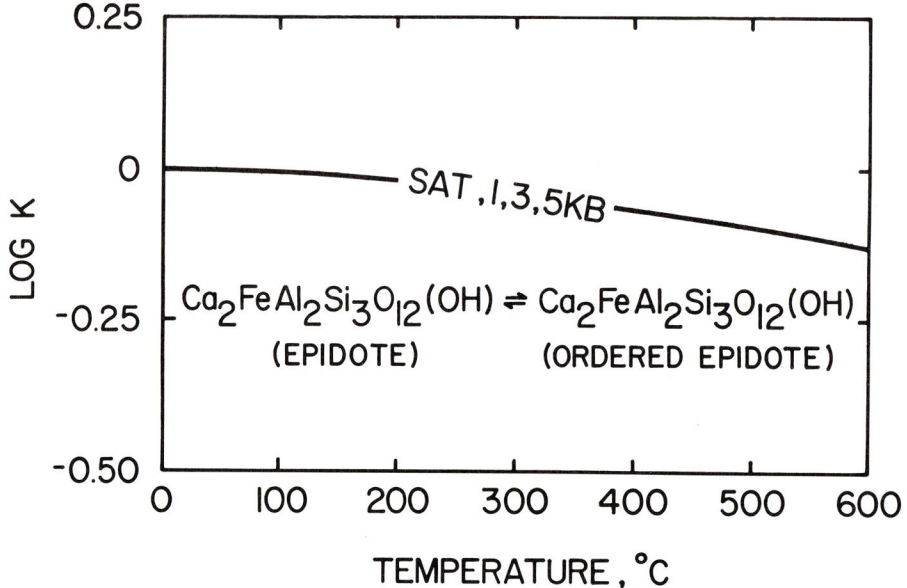

Fig. 12. Equilibrium constants for the metastable coexistence of completely ordered $Ca_2FeAl_2Si_3O_{12}(OH)$ with epidote in its stable state of substitutional order/disorder as a function of temperature at constant pressure (see caption of figure 2).

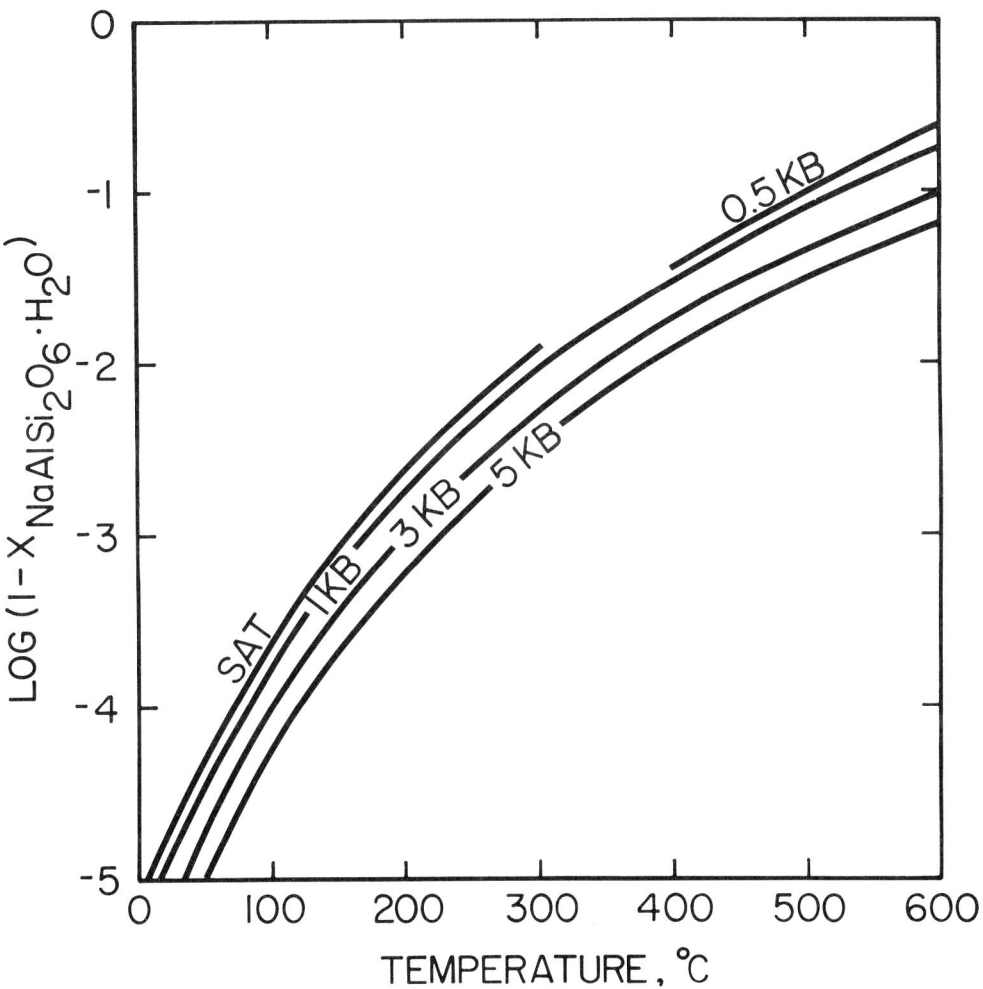

Fig. 13. Equilibrium composition of analcime computed from equation (27) as a function of temperature at constant pressure (see caption of figure 2).

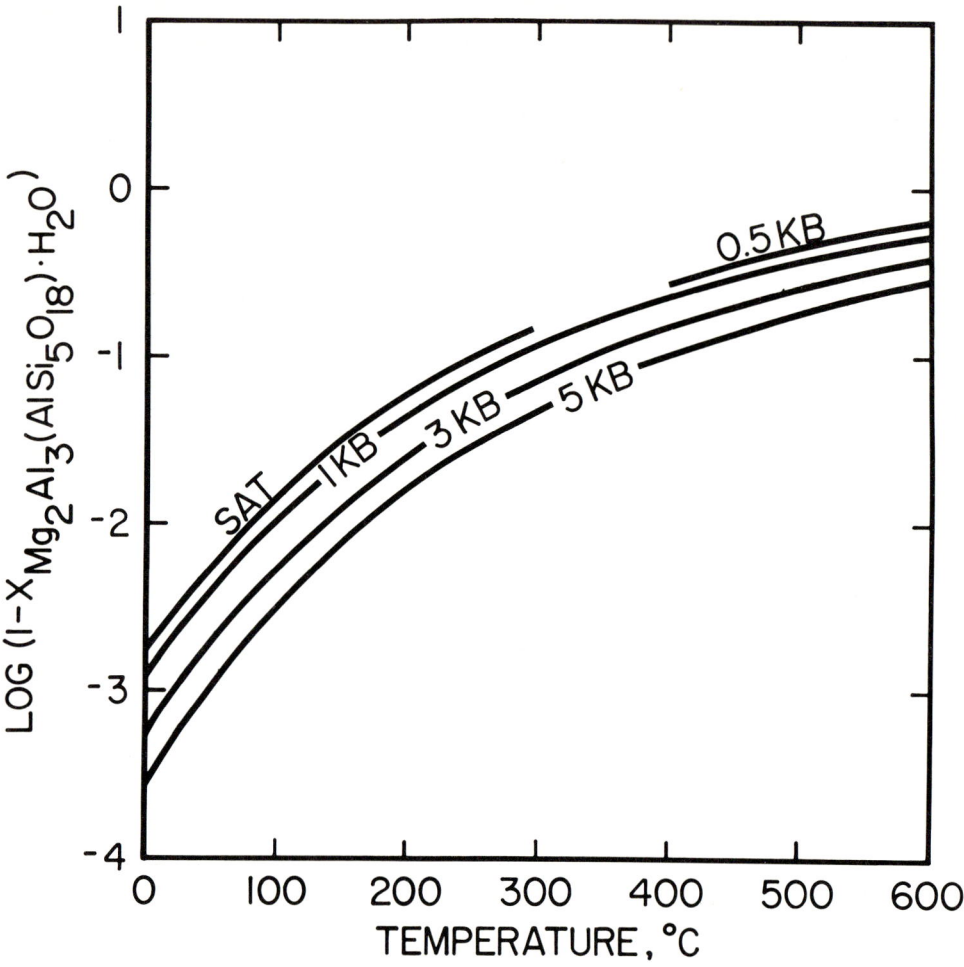

Fig. 14. Equilibrium composition of cordierite computed from equation (28) as a function of temperature at constant pressure (see caption of figure 2).

Table 1. Fugacity coefficients of H$_2$O (χ_{H_2O}) and CO$_2$ (χ_{CO_2}) in H$_2$O–CO$_2$ fluids computed from equations and parameters given by Flowers (1979) and Flowers and Helgeson (1983).

T,°C	χ_{H_2O}								
	PRESSURE = 0.5 KB								
	X_{CO_2}								
	0.01	0.05	0.10	0.30	0.50	0.70	0.90	0.95	0.99
400	0.394	0.399	0.413	0.487	0.529	0.558	0.583	0.588	0.592
500	0.665	0.666	0.667	0.670	0.670	0.667	0.663	0.661	0.660
600	0.789	0.789	0.788	0.779	0.764	0.746	0.726	0.721	0.717
	PRESSURE = 1.0 KB								
400	0.248	0.250	0.257	0.303	0.353	0.398	0.439	0.448	0.456
500	0.462	0.464	0.468	0.488	0.505	0.517	0.526	0.527	0.528
600	0.641	0.641	0.641	0.637	0.628	0.616	0.602	0.598	0.595
	PRESSURE = 3.0 KB								
400	0.182	0.184	0.189	0.223	0.271	0.327	0.386	0.402	0.414
500	0.350	0.351	0.354	0.377	0.405	0.433	0.458	0.463	0.468
600	0.520	0.521	0.522	0.526	0.529	0.529	0.527	0.526	0.526
	PRESSURE = 5.0 KB								
400	0.223	0.225	0.231	0.276	0.342	0.422	0.514	0.538	0.558
500	0.414	0.416	0.420	0.451	0.491	0.533	0.574	0.584	0.592
600	0.605	0.605	0.607	0.616	0.625	0.632	0.636	0.636	0.637

T,°C	χ_{CO_2}								
	PRESSURE = 0.5 KB								
	X_{CO_2}								
	0.01	0.05	0.10	0.30	0.50	0.70	0.90	0.95	0.99
400	6.125	4.128	2.714	1.248	1.092	1.051	1.039	1.038	1.038
500	1.167	1.133	1.110	1.084	1.084	1.086	1.088	1.088	1.088
600	0.977	0.990	1.006	1.056	1.087	1.105	1.112	1.113	1.113
	PRESSURE = 1.0 KB								
400	6.650	4.812	3.474	1.700	1.333	1.226	1.195	1.193	1.192
500	2.026	1.815	1.640	1.356	1.284	1.262	1.257	1.257	1.257
600	1.193	1.196	1.202	1.233	1.260	1.276	1.284	1.284	1.284
	PRESSURE = 3.0 KB								
400	17.035	12.667	9.366	4.572	3.369	2.966	2.839	2.830	2.828
500	5.354	4.772	4.250	3.244	2.903	2.775	2.735	2.732	2.731
600	2.849	2.789	2.733	2.629	2.604	2.603	2.606	2.606	2.606
	PRESSURE = 5.0 KB								
400	50.607	37.496	27.549	13.004	9.309	8.054	7.656	7.628	7.619
500	13.950	12.338	10.880	8.033	7.035	6.650	6.523	6.514	6.511
600	6.596	6.407	6.225	5.840	5.708	5.665	5.655	5.654	5.654

Table 2. Minerals, abbreviations, and formulas.

Index Number	Mineral Name	Abbreviation	Formula
1501	Acanthite	ACAN	Ag_2S
1044	Akermanite	AKER	$Ca_2MgSi_2O_7$
1092	Alabandite	ALA	MnS
1531	Albite	ALB	$NaAlSi_3O_8$
1560	Albite, High	HI-ALB	$NaAlSi_3O_8$
1025	Albite, Low	LO-ALB	$NaAlSi_3O_8$
1083	Alunite	ALUN	$KAl_3(OH)_6(SO_4)_2$
1125	Amorphous Silica	AMORPH SIL	$SiO_2 \cdot nH_2O$
1022	Analcime	ANAL	$NaAlSi_2O_6 \cdot H_2O$
1021	Analcime, Dehydrated	DEHYD-ANAL	$NaAlSi_2O_6$
1001	Andalusite	AND	Al_2SiO_5
1530	Andradite	ANDRA	$Ca_3Fe_2Si_3O_{12}$
1081	Anglesite	ANG	$PbSO_4$
1078	Anhydrite	ANHYD	$CaSO_4$
1015	Annite	ANNITE	$KFe_3(AlSi_3O_{10})(OH)_2$
1030	Anorthite	ANOR	$CaAl_2Si_2O_8$
1518	Anthophyllite	ANTH	$Mg_7Si_8O_{22}(OH)_2$
1542	Antigorite	ANTIG	$Mg_{48}Si_{34}O_{85}(OH)_{62}$
1072	Aragonite	ARAG	$CaCO_3$
1053	Artinite	ART	$Mg_2(OH)_2CO_3 \cdot 3H_2O$
1060	Azurite	AZUR	$Cu_3(OH)_2(CO_3)_2$
1080	Barite	BARITE	$BaSO_4$
1116	Boehmite	BOEH	$AlO(OH)$
1503	Bornite	BN	Cu_5FeS_4
1117	Brucite	BRUC	$Mg(OH)_2$
1040	Ca-Al Pyroxene	CA-AL PYX	$CaAl(AlSiO_6)$
1073	Calcite	CALC	$CaCO_3$
1082	Celestite	CELEST	$SrSO_4$
1068	Cerussite	CER	$PbCO_3$
1128	Chalcedony	CHALCED	SiO_2

Table 2 continued. Minerals, abbreviations, and formulas.

Index Number	Mineral Name	Abbreviation	Formula
1504	Chalcocite	CC	Cu_2S
1502	Chalcopyrite	CP	$CuFeS_2$
1007	Chrysotile	CHRYS	$Mg_3Si_2O_5(OH)_4$
1091	Cinnabar	CINN	HgS
1512	Clinochlore, 7-A	7-A CL	$Mg_5Al(AlSi_3O_{10})(OH)_8$
1513	Clinochlore, 14-A	14-A CL	$Mg_5Al(AlSi_3O_{10})(OH)_8$
1515	Clinozoisite	CLNZ	$Ca_2Al_3Si_3O_{12}(OH)$
1102	Copper, Native	COPPER	Cu
1065	Cordierite	ANH-CORD	$Mg_2Al_3(AlSi_5O_{18})$
1066	Cordierite, Hydrous	HYD-CORD	$Mg_2Al_3(AlSi_5O_{18}) \cdot H_2O$
1108	Corundum	COR	αAl_2O_3
1086	Covellite	COV	CuS
1126	Cristobalite, Alpha	A-CRIST	SiO_2
1127	Cristobalite, Beta	B-CRIST	SiO_2
1112	Cuprite	CUP	Cu_2O
1115	Diaspore	DIAS	$AlO(OH)$
1039	Diopside	DIOP	$CaMg(SiO_3)_2$
1075	Dolomite	DOL	$CaMg(CO_3)_2$
1070	Dolomite, Disordered	DIS-DOL	$CaMg(CO_3)_2$
1071	Dolomite, Ordered	ORD-DOL	$CaMg(CO_3)_2$
1537	Enstatite	ENST	$MgSiO_3$
1545	Epidote, Ordered	ORD-EP	$Ca_2FeAl_2Si_3O_{12}(OH)$
1559	Epidote	EPID	$Ca_2FeAl_2Si_3O_{12}(OH)$
1049	Fayalite	FAY	Fe_2SiO_4
1508	Ferrosilite	FERROSIL	$FeSiO_3$
1113	Ferrous Oxide	FE-OXIDE	FeO
1079	Fluorite	FLUORITE	CaF_2
1048	Forsterite	FORST	Mg_2SiO_4
1087	Galena	GALENA	PbS
1047	Gehlenite	GEHL	$Ca_2Al_2SiO_7$

Table 2 continued. Minerals, abbreviations, and formulas.

Index Number	Mineral Name	Abbreviation	Formula
1114	Gibbsite	GIBBS	$Al(OH)_3$
1100	Gold, Native	GOLD	Au
1103	Graphite	GRAPHITE	C
1529	Grossular	GROSS	$Ca_3Al_2Si_3O_{12}$
1106	Halite	HALITE	$NaCl$
1054	Hedenbergite	HED	$CaFe(SiO_3)_2$
1544	Hematite	HM	Fe_2O_3
1050	Huntite	HUNT	$CaMg_3(CO_3)_4$
1062	Hydromagnesite	HYDRO-MAG	$Mg_5(OH)_2(CO_3)_4 \cdot 4H_2O$
1067	Jadeite	JADEITE	$NaAl(SiO_3)_2$
1507	Kalsilite	KALS	$KAlSiO_4$
1004	Kaolinite	KAOL	$Al_2Si_2O_5(OH)_4$
1028	K-feldspar	K-SPAR	$KAlSi_3O_8$
1002	Kyanite	KYA	Al_2SiO_5
1132	Laumontite	LAUM	$Ca(Al_2Si_4O_{12}) \cdot 4H_2O$
1516	Lawsonite	LAWS	$CaAl_2Si_2O_7(OH)_2 \cdot H_2O$
1110	Lime	LIME	CaO
1074	Magnesite	MAG	$MgCO_3$
1506	Magnetite	MG	Fe_3O_4
1059	Malachite	MALACH	$Cu_2(OH)_2CO_3$
1119	Manganosite	MANGAN	MnO
1551	Margarite	MARG	$CaAl_2(Al_2Si_2O_{10})(OH)_2$
1045	Merwinite	MERW	$Ca_3Mg(SiO_4)_2$
1090	Metacinnabar	M-CINN	HgS
1027	Microcline, Maximum	MAX-MICRO	$KAlSi_3O_8$
1046	Monticellite	MONTI	$CaMgSiO_4$
1012	Muscovite	MUSC	$KAl_2(AlSi_3O_{10})(OH)_2$
1031	Nepheline	NEPH	$NaAlSiO_4$
1558	Nesquehonite	NESQ	$MgCO_3 \cdot 3H_2O$
1013	Paragonite	PARAG	$NaAl_2(AlSi_3O_{10})(OH)_2$

Table 2 continued. Minerals, abbreviations, and formulas.

Index Number	Mineral Name	Abbreviation	Formula
1538	Pargasite	PARG	$NaCa_2Mg_4Al(Al_2Si_6O_{22})(OH)_2$
1109	Periclase	PER	MgO
1014	Phlogopite	PHLOG	$KMg_3(AlSi_3O_{10})(OH)_2$
1123	Potassium Oxide	K-OXIDE	K_2O
1532	Prehnite	PREHN	$Ca_2Al(AlSi_3O_{10})(OH)_2$
1093	Pyrite	PYRITE	FeS_2
1509	Pyrophyllite	PYROPH	$Al_2Si_4O_{10}(OH)_2$
1555	Pyrrhotite	PO	FeS
1505	Quartz	QTZ	SiO_2
1552	Quicksilver	MERCURY	Hg
1085	Rhodochrosite	RHODO	$MnCO_3$
1029	Sanidine, High	HI-SAN	$KAlSi_3O_8$
1023	Sepiolite	SEPIO	$Mg_4Si_6O_{15}(OH)_2(OH_2)_2 \cdot (OH_2)_4$
1076	Siderite	SIDER	$FeCO_3$
1003	Sillimanite	SILL	Al_2SiO_5
1101	Silver, Native	SILVER	Ag
1064	Smithsonite	SMITH	$ZnCO_3$
1124	Sodium Oxide	NA-OXIDE	Na_2O
1088	Sphalerite	SL	ZnS
1120	Spinel	SPINEL	$MgAl_2O_4$
1069	Strontianite	STRONT	$SrCO_3$
1107	Sylvite	SYLVITE	KCl
1510	Talc	TALC	$Mg_3Si_4O_{10}(OH)_2$
1111	Tenorite	TO	CuO
1517	Tremolite	TREM	$Ca_2Mg_5Si_8O_{22}(OH)_2$
1130	Wairakite	WAIR	$Ca(Al_2Si_4O_{12}) \cdot 2H_2O$
1084	Witherite	WITH	$BaCO_3$
1035	Wollastonite	WOLL	$CaSiO_3$
1089	Wurtzite	WURT	ZnS
1519	Zoisite	ZOIS	$Ca_2Al_3Si_3O_{12}(OH)$

Table 3. Summary of hydrolysis reactions.

Index	Mineral	Reaction
1001	AND	$Al_2SiO_5 + 6H^+ = 2Al^{+++} + SiO_{2(aq)} + 3H_2O$
1002	KYA	$Al_2SiO_5 + 6H^+ = 2Al^{+++} + SiO_{2(aq)} + 3H_2O$
1003	SILL	$Al_2SiO_5 + 6H^+ = 2Al^{+++} + SiO_{2(aq)} + 3H_2O$
1004	KAOL	$Al_2Si_2O_5(OH)_4 + 6H^+ = 2Al^{+++} + 2SiO_{2(aq)} + 5H_2O$
1007	CHRYS	$Mg_3Si_2O_5(OH)_4 + 6H^+ = 3Mg^{++} + 2SiO_{2(aq)} + 5H_2O$
1012	MUSC	$KAl_2(AlSi_3O_{10})(OH)_2 + 10H^+ = K^+ + 3Al^{+++} + 3SiO_{2(aq)} + 6H_2O$
1013	PARAG	$NaAl_2(AlSi_3O_{10})(OH)_2 + 10H^+ = Na^+ + 3Al^{+++} + 3SiO_{2(aq)} + 6H_2O$
1014	PHLOG	$KMg_3(AlSi_3O_{10})(OH)_2 + 10H^+ = K^+ + 3Mg^{++} + Al^{+++} + 3SiO_{2(aq)} + 6H_2O$
1015	ANNITE	$KFe_3(AlSi_3O_{10})(OH)_2 + 10H^+ = K^+ + 3Fe^{++} + Al^{+++} + 3SiO_{2(aq)} + 6H_2O$
1021	DEHYD.-ANAL	$NaAlSi_2O_6 + 4H^+ = Na^+ + Al^{+++} + 2SiO_{2(aq)} + 2H_2O$
1022	ANAL	$NaAlSi_2O_6 \cdot H_2O + 4H^+ = Na^+ + Al^{+++} + 2SiO_{2(aq)} + 3H_2O$
1023	SEPIO	$Mg_4Si_6O_{15}(OH)_2(OH_2)_2 \cdot (OH_2)_4 + 8H^+ = 4Mg^{++} + 6SiO_{2(aq)} + 11H_2O$
1025	LO-ALB	$NaAlSi_3O_8 + 4H^+ = Na^+ + Al^{+++} + 3SiO_{2(aq)} + 2H_2O$
1027	MAX-MIC	$KAlSi_3O_8 + 4H^+ = K^+ + Al^{+++} + 3SiO_{2(aq)} + 2H_2O$
1028	K-SPAR	$KAlSi_3O_8 + 4H^+ = K^+ + Al^{+++} + 3SiO_{2(aq)} + 2H_2O$
1029	HI SAN	$KAlSi_3O_8 + 4H^+ = K^+ + Al^{+++} + 3SiO_{2(aq)} + 2H_2O$
1030	ANOR	$CaAl_2Si_2O_8 + 8H^+ = Ca^{++} + 2Al^{+++} + 2SiO_{2(aq)} + 4H_2O$
1031	NEPH	$NaAlSiO_4 + 4H^+ = Na^+ + Al^{+++} + SiO_{2(aq)} + 2H_2O$
1035	WOLL	$CaSiO_3 + 2H^+ = Ca^{++} + SiO_{2(aq)} + H_2O$
1039	DIOP	$CaMg(SiO_3)_2 + 4H^+ = Ca^{++} + Mg^{++} + 2SiO_{2(aq)} + 2H_2O$
1040	CA-AL PYX	$CaAl(AlSiO_6) + 8H^+ = Ca^{++} + 2Al^{+++} + SiO_{2(aq)} + 4H_2O$
1044	AKER	$Ca_2MgSi_2O_7 + 6H^+ = 2Ca^{++} + Mg^{++} + 2SiO_{2(aq)} + 3H_2O$
1045	MERW	$Ca_3Mg(SiO_4)_2 + 8H^+ = 3Ca^{++} + Mg^{++} + 2SiO_{2(aq)} + 4H_2O$
1046	MONTI	$CaMgSiO_4 + 4H^+ = Ca^{++} + Mg^{++} + SiO_{2(aq)} + 2H_2O$
1047	GEHL	$Ca_2Al_2SiO_7 + 10H^+ = 2Ca^{++} + 2Al^{+++} + SiO_{2(aq)} + 5H_2O$
1048	FORST	$Mg_2SiO_4 + 4H^+ = 2Mg^{++} + SiO_{2(aq)} + 2H_2O$
1049	FAY	$Fe_2SiO_4 + 4H^+ = 2Fe^{++} + SiO_{2(aq)} + 2H_2O$
1050	HUNT	$CaMg_3(CO_3)_4 + 4H^+ = Ca^{++} + 3Mg^{++} + 4HCO_3^-$
1053	ART	$Mg_2(OH)_2CO_3 \cdot 3H_2O + 3H^+ = 2Mg^{++} + HCO_3^- + 5H_2O$
1054	HED	$CaFe(SiO_3)_2 + 4H^+ = Ca^{++} + Fe^{++} + 2SiO_{2(aq)} + 2H_2O$
1059	MALACH	$Cu_2(OH)_2CO_3 + 3/4H^+ + 1/4HS^- = 2Cu^+ + HCO_3^- + H_2O + 1/4SO_4^=$

Table 3 continued. Summary of hydrolysis reactions.

Index	Mineral	Reaction
1060	AZUR	$Cu_3(OH)_2(CO_3)_2 + 5/8H^+ + 3/8HS^- = 3Cu^+ + 2HCO_3^- + 1/2H_2O + 3/8SO_4^=$
1062	HYDRO-MAG	$Mg_5(OH)_2(CO_3)_4 \cdot 4H_2O + 6H^+ = 5Mg^{++} + 4HCO_3^- + 6H_2O$
1064	SMITH	$ZnCO_3 + H^+ = Zn^{++} + HCO_3^-$
1065	ANH-CORD	$Mg_2Al_3(AlSi_5O_{18}) + 16H^+ = 2Mg^{++} + 4Al^{+++} + 5SiO_{2(aq)} + 8H_2O$
1066	HYD-CORD	$Mg_2Al_3(AlSi_5O_{18}) \cdot H_2O + 16H^+ = 2Mg^{++} + 4Al^{+++} + 5SiO_{2(aq)} + 9H_2O$
1067	JADEITE	$NaAl(SiO_3)_2 + 4H^+ = Na^+ + Al^{+++} + 2SiO_{2(aq)} + 2H_2O$
1068	CER	$PbCO_3 + H^+ = Pb^{++} + HCO_3^-$
1069	STRONT	$SrCO_3 + H^+ = Sr^{++} + HCO_3^-$
1070	DIS-DOL	$CaMg(CO_3)_2 + 2H^+ = Ca^{++} + Mg^{++} + 2HCO_3^-$
1071	ORD-DOL	$CaMg(CO_3)_2 + 2H^+ = Ca^{++} + Mg^{++} + 2HCO_3^-$
1072	ARAG	$CaCO_3 + H^+ = Ca^{++} + HCO_3^-$
1073	CALC	$CaCO_3 + H^+ = Ca^{++} + HCO_3^-$
1074	MAG	$MgCO_3 + H^+ = Mg^{++} + HCO_3^-$
1075	DOL	$CaMg(CO_3)_2 + 2H^+ = Ca^{++} + Mg^{++} + 2HCO_3^-$
1076	SIDER	$FeCO_3 + H^+ = Fe^{+++} + HCO_3^-$
1078	ANHYD	$CaSO_4 = Ca^{++} + SO_4^=$
1079	FLUORITE	$CaF_2 = Ca^{++} + 2F^-$
1080	BARITE	$BaSO_4 = Ba^{++} + SO_4^=$
1081	ANG	$PbSO_4 = Pb^{++} + SO_4^=$
1082	CELEST	$SrSO_4 = Sr^{++} + SO_4^=$
1083	ALUN	$KAl_3(OH)_6(SO_4)_2 + 6H^+ = K^+ + 3Al^{+++} + 2SO_4^= + 6H_2O$
1084	WITH	$BaCO_3 + H^+ = Ba^{++} + HCO_3^-$
1085	RHODO	$MnCO_3 + H^+ = Mn^{++} + HCO_3^-$
1086	COV	$CuS + 1/2H_2O = Cu^+ + 1/8SO_4^= + 7/8HS^- + 1/8H^+$
1087	GALENA	$PbS + H^+ = Pb^{++} + HS^-$
1088	SL	$ZnS + H^+ = Zn^{++} + HS^-$
1089	WURT	$ZnS + H^+ = Zn^{++} + HS^-$
1090	M-CINN	$HgS + H^+ = Hg^{++} + HS^-$
1091	CINN	$HgS + H^+ = Hg^{++} + HS^-$
1092	ALA	$MnS + H^+ = Mn^{++} + HS^-$
1093	PYRITE	$FeS_2 + H_2O = Fe^{+++} + 1/4SO_4^= + 7/4HS^- + 1/4H^+$

XXXVII

Table 3 continued. Summary of hydrolysis reactions.

Index	Mineral	Reaction
1100	GOLD	$Au + 9/8H^+ + 1/8SO_4^= = Au^+ + 1/2H_2O + 1/8HS^-$
1101	SILVER	$Ag + 9/8H^+ + 1/8SO_4^= = Ag^+ + 1/2H_2O + 1/8HS^-$
1102	COPPER	$Cu + 9/8H^+ + 1/8SO_4^= = Cu^+ + 1/2H_2O + 1/8HS^-$
1103	GRAPHITE	$C + H_2O + 1/2SO_4^= = HCO_3^- + 1/2HS^- + 1/2H^+$
1106	HALITE	$NaCl = Na^+ + Cl^-$
1107	SYLVITE	$KCl = K^+ + Cl^-$
1108	COR	$\alpha Al_2O_3 + 6H^+ = 2Al^{+++} + 3H_2O$
1109	PER	$MgO + 2H^+ = Mg^{++} + H_2O$
1110	LIME	$CaO + 2H^+ = Ca^{++} + H_2O$
1111	TO	$CuO + 7/8H^+ + 1/8HS^- = Cu^+ + 1/8SO_4^= + 1/2H_2O$
1112	CUP	$Cu_2O + 2H^+ = 2Cu^+ + H_2O$
1113	FE-OXIDE	$FeO + 2H^+ = Fe^{++} + H_2O$
1114	GIBBS	$Al(OH)_3 + 3H^+ = Al^{+++} + 3H_2O$
1115	DIAS	$AlO(OH) + 3H^+ = Al^{+++} + 2H_2O$
1116	BOEH	$AlO(OH) + 3H^+ = Al^{+++} + 2H_2O$
1117	BRUC	$Mg(OH)_2 + 2H^+ = Mg^{++} + 2H_2O$
1119	MANGAN	$MnO + 2H^+ = Mn^{++} + H_2O$
1120	SPINEL	$MgAl_2O_4 + 8H^+ = Mg^{++} + 2Al^{+++} + 4H_2O$
1123	K-OXIDE	$K_2O + 2H^+ = 2K^+ + H_2O$
1124	NA-OXIDE	$Na_2O + 2H^+ = 2Na^+ + H_2O$
1125	AMORPH SIL	$SiO_2 = SiO_{2(aq)}$
1126	A-CRIST	$SiO_2 = SiO_{2(aq)}$
1127	B-CRIST	$SiO_2 = SiO_{2(aq)}$
1128	CHALCED	$SiO_2 = SiO_{2(aq)}$
1130	WAIR	$Ca(Al_2Si_4O_{12}) \cdot 2H_2O + 8H^+ = Ca^{++} + 2Al^{+++} + 4SiO_{2(aq)} + 6H_2O$
1132	LAUM	$Ca(Al_2Si_4O_{12}) \cdot 4H_2O + 8H^+ = Ca^{++} + 2Al^{+++} + 4SiO_{2(aq)} + 8H_2O$
1501	ACAN	$Ag_2S + H^+ = 2Ag^+ + HS^-$
1502	CP	$CuFeS_2 + 7/8H^+ + 1/2H_2O = Cu^+ + Fe^{++} + 15/8HS^- + 1/8SO_4^=$
1503	BN	$Cu_5FeS_4 + 23/8H^+ + 1/2H_2O = 5Cu^+ + Fe^{++} + 31/8HS^- + 1/8SO_4^=$
1504	CC	$Cu_2S + H^+ = 2Cu^+ + HS^-$
1505	QTZ	$SiO_2 = SiO_{2(aq)}$

Table 3 continued. Summary of hydrolysis reactions.

Index	Mineral	Reaction
1506	MG	$Fe_3O_4 + 1/4HS^- + 23/4H^+ = 3Fe^{++} + 1/4SO_4^= + 3H_2O$
1507	KALS	$KAlSiO_4 + 4H^+ = K^+ + Al^{+++} + SiO_{2(aq)} + 2H_2O$
1508	FERROSIL	$FeSiO_3 + 2H^+ = Fe^{++} + SiO_{2(aq)} + H_2O$
1509	PYROPH	$Al_2Si_4O_{10}(OH)_2 + 6H^+ = 2Al^{+++} + 4SiO_{2(aq)} + 4H_2O$
1510	TALC	$Mg_3Si_4O_{10}(OH)_2 + 6H^+ = 3Mg^{++} + 4SiO_{2(aq)} + 4H_2O$
1512	7-A CL	$Mg_5Al(AlSi_3O_{10})(OH)_8 + 16H^+ = 5Mg^{++} + 2Al^{+++} + 3SiO_{2(aq)} + 12H_2O$
1513	14-A CL	$Mg_5Al(AlSi_3O_{10})(OH)_8 + 16H^+ = 5Mg^{++} + 2Al^{+++} + 3SiO_{2(aq)} + 12H_2O$
1515	CLNZ	$Ca_2Al_3Si_3O_{12}(OH) + 13H^+ = 2Ca^{++} + 3Al^{+++} + 3SiO_{2(aq)} + 7H_2O$
1516	LAWS	$CaAl_2Si_2O_7(OH)_2 \cdot H_2O + 8H^+ = Ca^{++} + 2Al^{+++} + 2SiO_{2(aq)} + 6H_2O$
1517	TREM	$Ca_2Mg_5Si_8O_{22}(OH)_2 + 14H^+ = 2Ca^{++} + 5Mg^{++} + 8SiO_{2(aq)} + 8H_2O$
1518	ANTH	$Mg_7Si_8O_{22}(OH)_2 + 14H^+ = 7Mg^{++} + 8SiO_{2(aq)} + 8H_2O$
1519	ZOIS	$Ca_2Al_3Si_3O_{12}(OH) + 13H^+ = 2Ca^{++} + 3Al^{+++} + 3SiO_{2(aq)} + 7H_2O$
1529	GROSS	$Ca_3Al_2Si_3O_{12} + 12H^+ = 3Ca^{++} + 2Al^{+++} + 3SiO_{2(aq)} + 6H_2O$
1530	ANDRA	$Ca_3Fe_2Si_3O_{12} + 1/4HS^- + 39/4H^+ =$ $3Ca^{++} + 2Fe^{++} + 3SiO_{2(aq)} + 1/4SO_4^= + 5H_2O$
1531	ALB	$NaAlSi_3O_8 + 4H^+ = Na^+ + Al^{+++} + 3SiO_{2(aq)} + 2H_2O$
1532	PREHN	$Ca_2Al(AlSi_3O_{10})(OH)_2 + 10H^+ = 2Ca^{++} + 2Al^{+++} + 3SiO_{2(aq)} + 6H_2O$
1537	ENST	$MgSiO_3 + 2H^+ = Mg^{++} + SiO_{2(aq)} + H_2O$
1538	PARG	$NaCa_2Mg_4Al(Al_2Si_6O_{22})(OH)_2 + 22H^+ =$ $Na^+ + 2Ca^{++} + 4Mg^{++} + 3Al^{+++} + 6SiO_{2(aq)} + 12H_2O$
1542	ANTIG	$Mg_{48}Si_{34}O_{85}(OH)_{62} + 96H^+ = 48Mg^{++} + 34SiO_{2(aq)} + 79H_2O$
1544	HM	$Fe_2O_3 + 15/4H^+ + 1/4HS^- = 2Fe^{++} + 1/4SO_4^= + 2H_2O$
1545	ORD-EP	$Ca_2FeAl_2Si_3O_{12}(OH) + 1/8HS^- + 95/8H^+ =$ $2Ca^{++} + Fe^{++} + 2Al^{+++} + 3SiO_{2(aq)} + 14H^+ = Ca^{++} + 4Al^{+++} + 2SiO_{2(aq)} + 8H_2O$
1551	MARG	$CaAl_2(Al_2Si_2O_{10})(OH)_2 + 14H^+ = Ca^{++} + 4Al^{+++} + 2SiO_{2(aq)} + 8H_2O$
1552	MERCURY	$Hg + 9/4H^+ + 1/4SO_4^= = Hg^+ + 1/4HS^- + H_2O$
1555	PO	$FeS + H^+ = Fe^{++} + HS^-$
1558	NESQ	$MgCO_3 \cdot 3H_2O + H^+ = Mg^{++} + HCO_3^- + 3H_2O$
1559	EPID	$Ca_2FeAl_2Si_3O_{12}(OH) + 1/8HS^- + 95/8H^+ =$ $2Ca^{++} + Fe^{++} + 2Al^{+++} + 3SiO_{2(aq)} + 1/8SO_4^= + 13/2H_2O$
1560	HI ALB	$NaAlSi_3O_8 + 4H^+ = Na^+ + Al^{+++} + 3SiO_{2(aq)} + 2H_2O$

REFERENCES

Aagaard, P., and Helgeson, H.C., 1983, Activity/composition relations among silicates and aqueous solutions: II. Chemical and thermodynamic consequences of ideal mixing of atoms on homological sites in montmorillonites, illites, and mixed-layer clays: Clays and Clay Minerals (in press).

Bird, D.K., and Helgeson, H.C., 1980, Chemical interaction of aqueous solutions with epidote-feldspar mineral assemblages in geologic systems: I. Thermodynamic analysis of phase relations in the system $CaO-FeO-Fe_2O_3-Al_2O_3-SiO_2-H_2O-CO_2$: Am. Jour. Sci., v. 280, p. 907-941.

Bird, D.K., and Helgeson, H.C., 1981, Chemical interaction of aqueous solutions with epidote-feldspar mineral assemblages in geologic systems: II. Equilibrium constraints in metamorphic/geothermal processes: Am. Jour. Sci., v. 281, p. 576-614.

Delany, J.M., and Helgeson, H.C., 1978, Calculation of the thermodynamic consequences of dehydration in subducting oceanic crust to 100 kb and $>$ 800°C: Am. Jour. Sci., v. 278, p. 638-686.

Flowers, G.C., 1979, Correction of Holloway's (1977) adaptation of the modified Redlich-Kwong equation of state for calculation of the fugacities of molecular species in supercritical fluids of geologic interest: Contr. Min. Pet., v. 69, p. 315-318.

Flowers, G.C., and Helgeson, H.C., 1983, Equilibrium and mass transfer during progressive metamorphism of siliceous dolomites: Am. Jour. Sci., v. 283, p. 230-286.

Franck, E.U., 1982, Survey of selected non-thermodynamic properties and chemical phenomena of fluids and fluid mixtures: in D. Rickard and F.E. Wickman, eds., Chemistry and Geochemistry of Solutions at High Temperatures and Pressures: Physics and Chemistry of the Earth, v. 13 and 14, Oxford, Pergamon Press, p. 65-88.

Frisch, C., and Helgeson, H.C., 1983, Metasomatic phase relations in dolomites of the Adamello Alps: Am. Jour. Sci., v. 283 (in press).

Garrels, R.M., and Christ, C.L., 1965, Solutions, Minerals, and Equilibria: New York, Harper and Row, 450 p.

Helgeson, H.C., 1968, Evaluation of irreversible reactions in geochemical processes involving minerals and aqueous solutions--I. Thermodynamic relations: Geochim. et Cosmochim. Acta, v. 32, p. 853-877.

Helgeson, H.C., 1969, Thermodynamics of hydrothermal systems at elevated temperatures and pressures: Am. Jour. Sci., v. 267, p. 729-804.

Helgeson, H.C., 1970a, Description and interpretation of phase relations in geochemical processes involving aqueous solutions: Am. Jour. Sci., v. 268, p. 415-438.

Helgeson, H.C., 1970b, A chemical and thermodynamic model of ore deposition in hydrothermal systems: in, B.A. Morgan, ed., Fiftieth Anniversary Symposia: Mineral. Soc. Am. Spec. Paper 3, p. 155-186.

Helgeson, H.C., 1979, Mass transfer among minerals and hydrothermal solutions: in, H.L. Barnes, ed., Geochemistry of Hydrothermal Ore Deposits, 2nd ed., New York, John Wiley and Sons, p. 568-610.

Helgeson, H.C., 1982a, Prediction of the thermodynamic properties of electrolytes at high temperatures: in, D. Rickard and F.E. Wickman, eds., Chemistry and Geochemistry of Solutions at High Temperatures and Presures: Physics and Chemistry of the Earth, v. 13 and 14, p. 133-177.

Helgeson, H.C., 1982b, Errata: Thermodynamics of minerals, reactions, and aqueous solutions at high pressures and temperatures: Am. Jour. Sci., v. 282, p. 1143-1149.

Helgeson, H.C., 1984, Errata II: Thermodynamics of minerals, reactions, and aqueous solutions at high pressures and temperatures: Am. Jour. Sci., v. 284 (in press).

Helgeson, H.C., and Aagaard, P., 1984, Activity/composition relations among silicates and aqueous solutions. I. Thermodynamics of intrasite mixing and substitutional order/disorder in minerals: Am. Jour. Sci., v. 284 (in press).

Helgeson, H.C., Brown, T.H., and Leeper, R.H., 1969, Handbook of Theoretical Activity Diagrams Depicting Chemical Equilibria in Geologic Systems Involving an Aqueous Phase at One Atm and 0° to 300°C: San Francisco, Freeman, Cooper and Co., 253 p.

Helgeson, H.C., Brown, T.H., Nigrini, A., and Jones, T.A., 1970, Calculation of mass transfer in geochemical processes involving aqueous solutions: Geochim. et Cosmochim. Acta, v. 34, p. 569-592.

Helgeson, H.C., Delany, J.M., Nesbitt, H.W., and Bird, D.K., 1978, Summary and critique of the thermodynamic properties of rock-forming minerals: Am. Jour. Sci., v. 278-A, p. 1-229.

Helgeson, H.C., Garrels, R.M., and MacKenzie, F.T., 1969, Evaluation of irreversible reactions in geochemical processes involving minerals and aqueous solutions--II. Applications: Geochim. et Cosmochim. Acta, v. 33, p. 455-481.

Helgeson, H.C., and Kirkham, D.H., 1974a, Theoretical prediction of the thermodynamic behavior of aqueous electrolytes at high pressures and temperatures: I. Summary of the thermodynamic/electrostatic properties of the solvent: Am. Jour. Sci., v. 274, p. 1089-1198.

Helgeson, H.C., and Kirkham, D.H., 1974b, Theoretical prediction of the thermodynamic behavior of aqueous electrolytes at high pressures and temperatures: II. Debye-Hückel parameters for activity coefficients and relative partial molal properties: Am. Jour. Sci., v. 274, p. 1199-1261.

Helgeson, H.C., and Kirkham, D.H., 1976, Theoretical prediction of the thermodynamic behavior of aqueous electrolytes at high pressures and temperatures: III. Equation of state for aqueous species at infinite dilution: Am. Jour. Sci., v. 276, p. 97-240.

Helgeson, H.C., Kirkham, D.H., and Flowers, G.C., 1981, Theoretical prediction of the thermodynamic behavior of aqueous electrolytes at high pressures and temperatures: IV. Calculation of activity and osmotic coefficients and apparent molal and standard and relative partial molal properties to 600°C and 5 kb: Am. Jour. Sci., v. 281, p. 1249-1493.

Helgeson, H.C., and Murphy, W.M., 1983, Calculation of mass transfer among minerals and aqueous solutions as a function of time and surface area in geochemical processes. I. Computational approach: Mathematical Geology, v. 15, p. 109-130.

Quist, A.S., and Marshall, W.L., 1968, Electrical conductances of aqueous sodium chloride solutions from 0-800° and at pressures to 4000 bars: Jour. Phys. Chem., v. 72, p. 684-703.

Walther, J.V., and Helgeson, H.C., 1977, Calculation of the thermodynamic properties of aqueous silica and the solubility of quartz and its polymorphs at high pressures and temperatures: Am. Jour. Sci., v. 277, p. 1315-1351.

Walther, J.V., and Helgeson, H.C., 1980, Description and interpretation of metasomatic phase relations at high pressures and temperatures: I. Equilibrium activities of ionic species in nonideal mixtures of CO_2 and H_2O: Am. Jour. Sci., v. 280, p. 575-606.

ACTIVITY DIAGRAMS

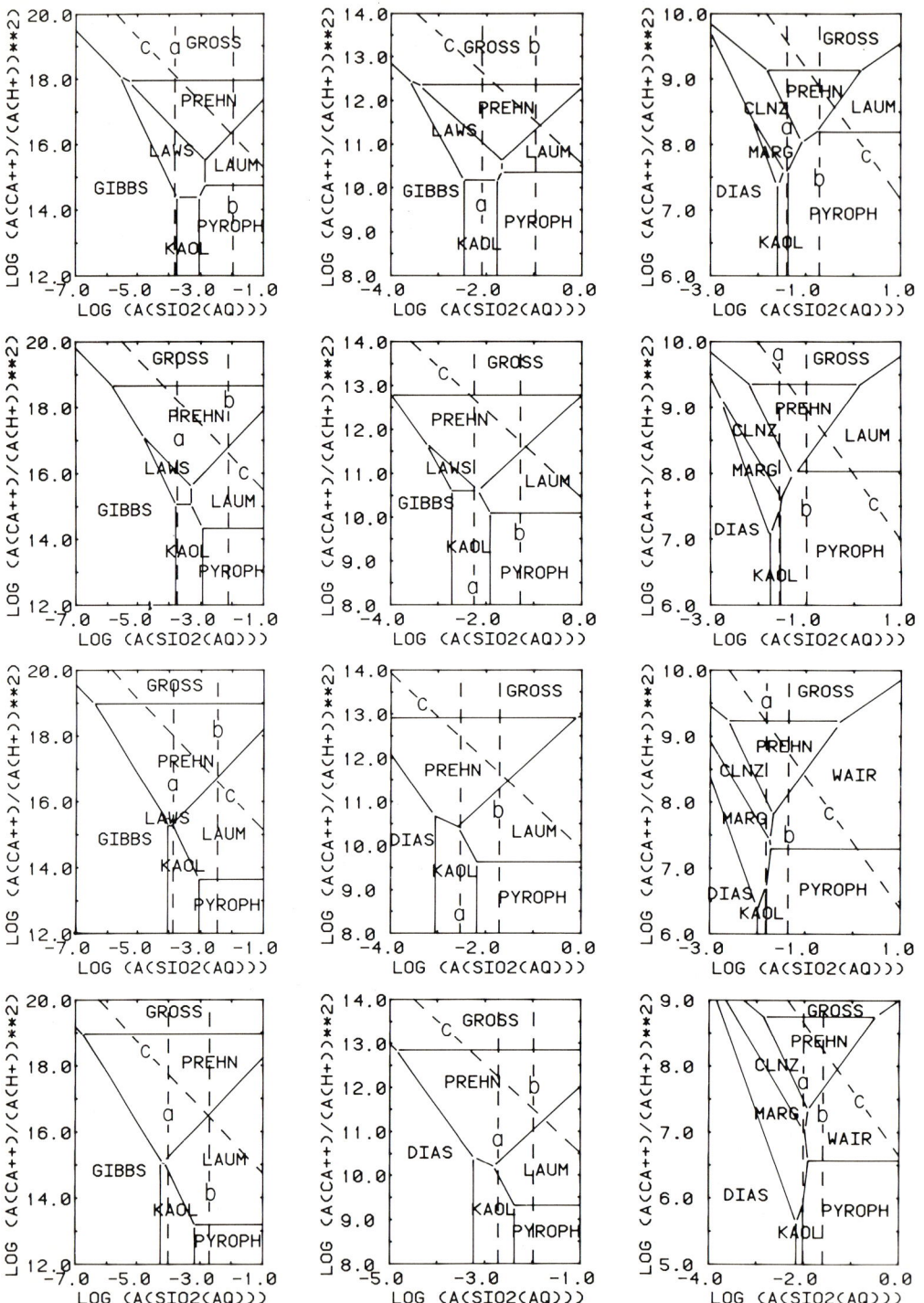

Phase relations in the system $HCl-H_2O-(Al_2O_3)-CaO-SiO_2$. Saturation limits: quartz (a), amorphous silica (b), wollastonite (c).

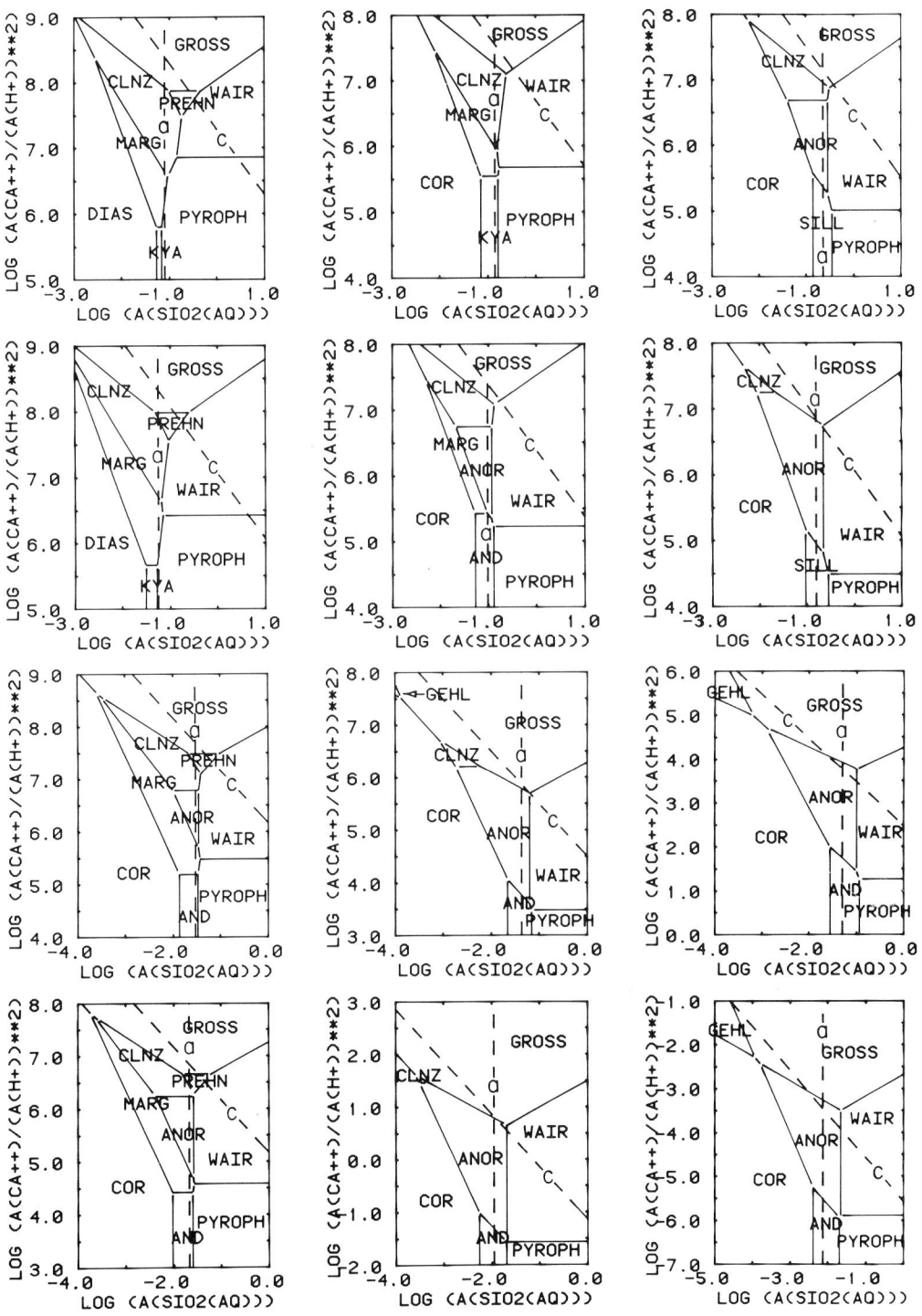

Phase relations in the system $HCl-H_2O-(Al_2O_3)-CaO-SiO_2$. Saturation limits: quartz (a), amorphous silica (b), wollastonite (c).

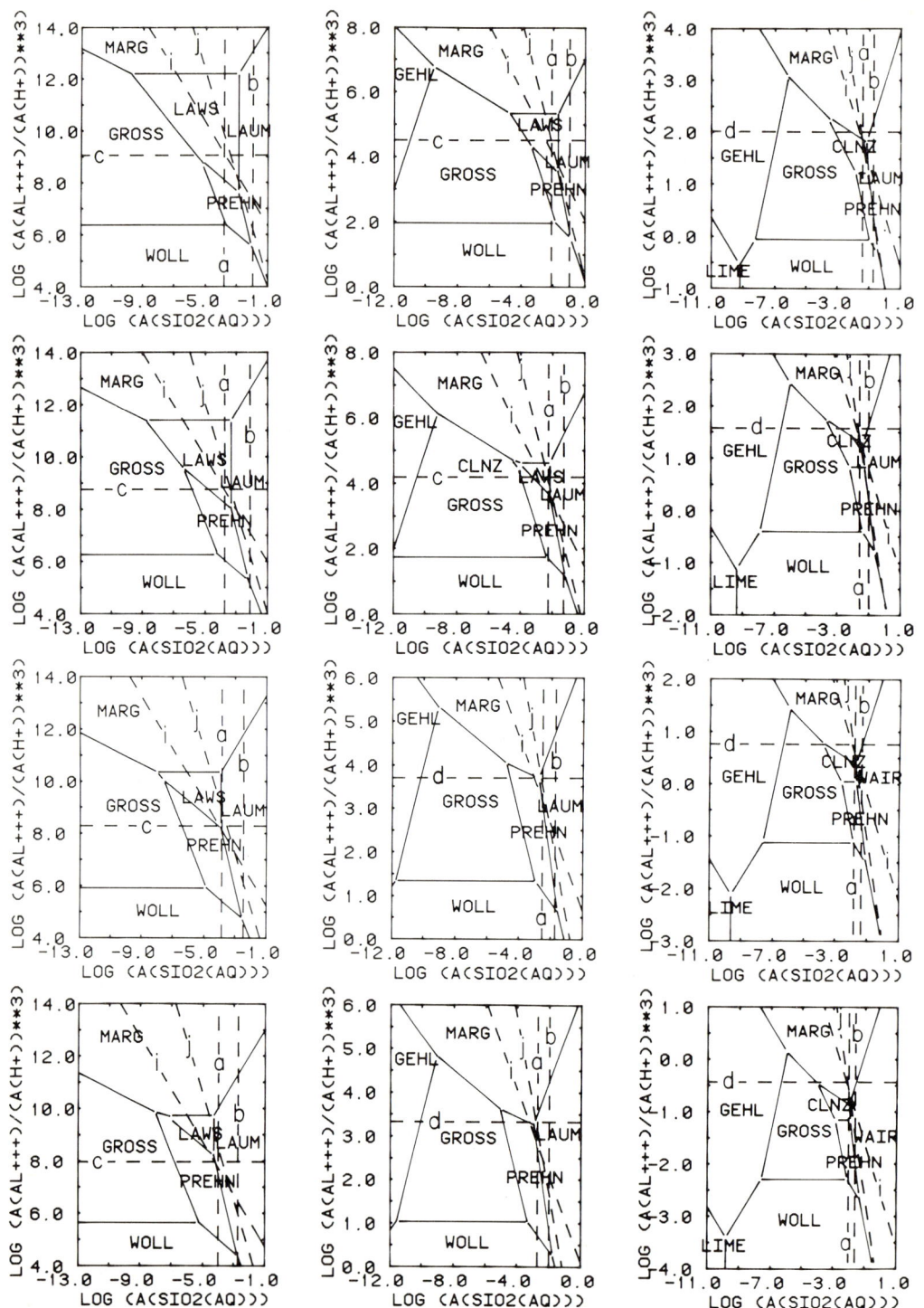

Phase relations in the system $HCl-H_2O-Al_2O_3-(CaO)-SiO_2$. Saturation limits: quartz (a), amorphous silica (b), gibbsite (c), diaspore (d), corundum (e), andalusite (f), kyanite (g), sillimanite (h), kaolinite (i), pyrophyllite (j).

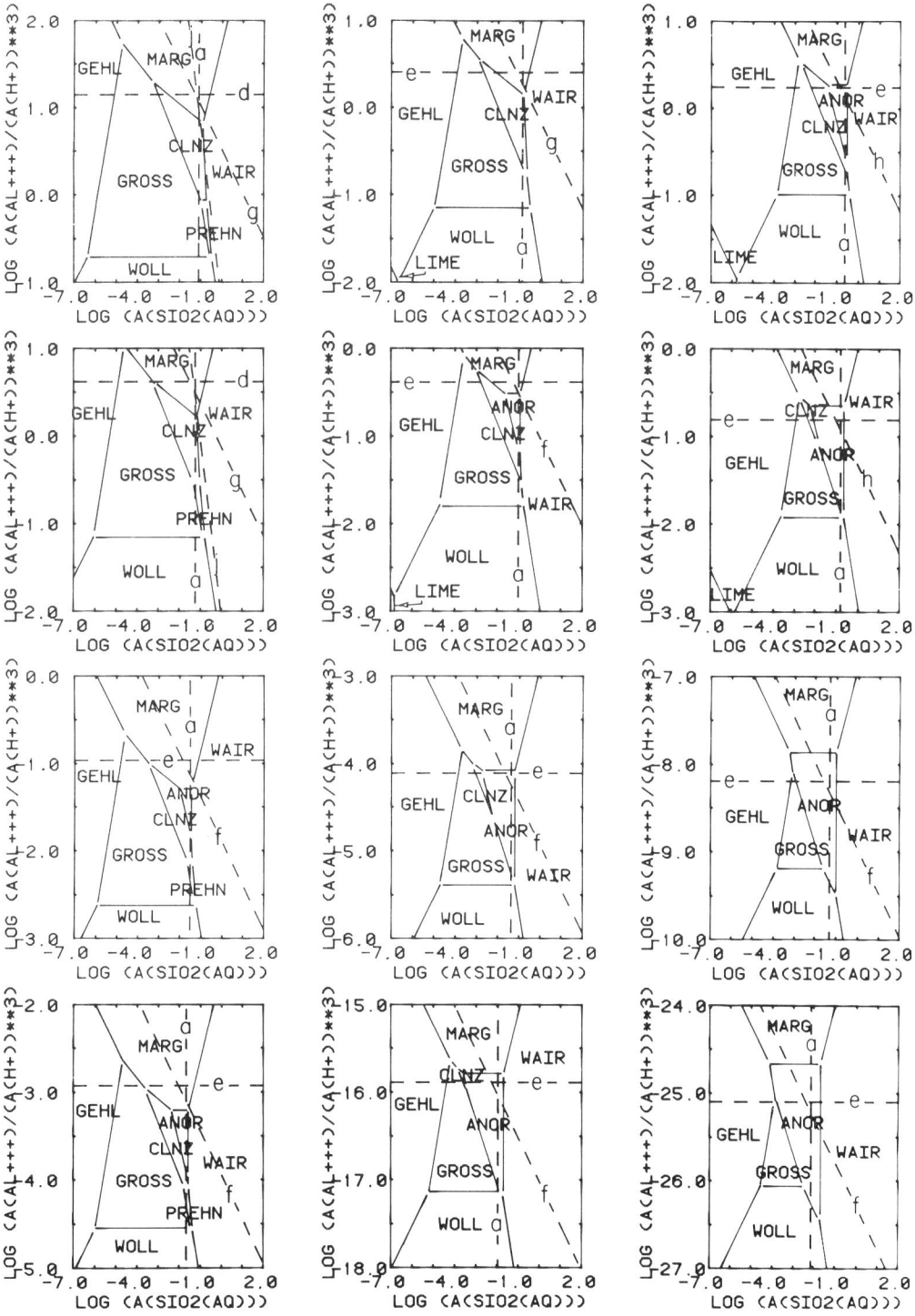

Phase relations in the system HCl−H$_2$O−Al$_2$O$_3$−(CaO)−SiO$_2$. Saturation limits: quartz (a), amorphous silica (b), gibbsite (c), diaspore (d), corundum (e), andalusite (f), kyanite (g), sillimanite (h), kaolinite (i), pyrophyllite (j).

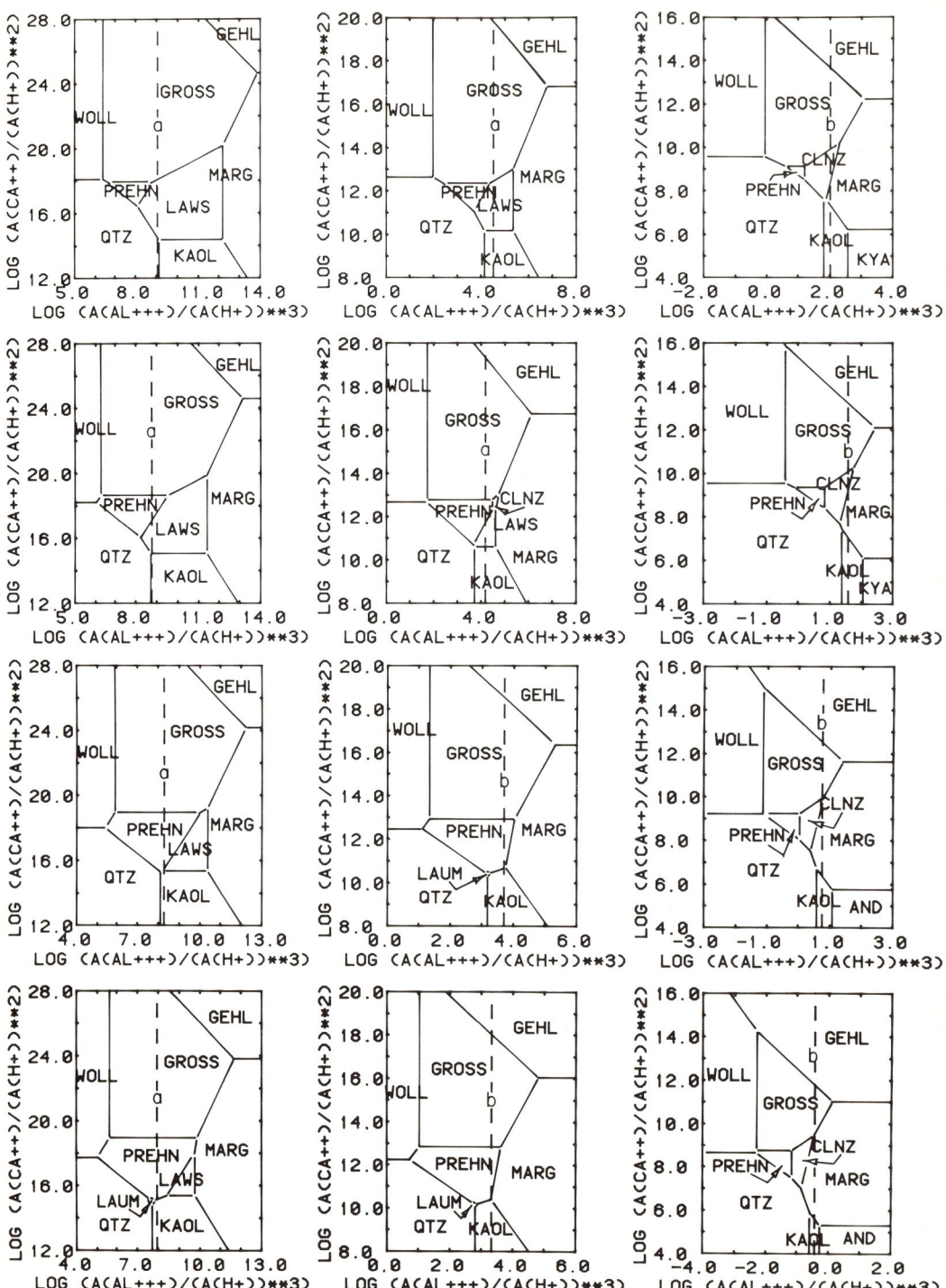

Phase relations in the system $HCl-H_2O-Al_2O_3-CaO-(SiO_2)$. Saturation limits: diaspore (a), gibbsite (b), corundum (c).

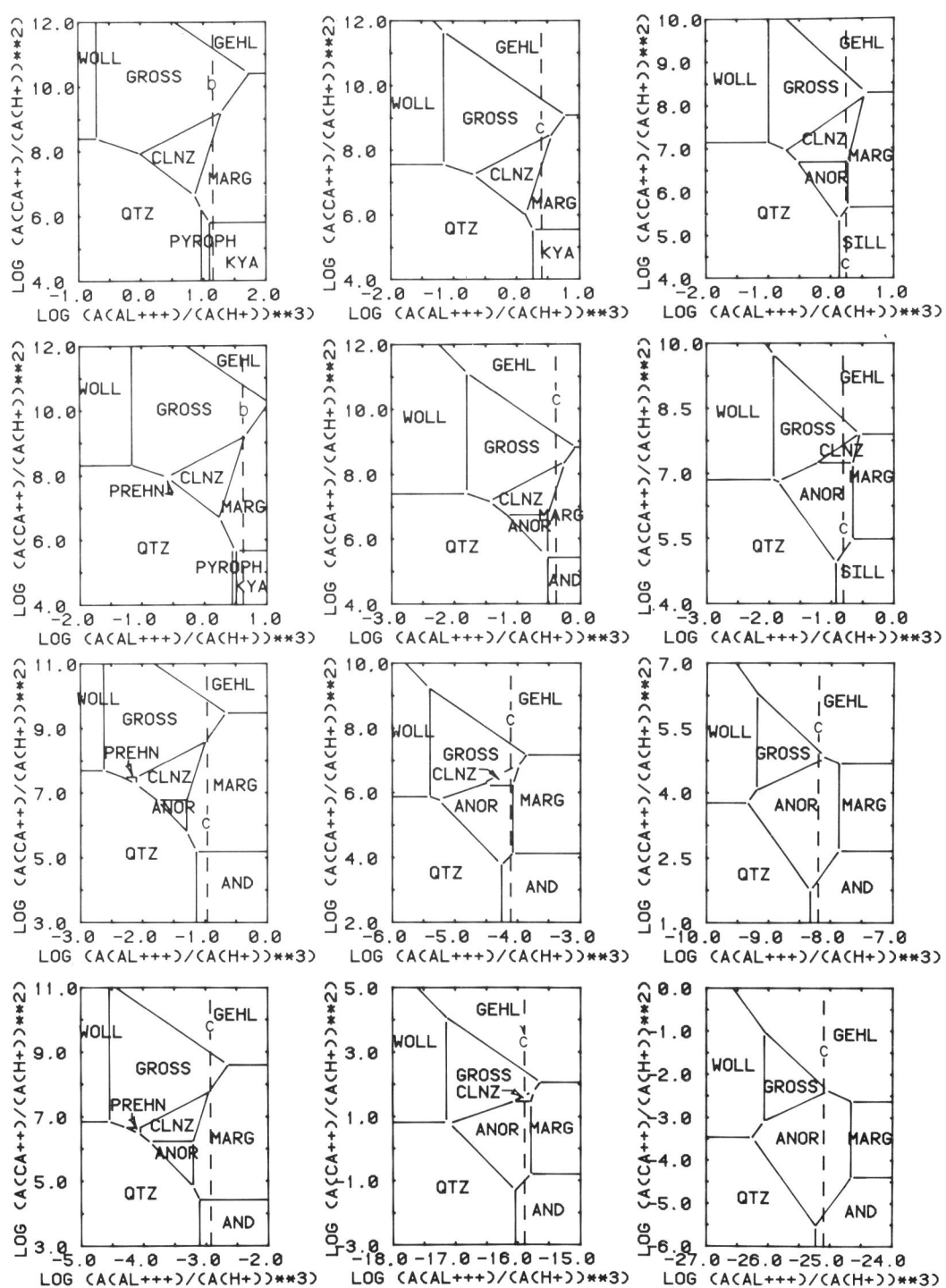

Phase relations in the system $HCl-H_2O-Al_2O_3-CaO-(SiO_2)$. Saturation limits: gibbsite (a), diaspore (b), corundum (c).

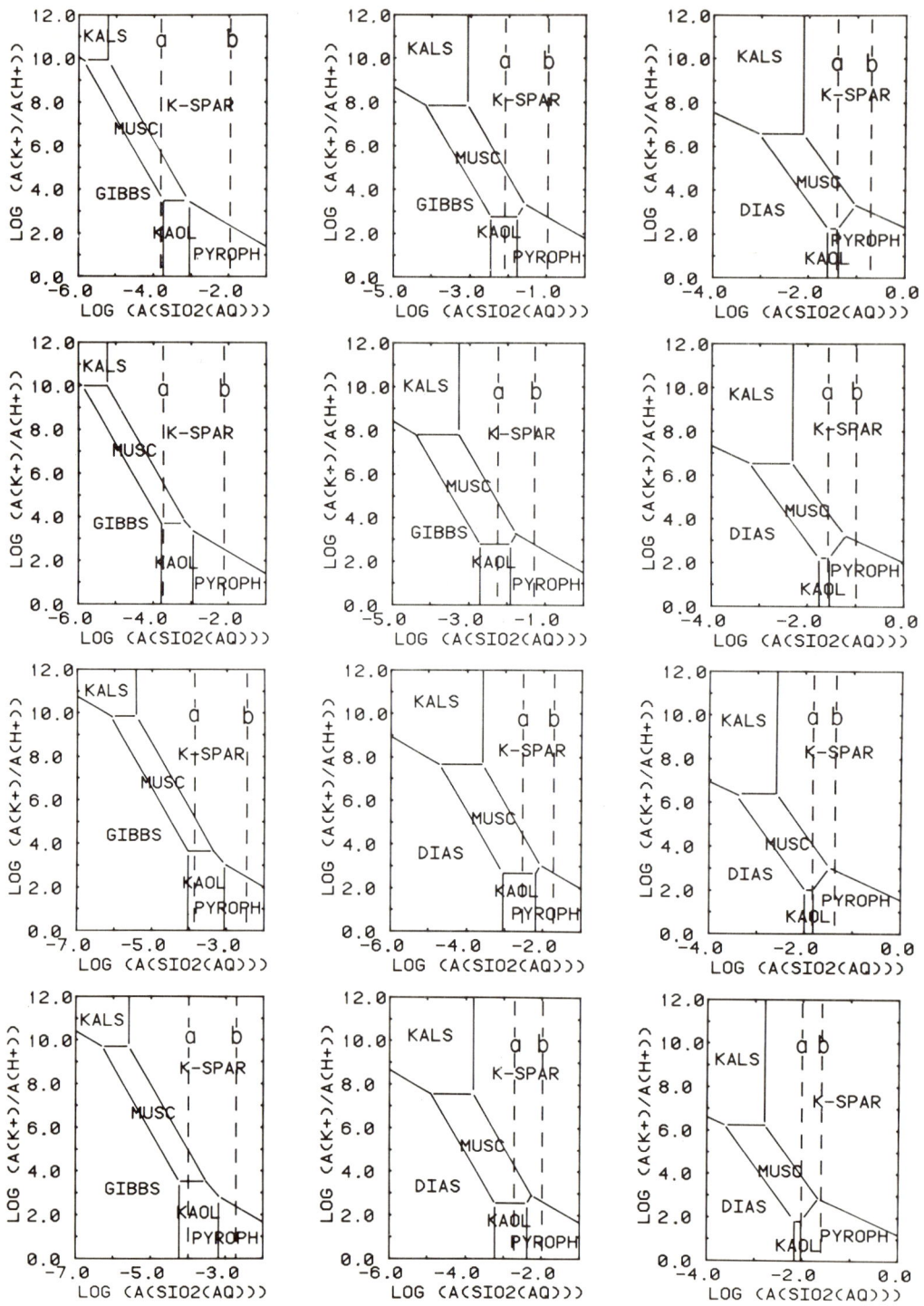

Phase relations in the system $HCl-H_2O-(Al_2O_3)-K_2O-SiO_2$. Saturation limits: quartz (a), amorphous silica (b).

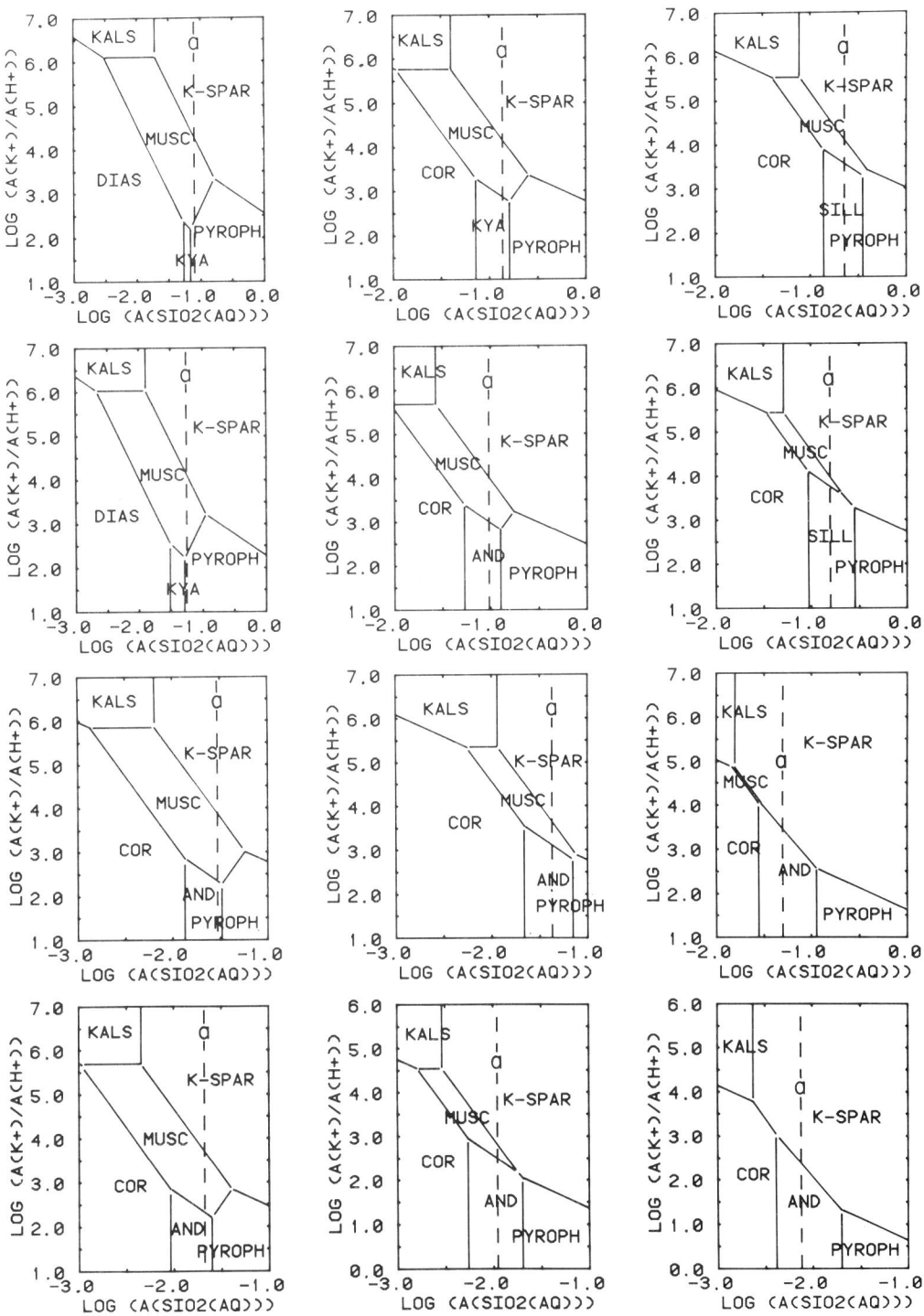

Phase relations in the system $HCl-H_2O-(Al_2O_3)-K_2O-SiO_2$. Saturation limits: quartz (a), amorphous silica (b).

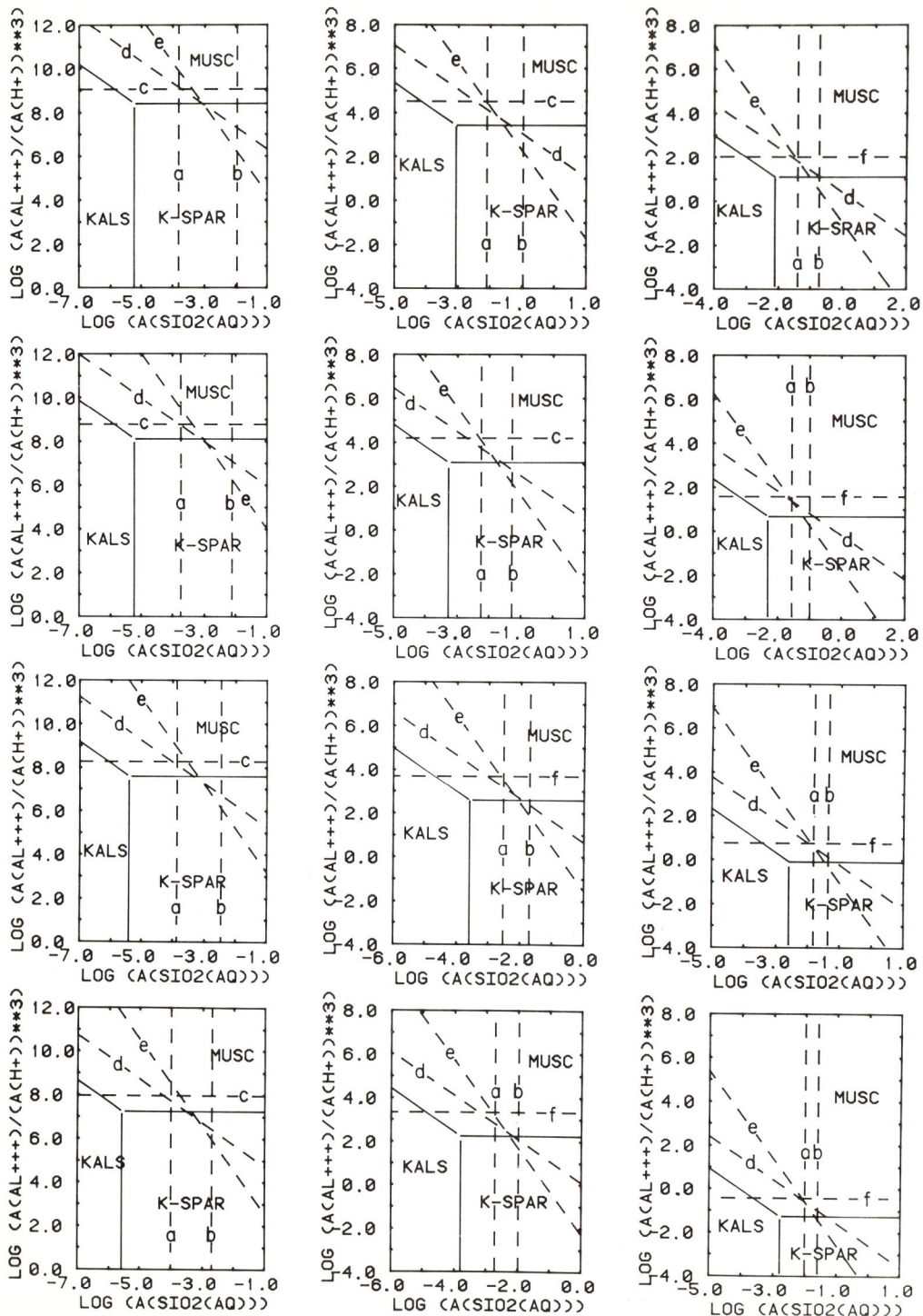

Phase relations in the system $HCl-H_2O-Al_2O_3-(K_2O)-SiO_2$. Saturation limits: quartz (a), amorphous silica (b), gibbsite (c), kaolinite (d), pyrophyllite (e), diaspore (f), kyanite (g), corundum (h), sillimanite (i), andalusite (j).

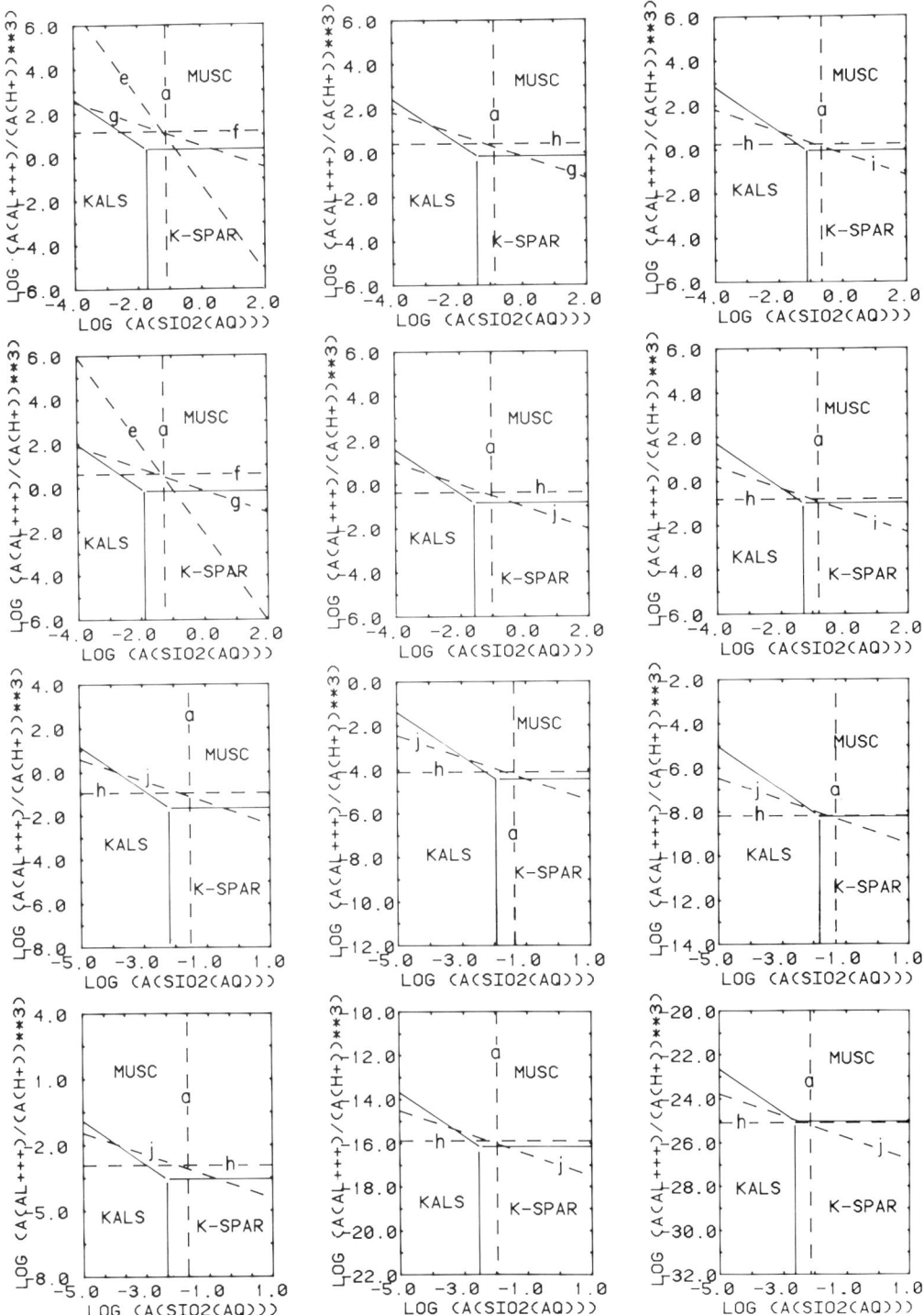

Phase relations in the system $HCl-H_2O-Al_2O_3-(K_2O)-SiO_2$. Saturation limits: quartz (a), amorphous silica (b), gibbsite (c), kaolinite (d), pyrophyllite (e), diaspore (f), kyanite (g), corundum (h), sillimanite (i), andalusite (j).

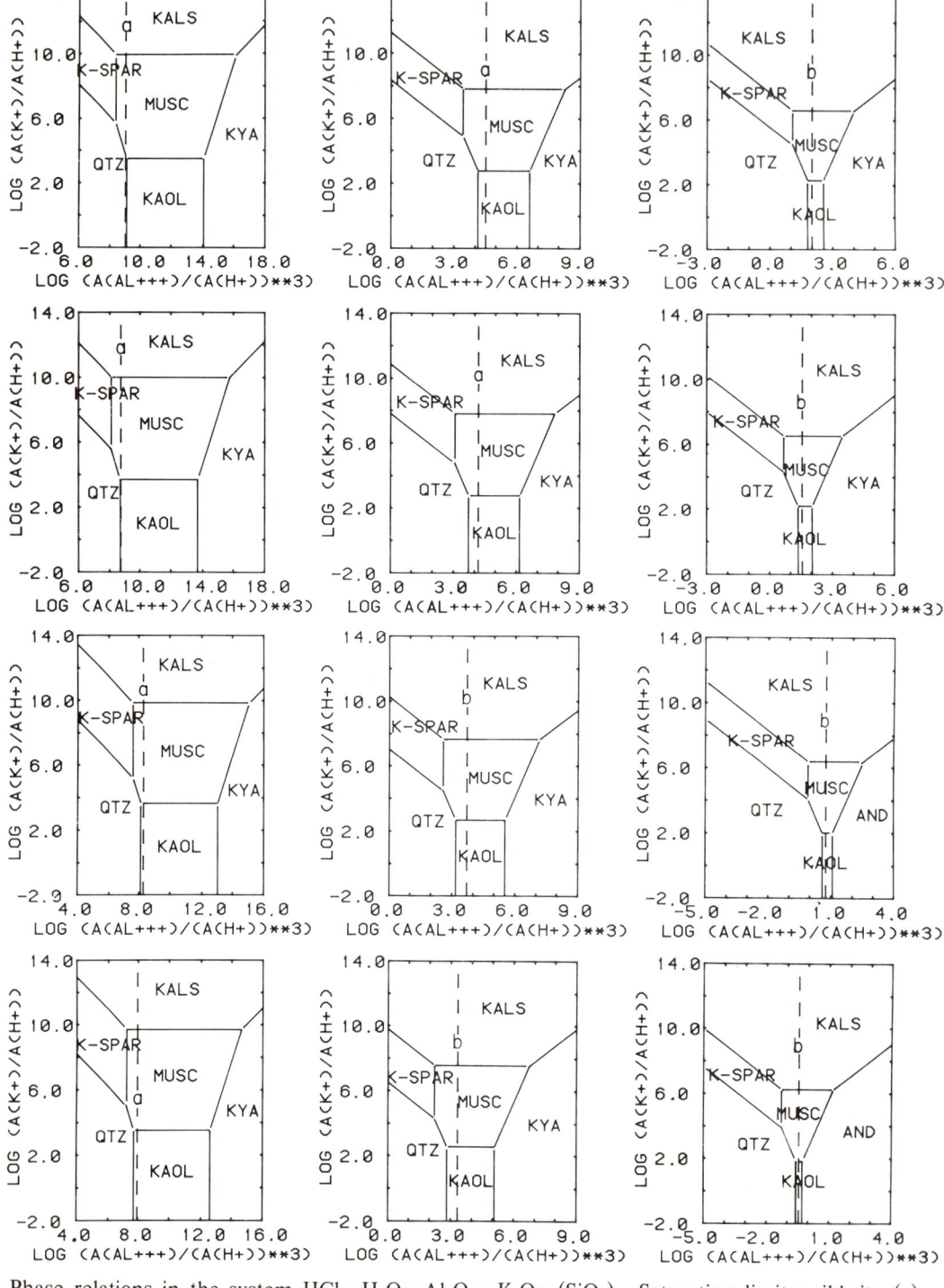

Phase relations in the system $HCl-H_2O-Al_2O_3-K_2O-(SiO_2)$. Saturation limits: gibbsite (a), diaspore (b), corundum (c).

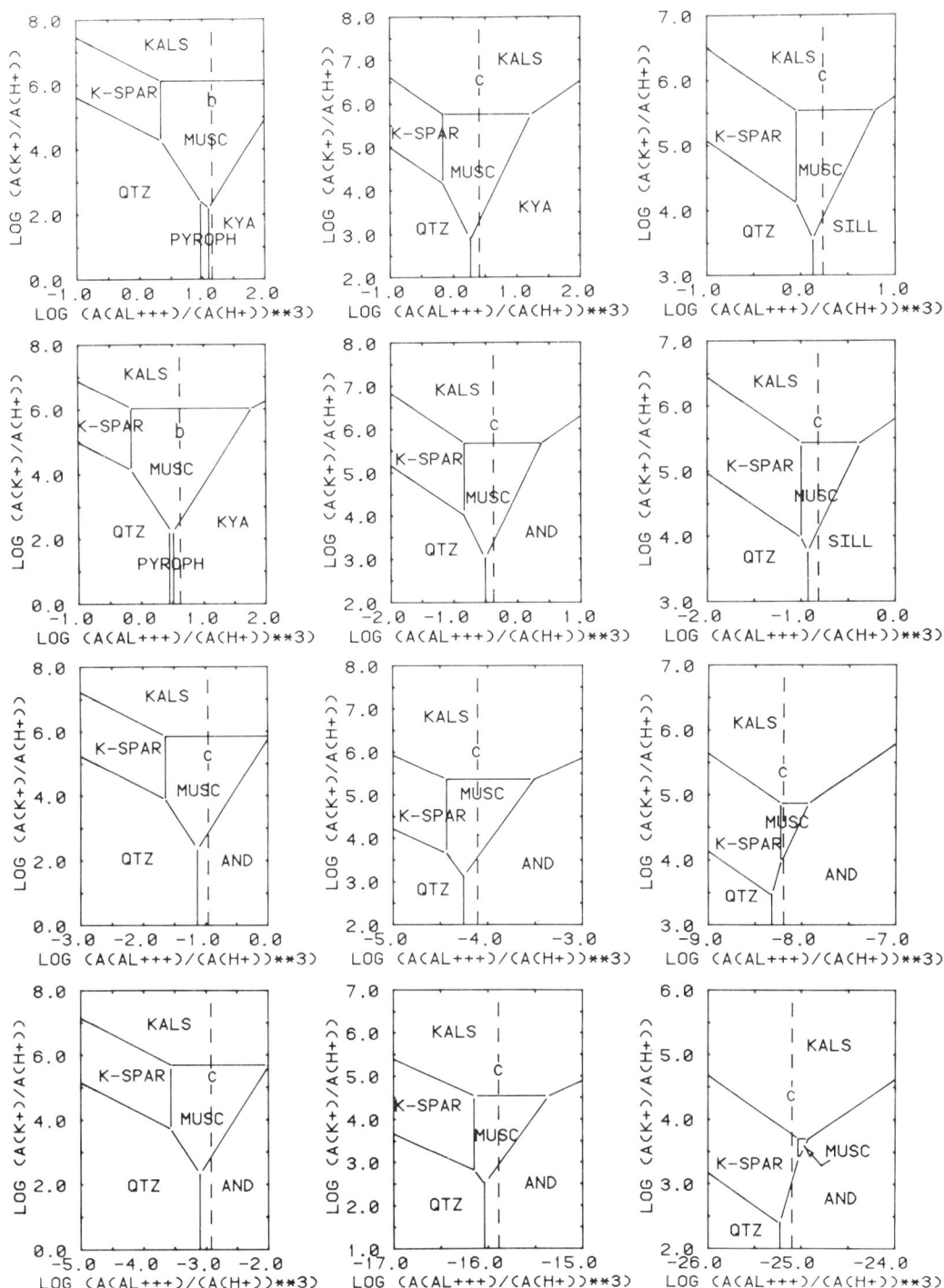

Phase relations in the system $HCl-H_2O-Al_2O_3-K_2O-(SiO_2)$. Saturation limits: gibbsite (a), diaspore (b), corundum (c).

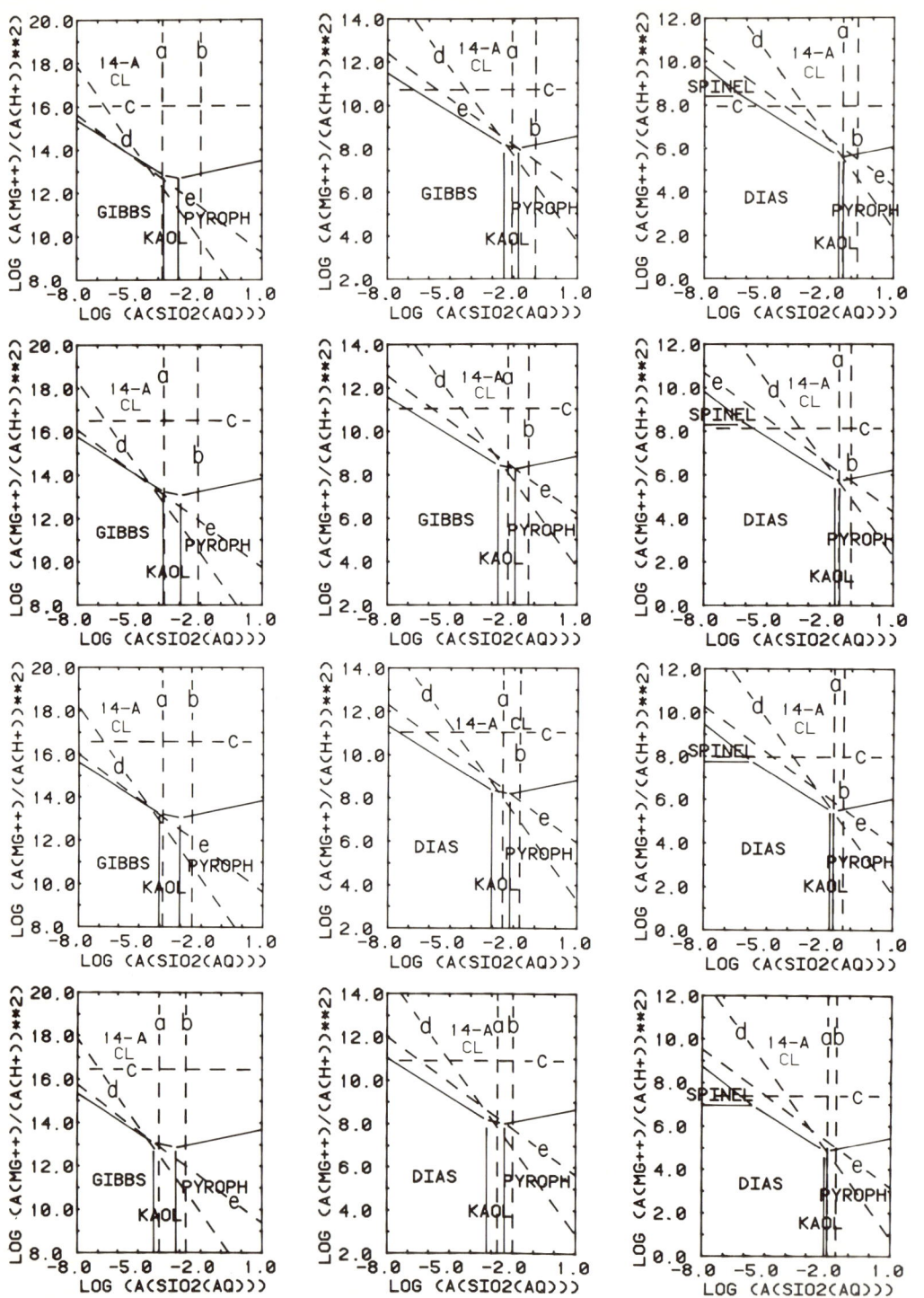

Phase relations in the system $HCl-H_2O-(Al_2O_3)-MgO-SiO_2$. Saturation limits: quartz (a), amorphous silica (b), brucite (c), talc (d), antigorite (e), forsterite (f), periclase (g).

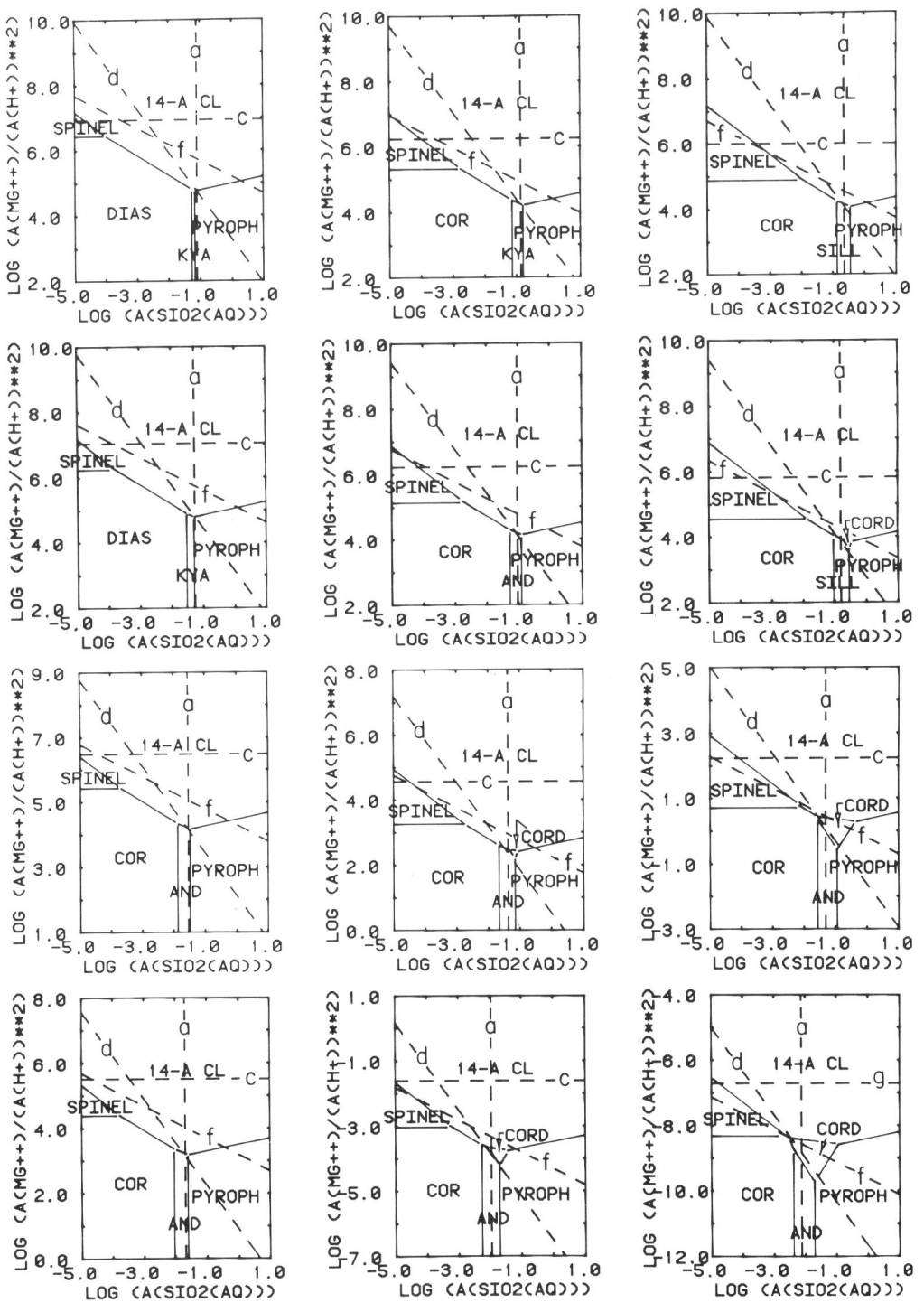

Phase relations in the system $HCl-H_2O-(Al_2O_3)-MgO-SiO_2$. Saturation limits: quartz (a), amorphous silica (b), brucite (c), talc (d), antigorite (e), forsterite (f), periclase (g).

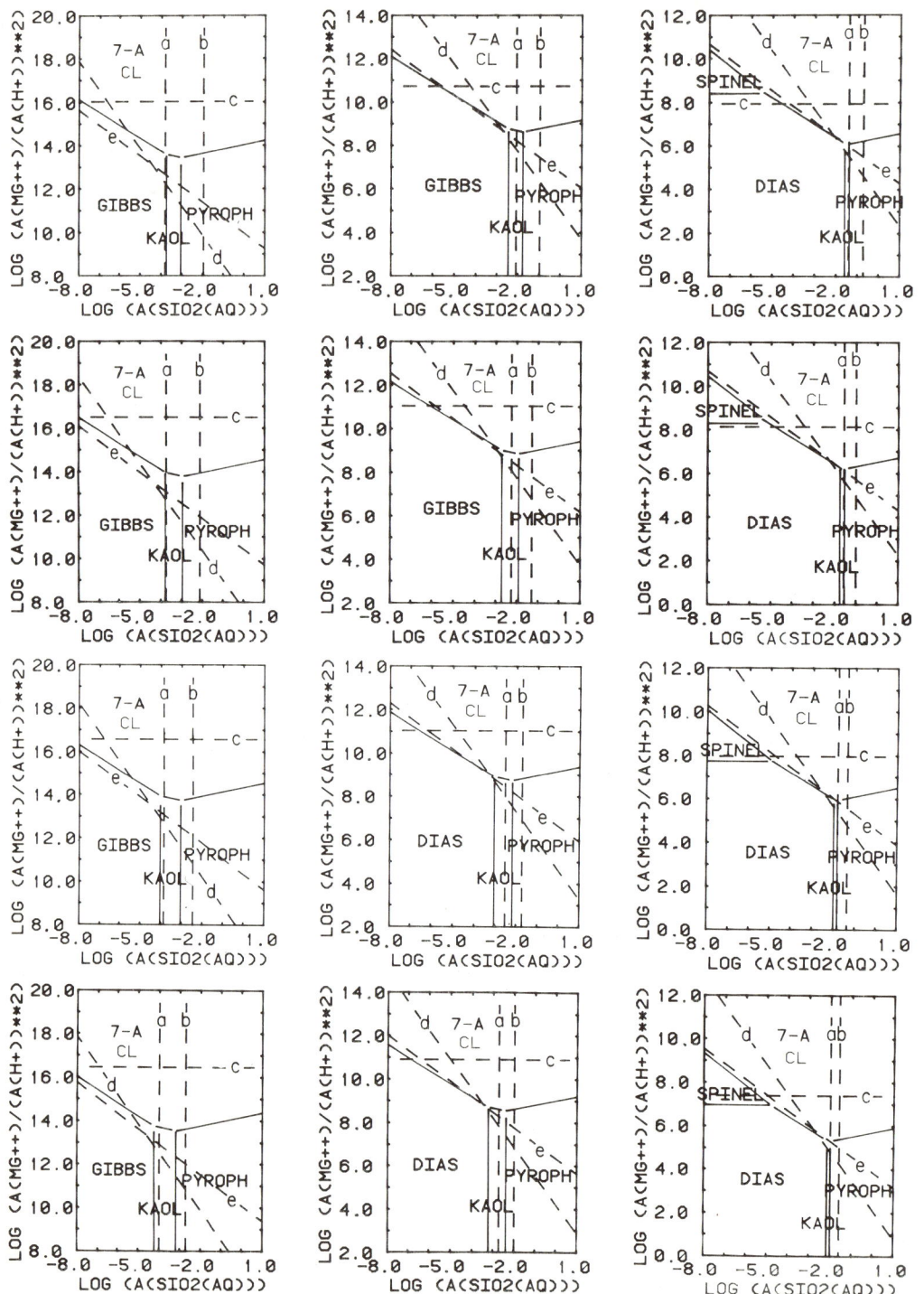

Phase relations in the system $HCl-H_2O-(Al_2O_3)-MgO-SiO_2$. Metastable 7-A clinochlore was considered instead of its stable counterpart, 14-A clinochlore. Saturation limits: quartz (a), amorphous silica (b), brucite (c), talc (d), antigorite (e), forsterite (f), periclase (g).

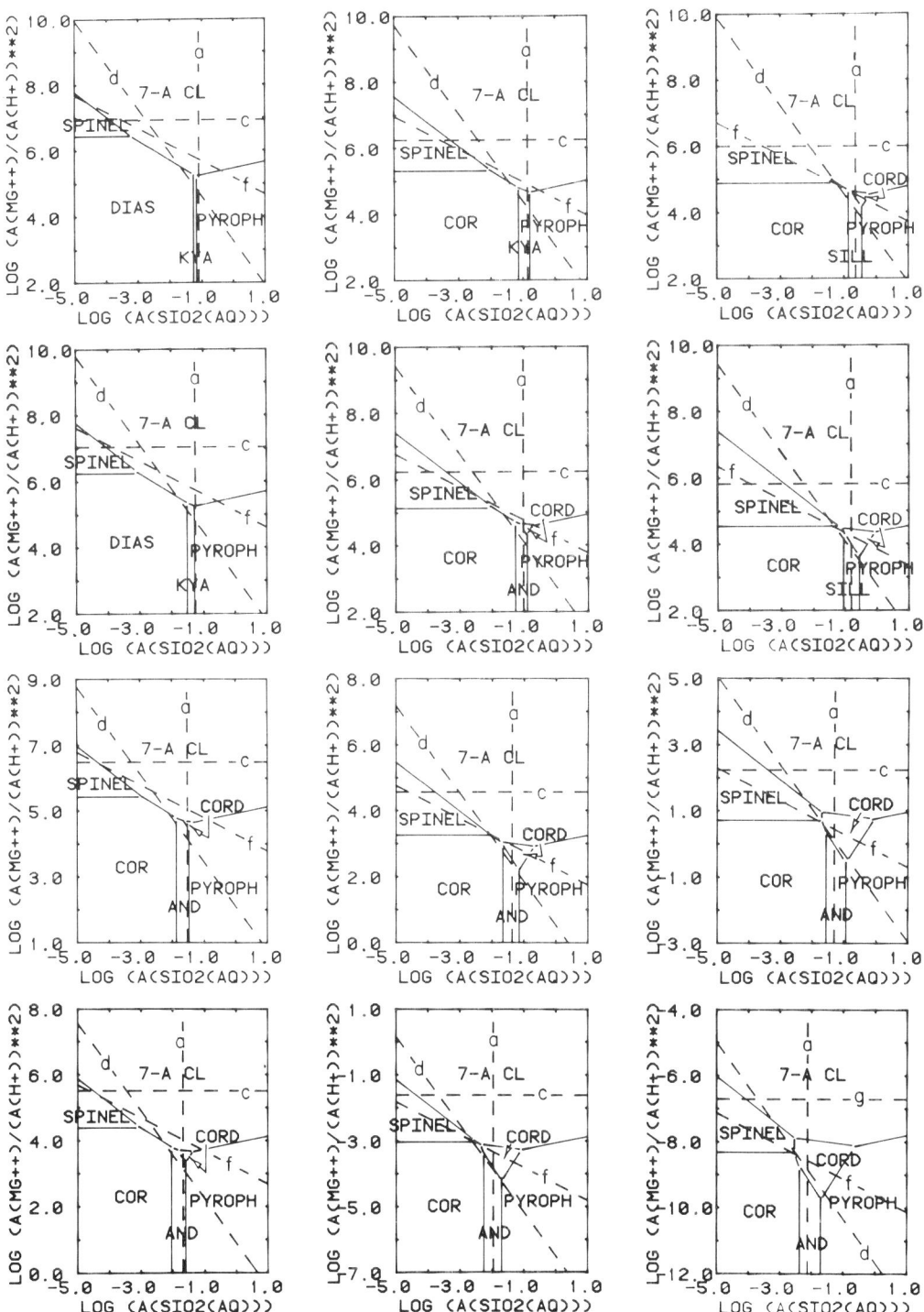

Phase relations in the system $HCl-H_2O-(Al_2O_3)-MgO-SiO_2$. Metastable 7-A clinochlore was considered instead of its stable counterpart, 14-A clinochlore. Saturation limits: quartz (a), amorphous silica (b), brucite (c), talc (d), antigorite (e), forsterite (f), periclase (g).

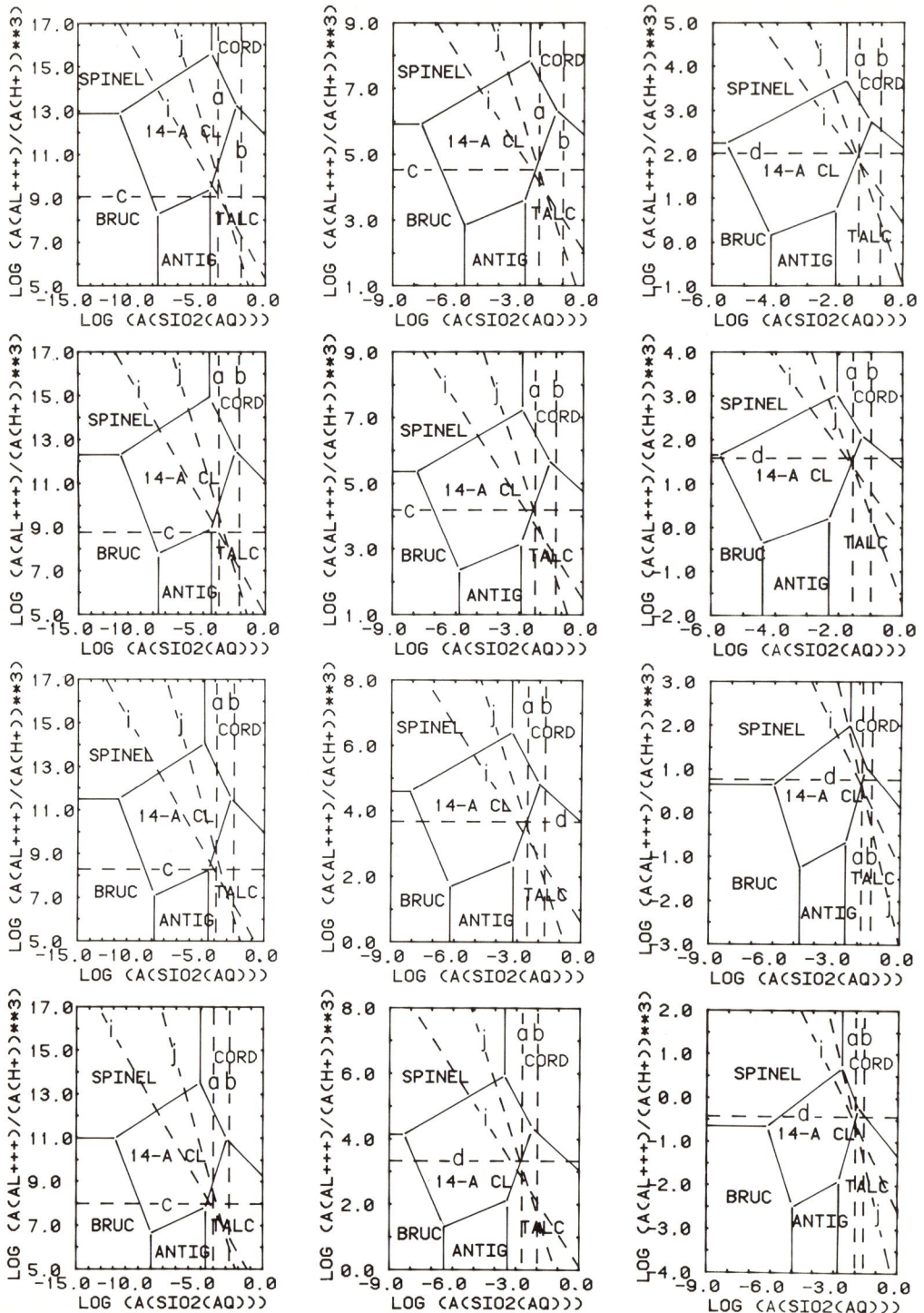

Phase relations in the system $HCl-H_2O-Al_2O_3-(MgO)-SiO_2$. Saturation limits: quartz (a), amorphous silica (b), gibbsite (c), diaspore (d), corundum (e), andalusite (f), kyanite (g), sillimanite (h), kaolinite (i), pyrophyllite (j).

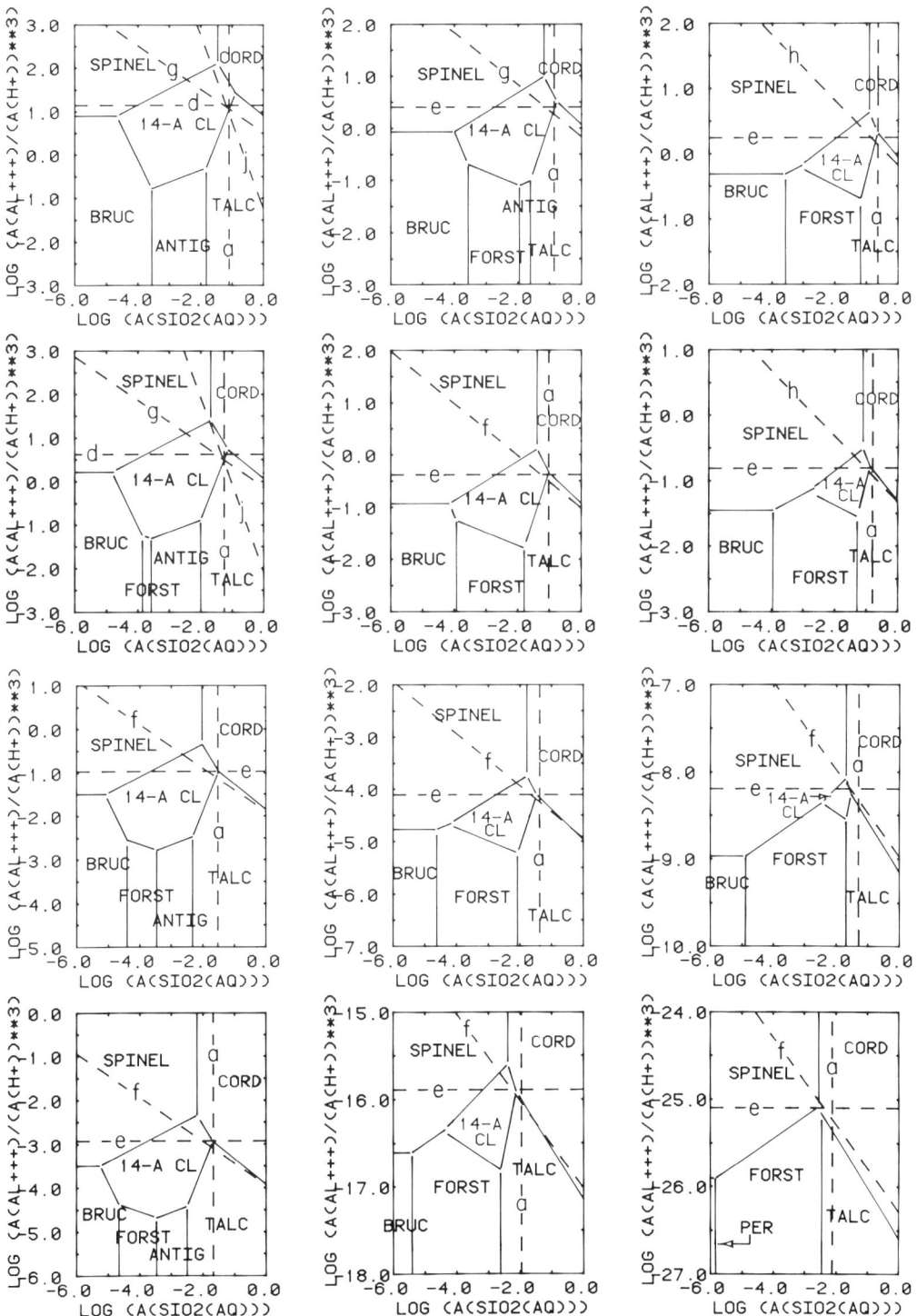

Phase relations in the system $HCl-H_2O-Al_2O_3-(MgO)-SiO_2$. Saturation limits: quartz (a), amorphous silica (b), gibbsite (c), diaspore (d), corundum (e), andalusite (f), kyanite (g), sillimanite (h), kaolinite (i), pyrophyllite (j).

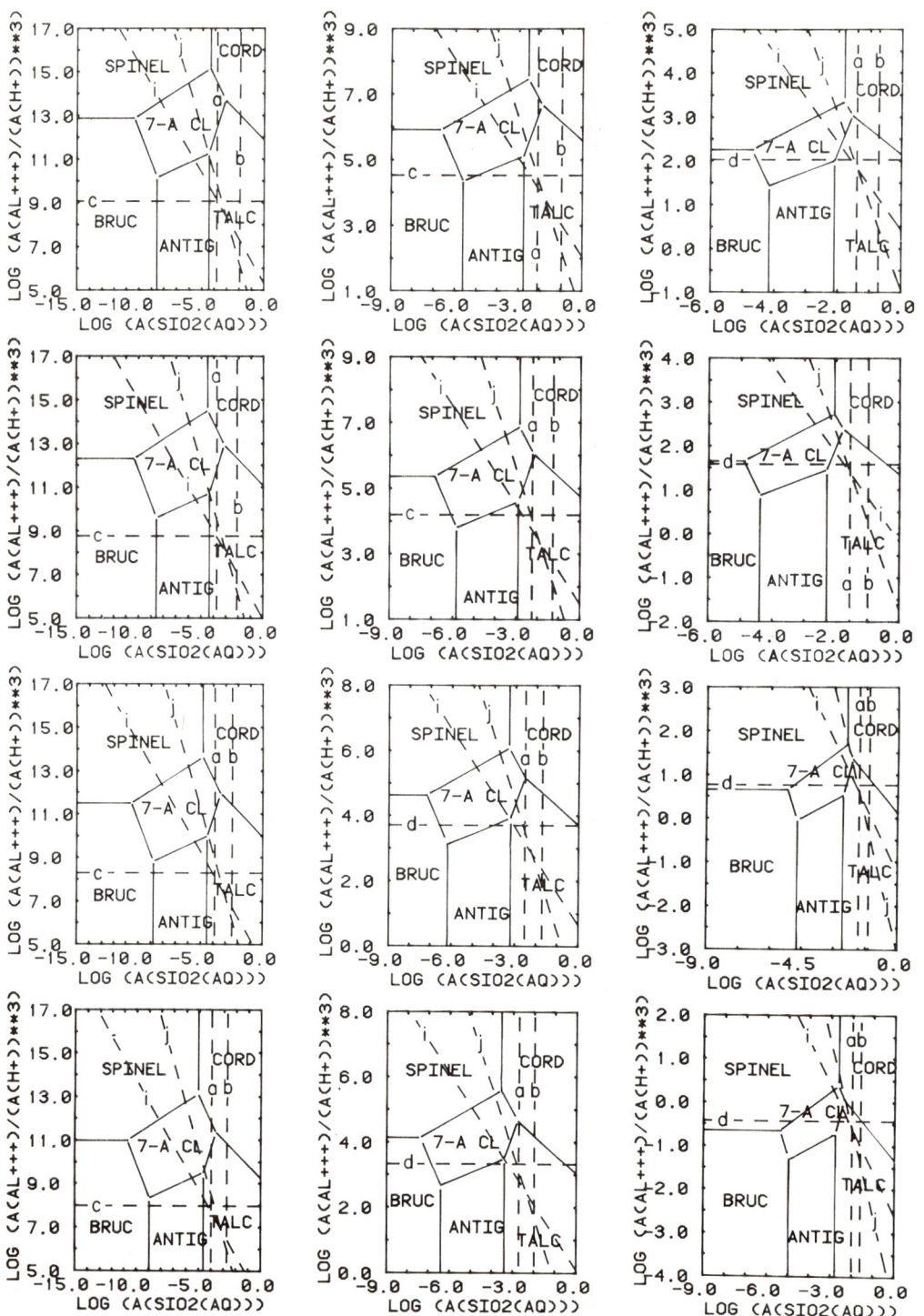

Phase relations in the system $HCl-H_2O-Al_2O_3-(MgO)-SiO_2$. Metastable 7-A clinochlore was considered instead of its stable counterpart, 14-A clinochlore. Saturation limits: quartz (a), amorphous silica (b), gibbsite (c), diaspore (d), corundum (e), andalusite (f), kyanite (g), sillimanite (h), kaolinite (i), pyrophyllite (j).

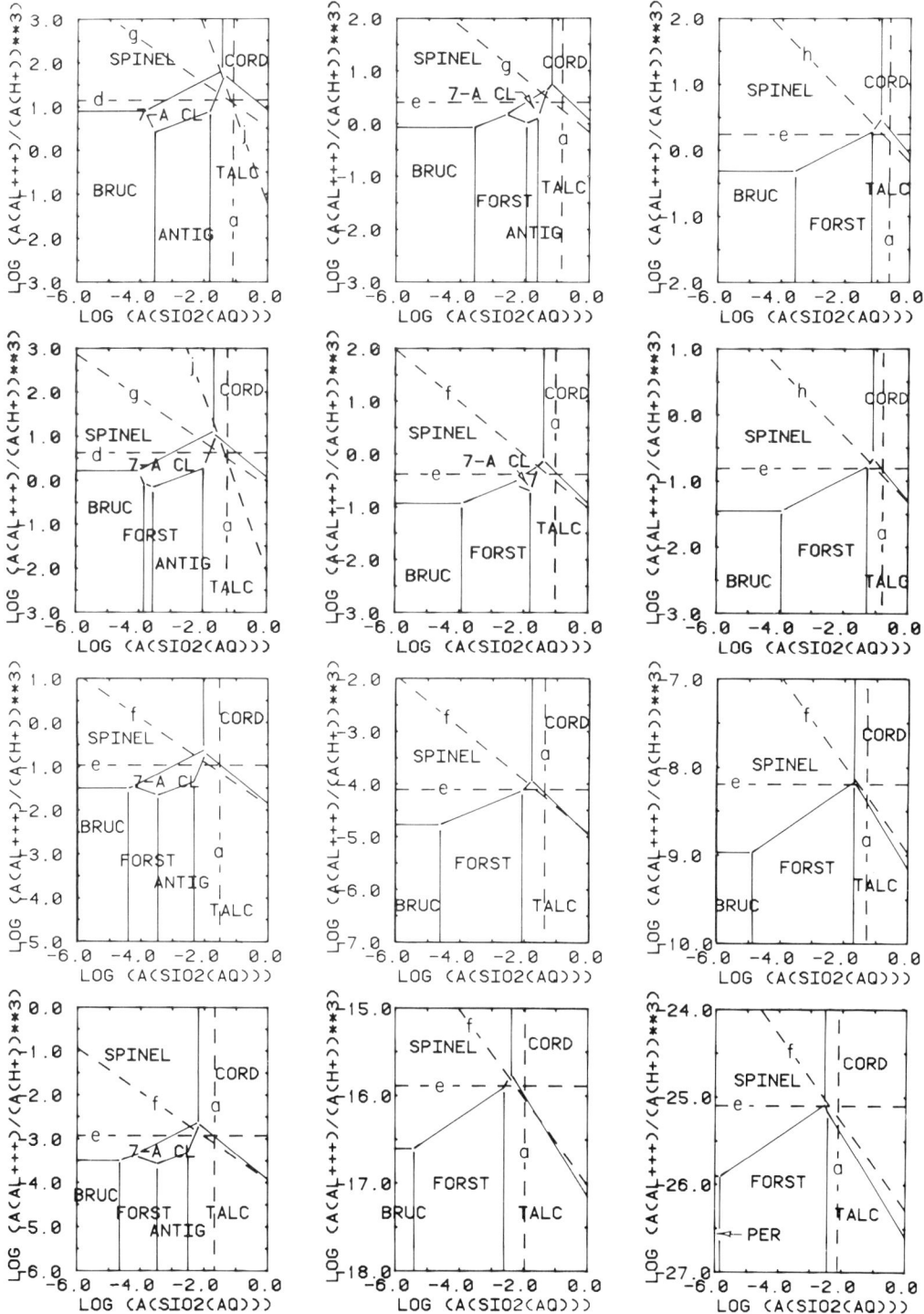

Phase relations in the system HCl−H$_2$O−Al$_2$O$_3$−(MgO)−SiO$_2$. Metastable 7-A clinochlore was considered instead of its stable counterpart, 14-A clinochlore. Saturation limits: quartz (a), amorphous silica (b), gibbsite (c), diaspore (d), corundum (e), andalusite (f), kyanite (g), sillimanite (h), kaolinite (i), pyrophyllite (j).

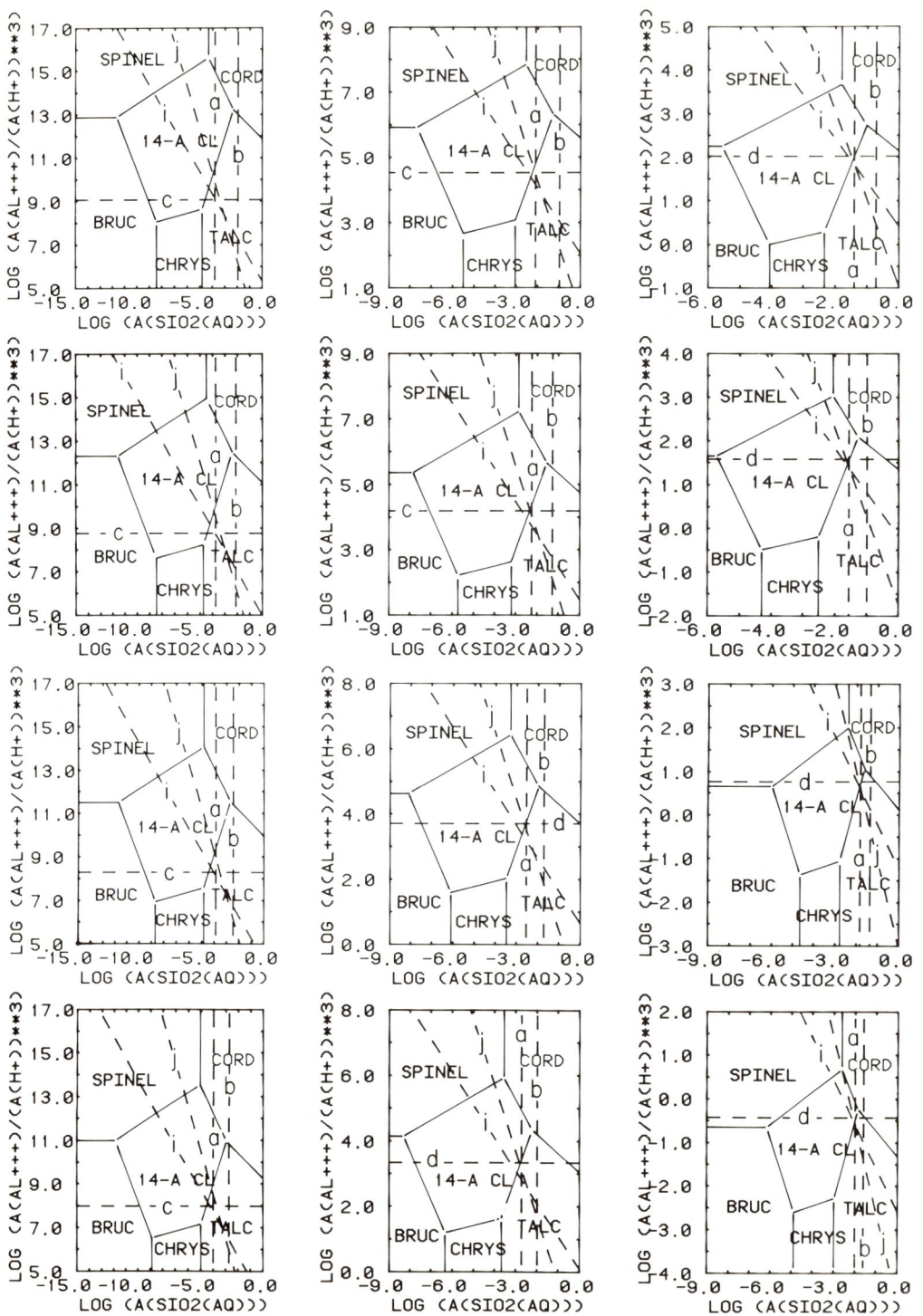

Phase relations in the system $HCl-H_2O-Al_2O_3-(MgO)-SiO_2$. Metastable chrysotile was considered instead of its stable counterpart, antigorite. Saturation limits: quartz (a), amorphous silica (b), gibbsite (c), diaspore (d), corundum (e), andalusite (f), kyanite (g), sillimanite (h), kaolinite (i), pyrophyllite (j).

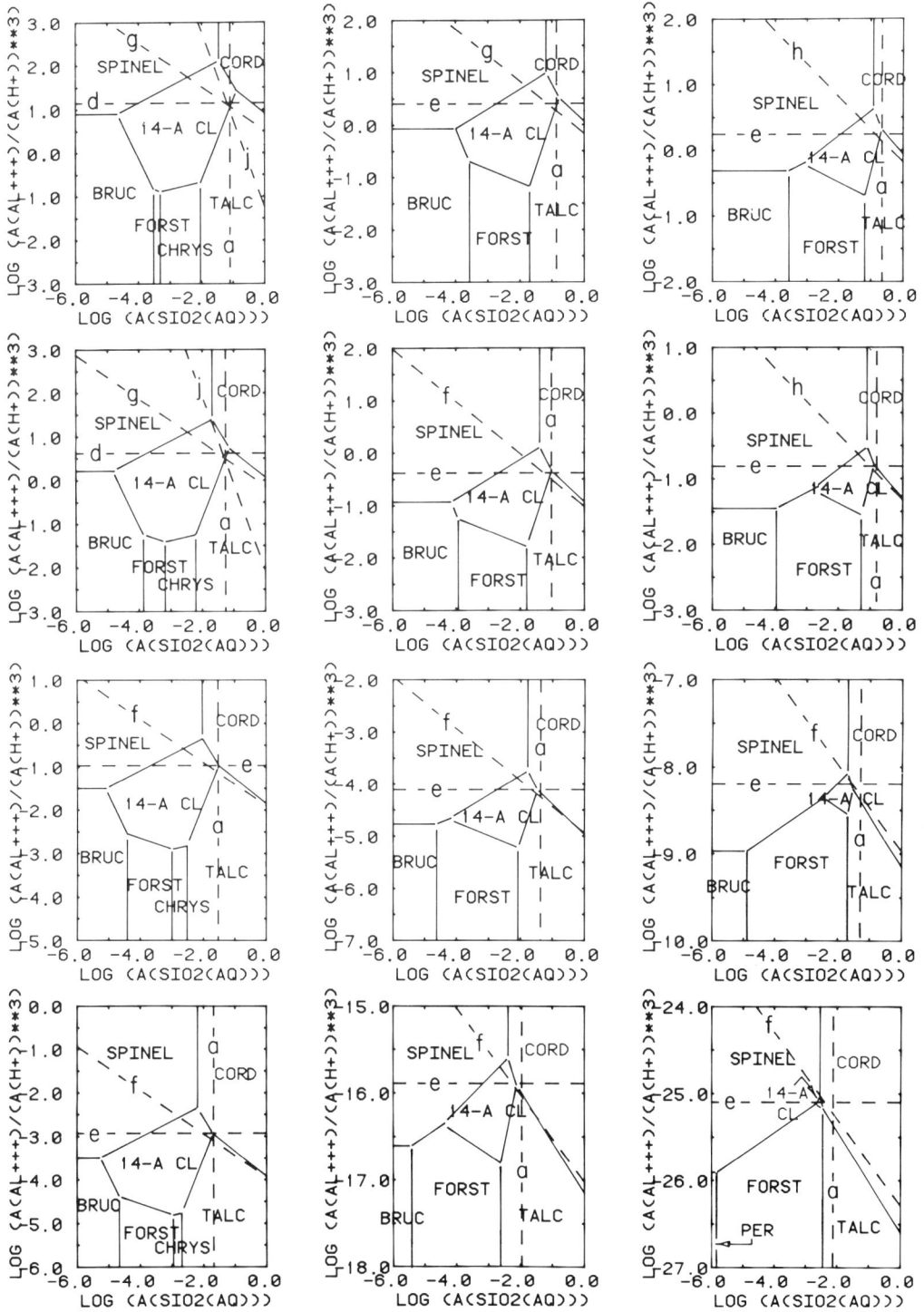

Phase relations in the system $HCl-H_2O-Al_2O_3-(MgO)-SiO_2$. Metastable chrysotile was considered instead of its stable counterpart, antigorite. Saturation limits: quartz (a), amorphous silica (b), gibbsite (c), diaspore (d), corundum (e), andalusite (f), kyanite (g), sillimanite (h), kaolinite (i), pyrophyllite (j).

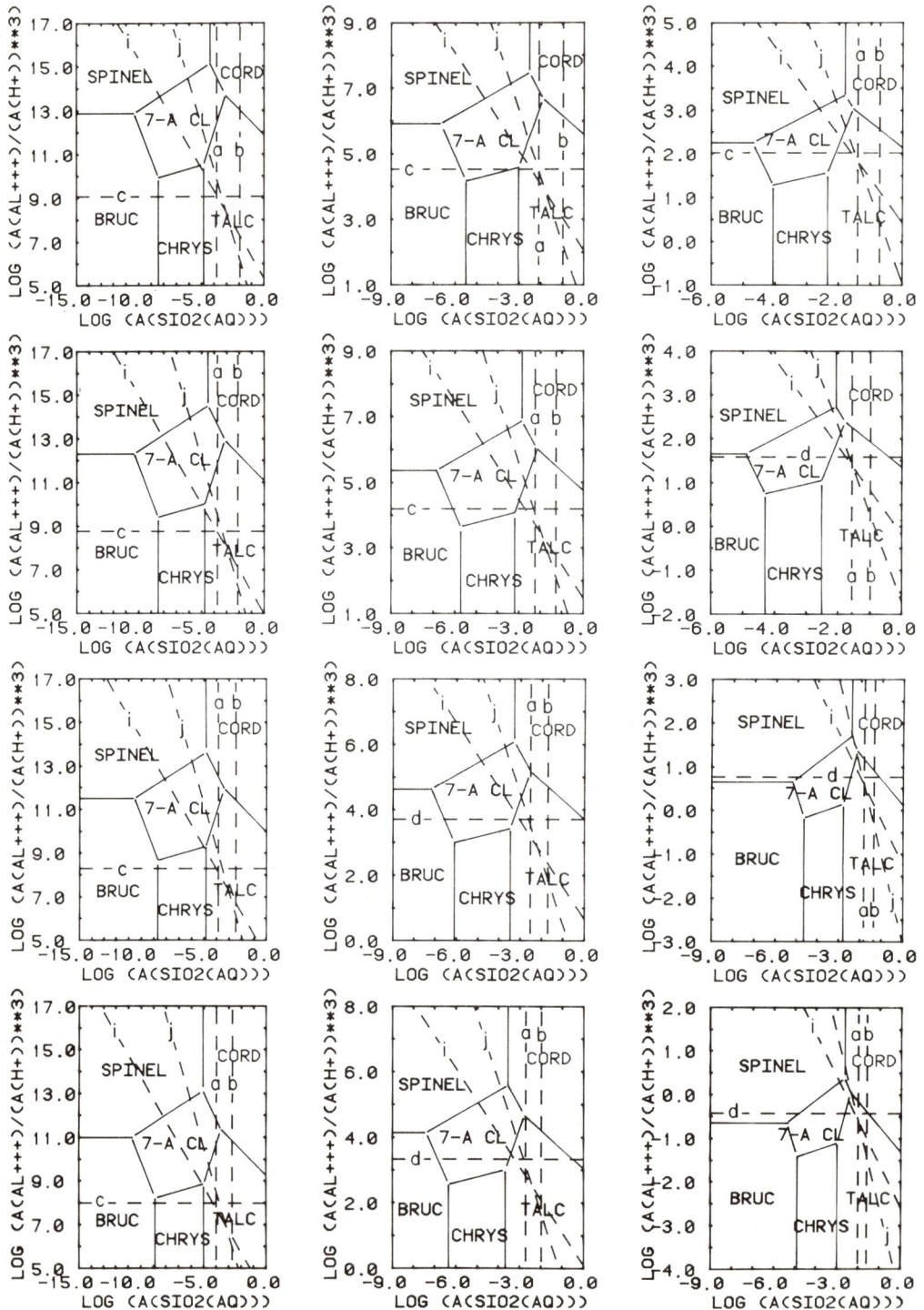

Phase relations in the system HCl−H$_2$O−Al$_2$O$_3$−(MgO)−SiO$_2$. Metastable 7-A clinochlore and chrysotile were considered instead of their stable counterparts, 14-A clinochlore and antigorite. Saturation limits: quartz (a), amorphous silica (b), gibbsite (c), diaspore (d), corundum (e), andalusite (f), kyanite (g), sillimanite (h), kaolinite (i), pyrophyllite (j).

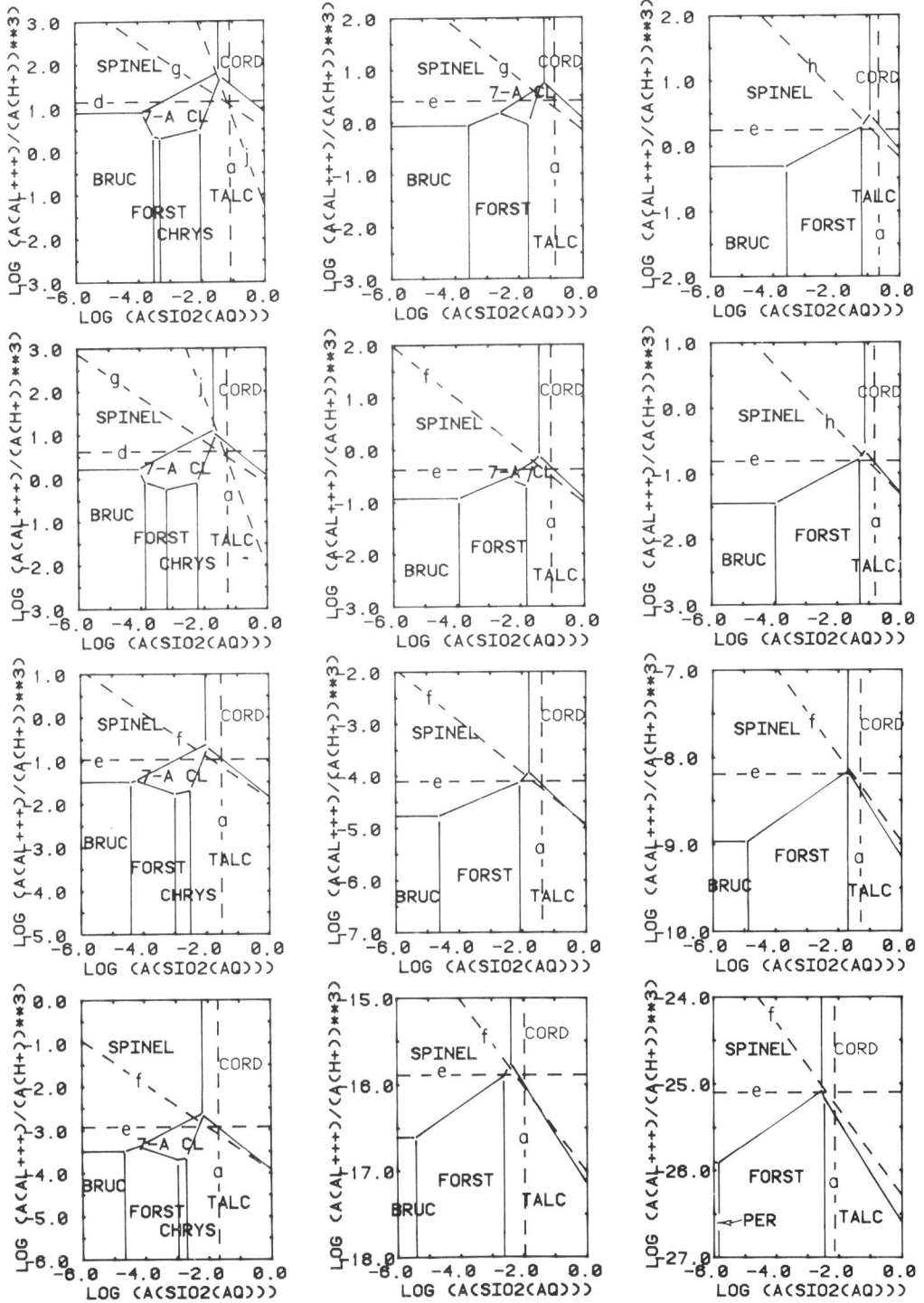

Phase relations in the system $HCl-H_2O-Al_2O_3-(MgO)-SiO_2$. Metastable 7-A clinochlore and chrysotile were considered instead of their stable counterparts, 14-A clinochlore and antigorite. Saturation limits: quartz (a), amorphous silica (b), gibbsite (c), diaspore (d), corundum (e), andalusite (f), kyanite (g), sillimanite (h), kaolinite (i), pyrophyllite (j).

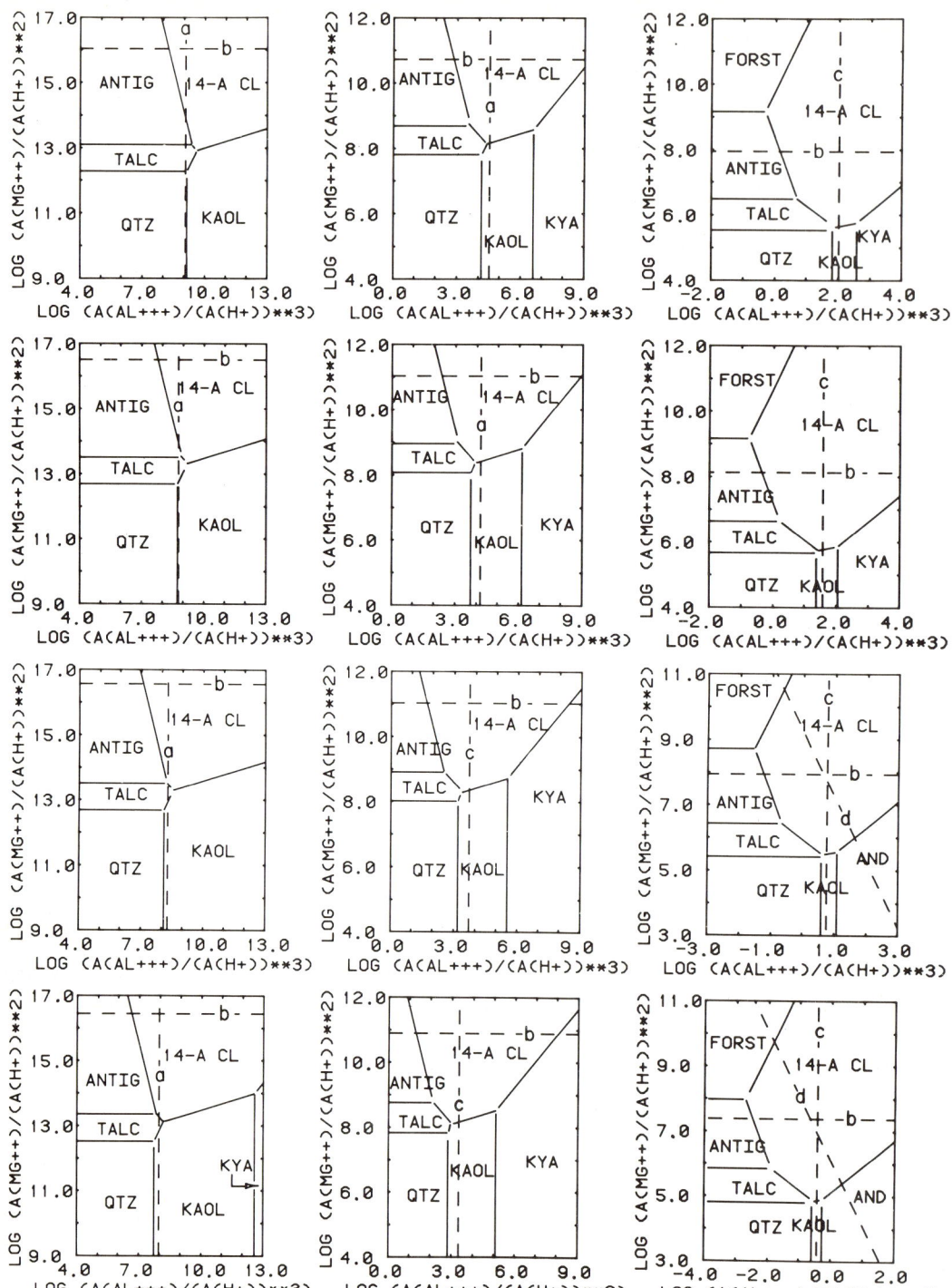

Phase relations in the system $HCl-H_2O-Al_2O_3-MgO-(SiO_2)$. Saturation limits: gibbsite (a), brucite (b), diaspore (c), spinel (d), corundum (e), periclase (f).

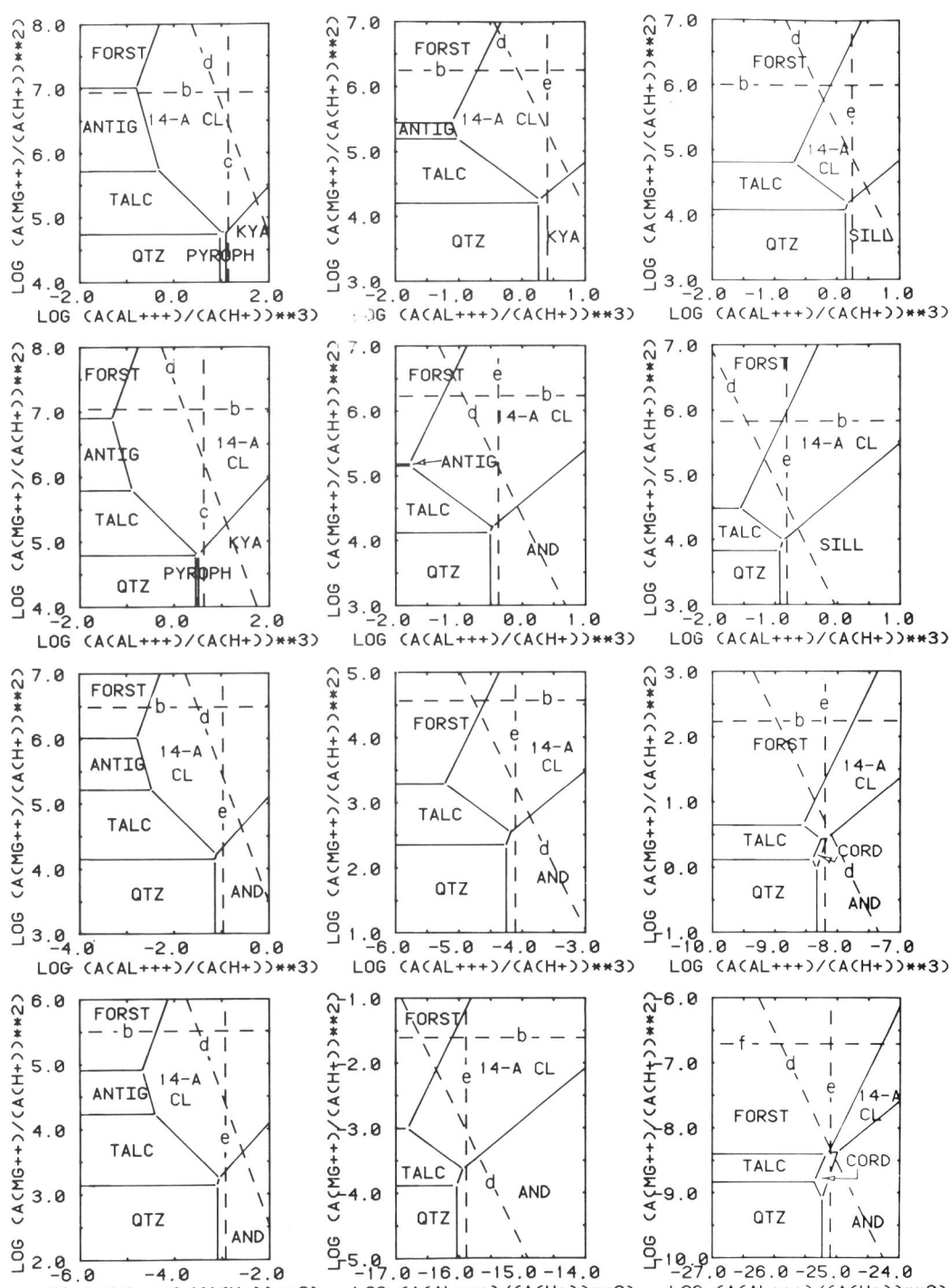

Phase relations in the system $HCl-H_2O-Al_2O_3-MgO-(SiO_2)$. Saturation limits: gibbsite (a), brucite (b), diaspore (c), spinel (d), corundum (e), periclase (f).

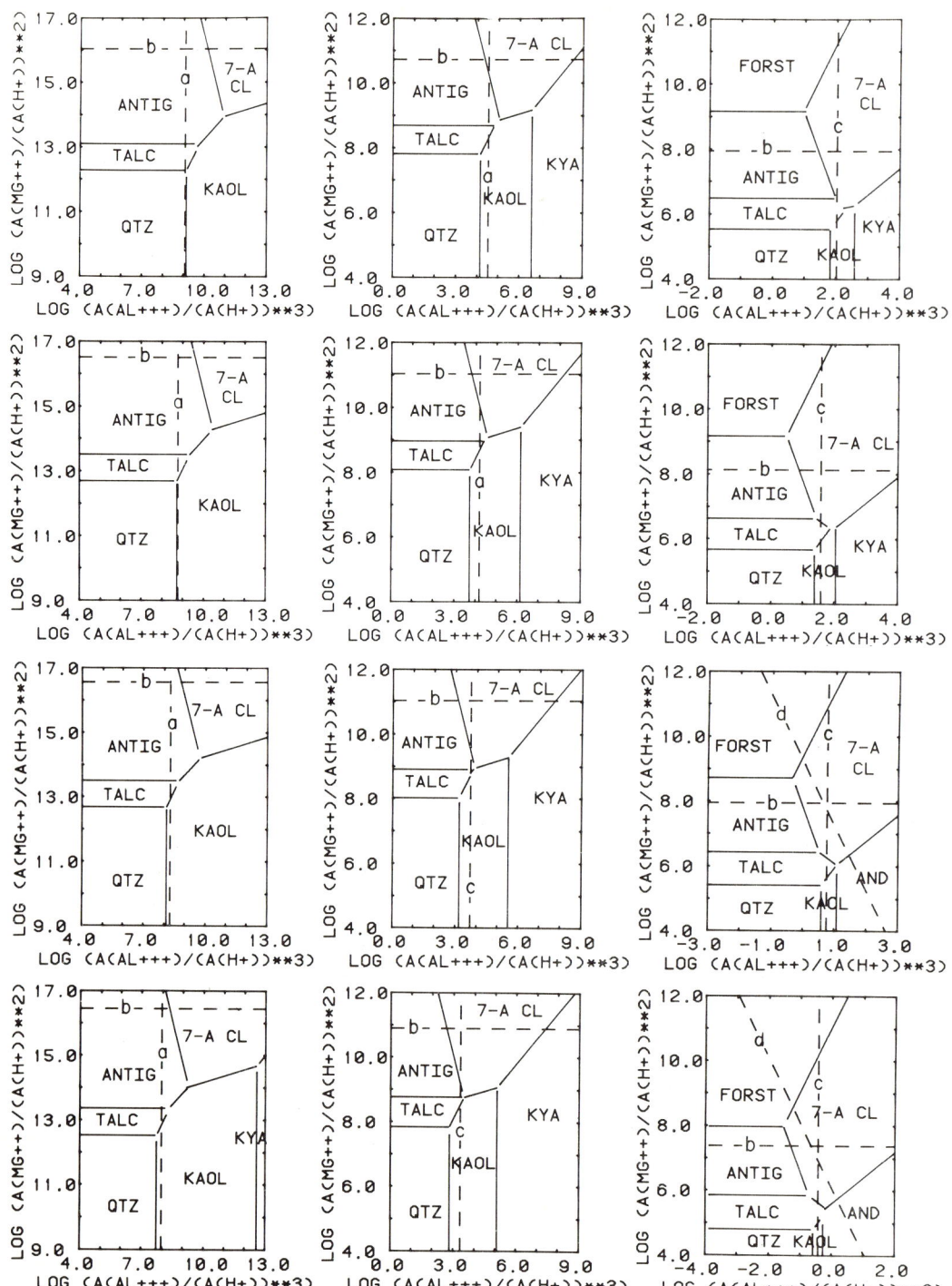

Phase relations in the system $HCl-H_2O-Al_2O_3-MgO-(SiO_2)$. Metastable 7-A clinochlore was considered instead of its stable counterpart, 14-A clinochlore. Saturation limits: gibbsite (a), brucite (b), diaspore (c), spinel (d), corundum (e), periclase (f).

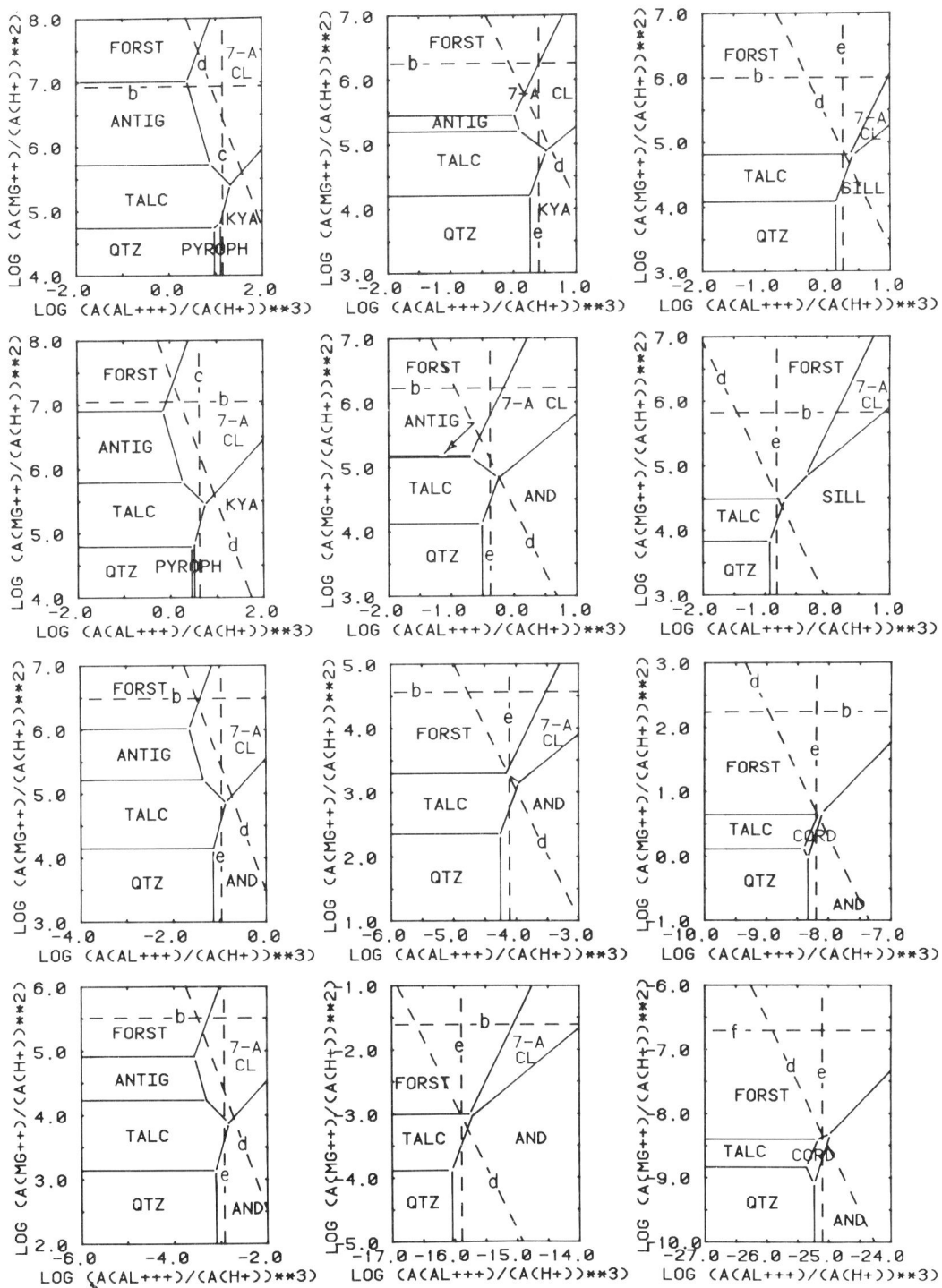

Phase relations in the system $HCl-H_2O-Al_2O_3-MgO-(SiO_2)$. Metastable 7-A clinochlore was considered instead of its stable counterpart, 14-A clinochlore. Saturation limits: gibbsite (a), brucite (b), diaspore (c), spinel (d), corundum (e), periclase (f).

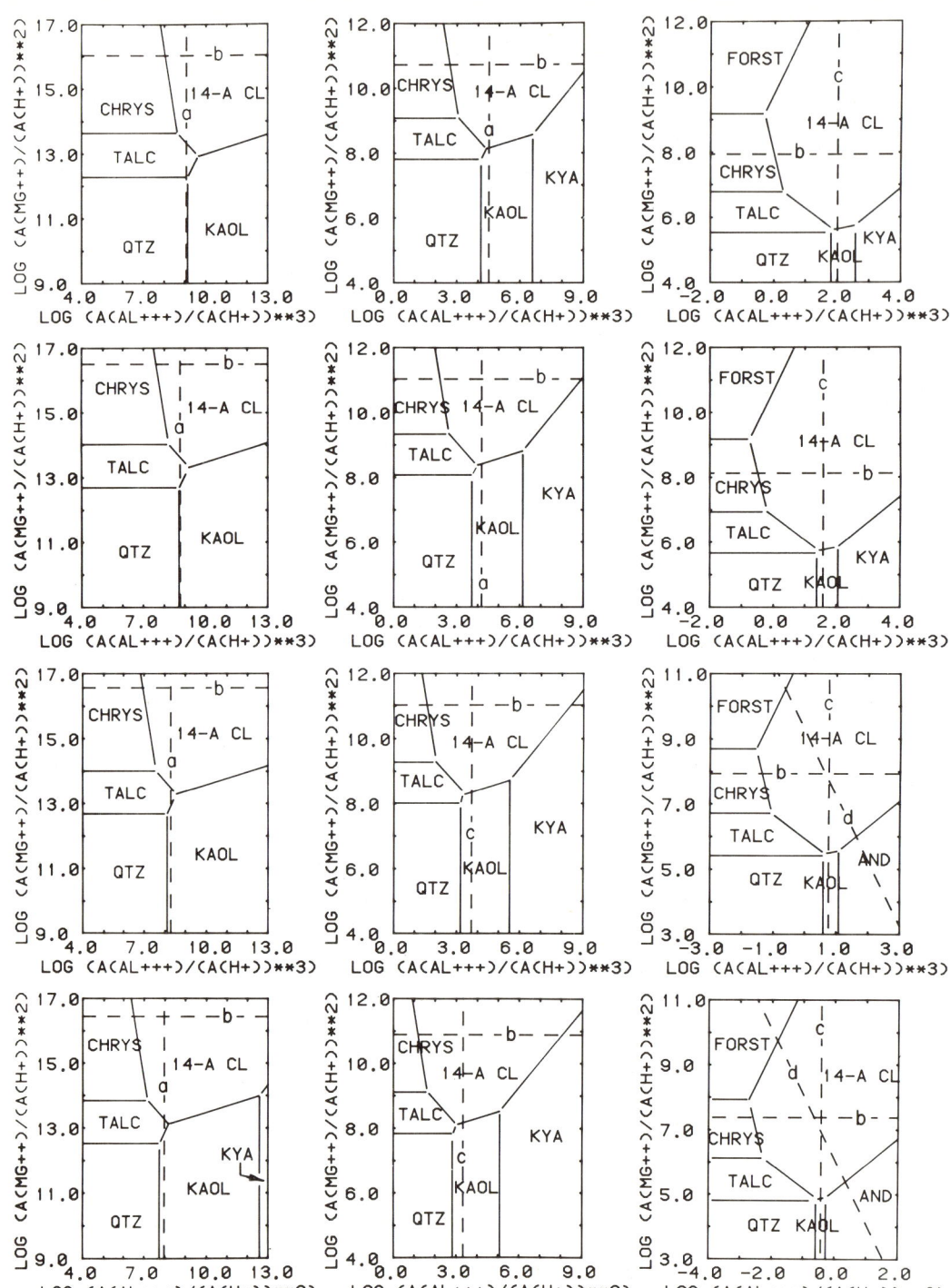

Phase relations in the system $HCl-H_2O-Al_2O_3-MgO-(SiO_2)$. Metastable chrysotile was considered instead of its stable counterpart, antigorite. Saturation limits: gibbsite (a), brucite (b), diaspore (c), spinel (d), corundum (e), periclase (f).

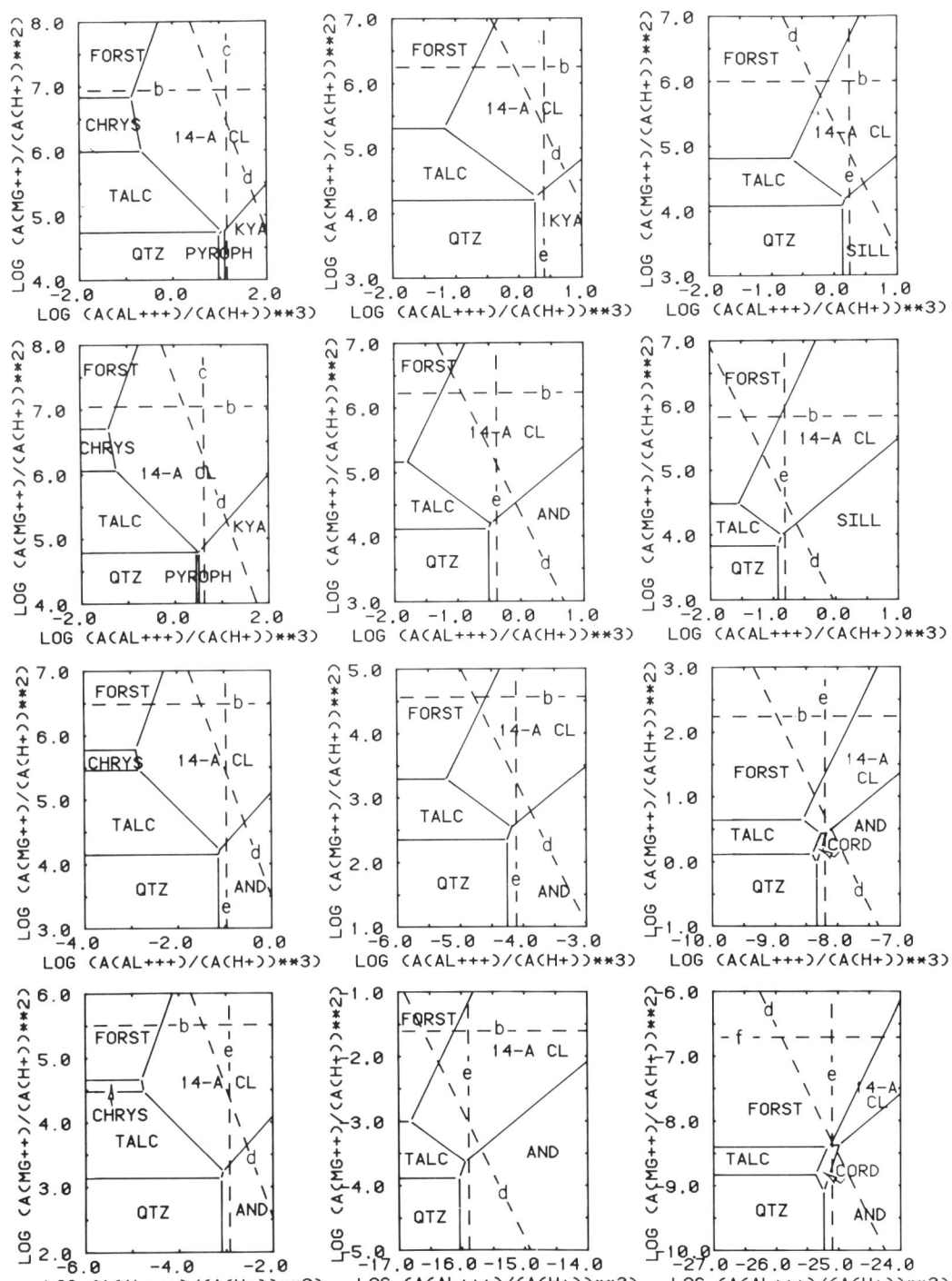

Phase relations in the system $HCl-H_2O-Al_2O_3-MgO-(SiO_2)$. Metastable chrysotile was considered instead of its stable counterpart, antigorite. Saturation limits: gibbsite (a), brucite (b), diaspore (c), spinel (d), corundum (e), periclase (f).

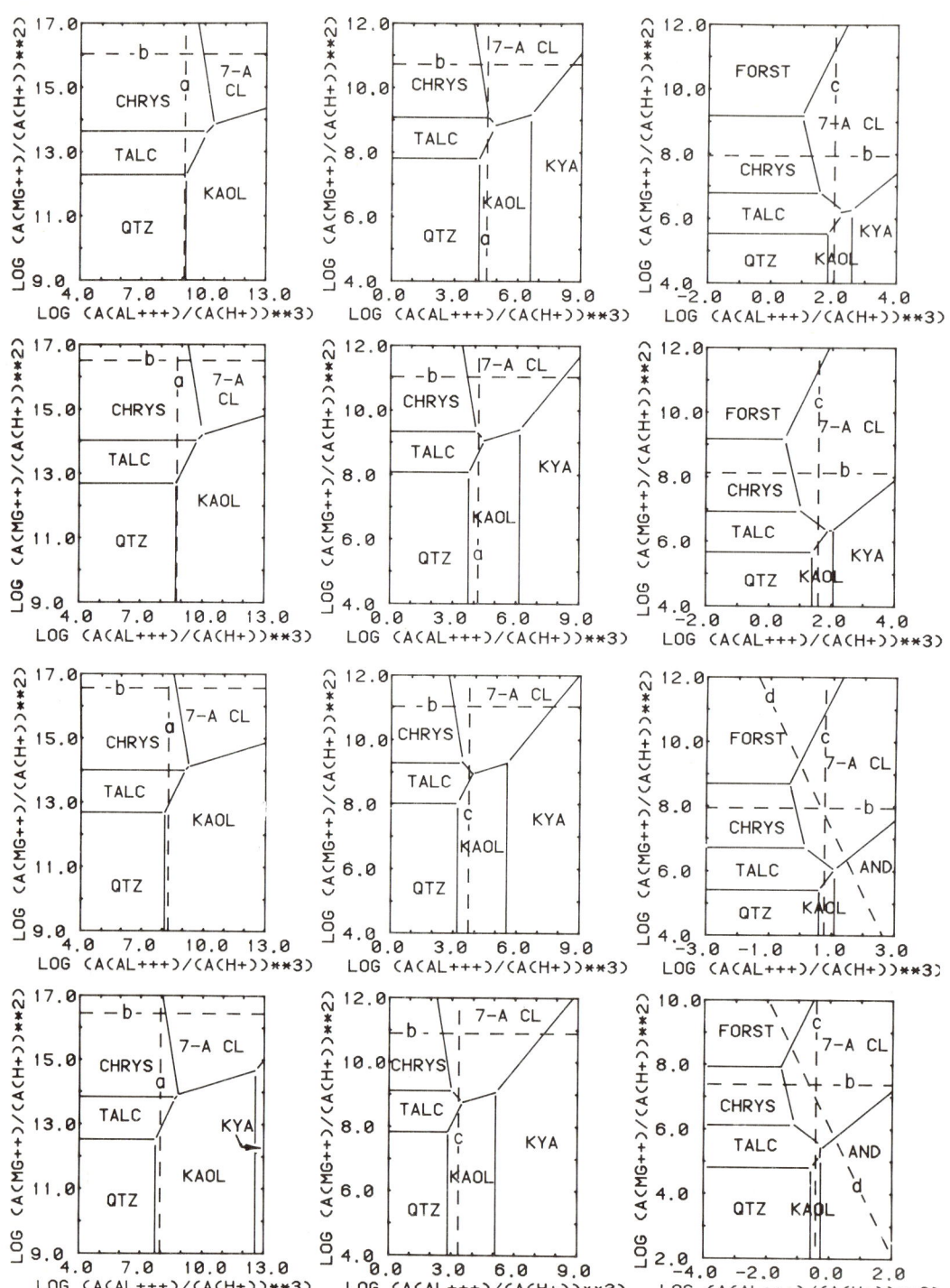

Phase relations in the system HCl−H$_2$O−Al$_2$O$_3$−MgO−(SiO$_2$). Metastable 7-A clinochlore and chrysotile were considered instead of their stable counterparts, 14-A clinochlore and antigorite. Saturation limits: gibbsite (a), brucite (b), diaspore (c), spinel (d), corundum (e), periclase (f).

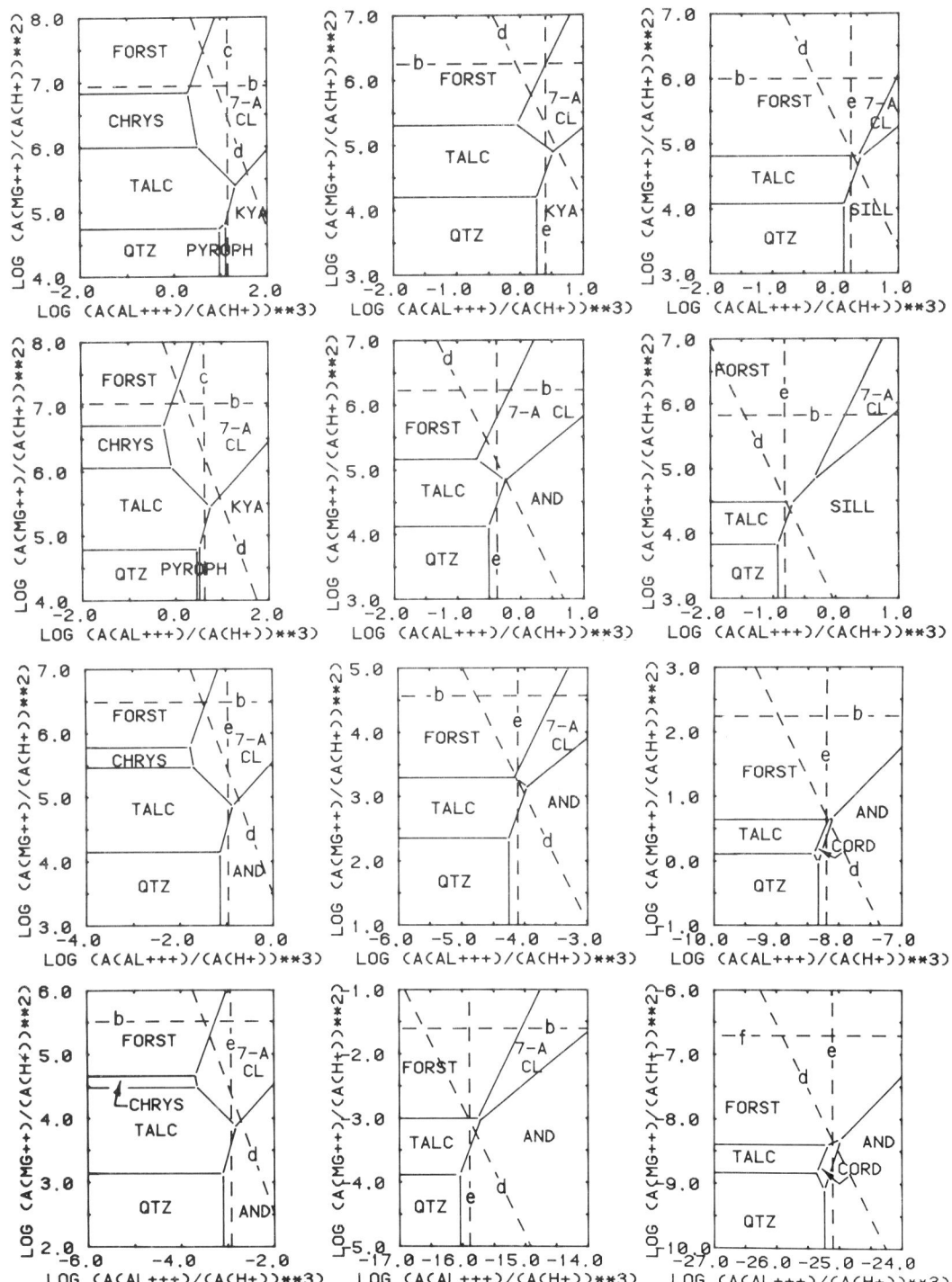

Phase relations in the system $HCl-H_2O-Al_2O_3-MgO-(SiO_2)$. Metastable 7-A clinochlore and chrysotile were considered instead of their stable counterparts, 14-A clinochlore and antigorite. Saturation limits: gibbsite (a), brucite (b), diaspore (c), spinel (d), corundum (e), periclase (f).

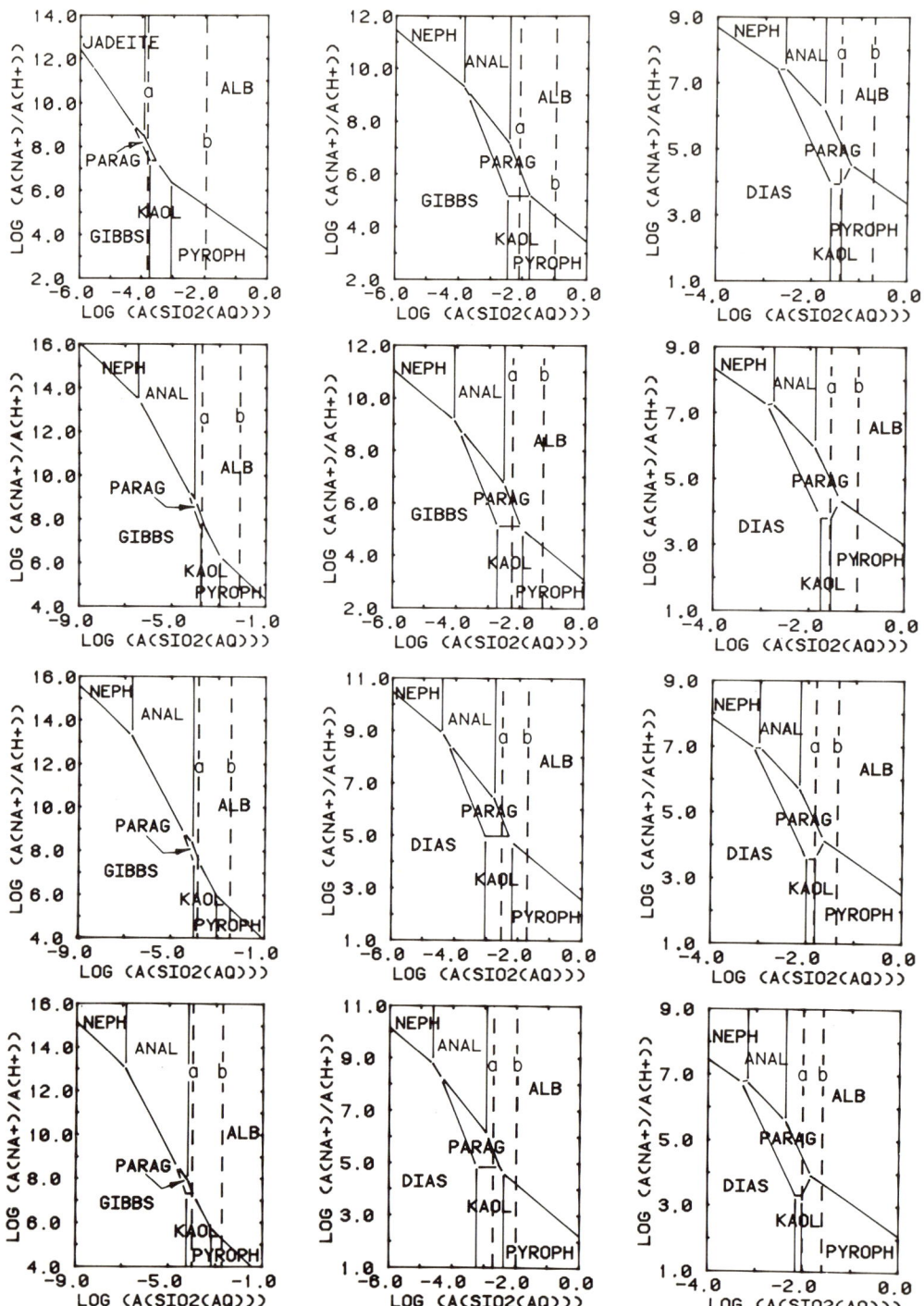

Phase relations in the system $HCl-H_2O-(Al_2O_3)-Na_2O-SiO_2$. Saturation limits: quartz (a), amorphous silica (b).

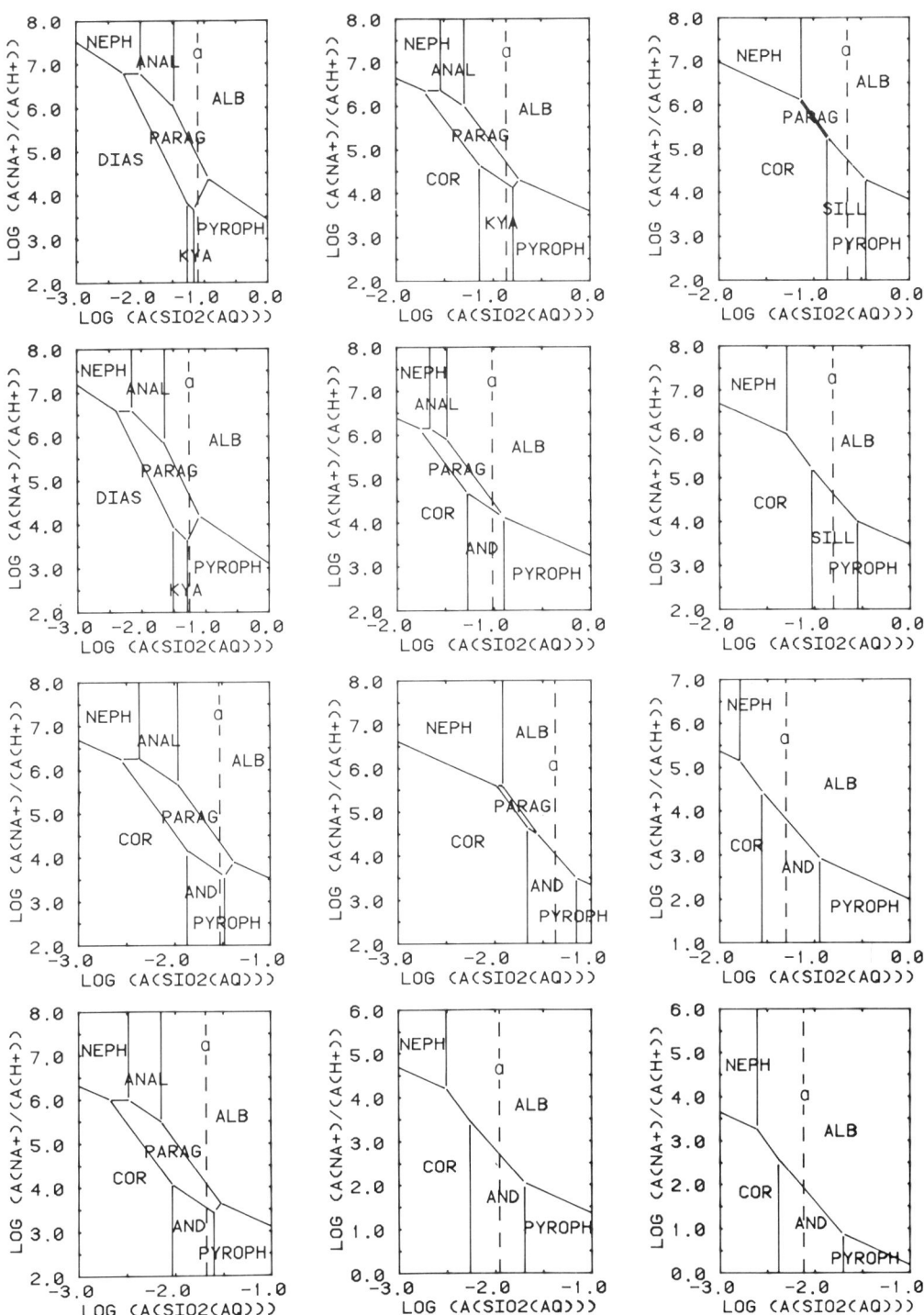

Phase relations in the system $HCl-H_2O-(Al_2O_3)-Na_2O-SiO_2$. Saturation limits: quartz (a), amorphous silica (b).

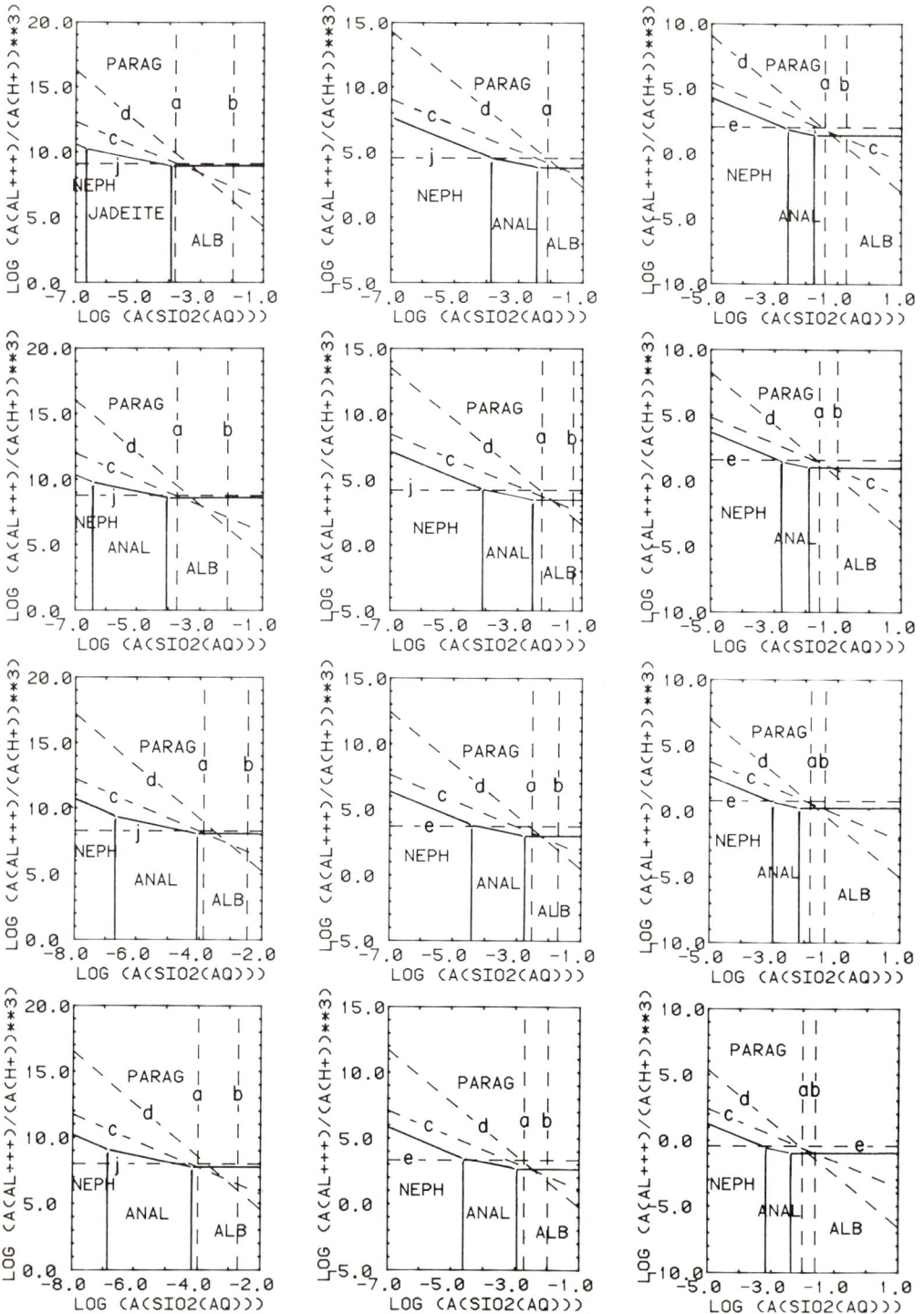

Phase relations in the system $HCl-H_2O-Al_2O_3-(Na_2O)-SiO_2$. Saturation limits: quartz (a), amorphous silica (b), kaolinite (c), pyrophyllite (d), diaspore (e), corundum (f), andalusite (g), kyanite (h), sillimanite (i), gibbsite (j).

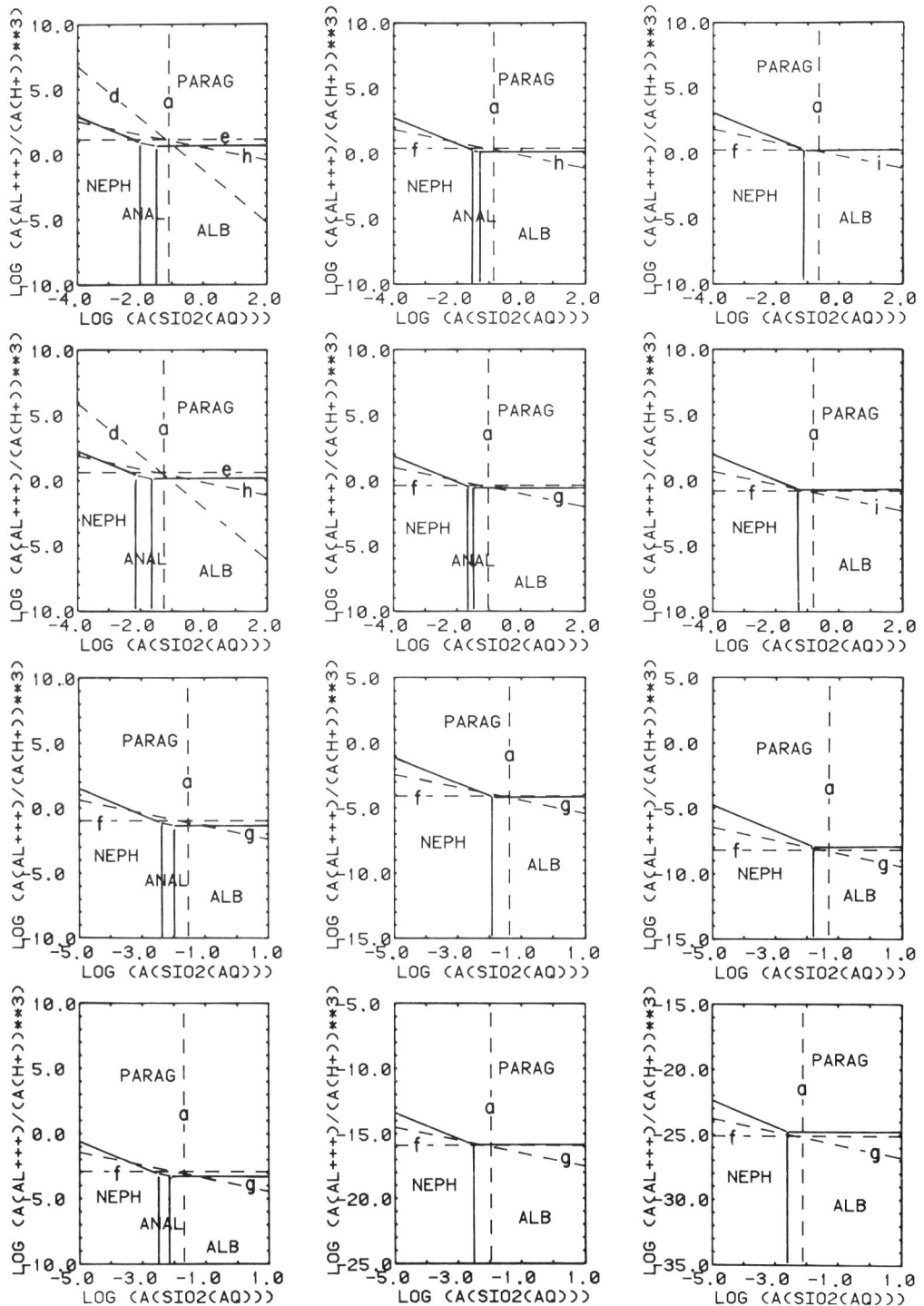

Phase relations in the system HCl−H$_2$O−Al$_2$O$_3$−(Na$_2$O)−SiO$_2$. Saturation limits: quartz (a), amorphous silica (b), kaolinite (c), pyrophyllite (d), diaspore (e), corundum (f), andalusite (g), kyanite (h), sillimanite (i), gibbsite (j).

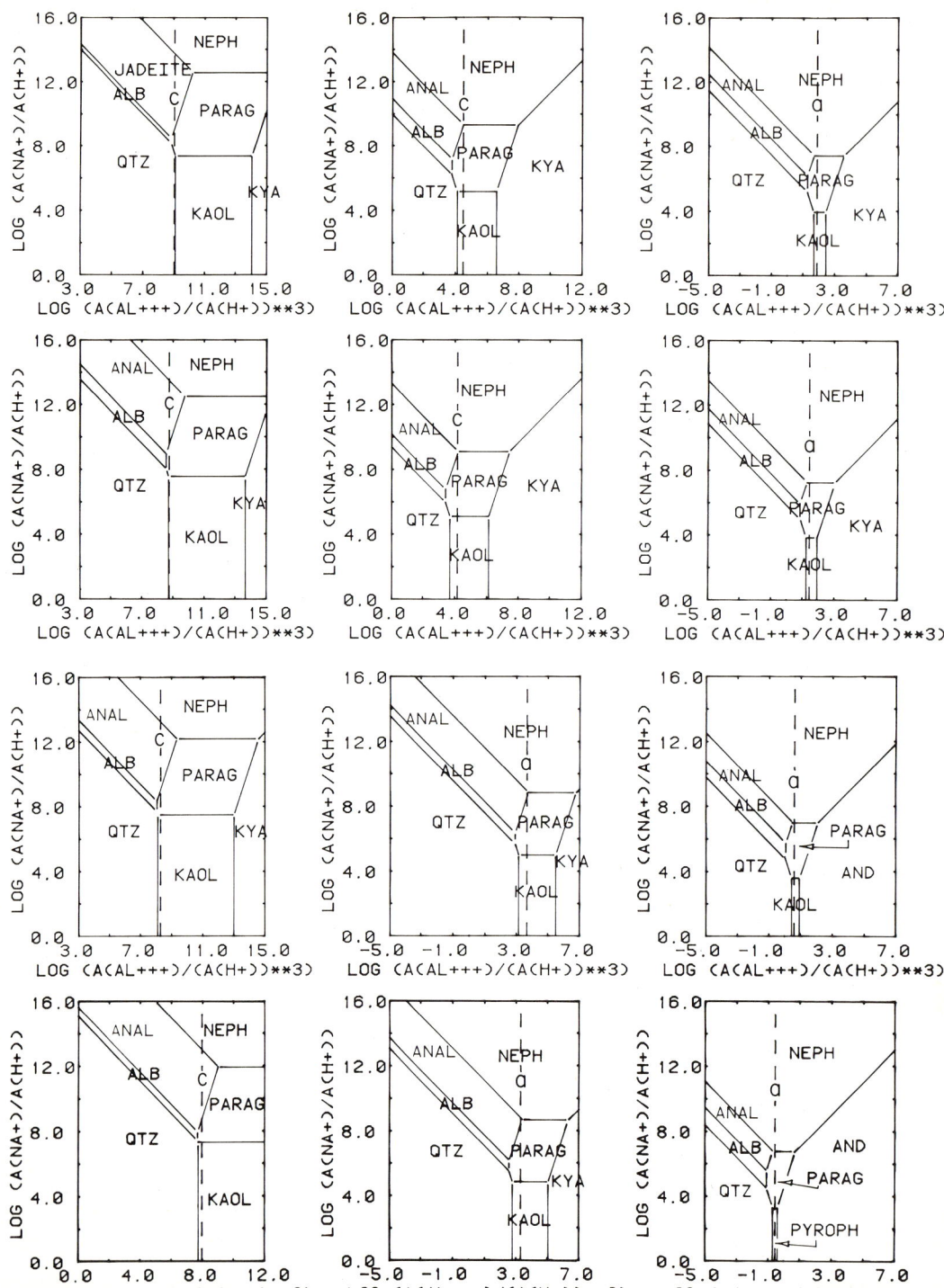

Phase relations in the system $HCl-H_2O-Al_2O_3-Na_2O-(SiO_2)$. Saturation limits: diaspore (a), corundum (b), gibbsite (c).

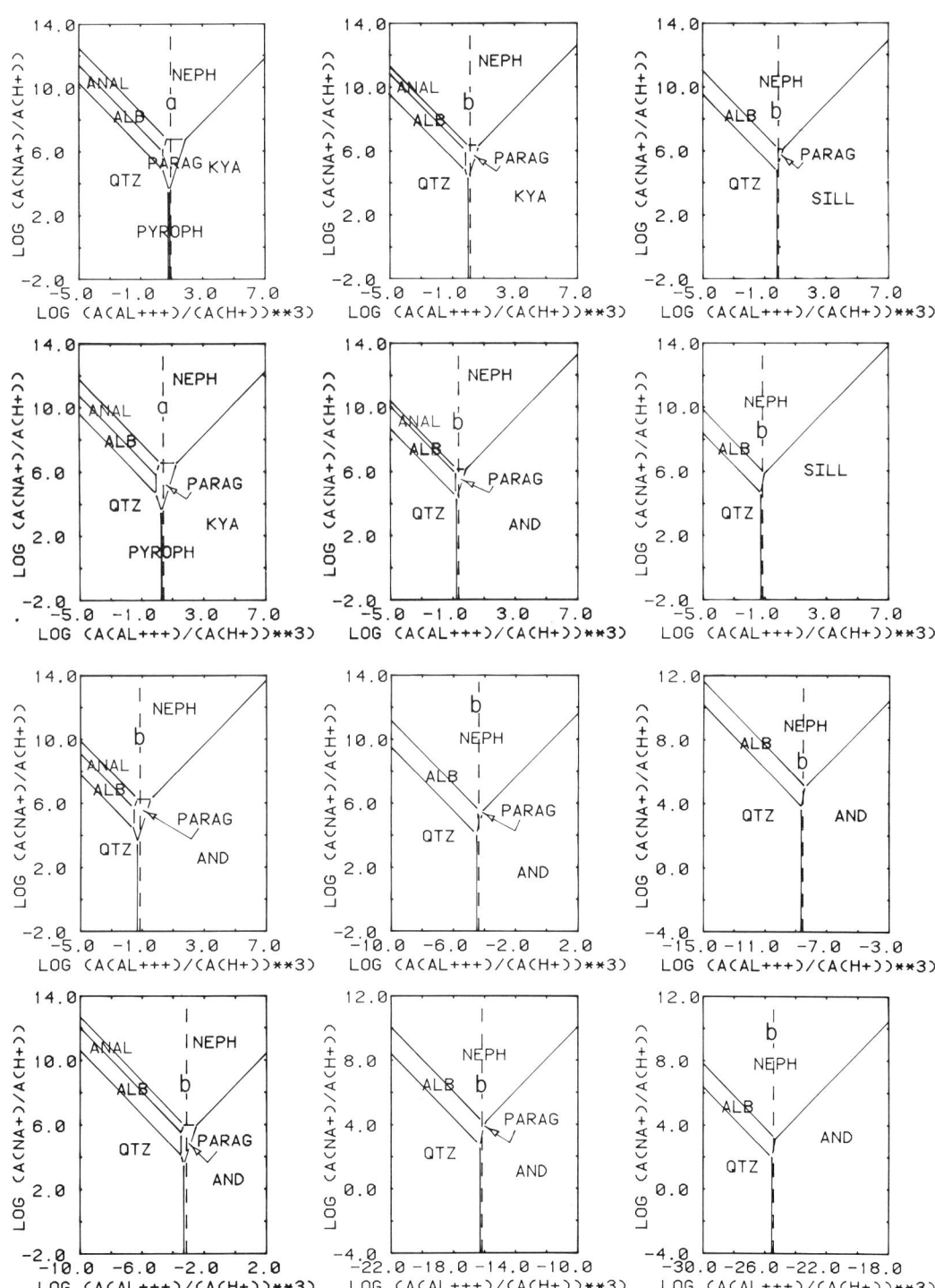

Phase relations in the system $HCl-H_2O-Al_2O_3-Na_2O-(SiO_2)$. Saturation limits: diaspore (a), corundum (b), gibbsite (c).

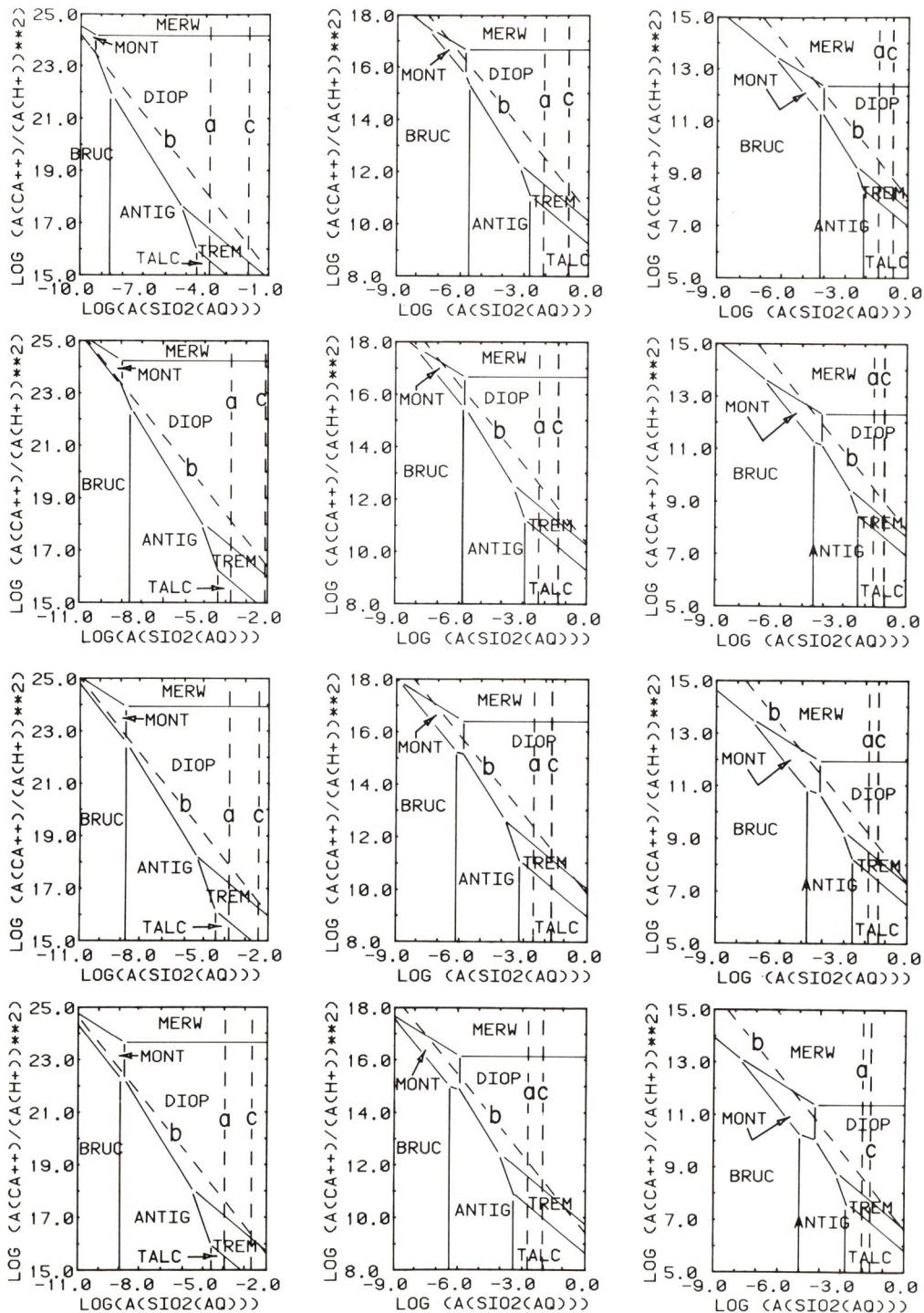

Phase relations in the system $HCl-H_2O-CaO-(MgO)-SiO_2$. Saturation limits: quartz (a), wollastonite (b), amorphous silica (c), lime (d).

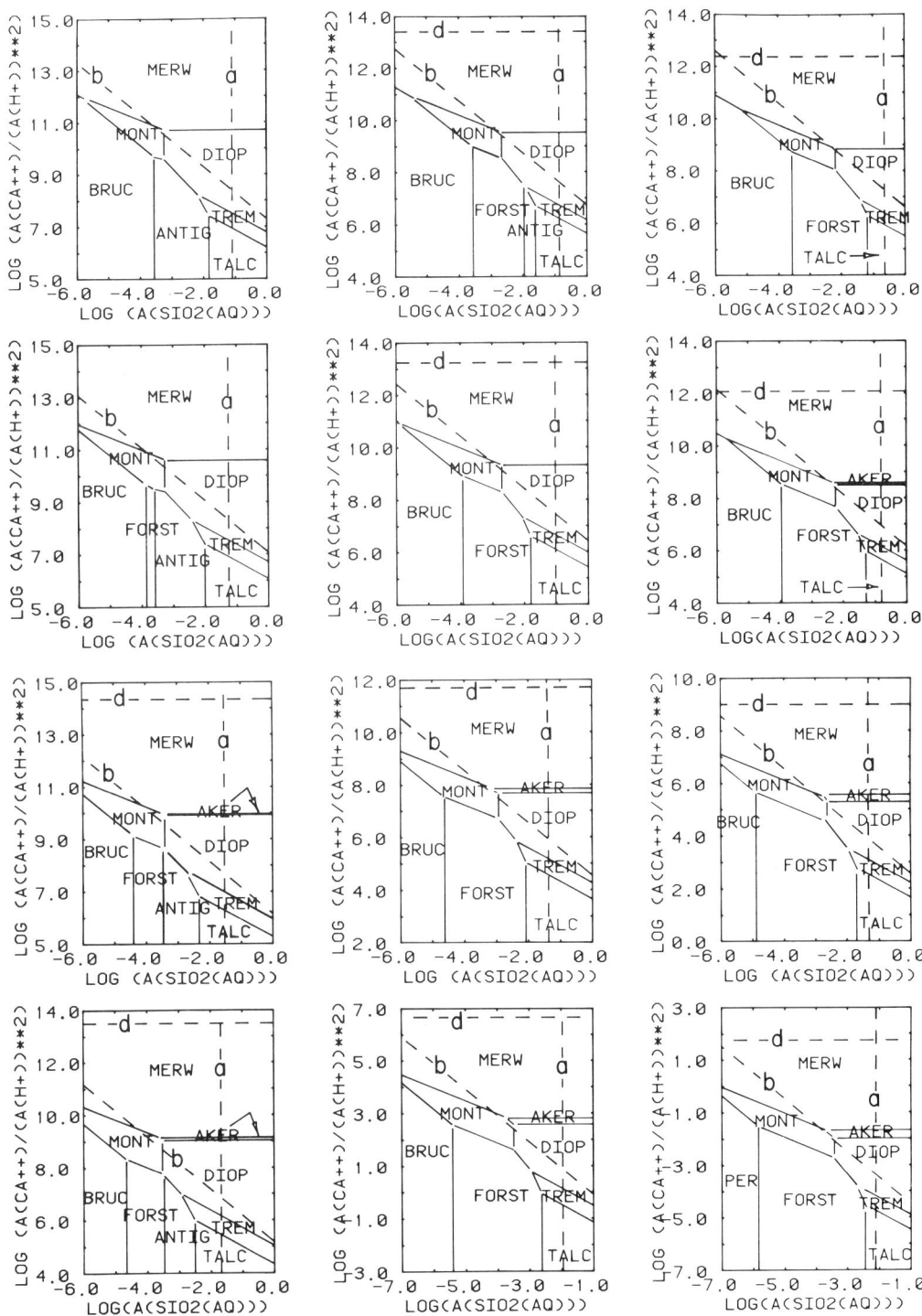

Phase relations in the system $HCl-H_2O-CaO-(MgO)-SiO_2$. Saturation limits: quartz (a), wollastonite (b), amorphous silica (c), lime (d).

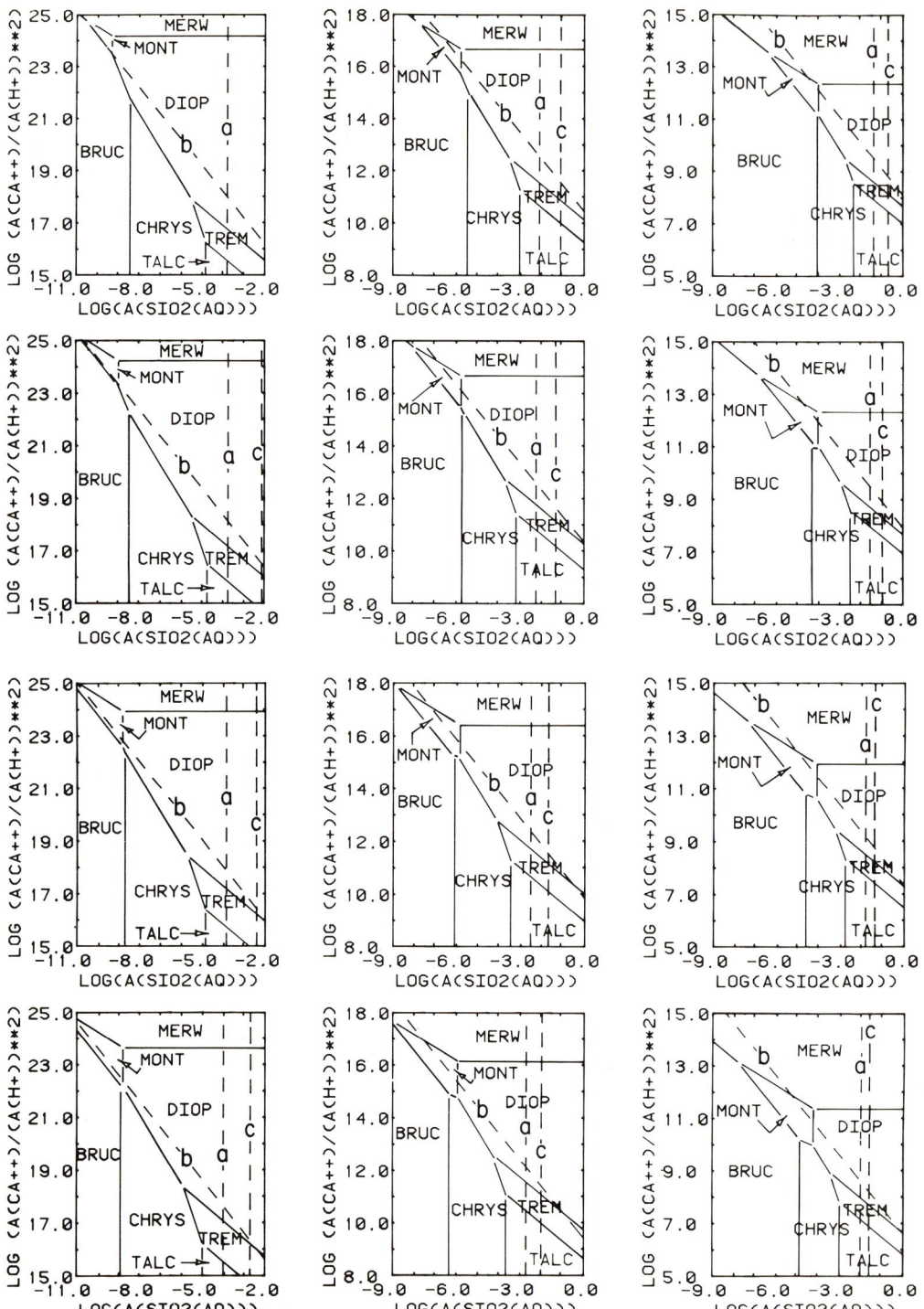

Phase relations in the system $HCl-H_2O-CaO-(MgO)-SiO_2$. Metastable chrysotile was considered instead of its stable counterpart, antigorite. Saturation limits: quartz (a), wollastonite (b), amorphous silica (c), lime (d).

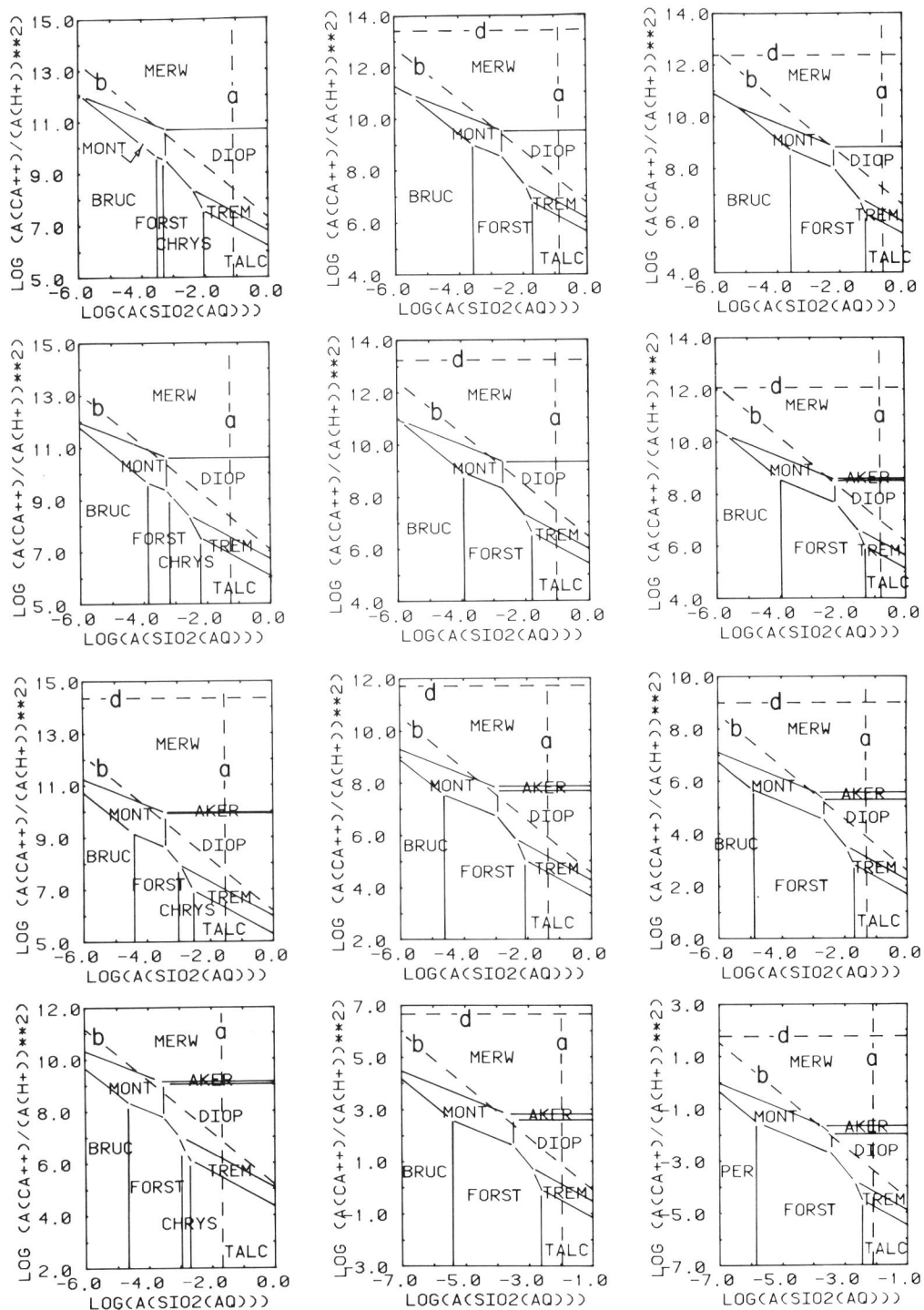

Phase relations in the system $HCl-H_2O-CaO-(MgO)-SiO_2$. Metastable chrysotile was considered instead of its stable counterpart, antigorite. Saturation limits: quartz (a), wollastonite (b), amorphous silica (c), lime (d).

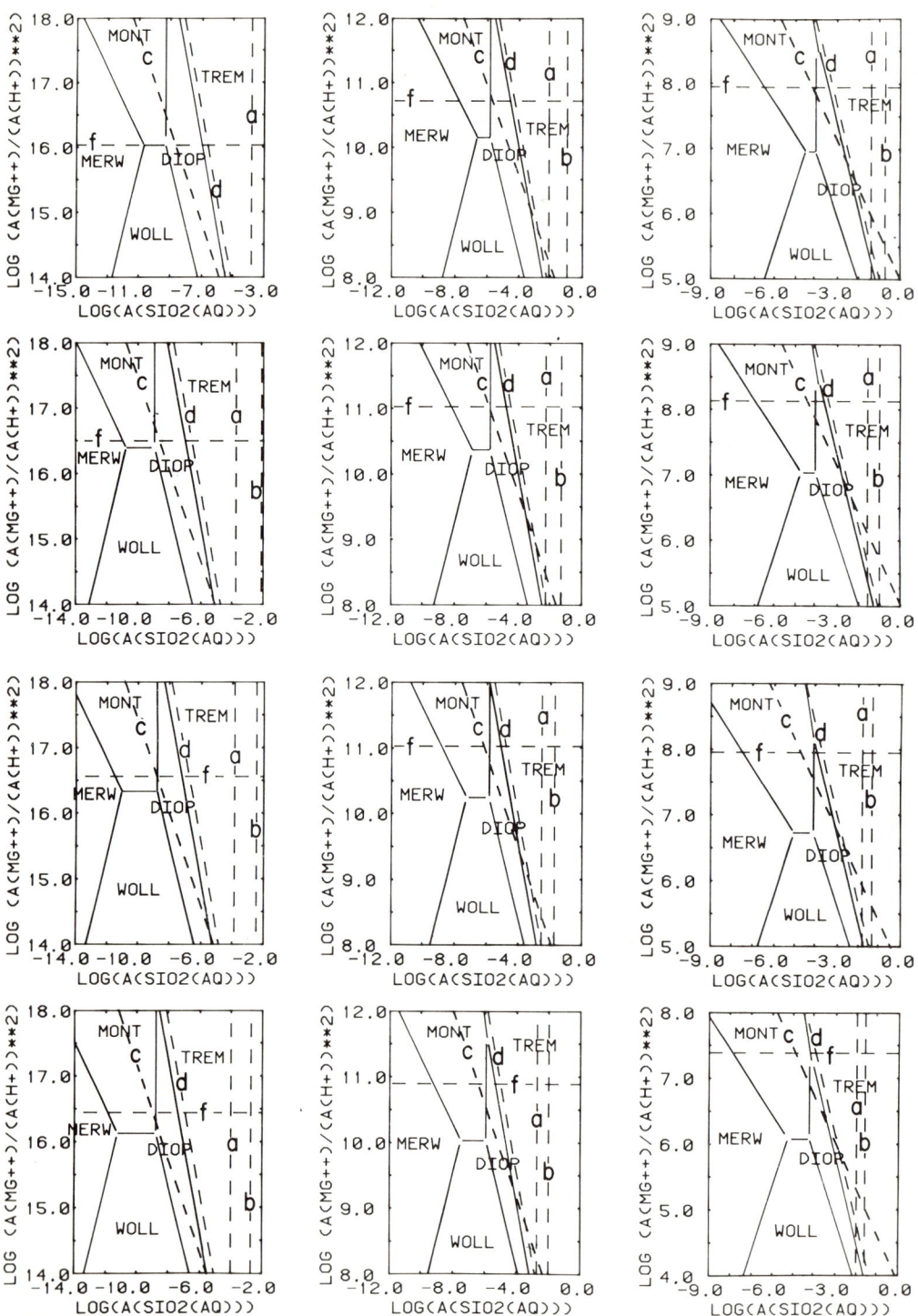

Phase relations in the system HCl−H₂O−(CaO)−MgO−SiO₂. Saturation limits: quartz (a), amorphous silica (b), antigorite (c), talc (d), forsterite (e), brucite (f), periclase (g).

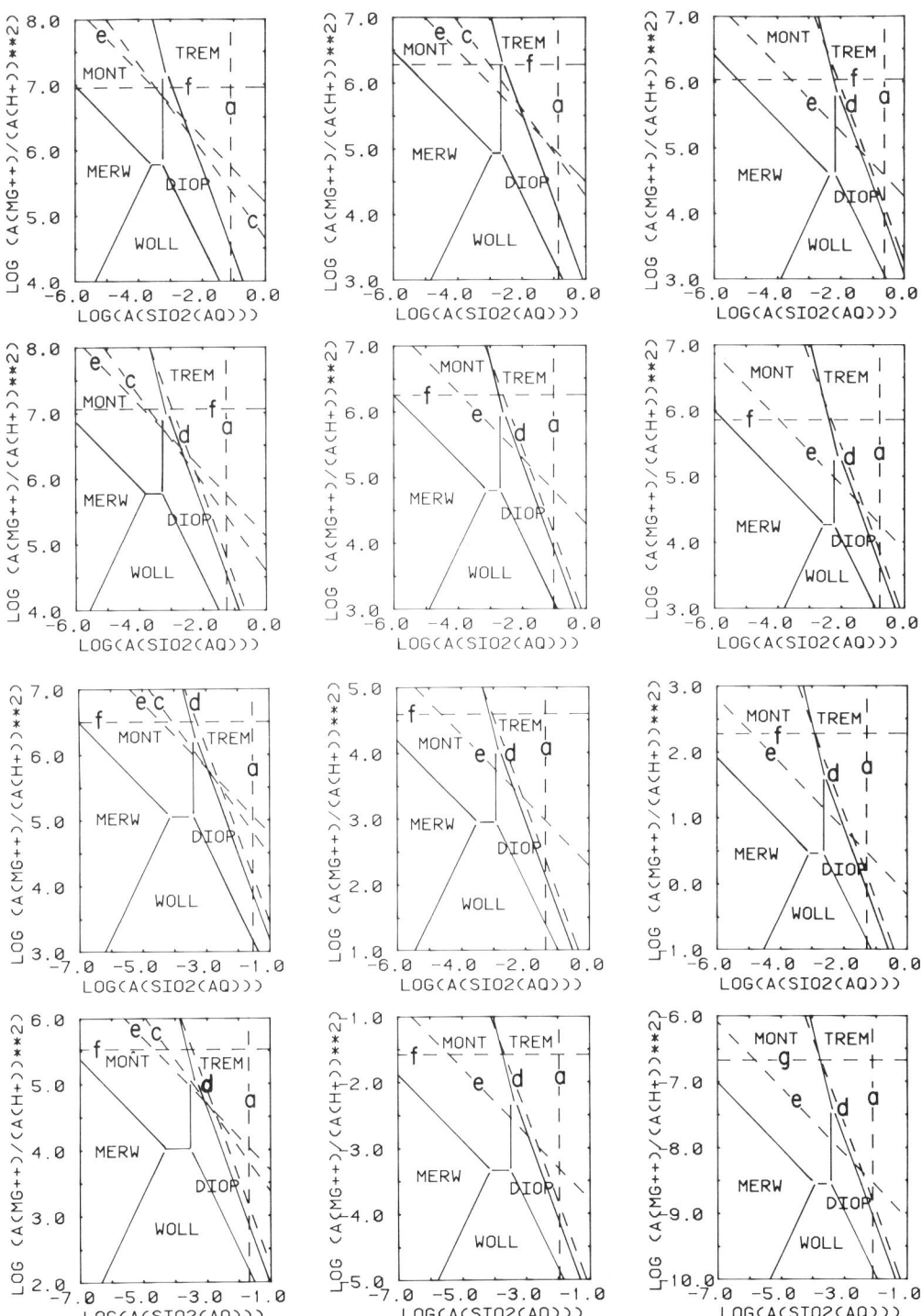

Phase relations in the system HCl−H$_2$O−(CaO)−MgO−SiO$_2$. Saturation limits: quartz (a), amorphous silica (b), antigorite (c), talc (d), forsterite (e), brucite (f), periclase (g).

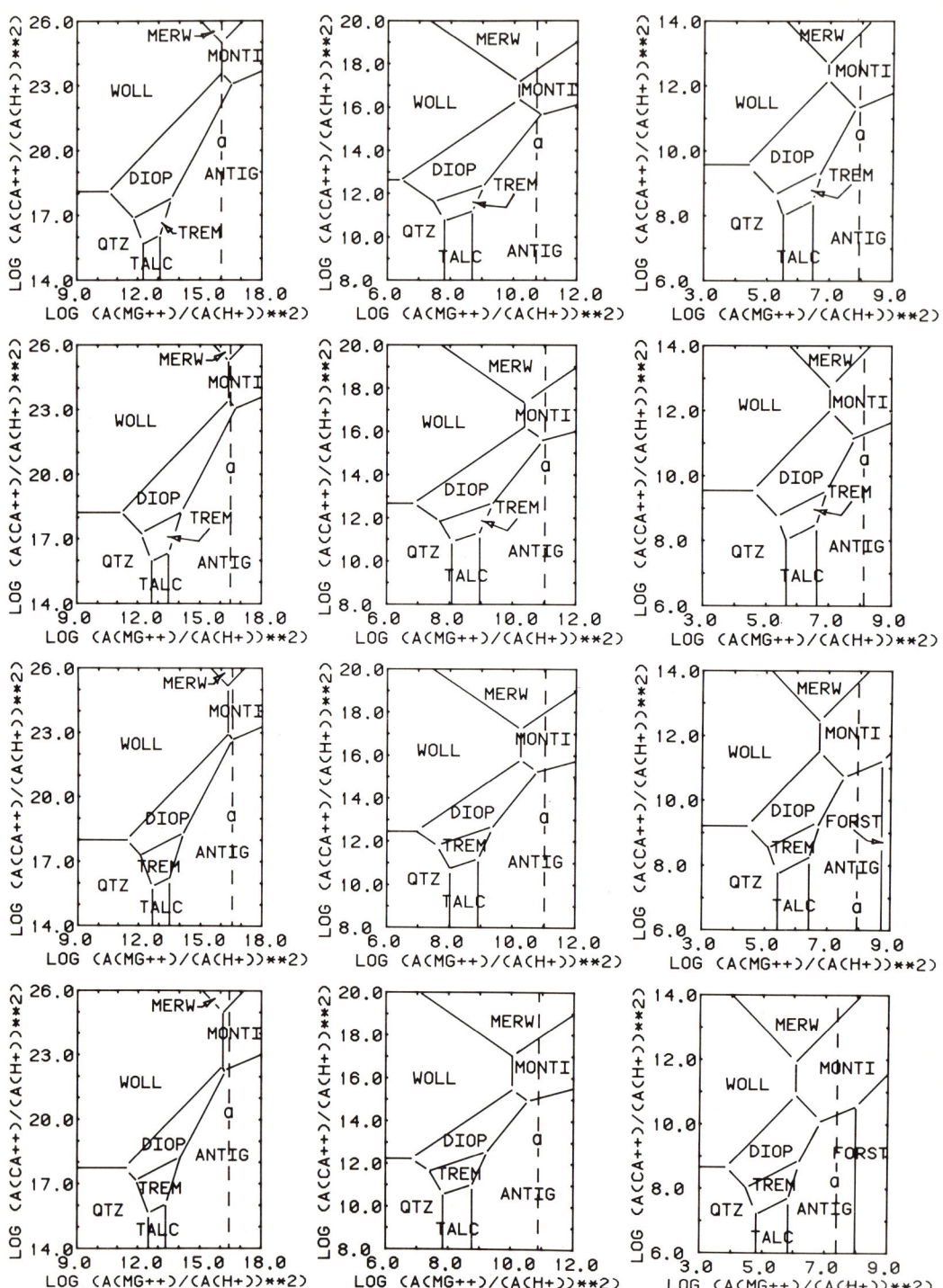

Phase relations in the system $HCl-H_2O-CaO-MgO-(SiO_2)$. Saturation limits: brucite (a), periclase (b).

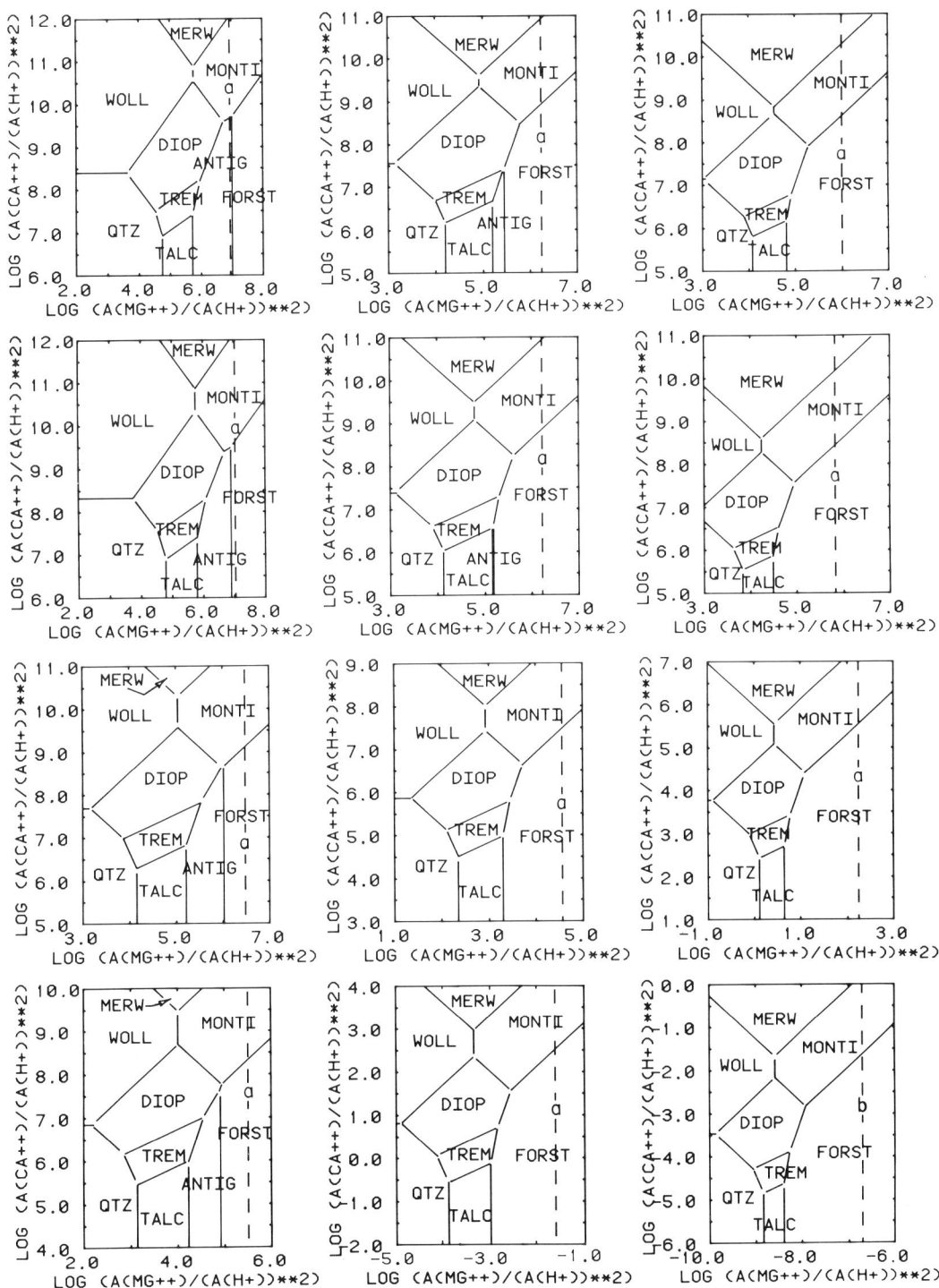

Phase relations in the system $HCl-H_2O-CaO-MgO-(SiO_2)$. Saturation limits: brucite (a), periclase (b).

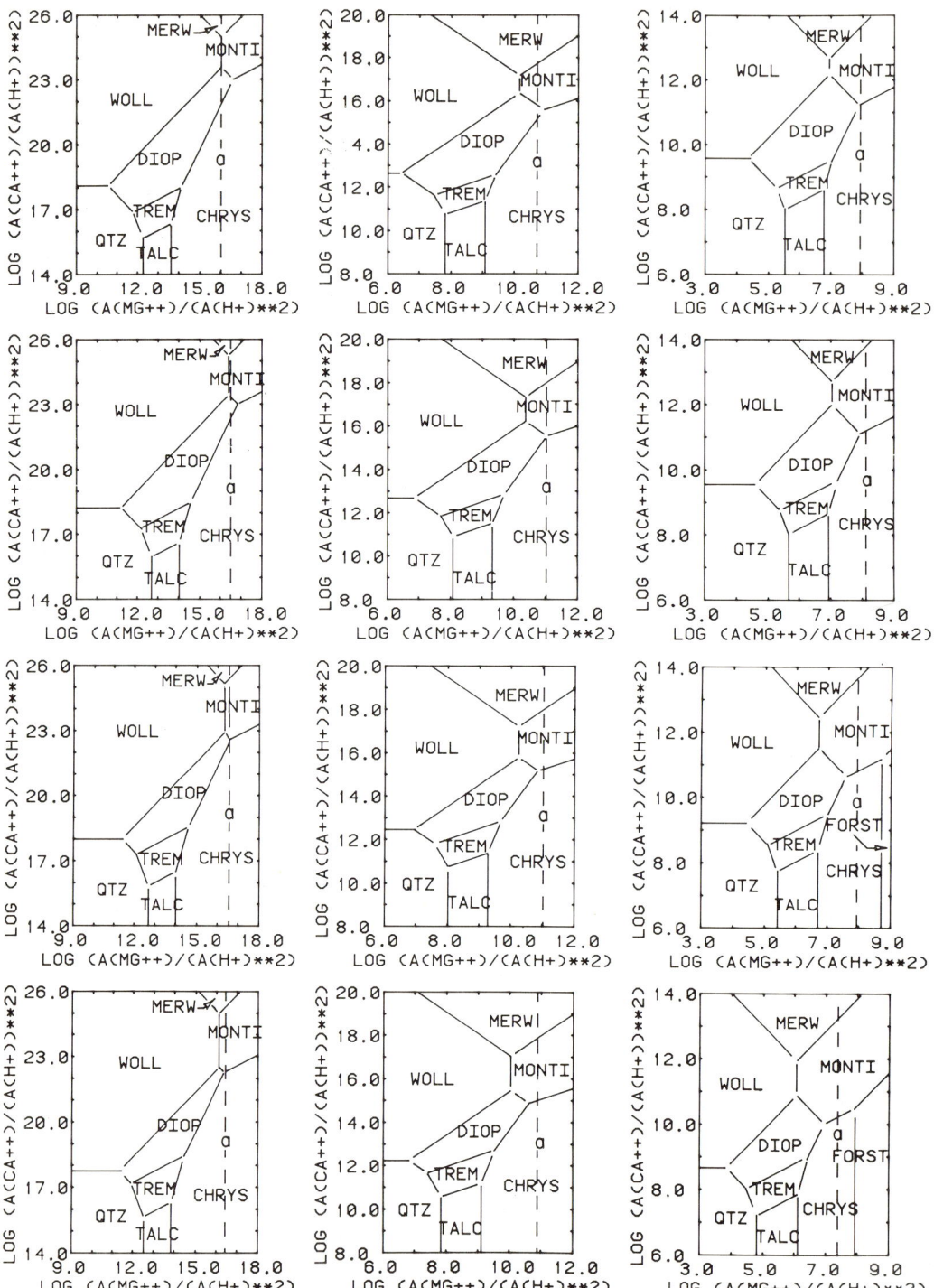

Phase relations in the system HCl−H₂O−CaO−MgO−(SiO₂). Metastable chrysotile was considered instead of its stable counterpart, antigorite. Saturation limits: brucite (a), periclase (b).

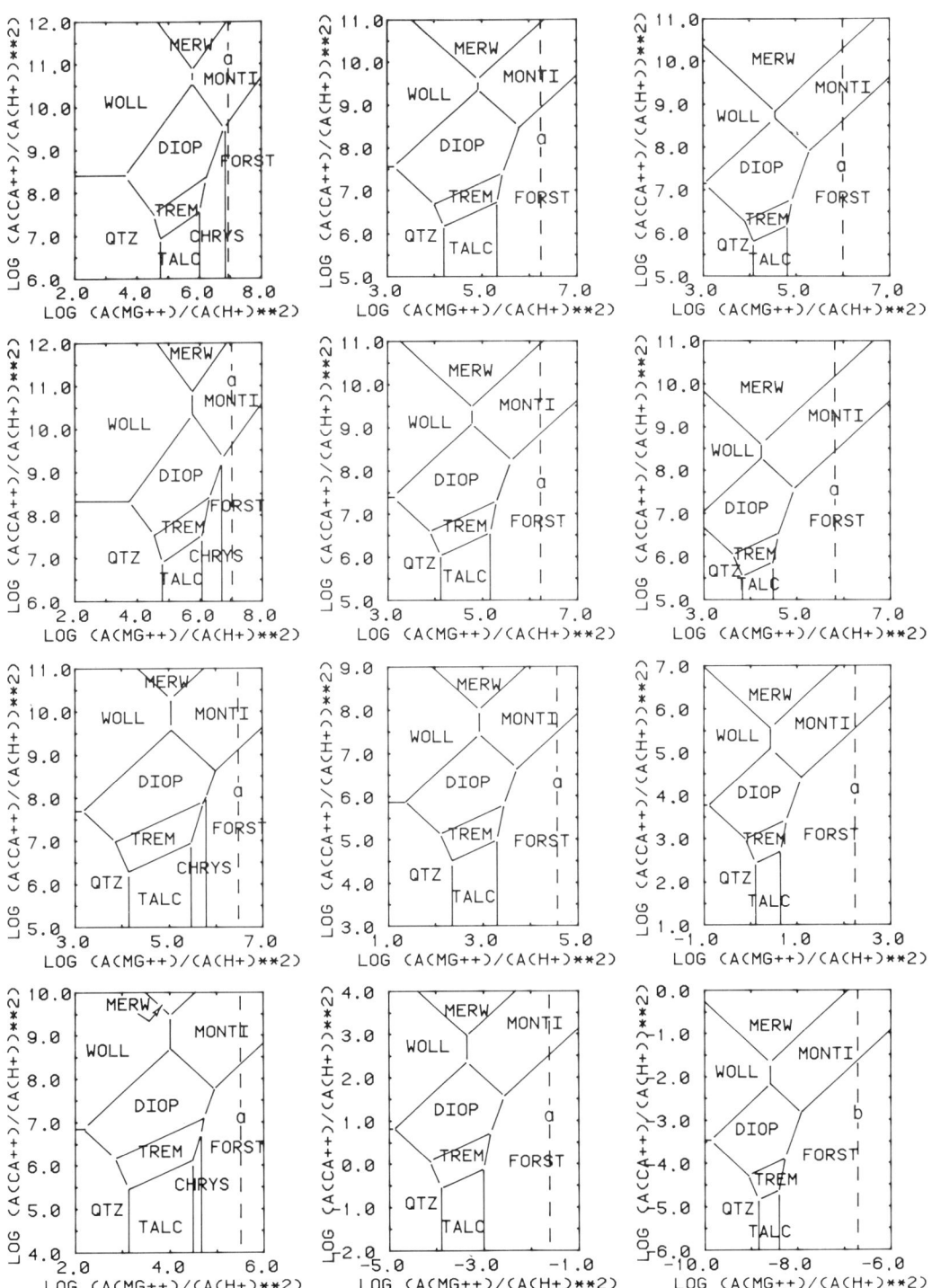

Phase relations in the system $HCl-H_2O-CaO-MgO-(SiO_2)$. Metastable chrysotile was considered instead of its stable counterpart, antigorite. Saturation limits: brucite (a), periclase (b).

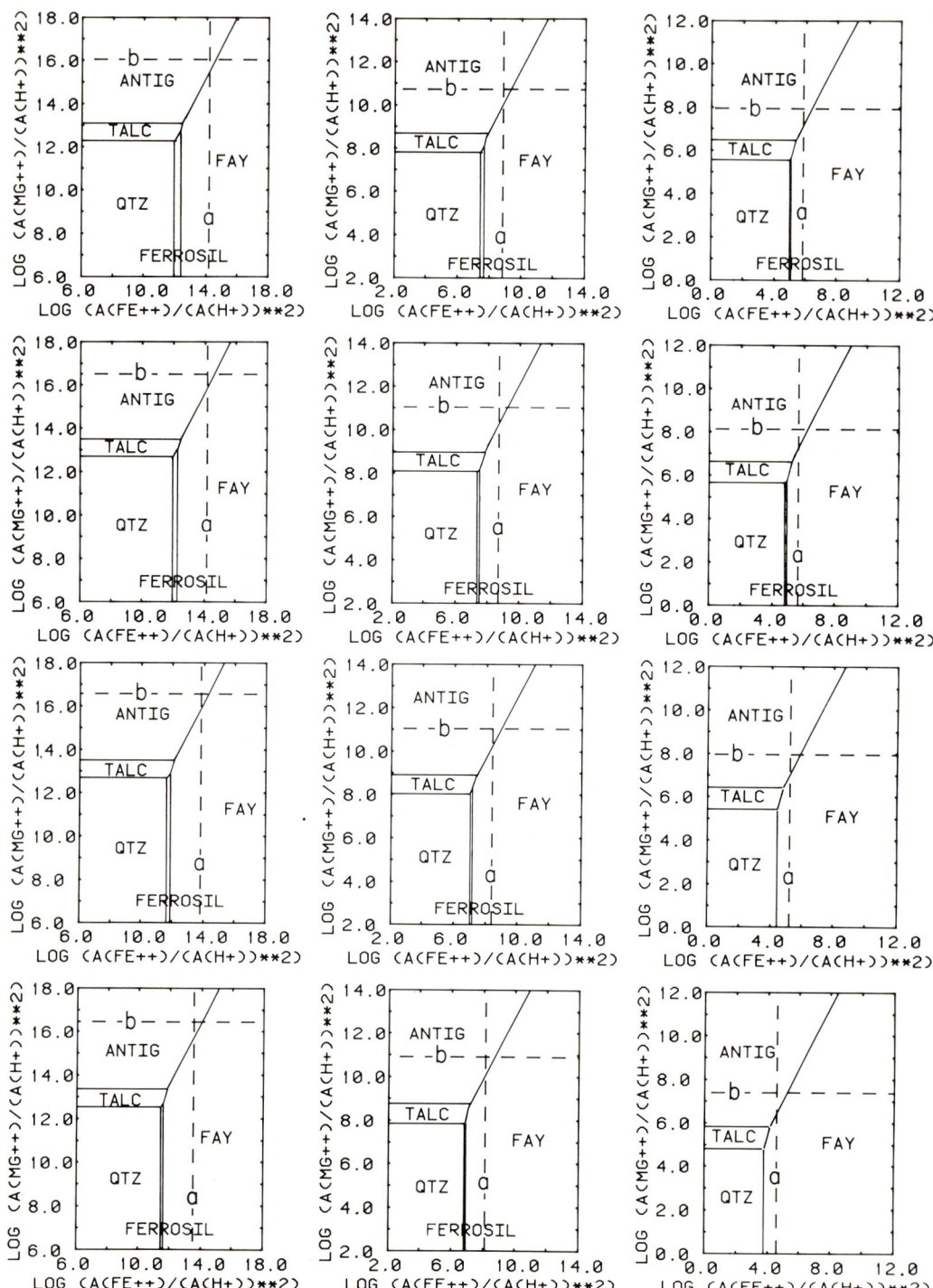

Phase relations in the system $HCl-H_2O-FeO-MgO-(SiO_2)$. Saturation limits: ferrous oxide (a), brucite (b), periclase (c).

FRANK J. MILLERO

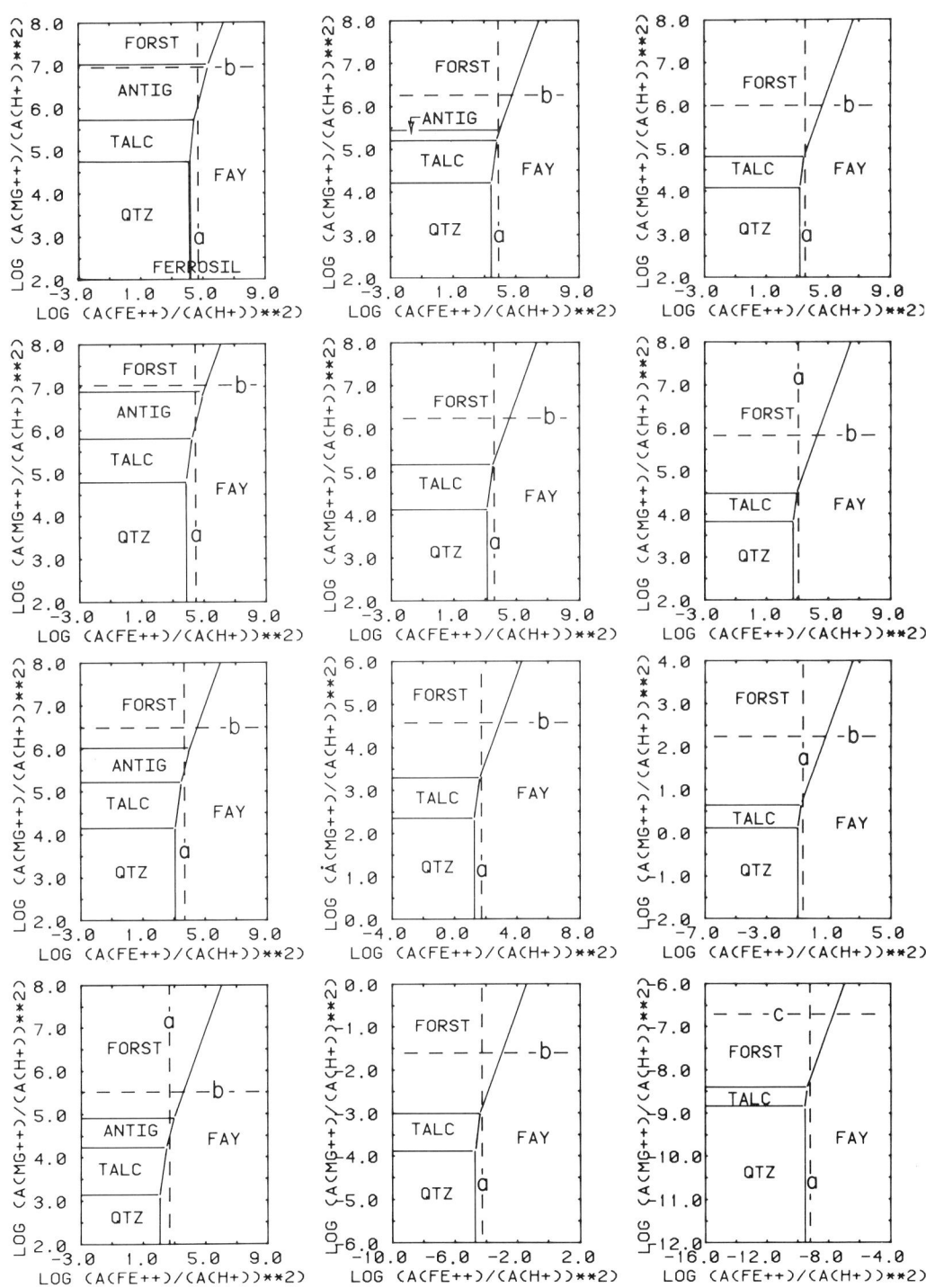

Phase relations in the system $HCl-H_2O-FeO-MgO-(SiO_2)$. Saturation limits: ferrous oxide (a), brucite (b), periclase (c).

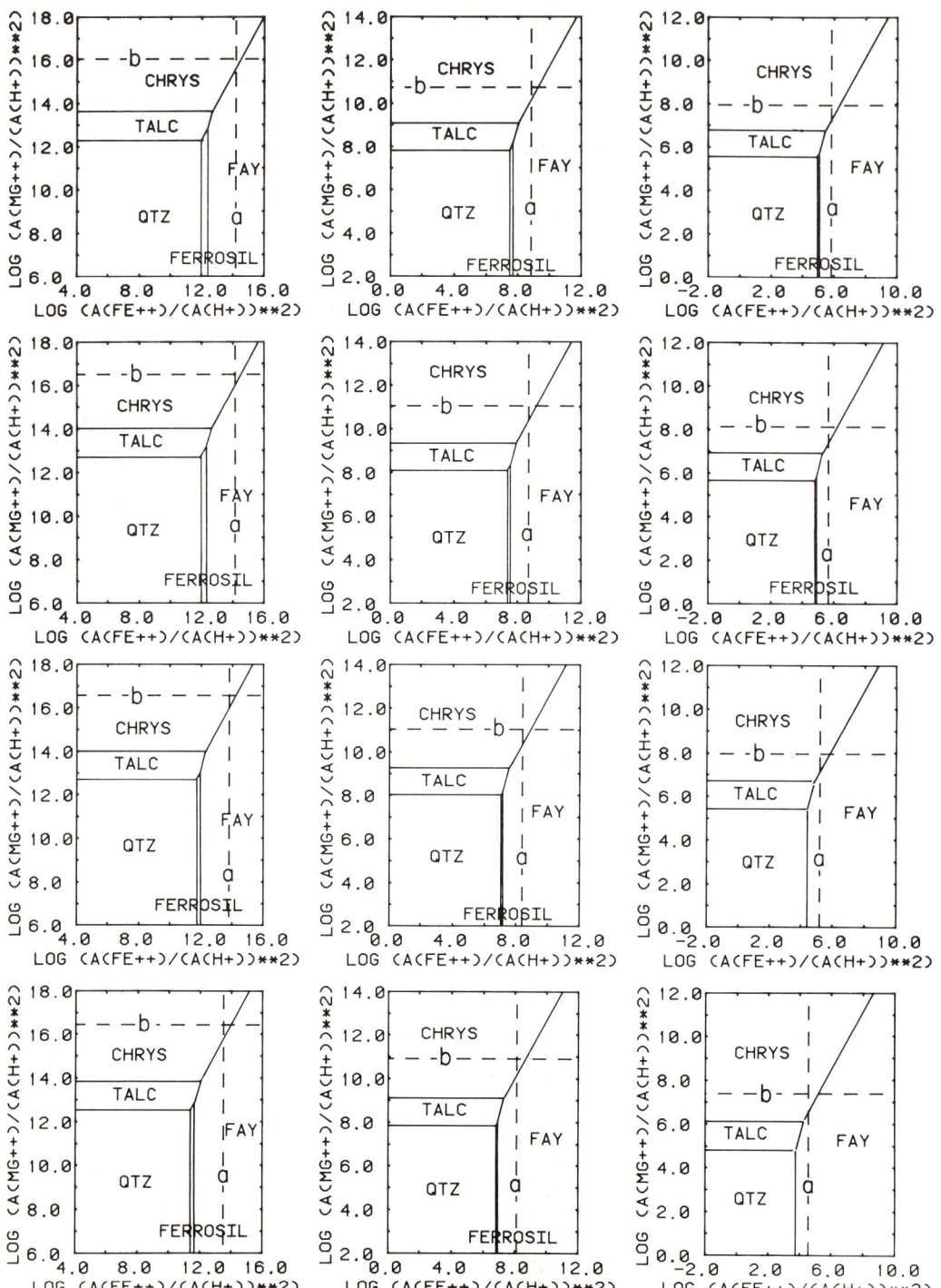

Phase relations in the system $HCl-H_2O-FeO-MgO-(SiO_2)$. Metastable chrysotile was considered instead of its stable counterpart, antigorite. Saturation limits: ferrous oxide (a), brucite (b), periclase (c).

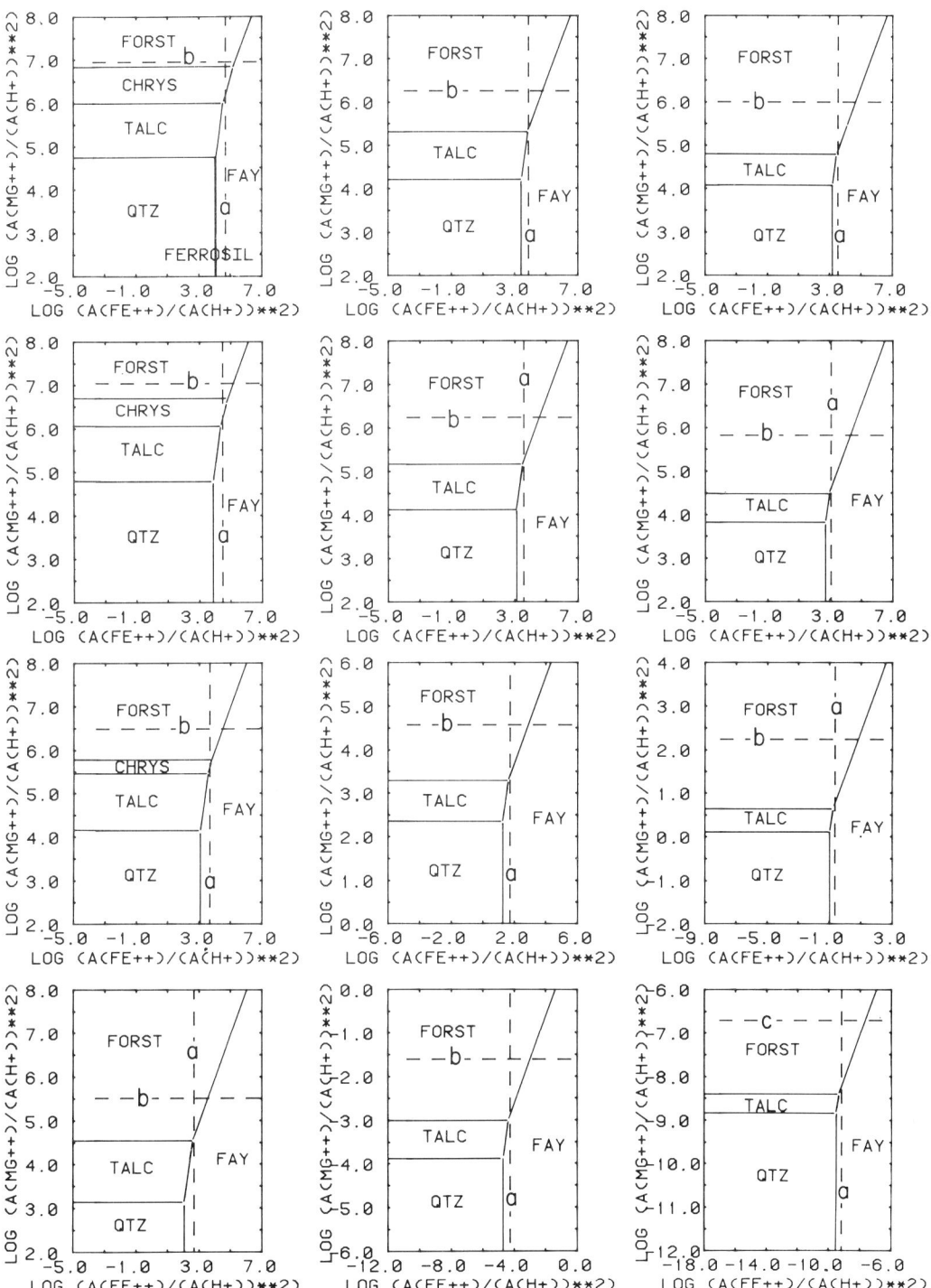

Phase relations in the system $HCl-H_2O-FeO-MgO-(SiO_2)$. Metastable chrysotile was considered instead of its stable counterpart, antigorite. Saturation limits: ferrous oxide (a), brucite (b), periclase (c).

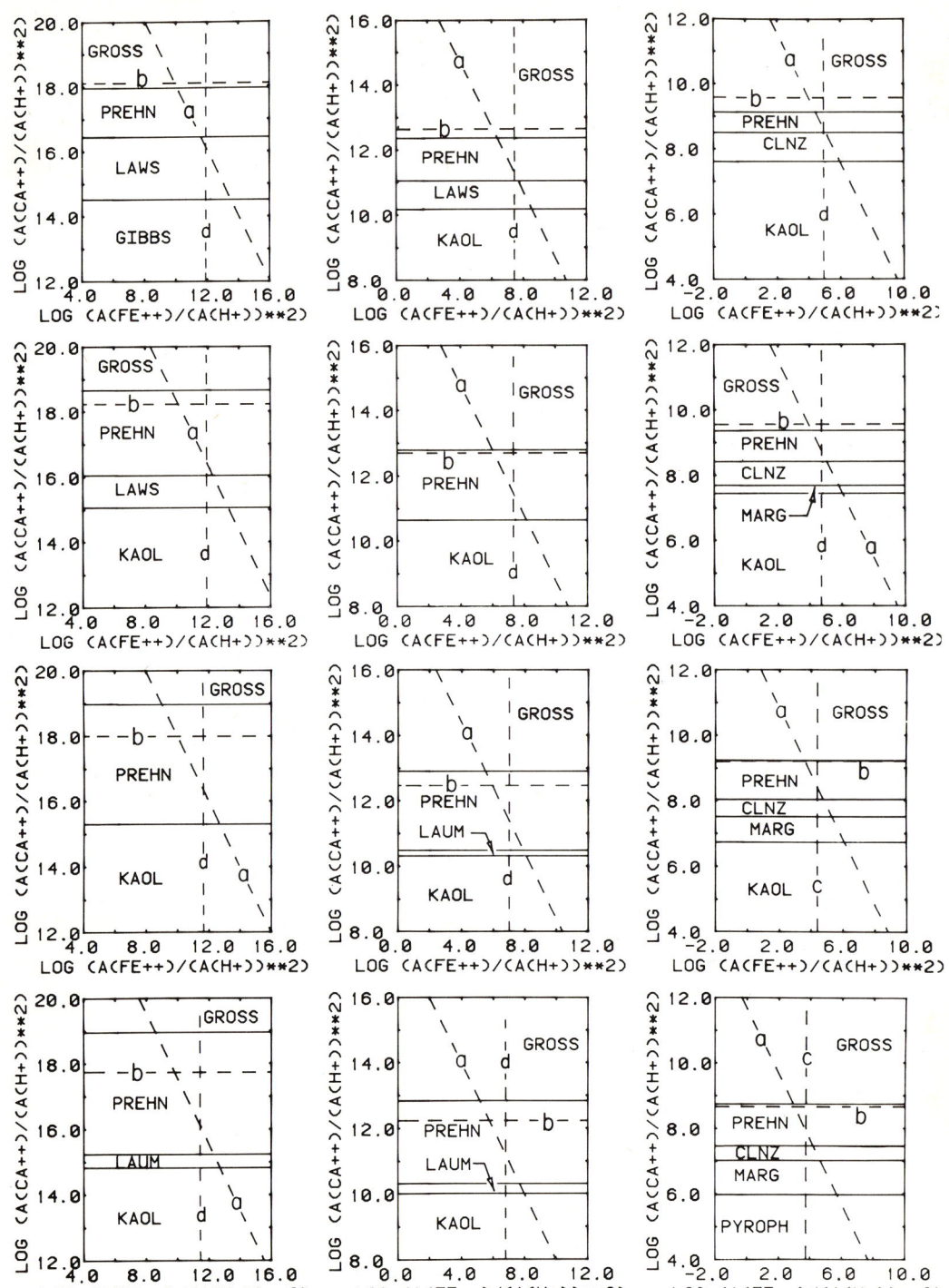

Phase relations in the system $HCl-H_2O-(Al_2O_3)-CaO-FeO-SiO_2$ in equilibrium with quartz. Saturation limits: hedenbergite (a), wollastonite (b), fayalite (c), ferrosilite (d).

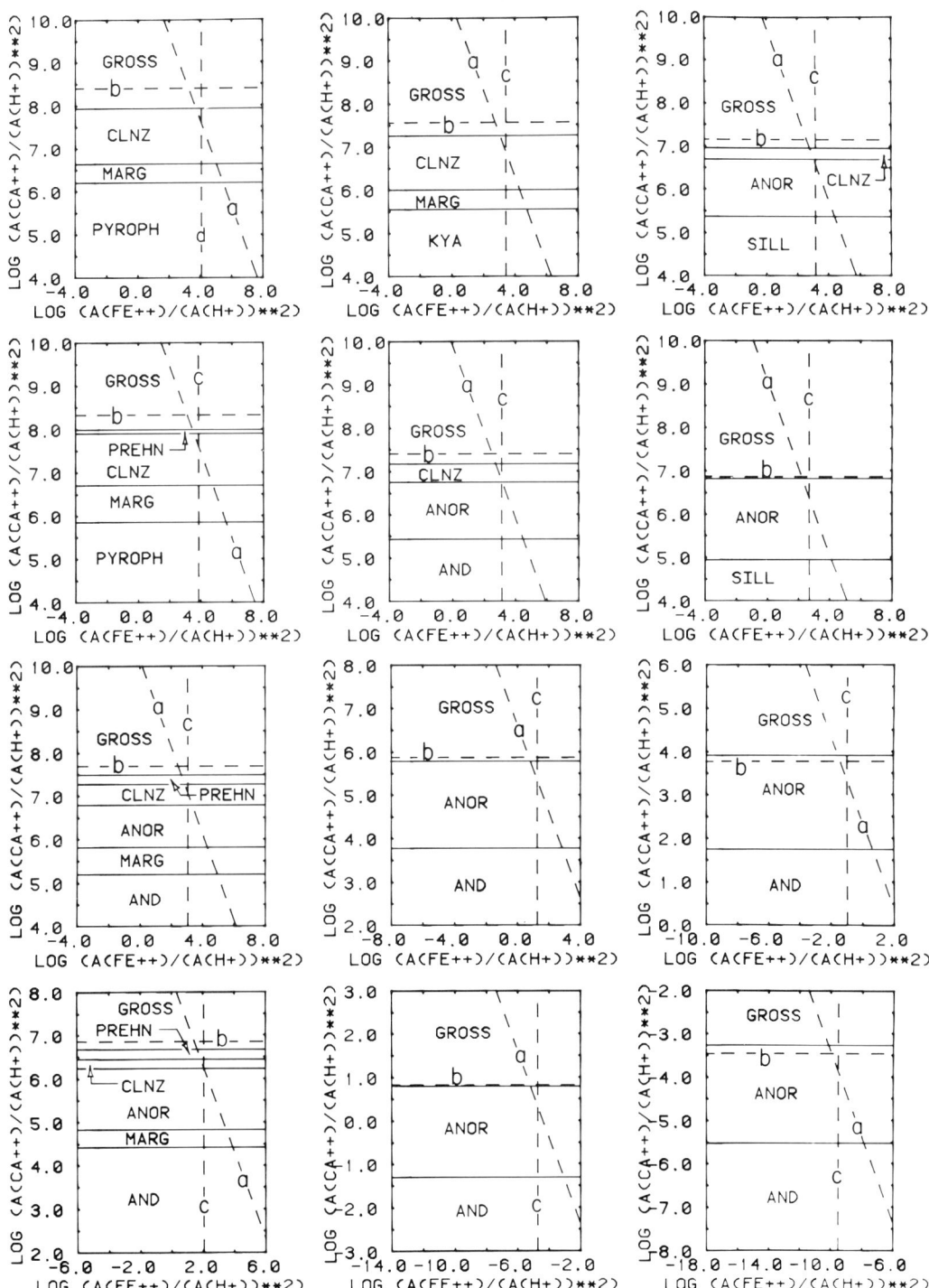

Phase relations in the system $HCl-H_2O-(Al_2O_3)-CaO-FeO-SiO_2$ in equilibrium with quartz. Saturation limits: hedenbergite (a), wollastonite (b), fayalite (c), ferrosilite (d).

56

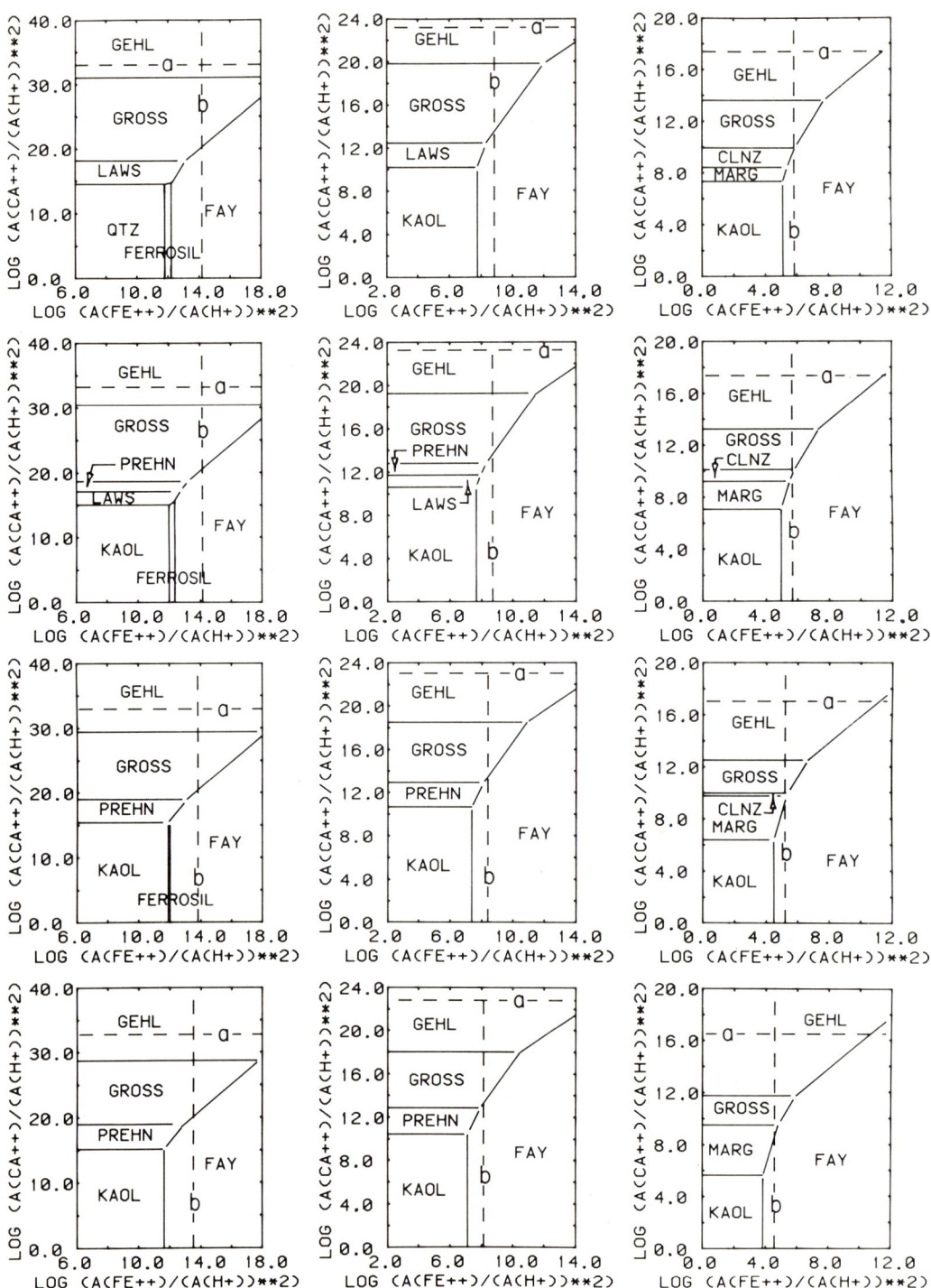

Phase relations in the system $HCl-H_2O-Al_2O_3-CaO-FeO-(SiO_2)$ in equilibrium with gibbsite, diaspore, or corundum, depending on which mineral is stable at each pressure and temperature. Saturation limits: lime (a), ferrous oxide (b).

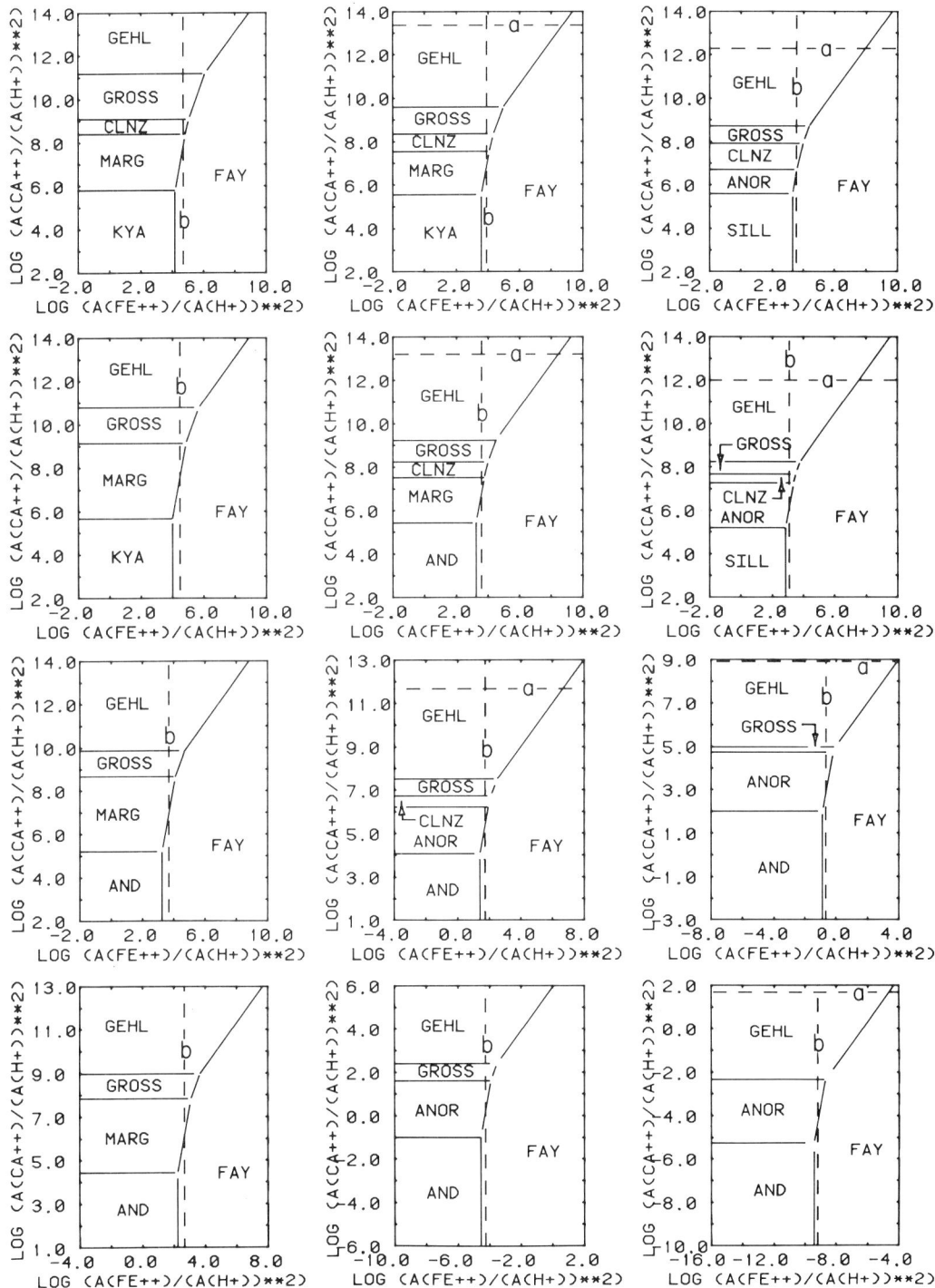

Phase relations in the system $HCl-H_2O-Al_2O_3-CaO-FeO-(SiO_2)$ in equilibrium with gibbsite, diaspore, or corundum, depending on which mineral is stable at each pressure and temperature. Saturation limits: lime (a), ferrous oxide (b).

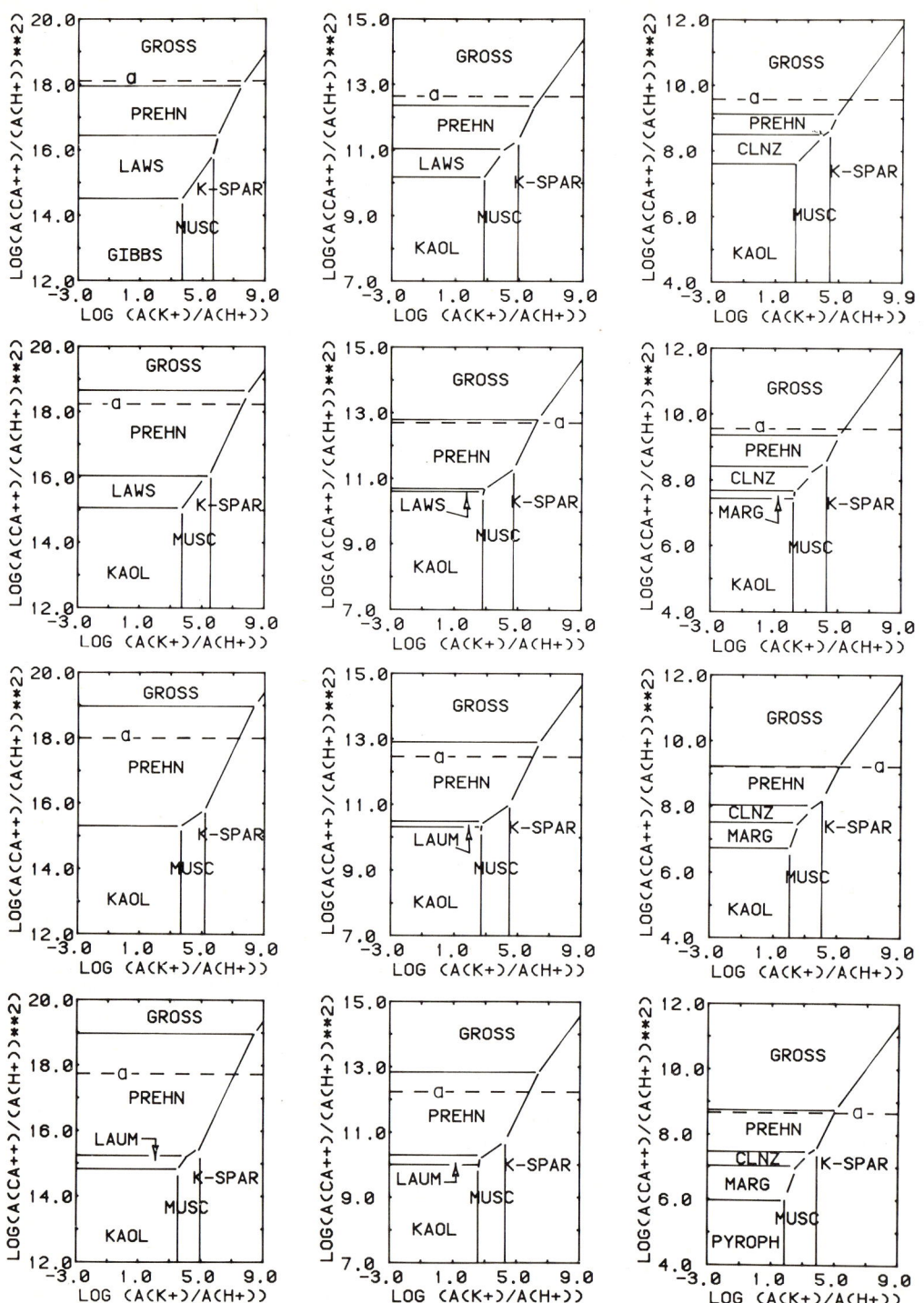

Phase relations in the system $HCl-H_2O-(Al_2O_3)-CaO-K_2O-SiO_2$ in equilibrium with quartz. Saturation limit: wollastonite (a).

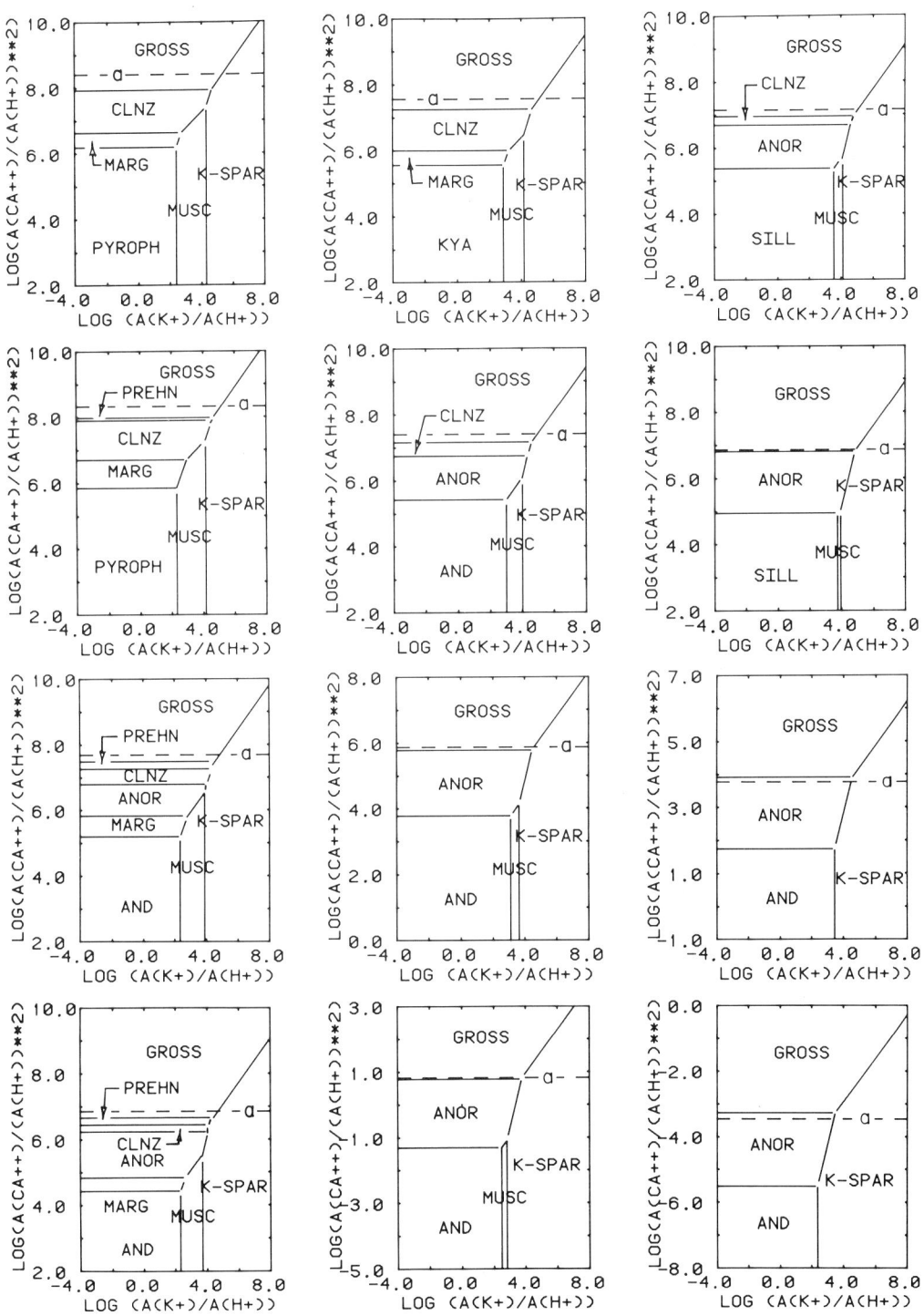

Phase relations in the system $HCl-H_2O-(Al_2O_3)-CaO-K_2O-SiO_2$ in equilibrium with quartz. Saturation limit: wollastonite (a).

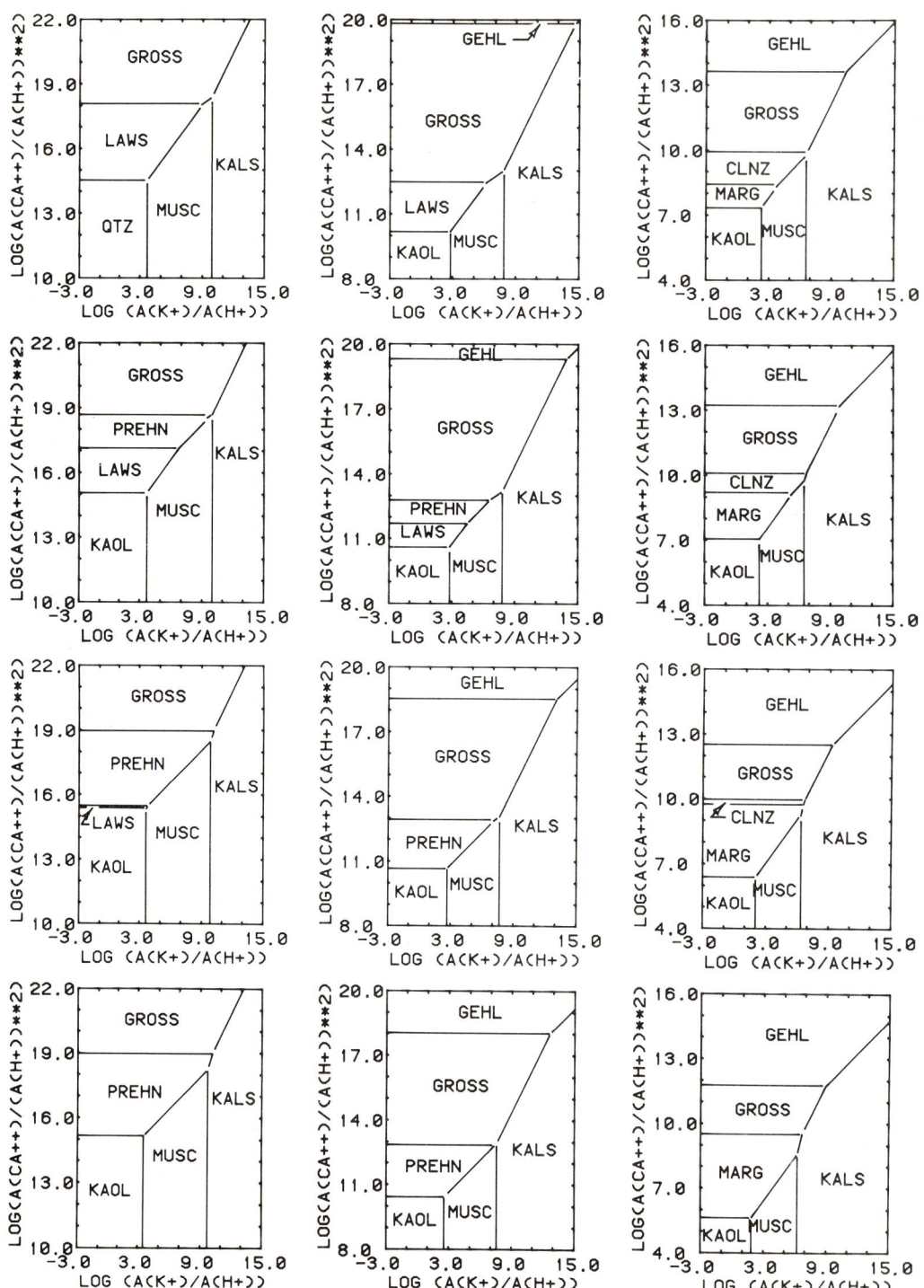

Phase relations in the system $HCl-H_2O-Al_2O_3-CaO-K_2O-(SiO_2)$ in equilibrium with gibbsite, diaspore, or corundum, depending on which mineral is stable at each pressure and temperature.

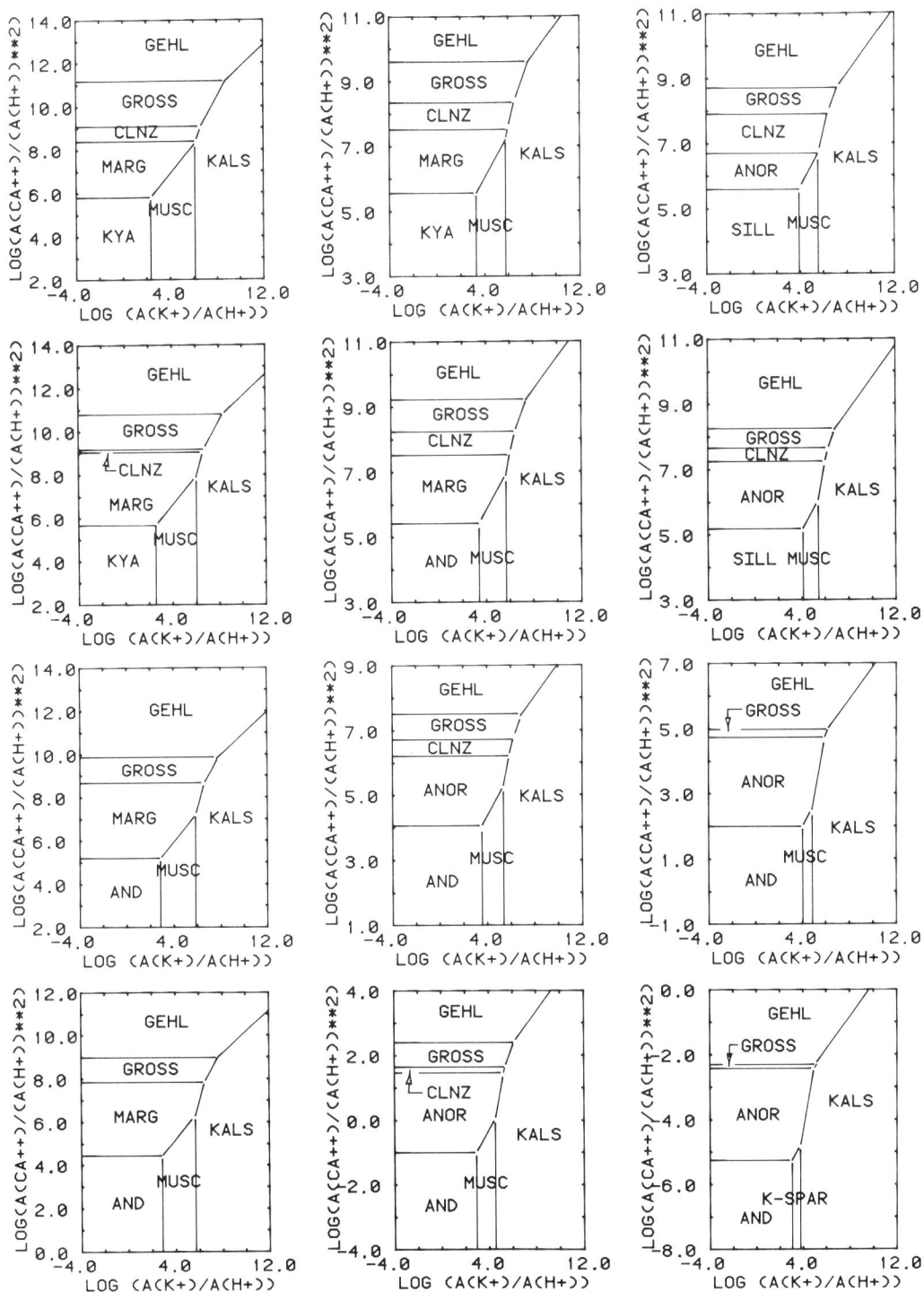

Phase relations in the system $HCl-H_2O-Al_2O_3-CaO-K_2O-(SiO_2)$ in equilibrium with gibbsite, diaspore, or corundum, depending on which mineral is stable at each pressure and temperature.

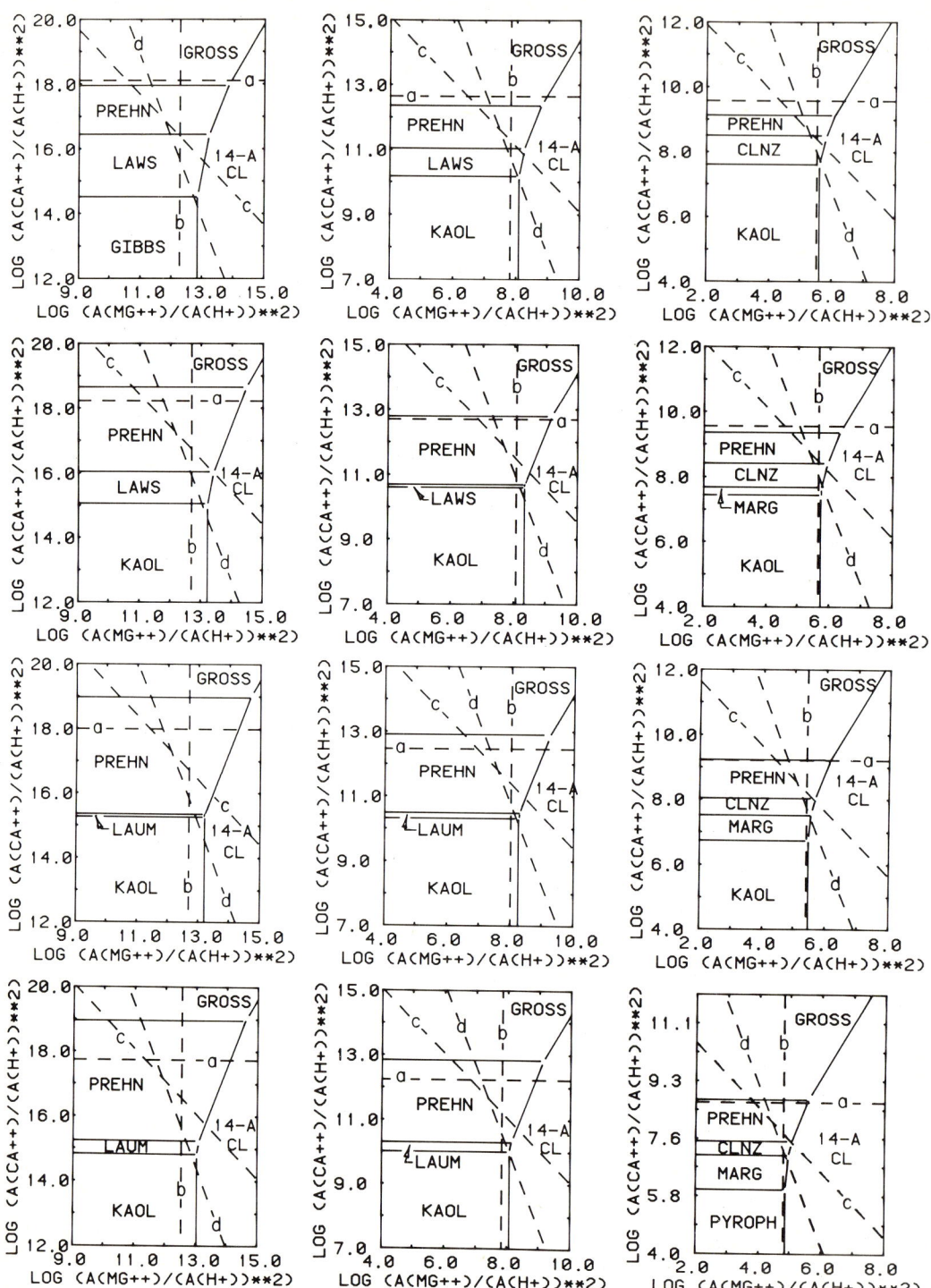

Phase relations in the system $HCl-H_2O-(Al_2O_3)-CaO-MgO-SiO_2$ in equilibrium with quartz. Saturation limits: wollastonite (a), talc (b), diopside (c), tremolite (d).

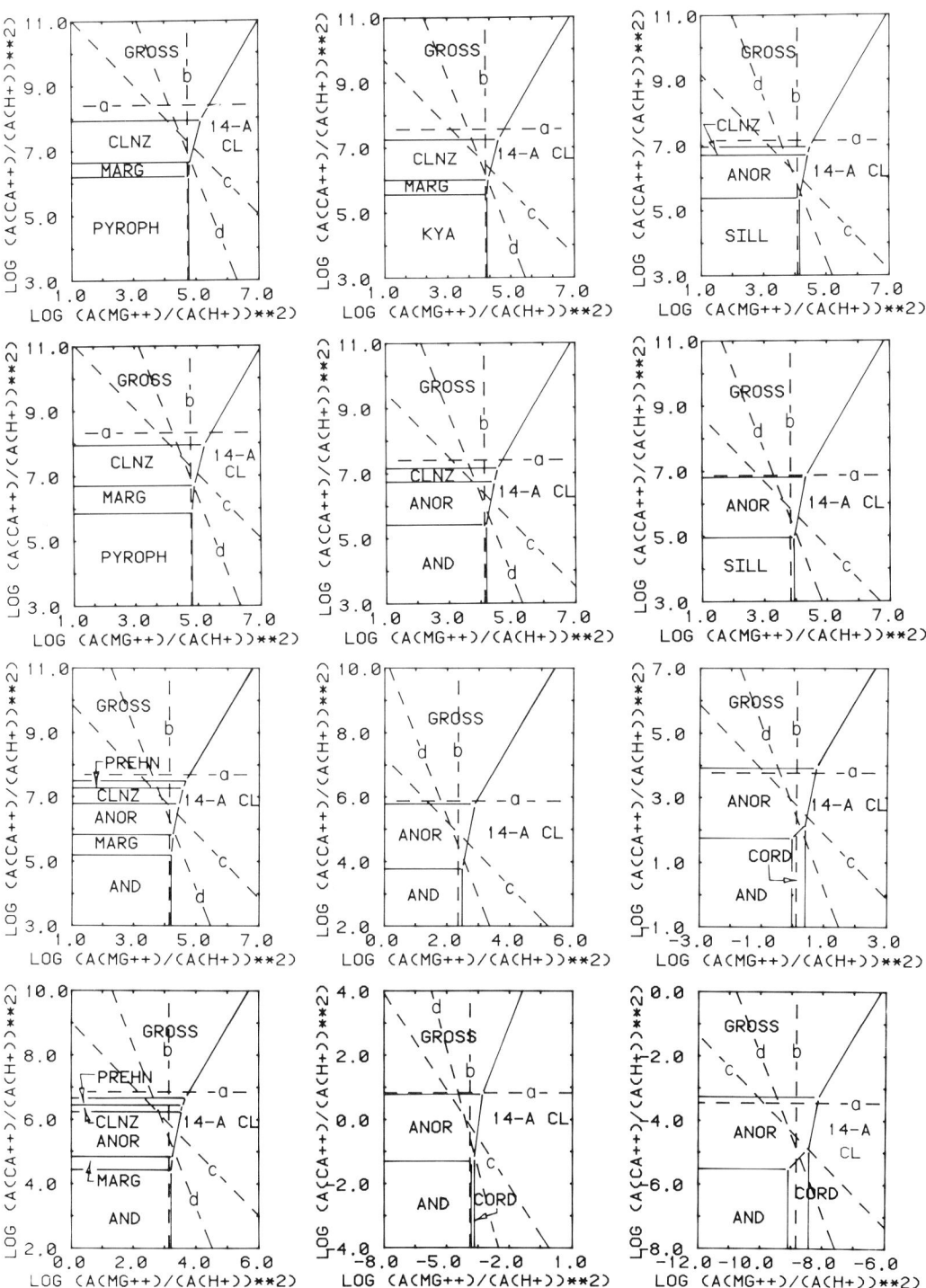

Phase relations in the system $HCl-H_2O-(Al_2O_3)-CaO-MgO-SiO_2$ in equilibrium with quartz. Saturation limits: wollastonite (a), talc (b), diopside (c), tremolite (d).

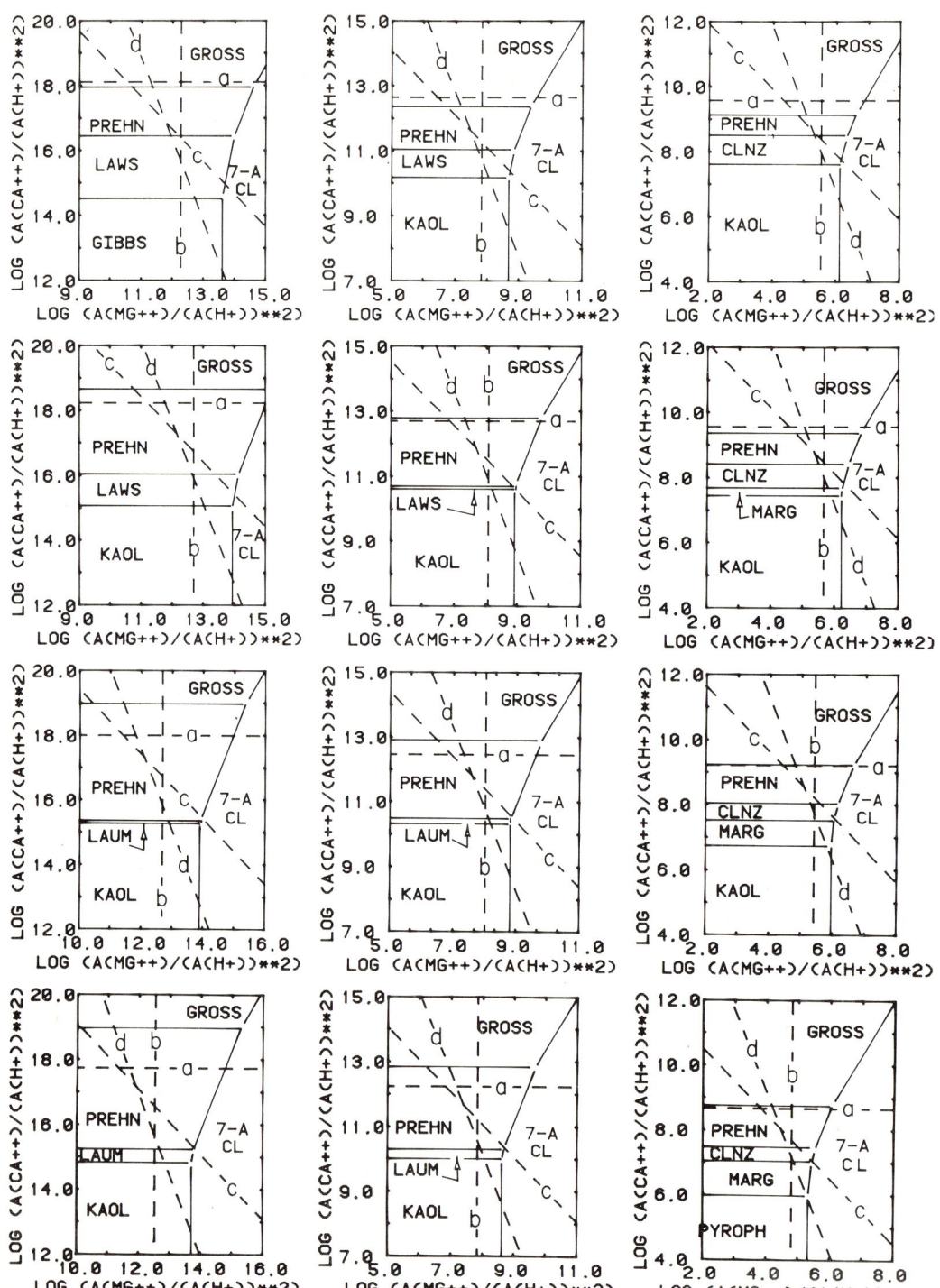

Phase relations in the system $HCl-H_2O-(Al_2O_3)-CaO-MgO-SiO_2$ in equilibrium with quartz. Metastable 7-A clinochlore was considered instead of its stable counterpart, 14-A clinochlore. Saturation limits: wollastonite (a), talc (b), diopside (c), tremolite (d).

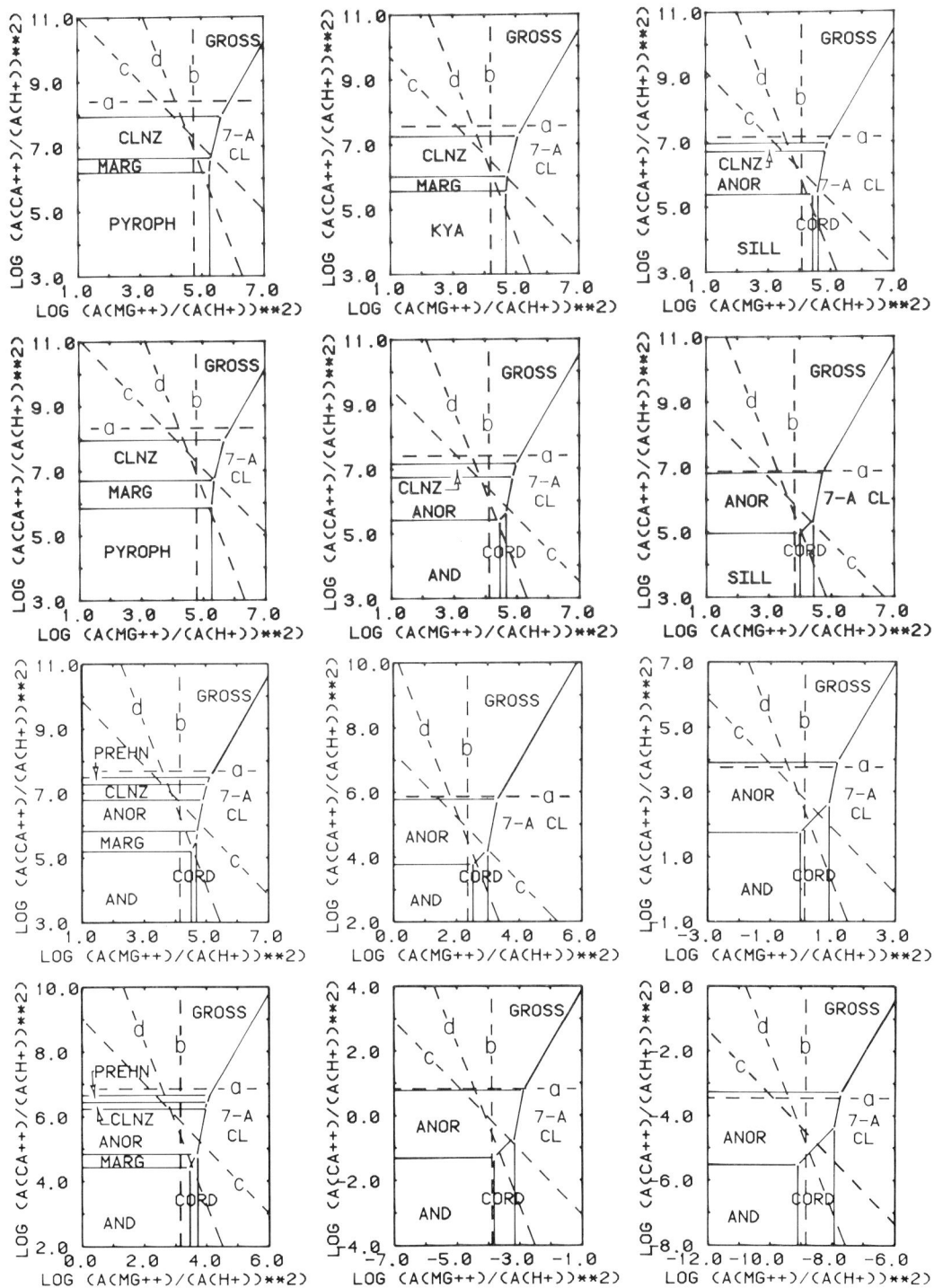

Phase relations in the system $HCl-H_2O-(Al_2O_3)-CaO-MgO-SiO_2$ in equilibrium with quartz. Metastable 7-A clinochlore was considered instead of its stable counterpart, 14-A clinochlore. Saturation limits: wollastonite (a), talc (b), diopside (c), tremolite (d).

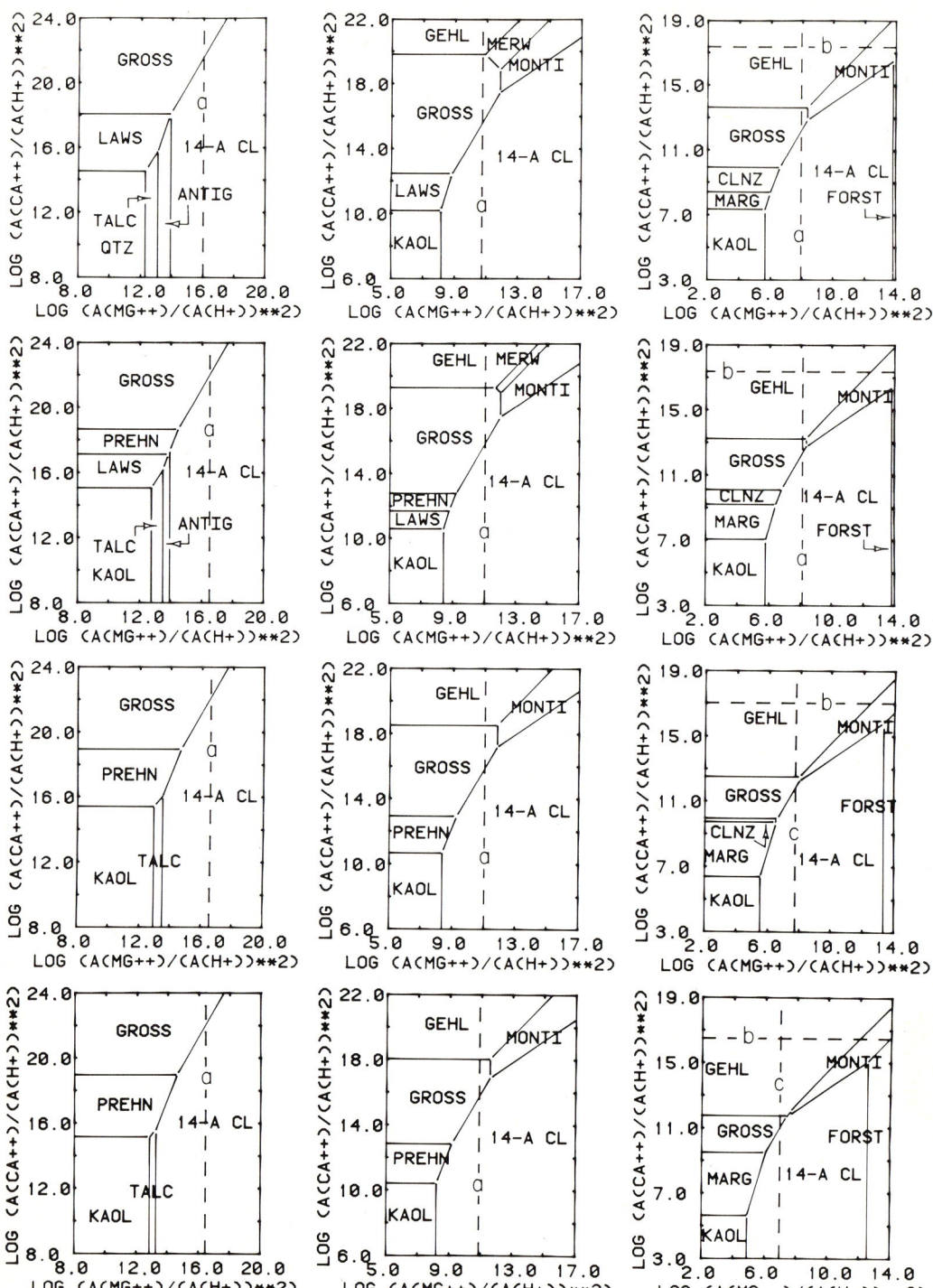

Phase relations in the system $HCl-H_2O-Al_2O_3-CaO-MgO-(SiO_2)$ in equilibrium with gibbsite, diaspore, or corundum, depending on which mineral is stable at each pressure and temperature. Saturation limits: brucite (a), lime (b), spinel (c).

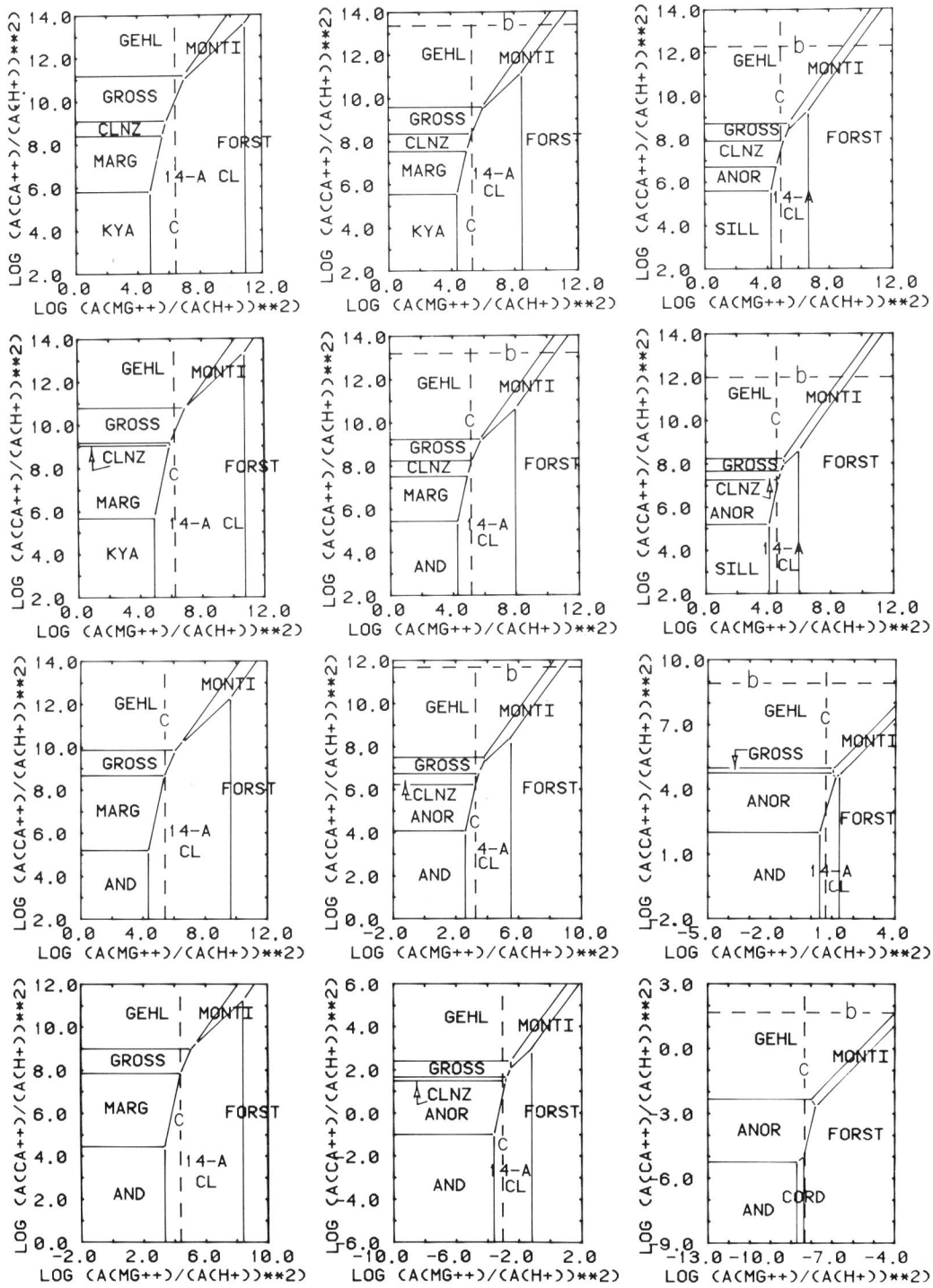

Phase relations in the system $HCl-H_2O-Al_2O_3-CaO-MgO-(SiO_2)$ in equilibrium with gibbsite, diaspore, or corundum, depending on which mineral is stable at each pressure and temperature. Saturation limits: brucite (a), lime (b), spinel (c).

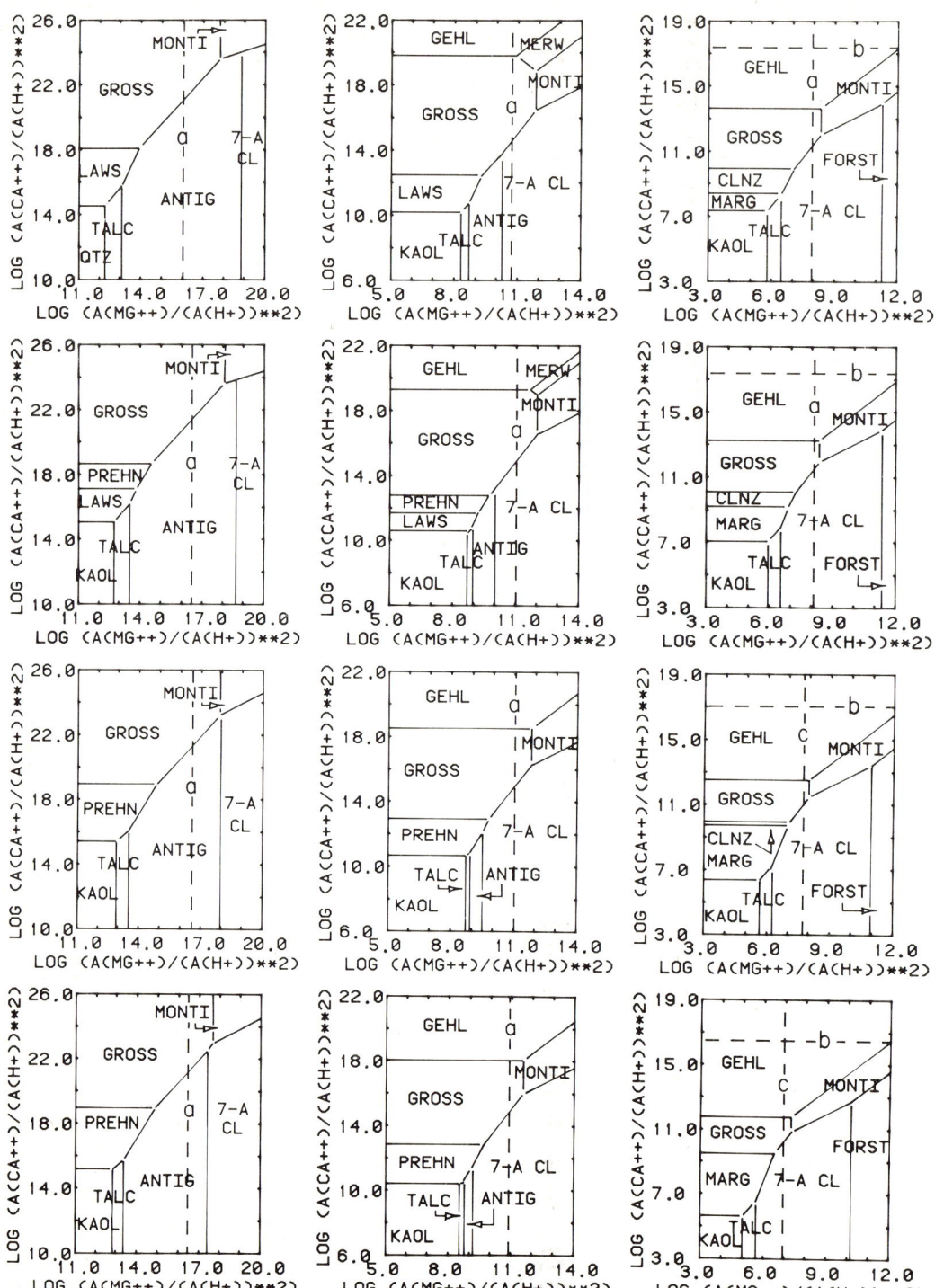

Phase relations in the system $HCl-H_2O-Al_2O_3-CaO-MgO-(SiO_2)$ in equilibrium with gibbsite, diaspore, or corundum, depending on which mineral is stable at each pressure and temperature. Metastable 7-A clinochlore was considered instead of its stable counterpart, 14-A clinochlore. Saturation limits: brucite (a), lime (b), spinel (c).

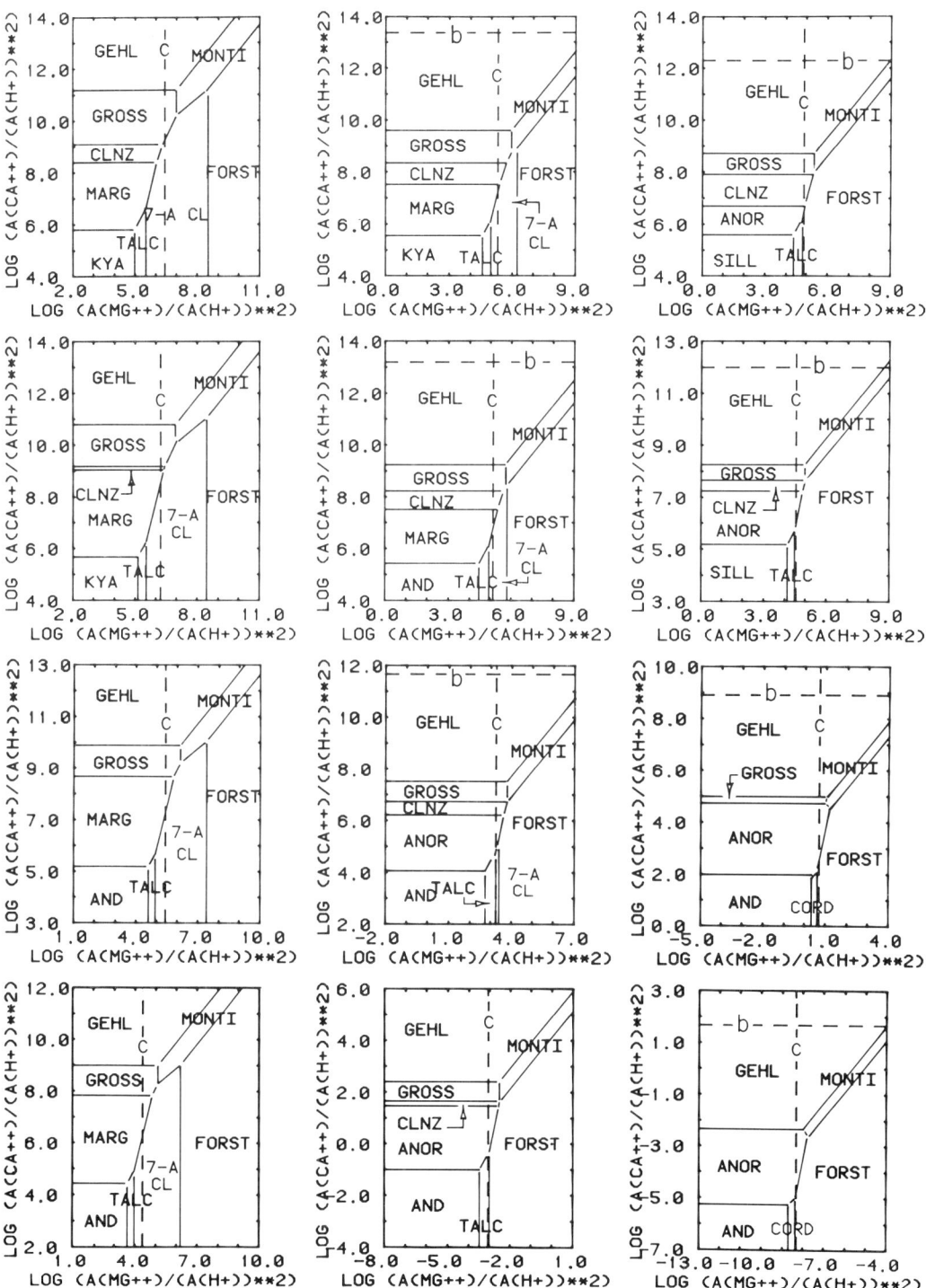

Phase relations in the system $HCl-H_2O-Al_2O_3-CaO-MgO-(SiO_2)$ in equilibrium with gibbsite, diaspore, or corundum, depending on which mineral is stable at each pressure and temperature. Metastable 7-A clinochlore was considered instead of its stable counterpart, 14-A clinochlore. Saturation limits: brucite (a), lime (b), spinel (c).

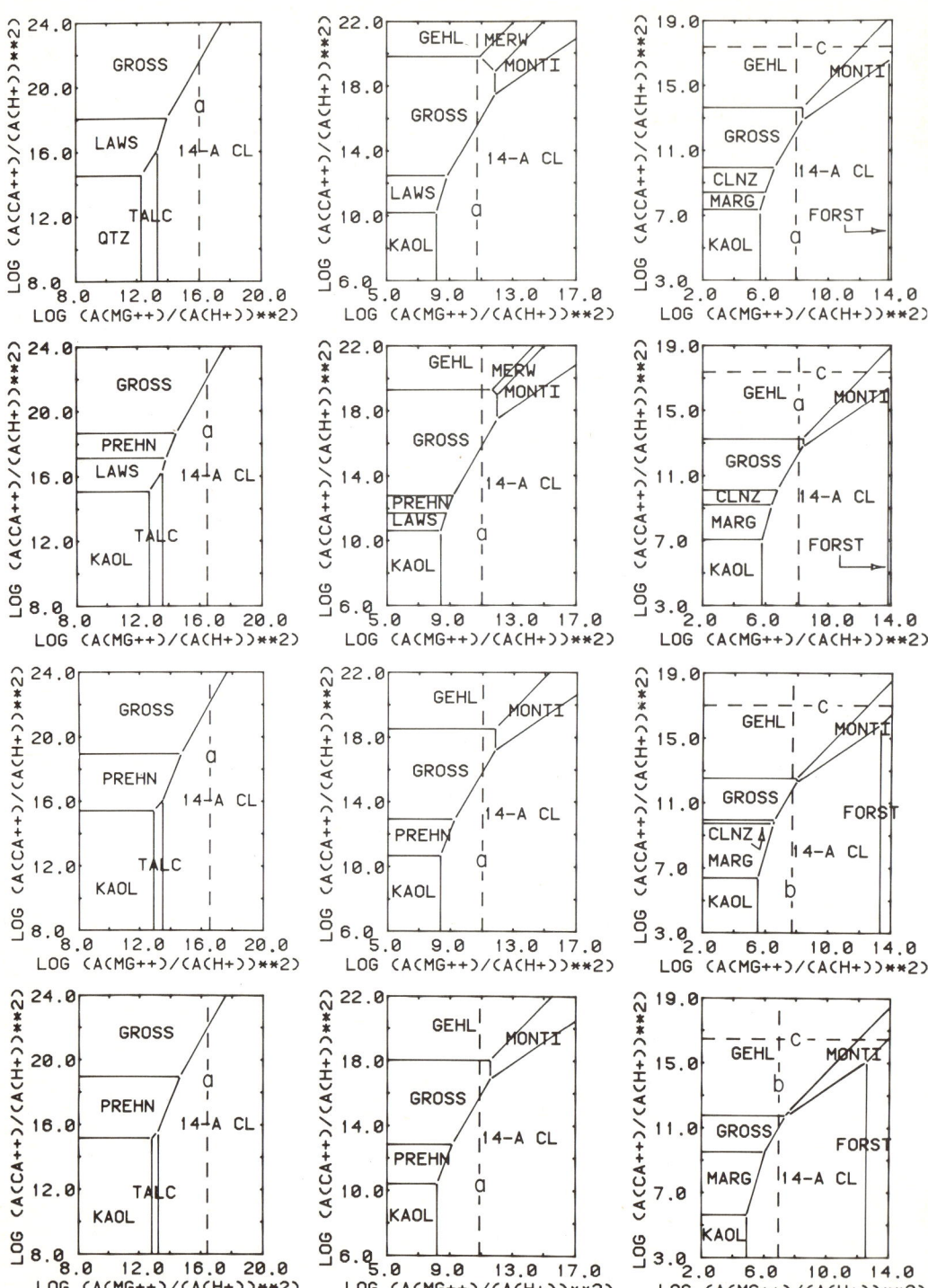

Phase relations in the system $HCl-H_2O-Al_2O_3-CaO-MgO-(SiO_2)$ in equilibrium with gibbsite, diaspore, or corundum, depending on which mineral is stable at each pressure and temperature. Metastable chrysotile was considered instead of its stable counterpart, antigorite. Saturation limits: brucite (a), spinel (b), lime (c).

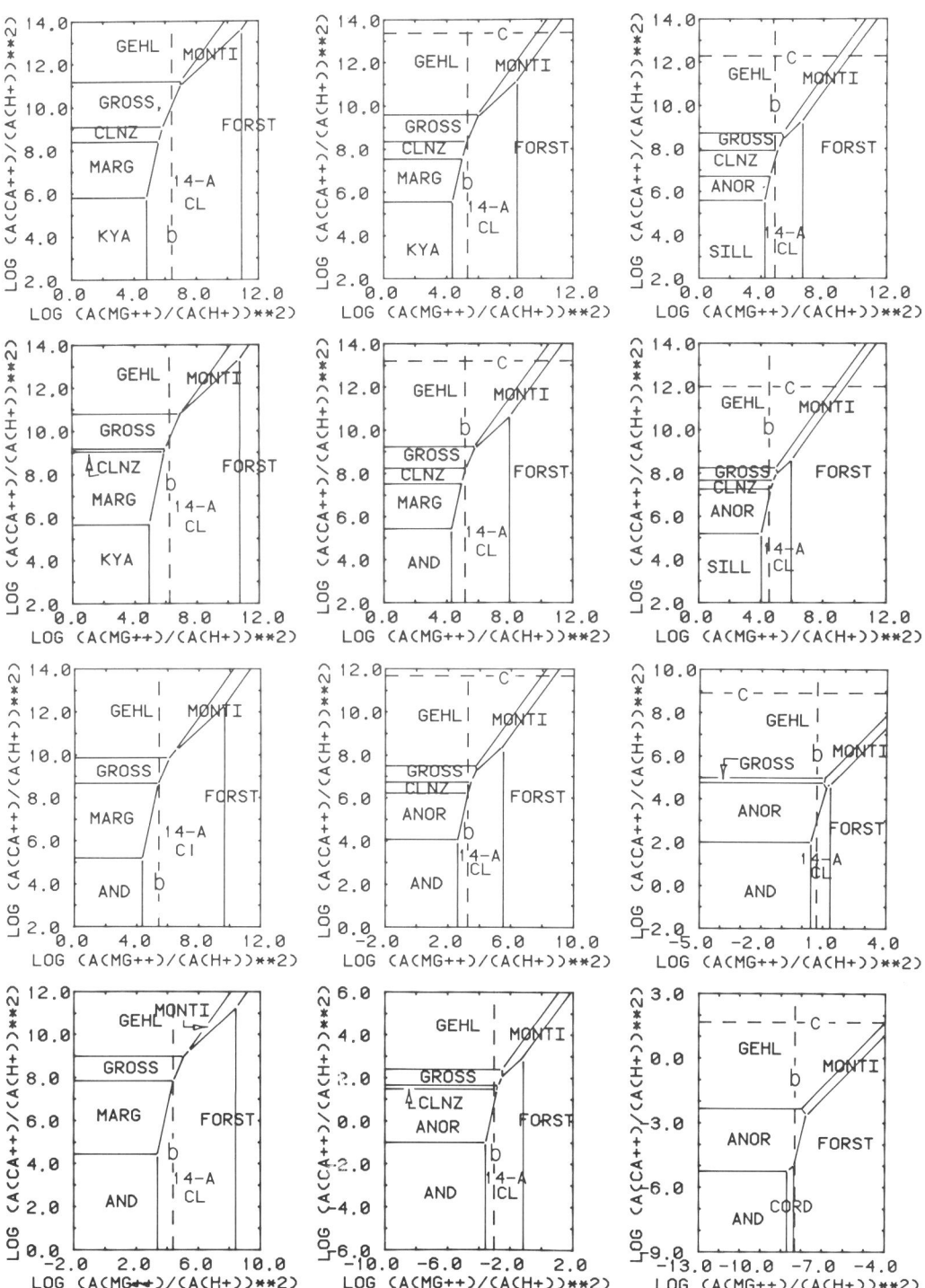

Phase relations in the system $HCl-H_2O-Al_2O_3-CaO-MgO-(SiO_2)$ in equilibrium with gibbsite, diaspore, or corundum, depending on which mineral is stable at each pressure and temperature. Metastable chrysotile was considered instead of its stable counterpart, antigorite. Saturation limits: brucite (a), spinel (b), lime (c).

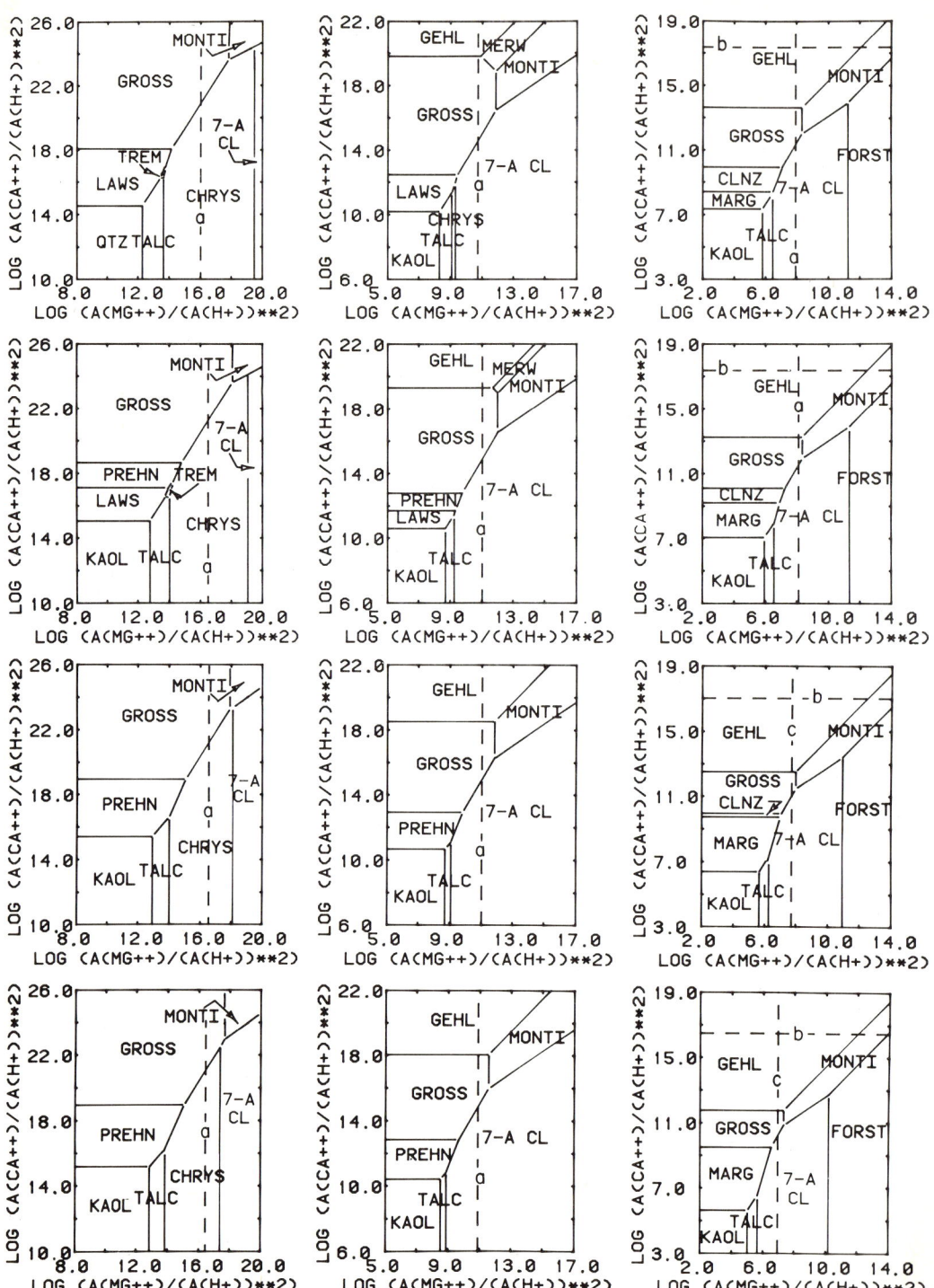

Phase relations in the system $HCl-H_2O-Al_2O_3-CaO-MgO-(SiO_2)$ in equilibrium with gibbsite, diaspore, or corundum, depending on which mineral is stable at each pressure and temperature. Metastable 7-A clinochlore and chrysotile were considered instead of their stable counterparts, 14-A clinochlore and antigorite. Saturation limits: brucite (a), lime (b), spinel (c).

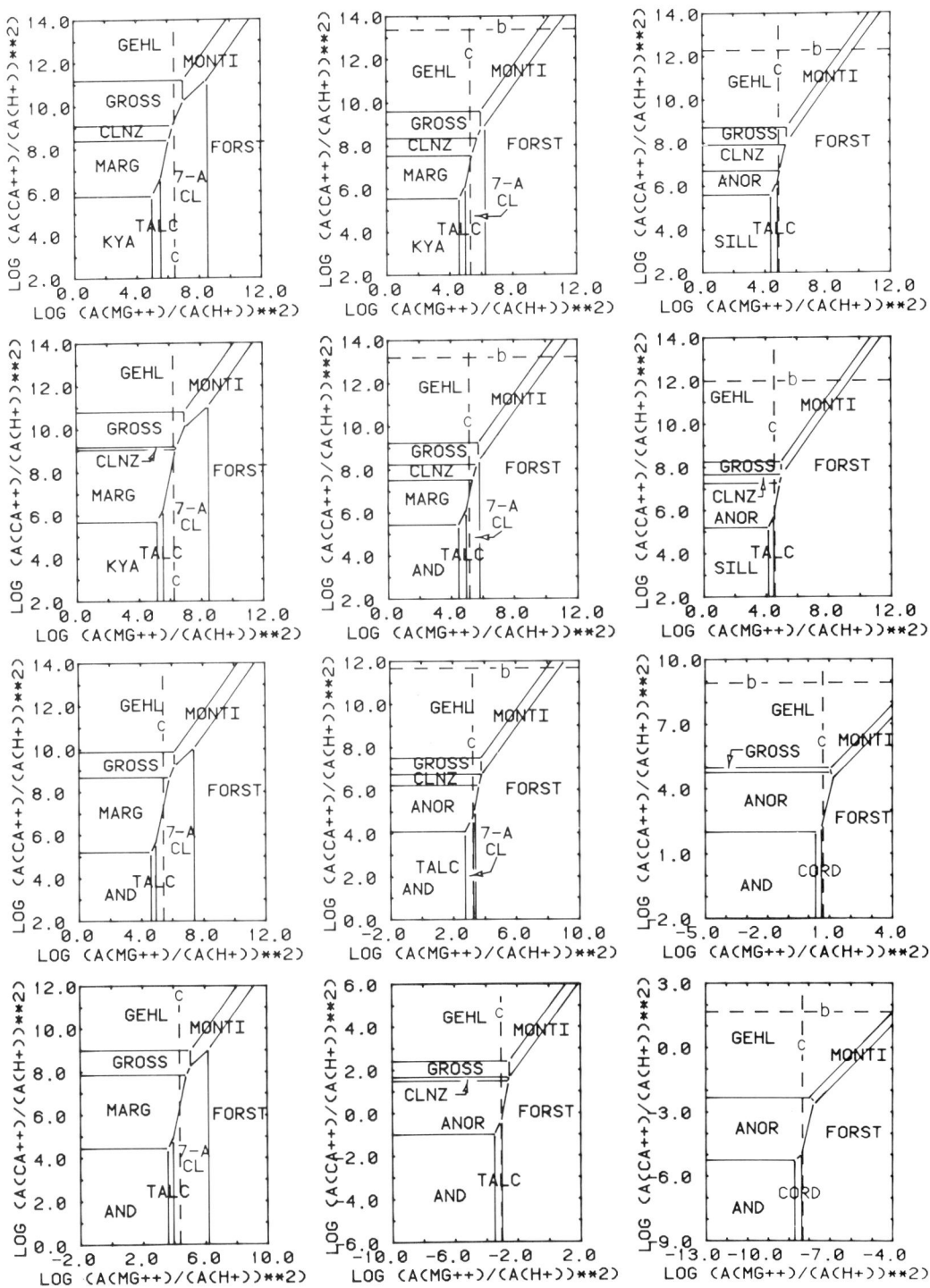

Phase relations in the system $HCl-H_2O-Al_2O_3-CaO-MgO-(SiO_2)$ in equilibrium with gibbsite, diaspore, or corundum, depending on which mineral is stable at each pressure and temperature. Metastable 7-A clinochlore and chrysotile were considered instead of their stable counterparts, 14-A clinochlore and antigorite. Saturation limits: brucite (a), lime (b), spinel (c).

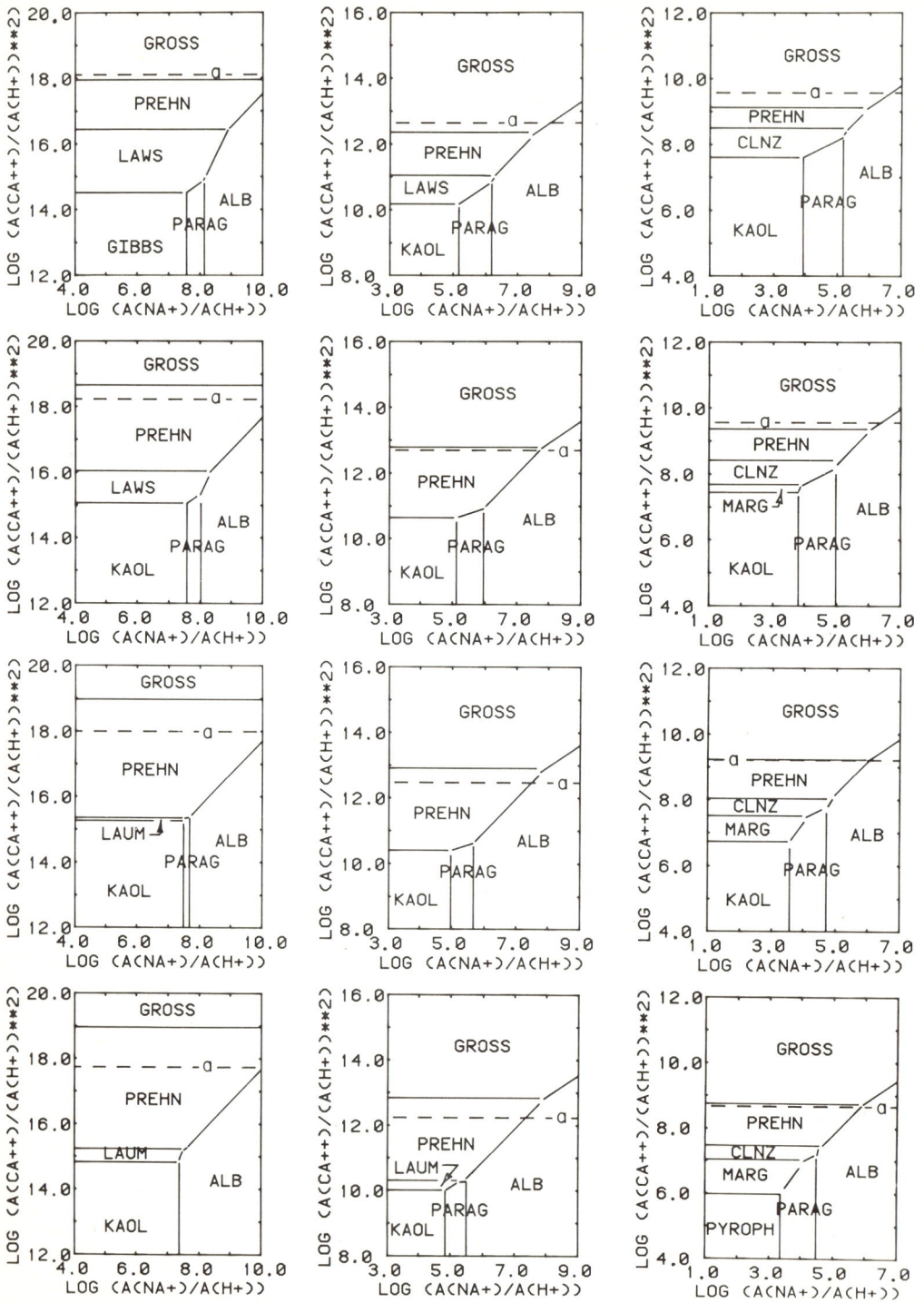

Phase relations in the system $HCl-H_2O-(Al_2O_3)-CaO-Na_2O-SiO_2$ in equilibrium with quartz. Saturation limit: wollastonite (a).

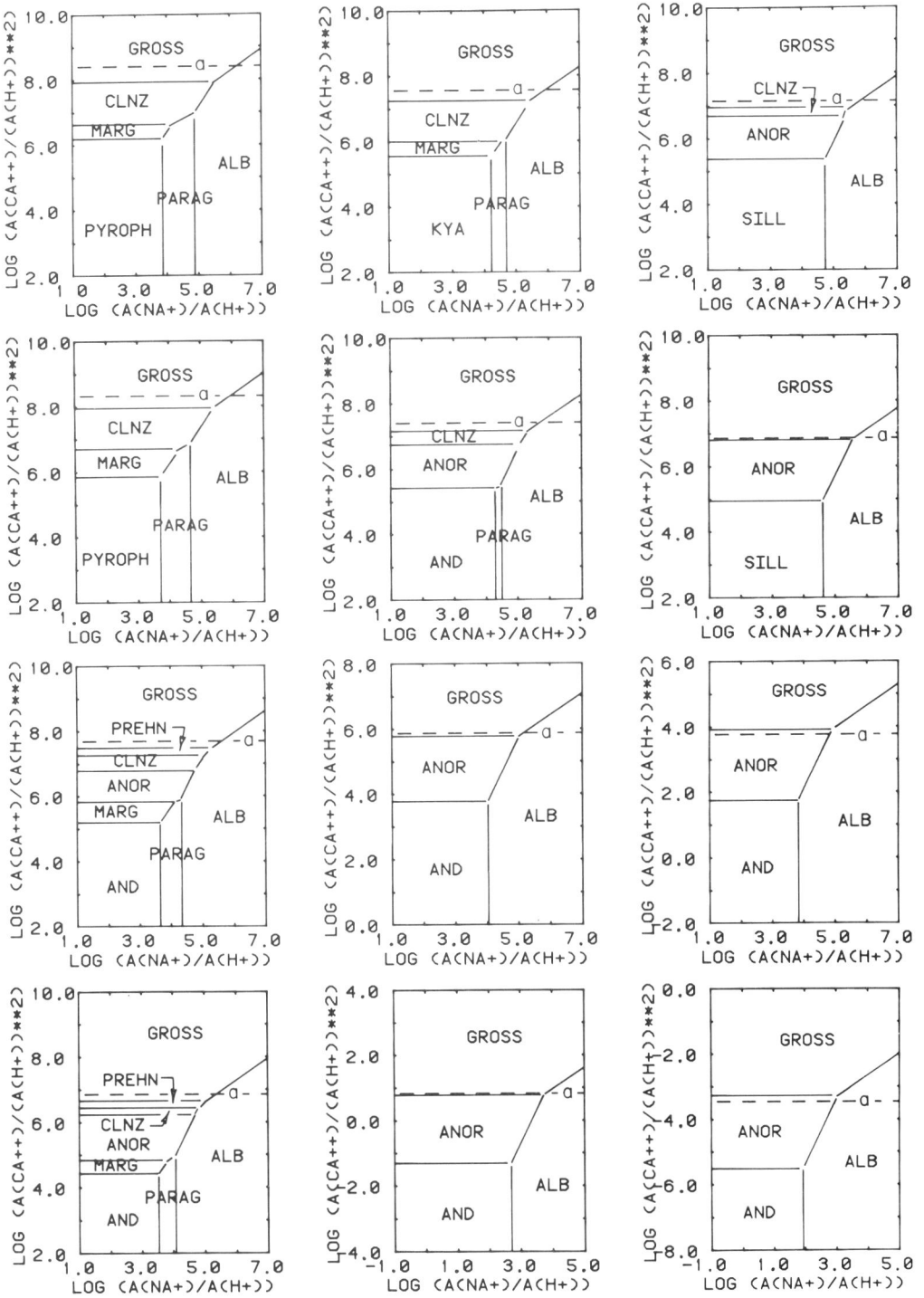

Phase relations in the system $HCl-H_2O-(Al_2O_3)-CaO-Na_2O-SiO_2$ in equilibrium with quartz. Saturation limit: wollastonite (a).

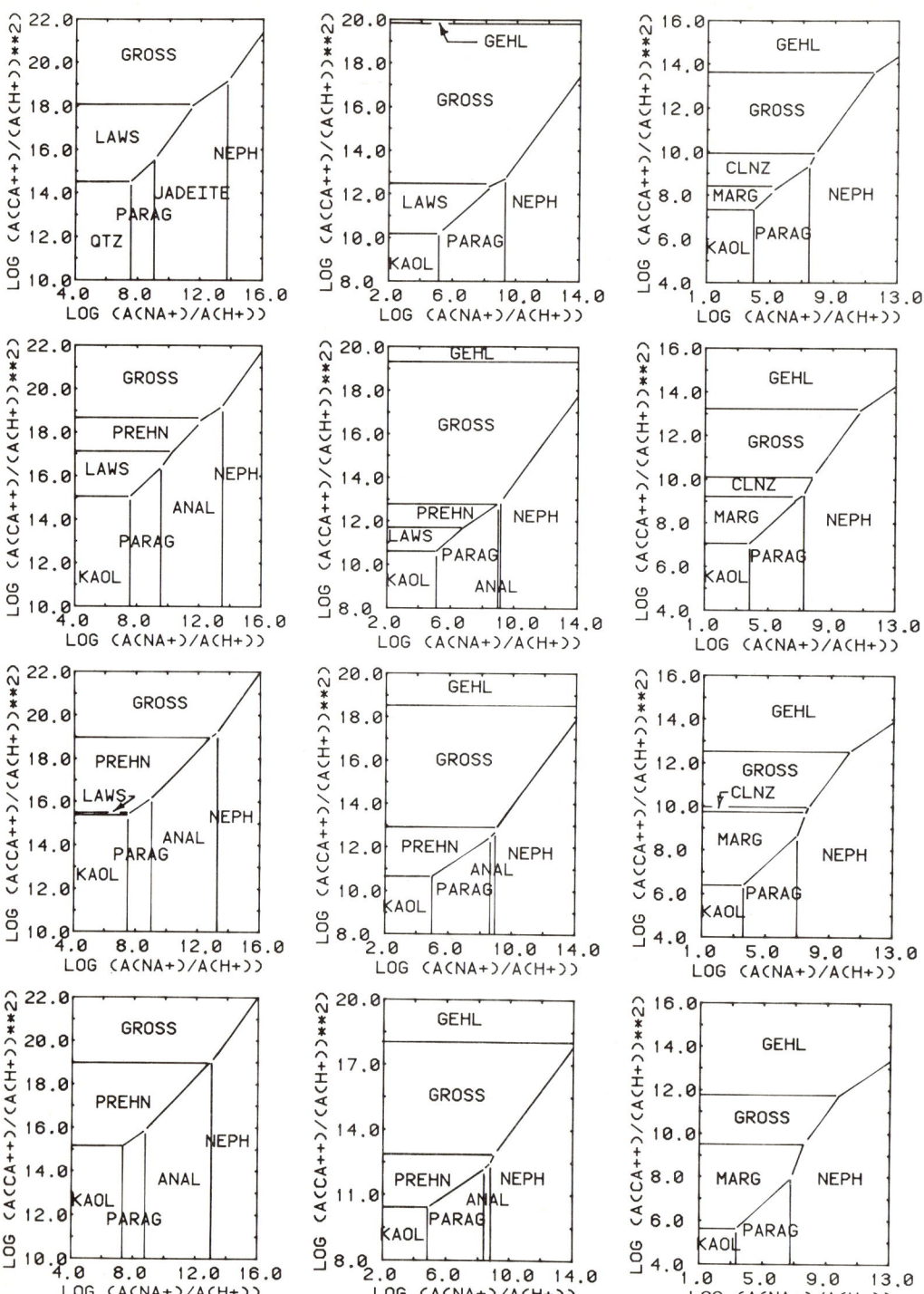

Phase relations in the system $HCl-H_2O-Al_2O_3-CaO-Na_2O-(SiO_2)$ in equilibrium with gibbsite, diaspore, or corundum, depending on which mineral is stable at each pressure and temperature.

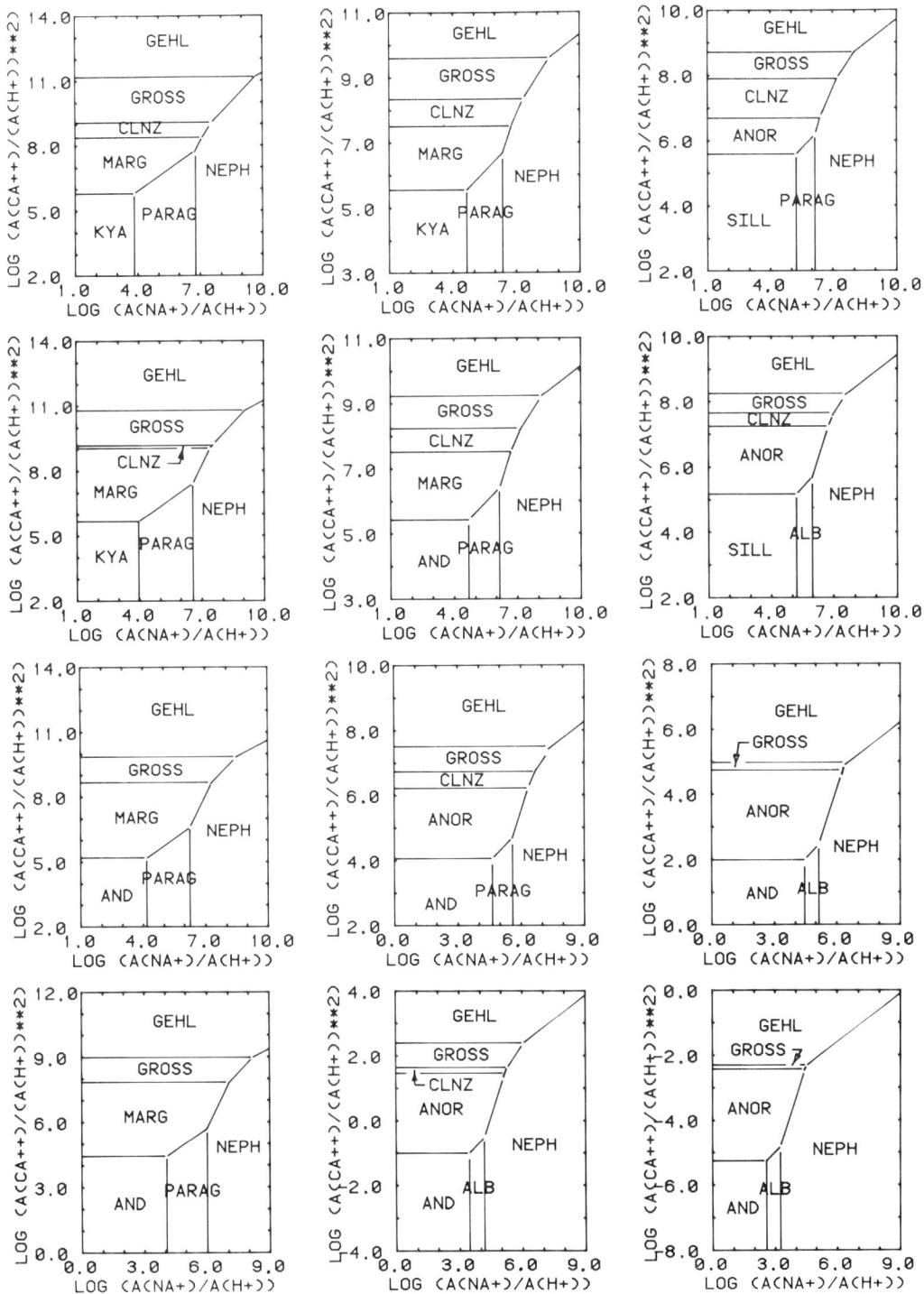

Phase relations in the system $HCl-H_2O-Al_2O_3-CaO-Na_2O-(SiO_2)$ in equilibrium with gibbsite, diaspore, or corundum, depending on which mineral is stable at each pressure and temperature.

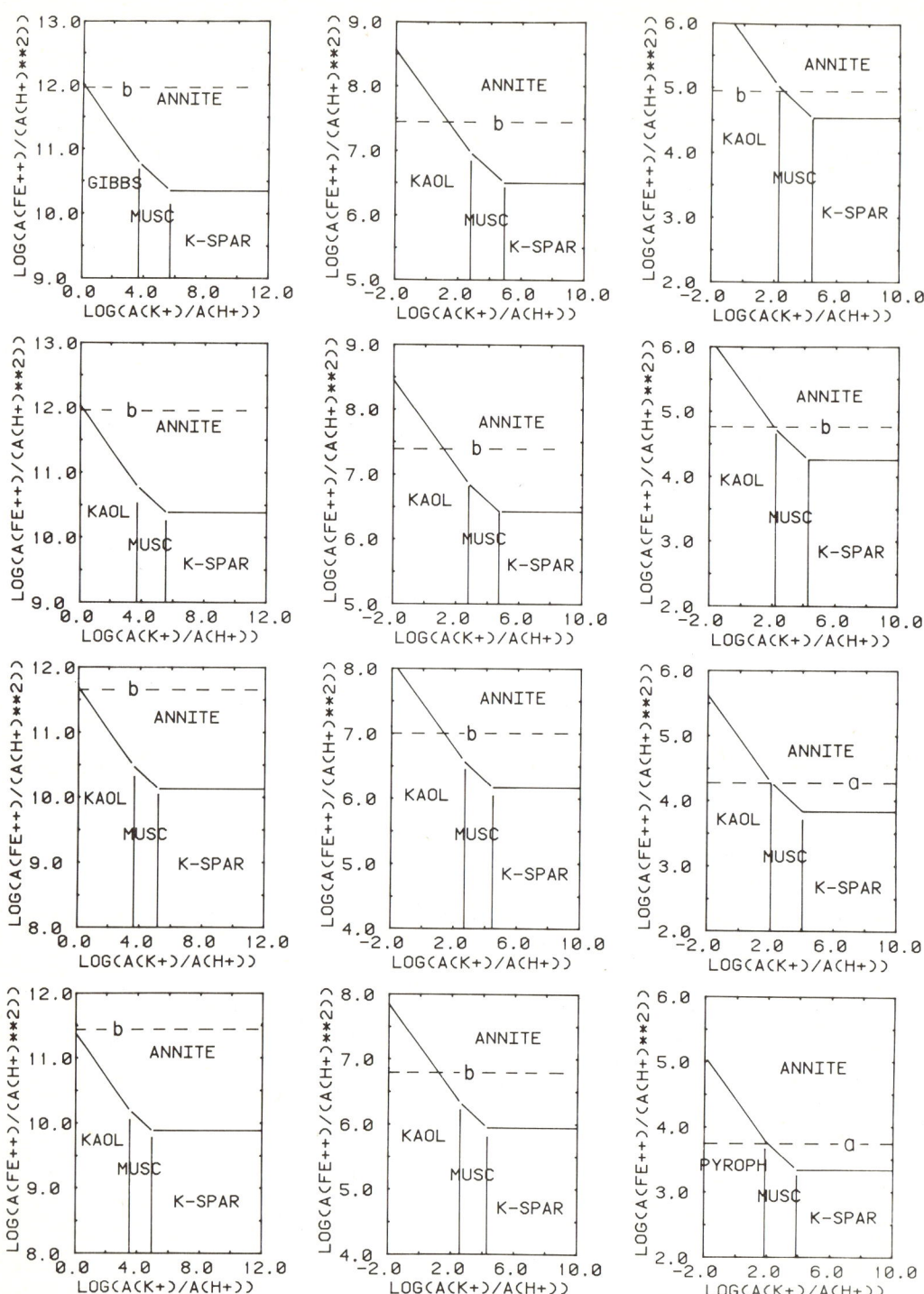

Phase relations in the system $HCl-H_2O-(Al_2O_3)-FeO-K_2O-SiO_2$ in equilibrium with quartz. Saturation limits: fayalite (a), ferrosilite (b).

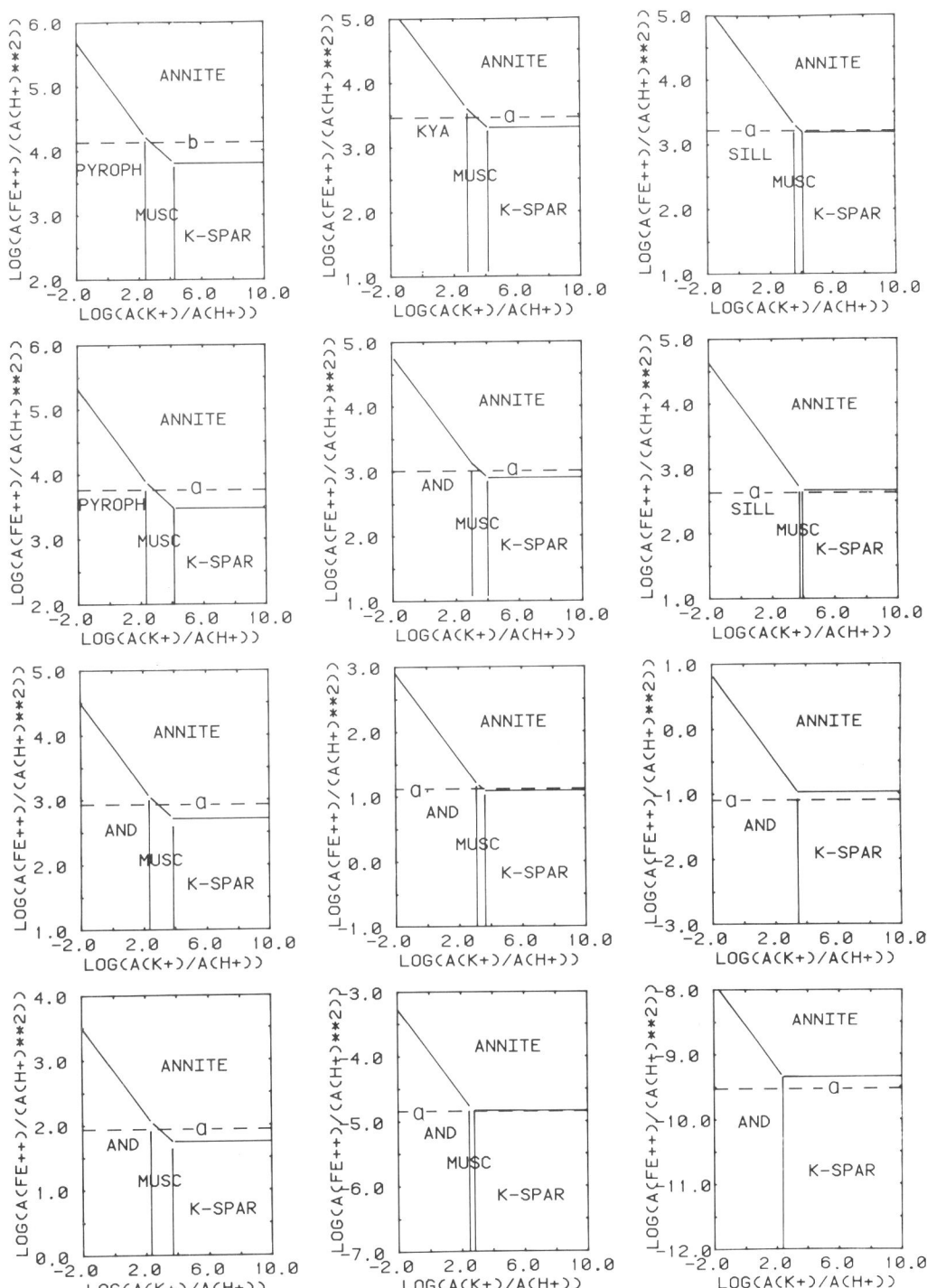

Phase relations in the system $HCl-H_2O-(Al_2O_3)-FeO-K_2O-SiO_2$ in equilibrium with quartz. Saturation limits: fayalite (a), ferrosilite (b).

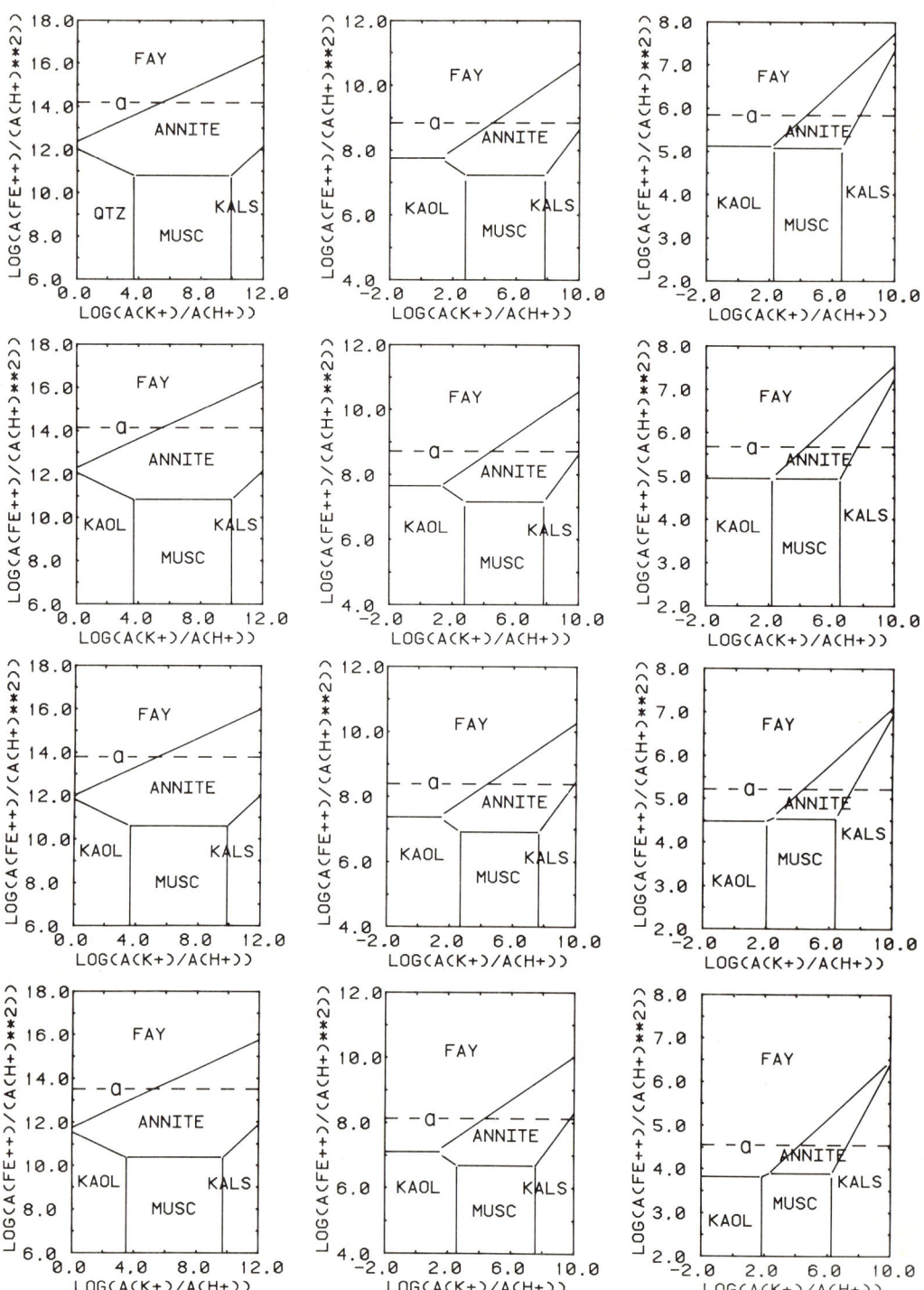

Phase relations in the system $HCl-H_2O-Al_2O_3-FeO-K_2O-(SiO_2)$ in equilibrium with gibbsite, diaspore, or corundum, depending on which mineral is stable at each pressure and temperature. Saturation limit: ferrous oxide (a).

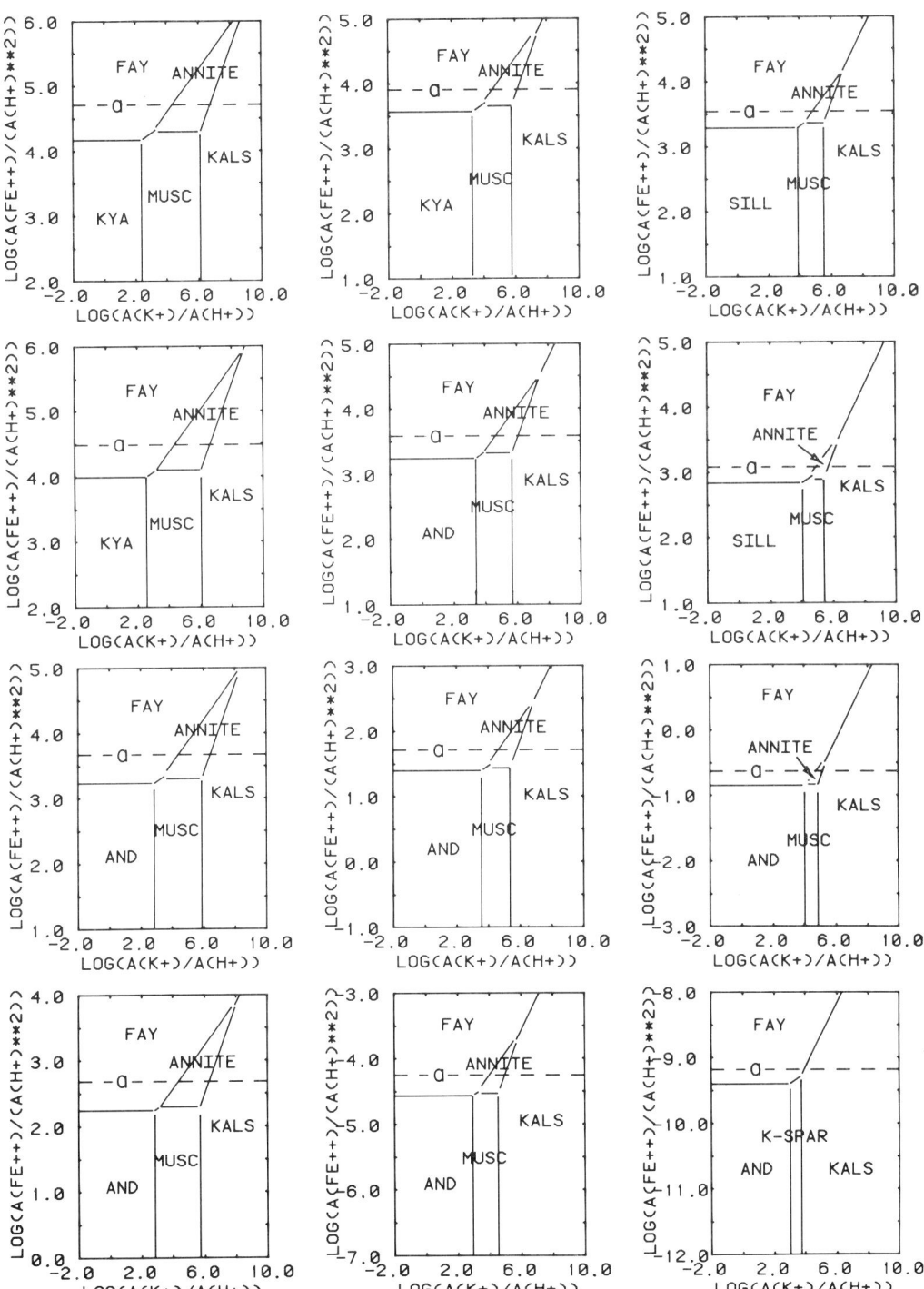

Phase relations in the system $HCl-H_2O-Al_2O_3-FeO-K_2O-(SiO_2)$ in equilibrium with gibbsite, diaspore, or corundum, depending on which mineral is stable at each pressure and temperature. Saturation limit: ferrous oxide (a).

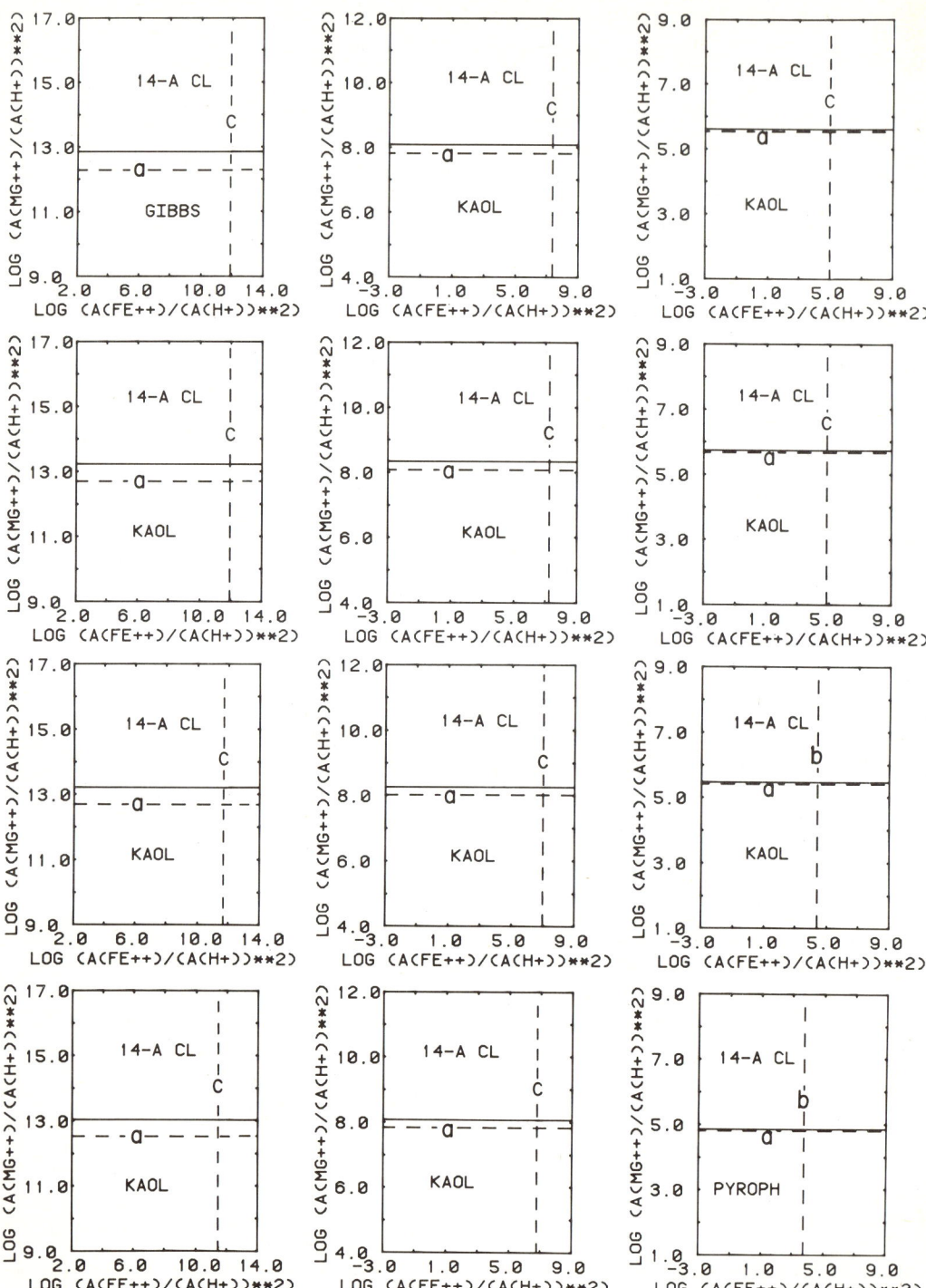

Phase relations in the system $HCl-H_2O-(Al_2O_3)-FeO-MgO-SiO_2$ in equilibrium with quartz. Saturation limits: talc (a), fayalite (b), ferrosilite (c).

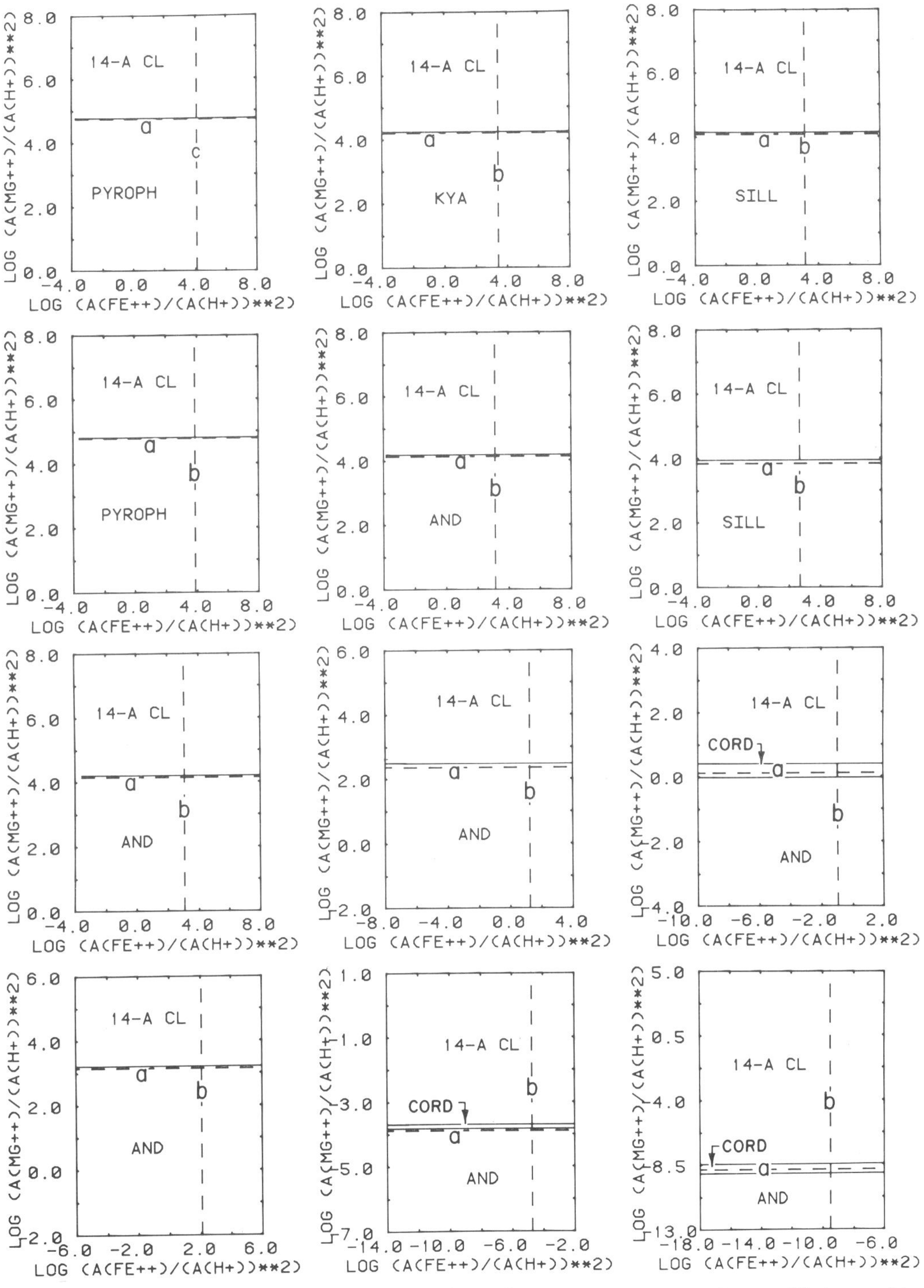

Phase relations in the system $HCl-H_2O-(Al_2O_3)-FeO-MgO-SiO_2$ in equilibrium with quartz. Saturation limits: talc (a), fayalite (b), ferrosilite (c).

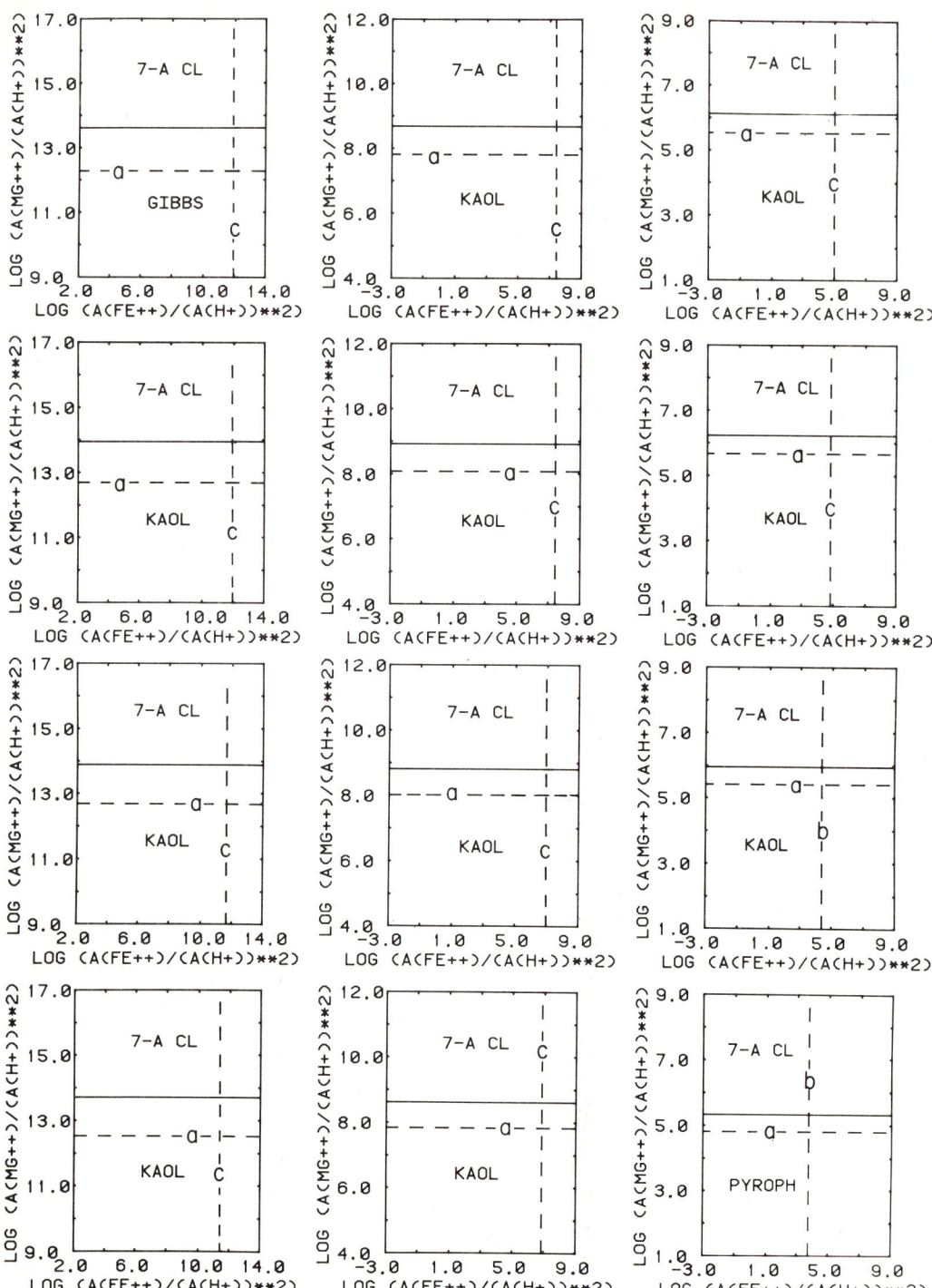

Phase relations in the system $HCl-H_2O-(Al_2O_3)-FeO-MgO-SiO_2$ in equilibrium with quartz. Metastable 7-A clinochlore was considered instead of its stable counterpart, 14-A clinochlore. Saturation limits: talc (a), fayalite (b), ferrosilite (c).

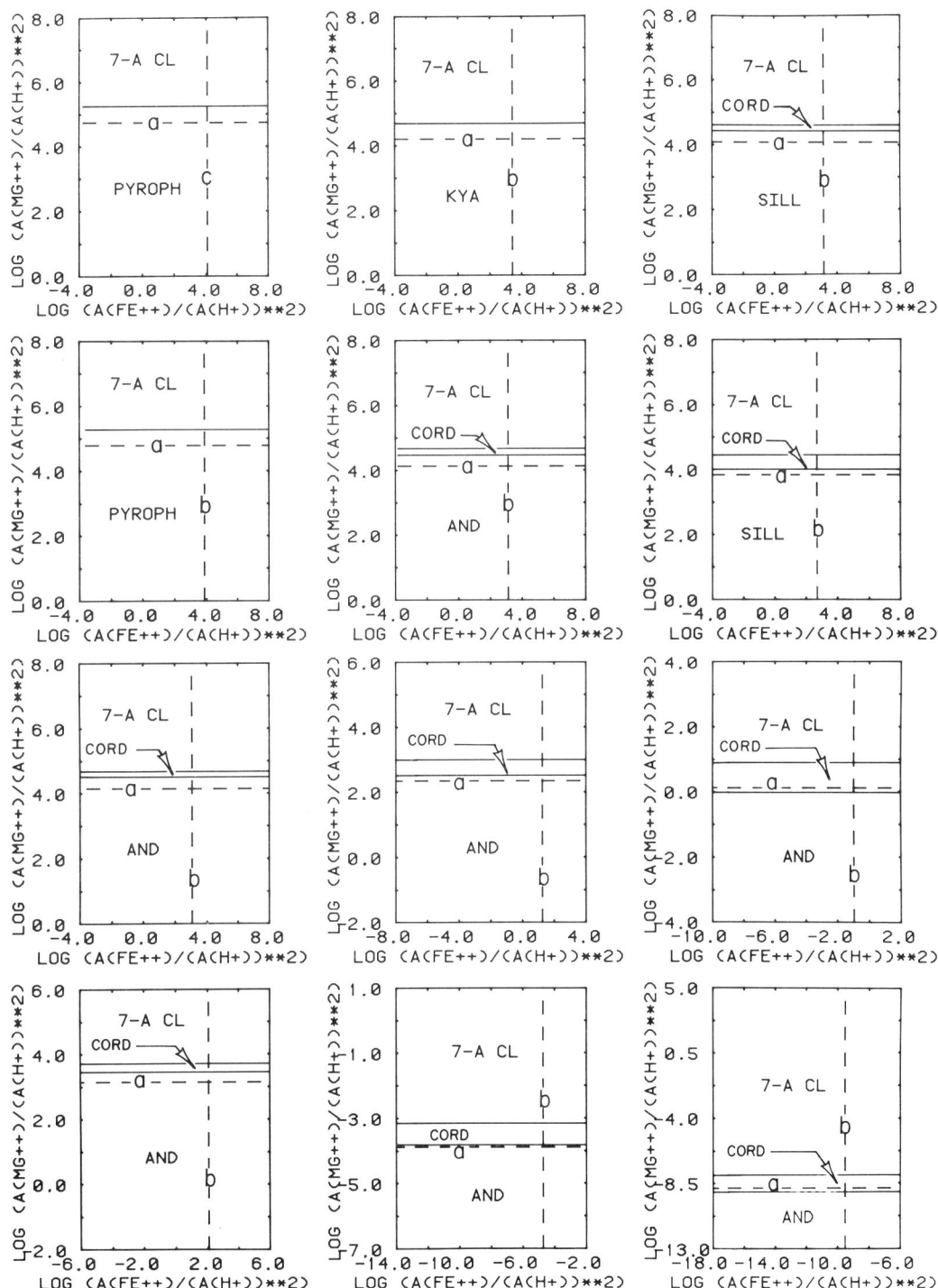

Phase relations in the system $HCl-H_2O-(Al_2O_3)-FeO-MgO-SiO_2$ in equilibrium with quartz. Metastable 7-A clinochlore was considered instead of its stable counterpart, 14-A clinochlore. Saturation limits: talc (a), fayalite (b), ferrosilite (c).

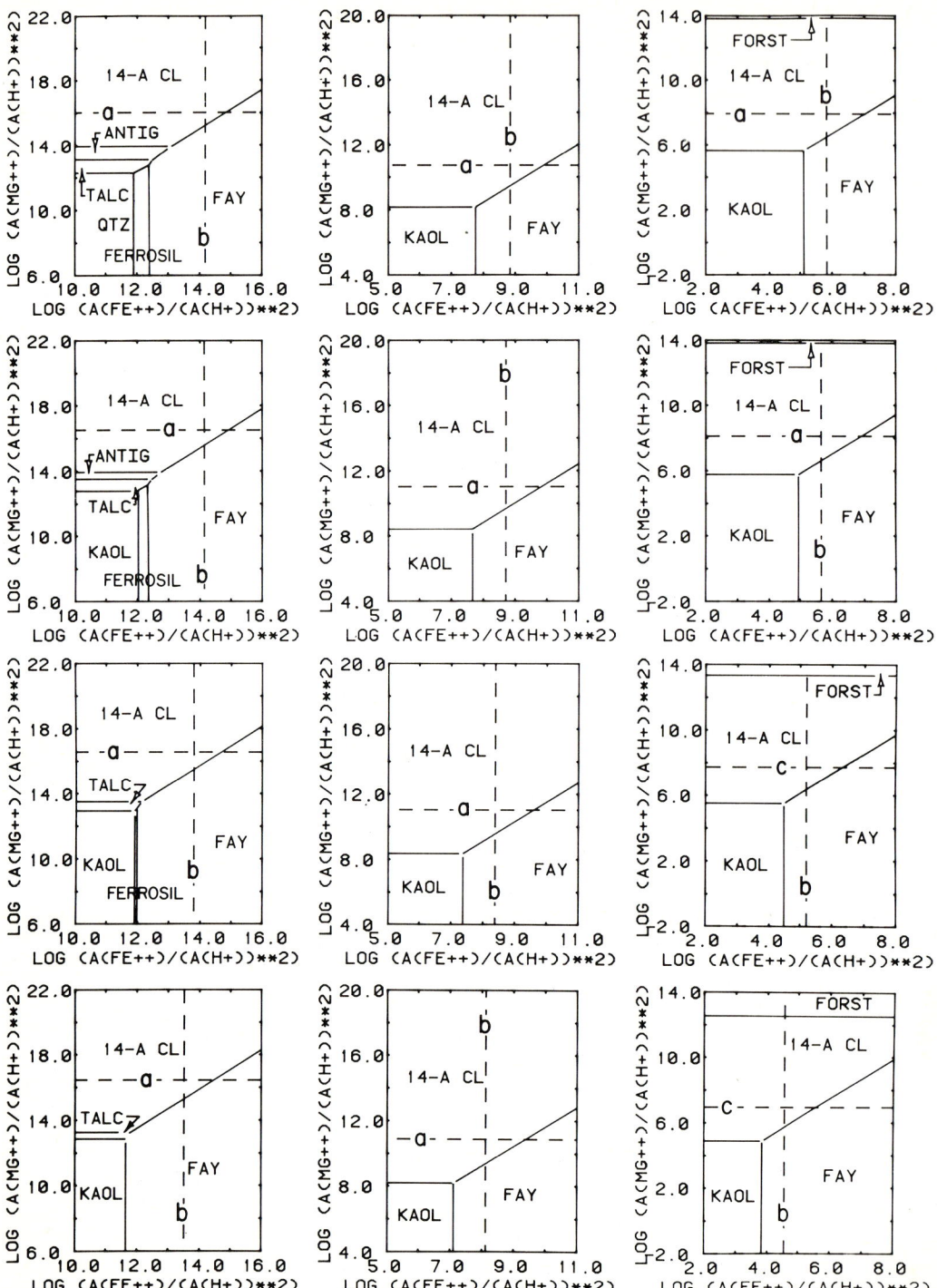

Phase relations in the system $HCl-H_2O-Al_2O_3-FeO-MgO-(SiO_2)$ in equilibrium with gibbsite, diaspore, or corundum, depending on which mineral is stable at each pressure and temperature. Saturation limits: brucite (a), ferrous oxide (b), spinel (c).

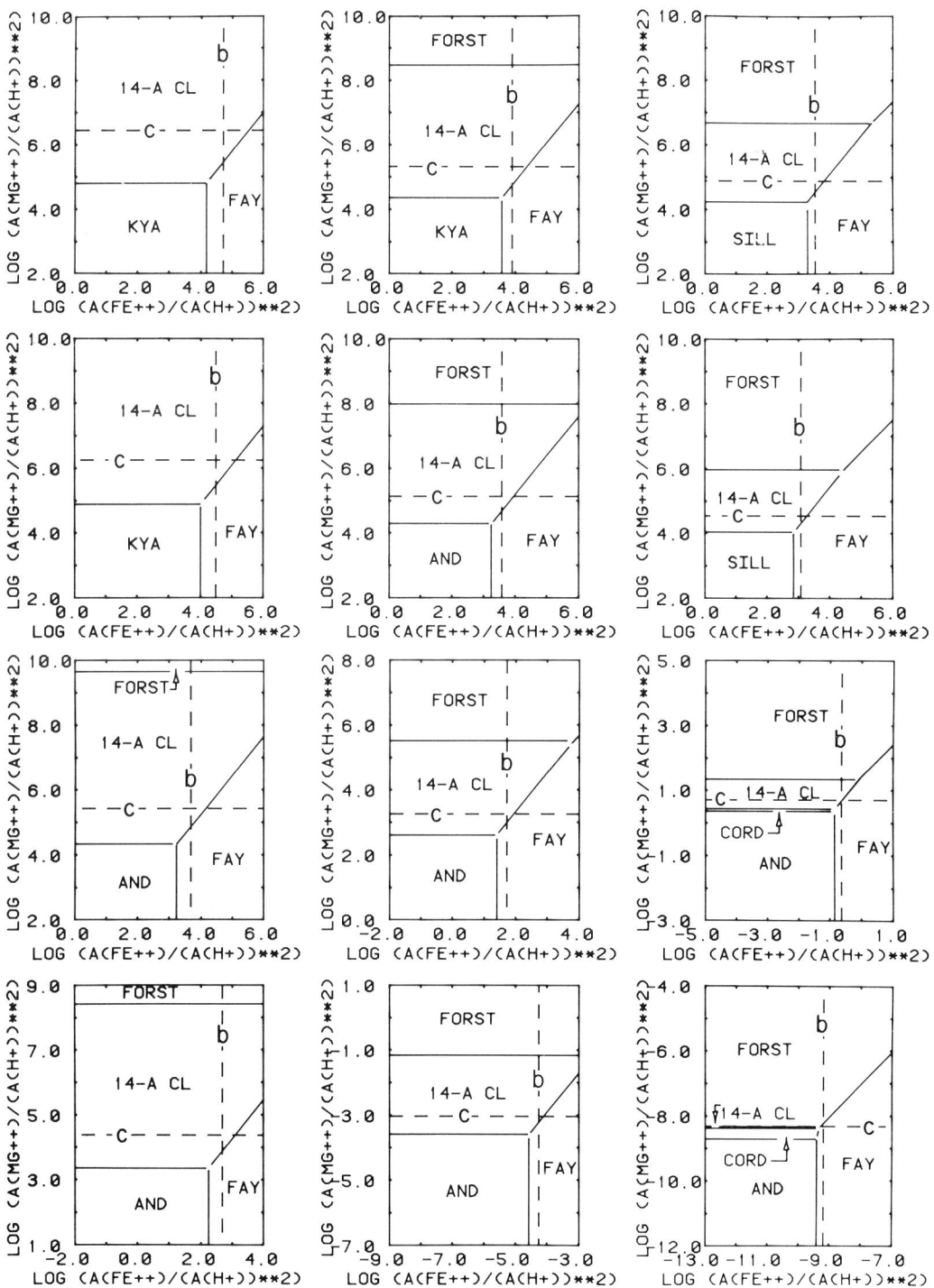

Phase relations in the system $HCl-H_2O-Al_2O_3-FeO-MgO-(SiO_2)$ in equilibrium with gibbsite, diaspore, or corundum, depending on which mineral is stable at each pressure and temperature. Saturation limits: brucite (a), ferrous oxide (b), spinel (c).

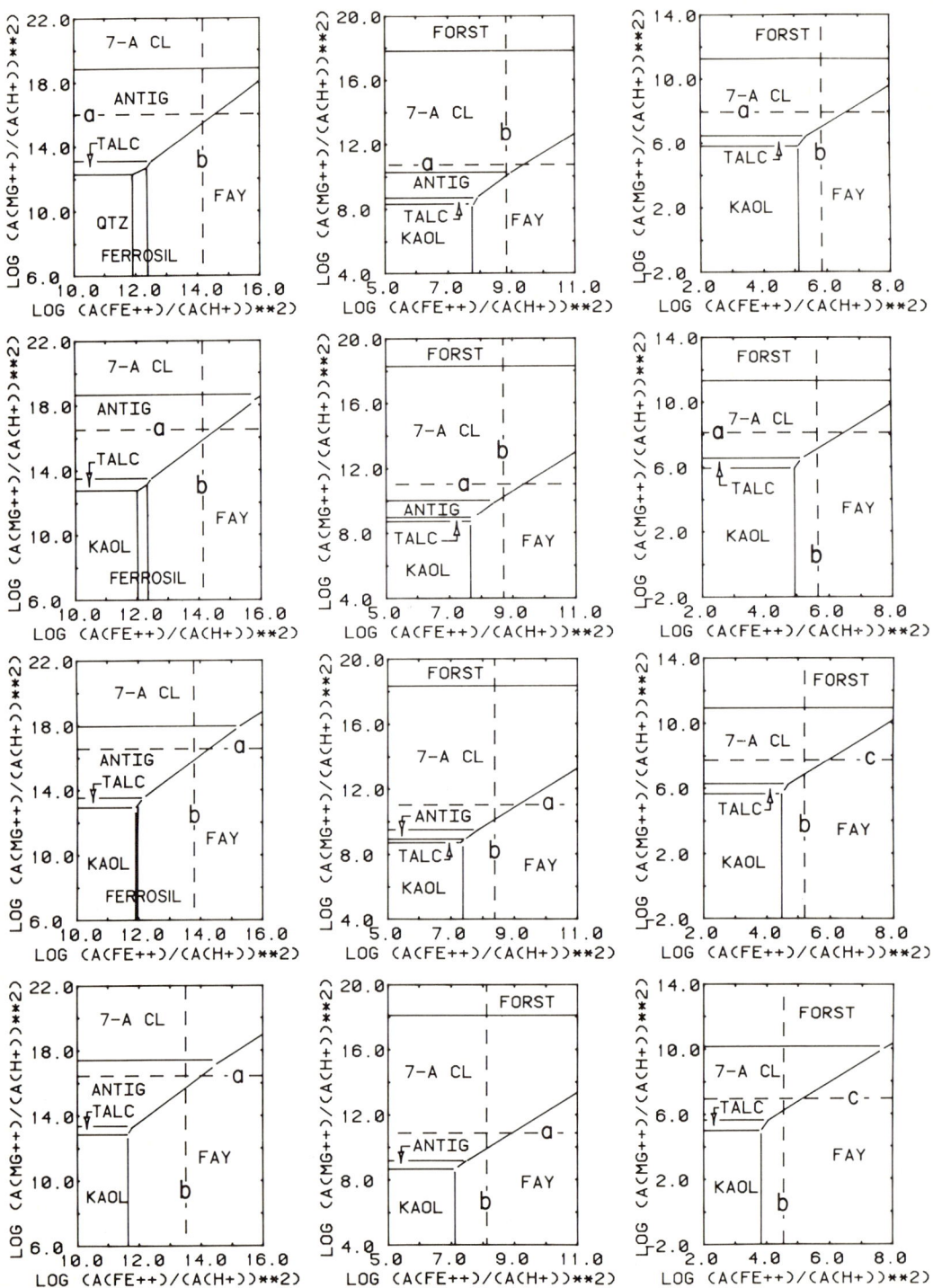

Phase relations in the system $HCl-H_2O-Al_2O_3-FeO-MgO-(SiO_2)$ in equilibrium with gibbsite, diaspore, or corundum, depending on which mineral is stable at each pressure and temperature. Metastable 7-A clinochlore was considered instead of its stable counterpart, 14-A clinochlore. Saturation limits: brucite (a), ferrous oxide (b), spinel (c).

89

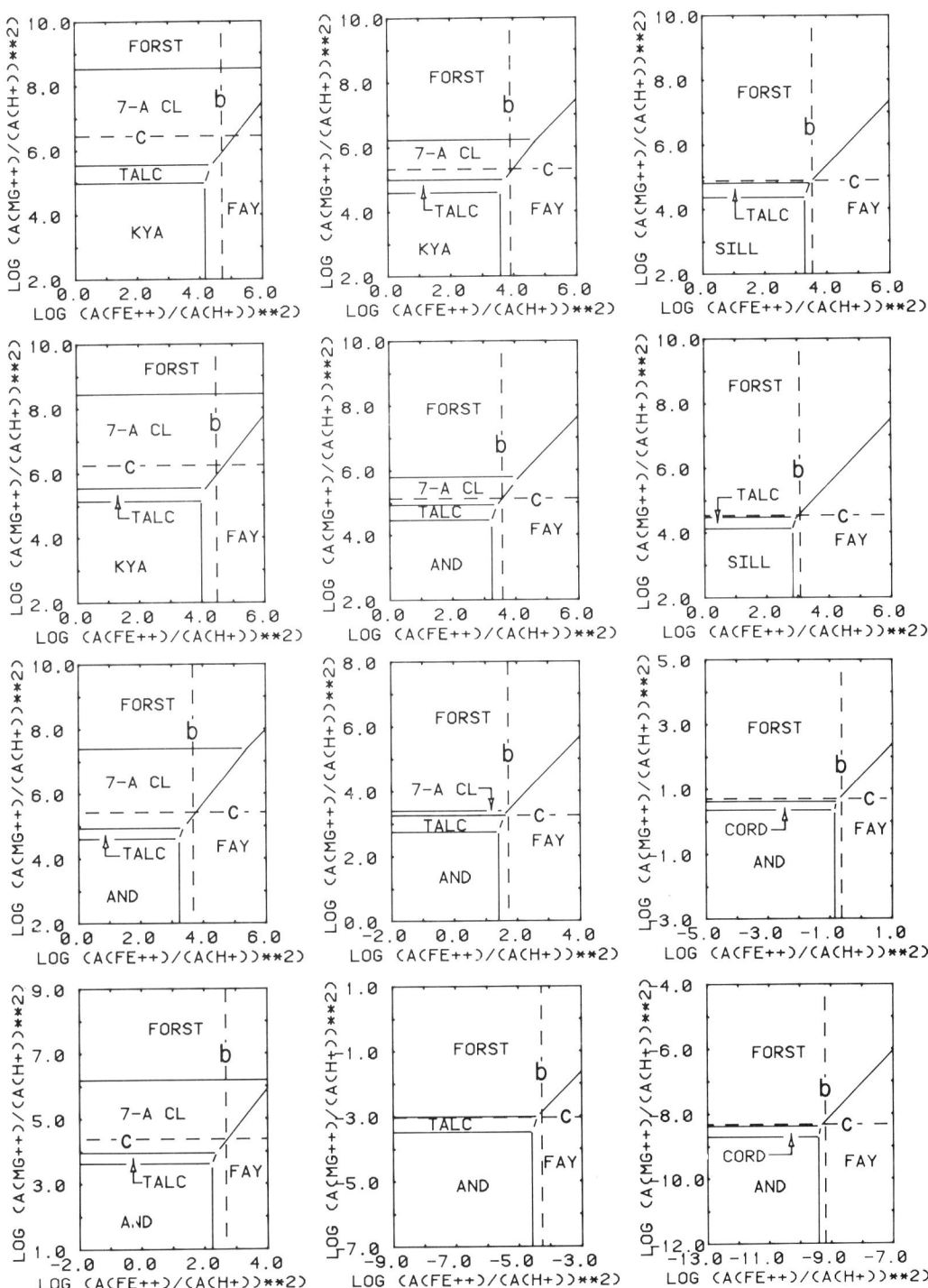

Phase relations in the system $HCl-H_2O-Al_2O_3-FeO-MgO-(SiO_2)$ in equilibrium with gibbsite, diaspore, or corundum, depending on which mineral is stable at each pressure and temperature. Metastable 7-A clinochlore was considered instead of its stable counterpart, 14-A clinochlore. Saturation limits: brucite (a), ferrous oxide (b), spinel (c).

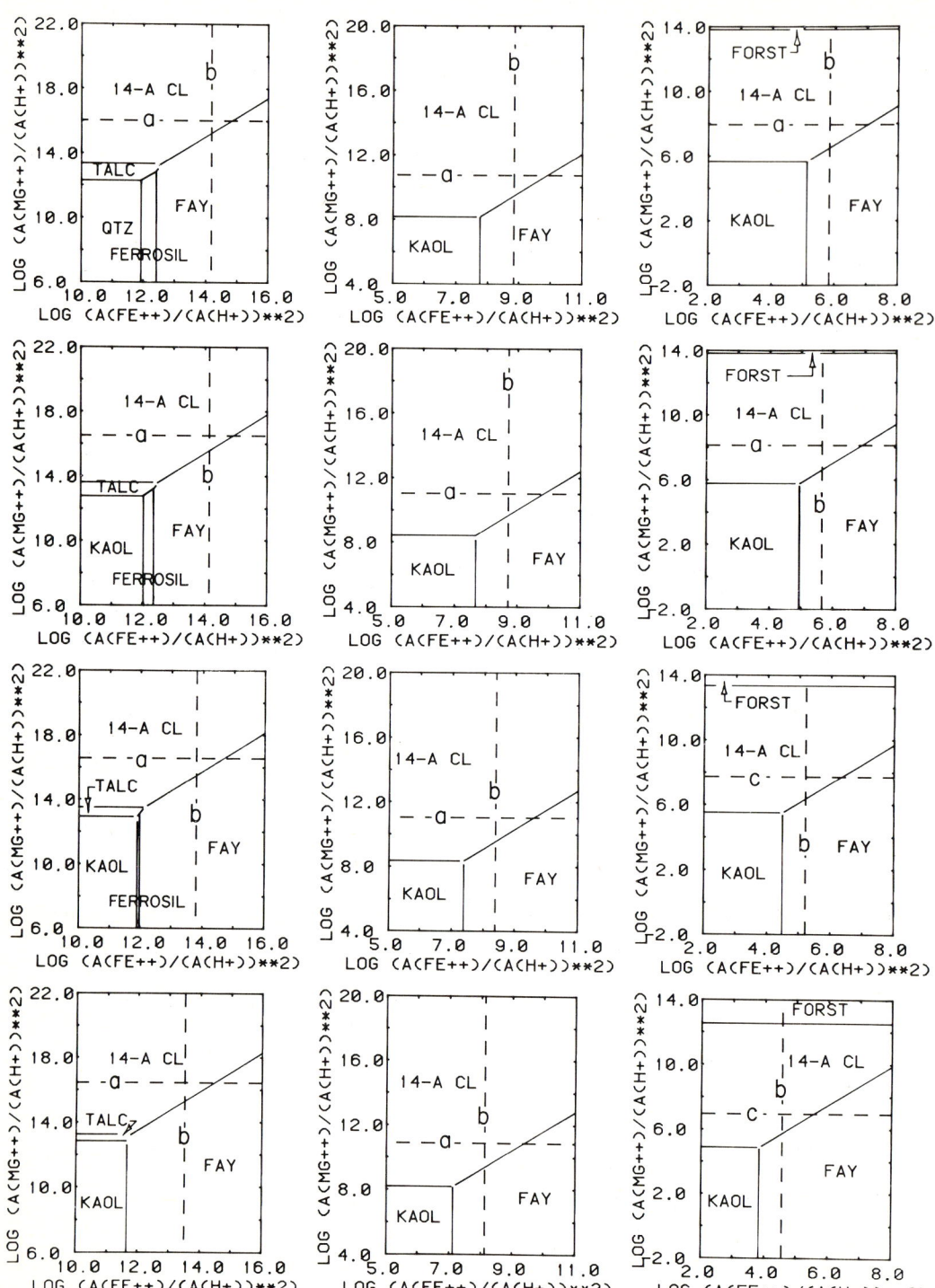

Phase relations in the system $HCl-H_2O-Al_2O_3-FeO-MgO-(SiO_2)$ in equilibrium with gibbsite, diaspore, or corundum, depending on which mineral is stable at each pressure and temperature. Metastable chrysotile was considered instead of its stable counterpart, antigorite. Saturation limits: brucite (a), ferrous oxide (b), spinel (c).

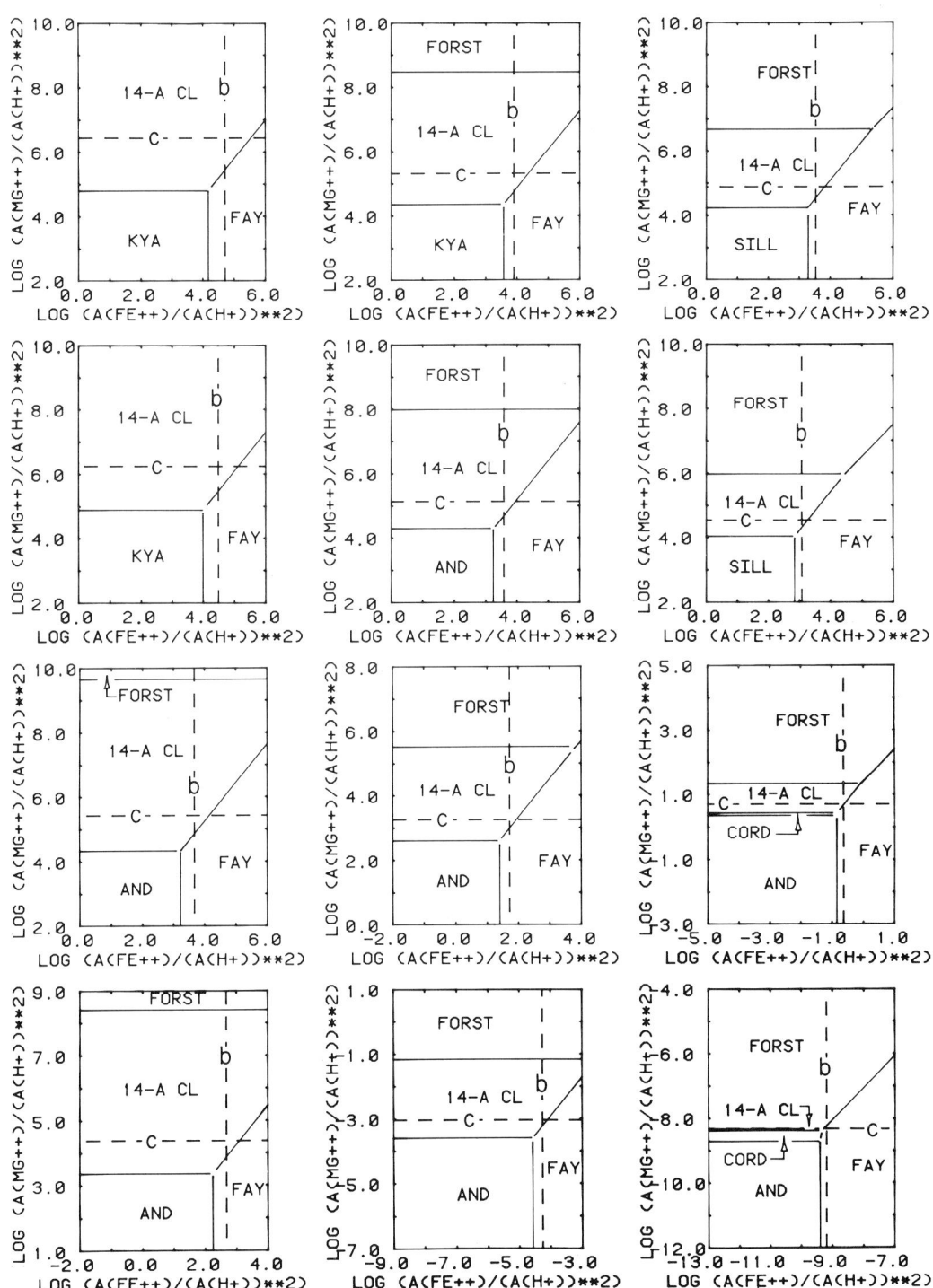

Phase relations in the system $HCl-H_2O-Al_2O_3-FeO-MgO-(SiO_2)$ in equilibrium with gibbsite, diaspore, or corundum, depending on which mineral is stable at each pressure and temperature. Metastable chrysotile was considered instead of its stable counterpart, antigorite. Saturation limits: brucite (a), ferrous oxide (b), spinel (c).

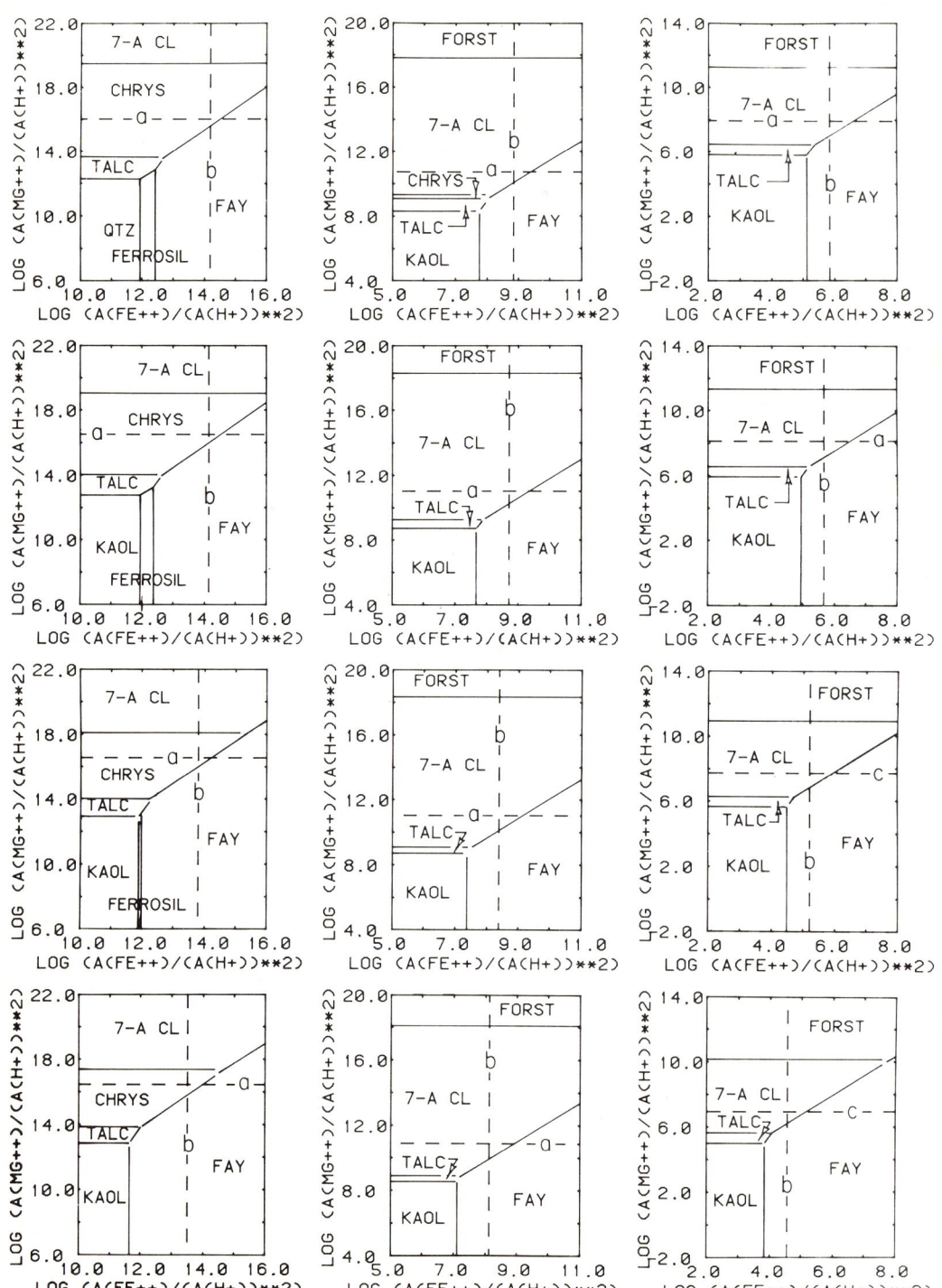

Phase relations in the system $HCl-H_2O-Al_2O_3-FeO-MgO-(SiO_2)$ in equilibrium with gibbsite, diaspore, or corundum, depending on which mineral is stable at each pressure and temperature. Metastable 7-A clinochlore and chrysotile were considered instead of their stable counterparts, 14-A clinochlore and antigorite. Saturation limits: brucite (a), ferrous oxide (b), spinel (c).

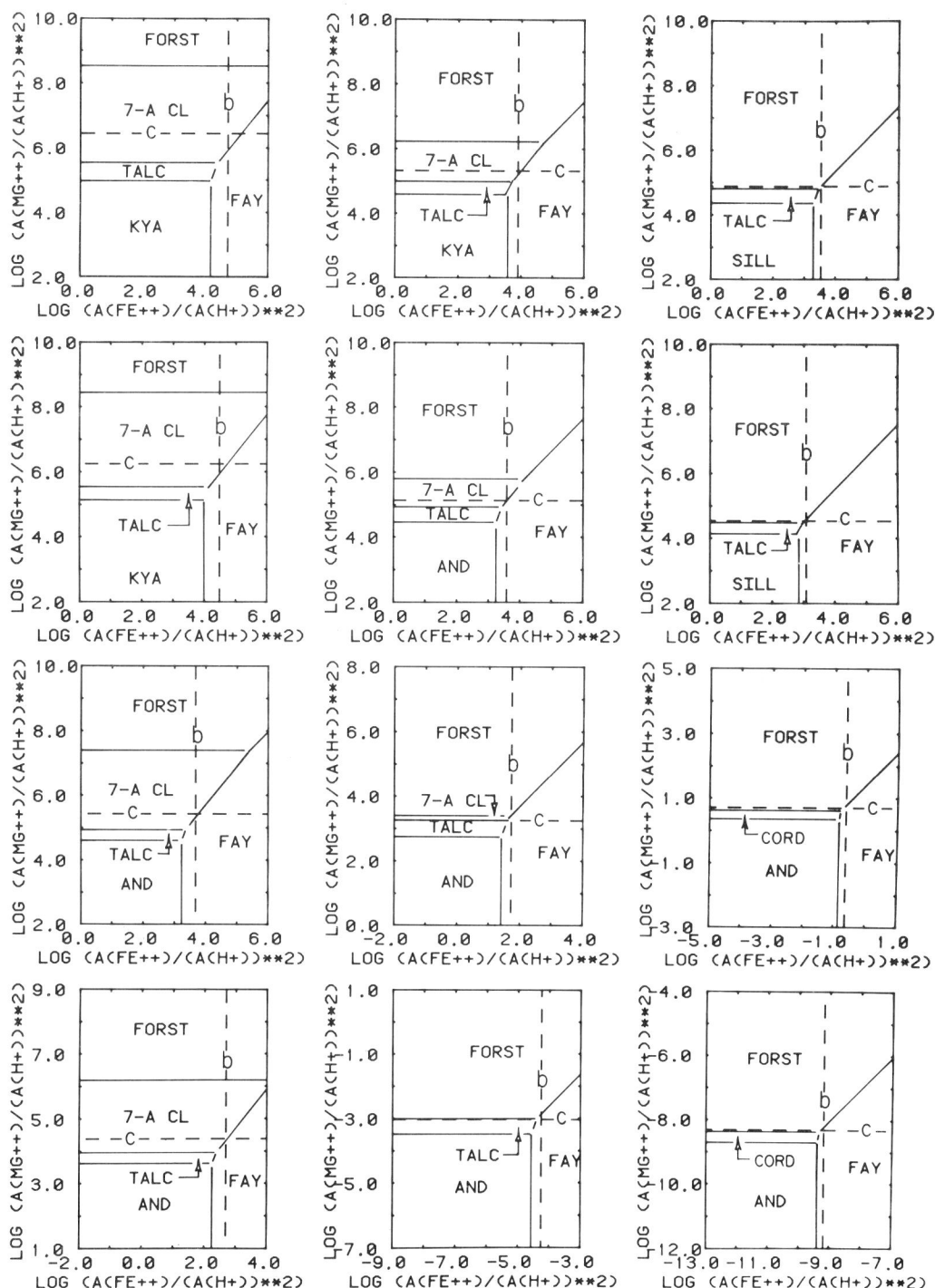

Phase relations in the system $HCl-H_2O-Al_2O_3-FeO-MgO-(SiO_2)$ in equilibrium with gibbsite, diaspore, or corundum, depending on which mineral is stable at each pressure and temperature. Metastable 7-A clinochlore and chrysotile were considered instead of their stable counterparts, 14-A clinochlore and antigorite. Saturation limits: brucite (a), ferrous oxide (b), spinel (c).

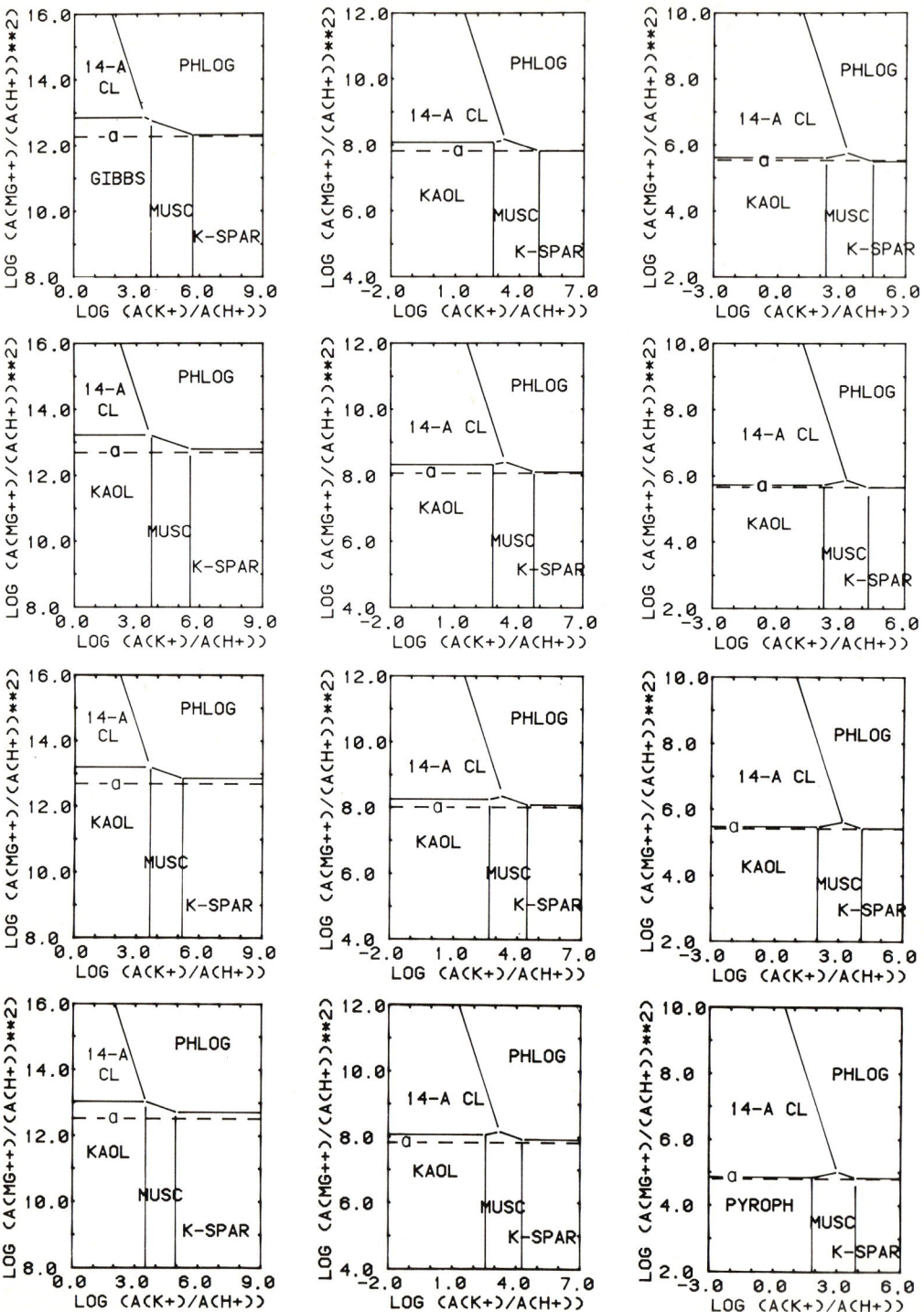

Phase relations in the system $HCl-H_2O-(Al_2O_3)-K_2O-MgO-SiO_2$ in equilibrium with quartz. Saturation limit: talc (a).

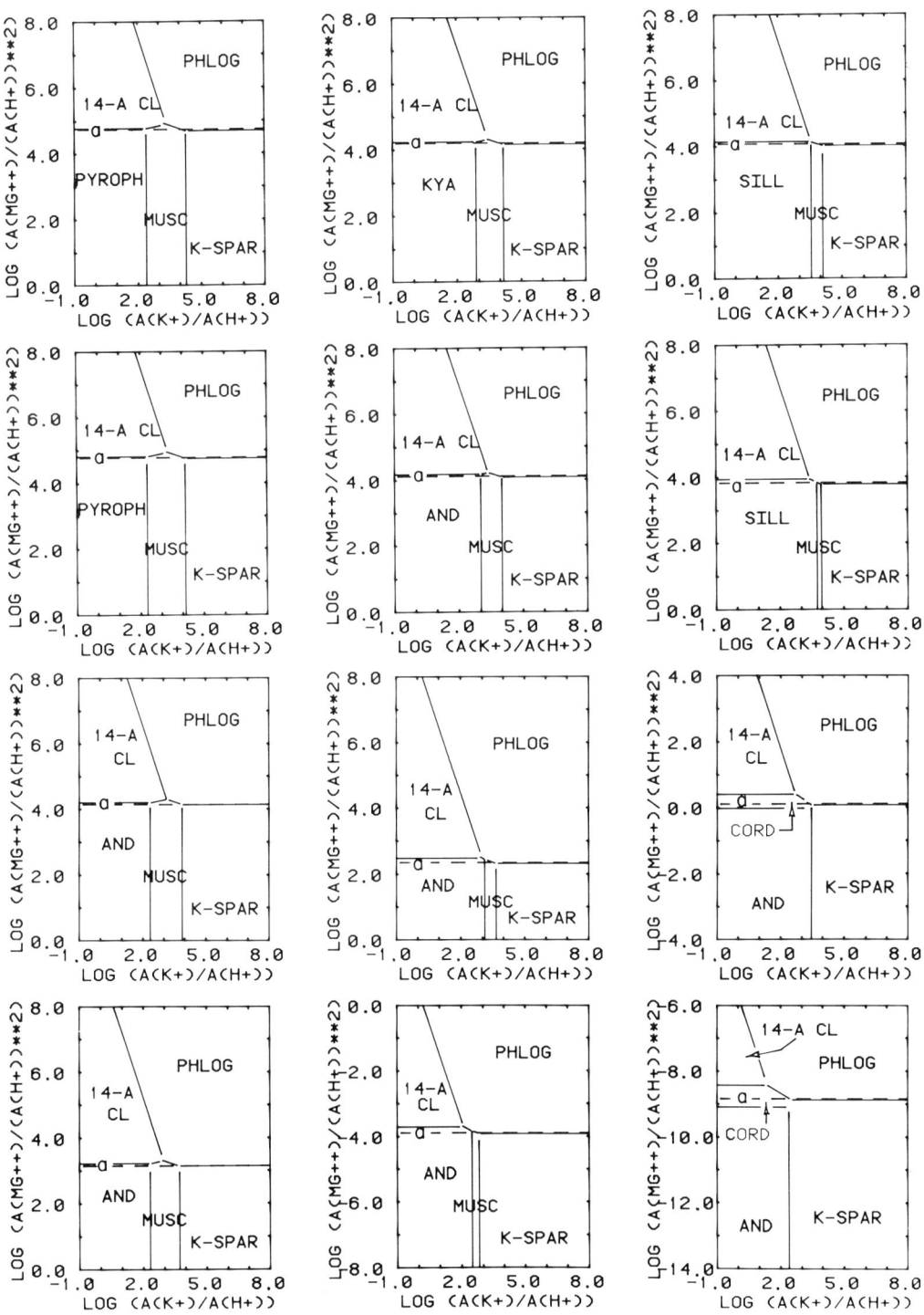

Phase relations in the system $HCl-H_2O-(Al_2O_3)-K_2O-MgO-SiO_2$ in equilibrium with quartz. Saturation limit: talc (a).

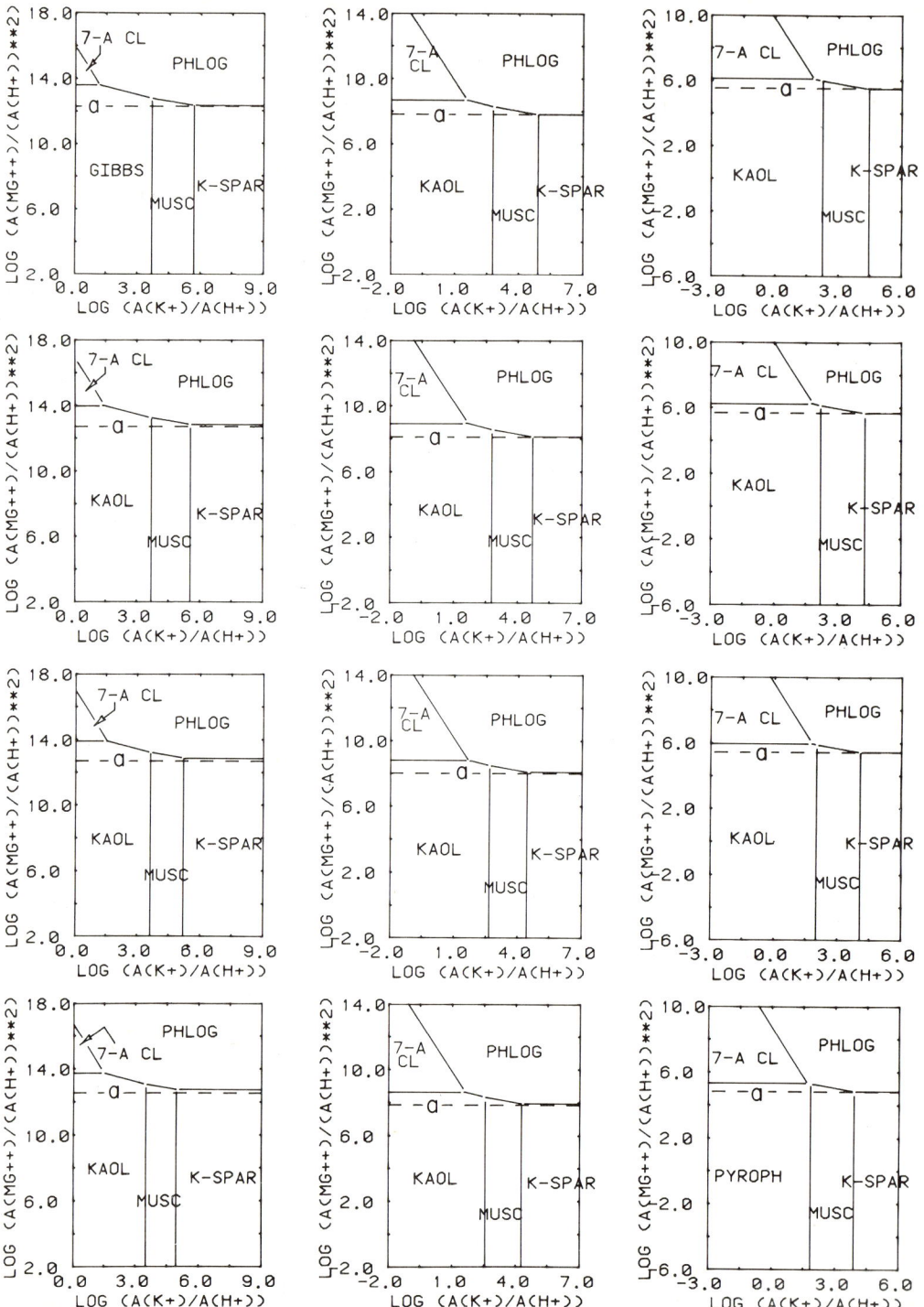

Phase relations in the system $HCl-H_2O-(Al_2O_3)-K_2O-MgO-SiO_2$ in equilibrium with quartz. Metastable 7-A clinochlore was considered instead of its stable counterpart, 14-A clinochlore. Saturation limit: talc (a).

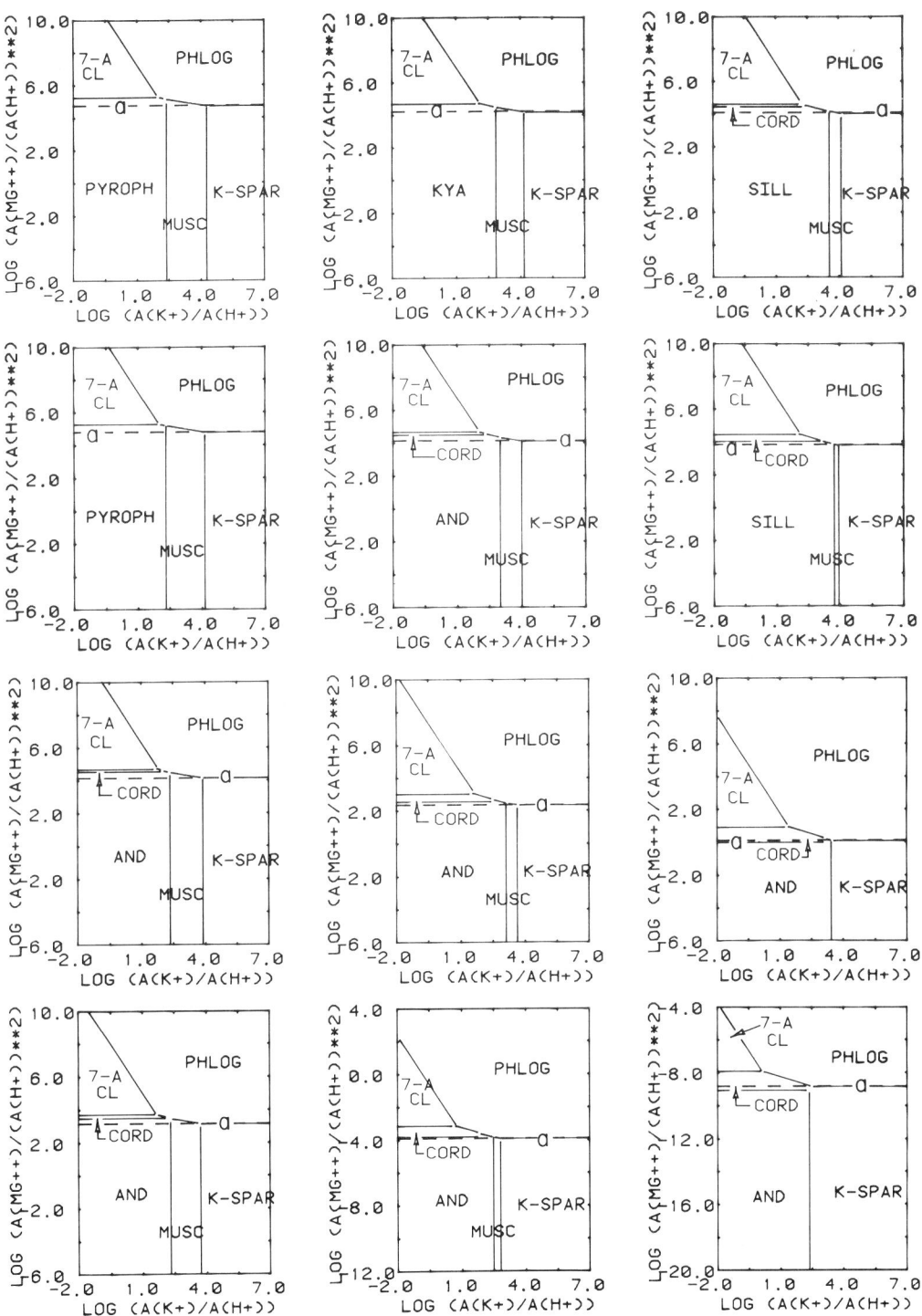

Phase relations in the system $HCl-H_2O-(Al_2O_3)-K_2O-MgO-SiO_2$ in equilibrium with quartz. Metastable 7-A clinochlore was considered instead of its stable counterpart, 14-A clinochlore. Saturation limit: talc (a).

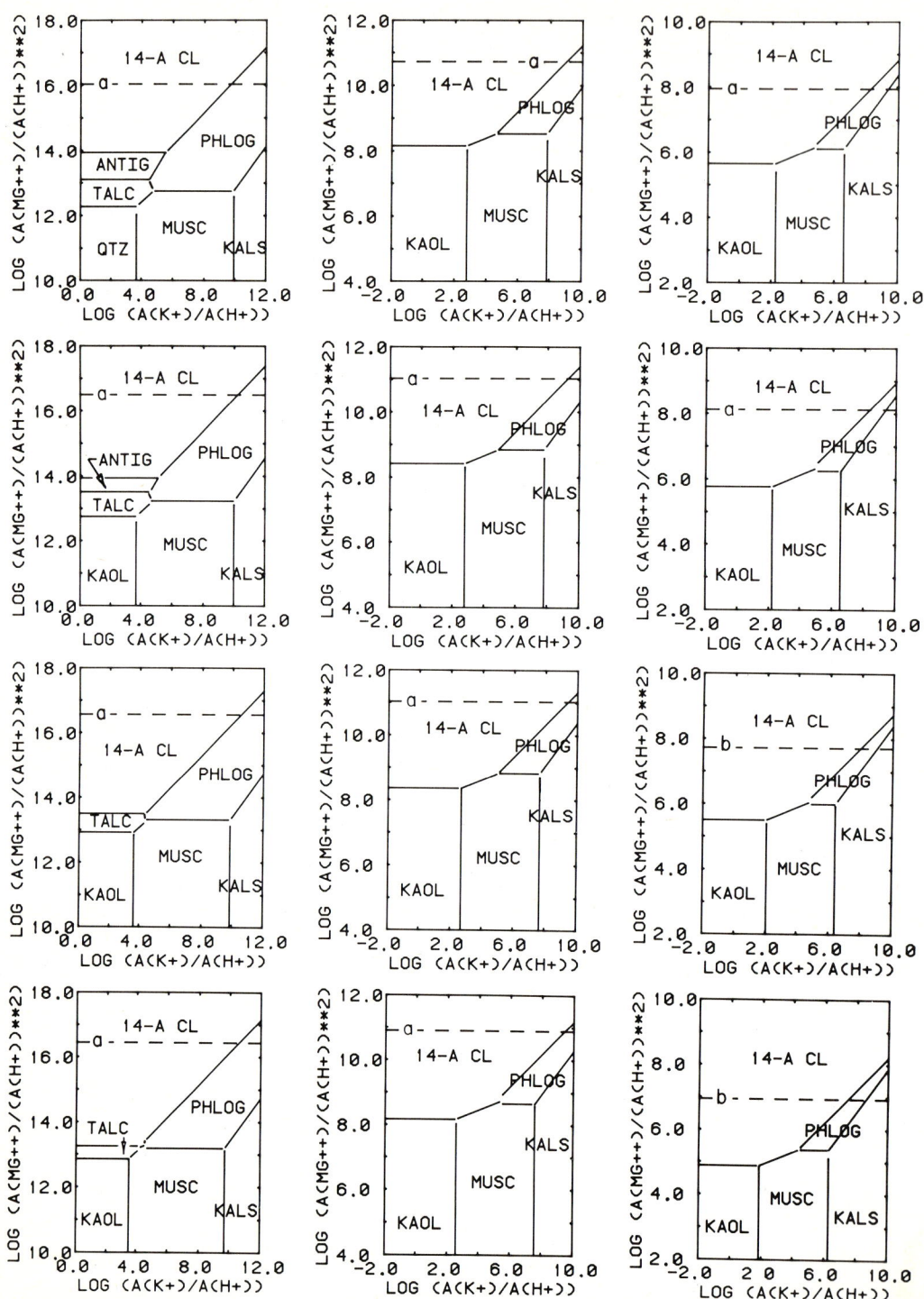

Phase relations in the system $HCl-H_2O-Al_2O_3-K_2O-MgO-(SiO_2)$ in equilibrium with gibbsite, diaspore, or corundum, depending on which mineral is stable at each pressure and temperature. Saturation limits: brucite (a), spinel (b).

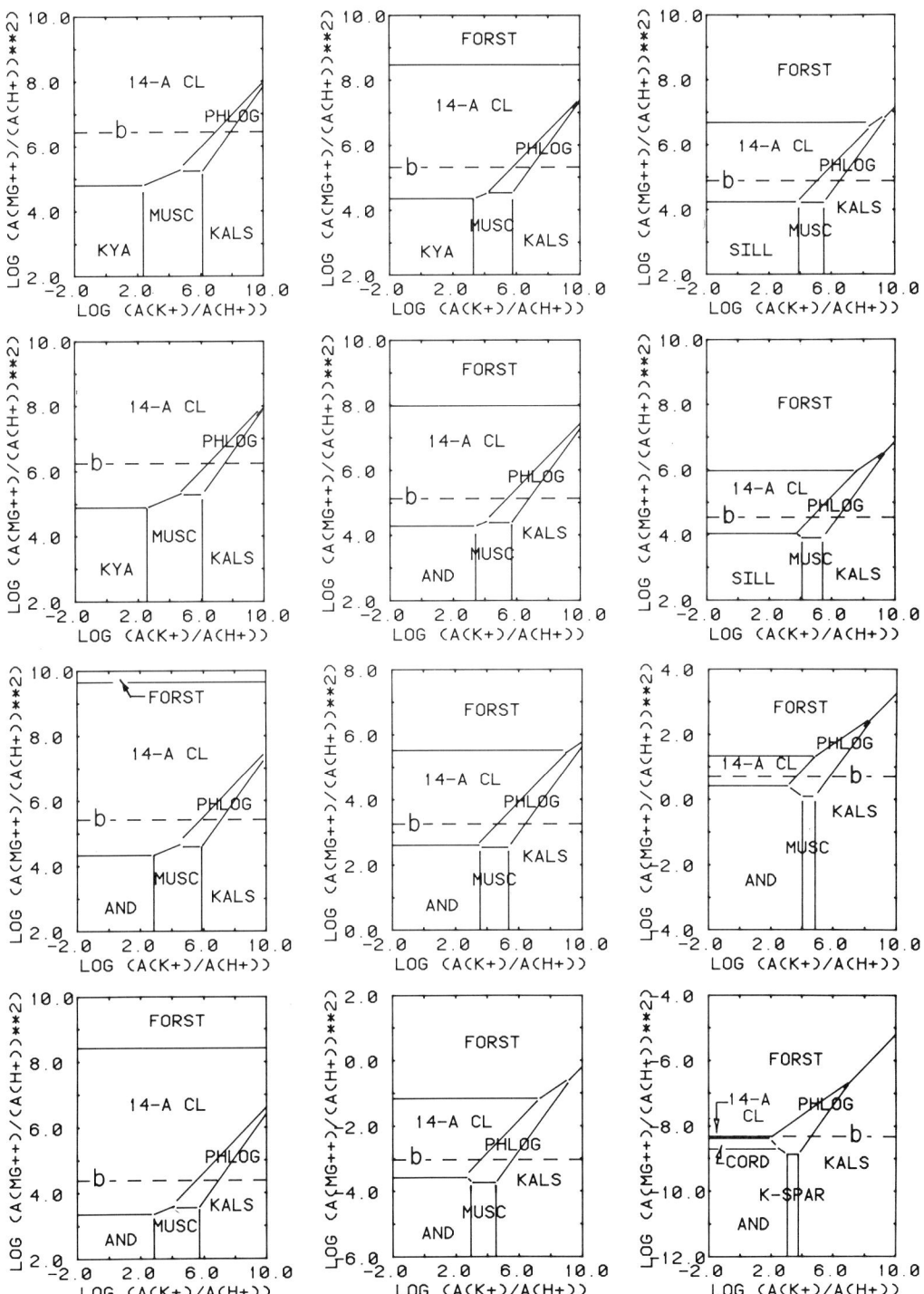

Phase relations in the system $HCl-H_2O-Al_2O_3-K_2O-MgO-(SiO_2)$ in equilibrium with gibbsite, diaspore, or corundum, depending on which mineral is stable at each pressure and temperature. Saturation limits: brucite (a), spinel (b).

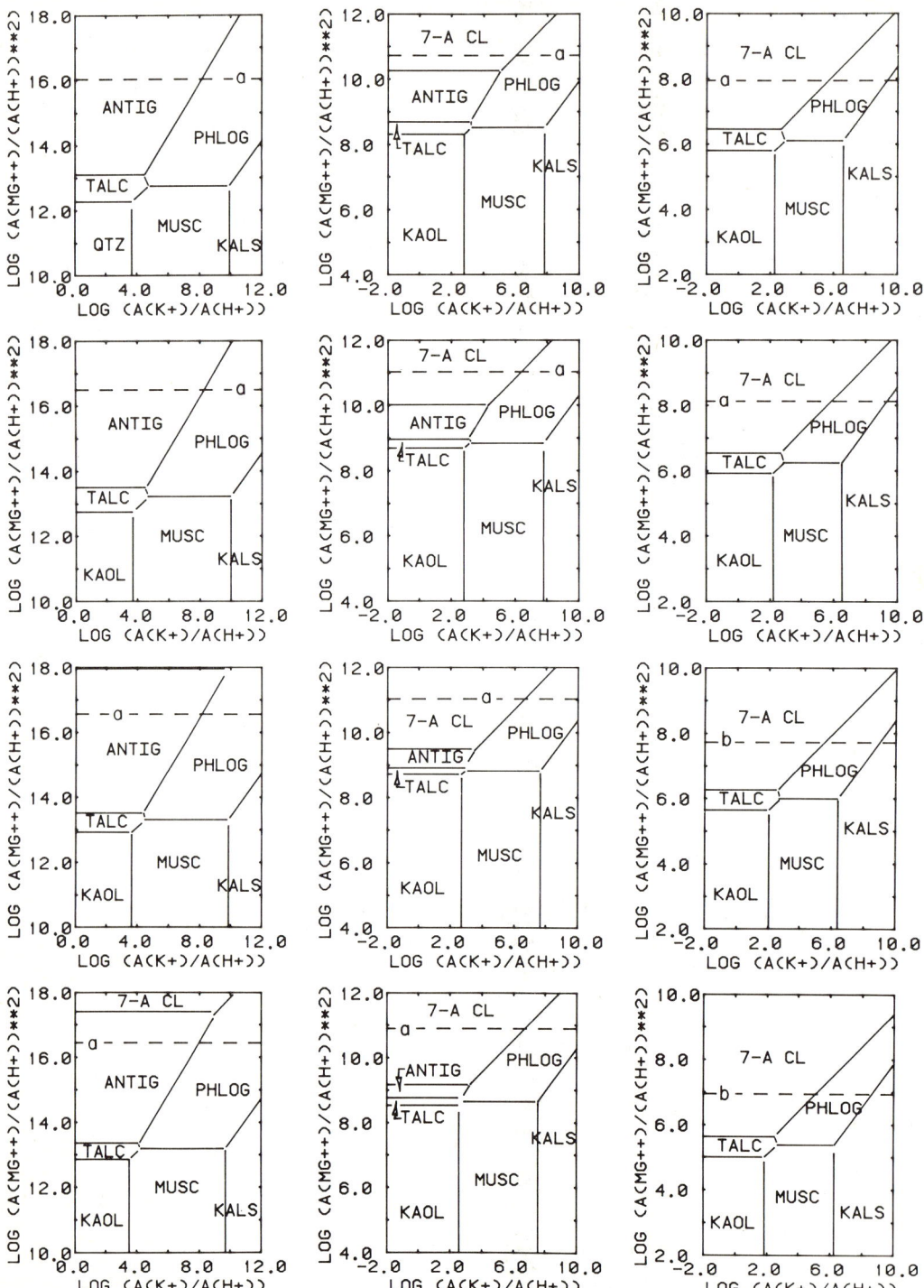

Phase relations in the system $HCl-H_2O-Al_2O_3-K_2O-MgO-(SiO_2)$ in equilibrium with gibbsite, diaspore, or corundum, depending on which mineral is stable at each pressure and temperature. Metastable 7-A clinochlore was considered instead of its stable counterpart, 14-A clinochlore. Saturation limits: brucite (a), spinel (b).

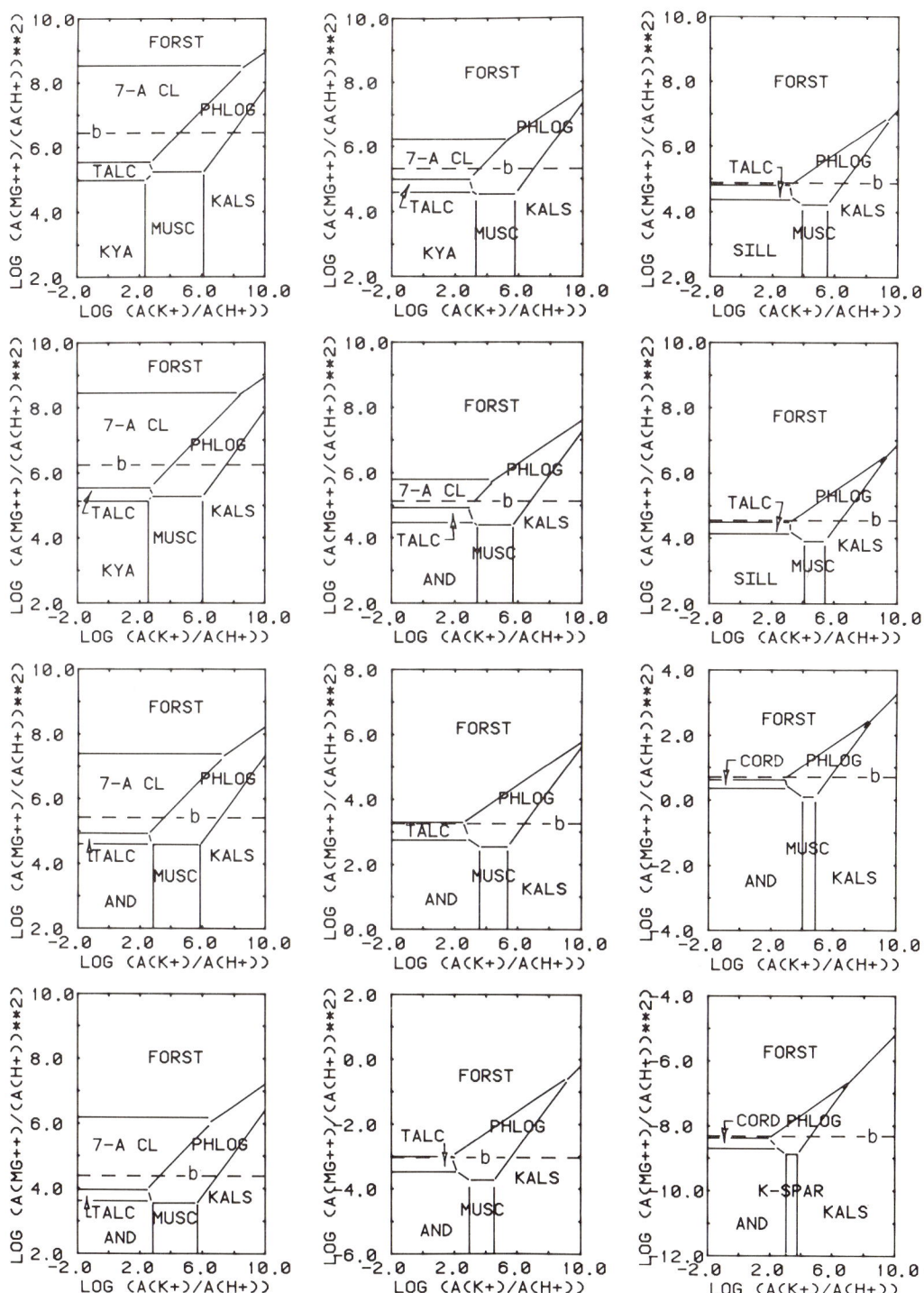

Phase relations in the system $HCl-H_2O-Al_2O_3-K_2O-MgO-(SiO_2)$ in equilibrium with gibbsite, diaspore, or corundum, depending on which mineral is stable at each pressure and temperature. Metastable 7-A clinochlore was considered instead of its stable counterpart, 14-A clinochlore. Saturation limits: brucite (a), spinel (b).

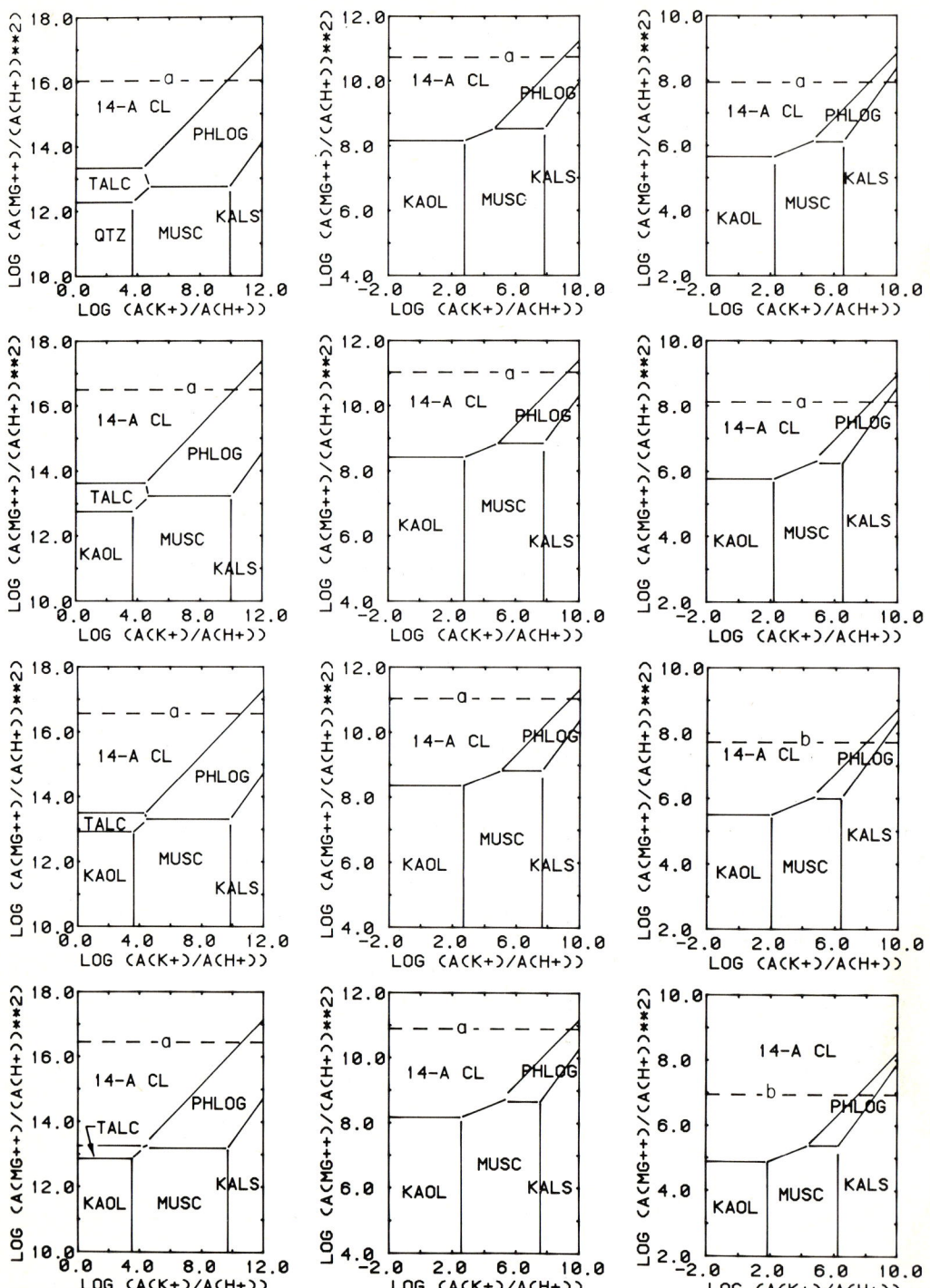

Phase relations in the system $HCl-H_2O-Al_2O_3-K_2O-MgO-(SiO_2)$ in equilibrium with gibbsite, diaspore, or corundum, depending on which mineral is stable at each pressure and temperature. Metastable chrysotile was considered instead of its stable counterpart, antigorite. Saturation limits: brucite (a), spinel (b).

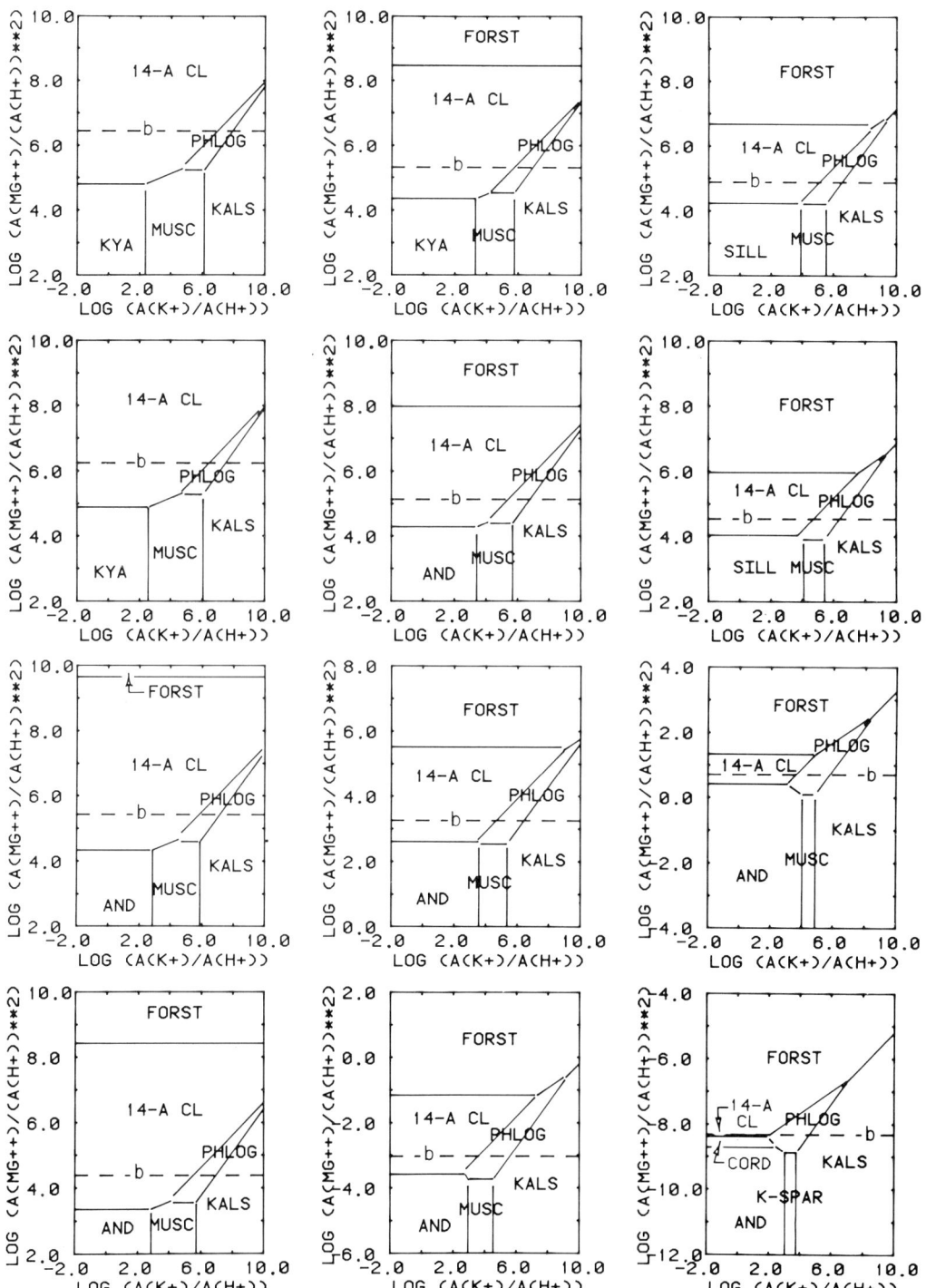

Phase relations in the system $HCl-H_2O-Al_2O_3-K_2O-MgO-(SiO_2)$ in equilibrium with gibbsite, diaspore, or corundum, depending on which mineral is stable at each pressure and temperature. Metastable chrysotile was considered instead of its stable counterpart, antigorite. Saturation limits: brucite (a), spinel (b).

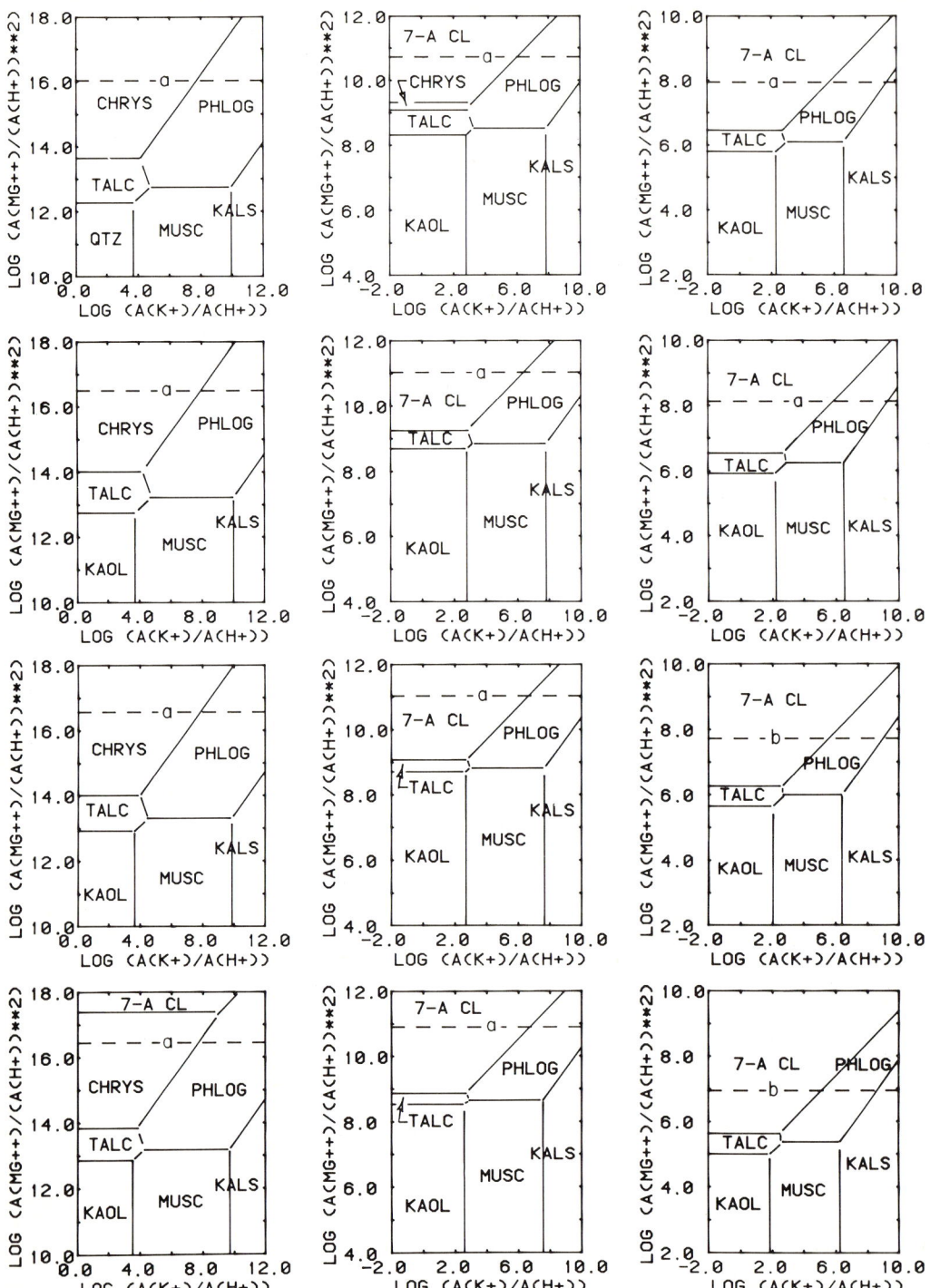

Phase relations in the system $HCl-H_2O-Al_2O_3-K_2O-MgO-(SiO_2)$ in equilibrium with gibbsite, diaspore, or corundum, depending on which mineral is stable at each pressure and temperature. Metastable 7-A clinochlore and chrysotile were considered instead of their stable counterparts, 14-A clinochlore and antigorite. Saturation limits: brucite (a), spinel (b).

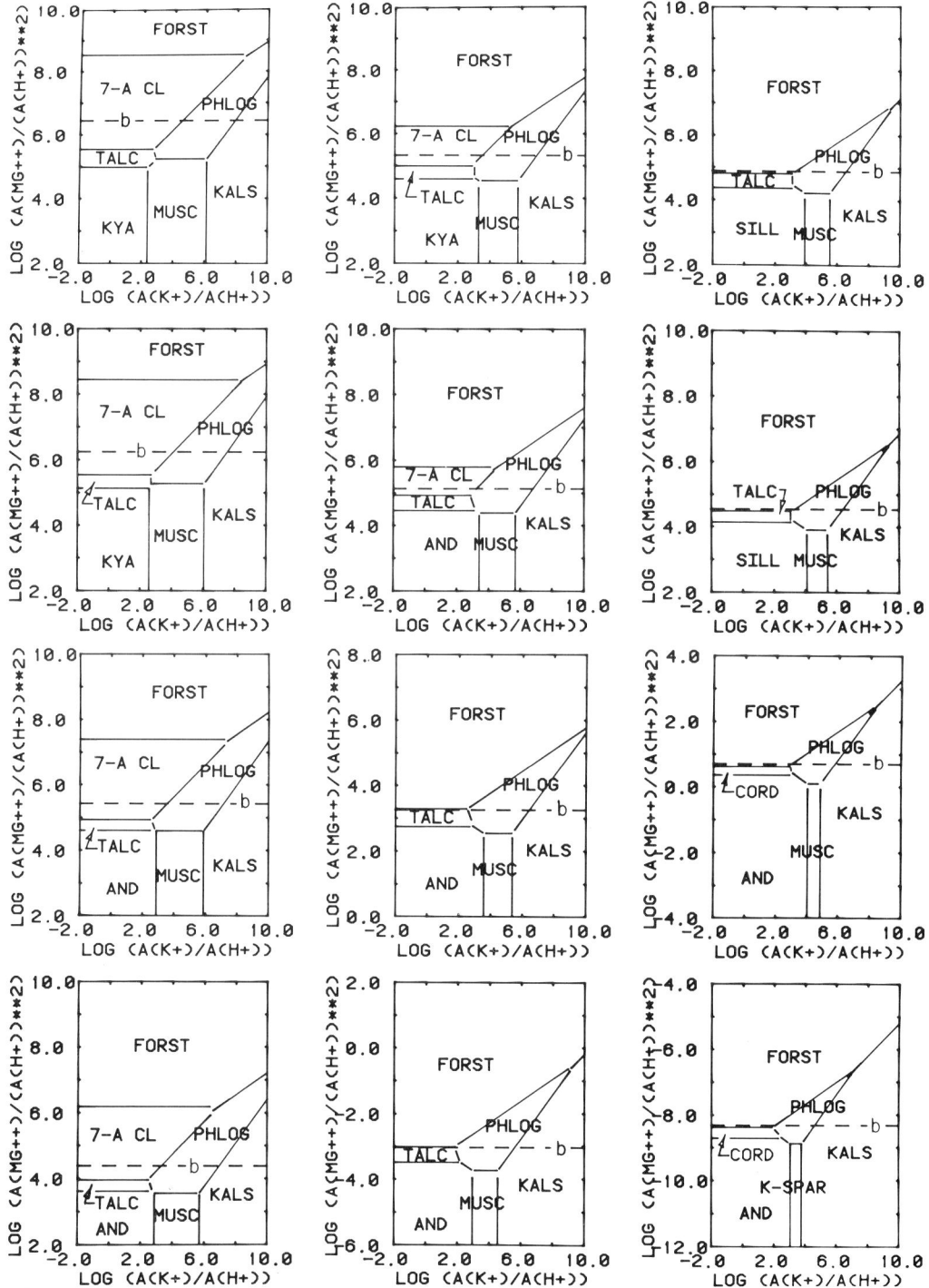

Phase relations in the system $HCl-H_2O-Al_2O_3-K_2O-MgO-(SiO_2)$ in equilibrium with gibbsite, diaspore, or corundum, depending on which mineral is stable at each pressure and temperature. Metastable 7-A clinochlore and chrysotile were considered instead of their stable counterparts, 14-A clinochlore and antigorite. Saturation limits: brucite (a), spinel (b).

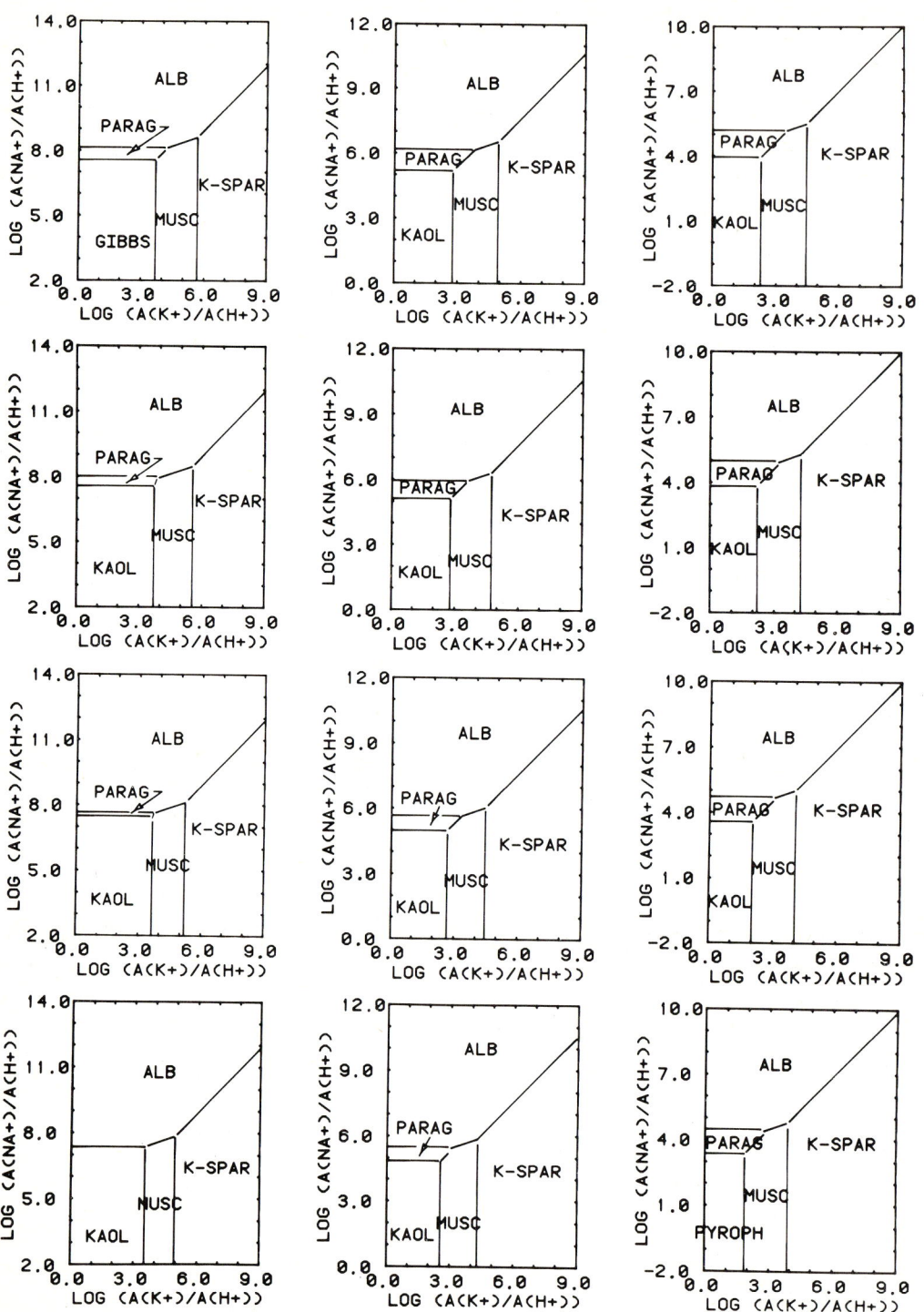

Phase relations in the system $HCl-H_2O-(Al_2O_3)-K_2O-Na_2O-SiO_2$ in equilibrium with quartz.

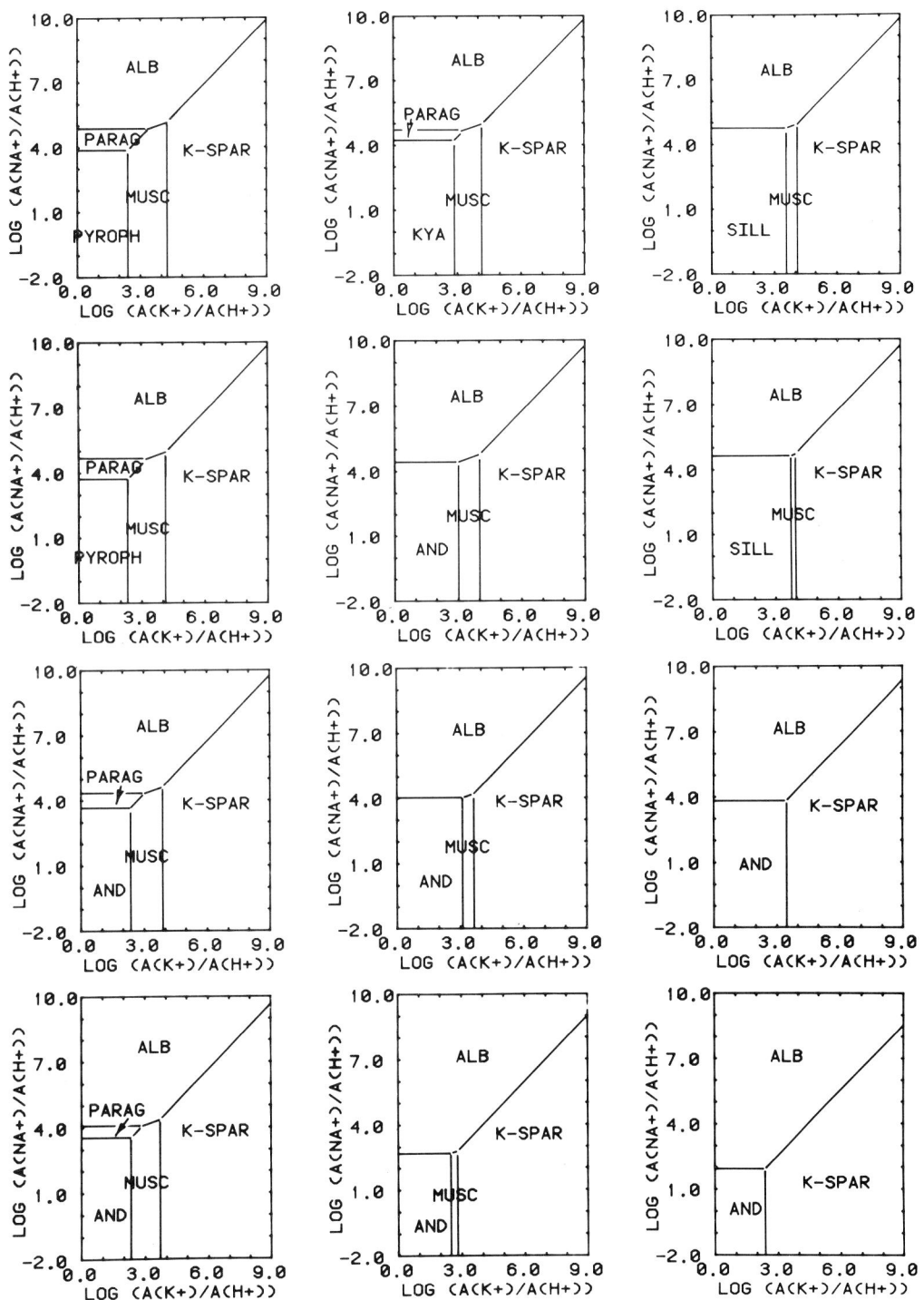

Phase relations in the system $HCl-H_2O-(Al_2O_3)-K_2O-Na_2O-SiO_2$ in equilibrium with quartz.

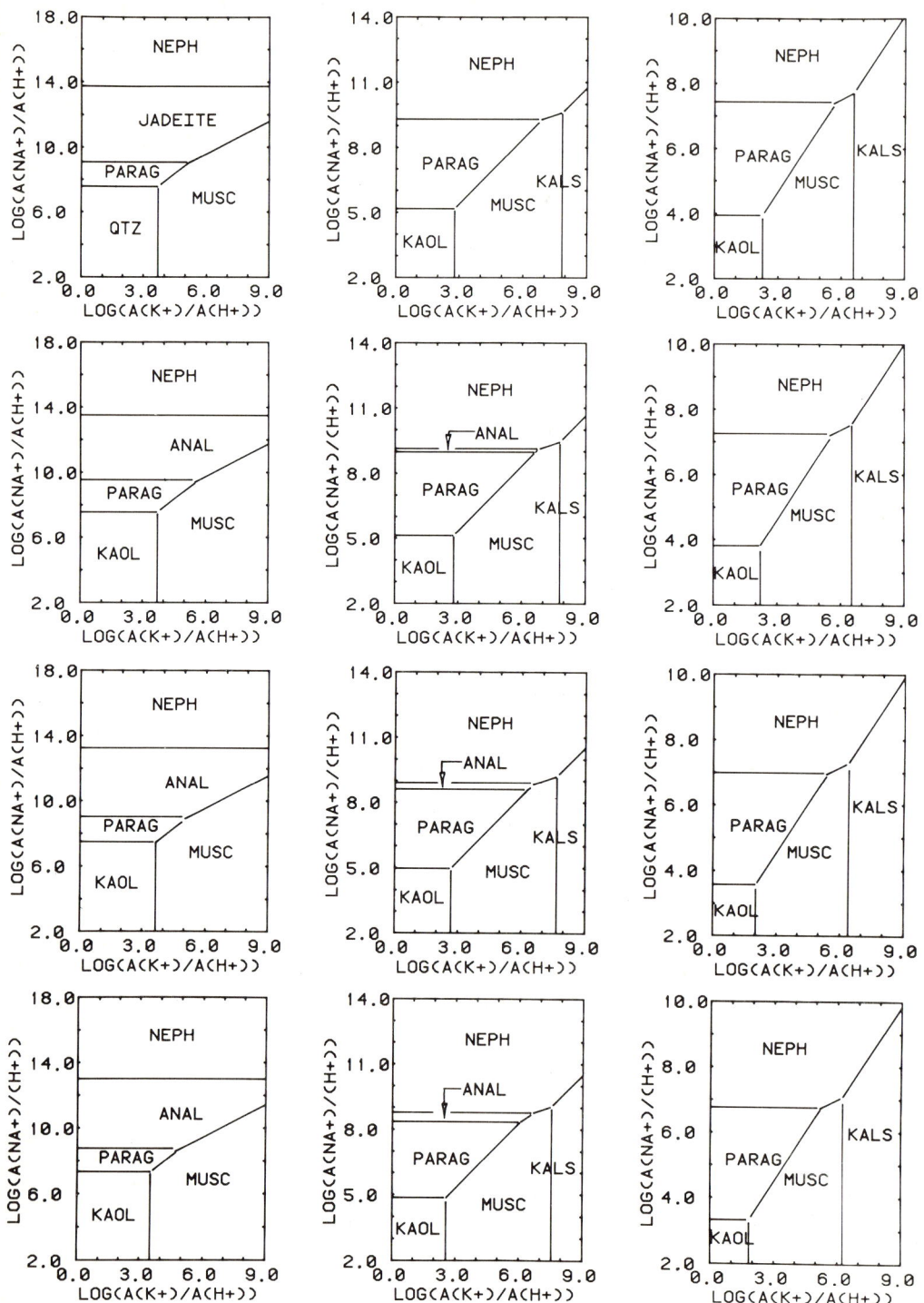

Phase relations in the system $HCl-H_2O-Al_2O_3-K_2O-Na_2O-(SiO_2)$ in equilibrium with gibbsite, diaspore, or corundum, depending on which mineral is stable at each pressure and temperature.

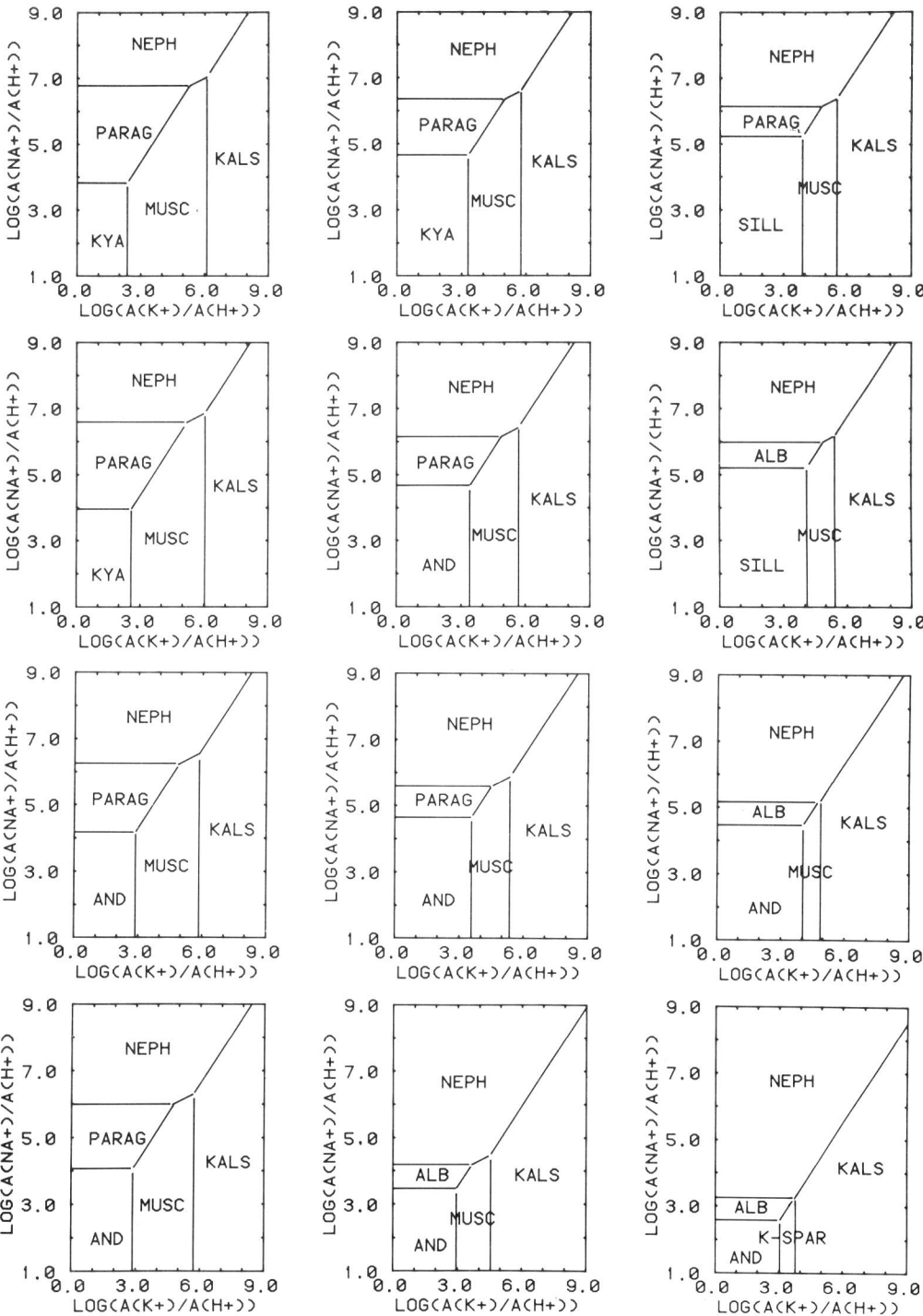

Phase relations in the system $HCl-H_2O-Al_2O_3-K_2O-Na_2O-(SiO_2)$ in equilibrium with gibbsite, diaspore, or corundum, depending on which mineral is stable at each pressure and temperature.

110

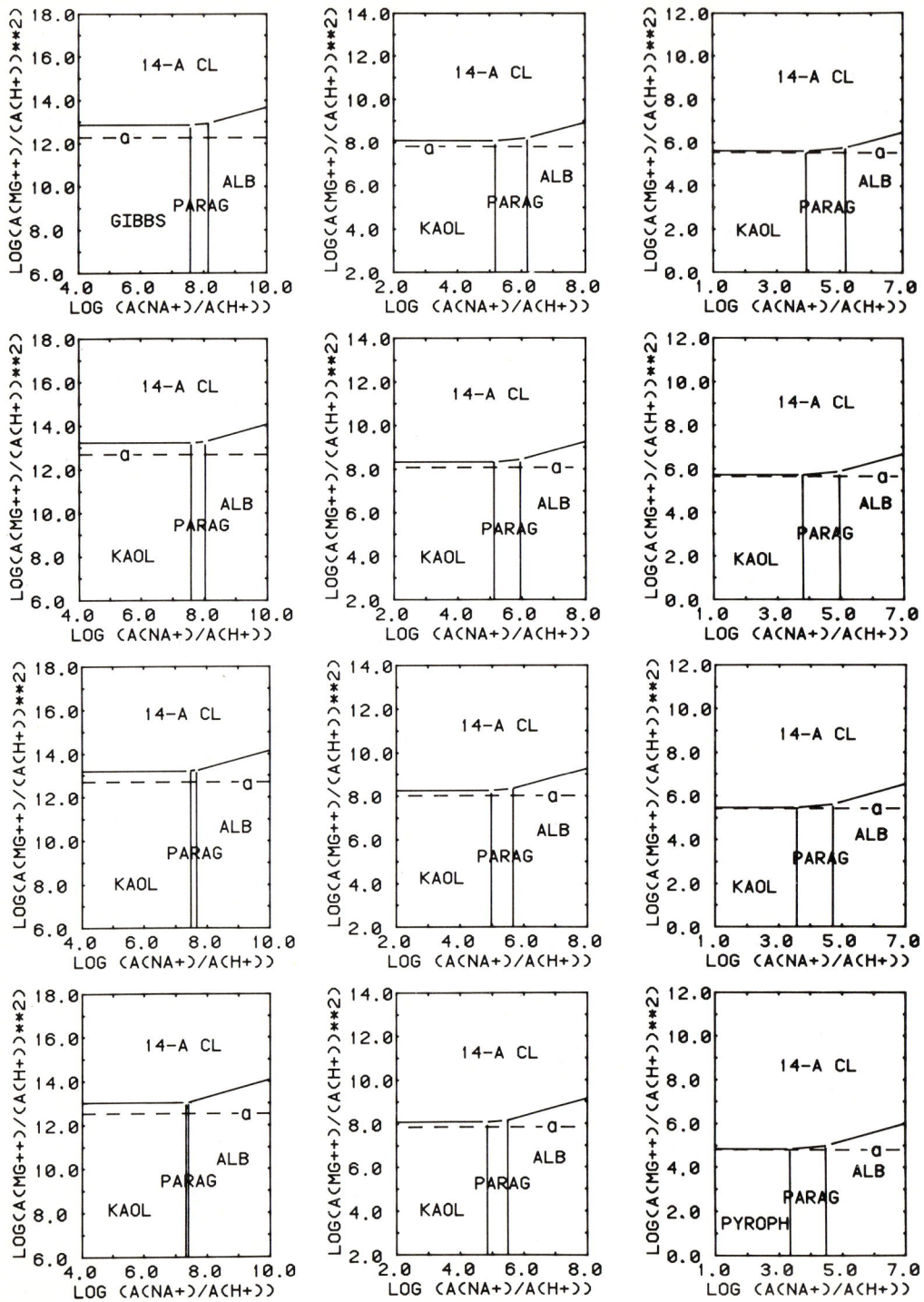

Phase relations in the system $HCl-H_2O-(Al_2O_3)-MgO-Na_2O-SiO_2$ in equilibrium with quartz. Saturation limit: talc.

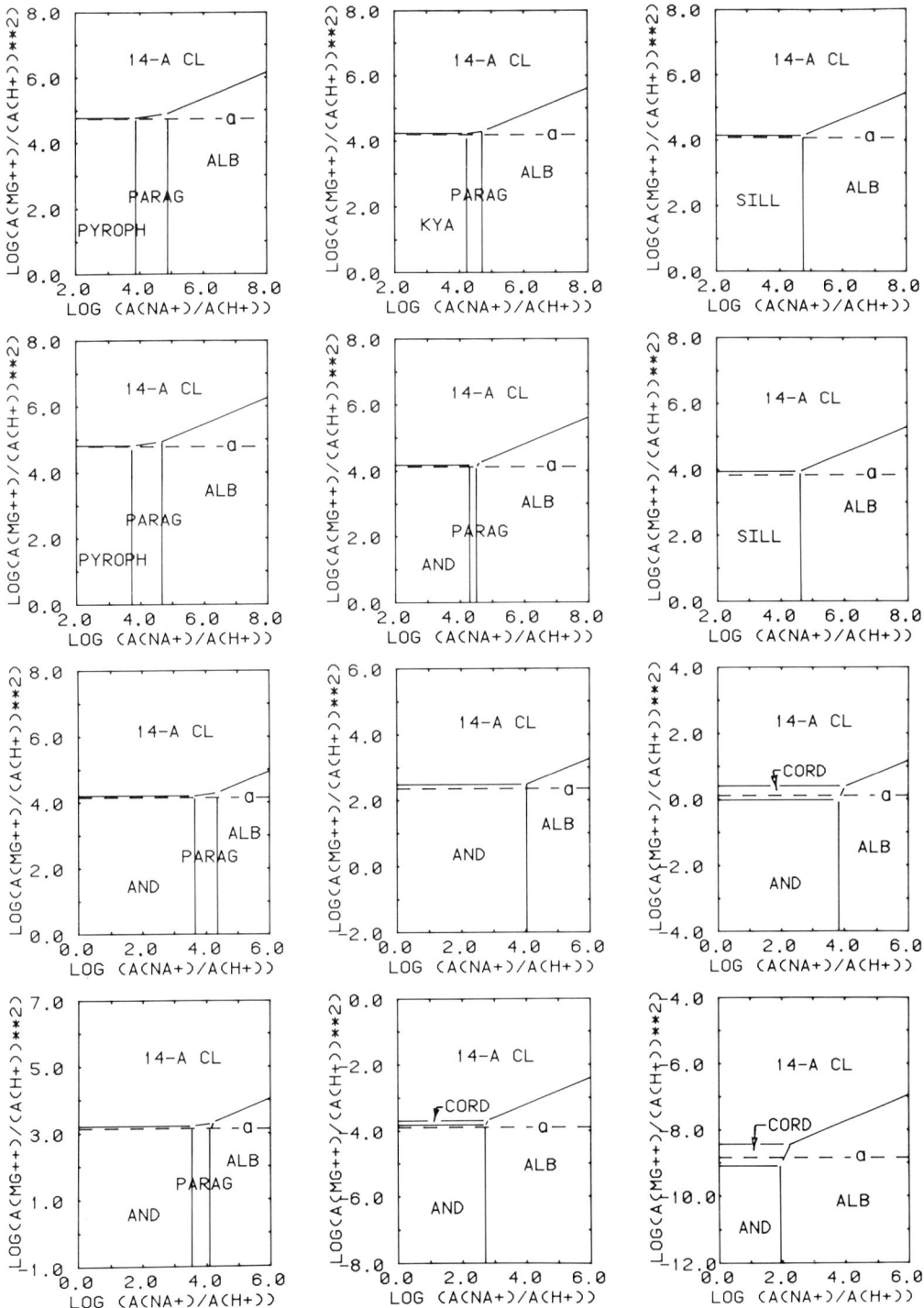

Phase relations in the system $HCl-H_2O-(Al_2O_3)-MgO-Na_2O-SiO_2$ in equilibrium with quartz. Saturation limit: talc.

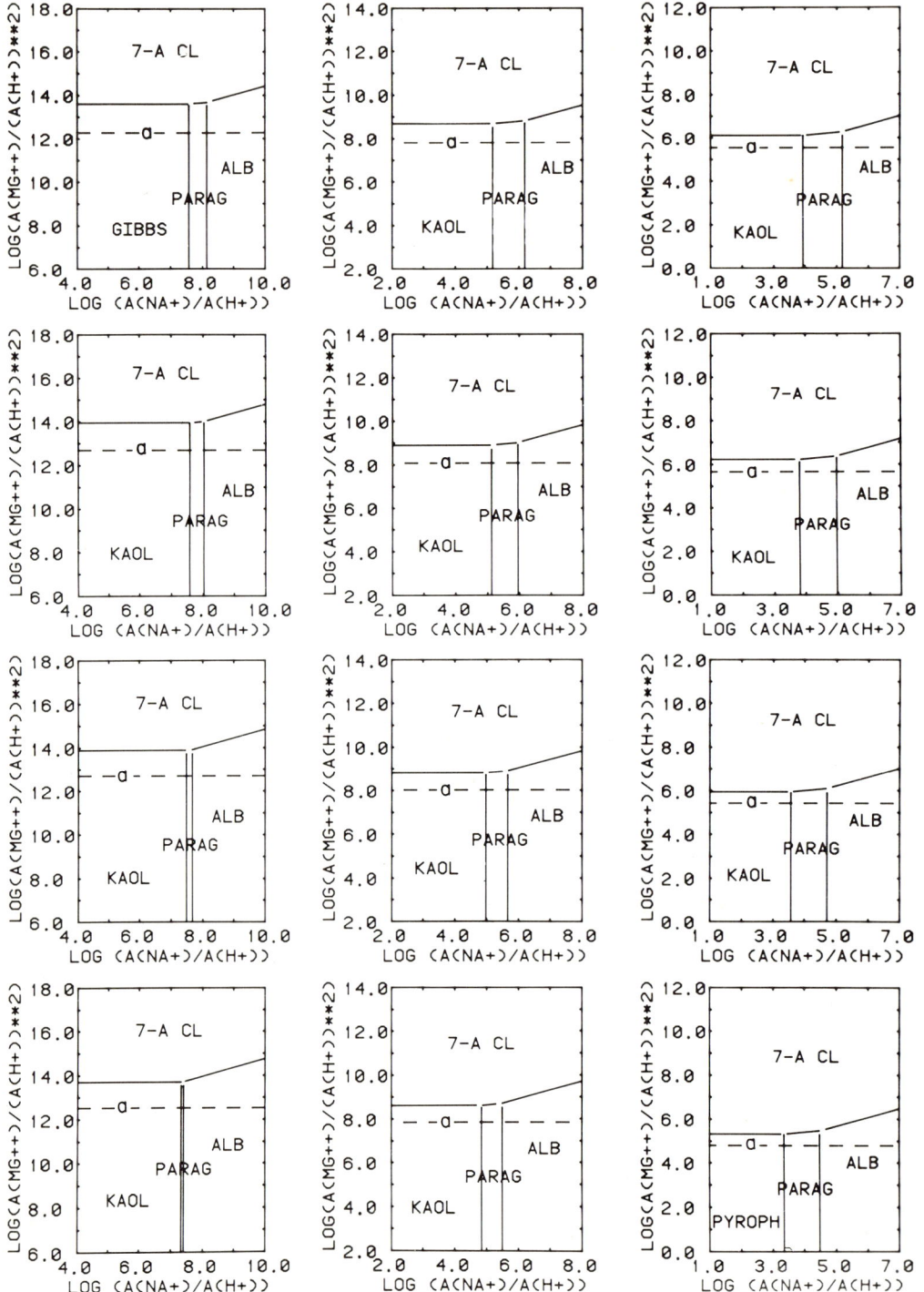

Phase relations in the system $HCl-H_2O-(Al_2O_3)-MgO-Na_2O-SiO_2$ in equilibrium with quartz. Metastable 7-A clinochlore was considered instead of its stable counterpart, 14-A clinochlore. Saturation limit: talc.

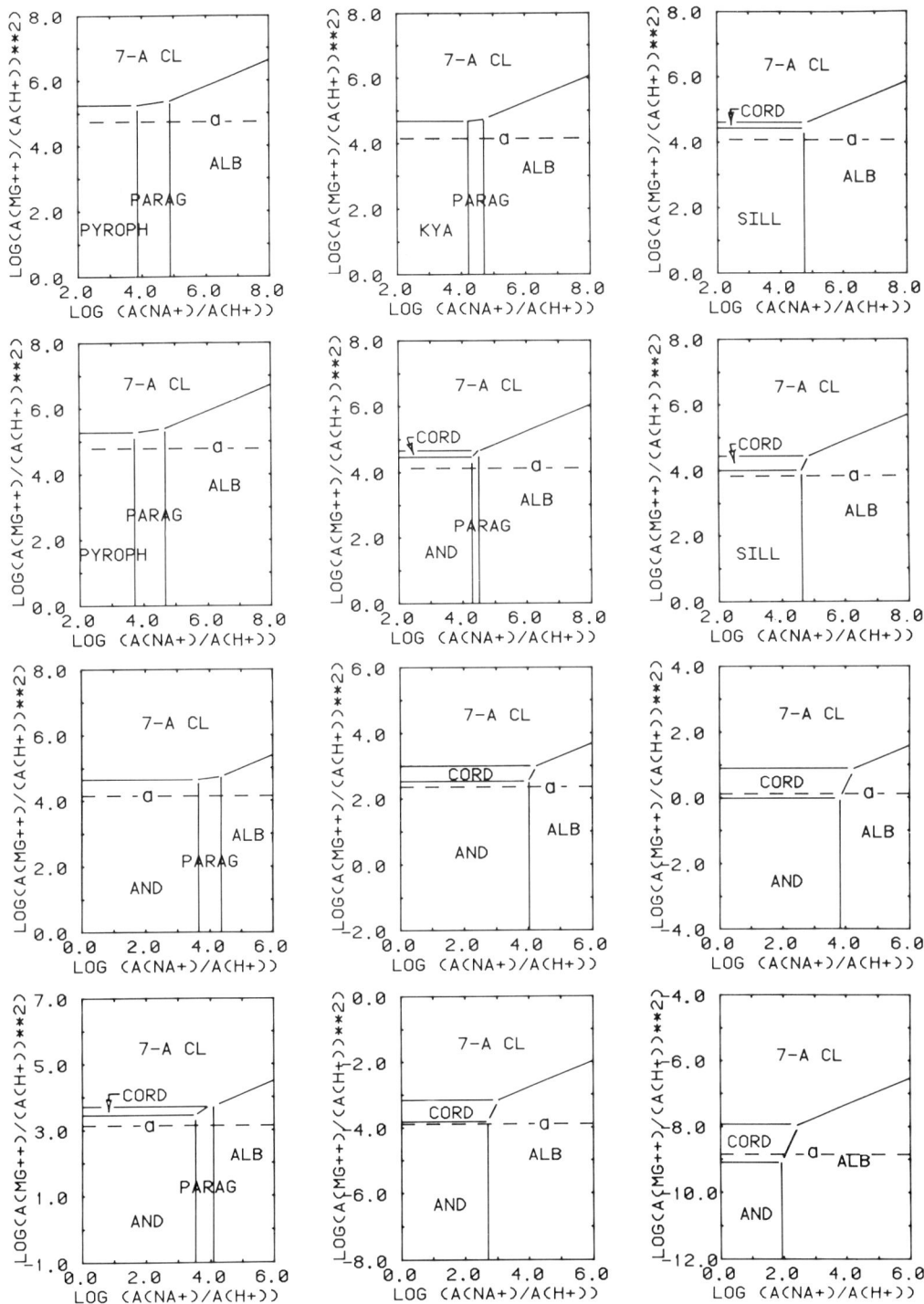

Phase relations in the system $HCl-H_2O-(Al_2O_3)-MgO-Na_2O-SiO_2$ in equilibrium with quartz. Metastable 7-A clinochlore was considered instead of its stable counterpart, 14-A clinochlore. Saturation limit: talc.

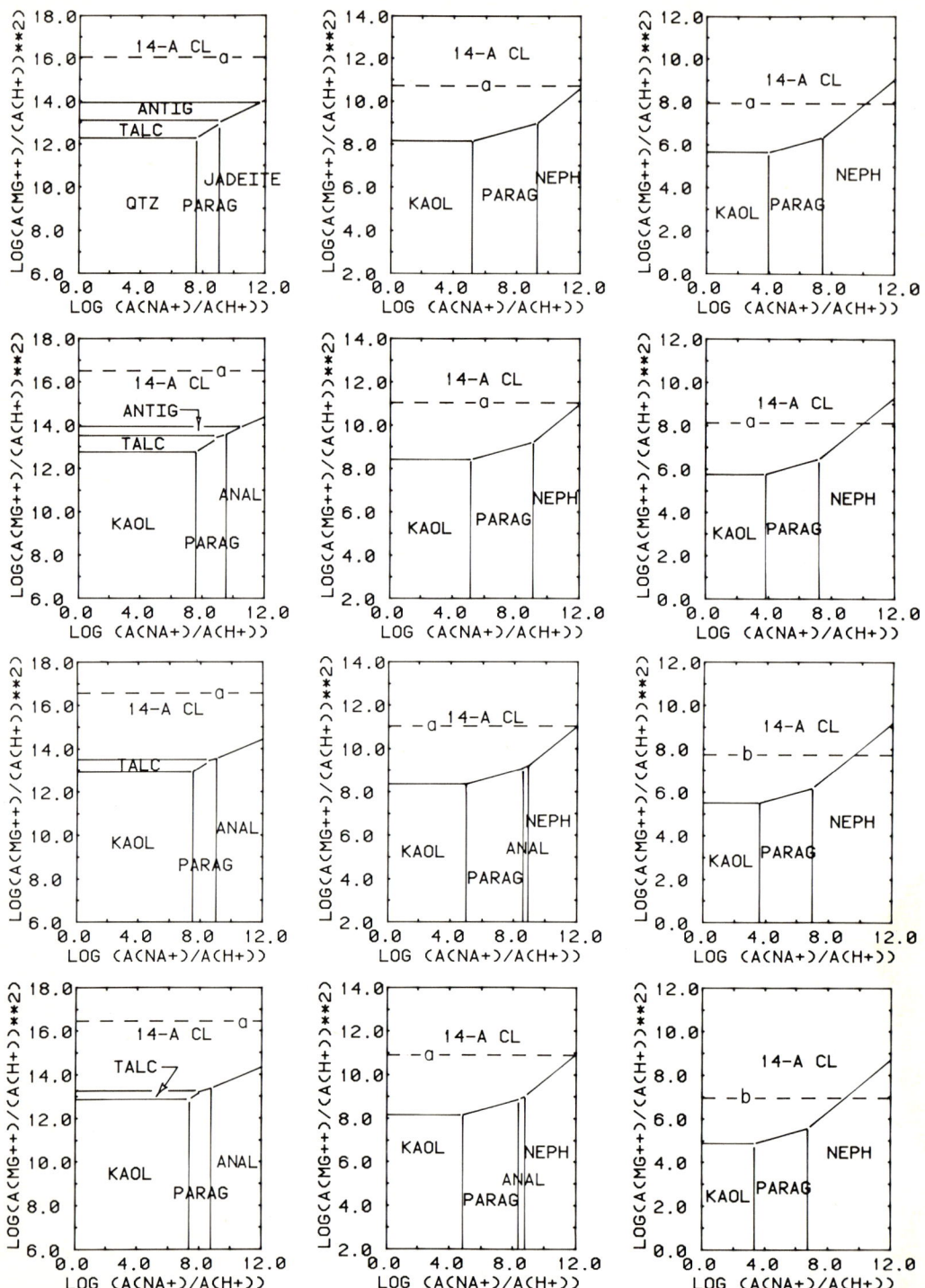

Phase relations in the system $HCl-H_2O-Al_2O_3-MgO-Na_2O-(SiO_2)$ in equilibrium with gibbsite, diaspore, or corundum, depending on which mineral is stable at each pressure and temperature. Saturation limits: brucite (a), spinel (b).

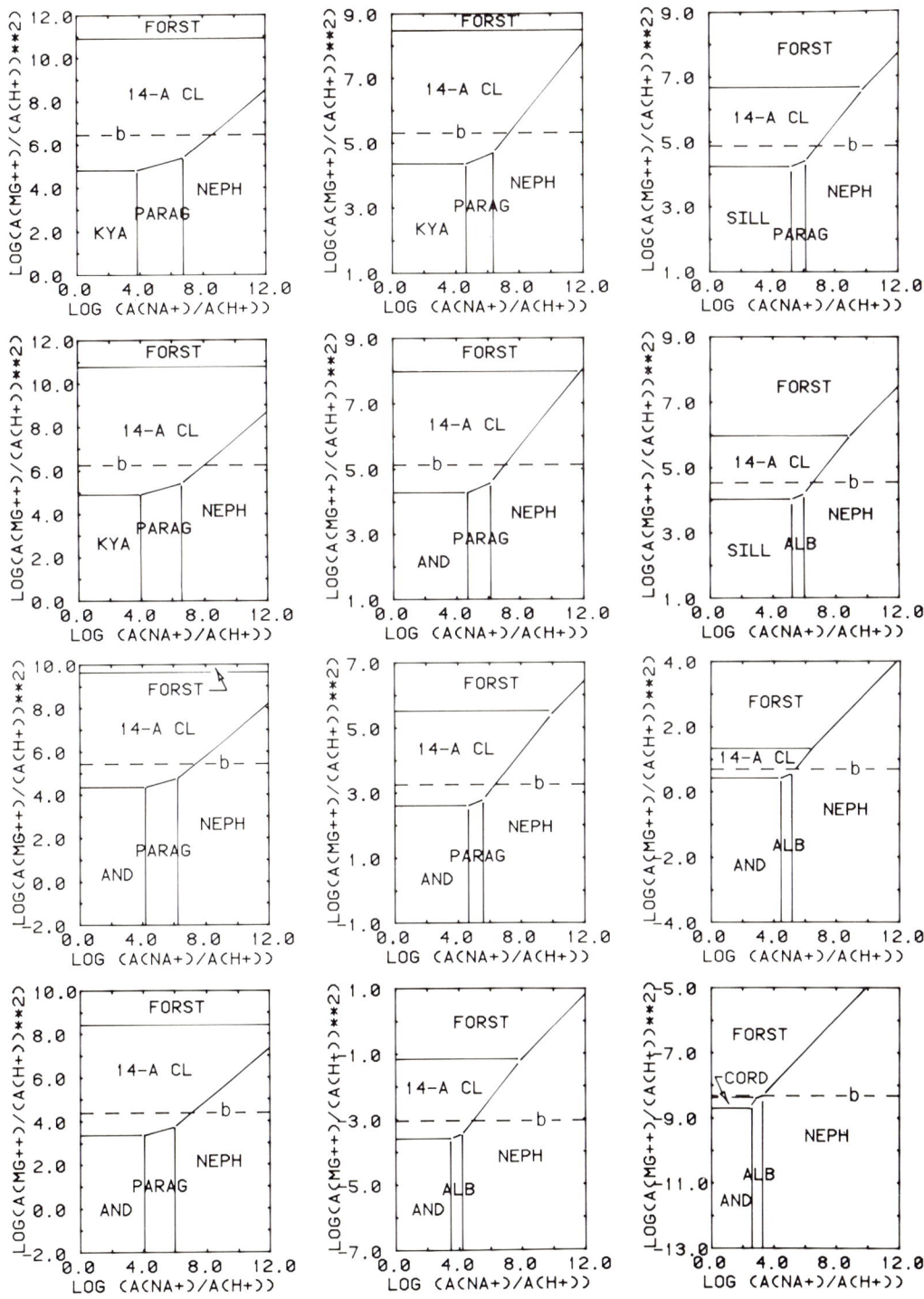

Phase relations in the system $HCl-H_2O-Al_2O_3-MgO-Na_2O-(SiO_2)$ in equilibrium with gibbsite, diaspore, or corundum, depending on which mineral is stable at each pressure and temperature. Saturation limits: brucite (a), spinel (b).

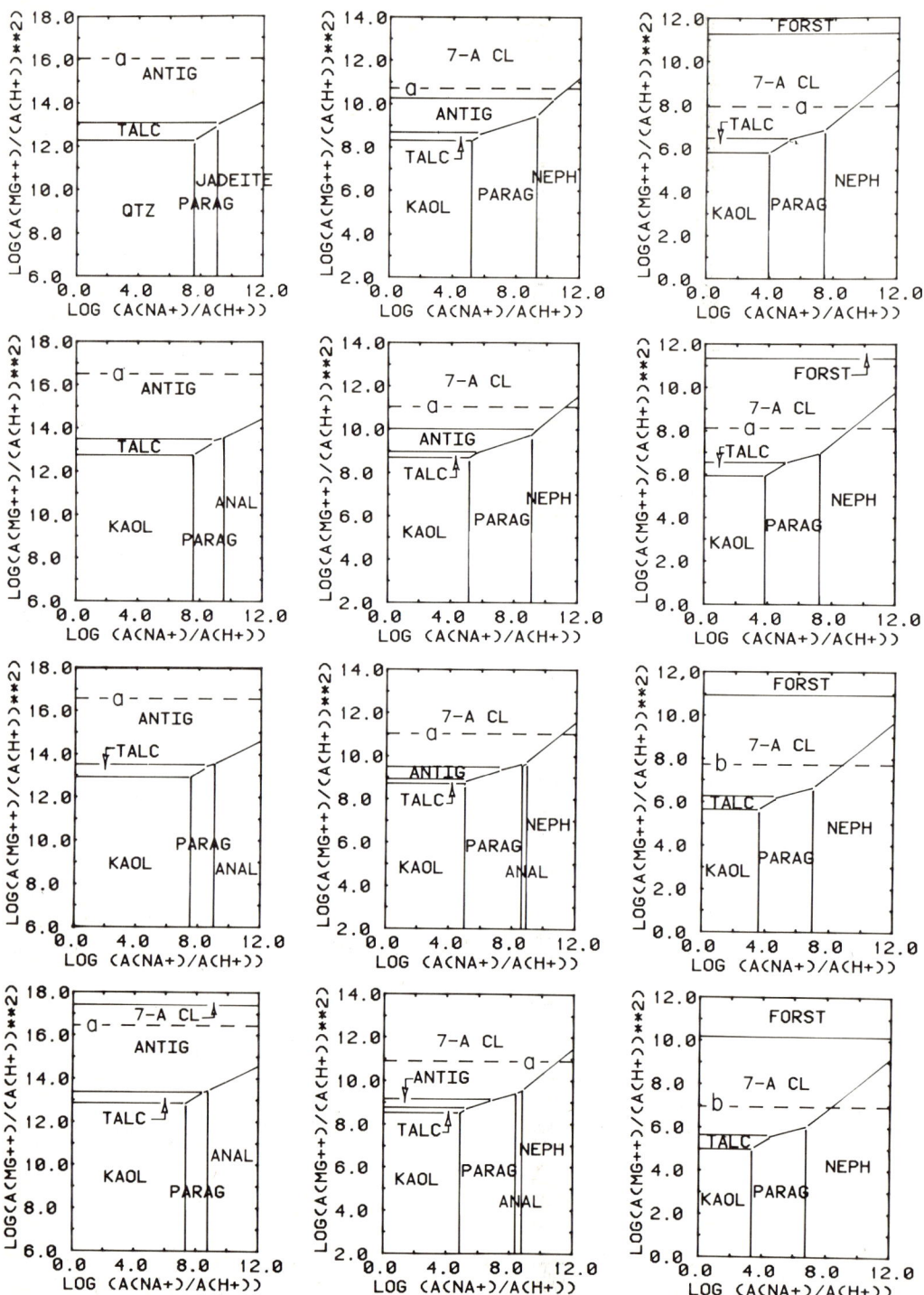

Phase relations in the system $HCl-H_2O-Al_2O_3-MgO-Na_2O-(SiO_2)$ in equilibrium with gibbsite, diaspore, or corundum, depending on which mineral is stable at each pressure and temperature. Metastable 7-Å clinochlore was considered instead of its stable counterpart, 14-Å clinochlore. Saturation limits: brucite (a), spinel (b).

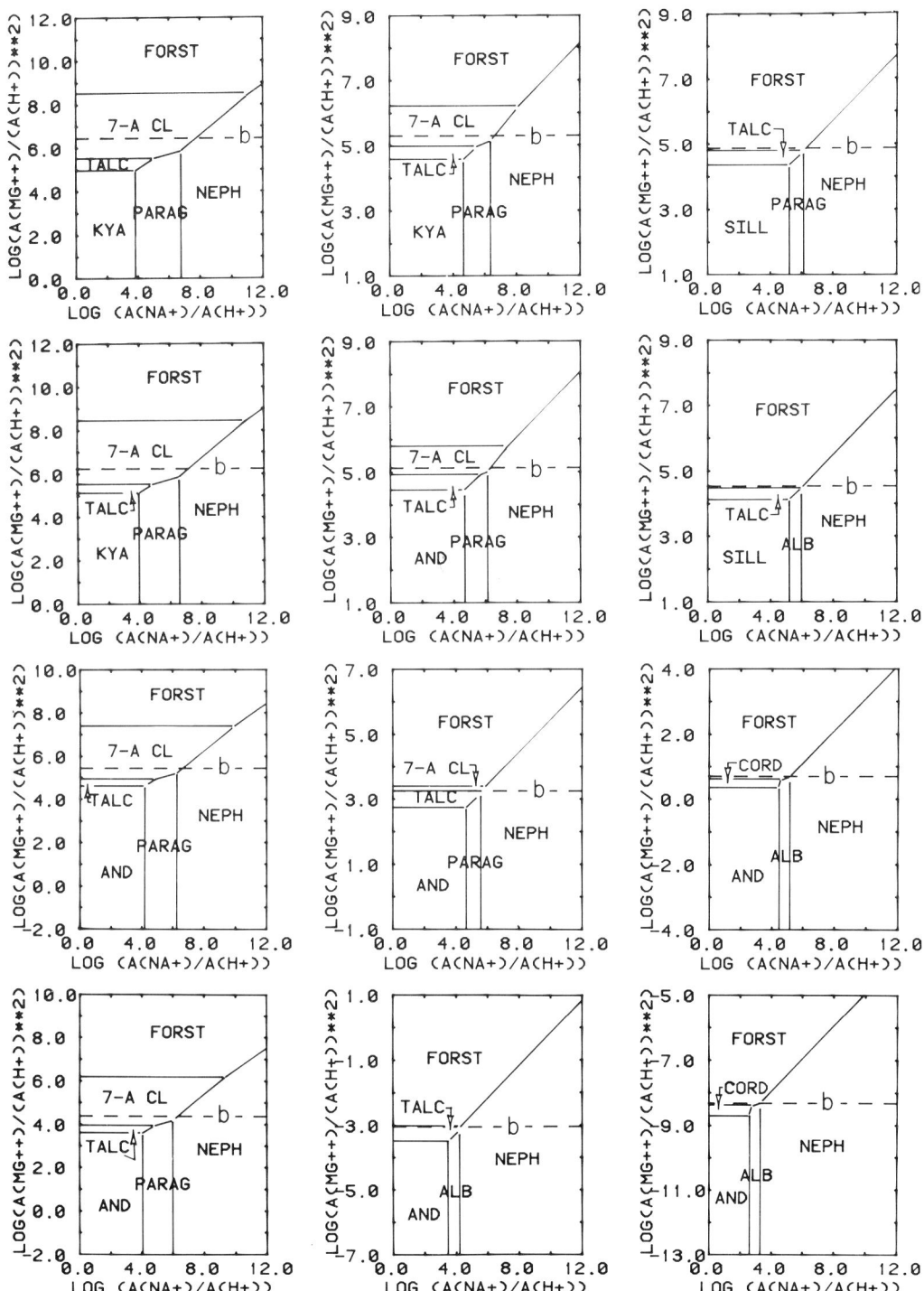

Phase relations in the system $HCl-H_2O-Al_2O_3-MgO-Na_2O-(SiO_2)$ in equilibrium with gibbsite, diaspore, or corundum, depending on which mineral is stable at each pressure and temperature. Metastable 7-A clinochlore was considered instead of its stable counterpart, 14-A clinochlore. Saturation limits: brucite (a), spinel (b).

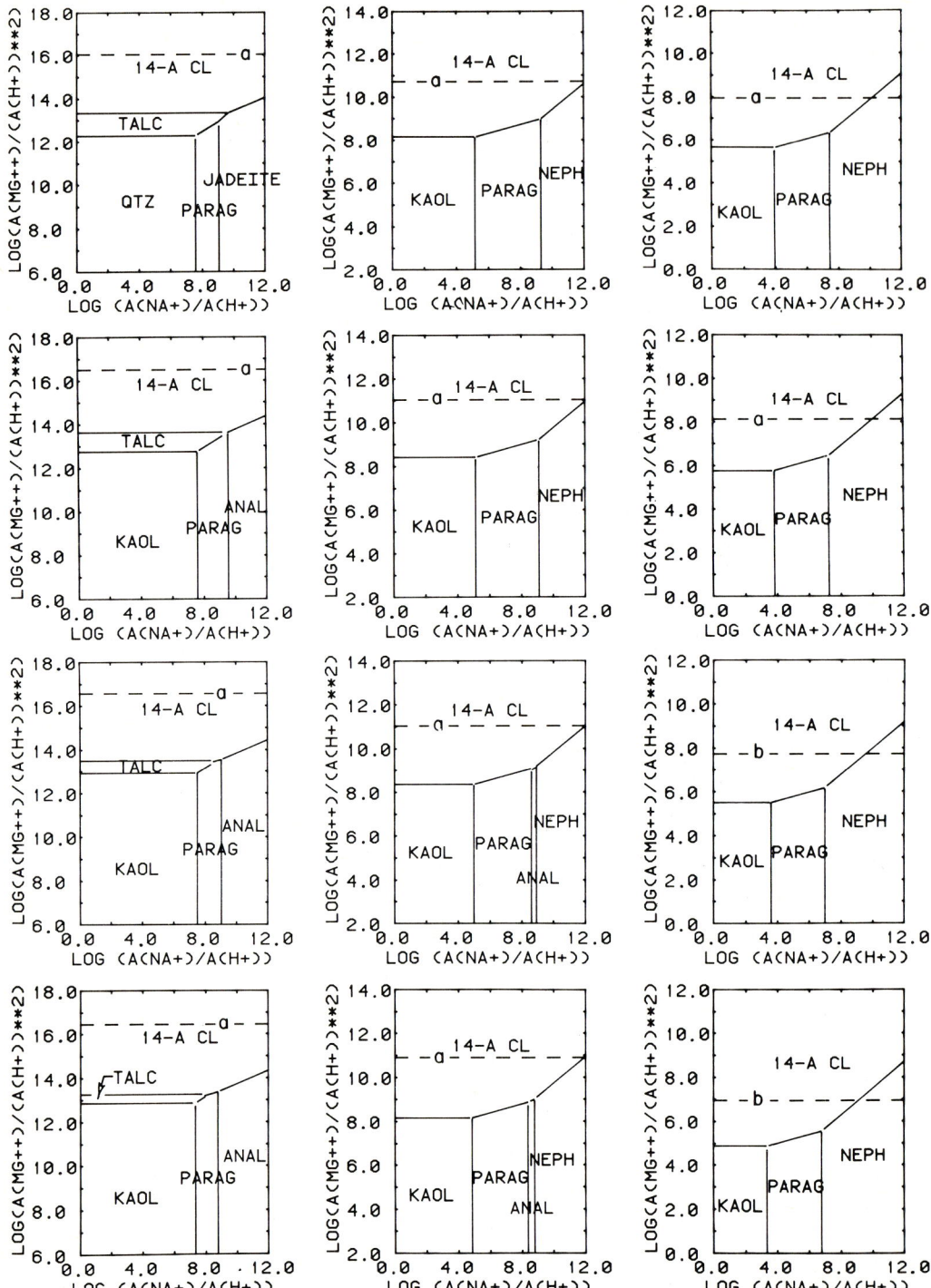

Phase relations in the system $HCl-H_2O-Al_2O_3-MgO-Na_2O-(SiO_2)$ in equilibrium with gibbsite, diaspore, or corundum, depending on which mineral is stable at each pressure and temperature. Metastable chrysotile was considered instead of its stable counterpart, antigorite. Saturation limits: brucite (a), spinel (b).

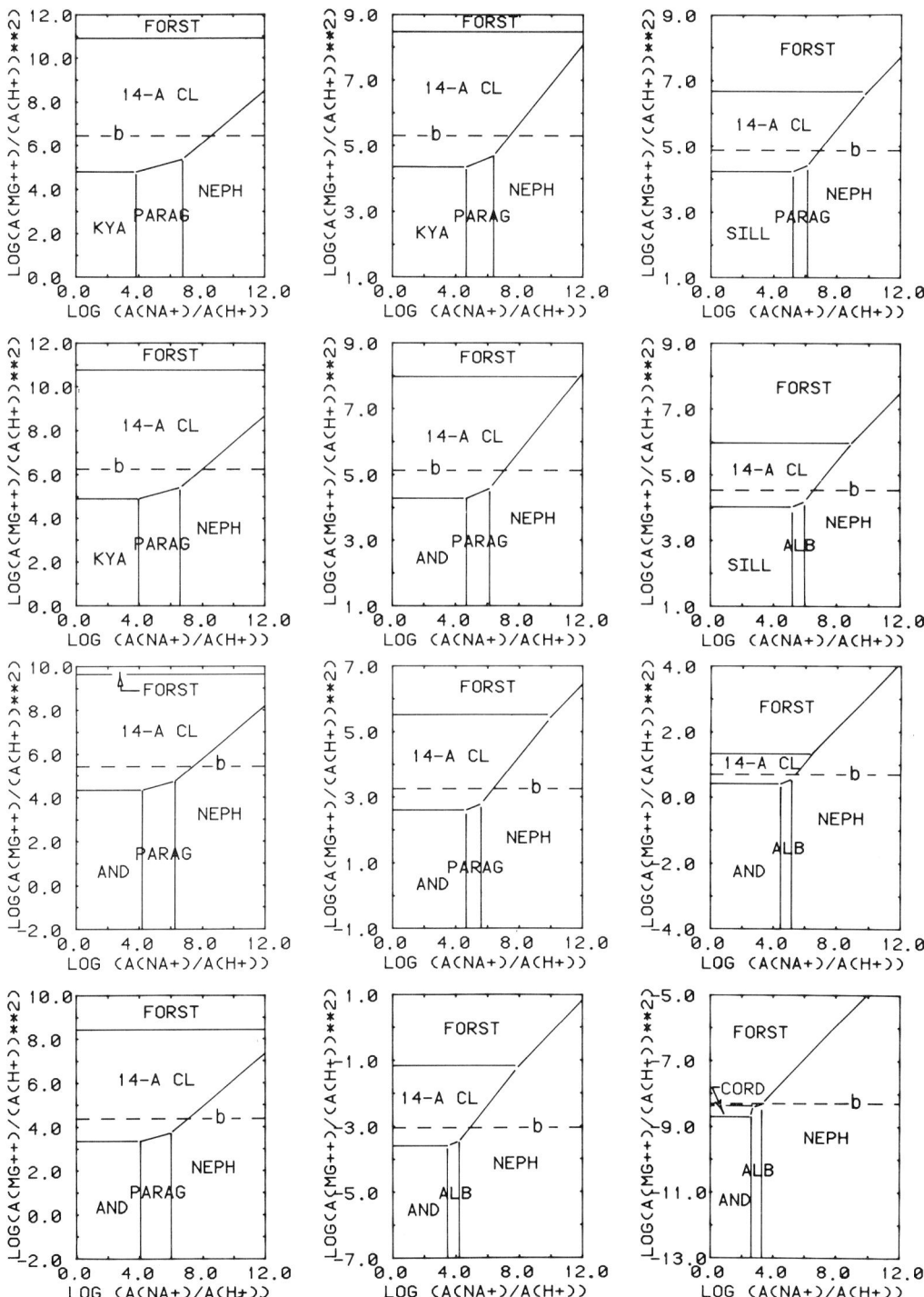

Phase relations in the system $HCl-H_2O-Al_2O_3-MgO-Na_2O-(SiO_2)$ in equilibrium with gibbsite, diaspore, or corundum, depending on which mineral is stable at each pressure and temperature. Metastable chrysotile was considered instead of its stable counterpart, antigorite. Saturation limits: brucite (a), spinel (b).

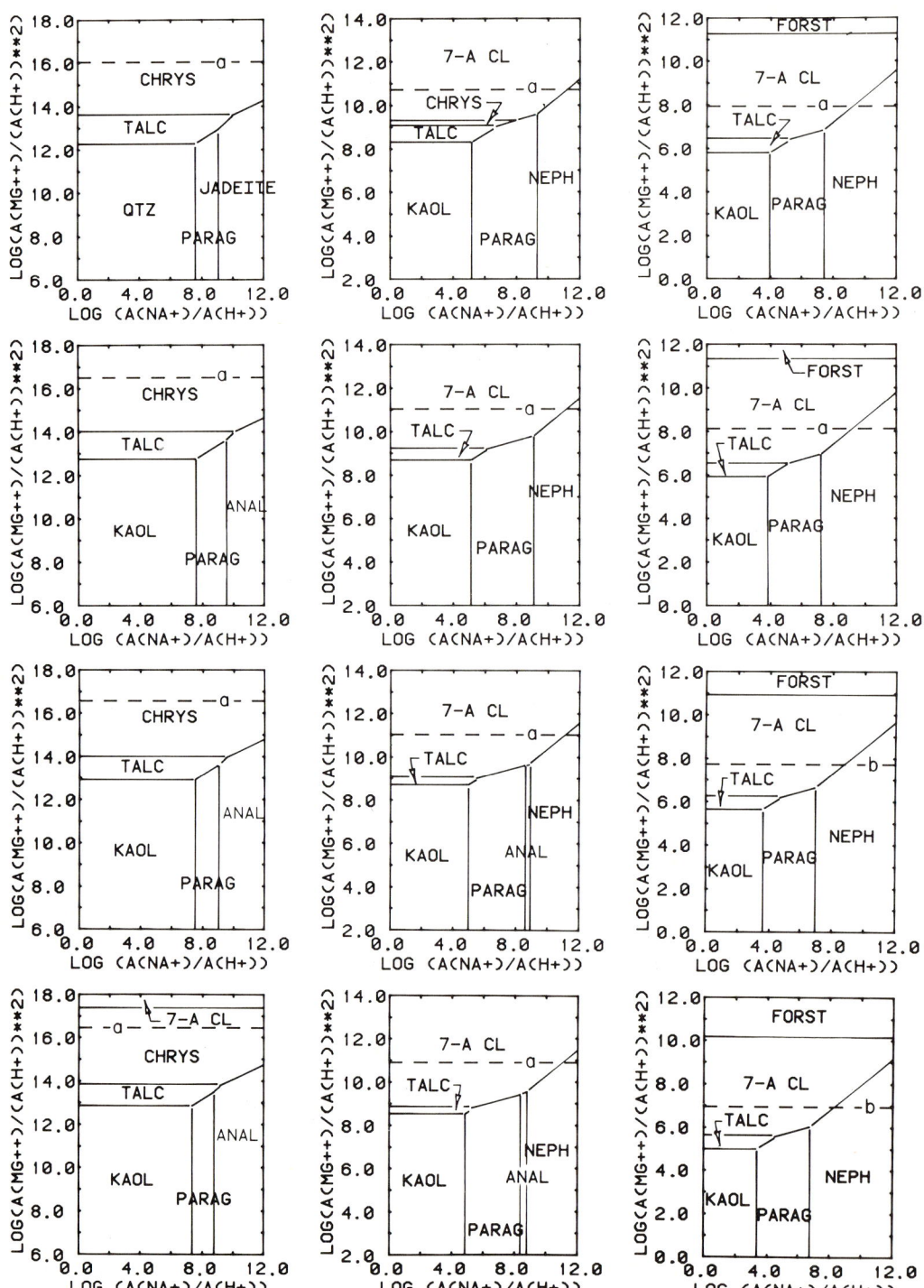

Phase relations in the system HCl−H₂O−Al₂O₃−MgO−Na₂O−(SiO₂) in equilibrium with gibbsite, diaspore, or corundum, depending on which mineral is stable at each pressure and temperature. Metastable 7-A clinochlore and chrysotile were considered instead of their stable counterparts, 14-A clinochlore and antigorite. Saturation limits: brucite (a), spinel (b).

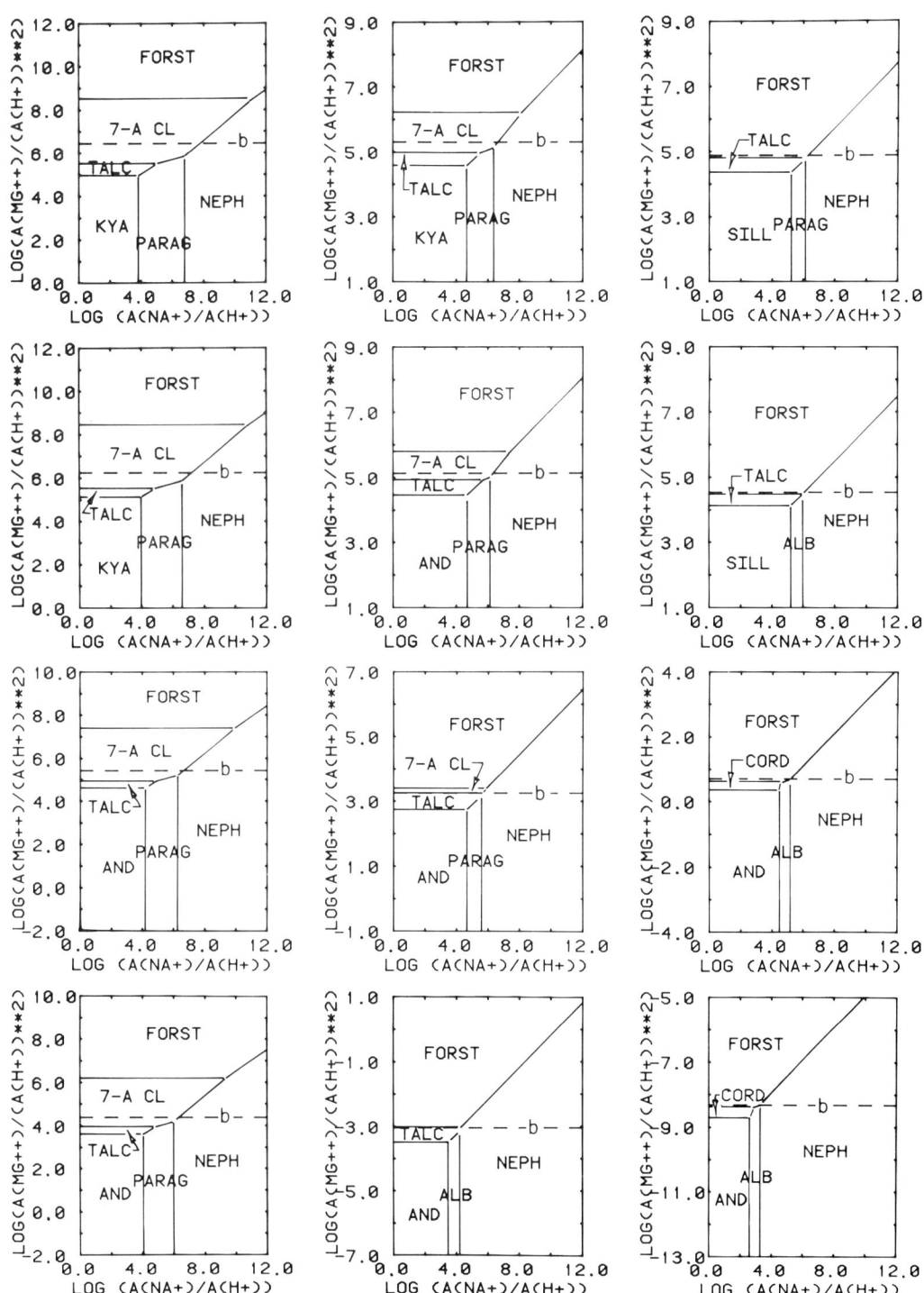

Phase relations in the system $HCl-H_2O-Al_2O_3-MgO-Na_2O-(SiO_2)$ in equilibrium with gibbsite, diaspore, or corundum, depending on which mineral is stable at each pressure and temperature. Metastable 7-A clinochlore and chrysotile were considered instead of their stable counterparts, 14-A clinochlore and antigorite. Saturation limits: brucite (a), spinel (b).

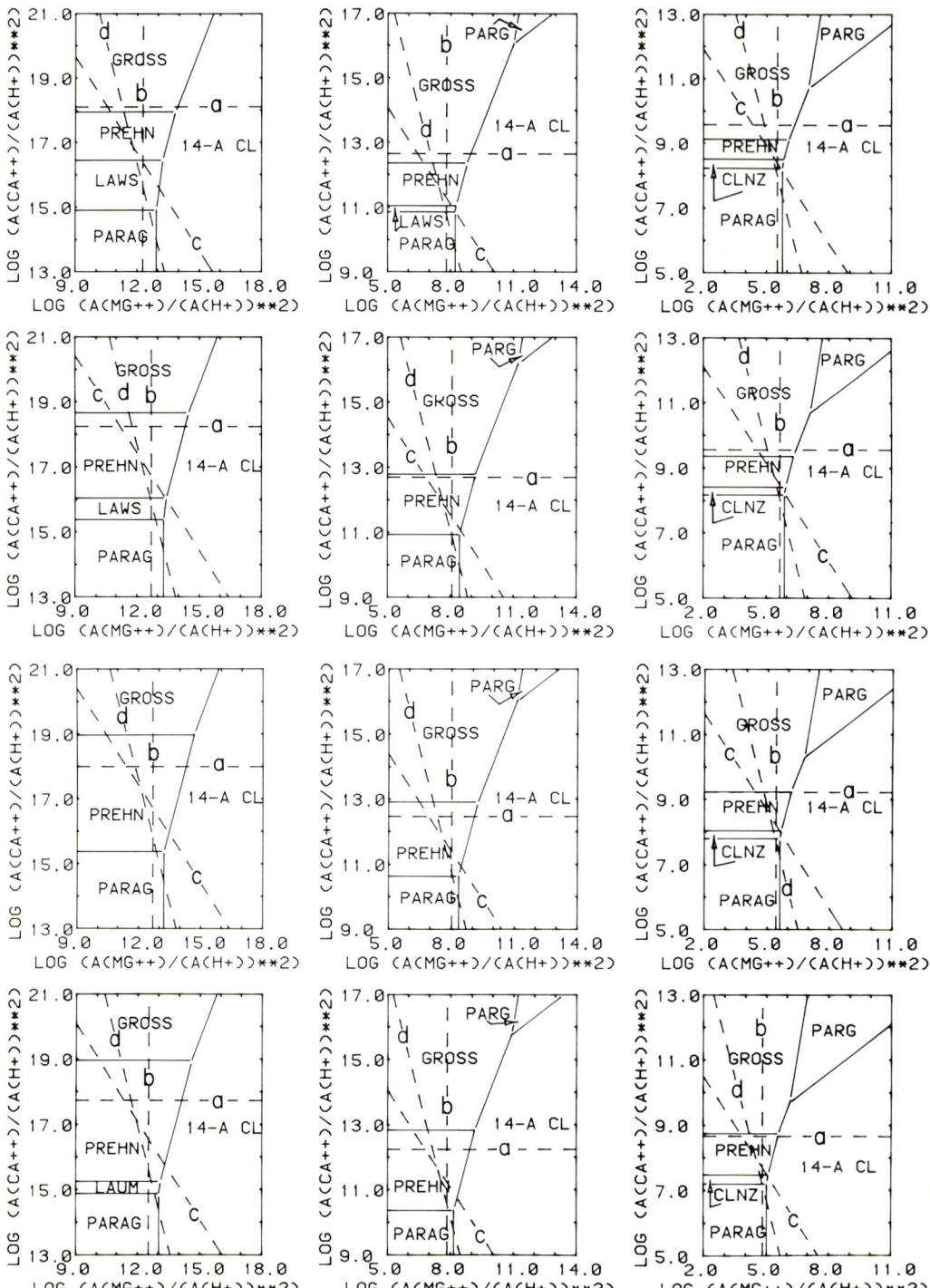

Phase relations in the system $HCl-H_2O-(Al_2O_3)-CaO-MgO-Na_2O-SiO_2$ in equilibrium with quartz and albite. Saturation limits: wollastonite (a), talc (b), diopside (c), tremolite (d).

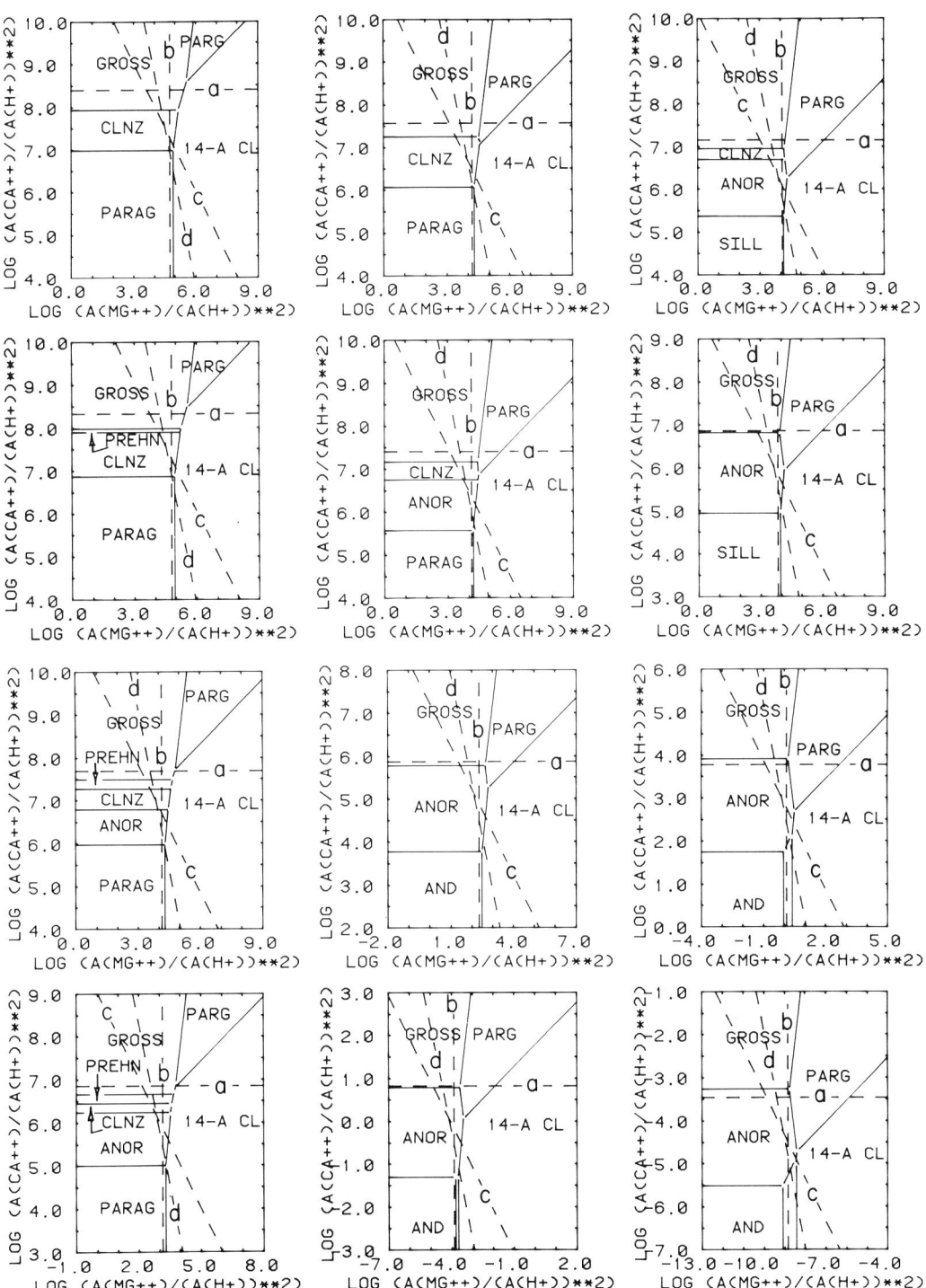

Phase relations in the system $HCl-H_2O-(Al_2O_3)-CaO-MgO-Na_2O-SiO_2$ in equilibrium with quartz and albite. Saturation limits: wollastonite (a), talc (b), diopside (c), tremolite (d).

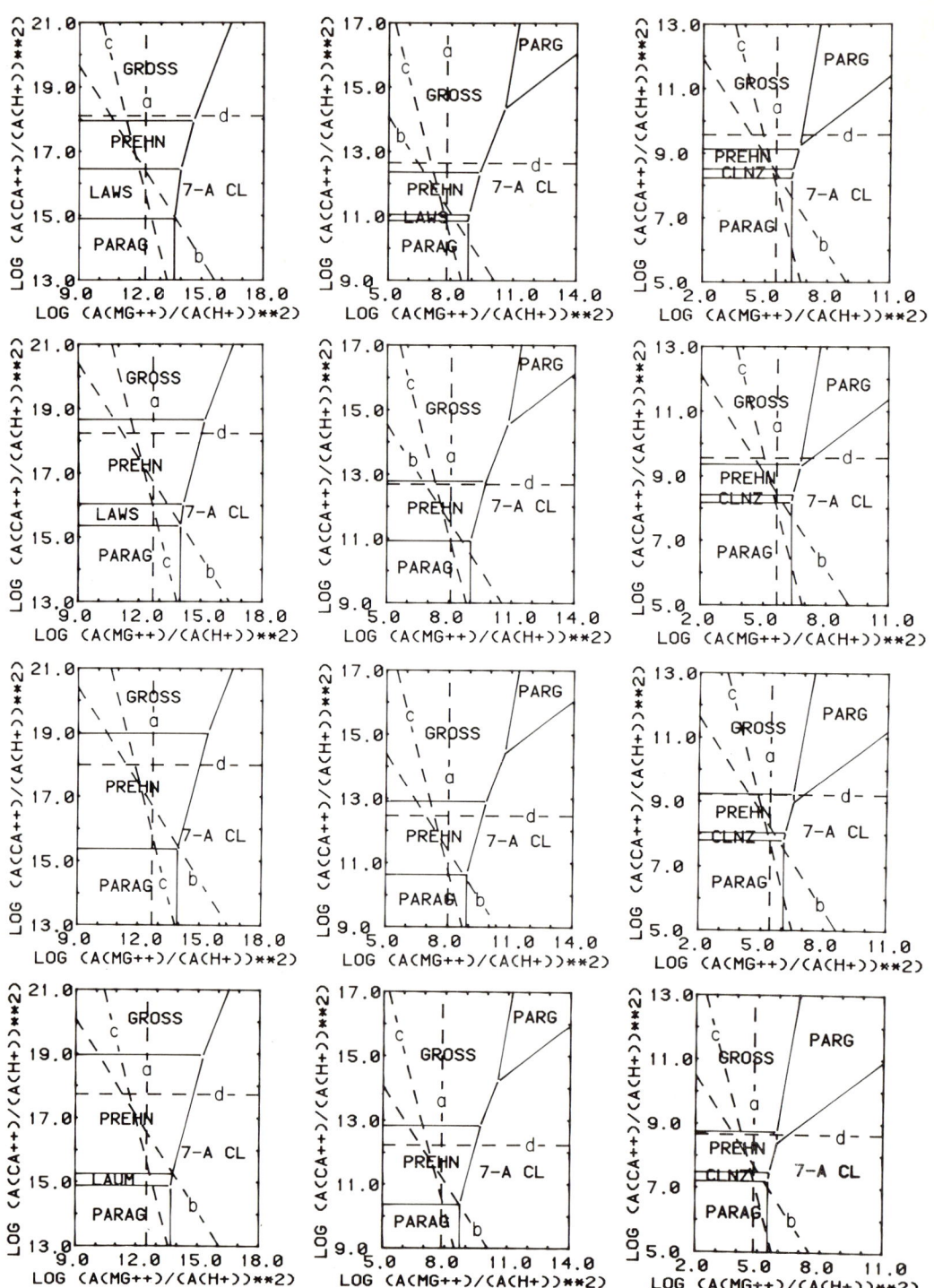

Phase relations in the system $HCl-H_2O-(Al_2O_3)-CaO-MgO-Na_2O-SiO_2$ in equilibrium with quartz and albite. Metastable 7-A clinochlore was considered instead of its stable counterpart, 14-A clinochlore. Saturation limits: talc (a), diopside (b), tremolite (c), wollastonite (d).

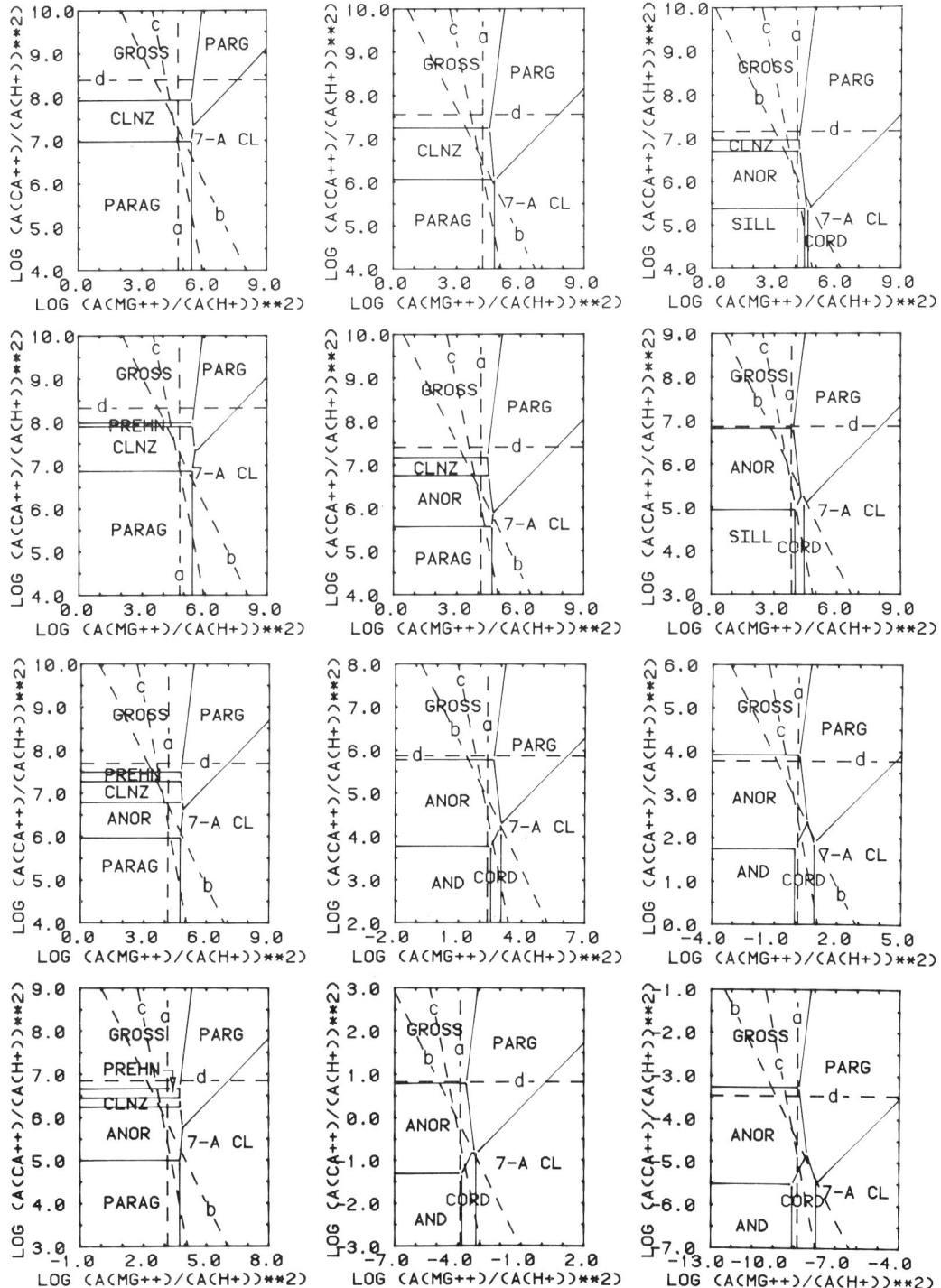

Phase relations in the system $HCl-H_2O-(Al_2O_3)-CaO-MgO-Na_2O-SiO_2$ in equilibrium with quartz and albite. Metastable 7-A clinochlore was considered instead of its stable counterpart, 14-A clinochlore. Saturation limits: talc (a), diopside (b), tremolite (c), wollastonite (d).

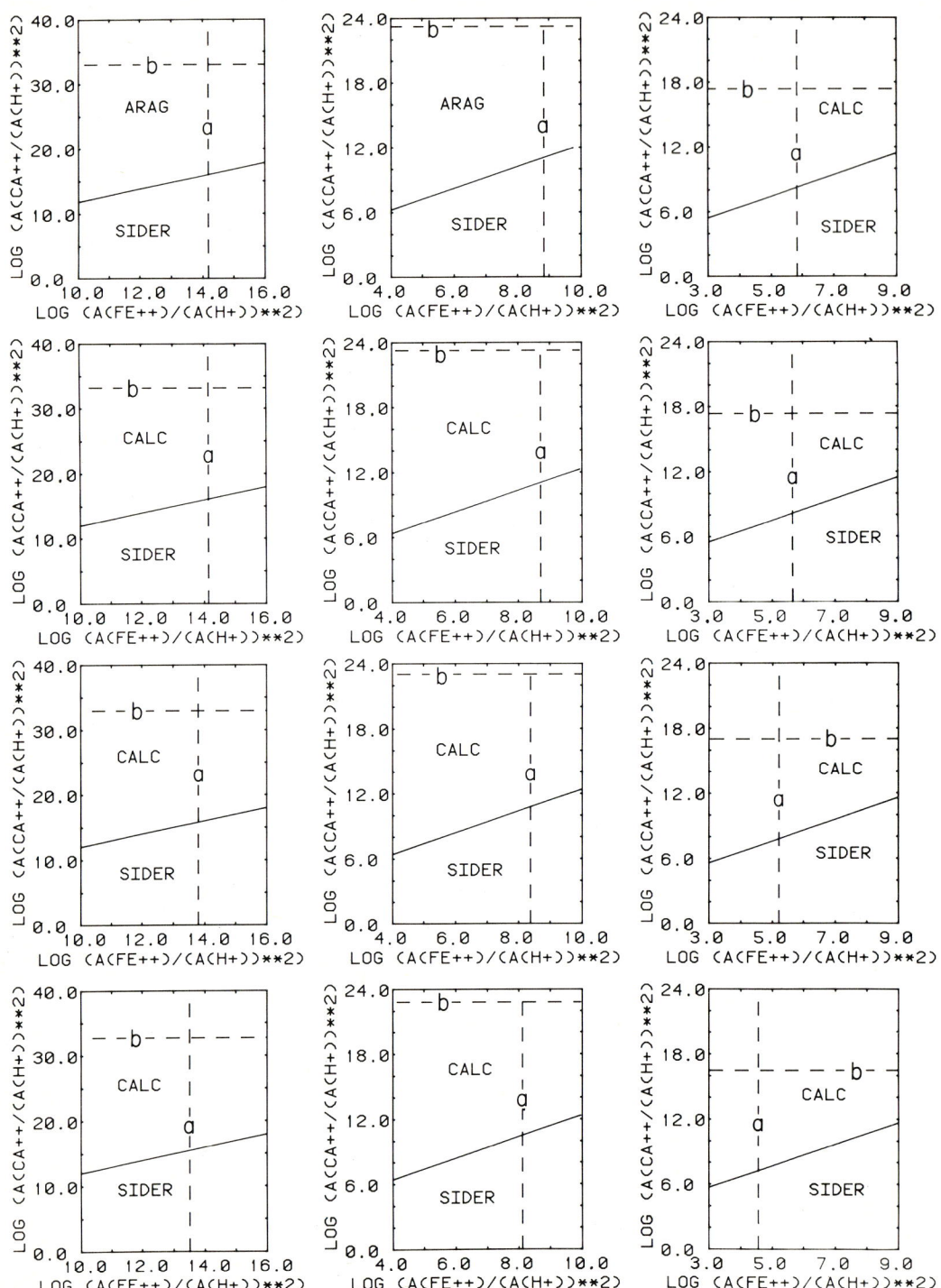

Phase relations in the system $HCl-H_2O-CaO-(CO_2)-FeO$. Saturation limits: ferrous oxide (a), lime (b).

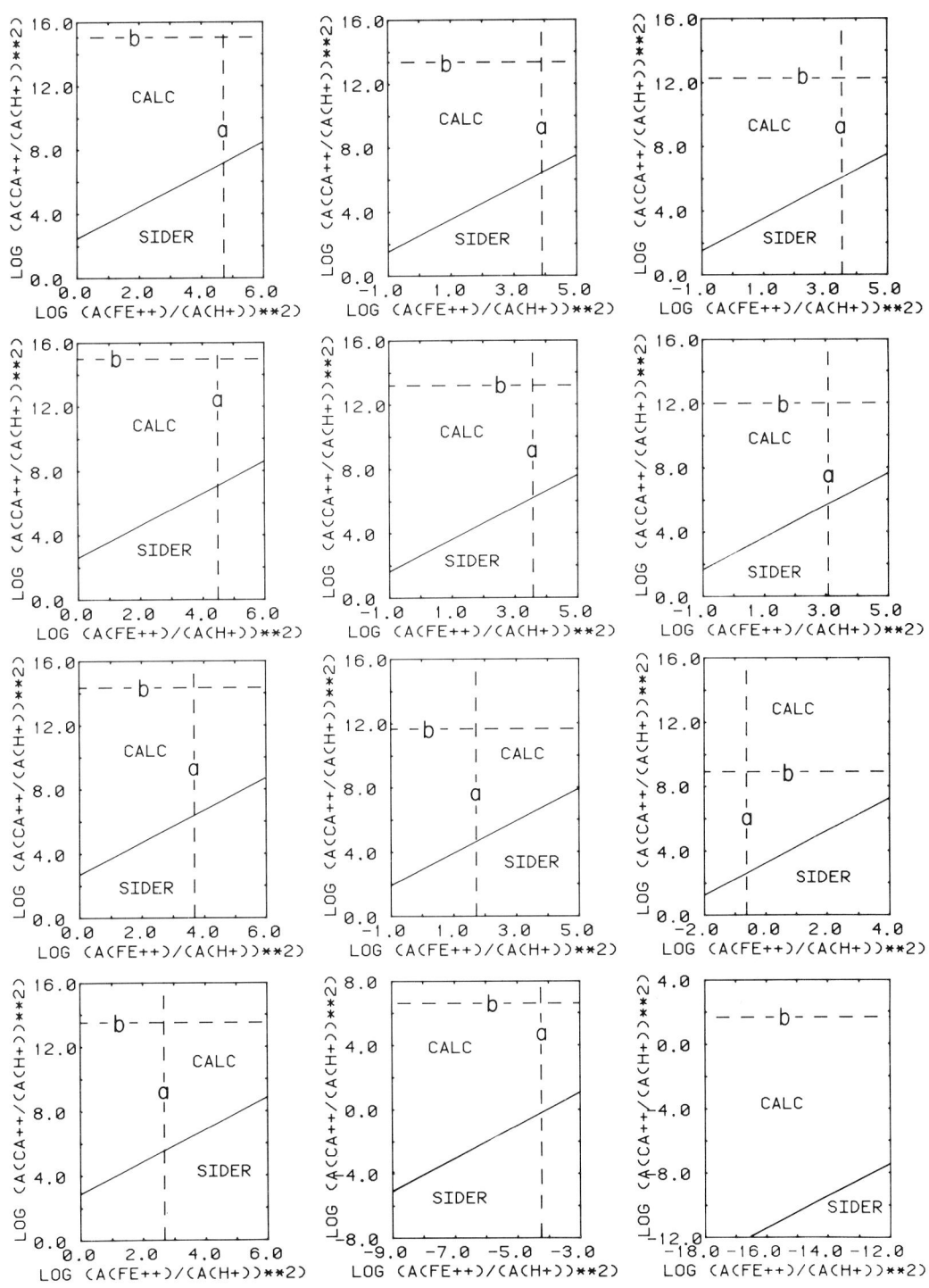

Phase relations in the system $HCl-H_2O-CaO-(CO_2)-FeO$. Saturation limits: ferrous oxide (a), lime (b).

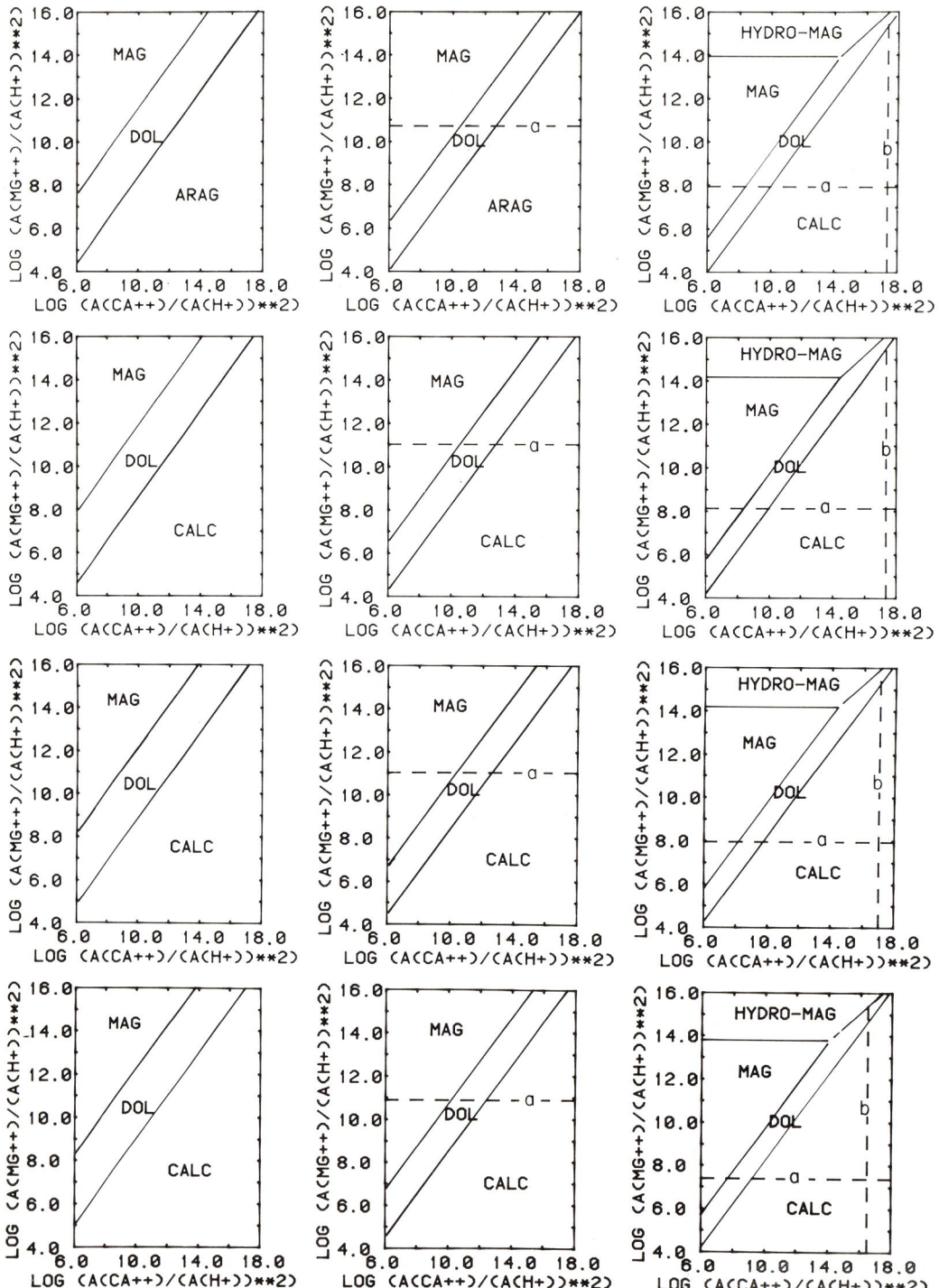

Phase relations in the system $HCl-H_2O-CaO-(CO_2)-MgO$. Saturation limits: brucite (a), lime (b), periclase (c).

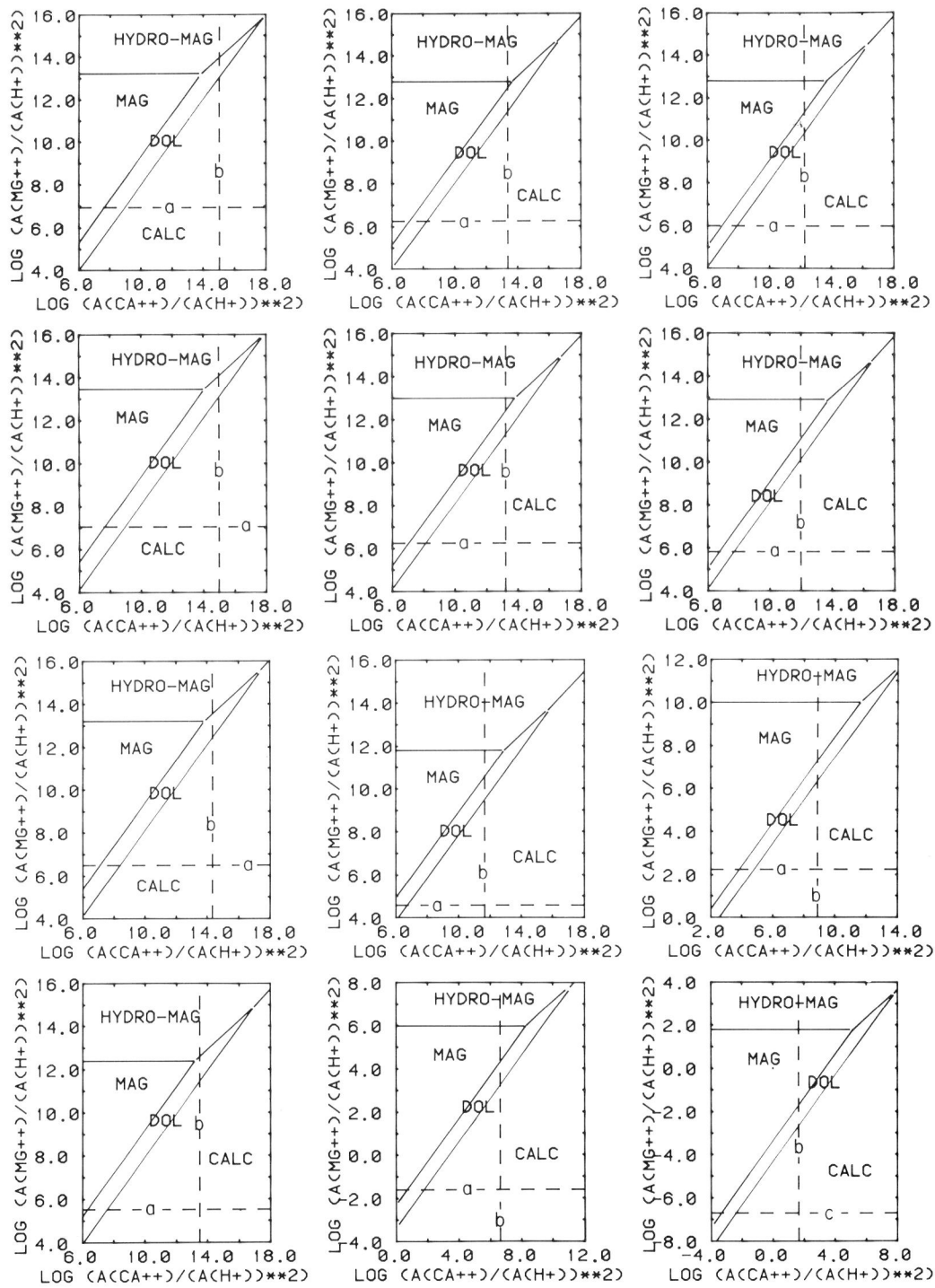

Phase relations in the system $HCl-H_2O-CaO-(CO_2)-MgO$. Saturation limits: brucite (a), lime (b), periclase (c).

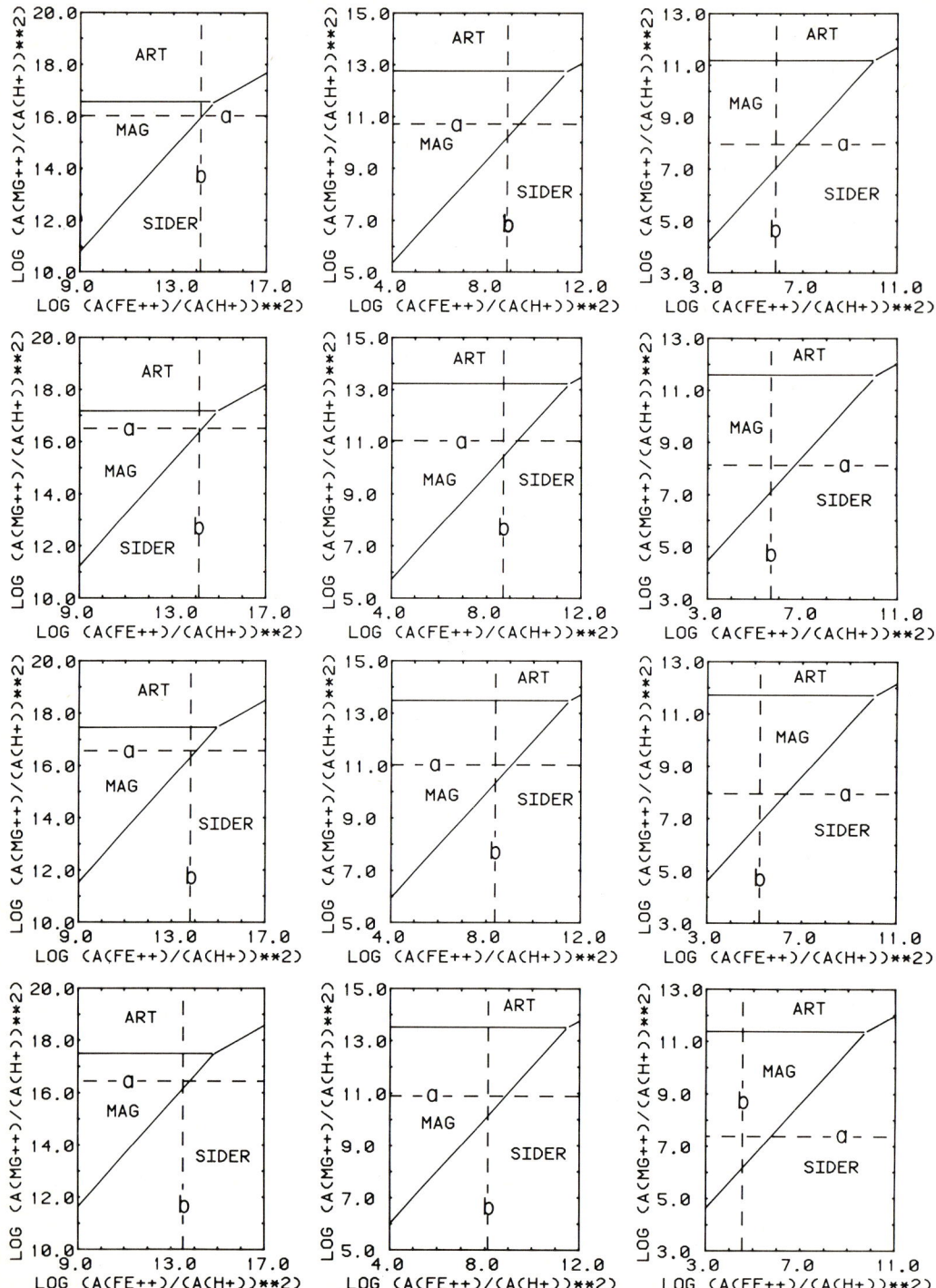

Phase relations in the system $HCl-H_2O-(CO_2)-FeO-MgO$. Saturation limits: brucite (a), ferrous oxide (b), periclase (c).

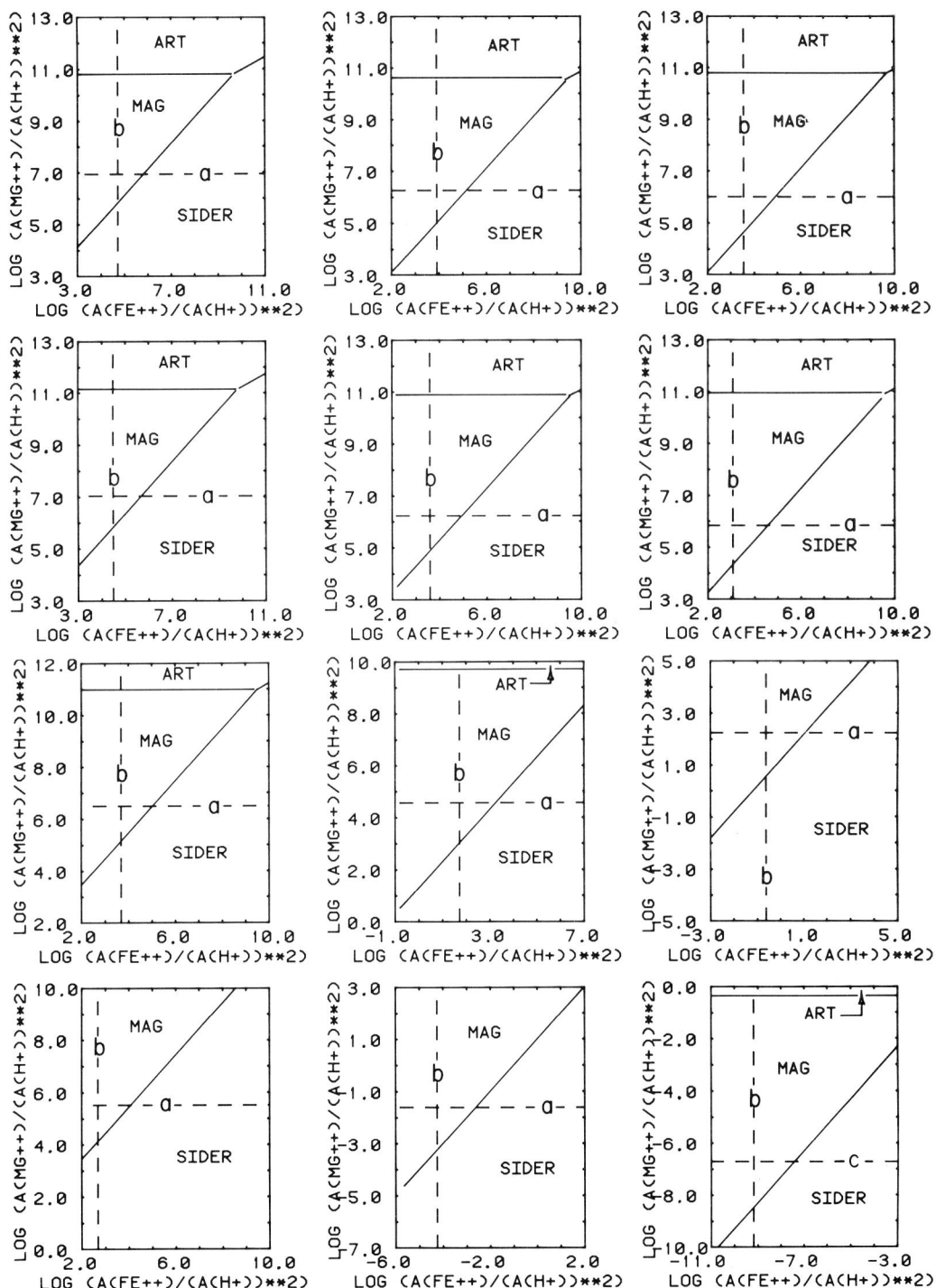

Phase relations in the system $HCl-H_2O-(CO_2)-FeO-MgO$. Saturation limits: brucite (a), ferrous oxide (b), periclase (c).

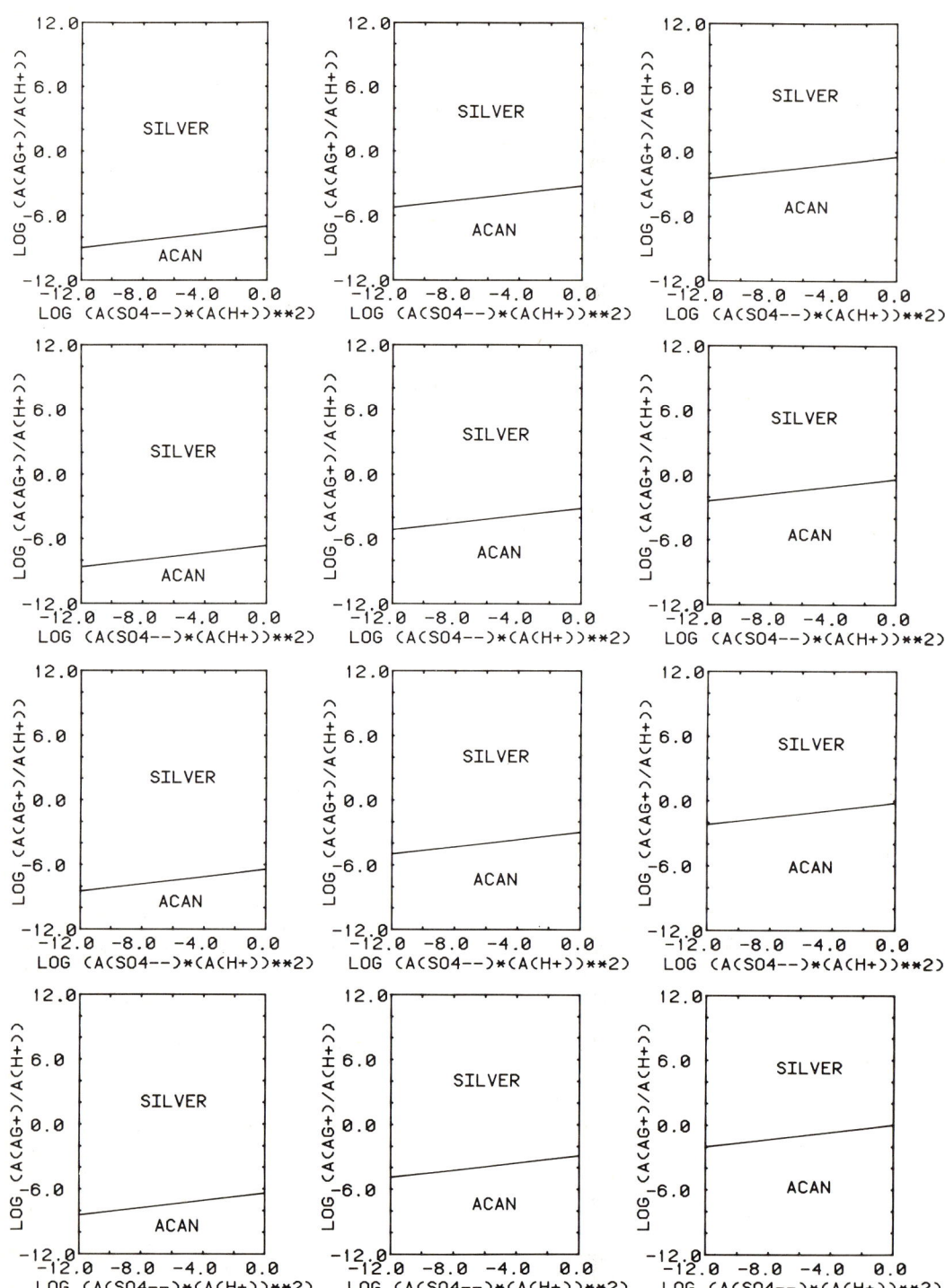

Phase relations in the system $HCl-H_2O-Ag_2O-(H_2S)-H_2SO_4$.

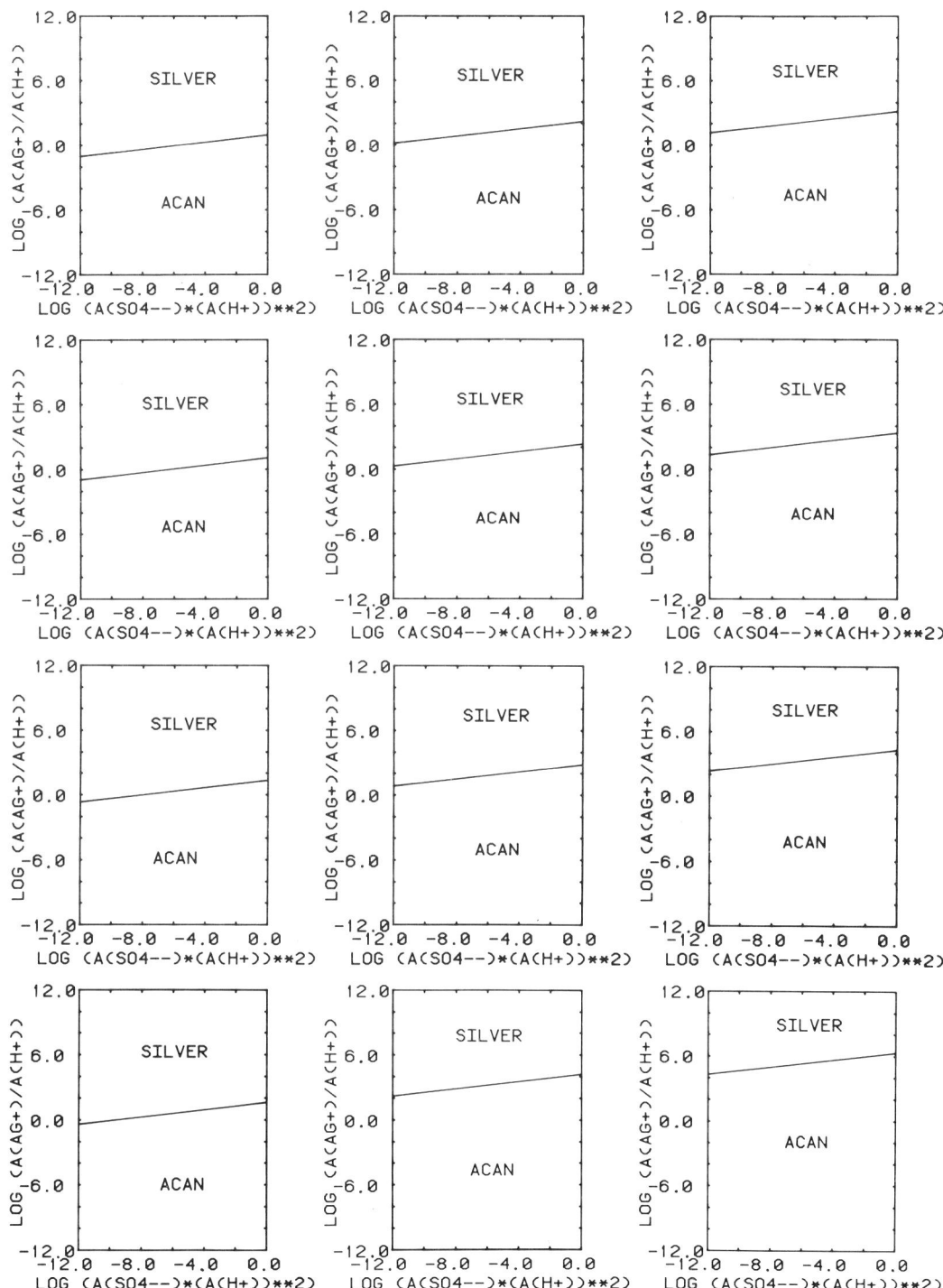

Phase relations in the system $HCl-H_2O-Ag_2O-(H_2S)-H_2SO_4$.

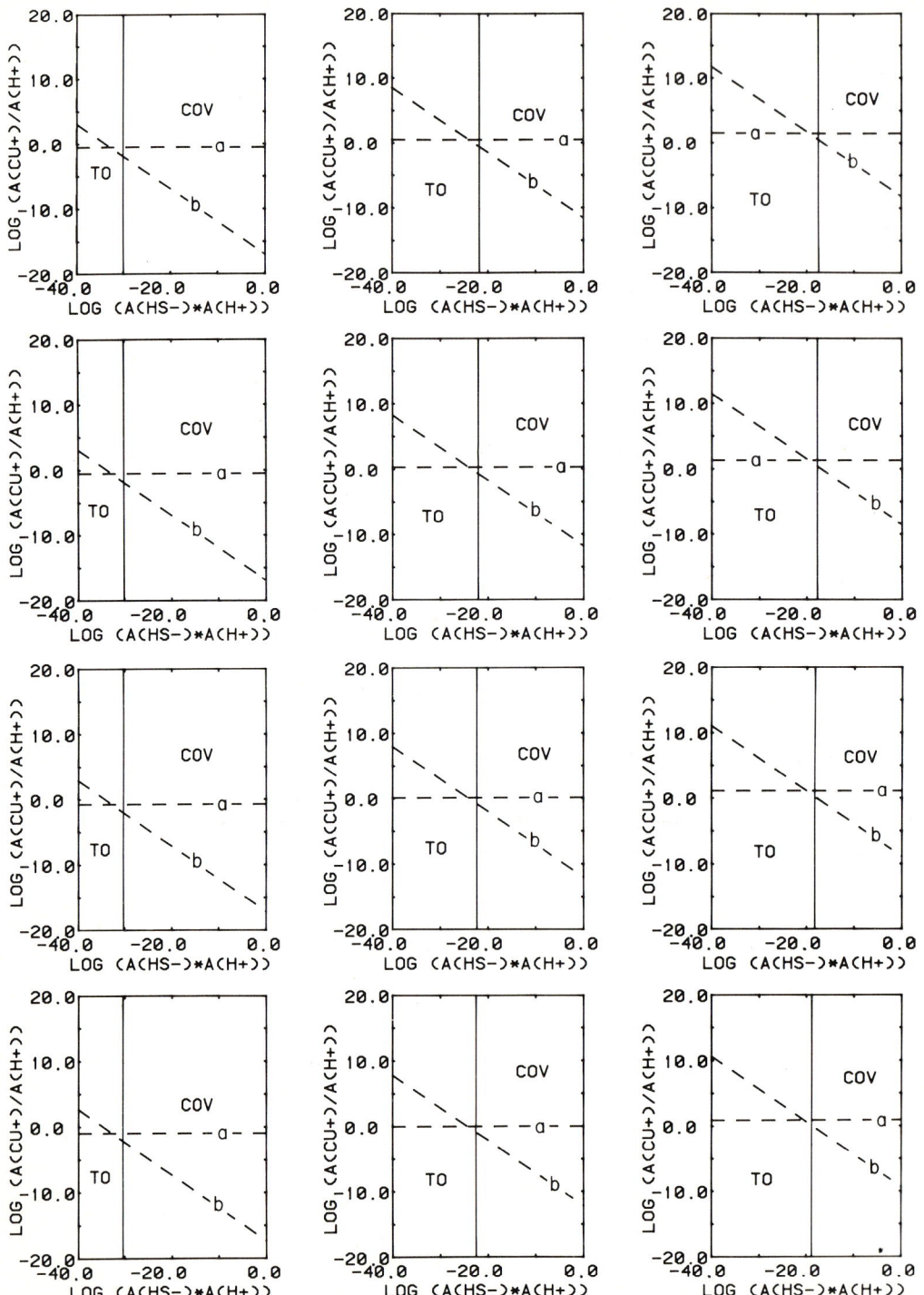

Phase relations in the system $HCl-H_2O-CuO-H_2S-(H_2SO_4)$. Saturation limits: cuprite (a), chalcocite (b).

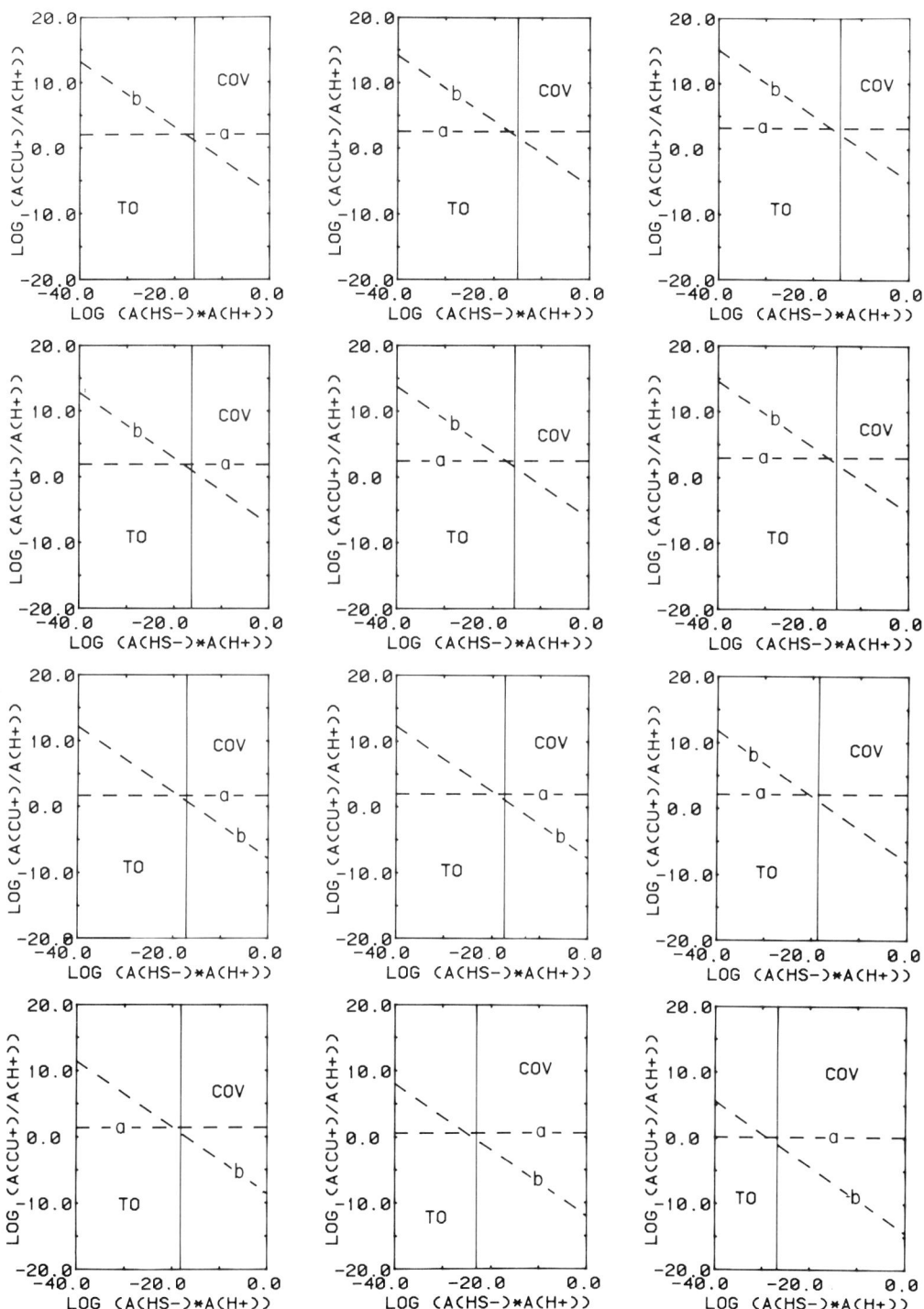

Phase relations in the system $HCl-H_2O-CuO-H_2S-(H_2SO_4)$. Saturation limits: cuprite (a), chalcocite (b).

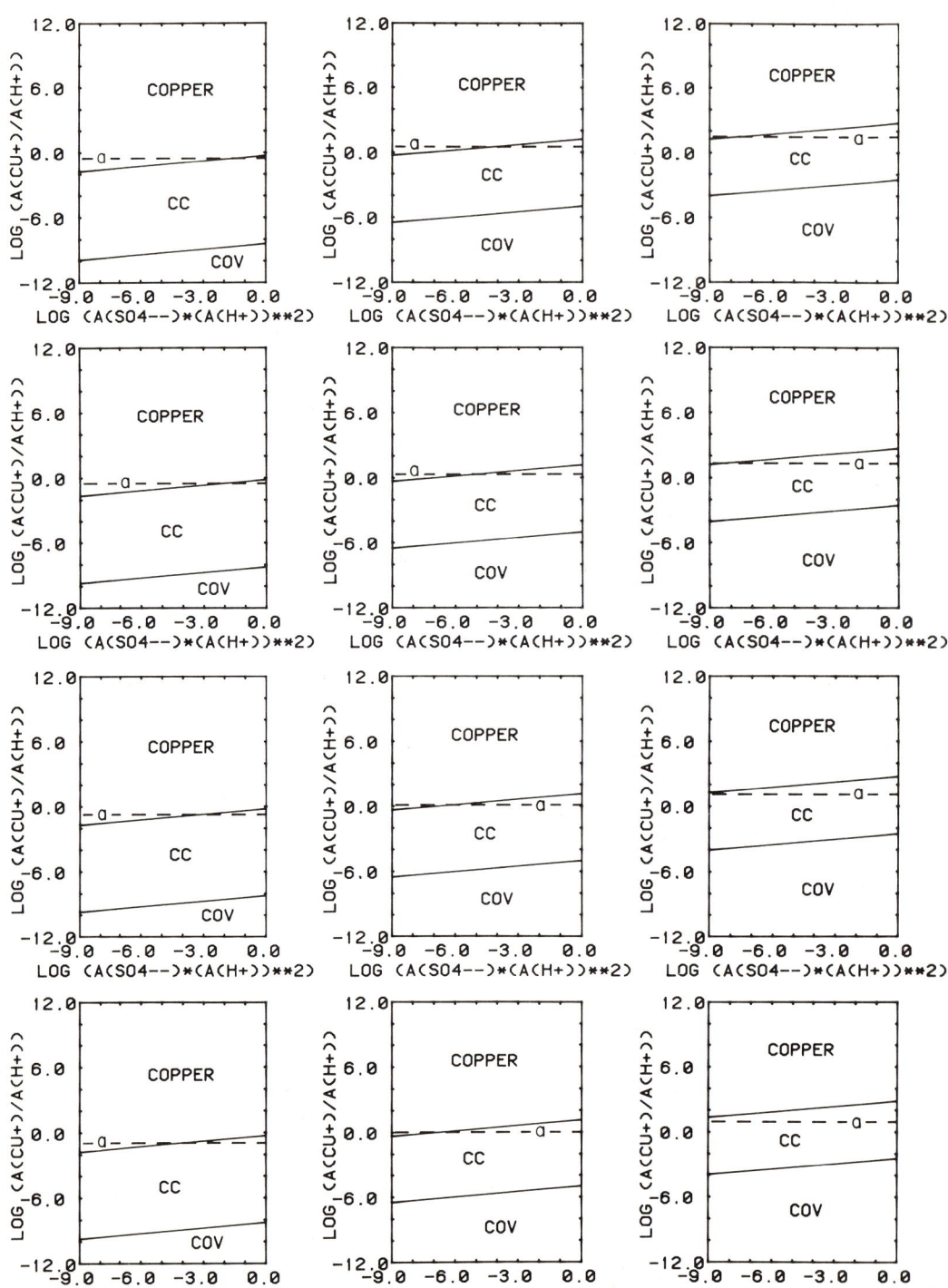

Phase relations in the system $HCl-H_2O-CuO-(H_2S)-H_2SO_4$. Saturation limit: cuprite (a).

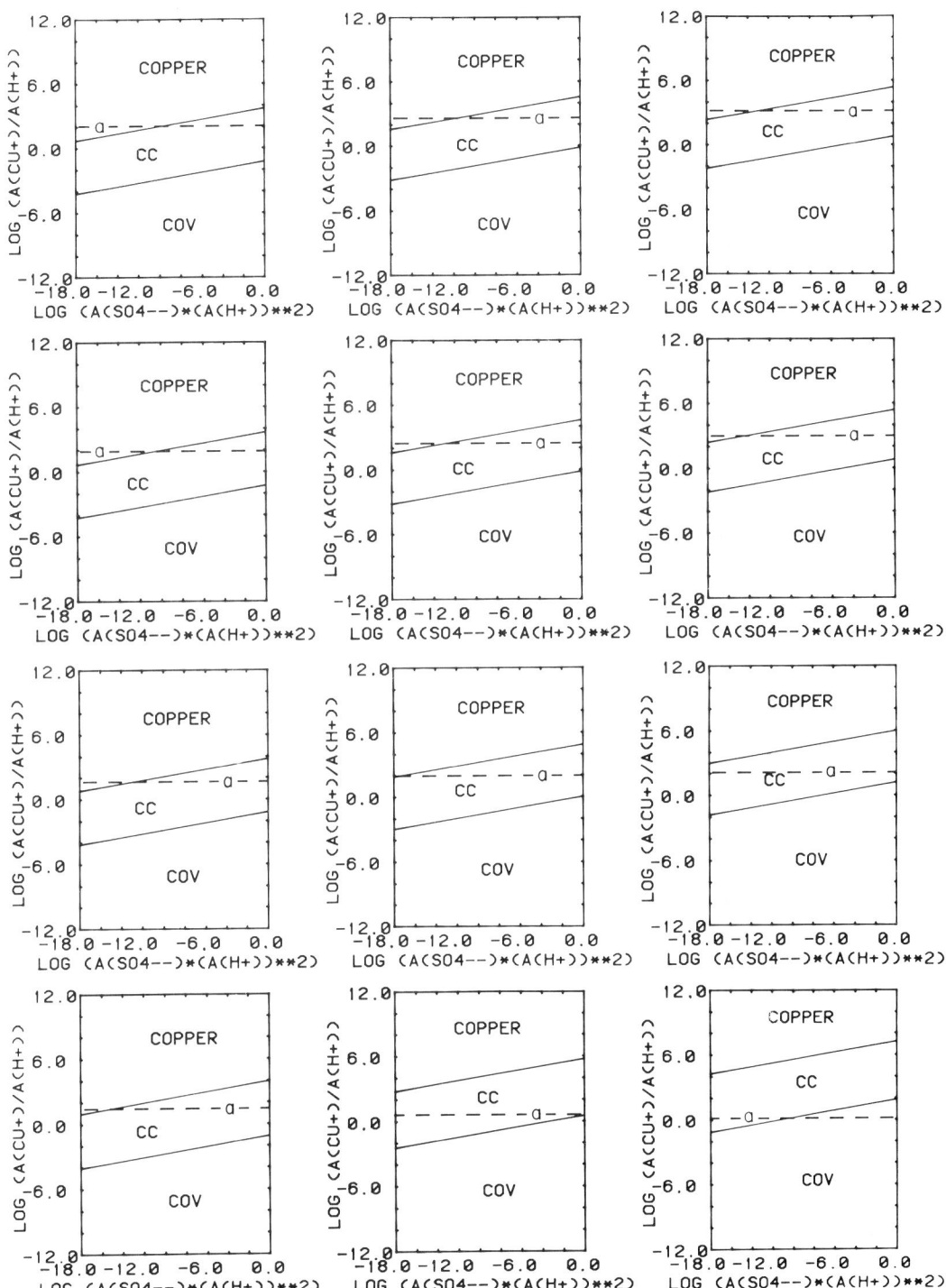

Phase relations in the system $HCl-H_2O-CuO-(H_2S)-H_2SO_4$. Saturation limit: cuprite (a).

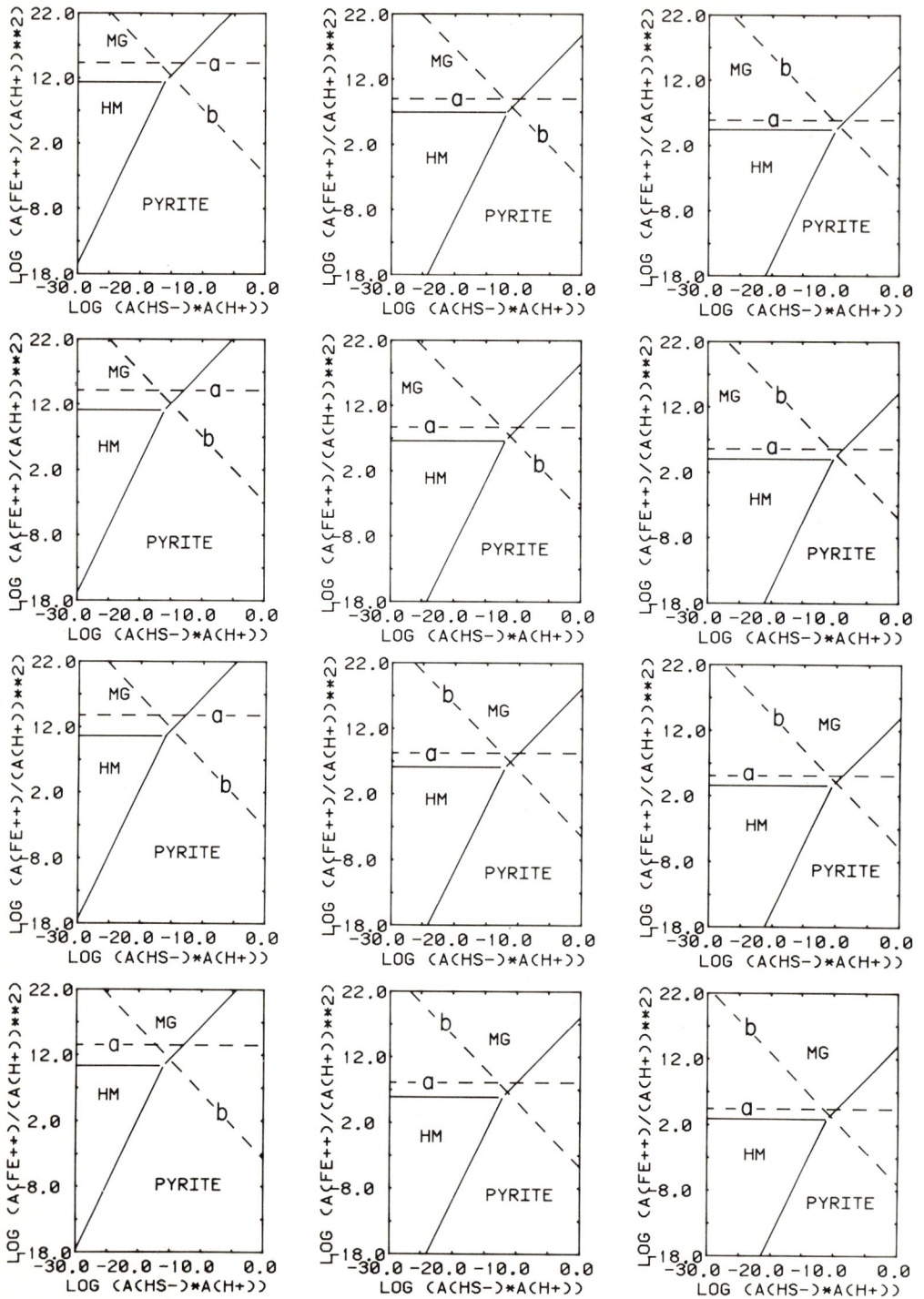

Phase relations in the system $HCl-H_2O-FeO-H_2S-(H_2SO_4)$. Saturation limits: ferrous oxide (a), pyrrhotite (b).

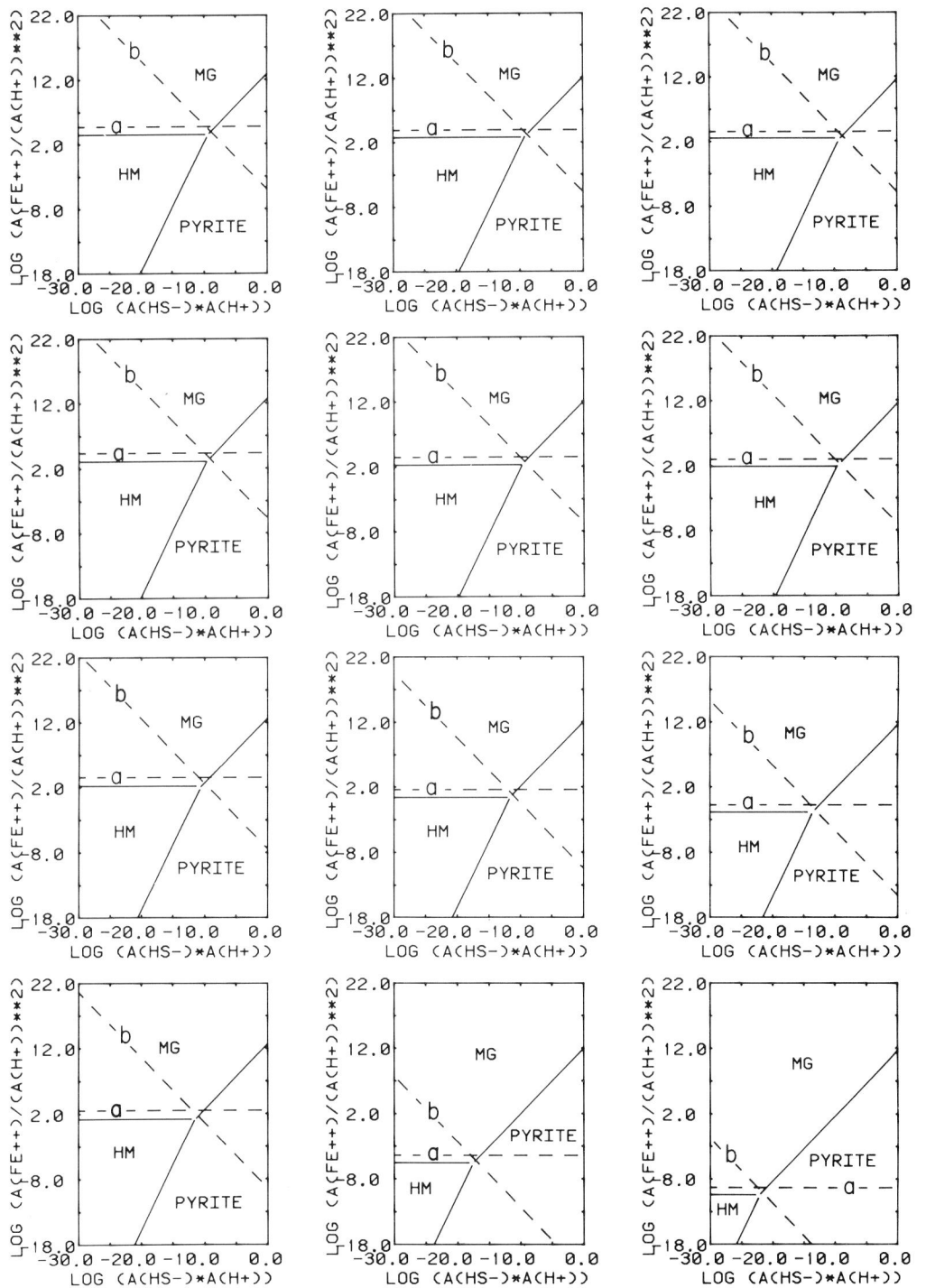

Phase relations in the system $HCl-H_2O-FeO-H_2S-(H_2SO_4)$. Saturation limits: ferrous oxide (a), pyrrhotite (b).

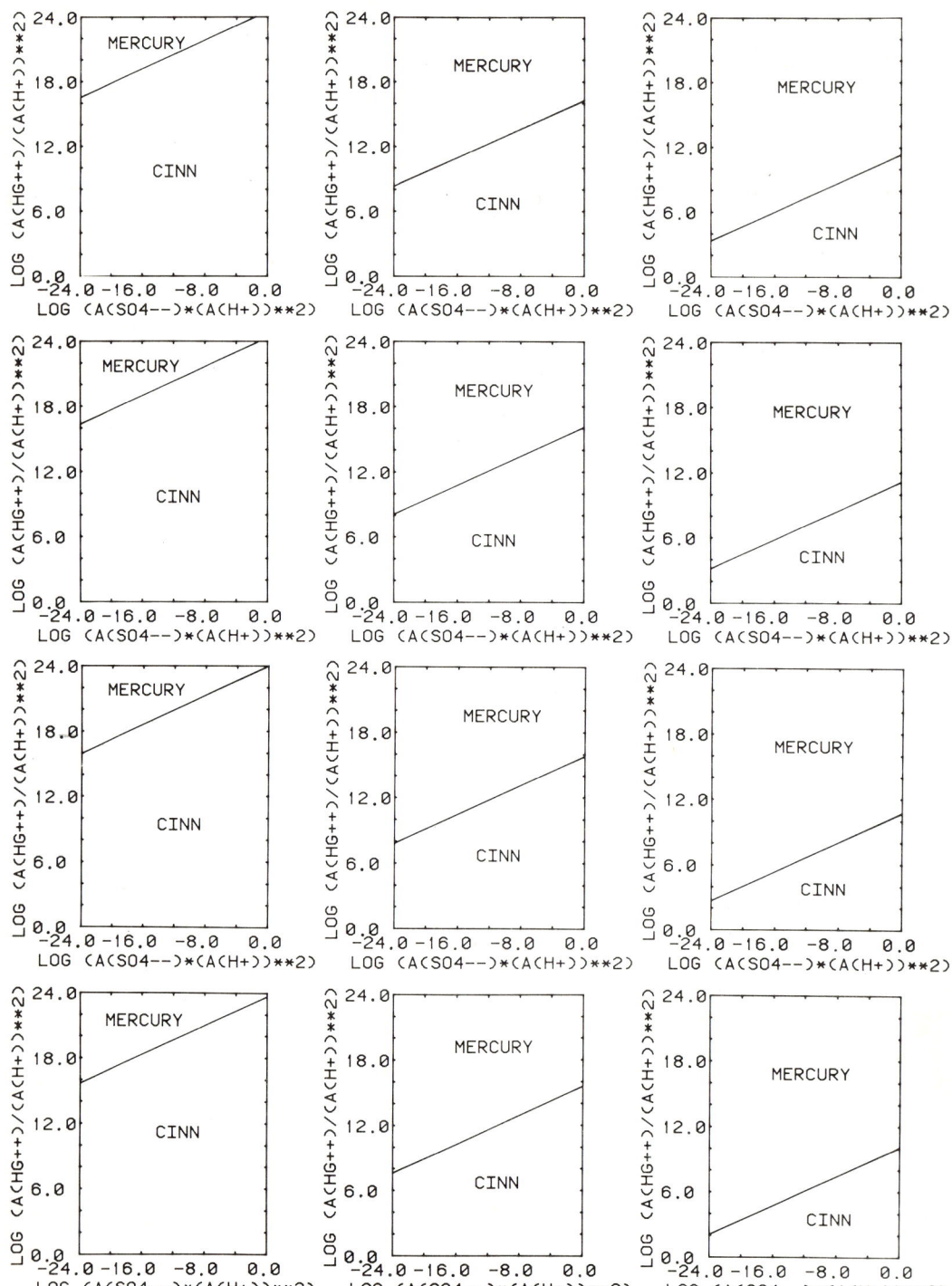

Phase relations in the system $HCl-H_2O-HgS-(H_2S)-H_2SO_4$.

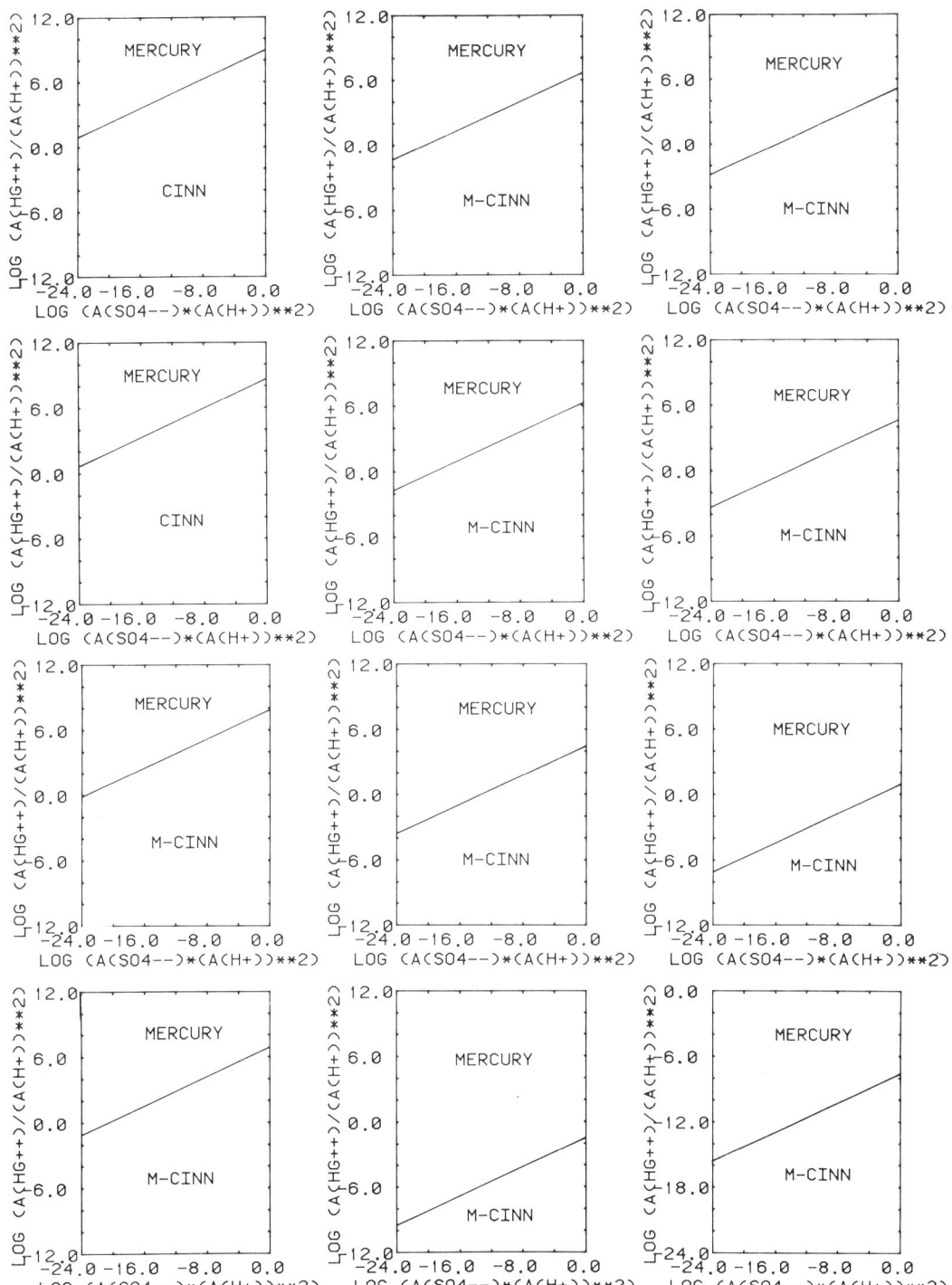

Phase relations in the system $HCl-H_2O-HgS-(H_2S)-H_2SO_4$.

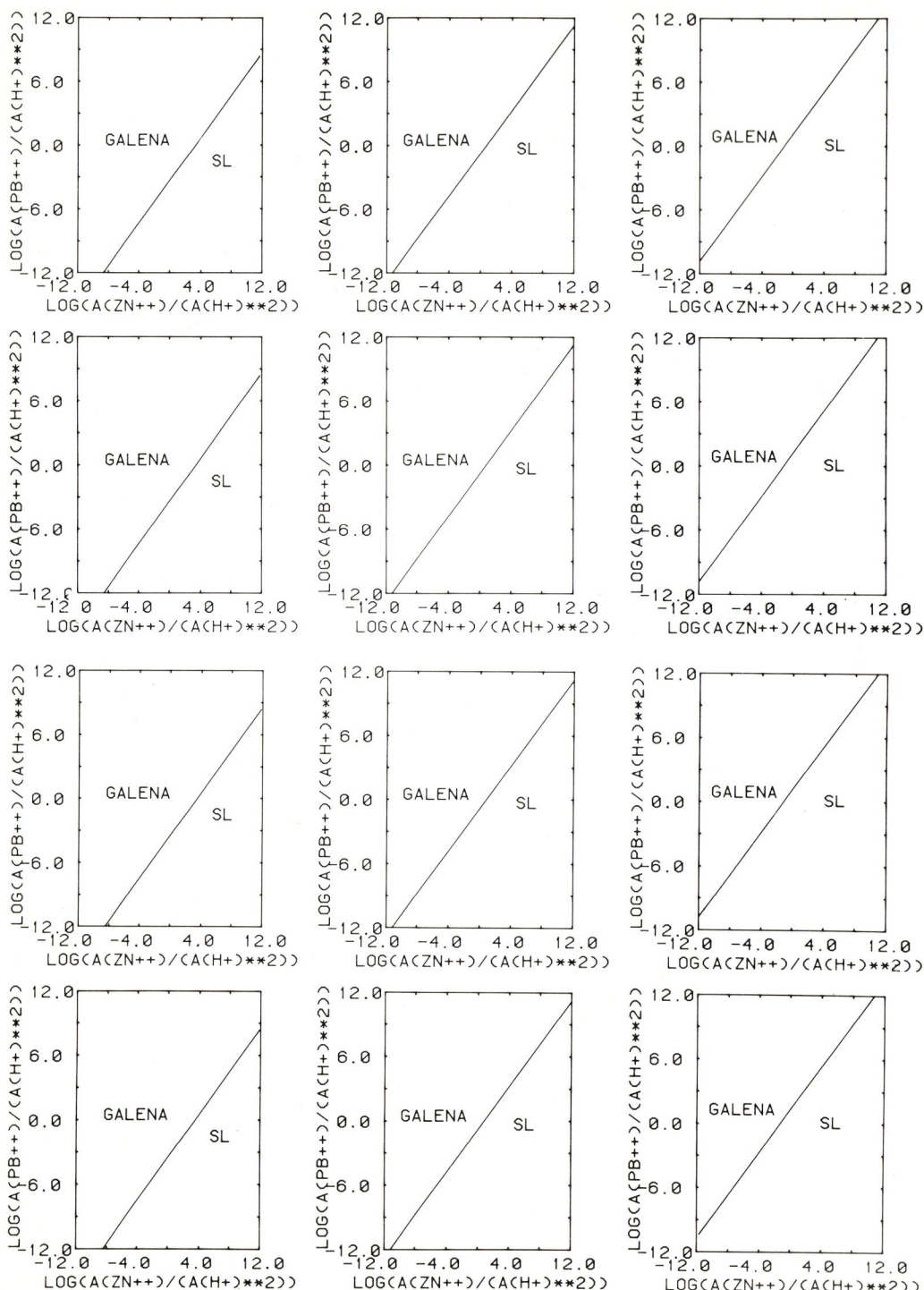

Phase relations in the system $HCl-H_2O-PbS-(H_2S)-ZnS$.

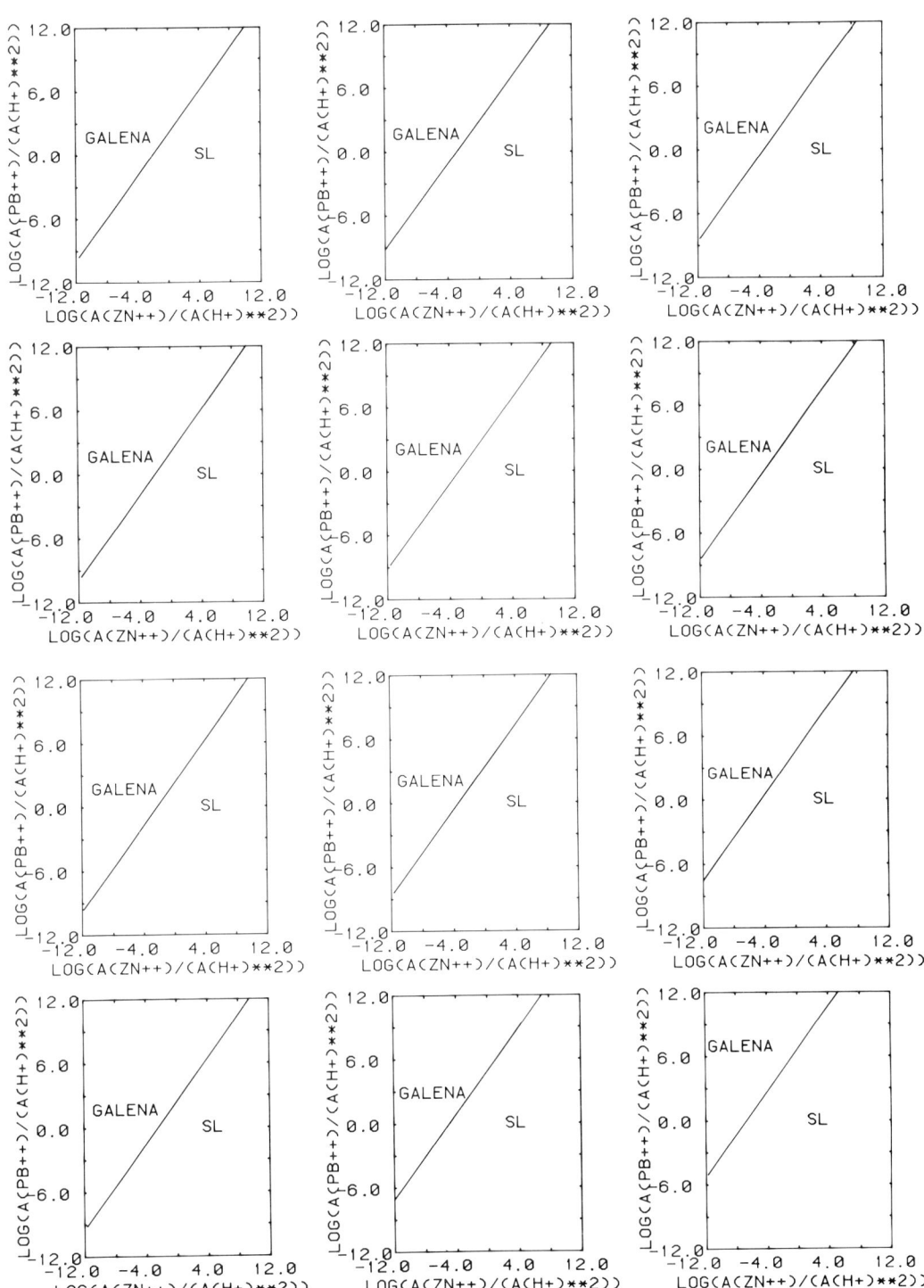

Phase relations in the system $HCl-H_2O-PbS-(H_2S)-ZnS$.

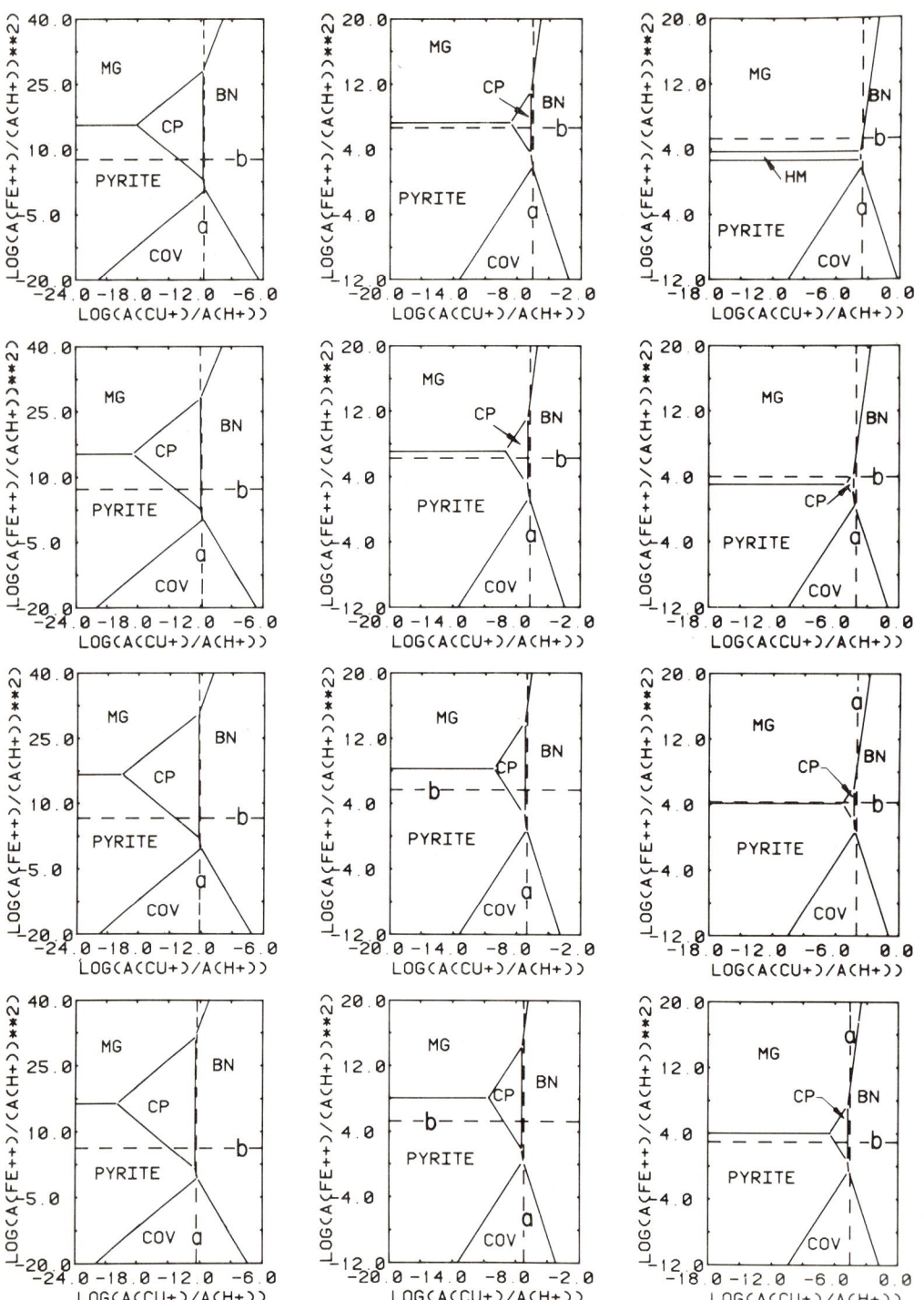

Phase relations in the system $HCl-H_2O-Cu_2O-FeO-H_2S-(H_2SO_4)$, at $\log(a_{H^+}a_{HS^-}) = -10.0$. Saturation limits: chalcocite (a), pyrrhotite (b).

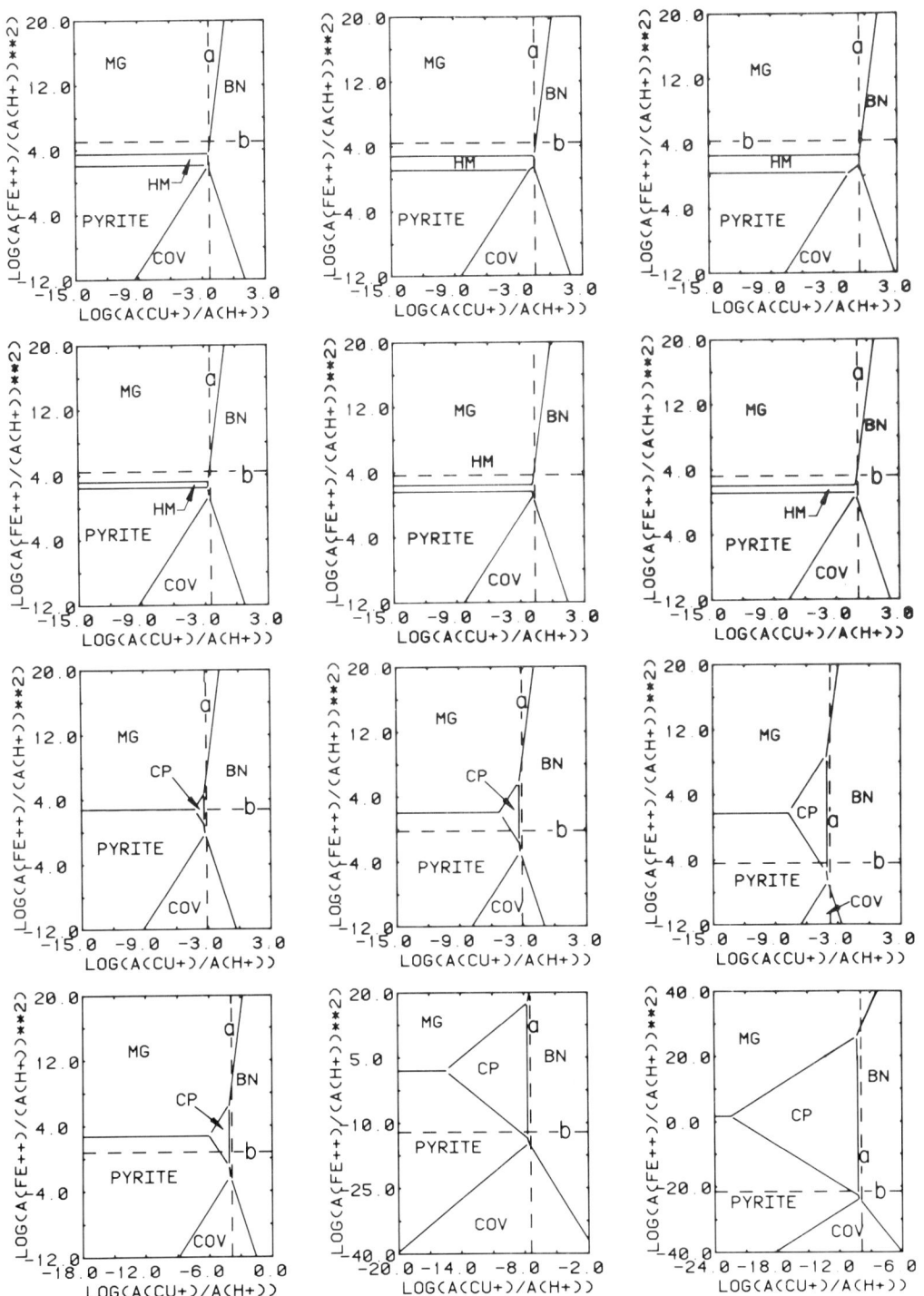

Phase relations in the system $HCl-H_2O-Cu_2O-FeO-H_2S-(H_2SO_4)$, at $\log(a_{H^+}a_{HS^-}) = -10.0$. Saturation limits: chalcocite (a), pyrrhotite (b).

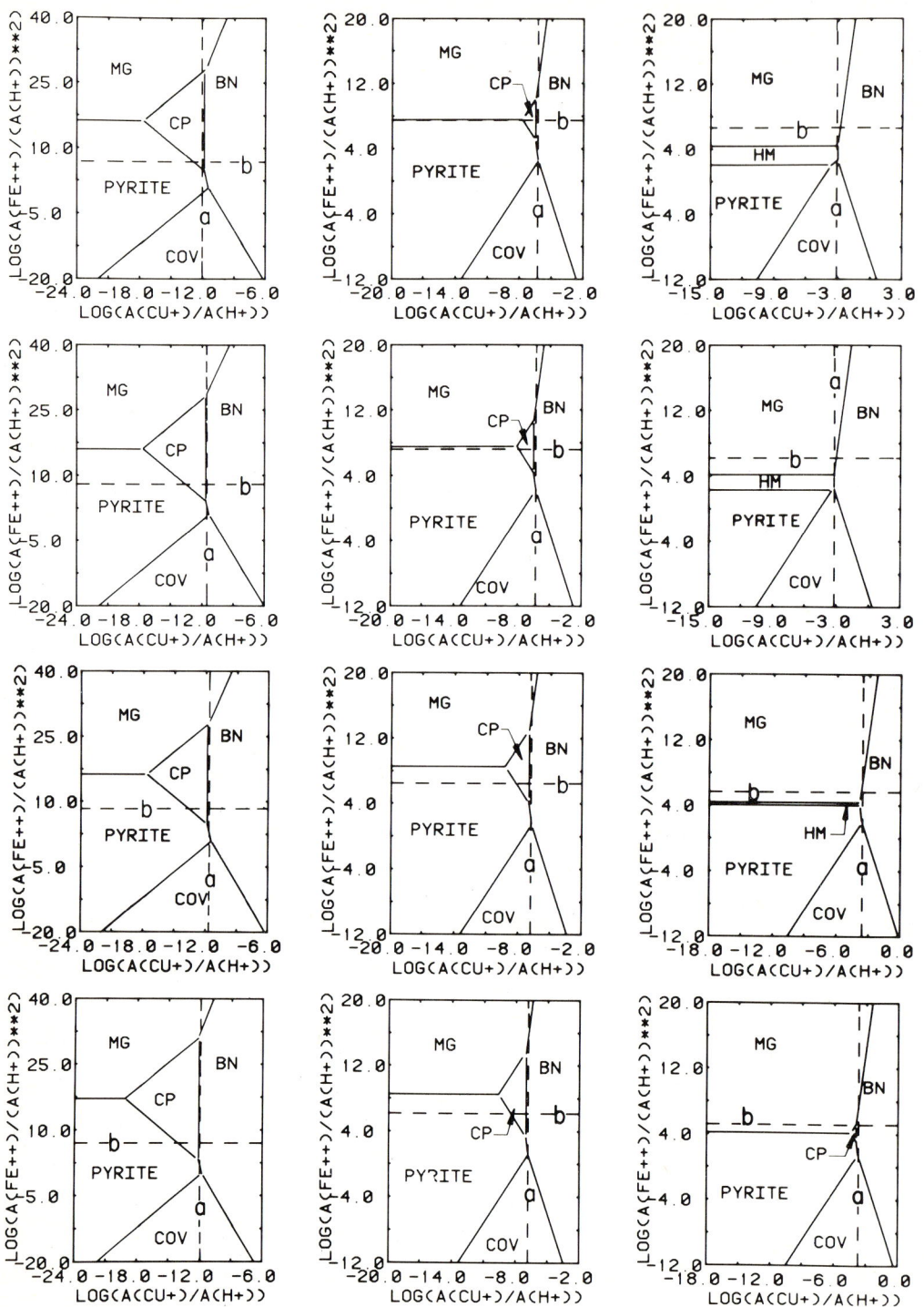

Phase relations in the system $HCl-H_2O-Cu_2O-FeO-H_2S-(H_2SO_4)$, at $\log(a_{H^+}a_{HS^-}) = -11.0$. Saturation limits: chalcocite (a), pyrrhotite (b).

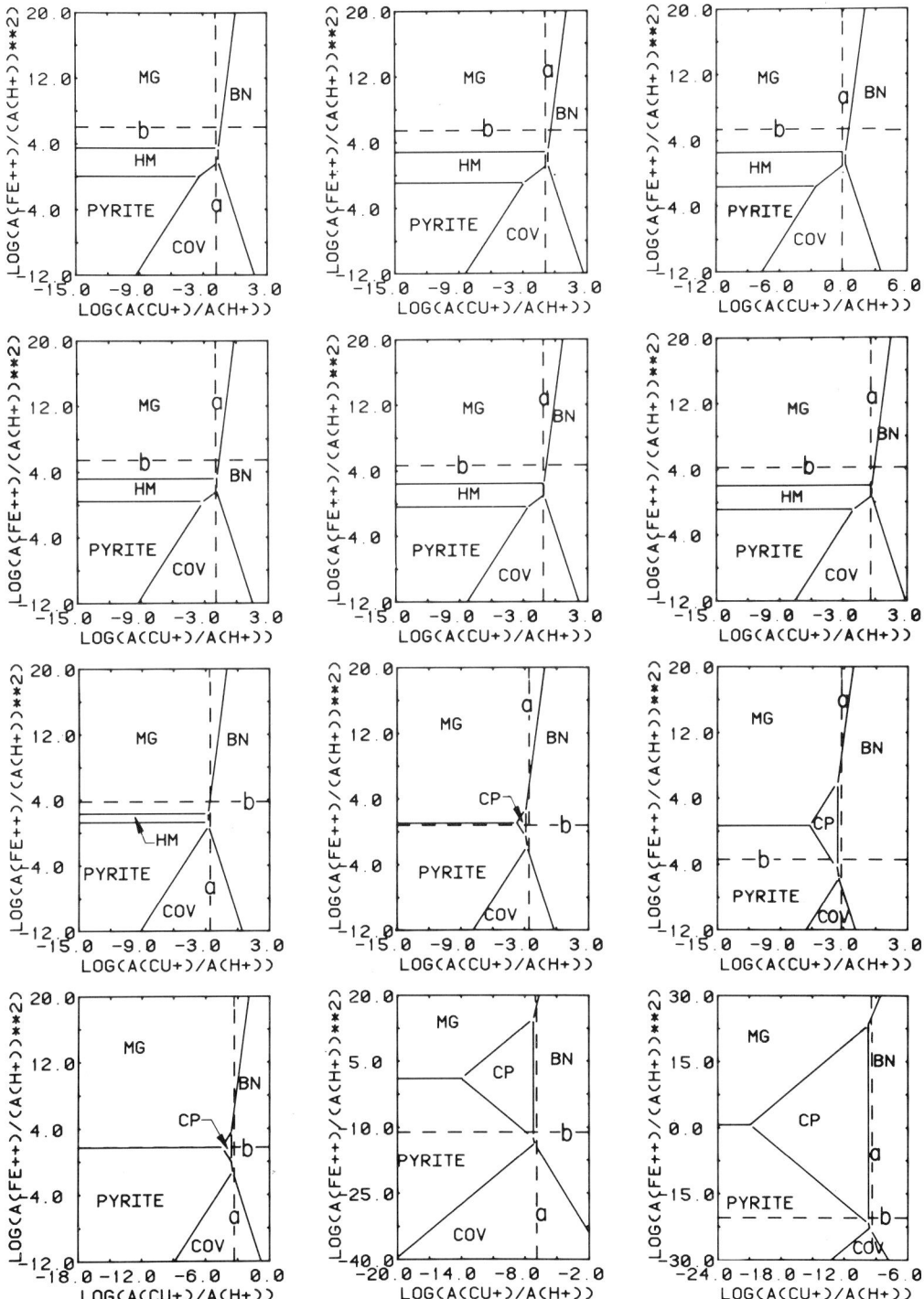

Phase relations in the system $HCl-H_2O-Cu_2O-FeO-H_2S-(H_2SO_4)$, at $\log(a_{H^+}a_{HS^-}) = -11.0$. Saturation limits: chalcocite (a), pyrrhotite (b).

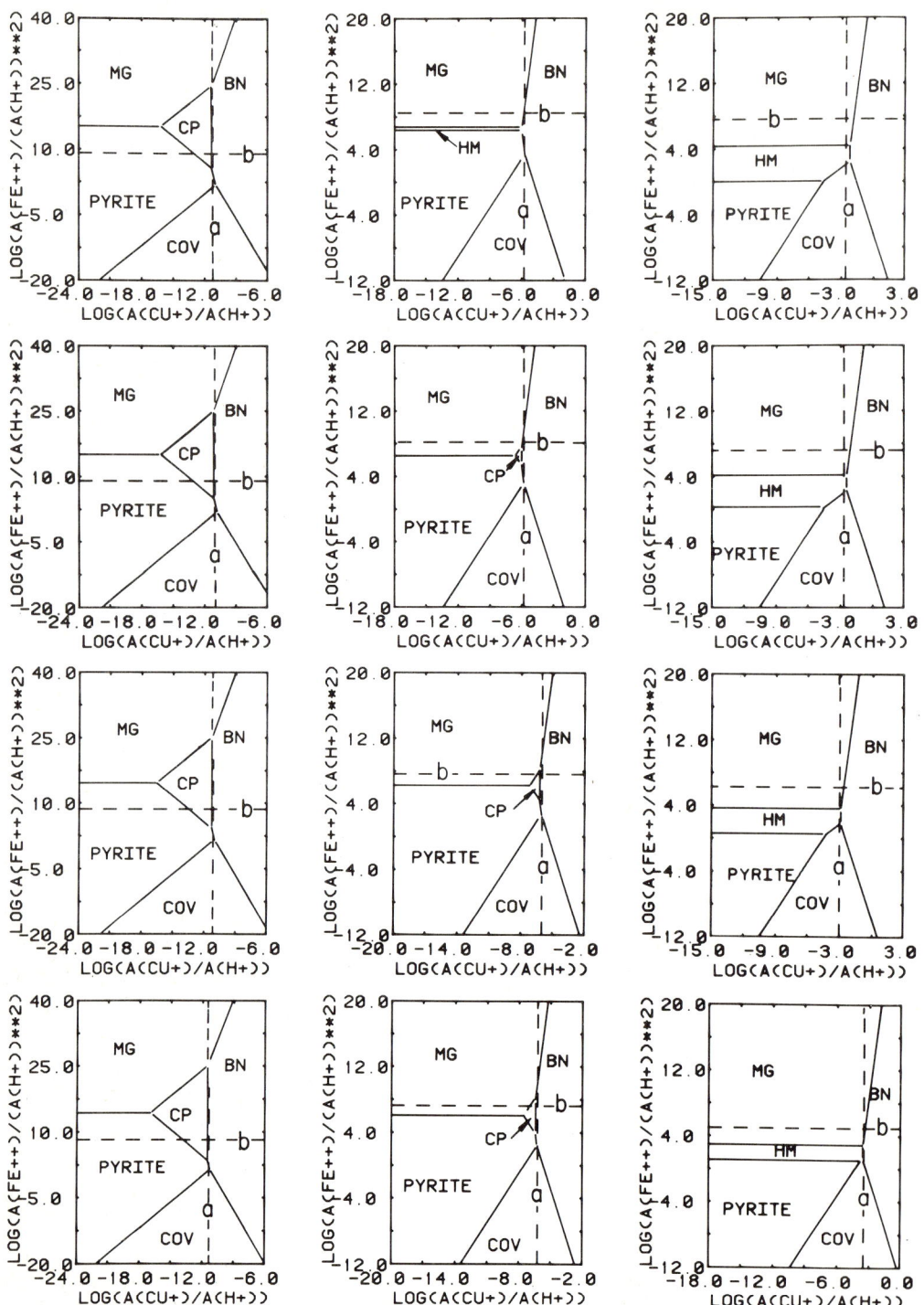

Phase relations in the system $HCl-H_2O-Cu_2O-FeO-H_2S-(H_2SO_4)$, at $\log(a_{H^+}a_{HS^-}) = -12.0$. Saturation limits: chalcocite (a), pyrrhotite (b).

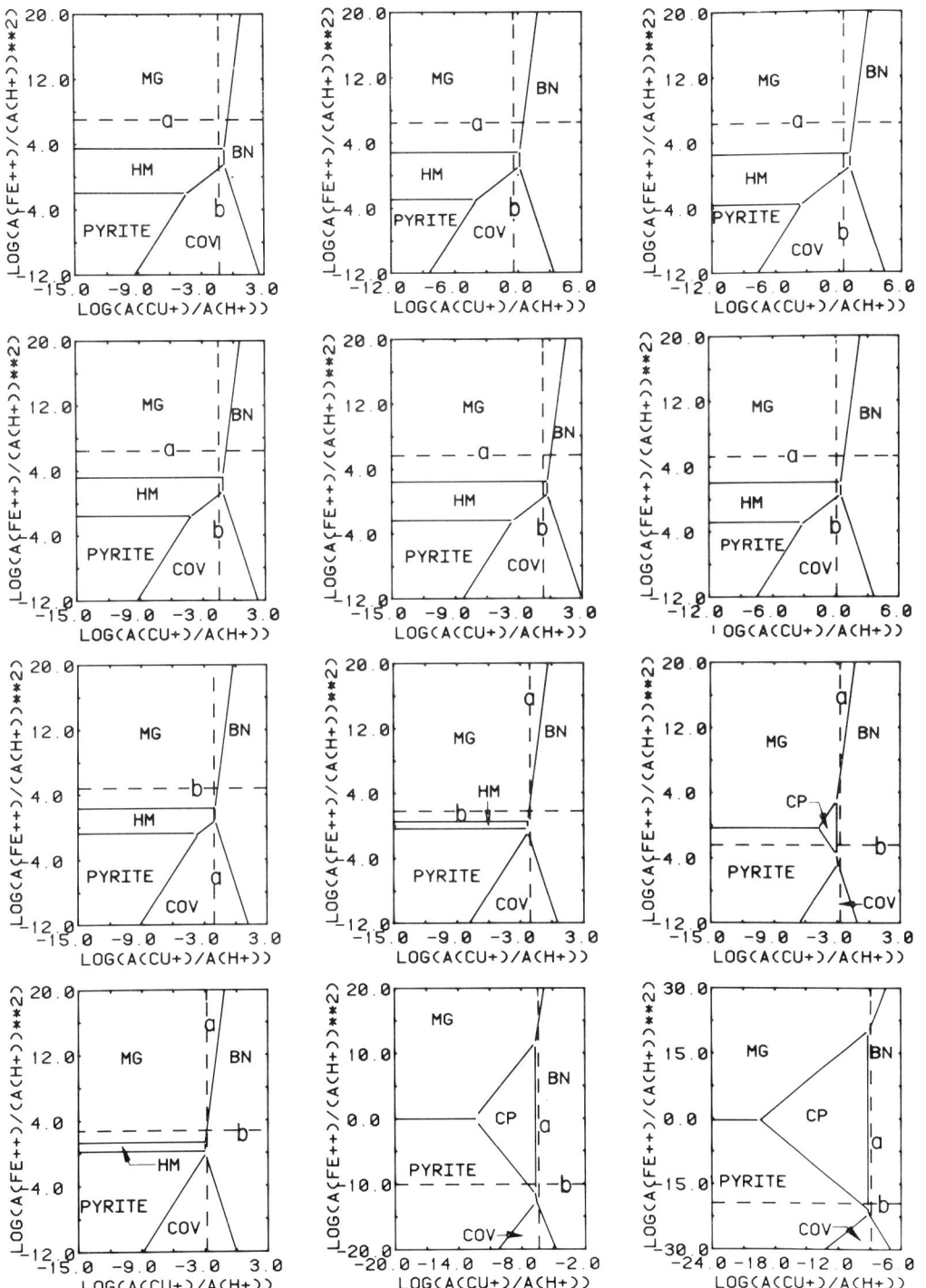

Phase relations in the system $HCl-H_2O-Cu_2O-FeO-H_2S-(H_2SO_4)$, at $\log(a_{H^+}a_{HS^-}) = -12.0$. Saturation limits: chalcocite (a), pyrrhotite (b).

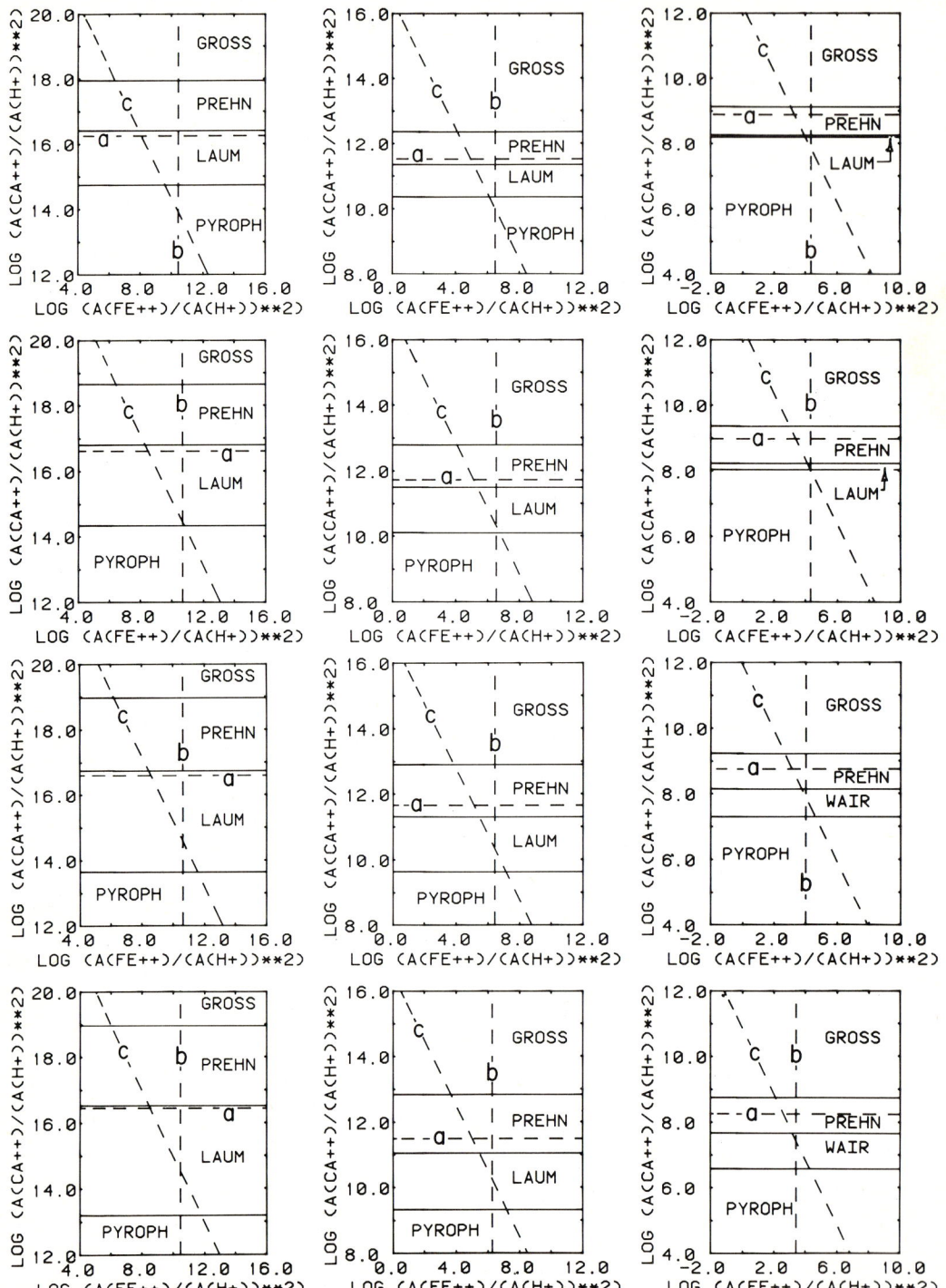

Phase relations in the system $HCl-H_2O-(Al_2O_3)-CaO-FeO-SiO_2$ in equilibrium with amorphous silica. Saturation limits: wollastonite (a), ferrosilite (b), hedenbergite (c).

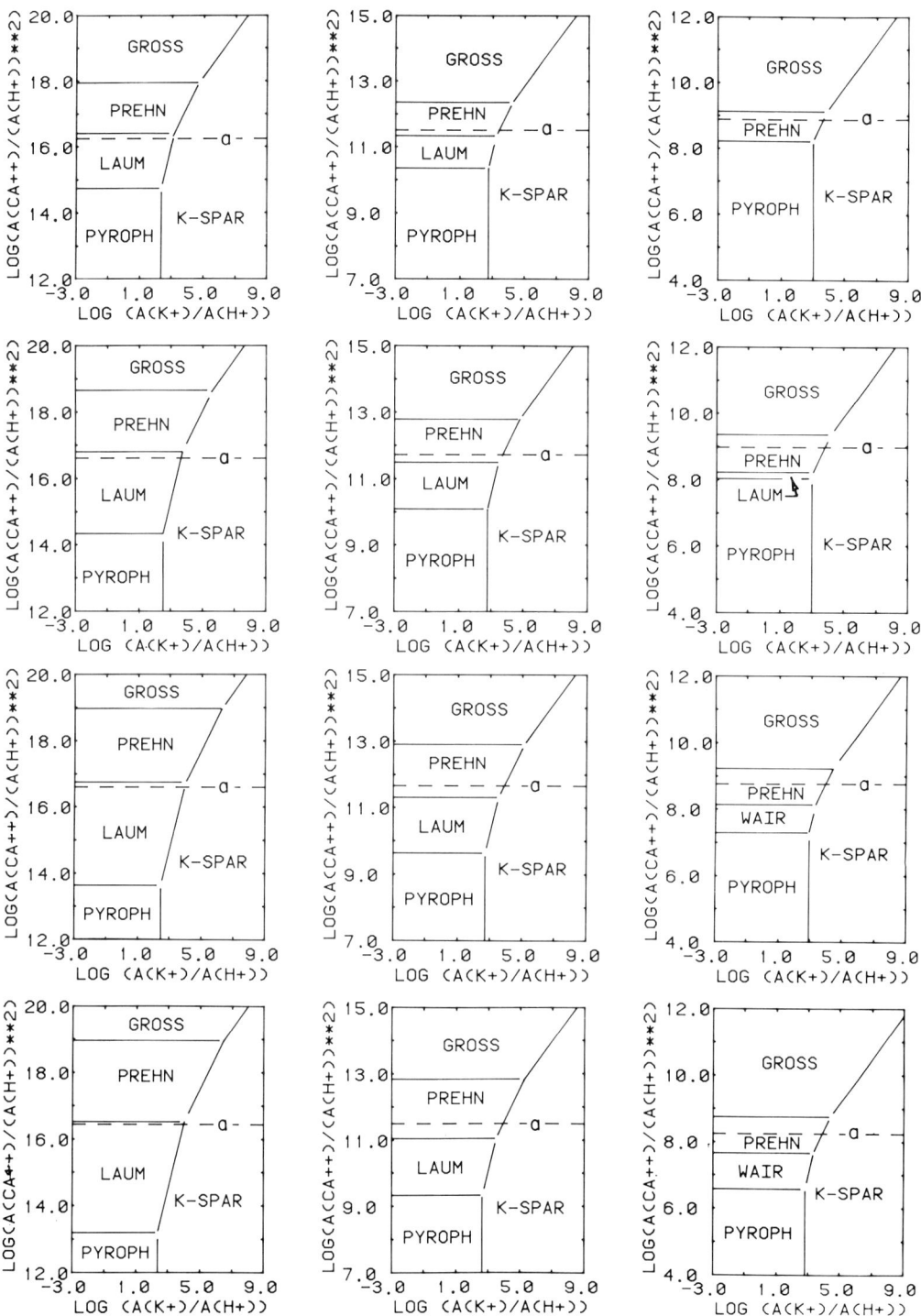

Phase relations in the system $HCl-H_2O-(Al_2O_3)-CaO-K_2O-SiO_2$ in equilibrium with amorphous silica. Saturation limit: wollastonite (a).

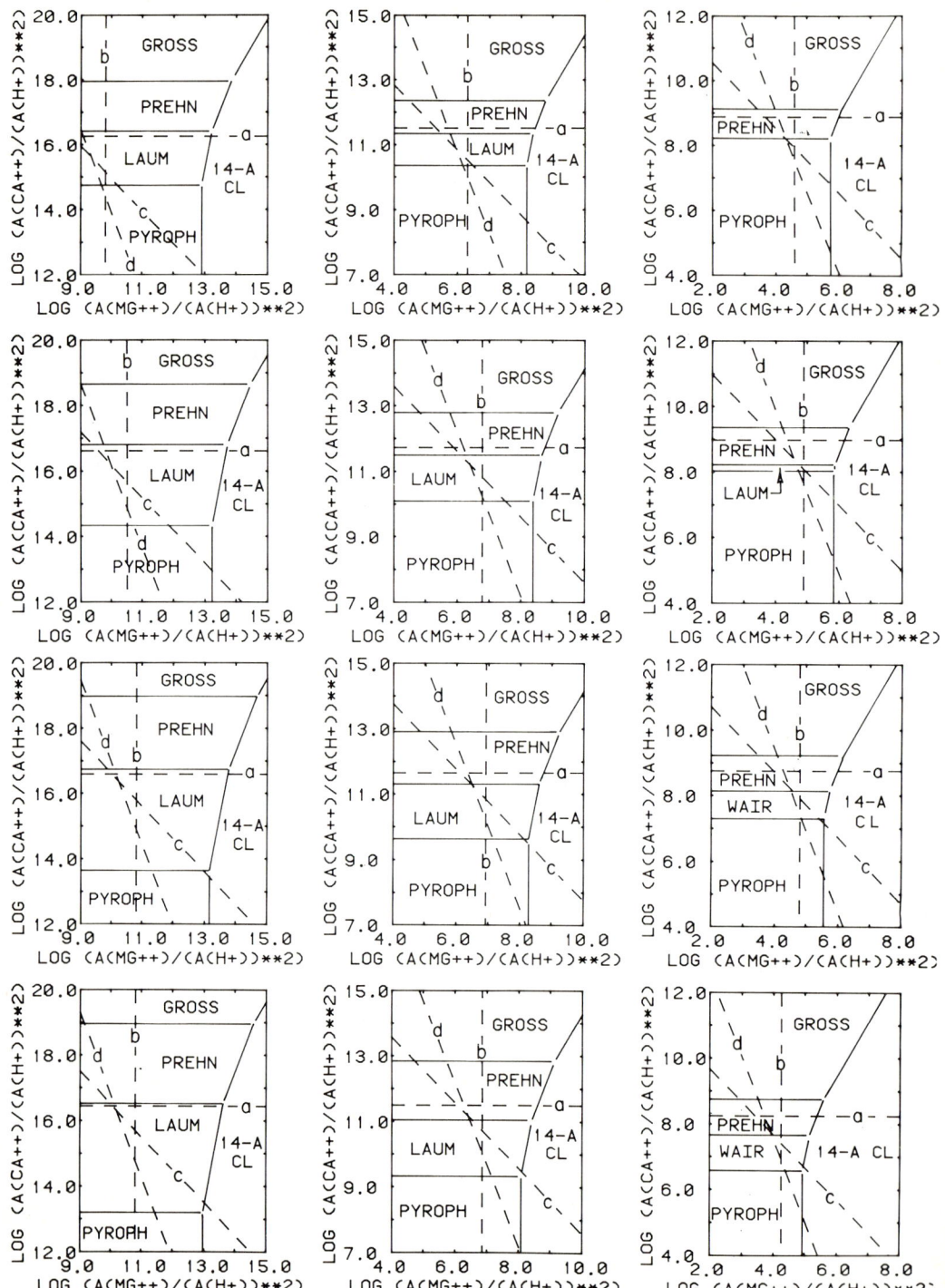

Phase relations in the system $HCl-H_2O-(Al_2O_3)-CaO-MgO-SiO_2$ in equilibrium with amorphous silica. Saturation limits: wollastonite (a), talc (b), diopside (c), tremolite (d).

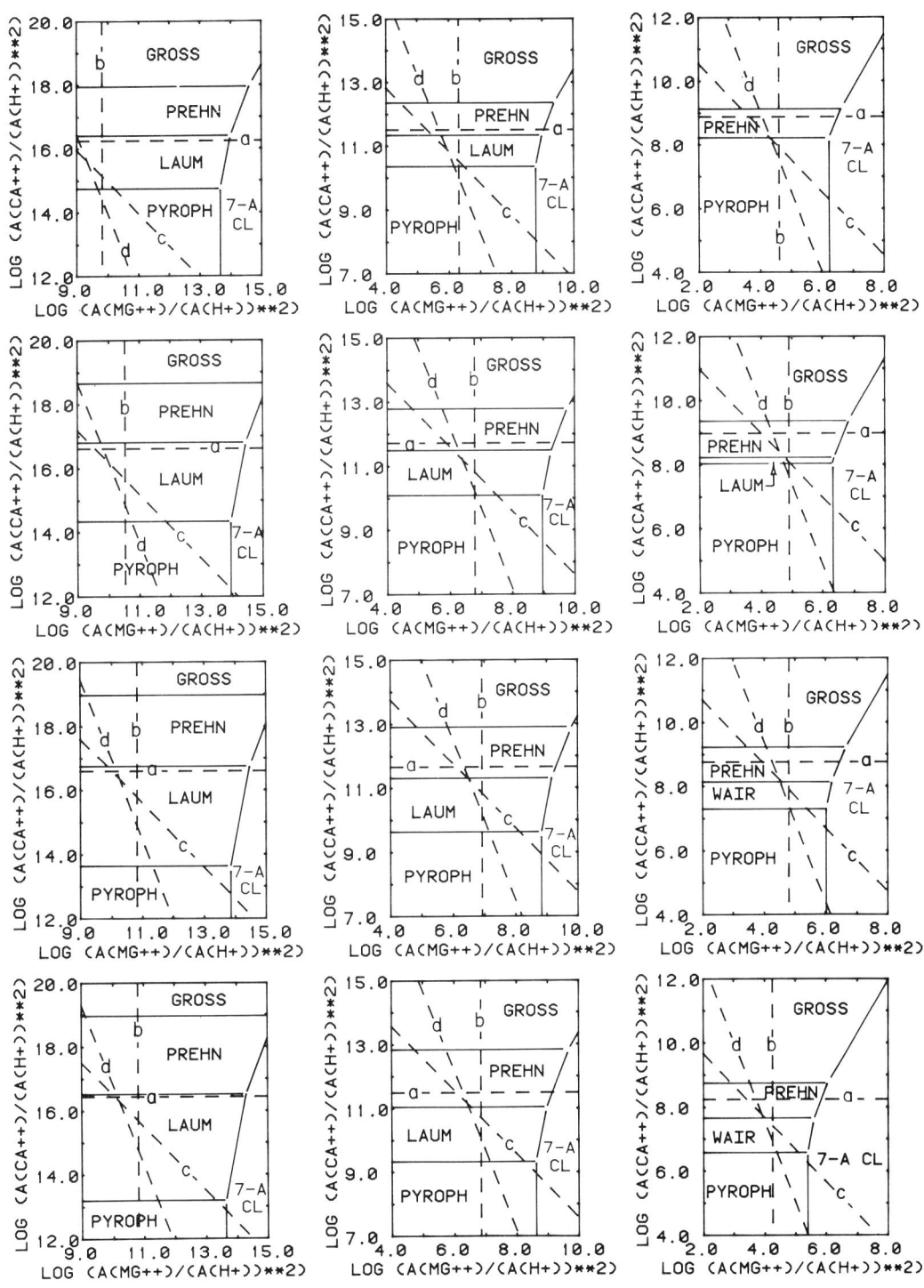

Phase relations in the system $HCl-H_2O-(Al_2O_3)-CaO-MgO-SiO_2$ in equilibrium with amorphous silica. Metastable 7-A clinochlore was considered instead of its stable counterpart, 14-A clinochlore. Saturation limits: wollastonite (a), talc (b), diopside (c), tremolite (d).

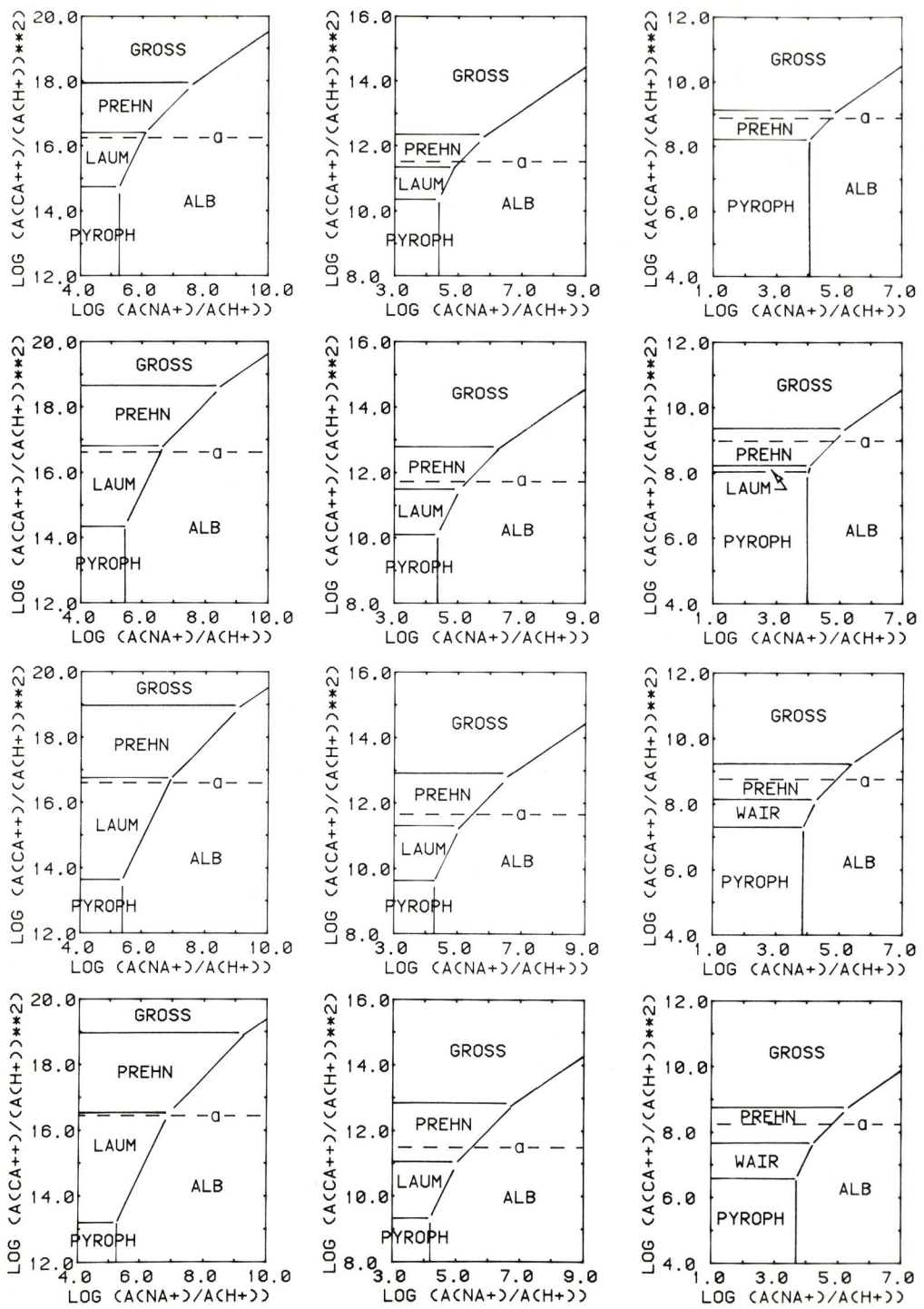

Phase relations in the system $HCl-H_2O-(Al_2O_3)-CaO-Na_2O-SiO_2$ in equilibrium with amorphous silica. Saturation limit: wollastonite (a).

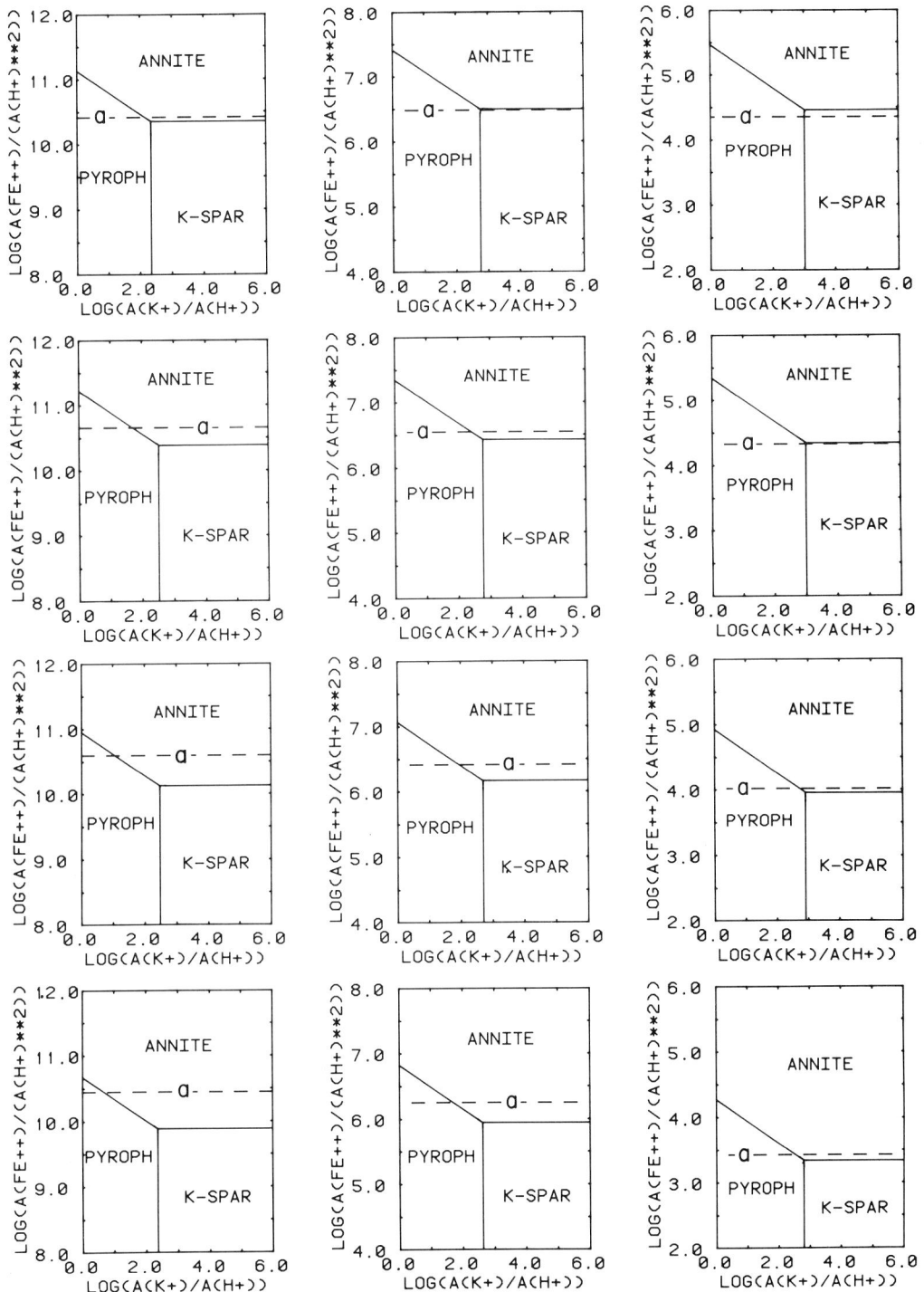

Phase relations in the system $HCl-H_2O-(Al_2O_3)-FeO-K_2O-SiO_2$ in equilibrium with amorphous silica. Saturation limit: ferrosilite (a).

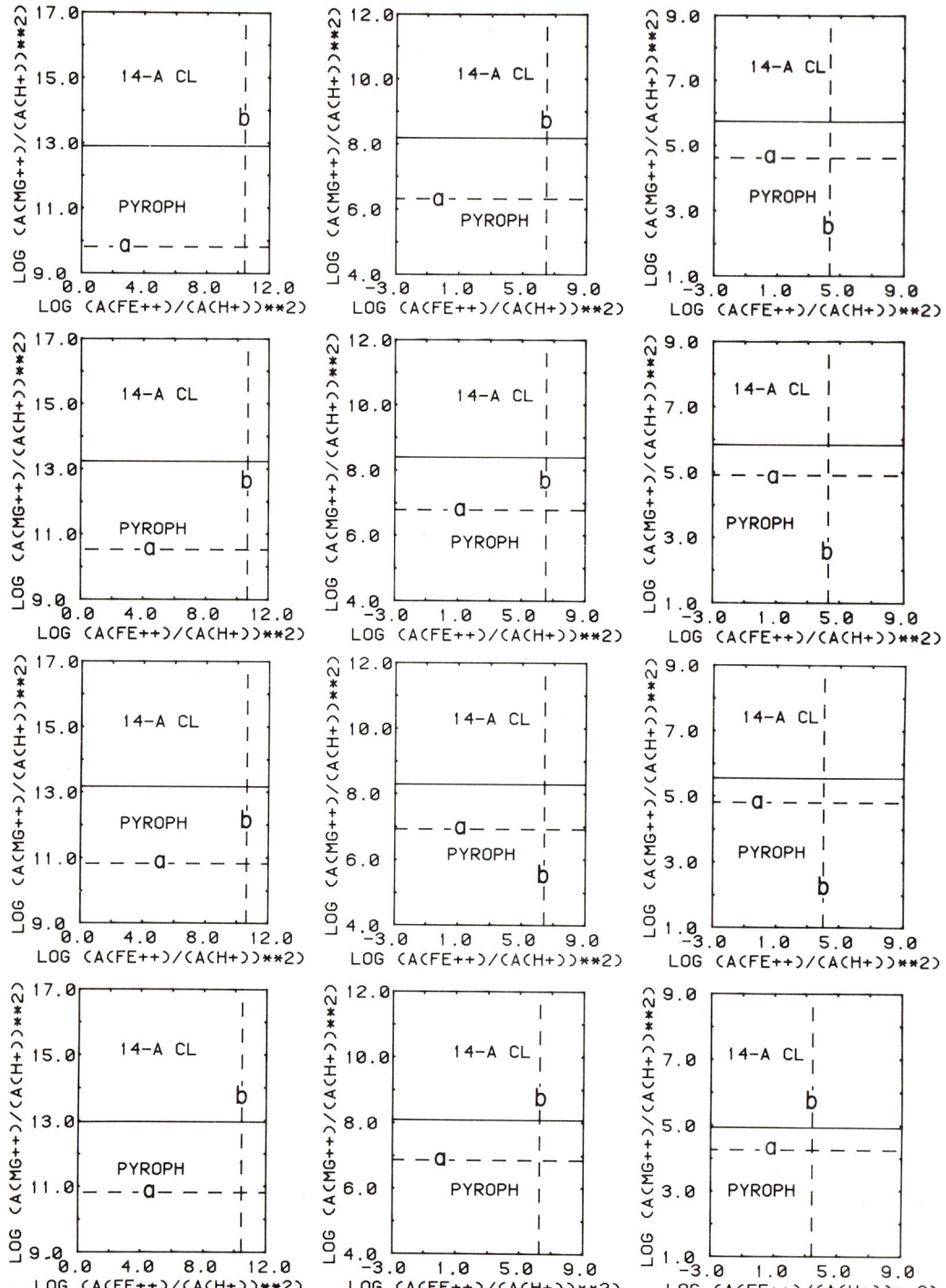

HCl−H$_2$O−(Al$_2$O$_3$)−FeO−MgO−SiO$_2$ in equilibrium with amorphous silica. Saturation limits: talc (a), ferrosilite (b).

157

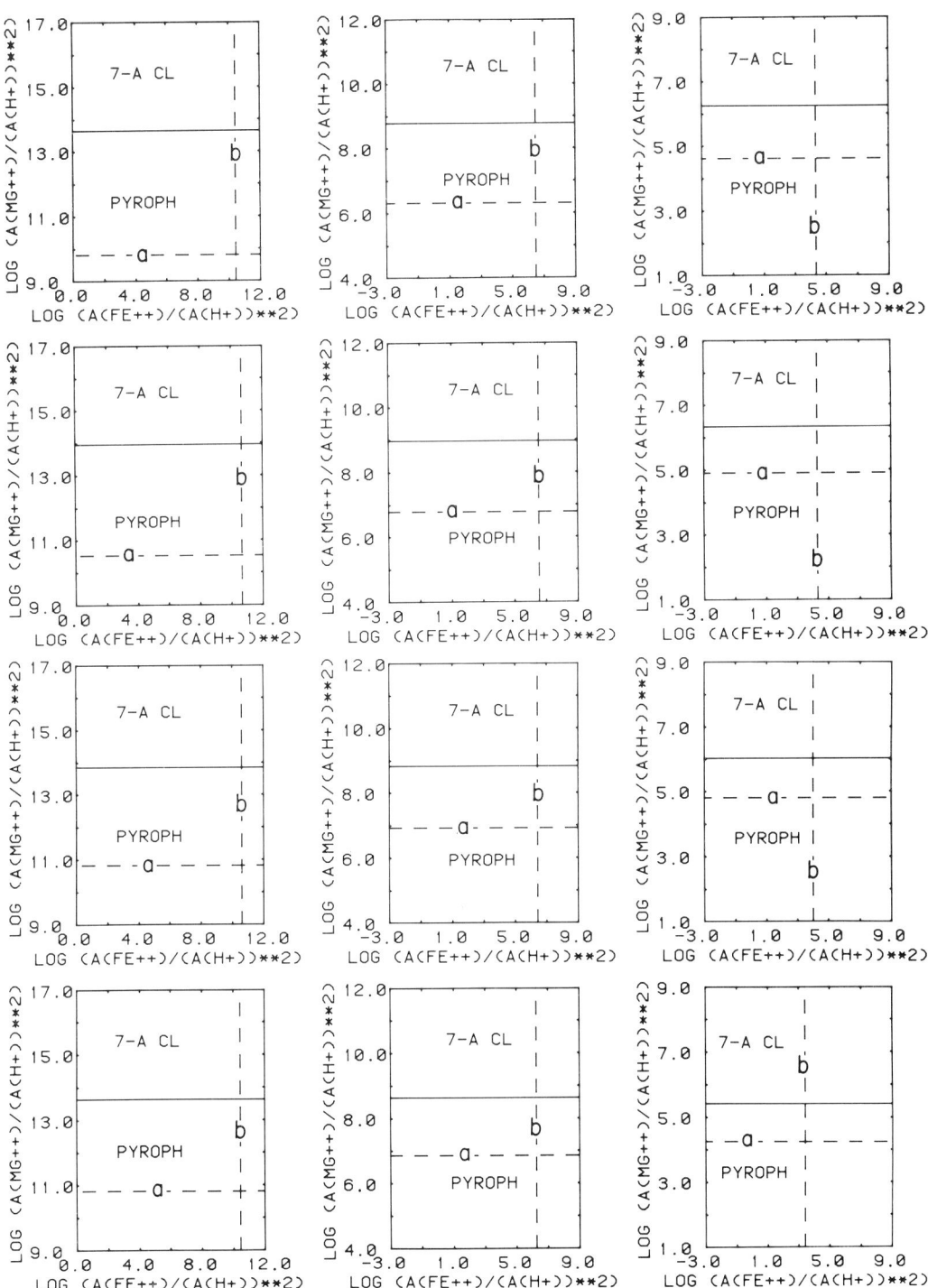

Phase relations in the system HCl—H$_2$O—(Al$_2$O$_3$)—FeO—MgO—SiO$_2$ in equilibrium with amorphous silica. Metastable 7-A clinochlore was considered instead of its stable counterpart, 14-A clinochlore. Saturation limits: talc (a), ferrosilite (b).

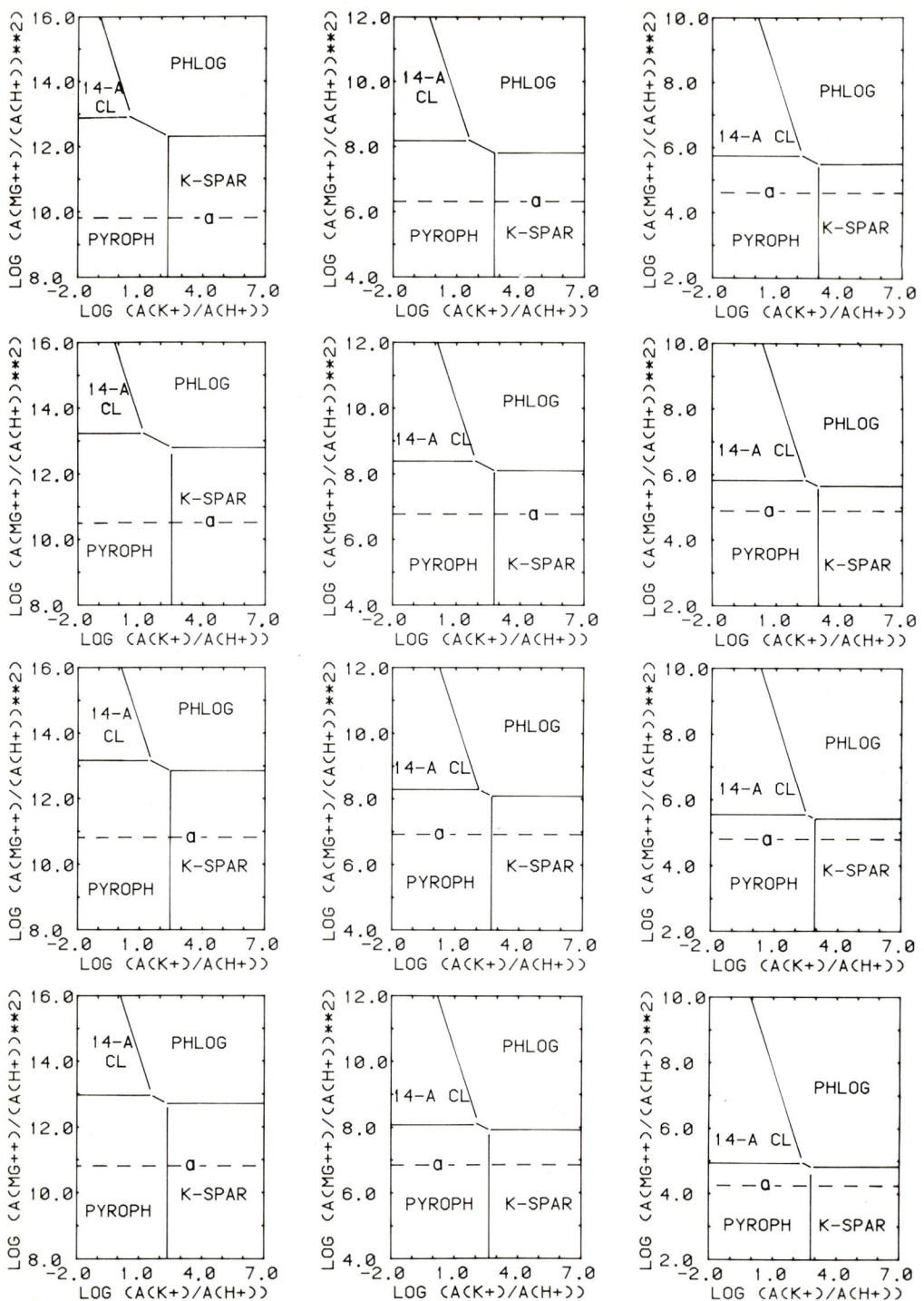

Phase relations in the system $HCl-H_2O-(Al_2O_3)-K_2O-MgO-SiO_2$ in equilibrium with amorphous silica. Saturation limit: talc (a).

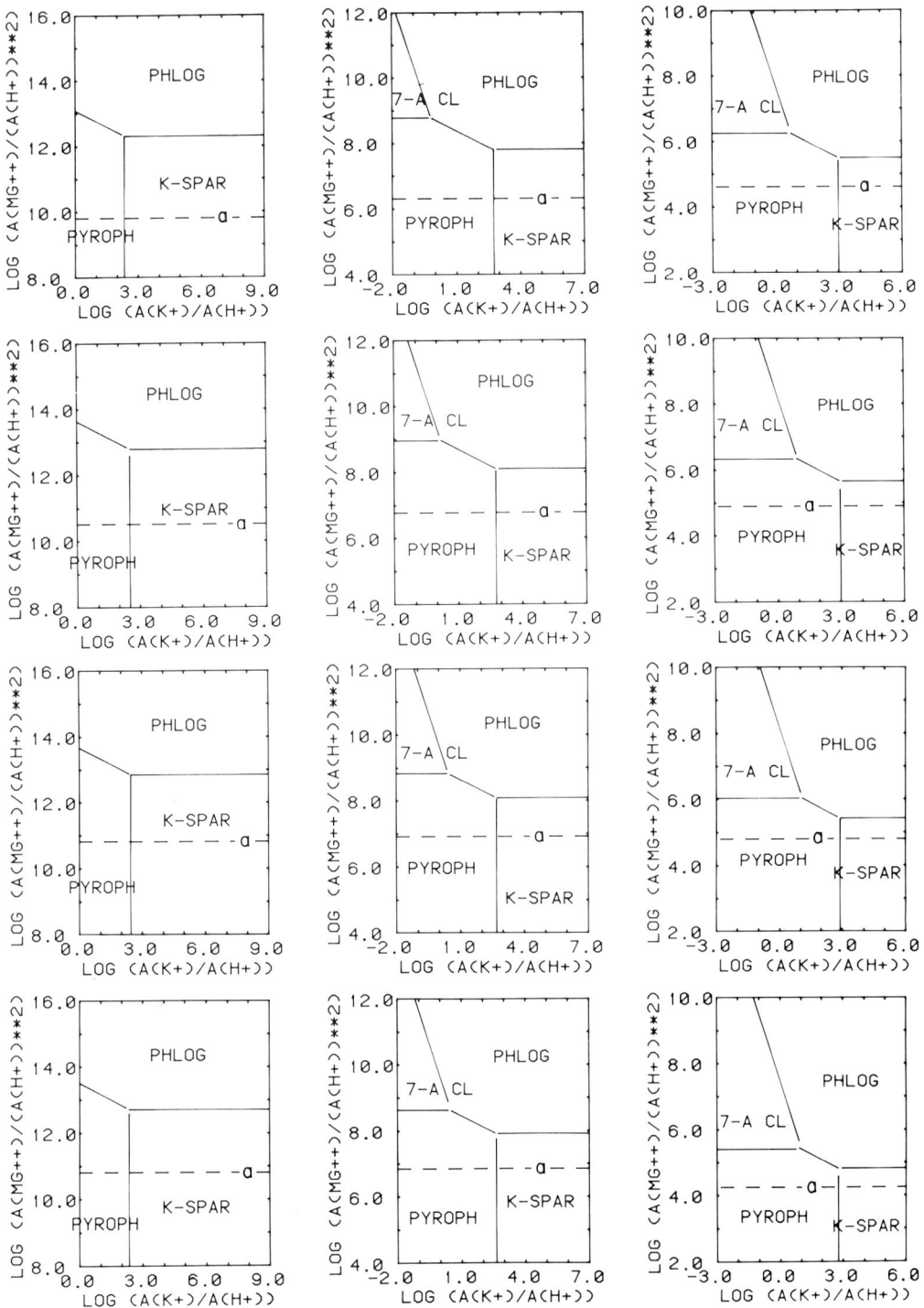

Phase relations in the system $HCl-H_2O-(Al_2O_3)-K_2O-MgO-SiO_2$ in equilibrium with amorphous silica. Metastable 7-Å clinochlore was considered instead of its stable counterpart, 14-Å clinochlore. Saturation limit: talc (a).

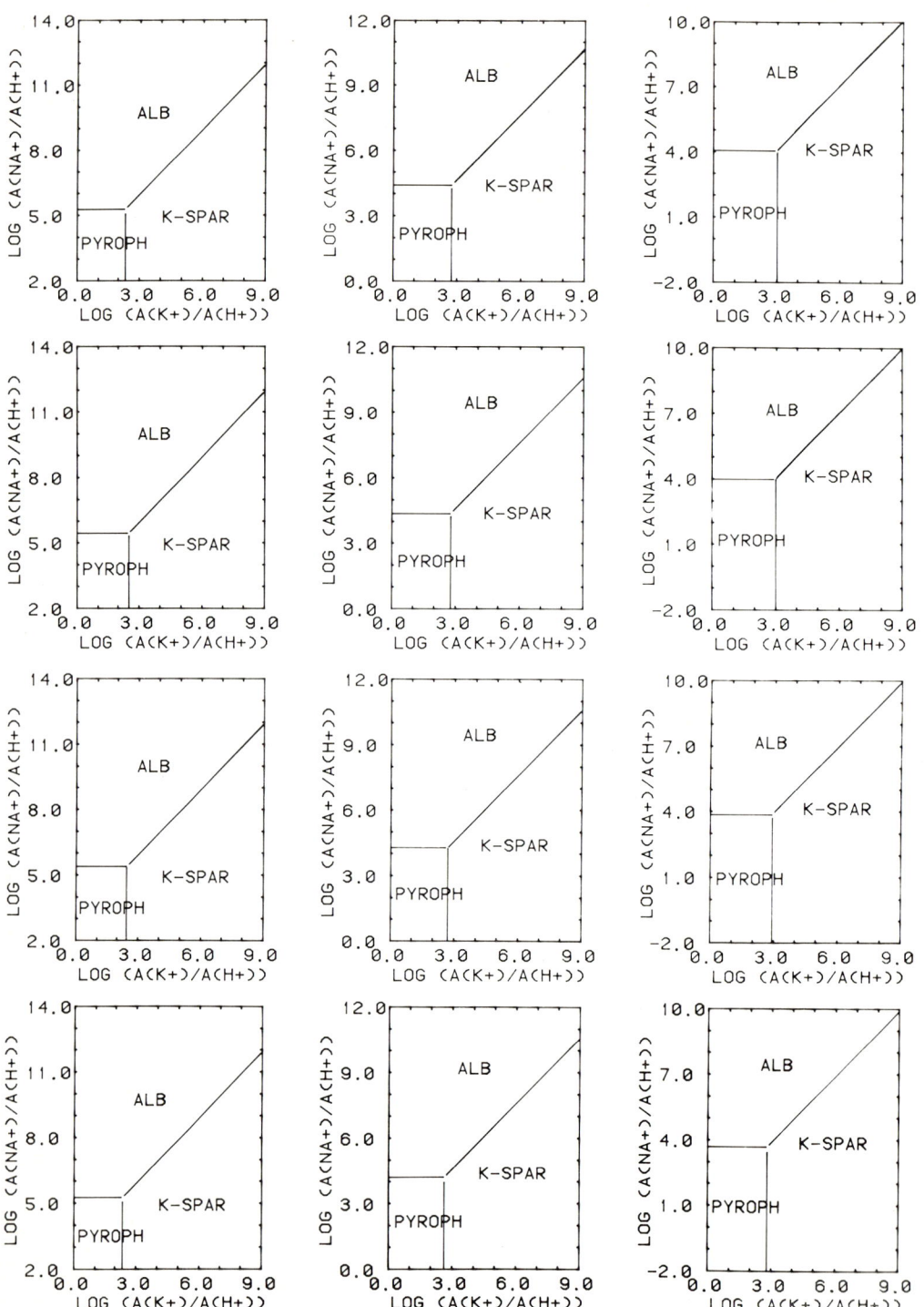

Phase relations in the system $HCl-H_2O-(Al_2O_3)-K_2O-Na_2O-SiO_2$ in equilibrium with amorphous silica.

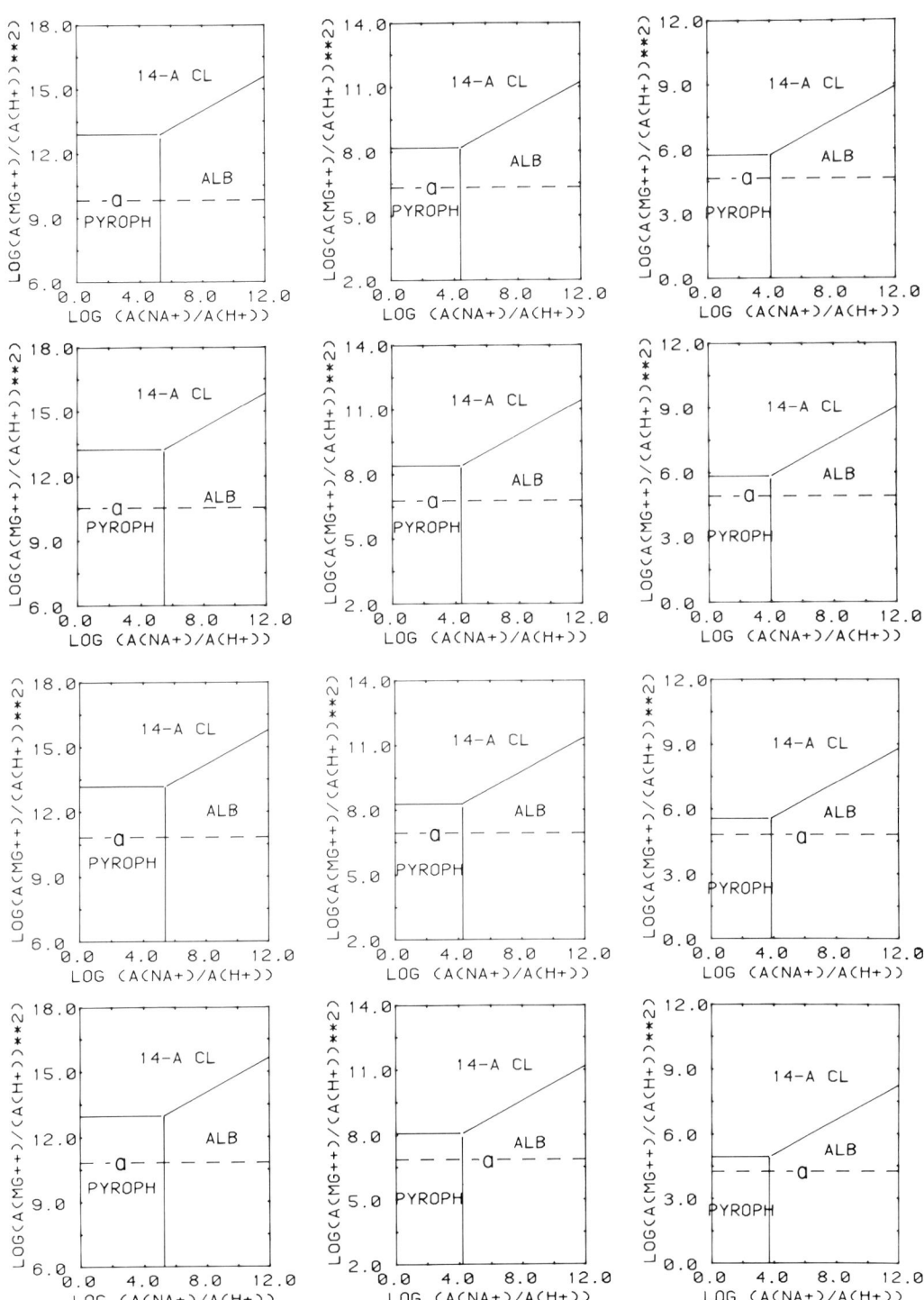

Phase relations in the system $HCl-H_2O-(Al_2O_3)-MgO-Na_2O-SiO_2$ in equilibrium with amorphous silica. Saturation limit: talc (a).

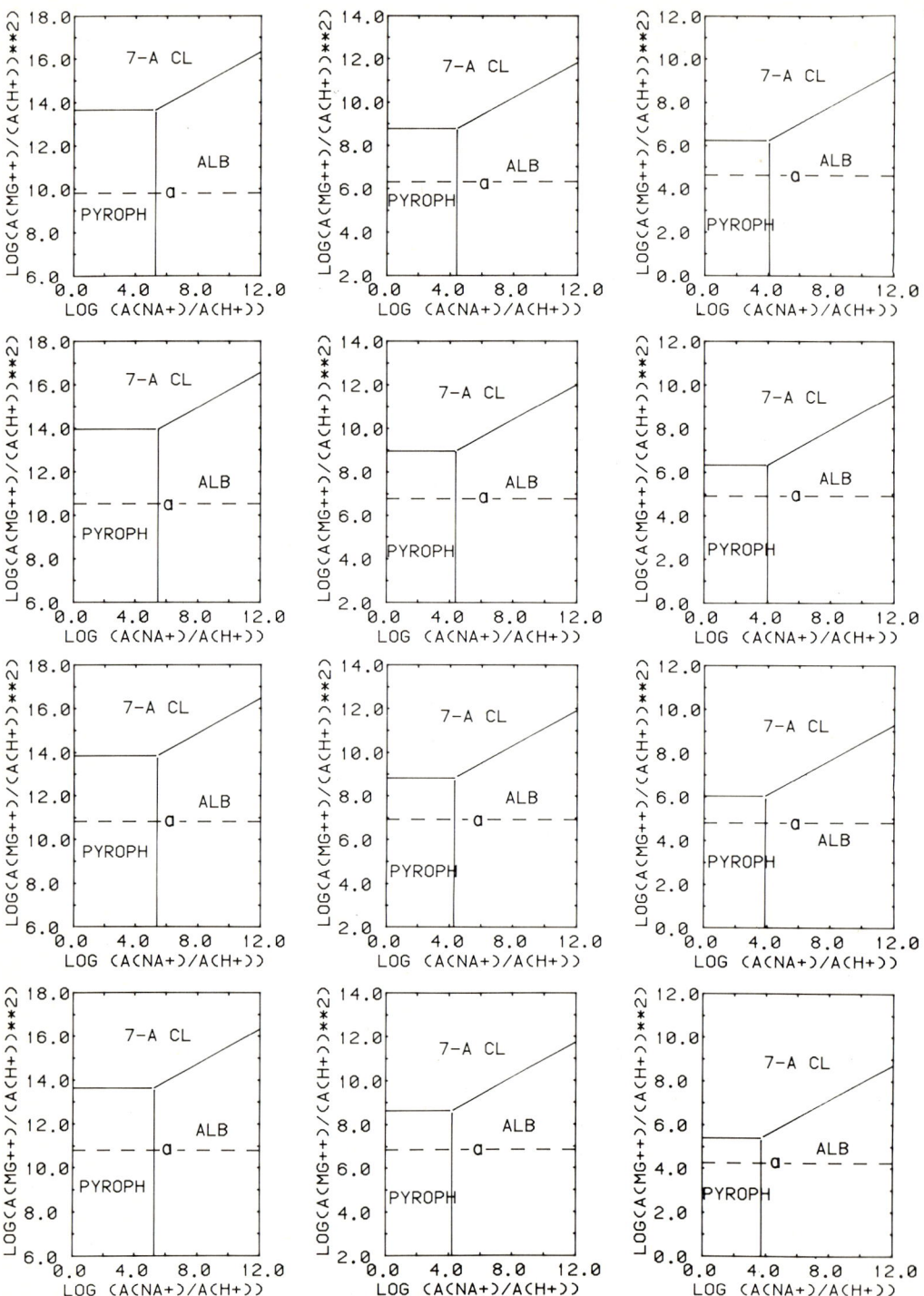

Phase relations in the system $HCl-H_2O-(Al_2O_3)-MgO-Na_2O-SiO_2$ in equilibrium with amorphous silica. Metastable 7-A clinochlore was considered instead of its stable counterpart, 14-A clinochlore. Saturation limit: talc (a).

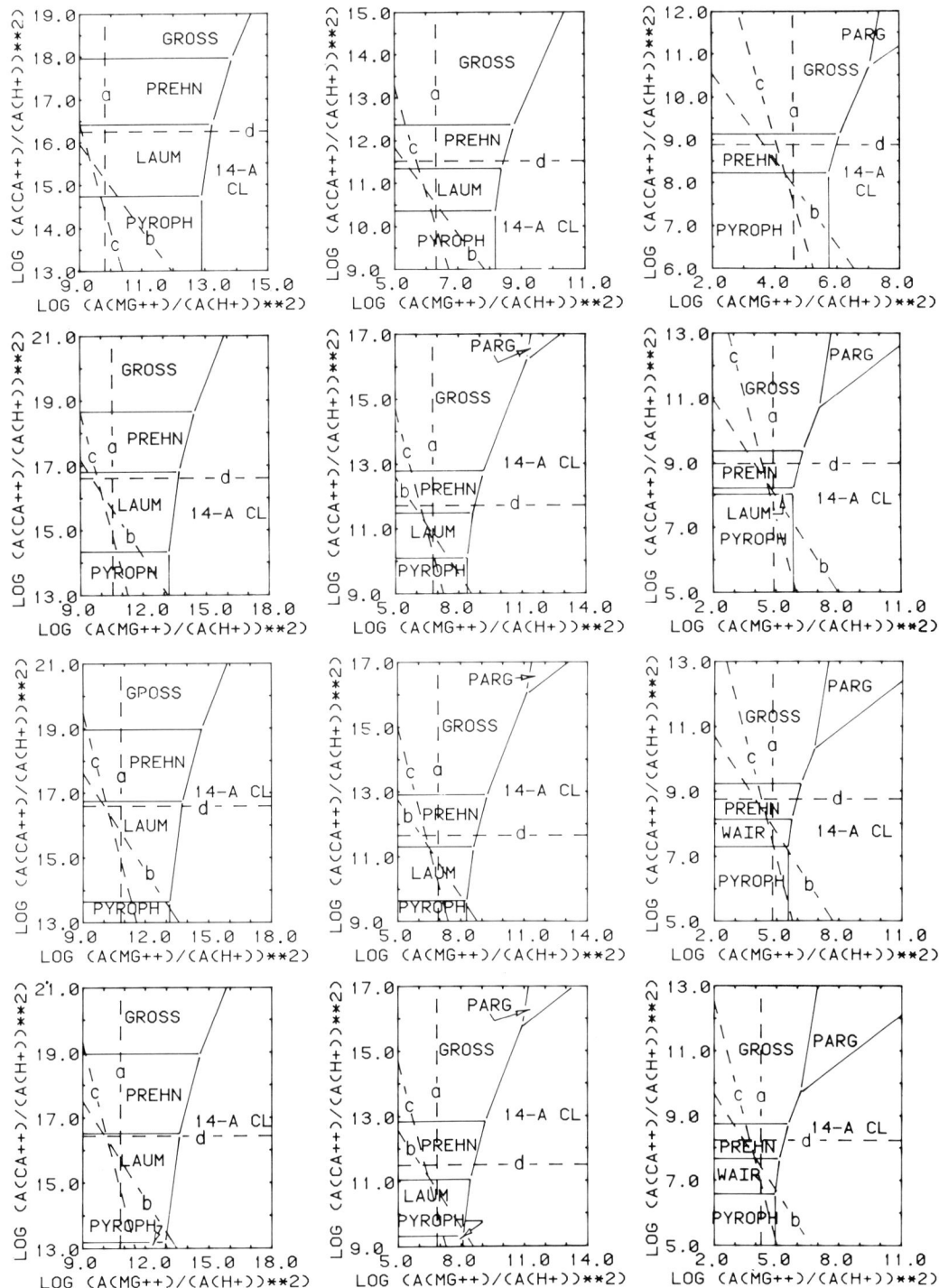

Phase relations in the system $HCl-H_2O-(Al_2O_3)-CaO-MgO-Na_2O-SiO_2$ in equilibrium with albite and amorphous silica. Saturation limits: talc (a), diopside (b), tremolite (c), wollastonite (d).

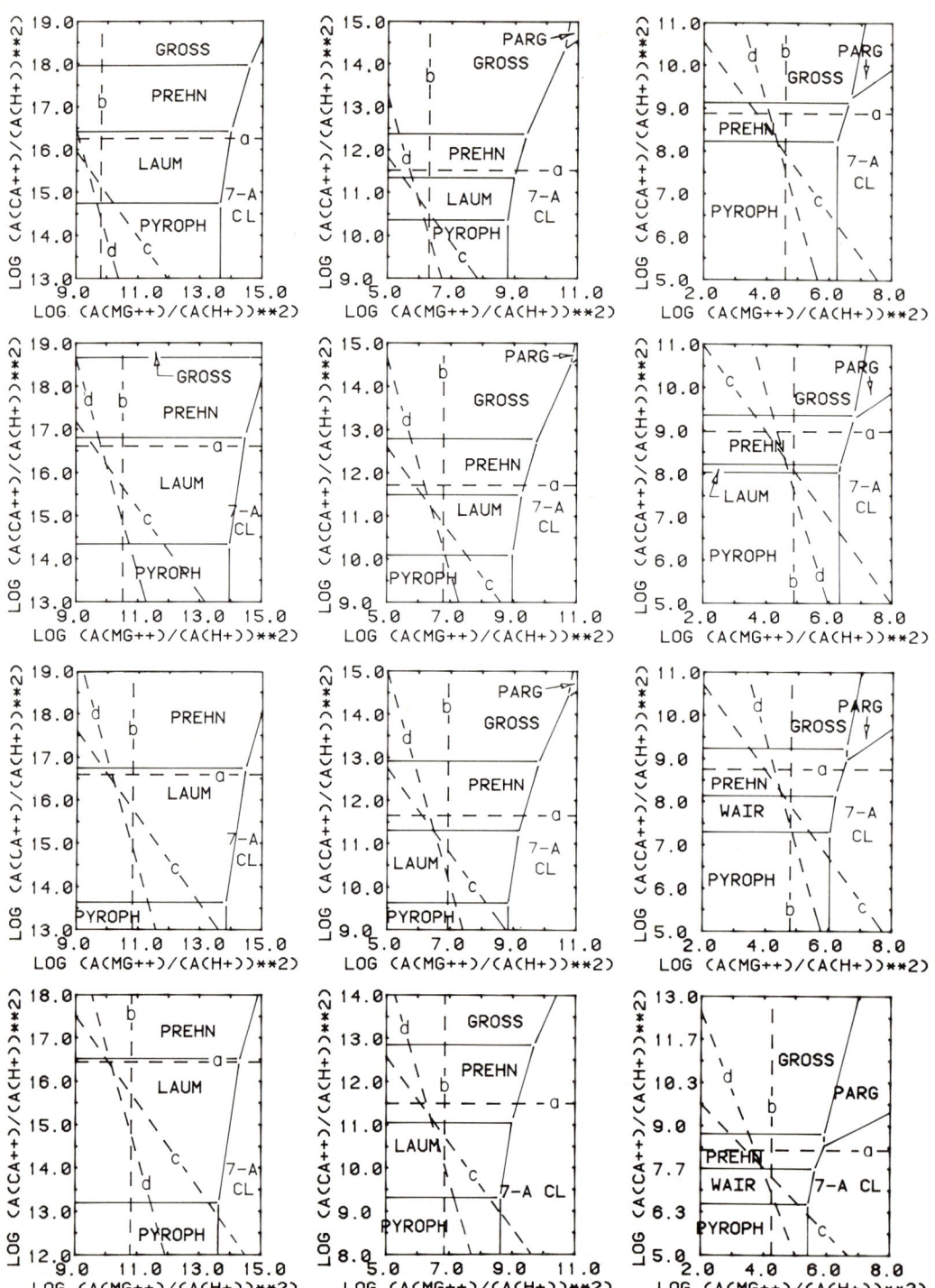

Phase relations in the system $HCl-H_2O-(Al_2O_3)-CaO-MgO-Na_2O-SiO_2$ in equilibrium with albite and amorphous silica. Metastable 7-A clinochlore was considered instead of its stable counterpart, 14-A clinochlore. Saturation limits: wollastonite (a), talc (b), diopside (c), tremolite (d).

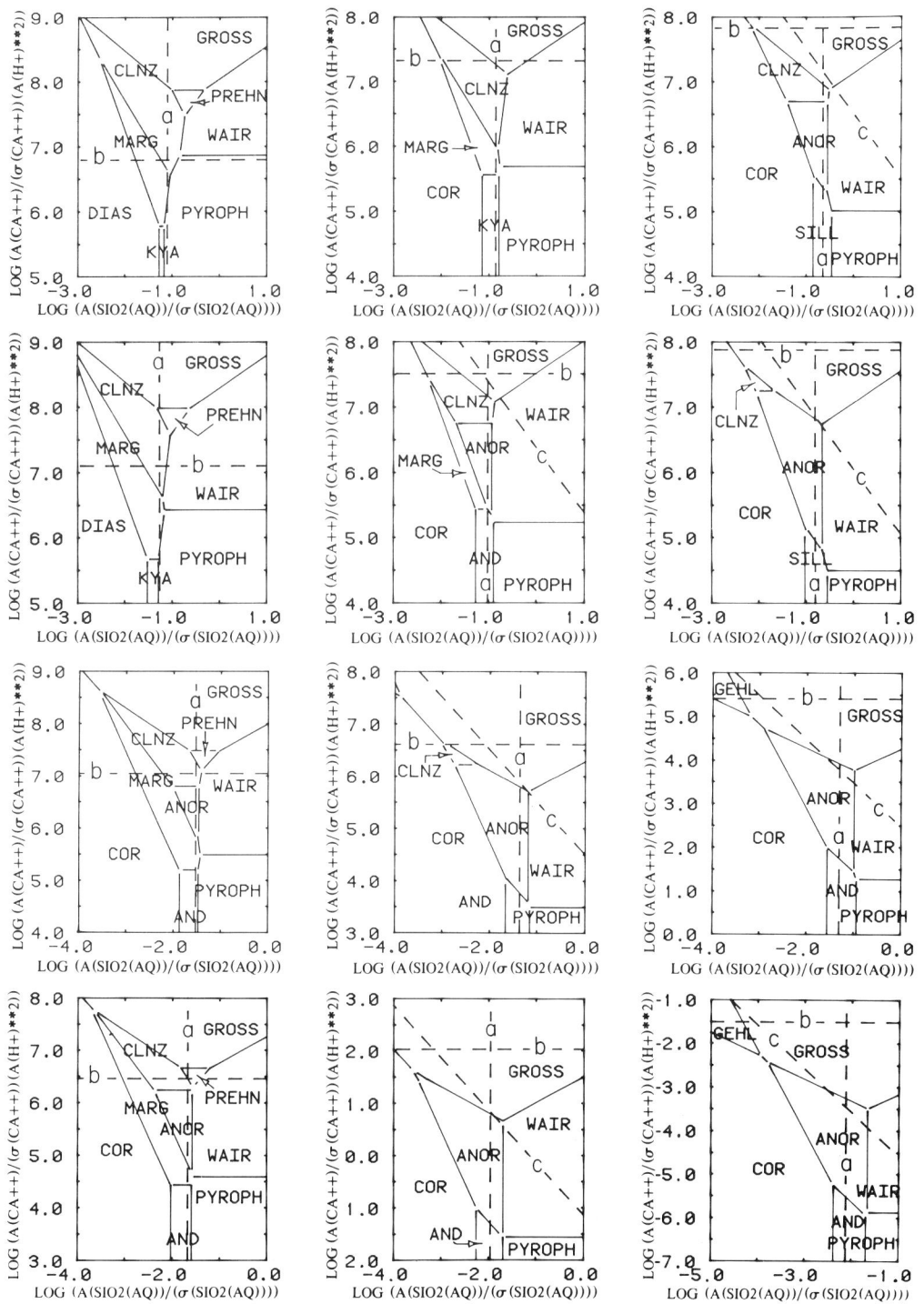

Phase relations in the system $HCl-H_2O-(Al_2O_3)-CaO-CO_2-SiO_2$ at $X_{CO_2} = 0.01$. Saturation limits: quartz (a), calcite (b), wollastonite (c).

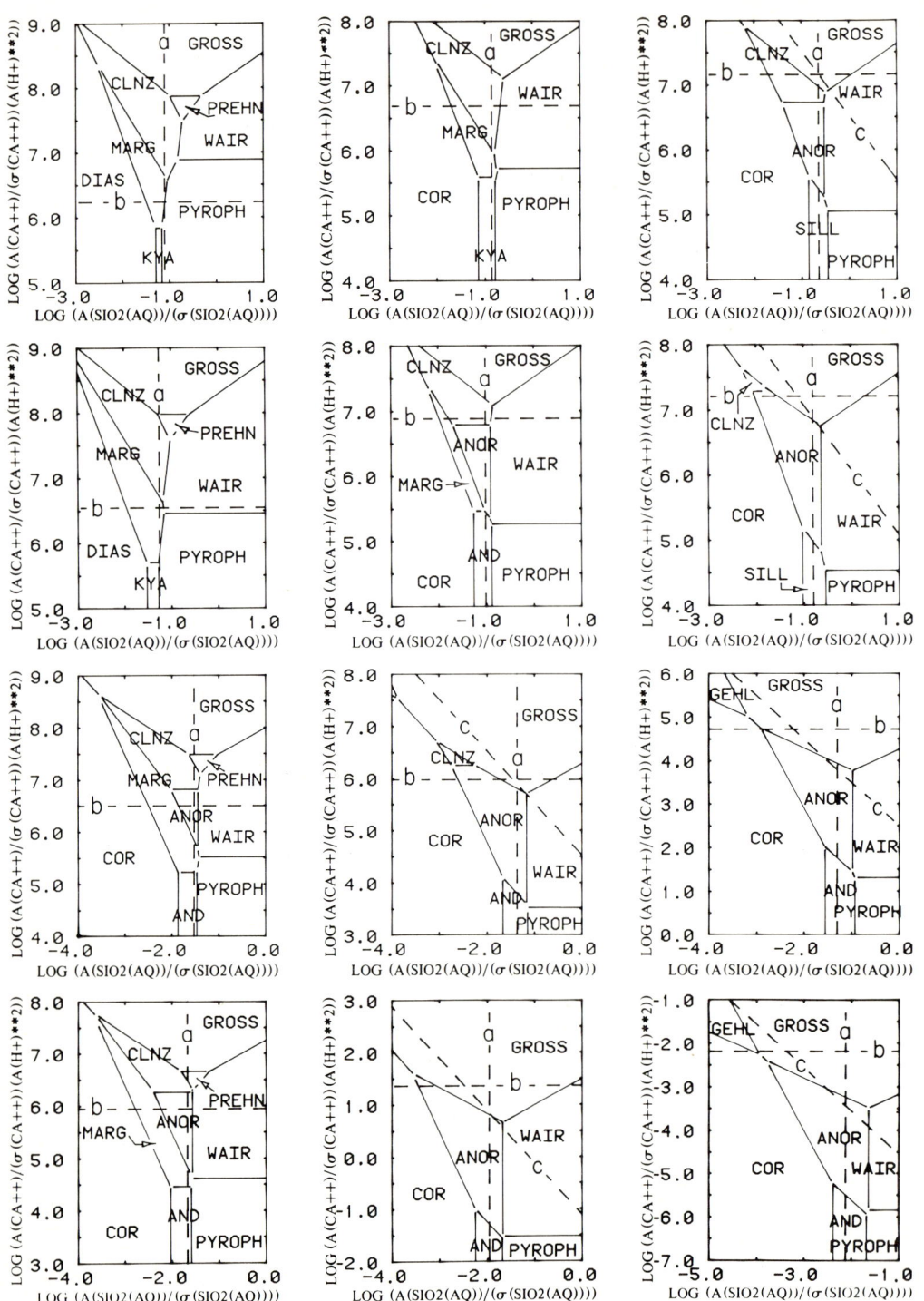

Phase relations in the system $HCl-H_2O-(Al_2O_3)-CaO-CO_2-SiO_2$ at $X_{CO_2} = 0.05$. Saturation limits: quartz (a), calcite (b), wollastonite (c).

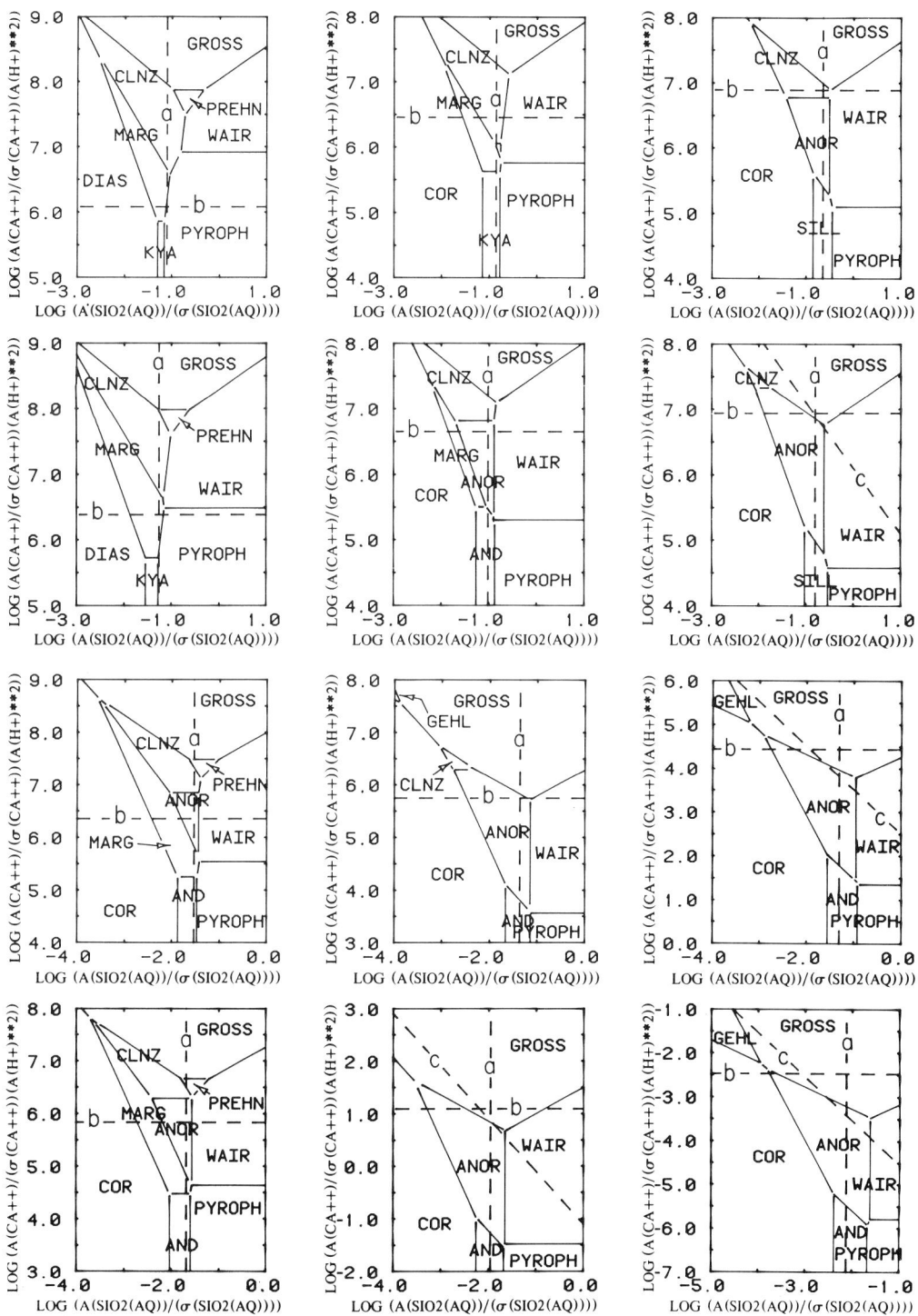

Phase relations in the system $HCl-H_2O-(Al_2O_3)-CaO-CO_2-SiO_2$ at $X_{CO_2} = 0.10$. Saturation limits: quartz (a), calcite (b), wollastonite (c).

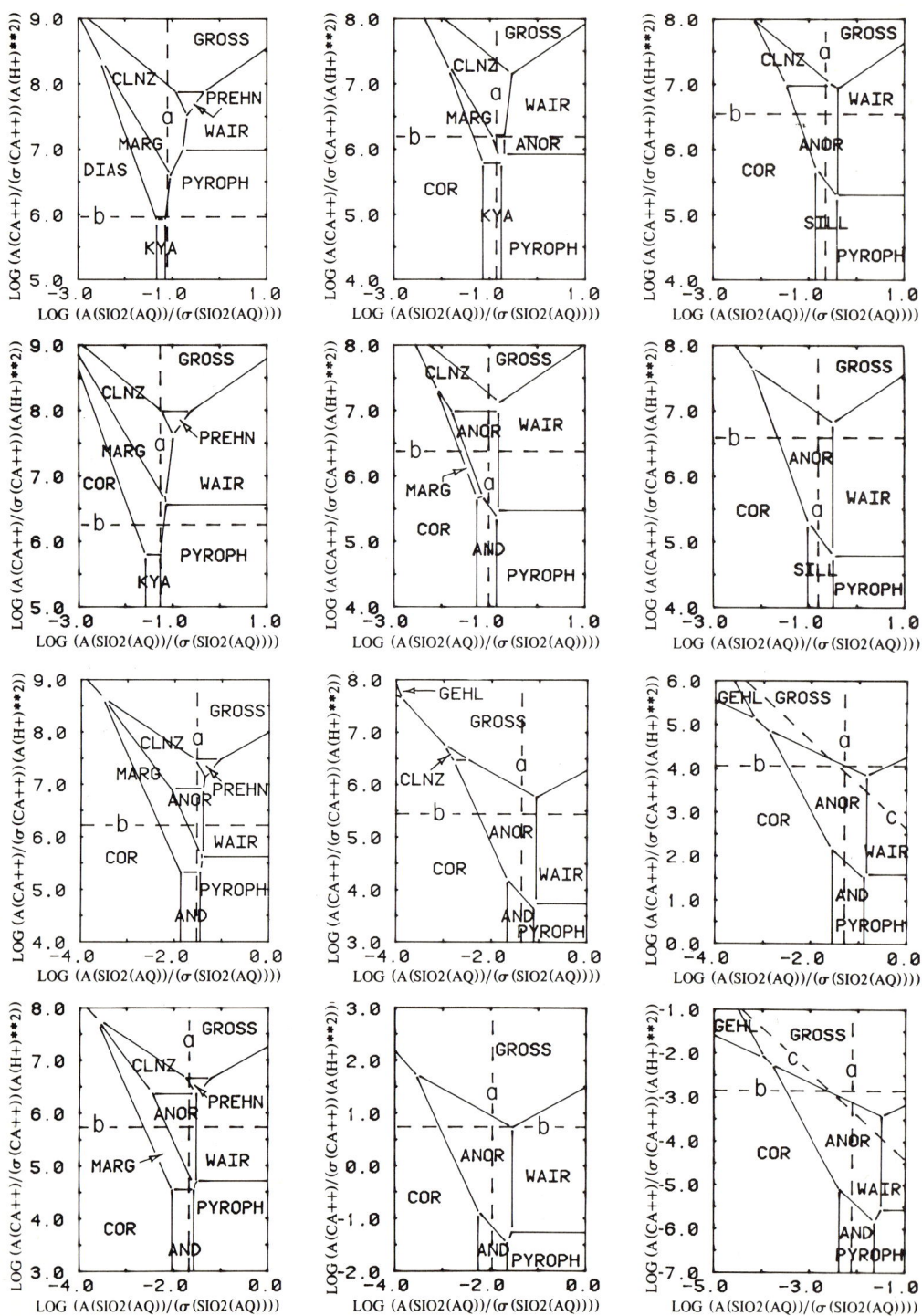

Phase relations in the system $HCl-H_2O-(Al_2O_3)-CaO-CO_2-SiO_2$ at $X_{CO_2} = 0.30$. Saturation limits: quartz (a), calcite (b), wollastonite (c).

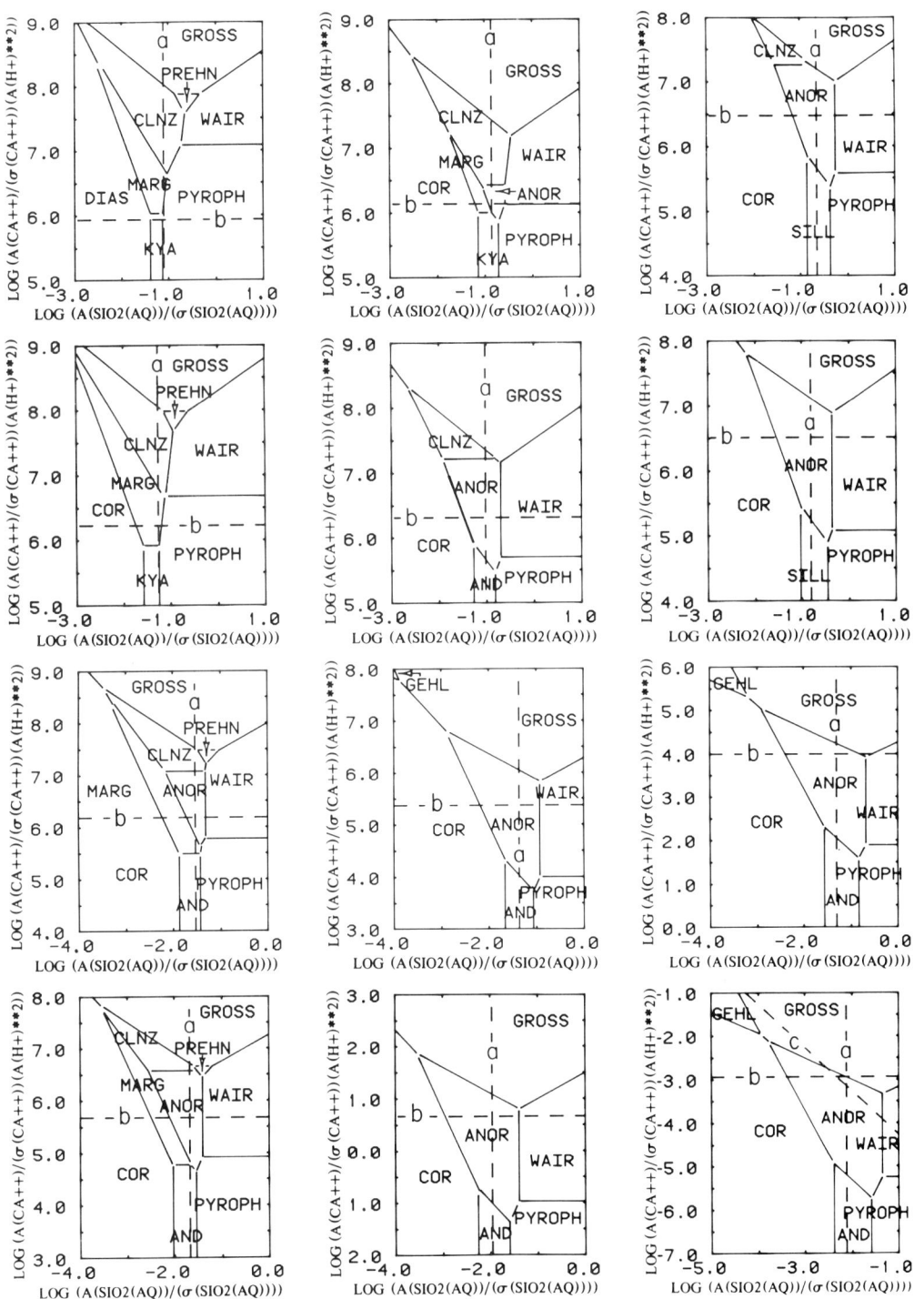

Phase relations in the system $HCl-H_2O-(Al_2O_3)-CaO-CO_2-SiO_2$ at $X_{CO_2} = 0.50$. Saturation limits: quartz (a), calcite (b), wollastonite (c).

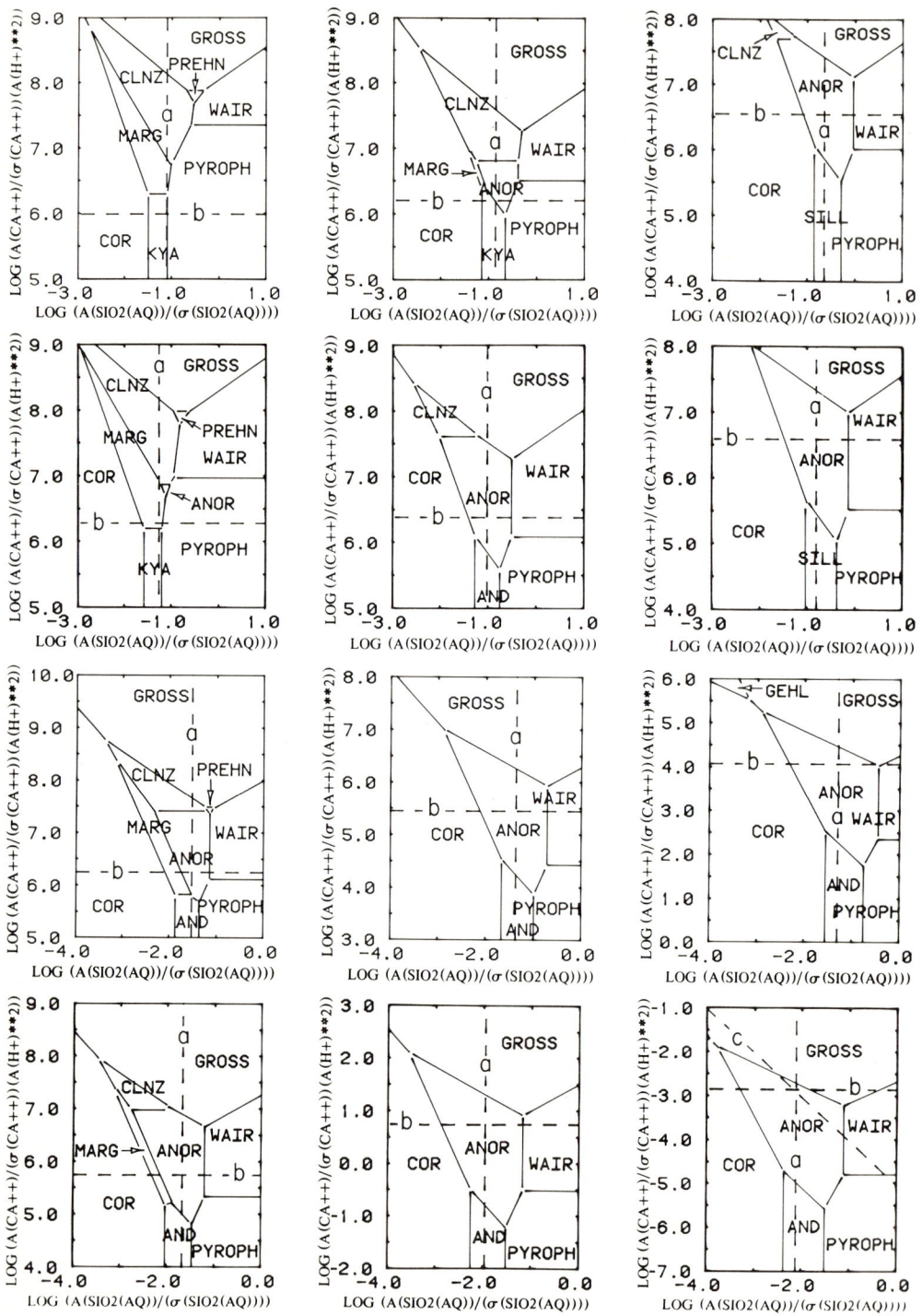

Phase relations in the system $HCl-H_2O-(Al_2O_3)-CaO-CO_2-SiO_2$ at $X_{CO_2} = 0.70$. Saturation limits: quartz (a), calcite (b), wollastonite (c).

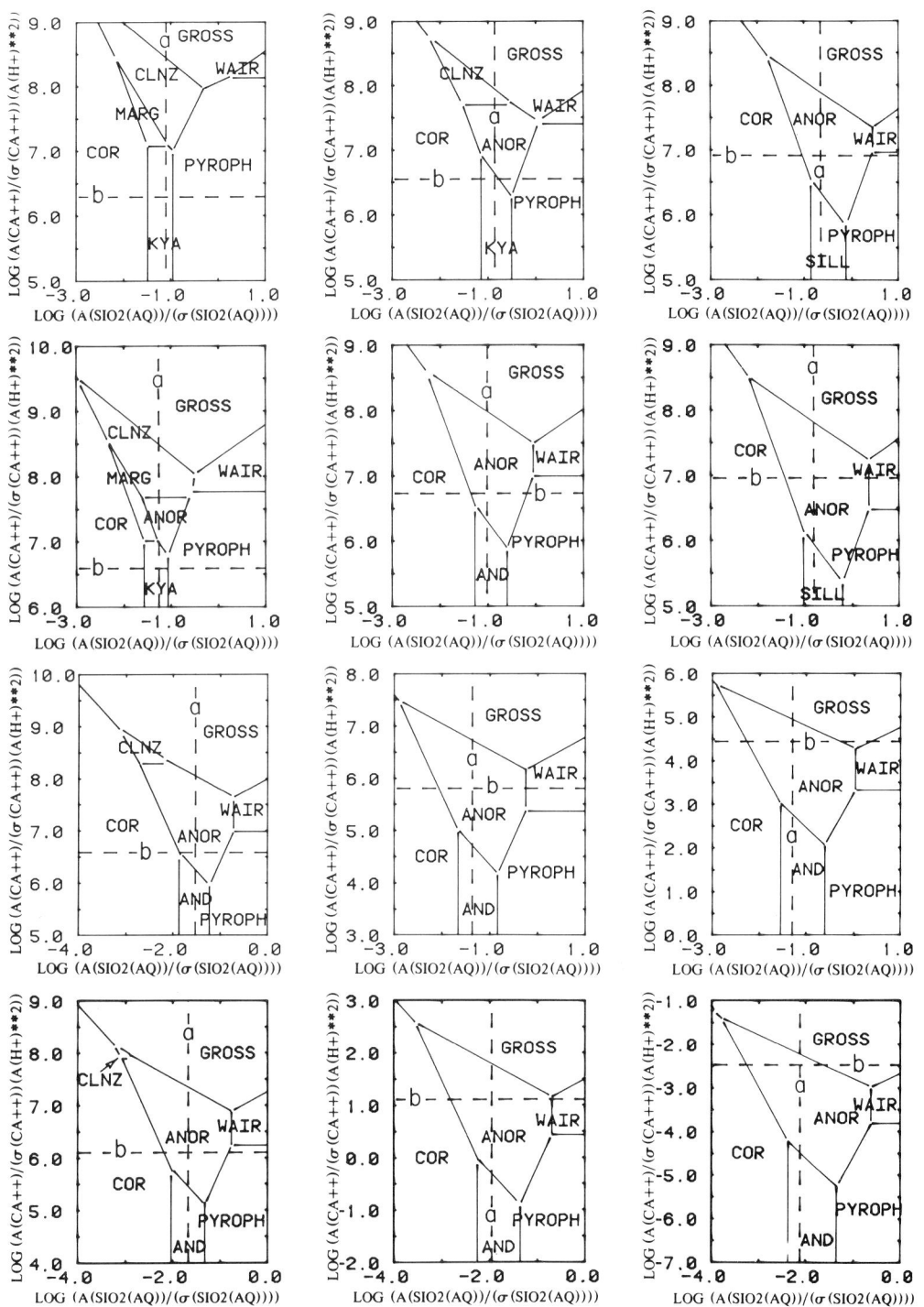

Phase relations in the system $HCl-H_2O-(Al_2O_3)-CaO-CO_2-SiO_2$ at $X_{CO_2} = 0.90$. Saturation limits: quartz (a), calcite (b).

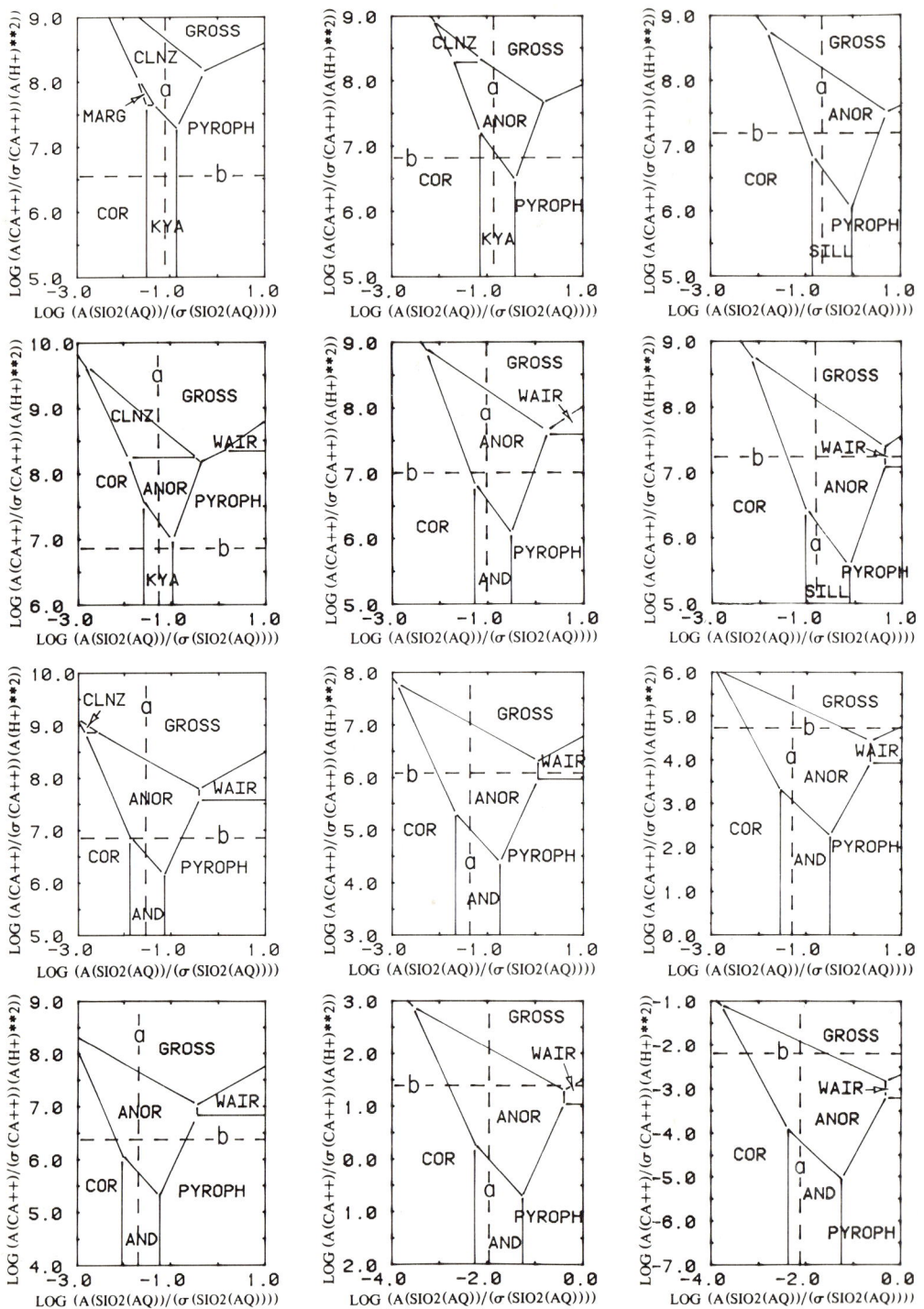

Phase relations in the system $HCl-H_2O-(Al_2O_3)-CaO-CO_2-SiO_2$ at $X_{CO_2} = 0.95$. Saturation limits: quartz (a), calcite (b).

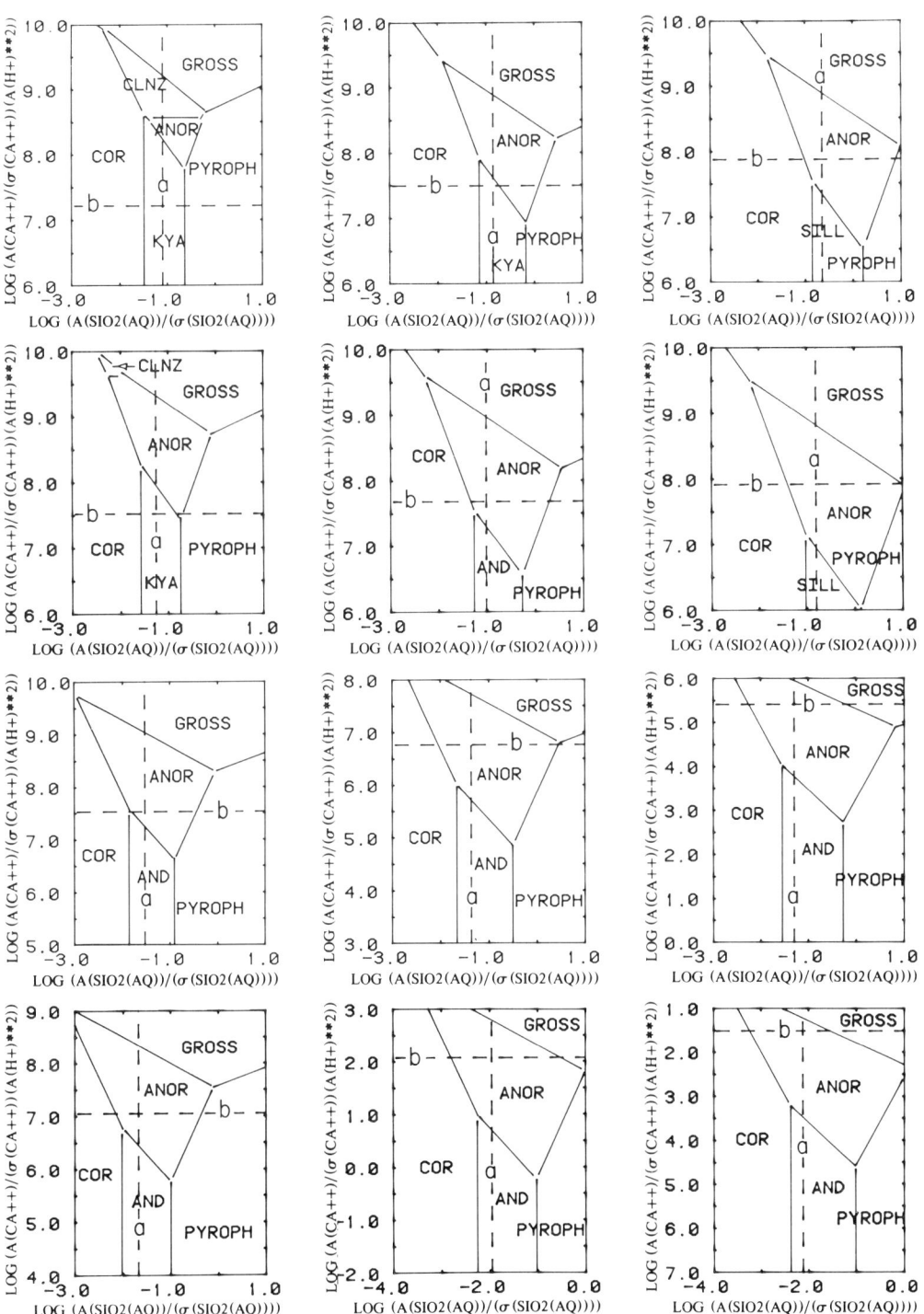

Phase relations in the system $HCl-H_2O-(Al_2O_3)-CaO-CO_2-SiO_2$ at $X_{CO_2} = 0.99$. Saturation limits: quartz (a), calcite (b).

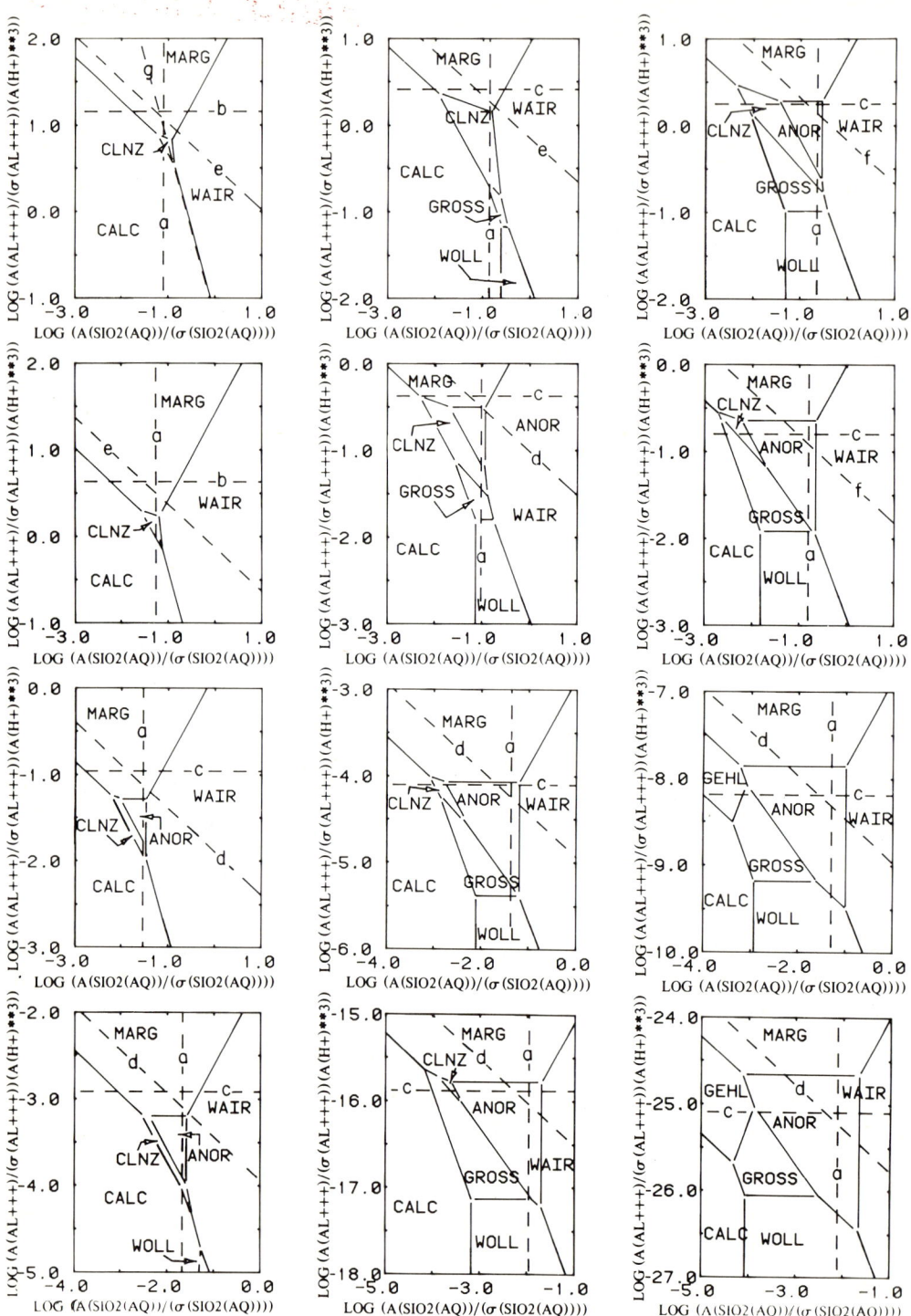

Phase relations in the system $HCl-H_2O-Al_2O_3-(CaO)-CO_2-SiO_2$ at $X_{CO_2} = 0.01$. Saturation limits: quartz (a), diaspore (b), corundum (c), andalusite (d), kyanite (e), sillimanite (f), pyrophyllite (g).

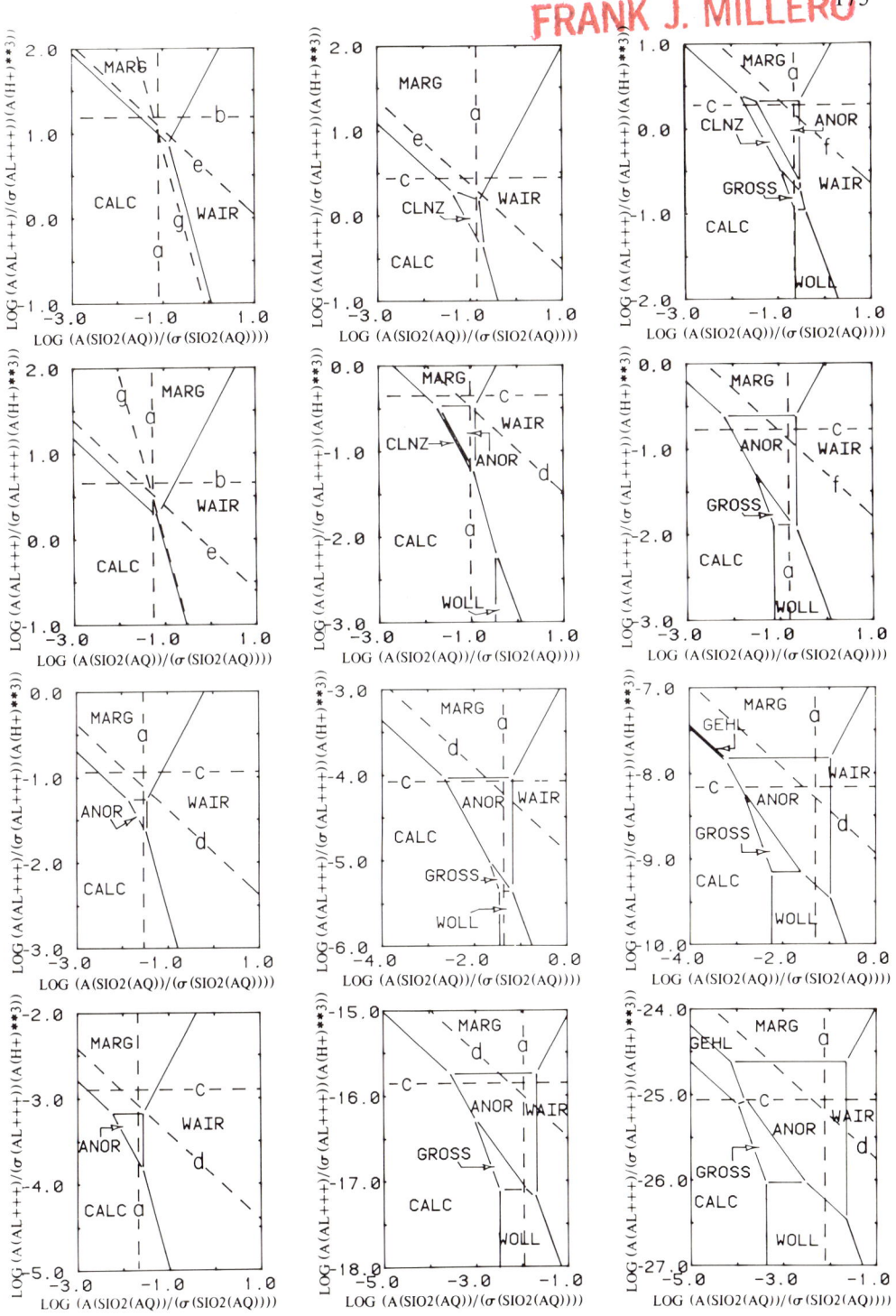

Phase relations in the system $HCl-H_2O-Al_2O_3-(CaO)-CO_2-SiO_2$ at $X_{CO_2} = 0.05$. Saturation limits: quartz (a), diaspore (b), corundum (c), andalusite (d), kyanite (e), sillimanite (f), pyrophyllite (g).

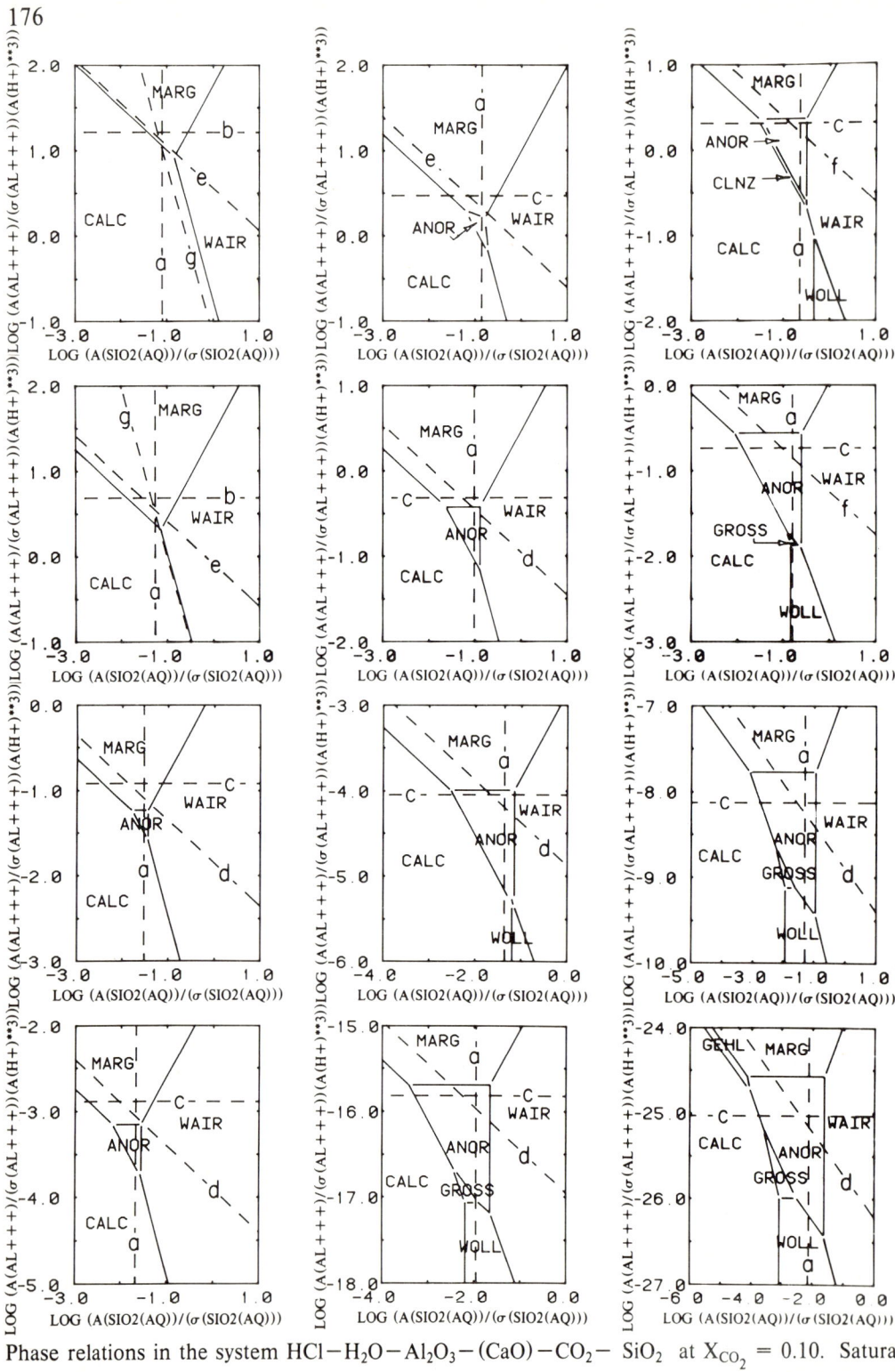

Phase relations in the system $HCl-H_2O-Al_2O_3-(CaO)-CO_2-SiO_2$ at $X_{CO_2} = 0.10$. Saturation limits: quartz (a), diaspore (b), corundum (c), andalusite (d), kyanite (e), sillimanite (f), pyrophyllite (g).

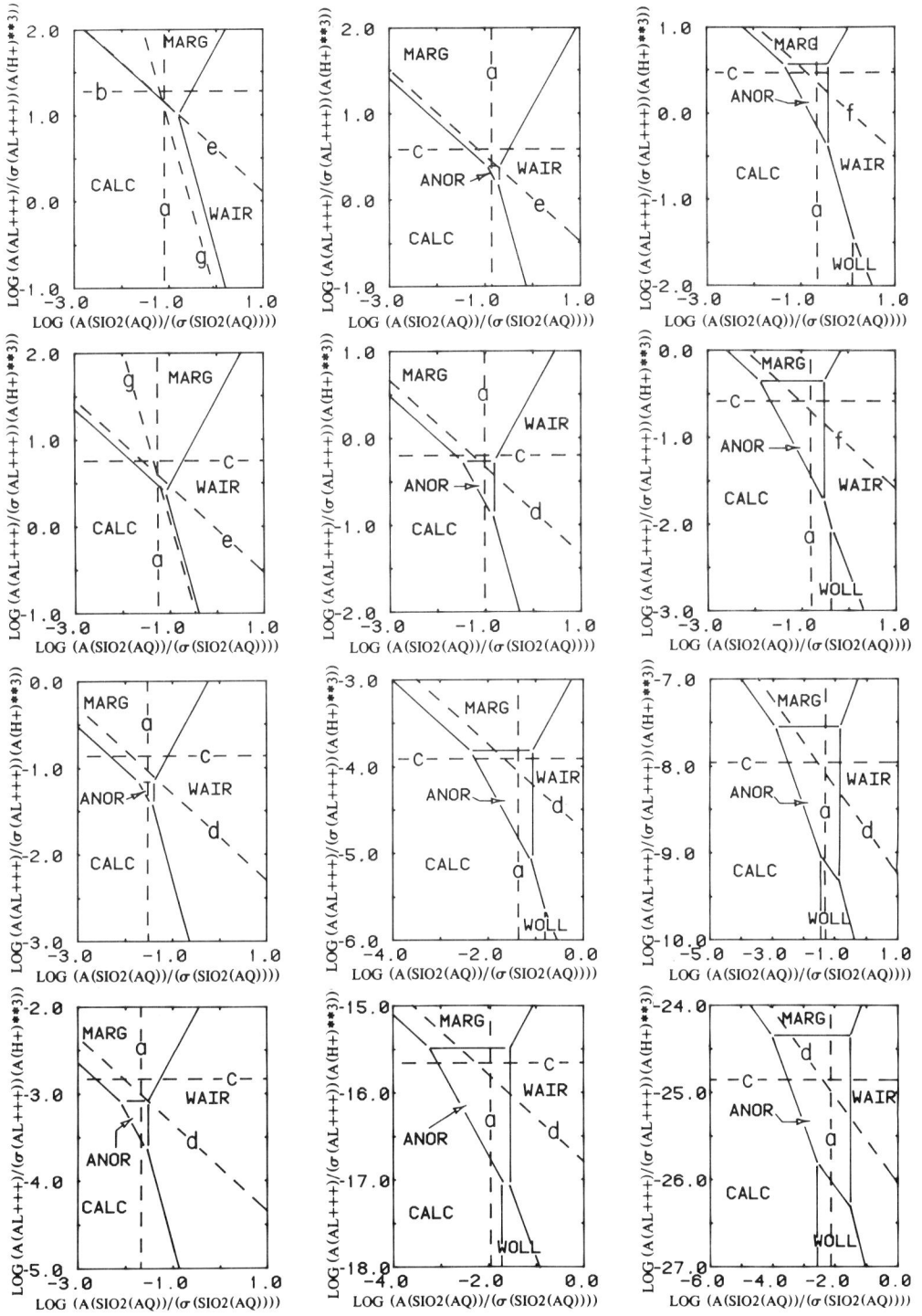

Phase relations in the system $HCl-H_2O-Al_2O_3-(CaO)-CO_2-SiO_2$ at $X_{CO_2} = 0.30$. Saturation limits: quartz (a), diaspore (b), corundum (c), andalusite (d), kyanite (e), sillimanite (f), pyrophyllite (g).

Phase relations in the system $HCl-H_2O-Al_2O_3-(CaO)-CO_2-SiO_2$ at $X_{CO_2} = 0.50$. Saturation limits: quartz (a), corundum (b), andalusite (c), kyanite (d), sillimanite (e), pyrophyllite (f), diaspore (g).

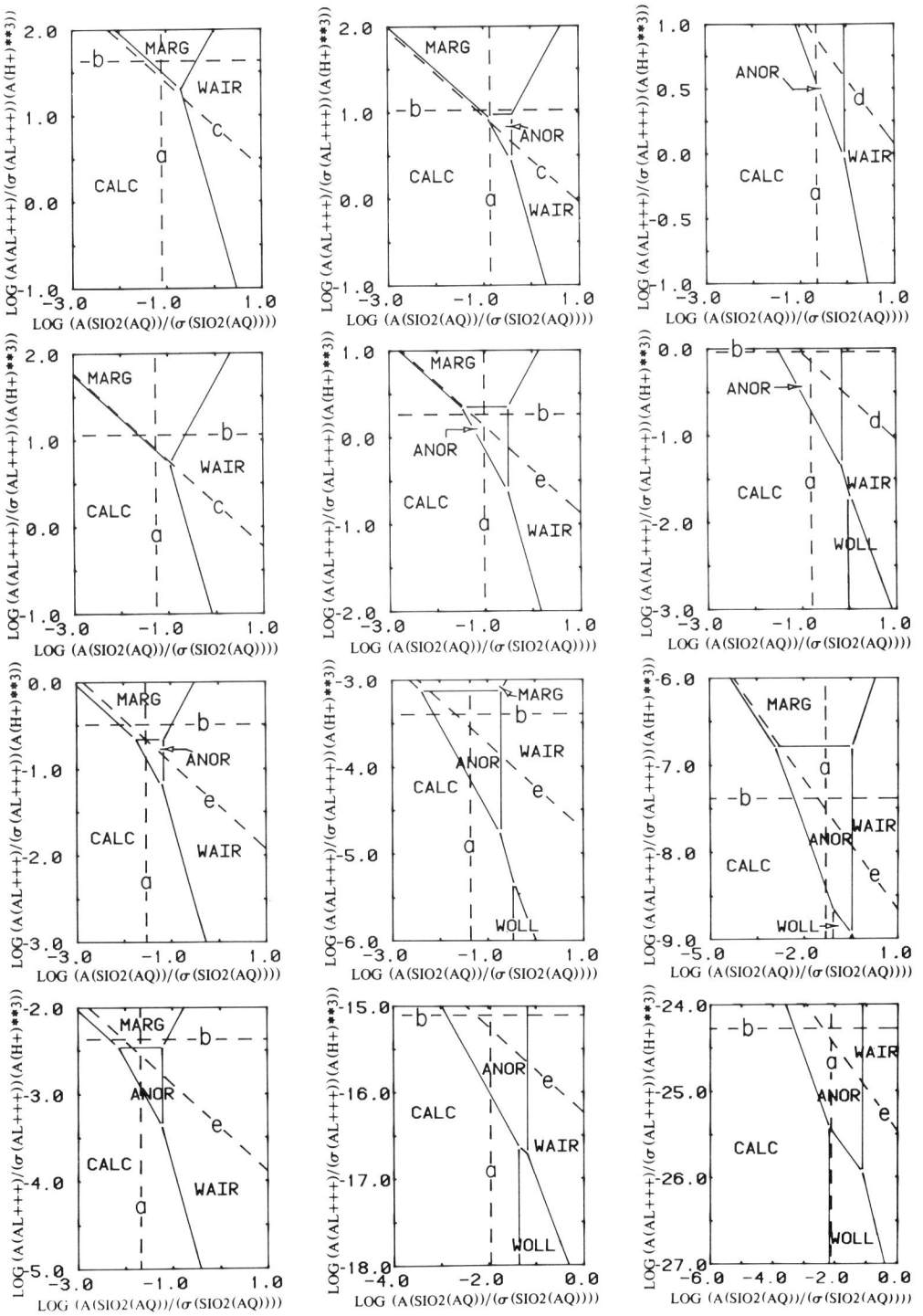

Phase relations in the system $HCl-H_2O-Al_2O_3-(CaO)-CO_2-SiO_2$ at $X_{CO_2} = 0.70$. Saturation limits: quartz (a), corundum (b), kyanite (c), sillimanite (d), andalusite (e).

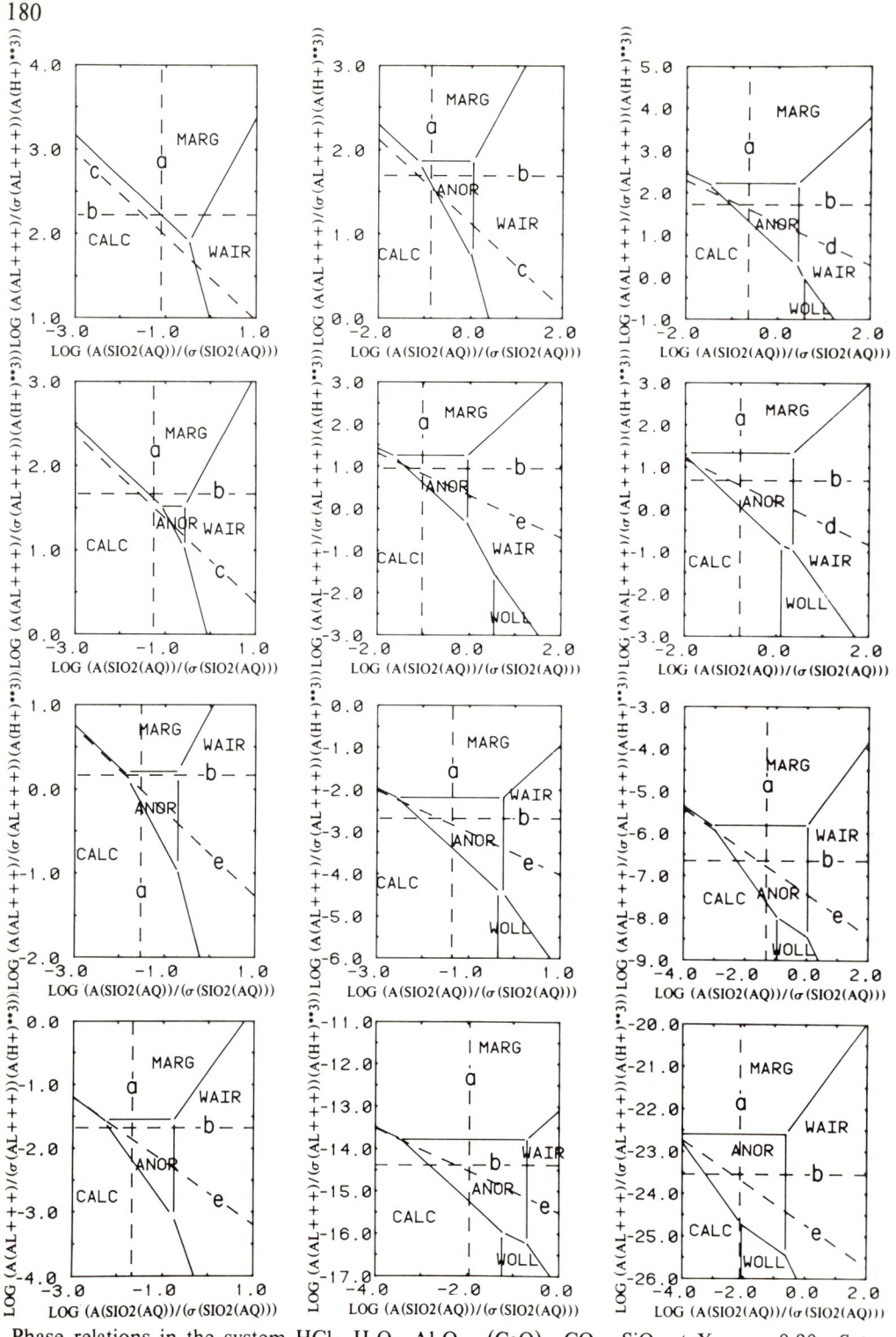

Phase relations in the system $HCl-H_2O-Al_2O_3-(CaO)-CO_2-SiO_2$ at $X_{CO_2} = 0.90$. Saturation limits: quartz (a), corundum (b), kyanite (c), sillimanite (d), andalusite (e).

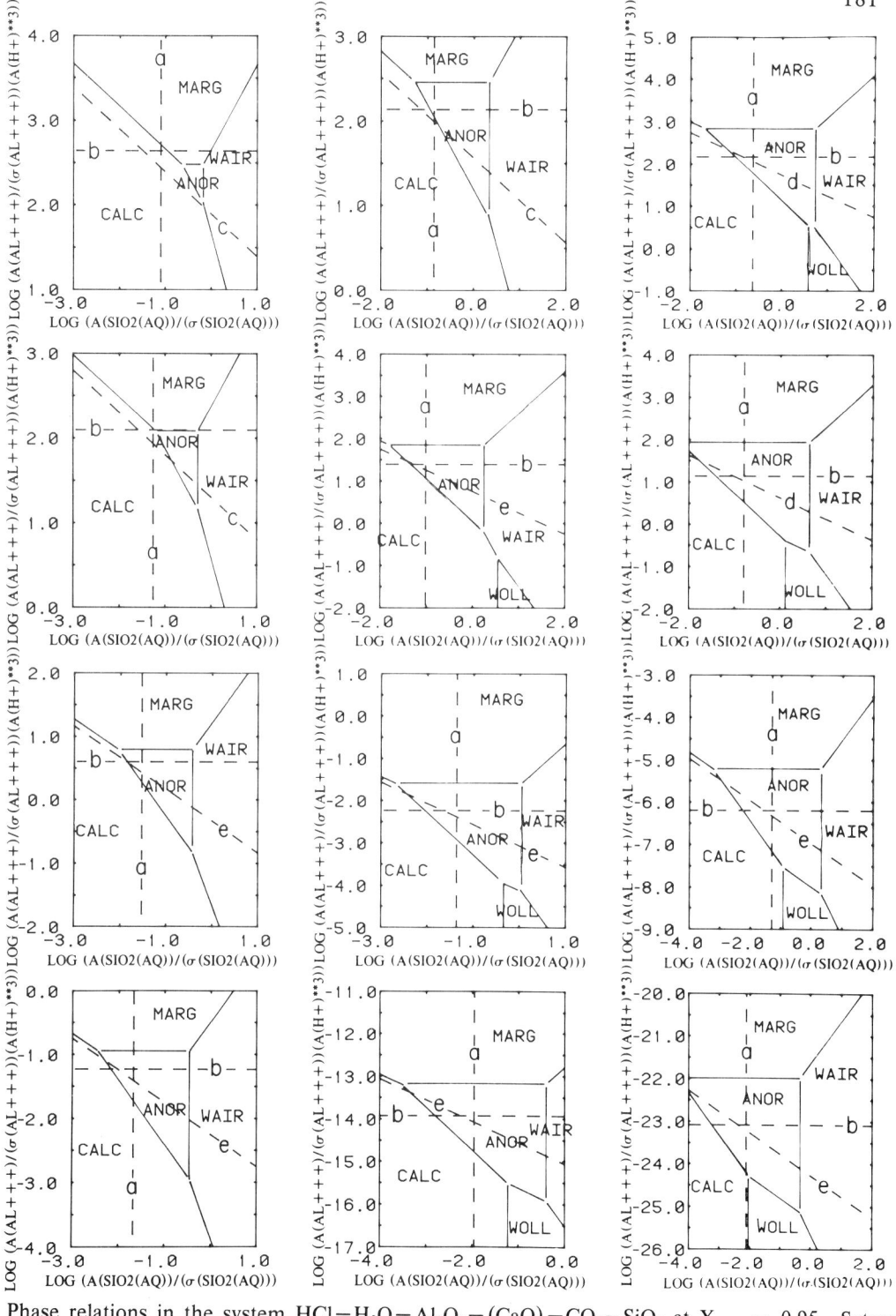

Phase relations in the system $HCl-H_2O-Al_2O_3-(CaO)-CO_2-SiO_2$ at $X_{CO_2} = 0.95$. Saturation limits: quartz (a), corundum (b), kyanite (c), sillimanite (d), andalusite (e).

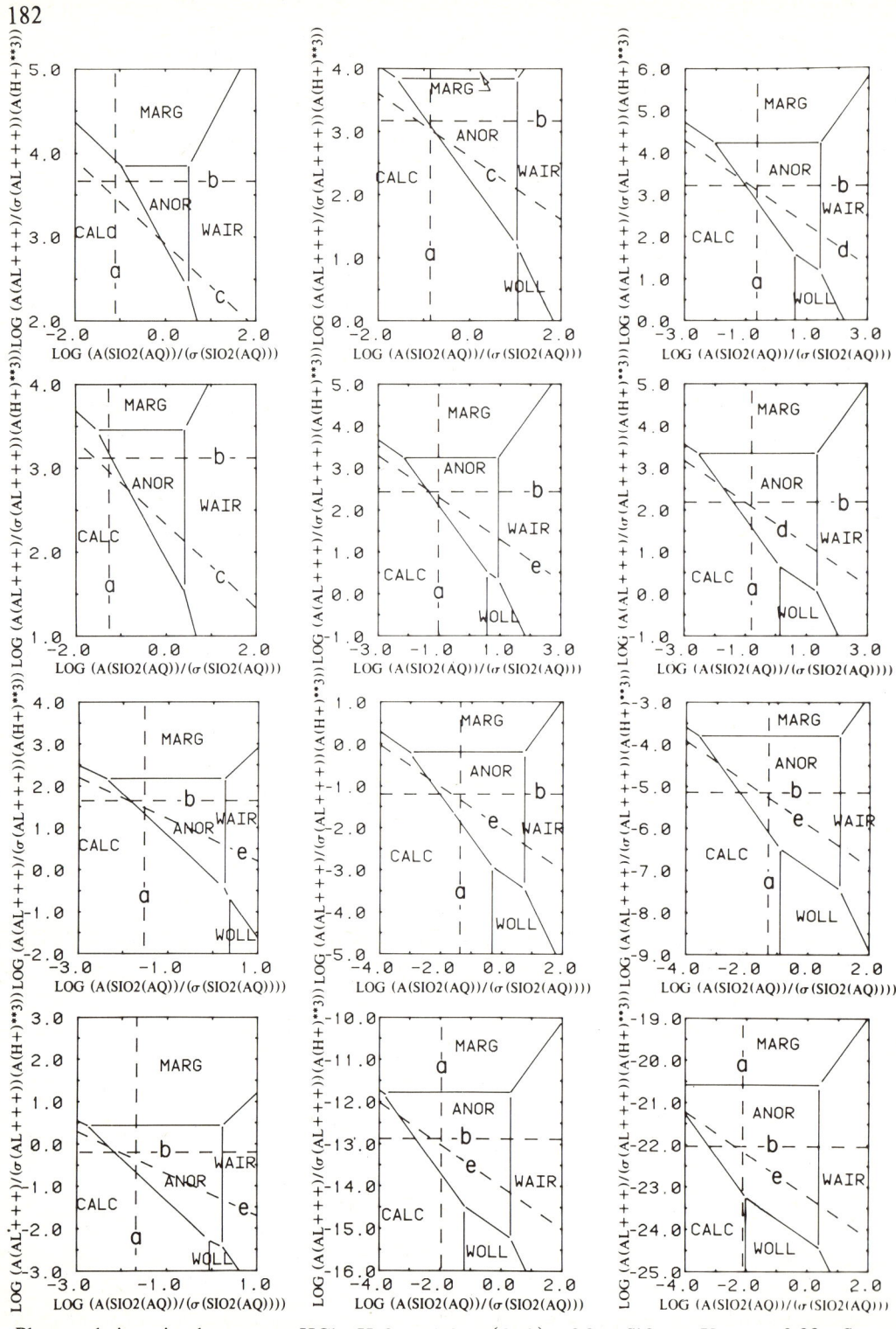

Phase relations in the system $HCl-H_2O-Al_2O_3-(CaO)-CO_2-SiO_2$ at $X_{CO_2} = 0.99$. Saturation limits: quartz (a), corundum (b), kyanite (c), sillimanite (d), andalusite (e).

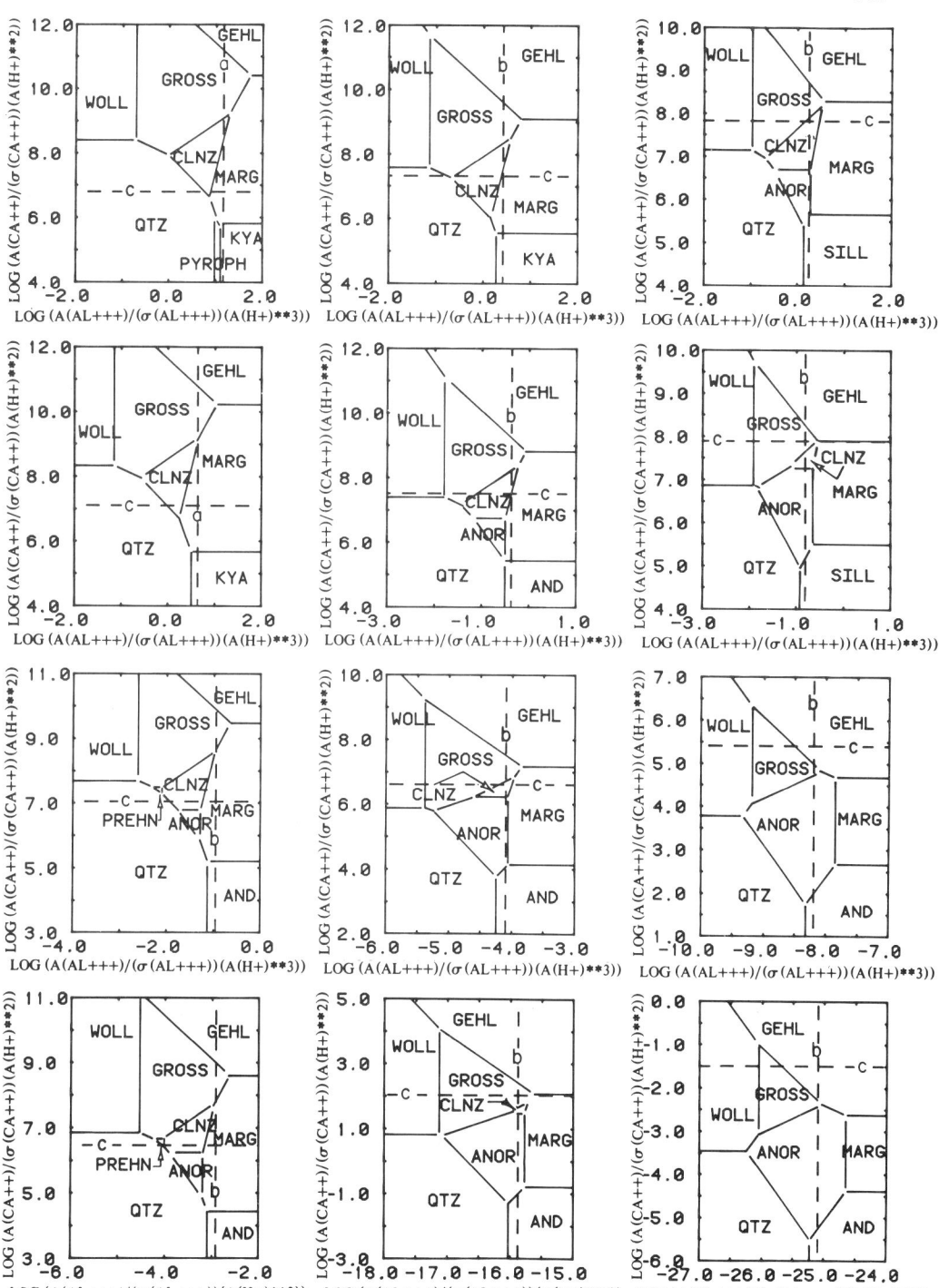

Phase relations in the system $HCl-H_2O-Al_2O_3-CaO-CO_2-(SiO_2)$ at $X_{CO_2} = 0.01$. Saturation limits: diaspore (a), corundum (b), calcite (c).

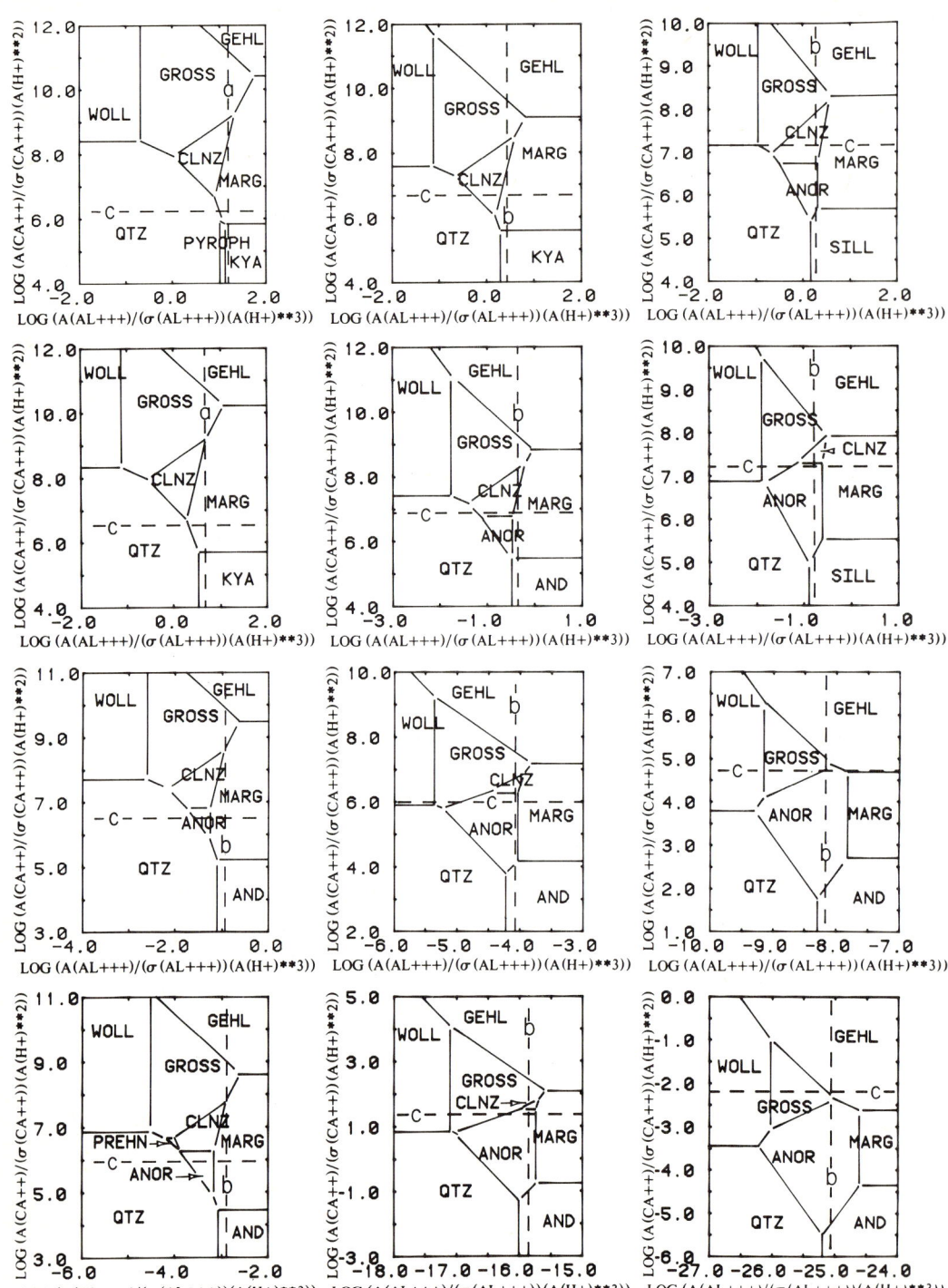

Phase relations in the system $HCl-H_2O-Al_2O_3-CaO-CO_2-(SiO_2)$ at $X_{CO_2} = 0.05$. Saturation limits: diaspore (a), corundum (b), calcite (c).

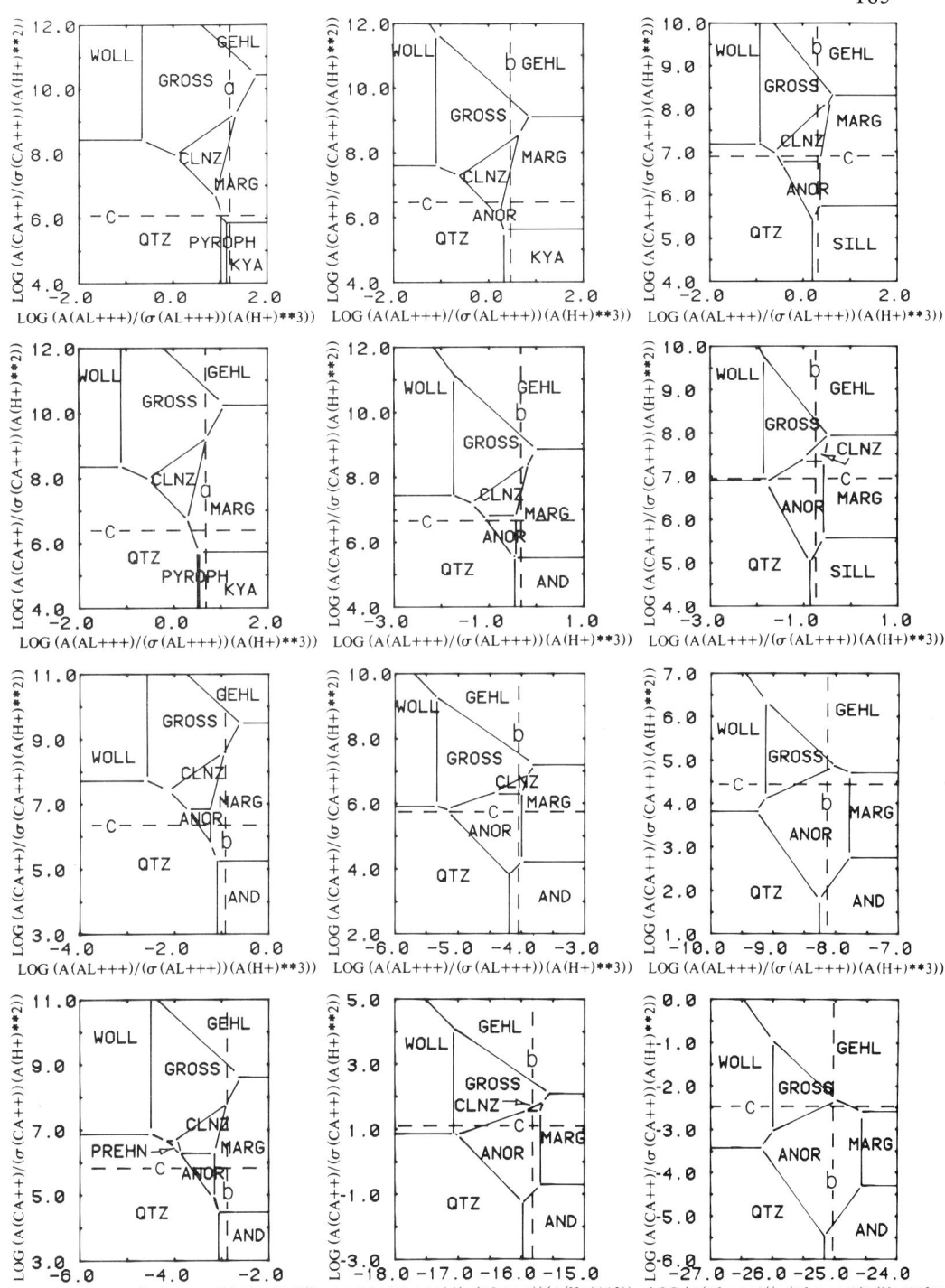

Phase relations in the system $HCl-H_2O-Al_2O_3-CaO-CO_2-(SiO_2)$ at $X_{CO_2} = 0.10$. Saturation limits: diaspore (a), corundum (b), calcite (c).

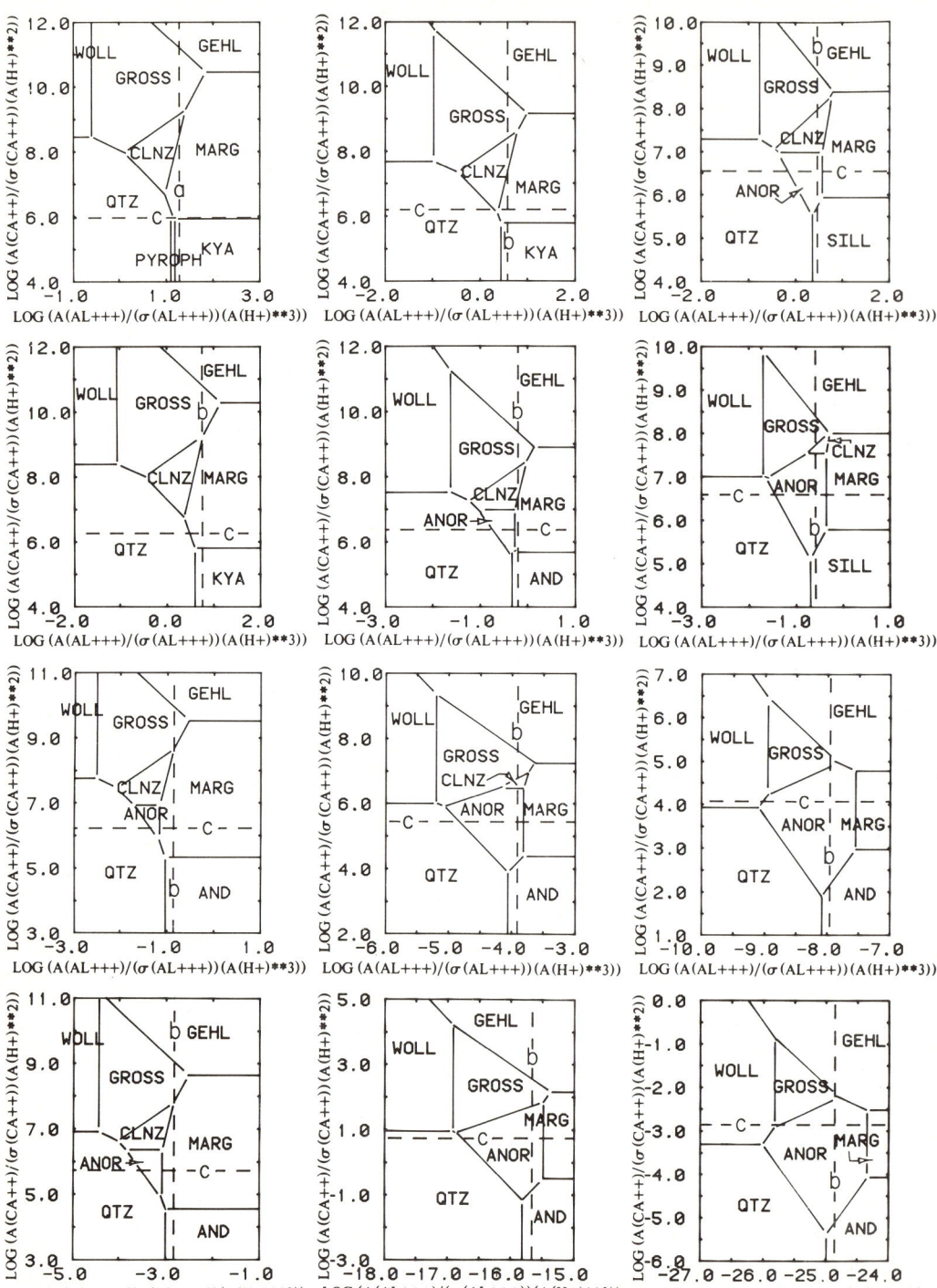

Phase relations in the system $HCl-H_2O-Al_2O_3-CaO-CO_2-(SiO_2)$ at $X_{CO_2} = 0.30$. Saturation limits: diaspore (a), corundum (b), calcite (c).

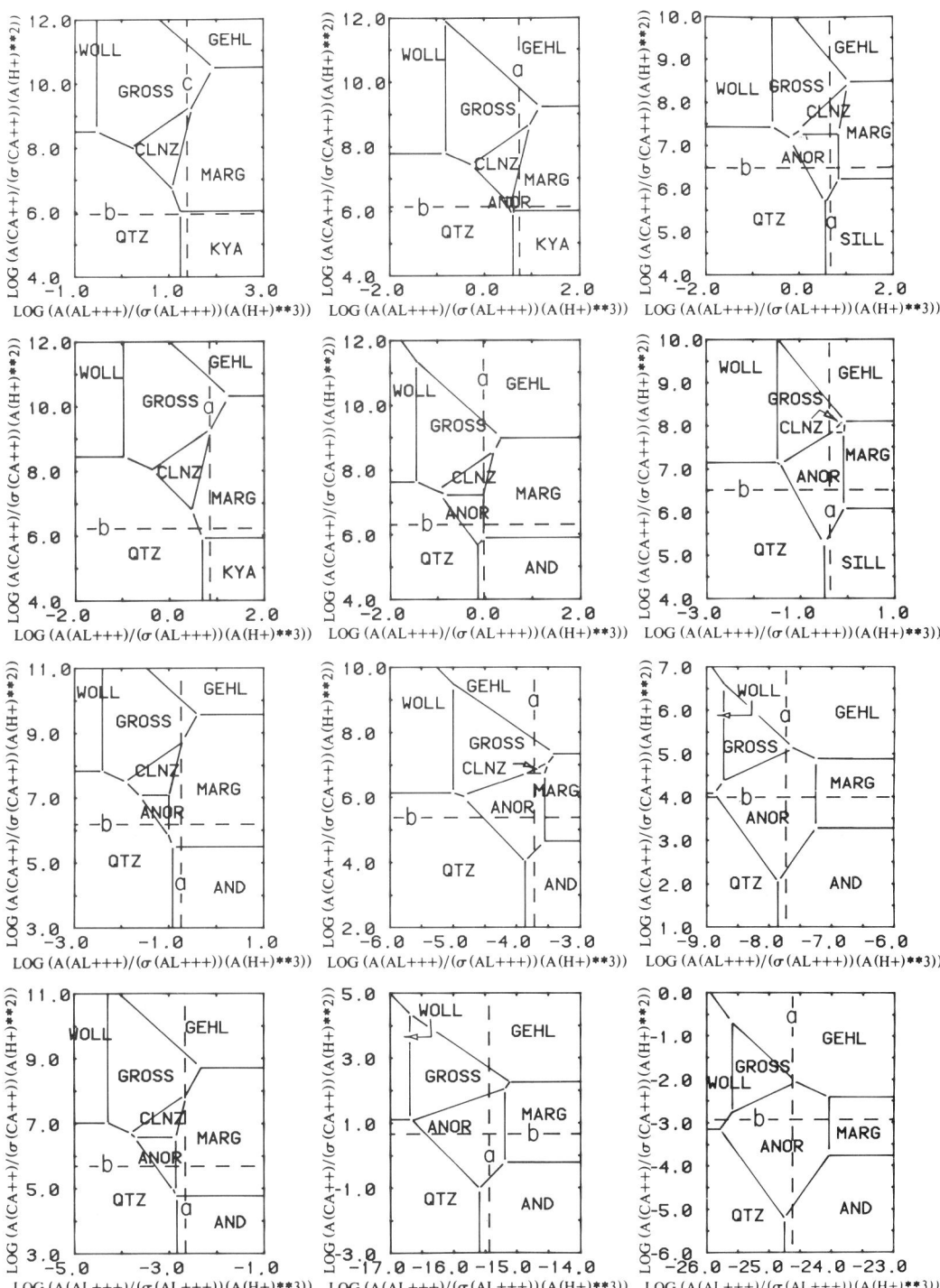

Phase relations in the system $HCl-H_2O-Al_2O_3-CaO-CO_2-(SiO_2)$ at $X_{CO_2} = 0.50$. Saturation limits: corundum (a), calcite (b), diaspore (c).

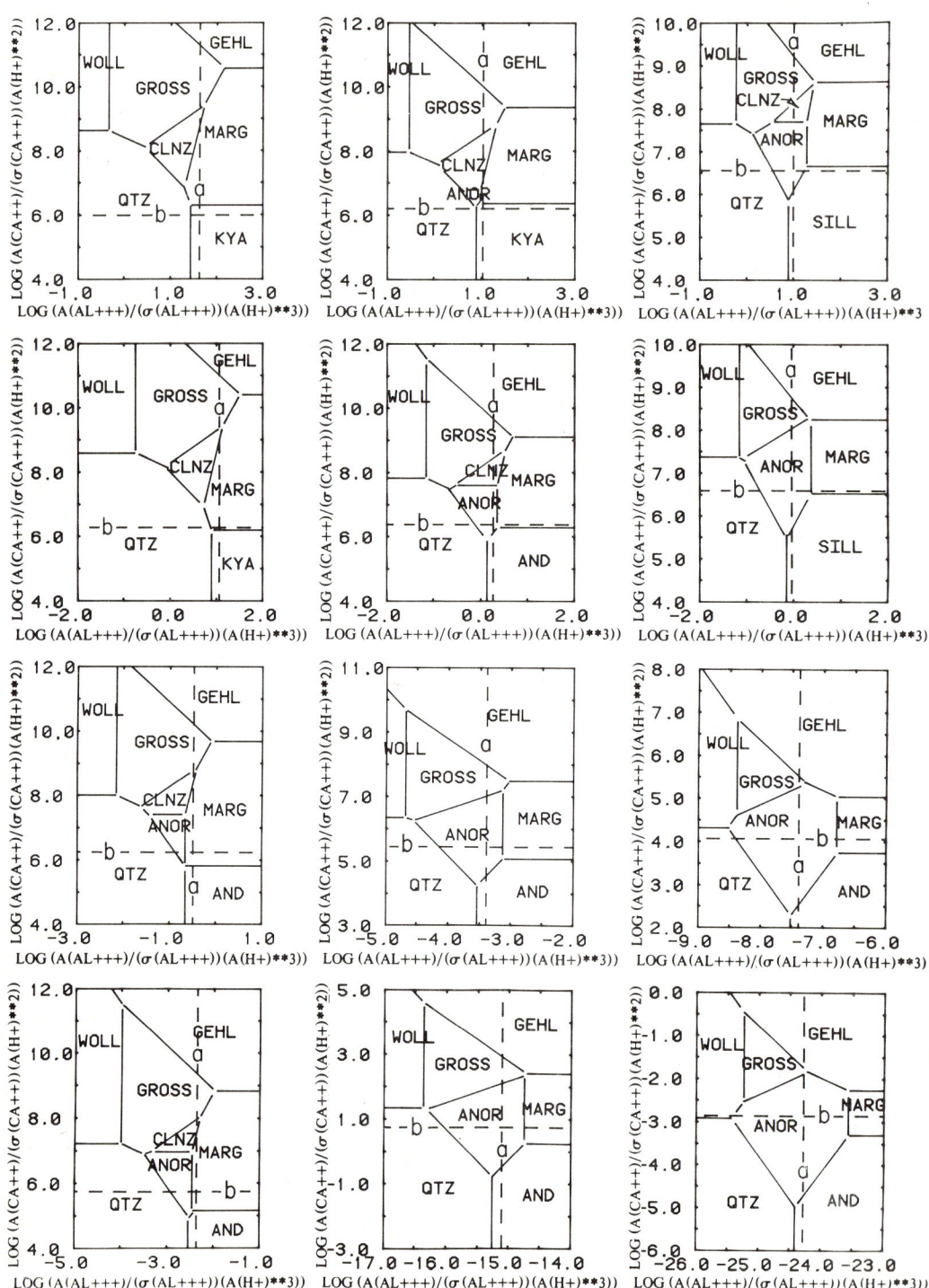

Phase relations in the system $HCl-H_2O-Al_2O_3-CaO-CO_2-(SiO_2)$ at $X_{CO_2} = 0.70$. Saturation limits: corundum (a), calcite (b).

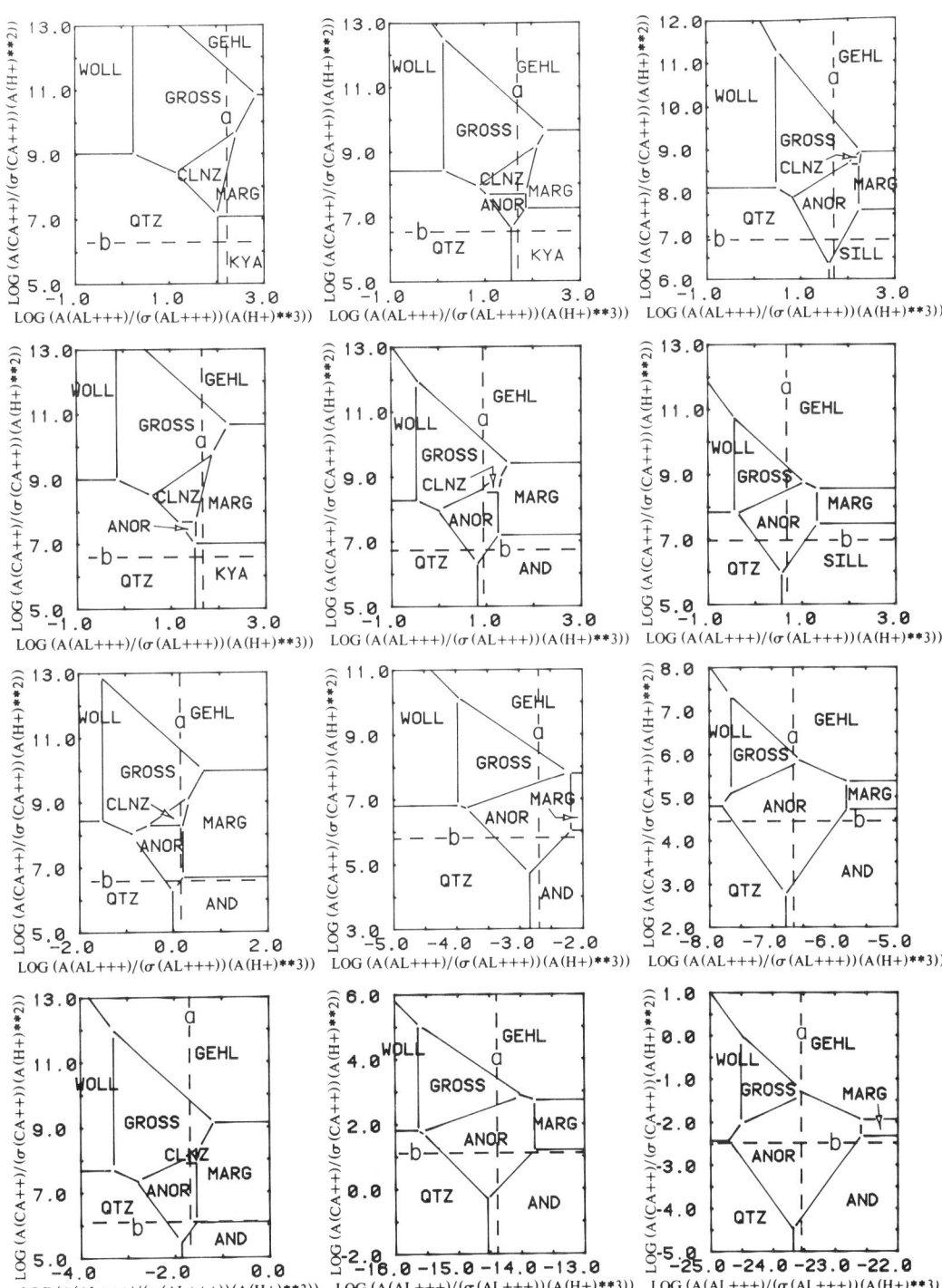

Phase relations in the system $HCl-H_2O-Al_2O_3-CaO-CO_2-(SiO_2)$ at $X_{CO_2} = 0.90$. Saturation limits: corundum (a), calcite (b).

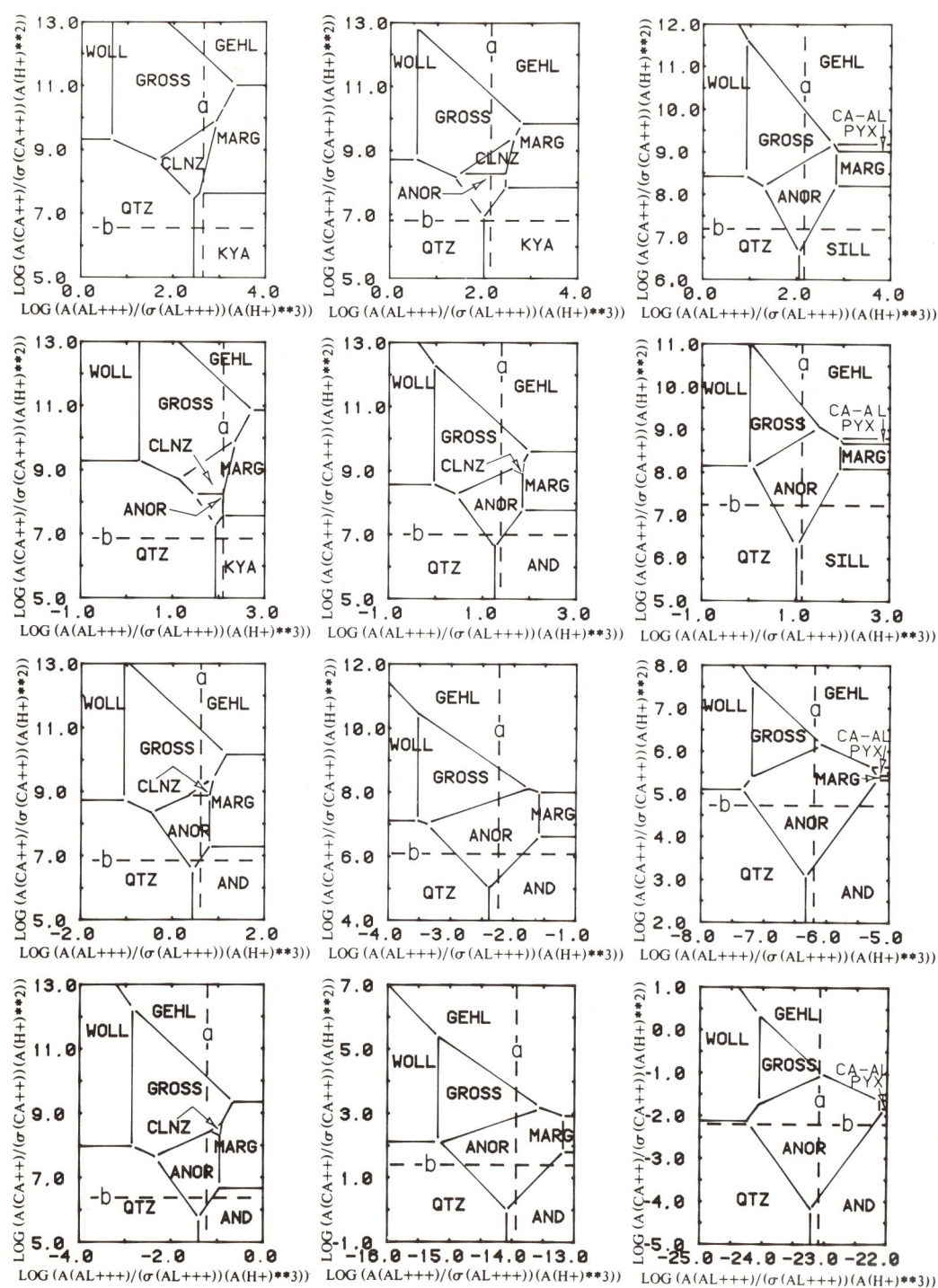

Phase relations in the system $HCl-H_2O-Al_2O_3-CaO-CO_2-(SiO_2)$ at $X_{CO_2} = 0.95$. Saturation limits: corundum (a), calcite (b).

191

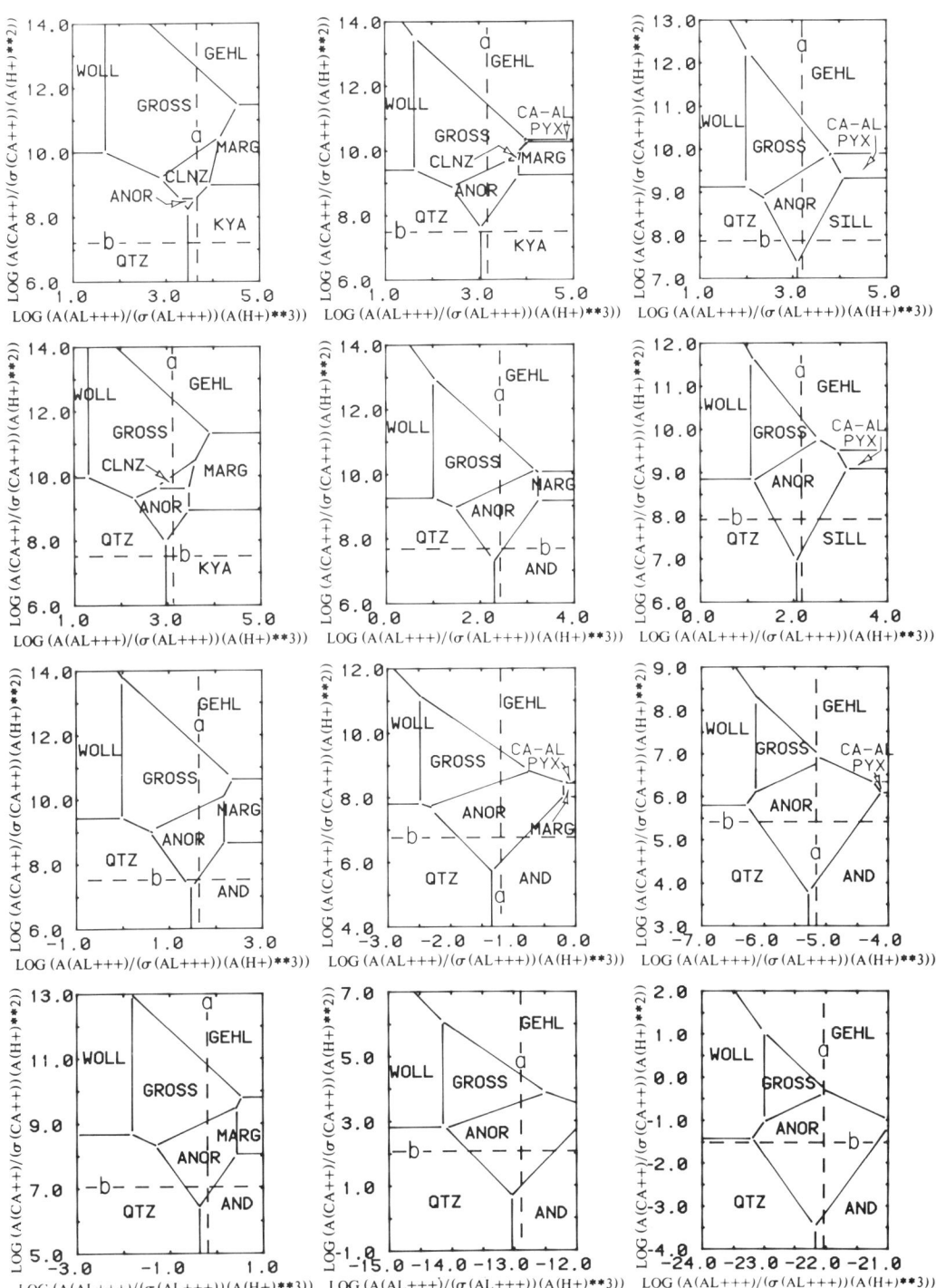

Phase relations in the system $HCl-H_2O-Al_2O_3-CaO-CO_2-(SiO_2)$ at $X_{CO_2} = 0.99$. Saturation limits: corundum (a), calcite (b).

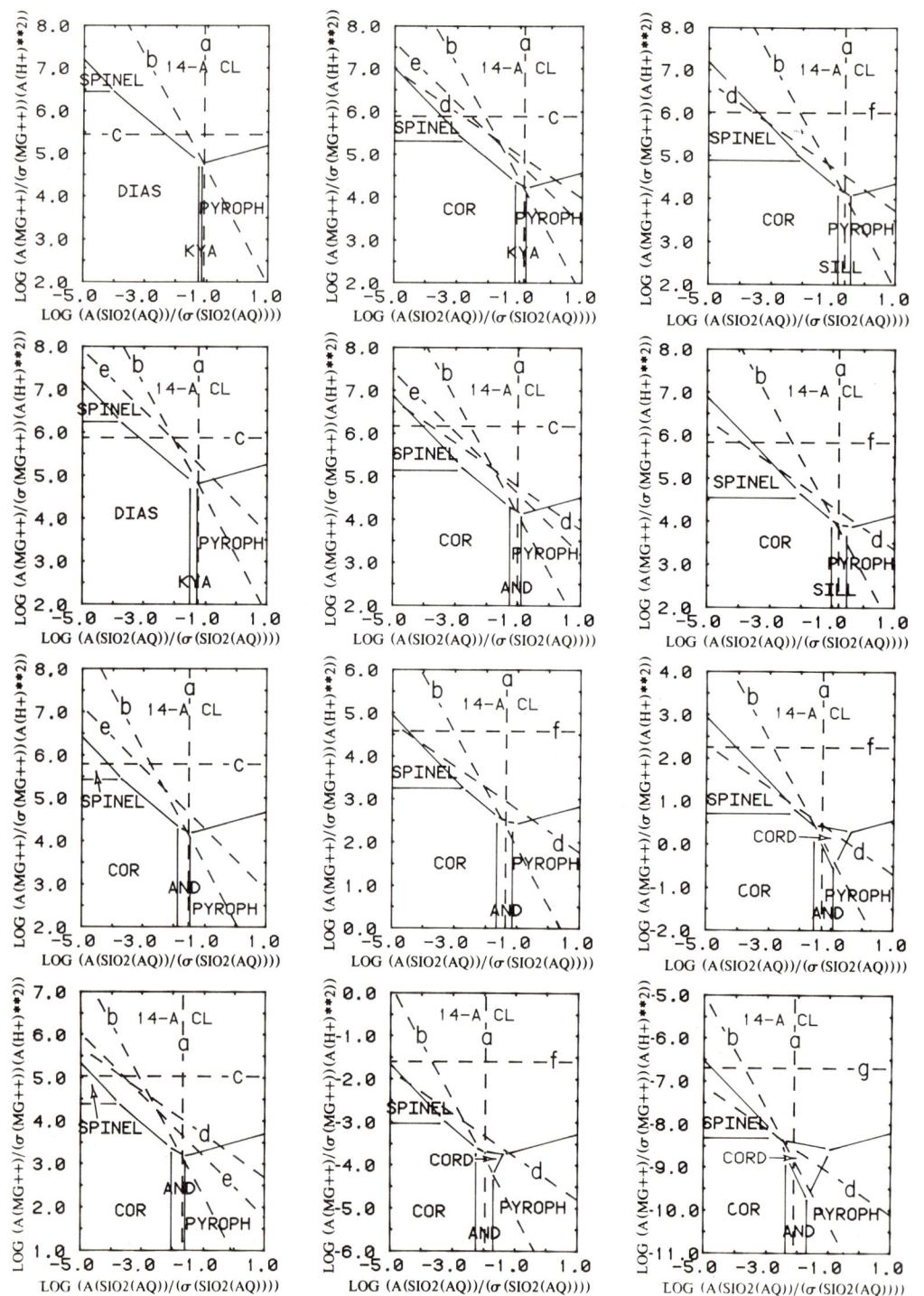

Phase relations in the system $HCl-H_2O-(Al_2O_3)-CO_2-MgO-SiO_2$ at $X_{CO_2} = 0.01$. Saturation limits: quartz (a), talc (b), magnesite (c), forsterite (d), antigorite (e), brucite (f), periclase (g).

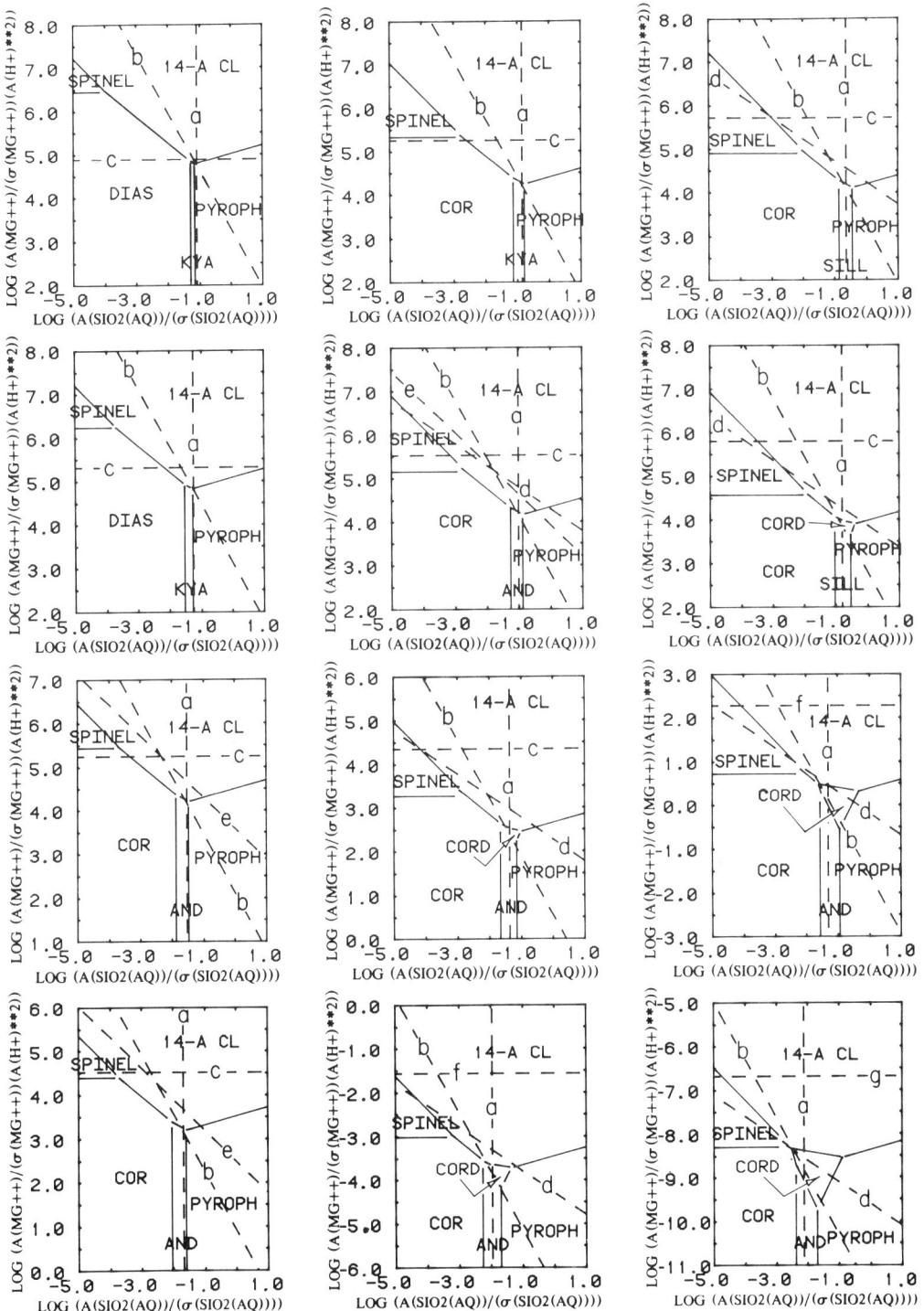

Phase relations in the system $HCl-H_2O-(Al_2O_3)-CO_2-MgO-SiO_2$ at $X_{CO_2} = 0.05$. Saturation limits: quartz (a), talc (b), magnesite (c), forsterite (d), antigorite (e), brucite (f), periclase (g).

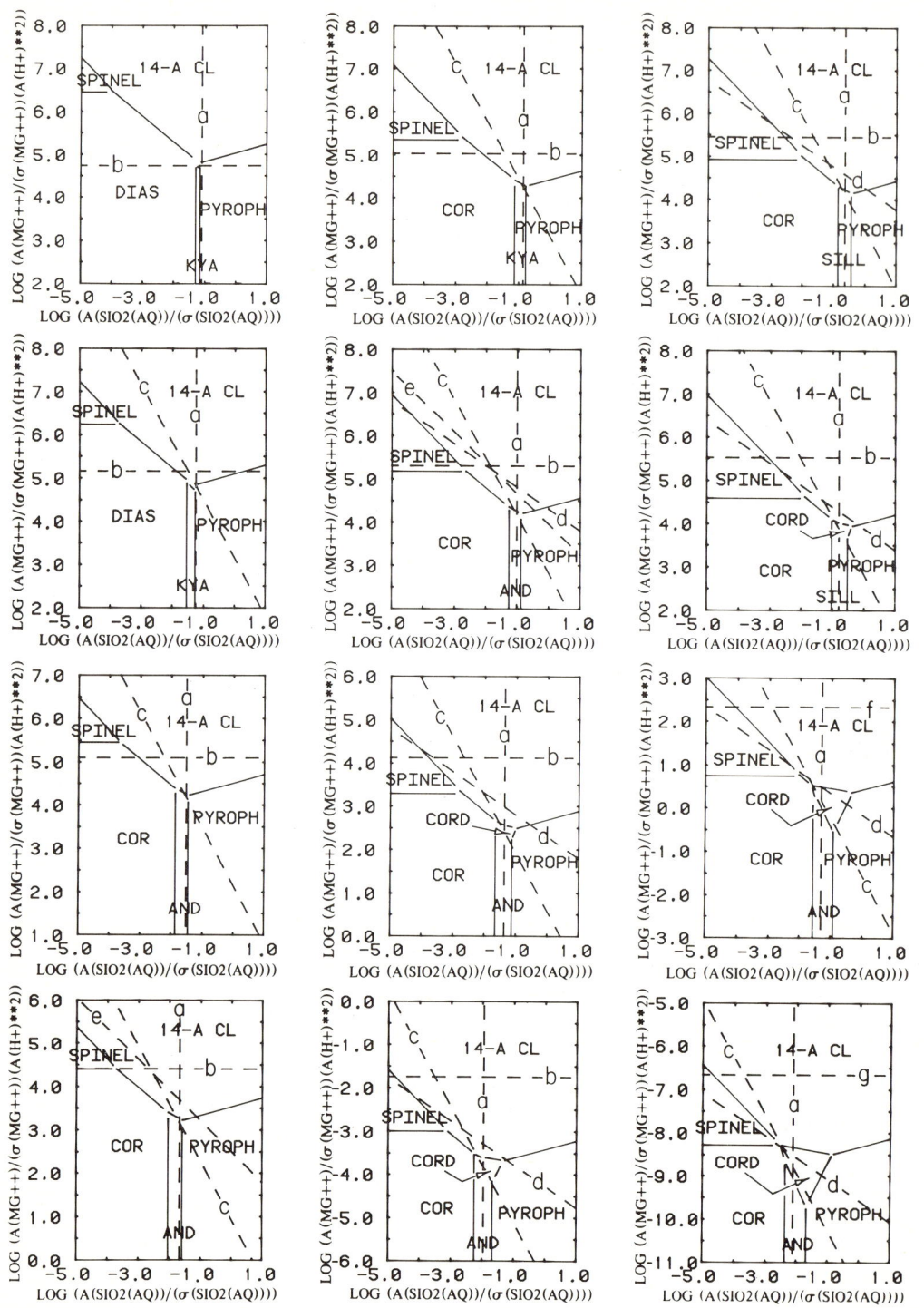

Phase relations in the system $HCl-H_2O-(Al_2O_3)-CO_2-MgO-SiO_2$ at $X_{CO_2} = 0.10$. Saturation limits: quartz (a), magnesite (b), talc (c), forsterite (d), antigorite (e), brucite (f), periclase (g).

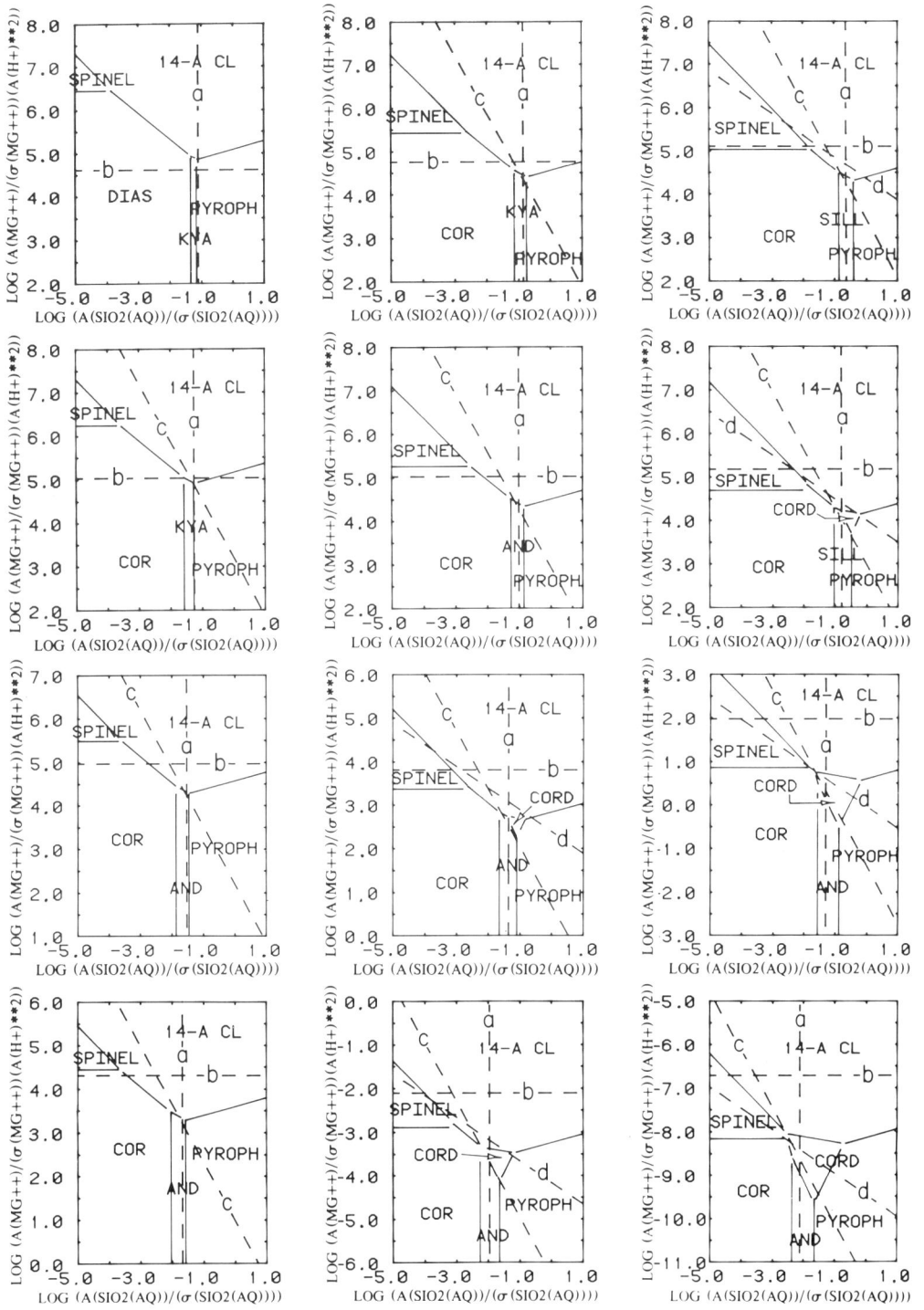

Phase relations in the system $HCl-H_2O-(Al_2O_3)-CO_2-MgO-SiO_2$ at $X_{CO_2} = 0.30$. Saturation limits: quartz (a), magnesite (b), talc (c), forsterite (d).

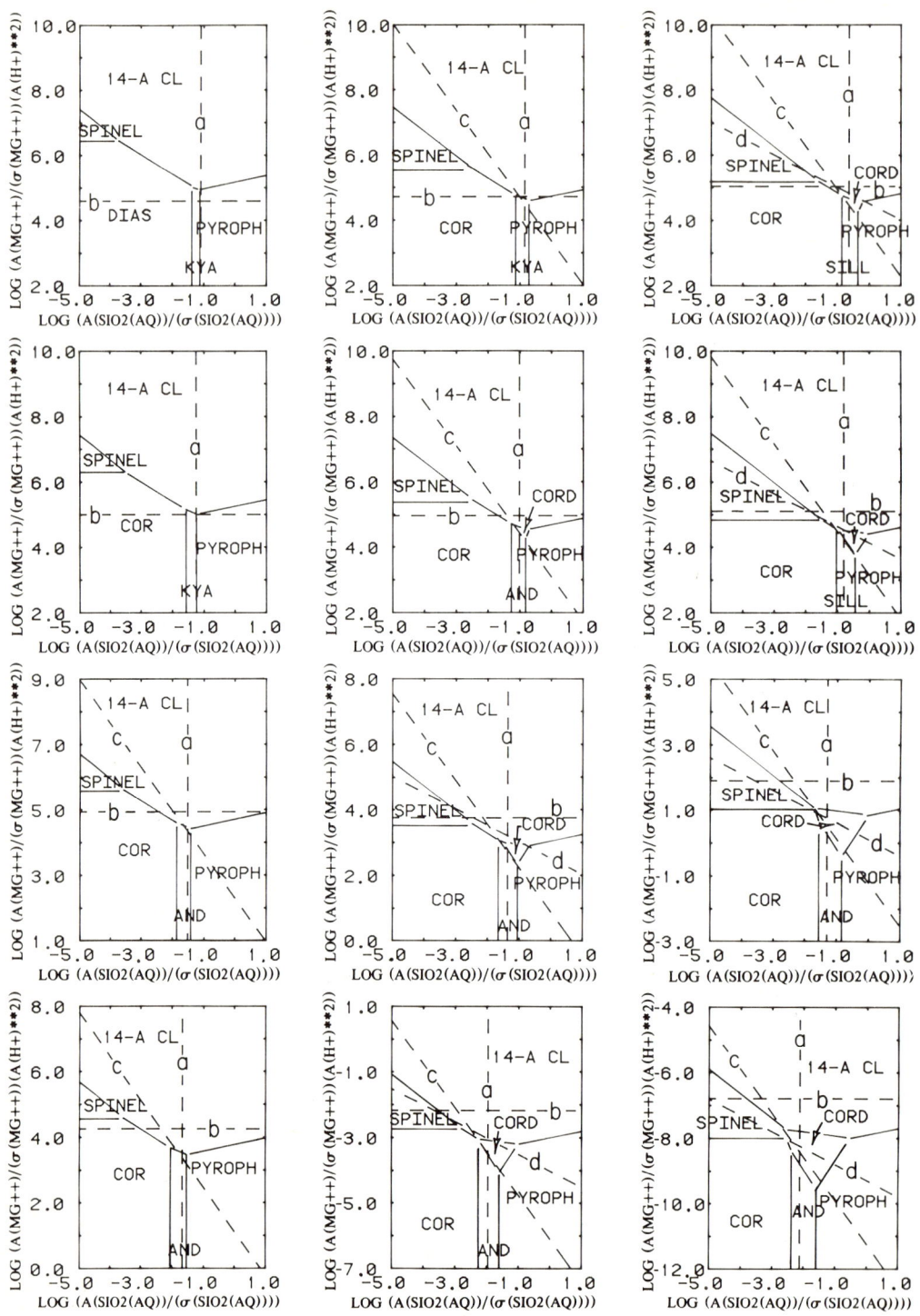

Phase relations in the system $HCl-H_2O-(Al_2O_3)-CO_2-MgO-SiO_2$ at $X_{CO_2} = 0.50$. Saturation limits: quartz (a), magnesite (b), talc (c), forsterite (d).

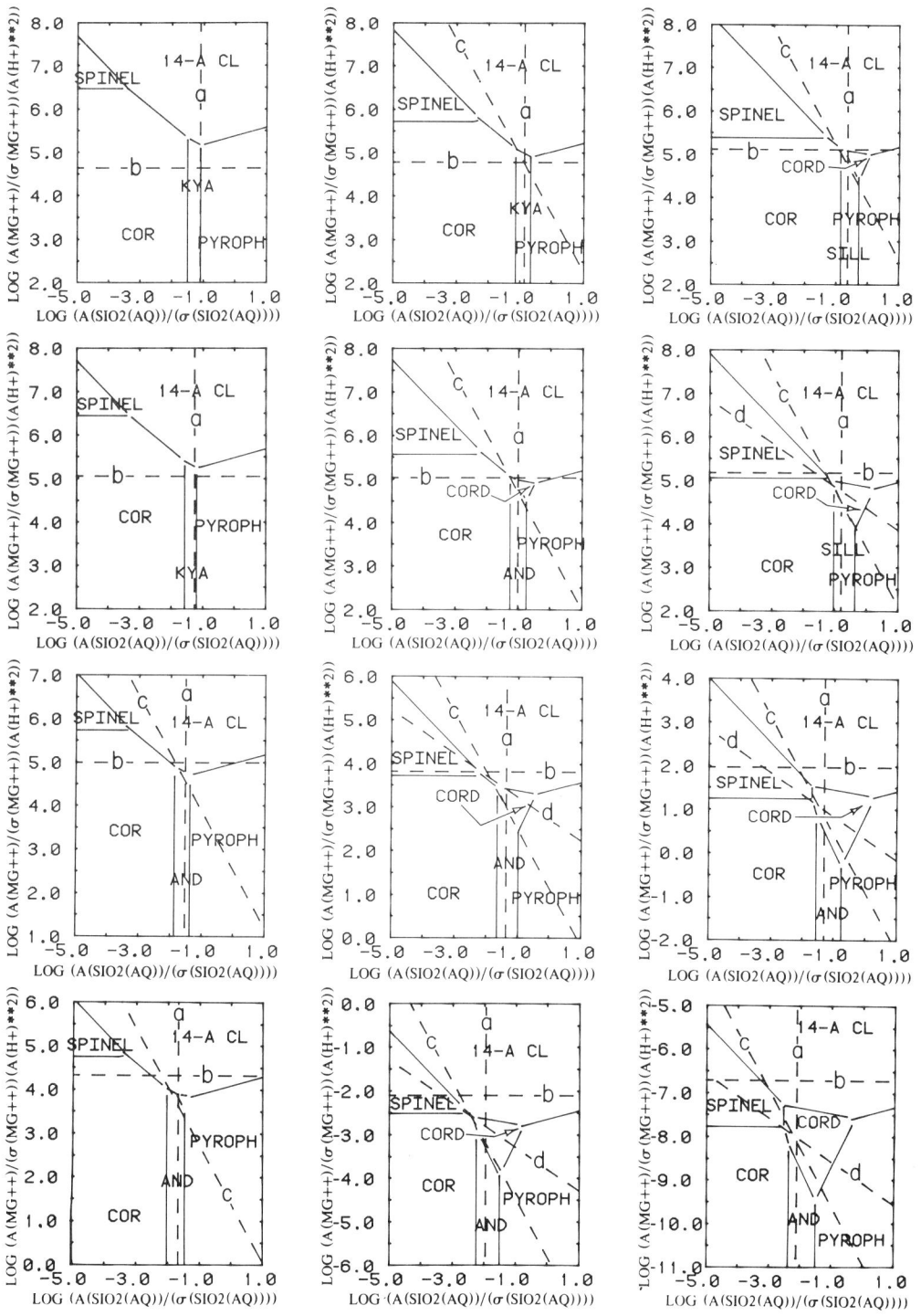

Phase relations in the system $HCl-H_2O-(Al_2O_3)-CO_2-MgO-SiO_2$ at $X_{CO_2} = 0.70$. Saturation limits: quartz (a), magnesite (b), talc (c), forsterite (d).

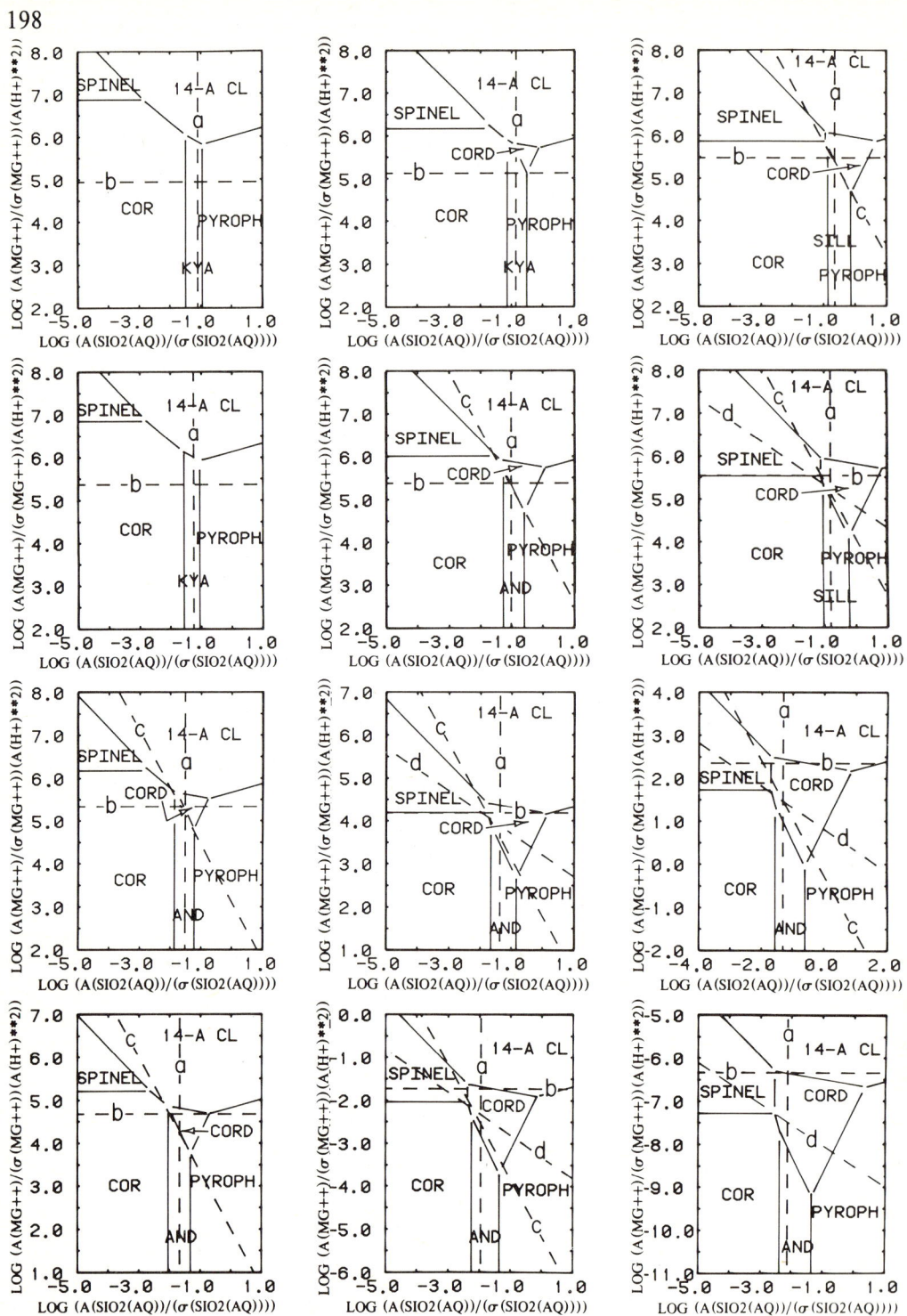

Phase relations in the system $HCl-H_2O-(Al_2O_3)-CO_2-MgO-SiO_2$ at $X_{CO_2} = 0.90$. Saturation limits: quartz (a), magnesite (b), talc (c), forsterite (d).

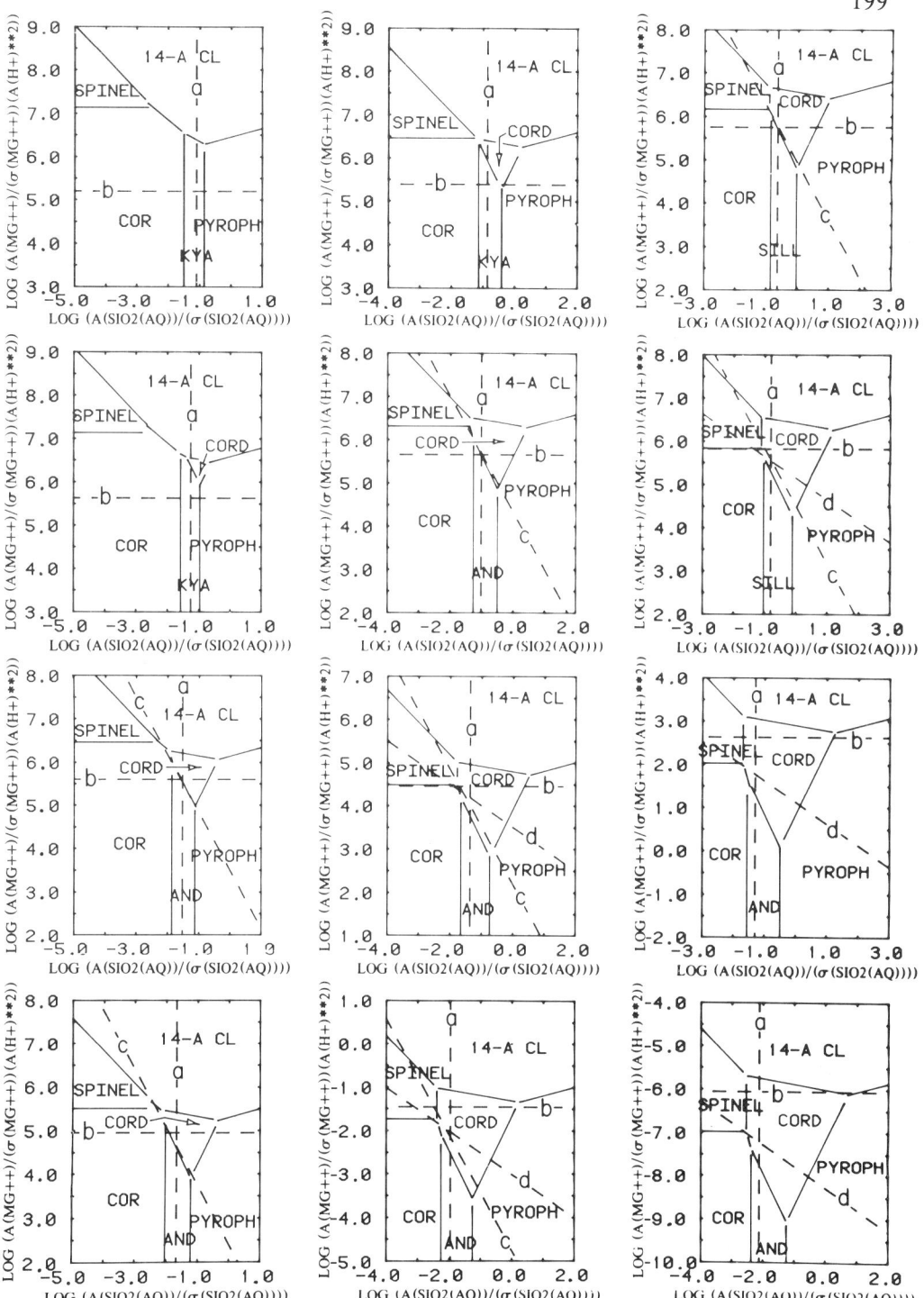

Phase relations in the system $HCl-H_2O-(Al_2O_3)-CO_2-MgO-SiO_2$ at $X_{CO_2} = 0.95$. Saturation limits: quartz (a), magnesite (b), talc (c), forsterite (d).

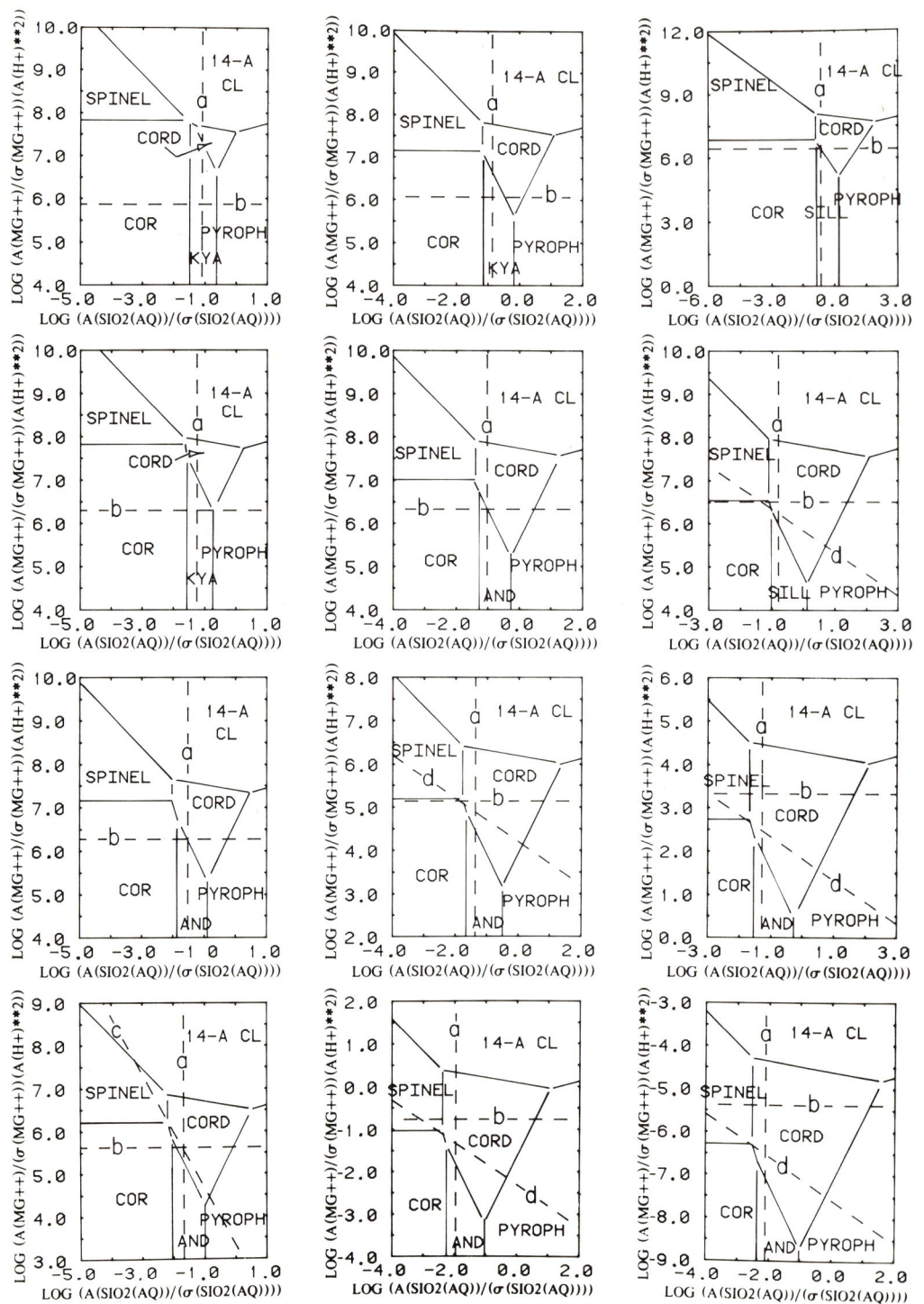

Phase relations in the system $HCl-H_2O-(Al_2O_3)-CO_2-MgO-SiO_2$ at $X_{CO_2} = 0.99$. Saturation limits: quartz (a), magnesite (b), talc (c), forsterite (d).

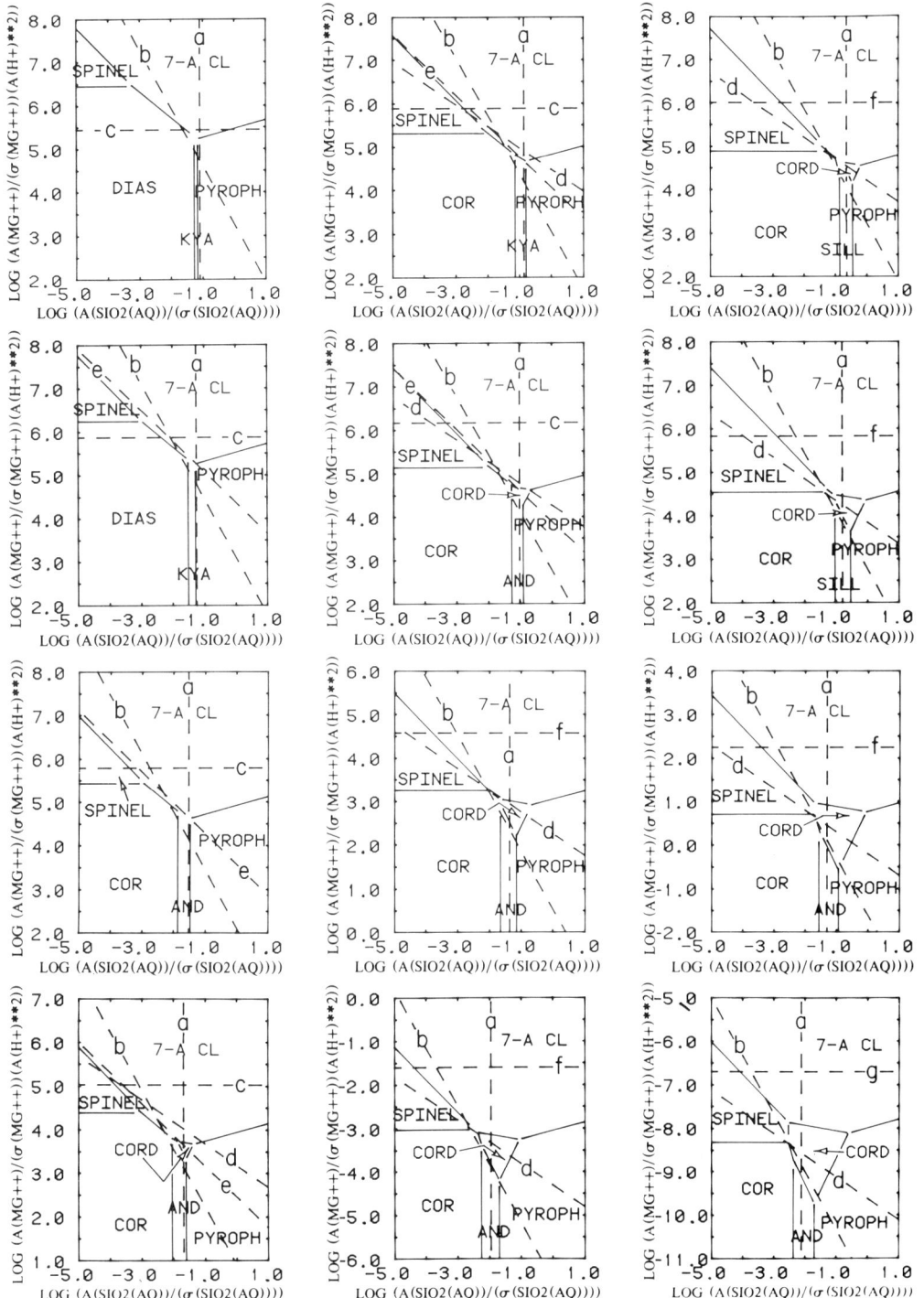

Phase relations in the system $HCl-H_2O-(Al_2O_3)-CO_2-MgO-SiO_2$ at $X_{CO_2} = 0.01$. Metastable 7-A clinochlore was considered instead of its stable counterpart, 14-A clinochlore. Saturation limits: quartz (a), talc (b), magnesite (c), forsterite (d), antigorite (e), brucite (f), periclase (g).

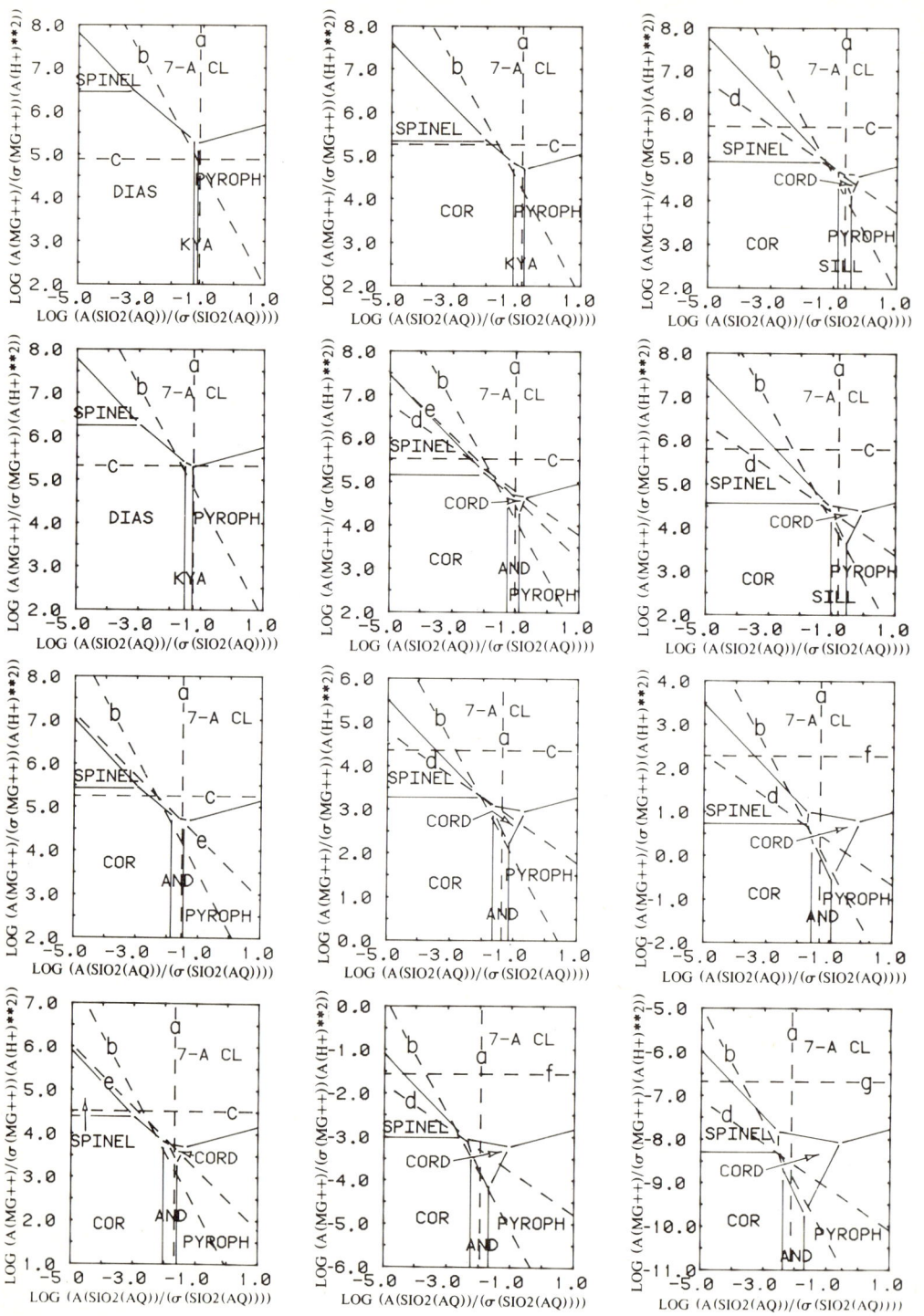

Phase relations in the system $HCl-H_2O-(Al_2O_3)-CO_2-MgO-SiO_2$ at $X_{CO_2} = 0.05$. Metastable 7-A clinochlore was considered instead of its stable counterpart, 14-A clinochlore. Saturation limits: quartz (a), talc (b), magnesite (c), forsterite (d), antigorite (e), brucite (f), periclase (g).

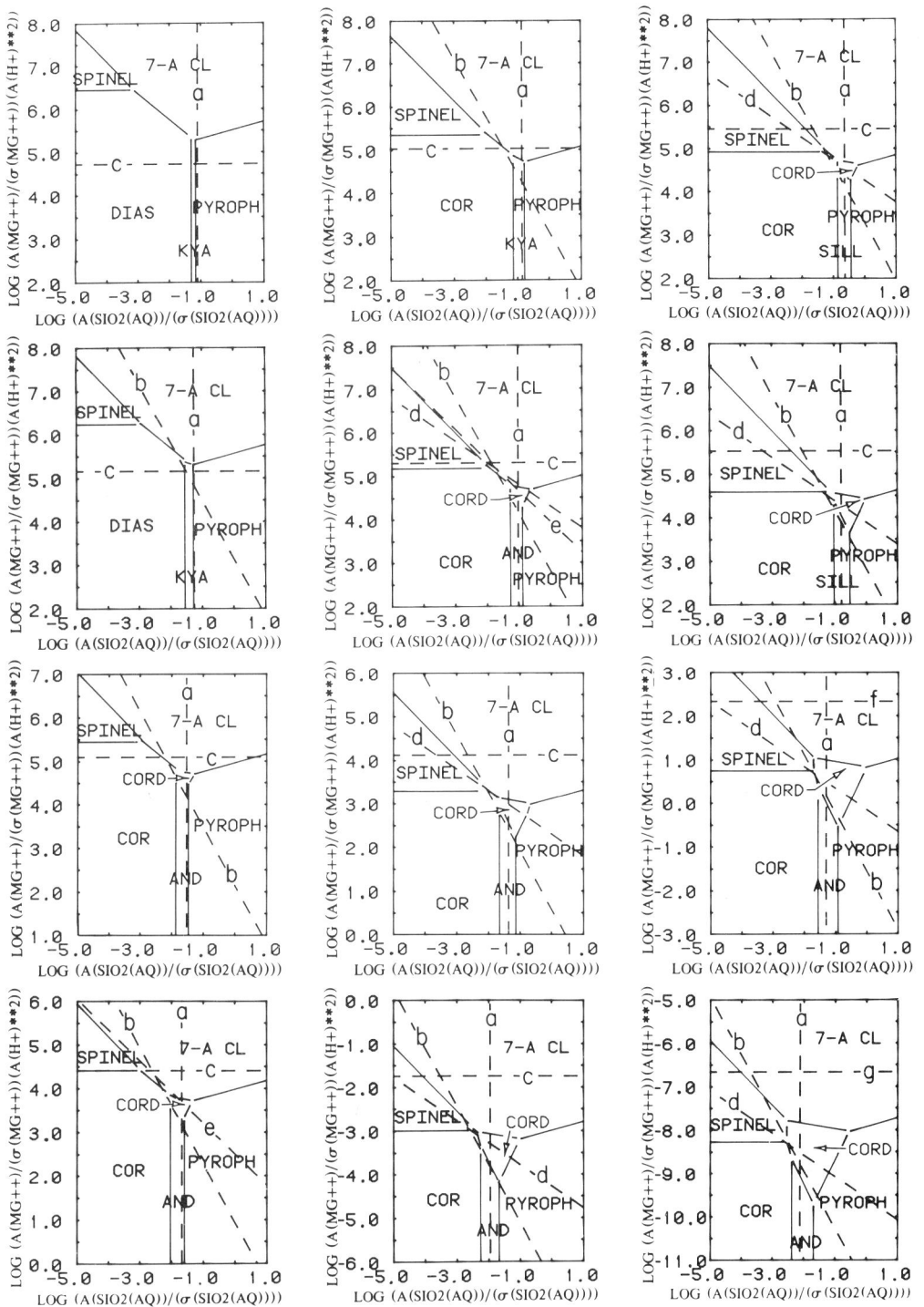

Phase relations in the system $HCl-H_2O-(Al_2O_3)-CO_2-MgO-SiO_2$ at $X_{CO_2} = 0.10$. Metastable 7-A clinochlore was considered instead of its stable counterpart, 14-A clinochlore. Saturation limits: quartz (a), talc (b), magnesite (c), forsterite (d), antigorite (e), brucite (f), periclase (g).

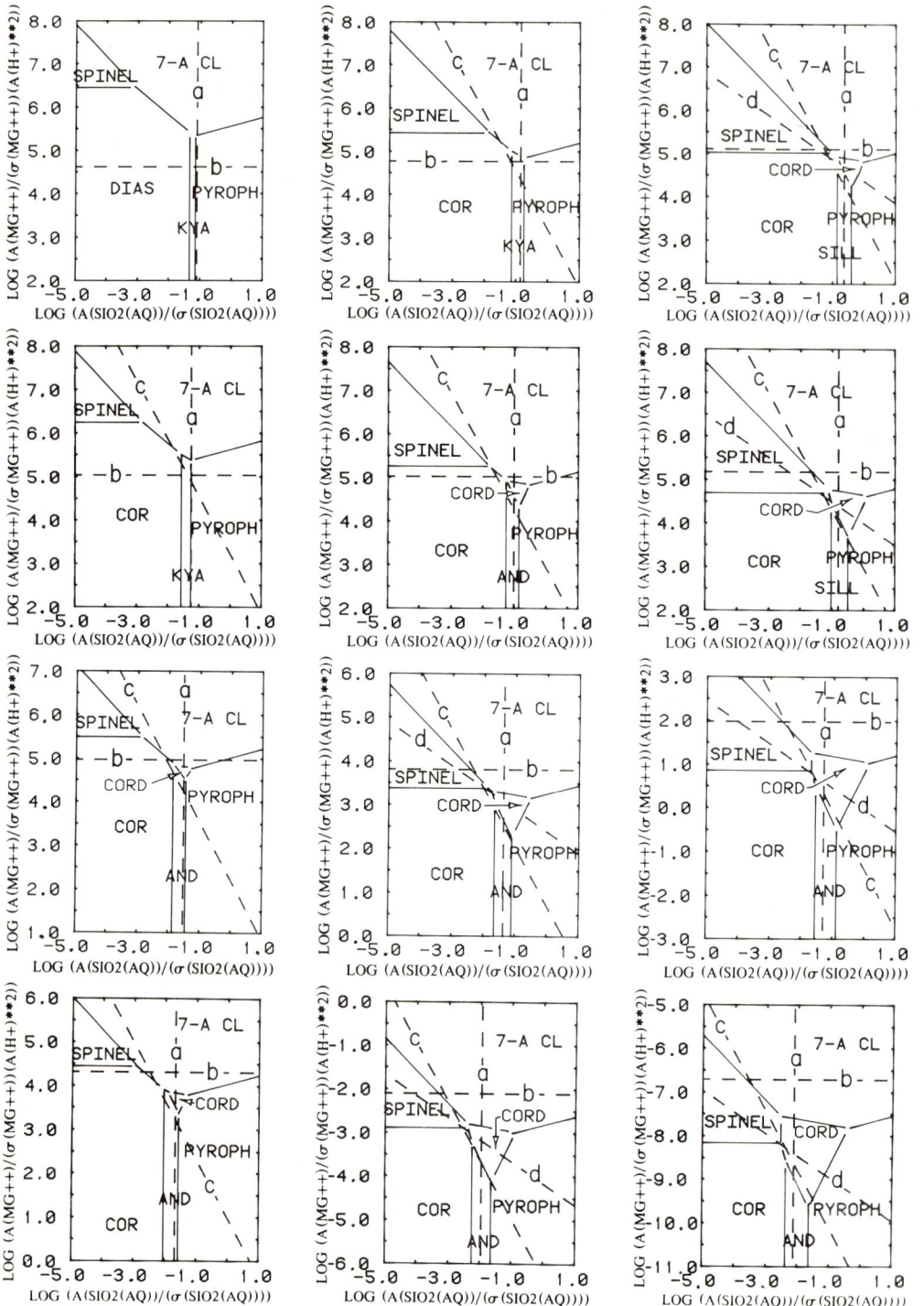

Phase relations in the system $HCl-H_2O-(Al_2O_3)-CO_2-MgO-SiO_2$ at $X_{CO_2} = 0.30$. Metastable 7-A clinochlore was considered instead of its stable counterpart, 14-A clinochlore. Saturation limits: quartz (a), magnesite (b), talc (c), forsterite (d).

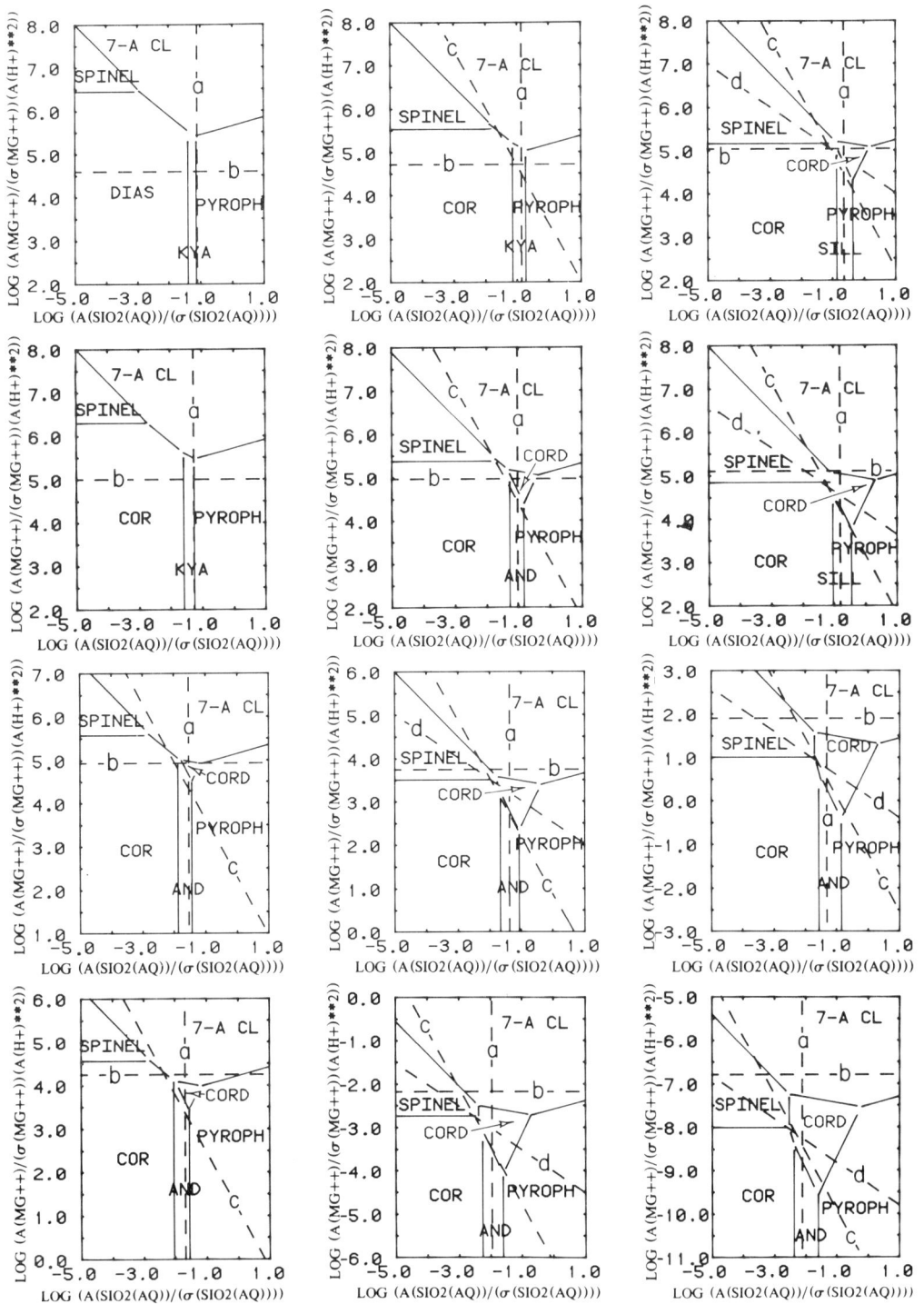

Phase relations in the system $HCl-H_2O-(Al_2O_3)-CO_2-MgO-SiO_2$ at $X_{CO_2} = 0.50$. Metastable 7-A clinochlore was considered instead of its stable counterpart, 14-A clinochlore. Saturation limits: quartz (a), magnesite (b), talc (c), forsterite (d).

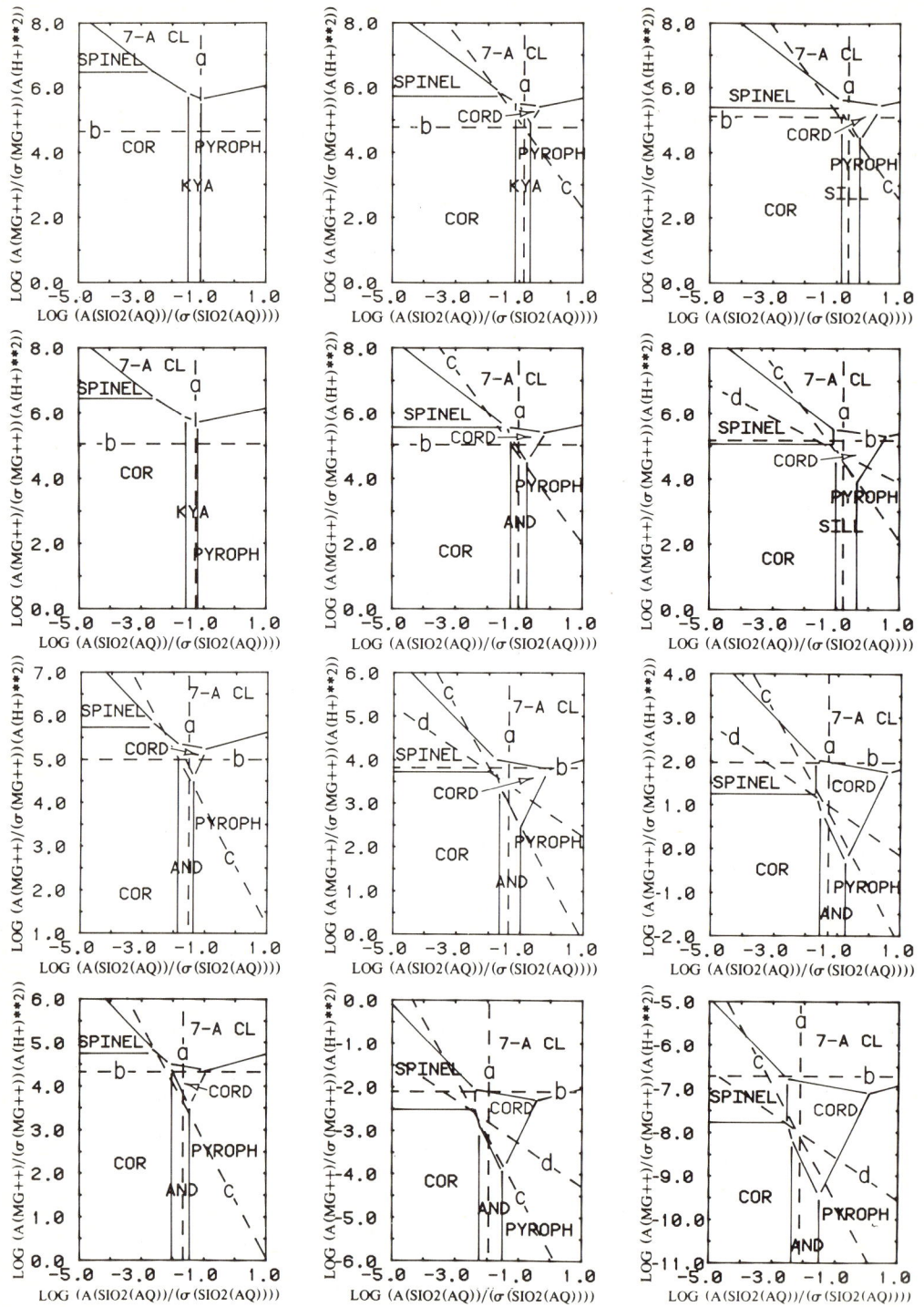

Phase relations in the system $HCl-H_2O-(Al_2O_3)-CO_2-MgO-SiO_2$ at $X_{CO_2} = 0.70$. Metastable 7-Å clinochlore was considered instead of its stable counterpart, 14-Å clinochlore. Saturation limits: quartz (a), magnesite (b), talc (c), forsterite (d).

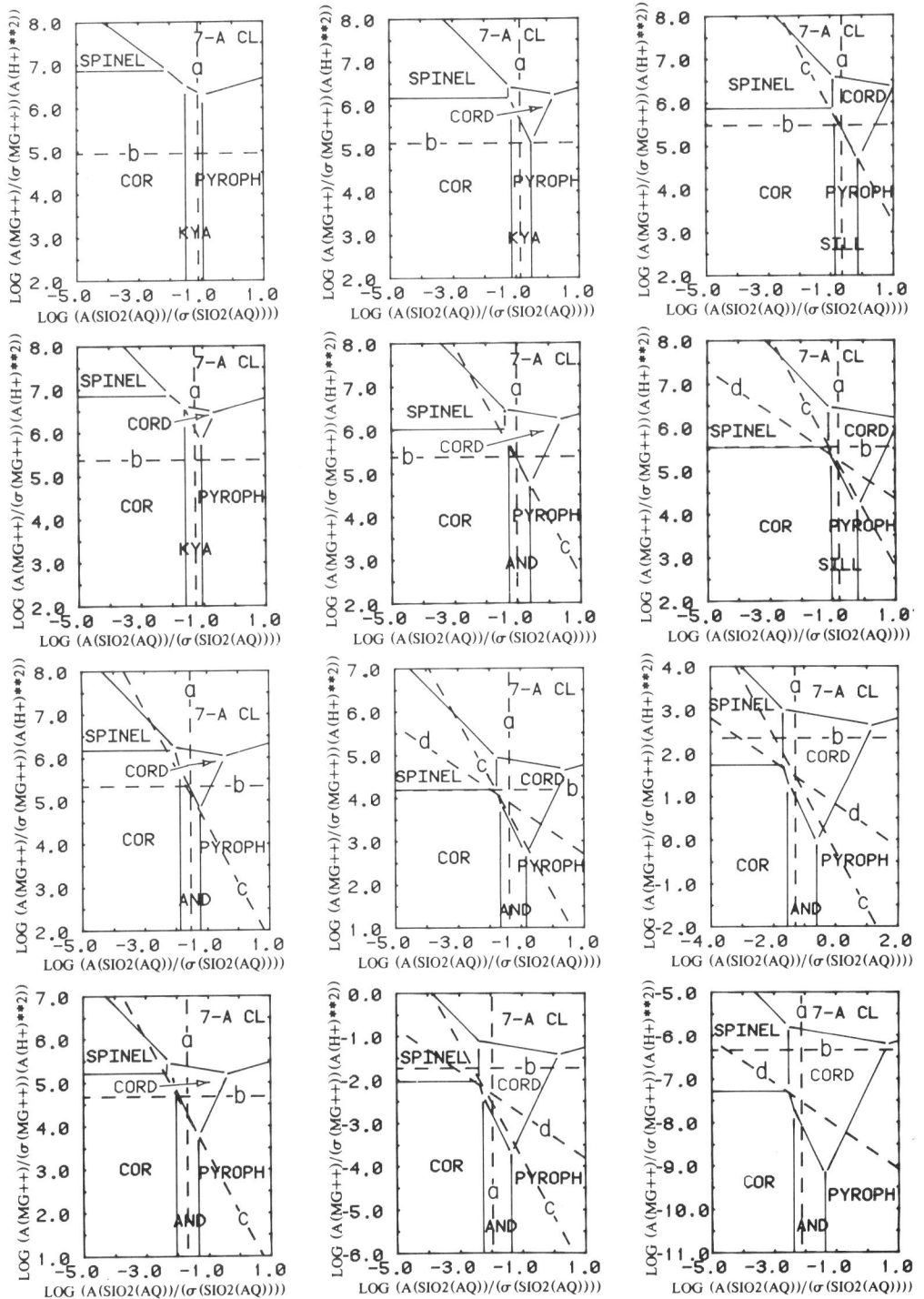

Phase relations in the system $HCl-H_2O-(Al_2O_3)-CO_2-MgO-SiO_2$ at $X_{CO_2} = 0.90$. Metastable 7-A clinochlore was considered instead of its stable counterpart, 14-A clinochlore. Saturation limits: quartz (a), magnesite (b), talc (c), forsterite (d).

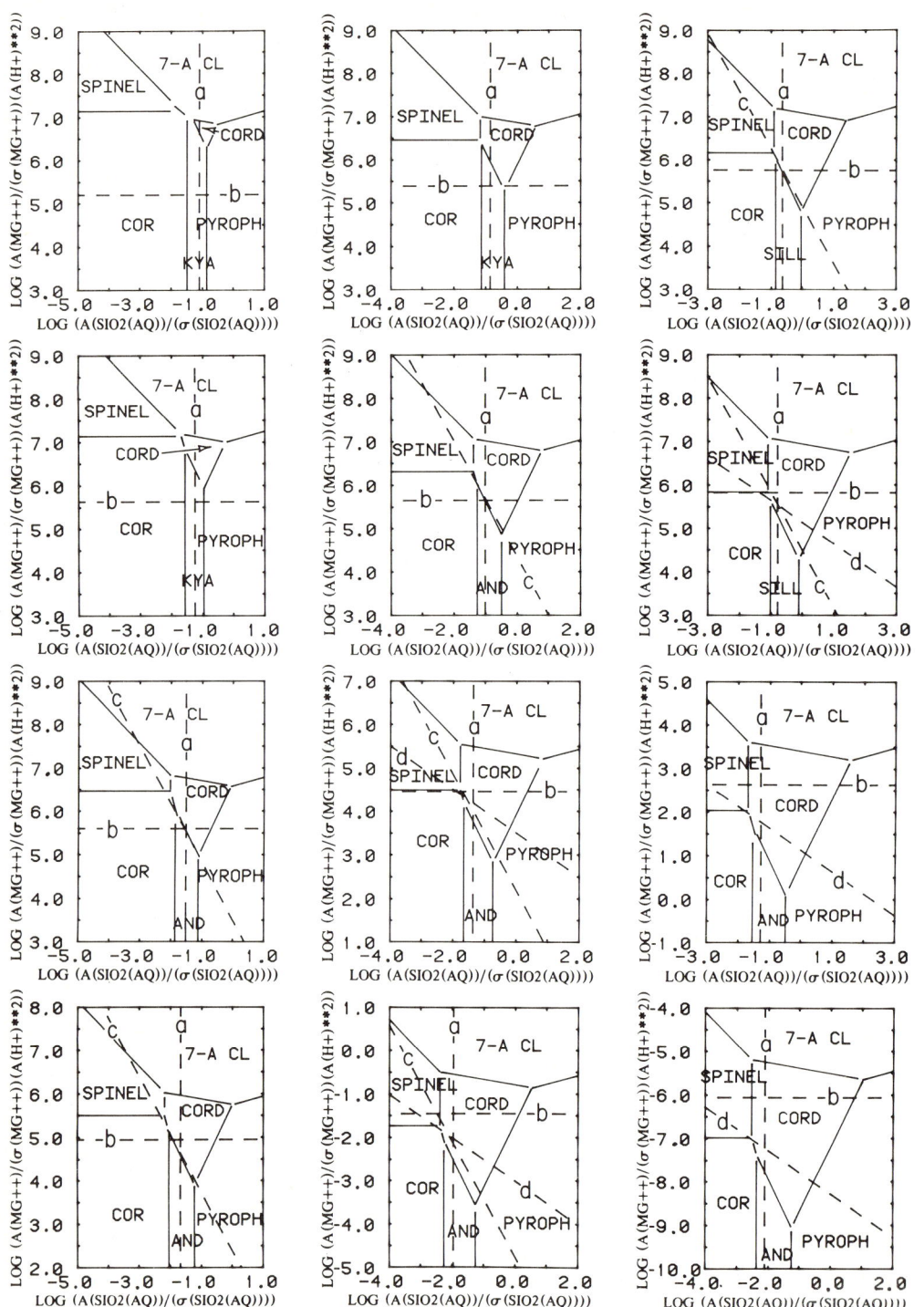

Phase relations in the system $HCl-H_2O-(Al_2O_3)-CO_2-MgO-SiO_2$ at $X_{CO_2} = 0.95$. Metastable 7-A clinochlore was considered instead of its stable counterpart, 14-A clinochlore. Saturation limits: quartz (a), magnesite (b), talc (c), forsterite (d).

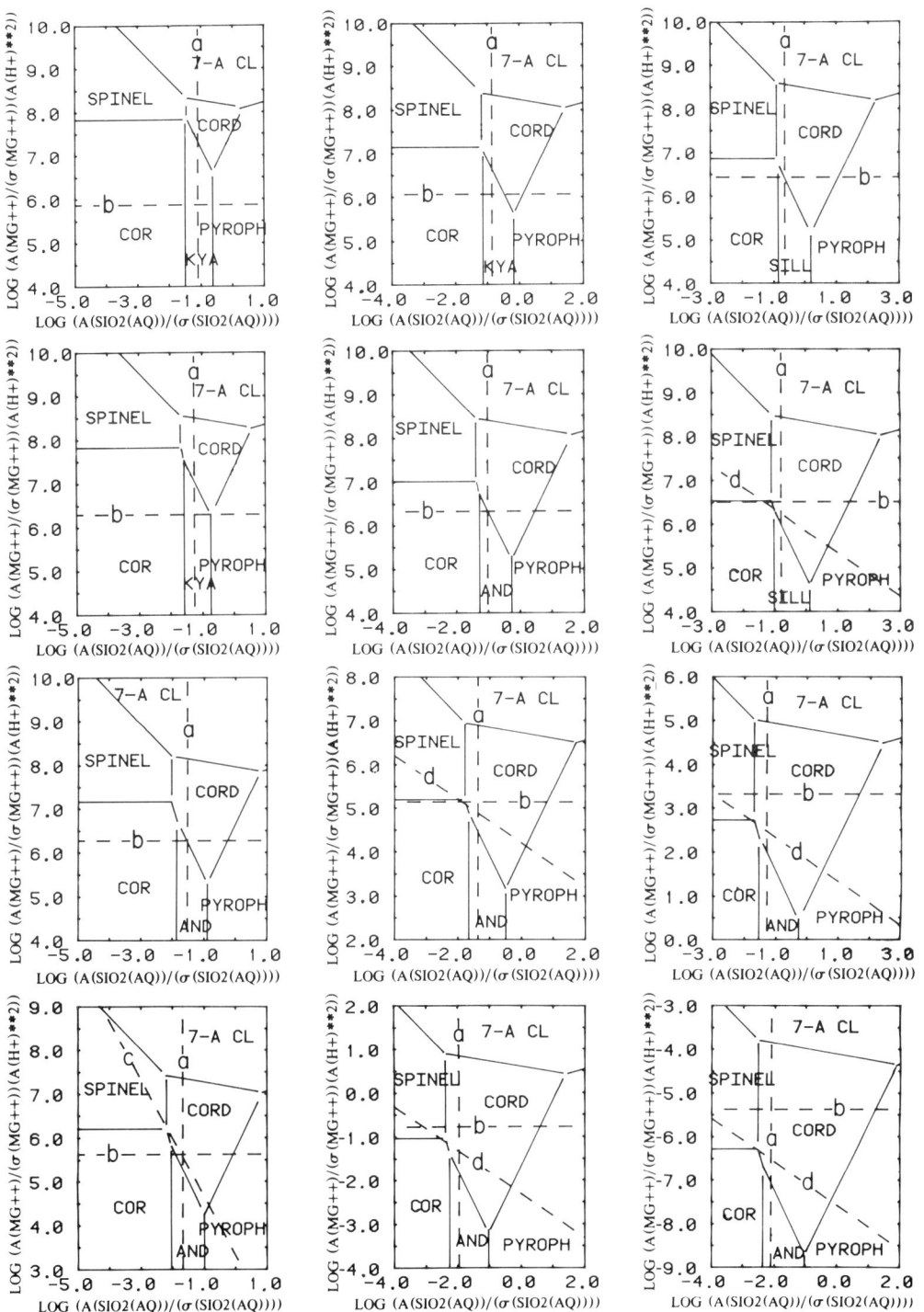

Phase relations in the system $HCl-H_2O-(Al_2O_3)-CO_2-MgO-SiO_2$ at $X_{CO_2} = 0.99$. Metastable 7-A clinochlore was considered instead of its stable counterpart, 14-A clinochlore. Saturation limits: quartz (a), magnesite (b), talc (c), forsterite (d).

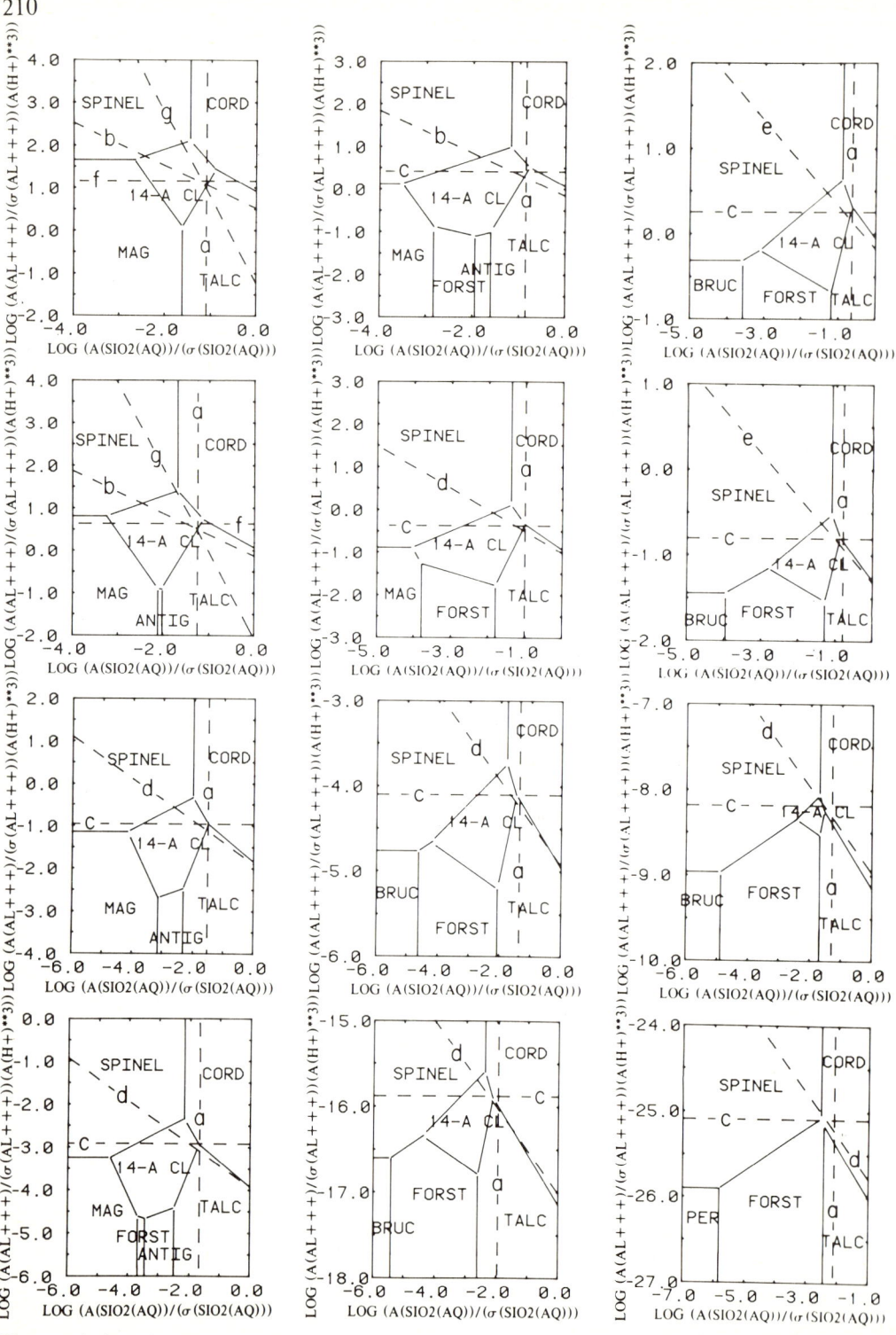

Phase relations in the system $HCl-H_2O-Al_2O_3-CO_2-(MgO)-SiO_2$ at $X_{CO_2} = 0.01$. Saturation limits: quartz (a), kyanite (b), corundum (c), andalusite (d), sillimanite (e), diaspore (f), pyrophyllite (g).

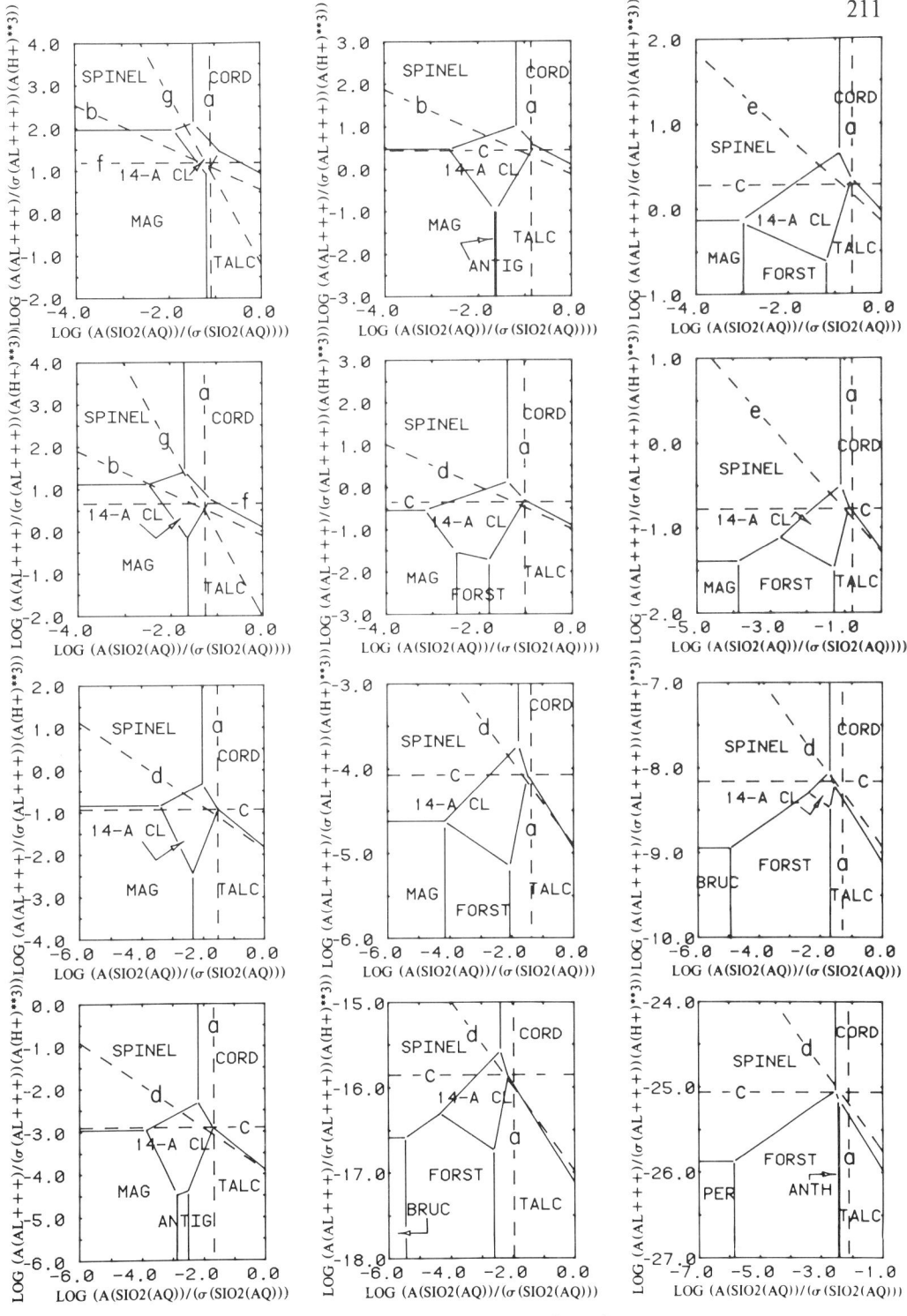

Phase relations in the system $HCl-H_2O-Al_2O_3-CO_2-(MgO)-SiO_2$ at $X_{CO_2} = 0.05$. Saturation limits: quartz (a), kyanite (b), corundum (c), andalusite (d), sillimanite (e), diaspore (f), pyrophyllite (g).

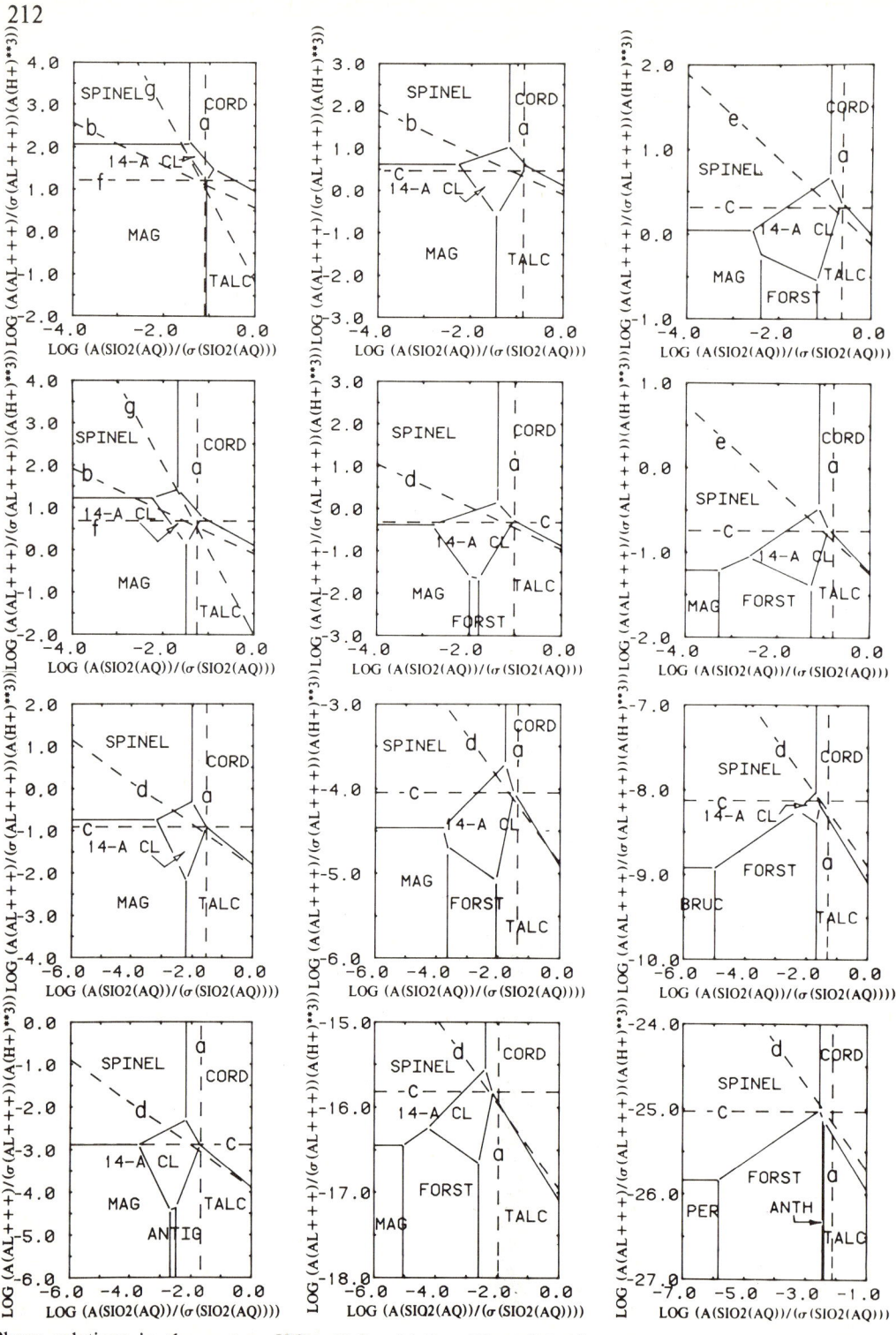

Phase relations in the system $HCl-H_2O-Al_2O_3-CO_2-(MgO)-SiO_2$ at $X_{CO_2} = 0.10$. Saturation limits: quartz (a), kyanite (b), corundum (c), andalusite (d), sillimanite (e), diaspore (f), pyrophyllite (g).

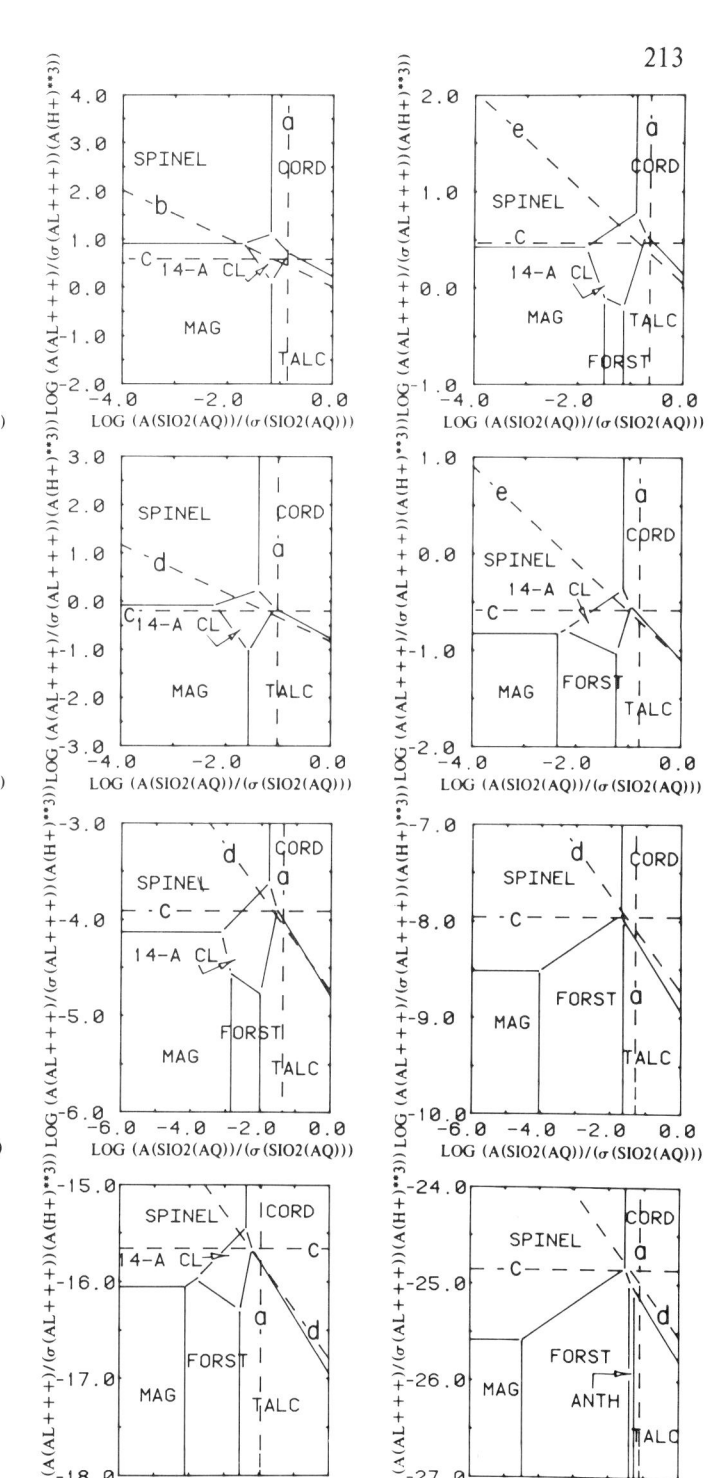

Phase relations in the system $HCl-H_2O-Al_2O_3-CO_2-(MgO)-SiO_2$ at $X_{CO_2} = 0.30$. Saturation limits: quartz (a), kyanite (b), corundum (c), andalusite (d), sillimanite (e), diaspore (f), pyrophyllite (g).

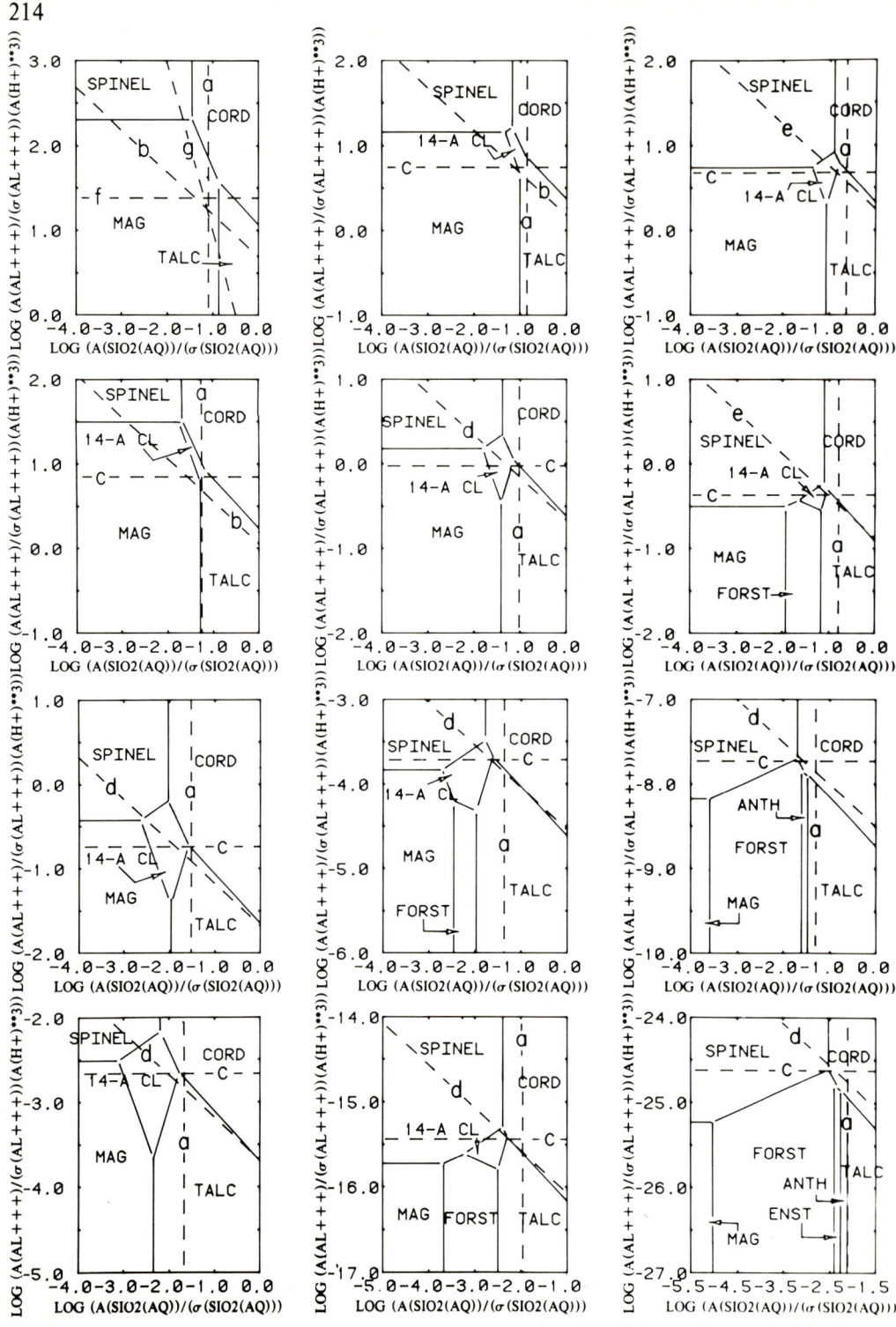

Phase relations in the system $HCl-H_2O-Al_2O_3-CO_2-(MgO)-SiO_2$ at $X_{CO_2} = 0.50$. Saturation limits: quartz (a), kyanite (b), corundum (c), andalusite (d), sillimanite (e), diaspore (f), pyrophyllite (g).

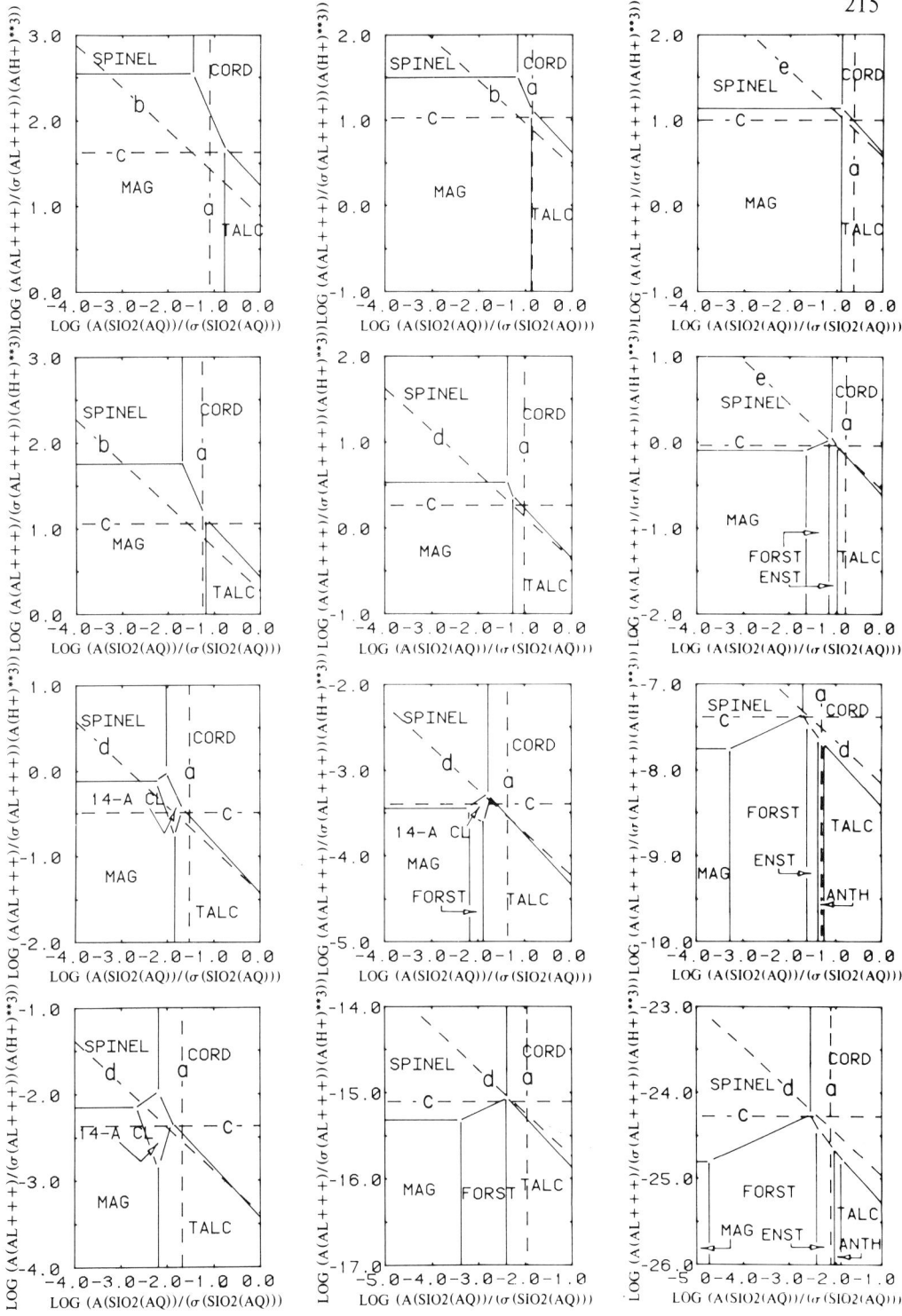

Phase relations in the system $HCl-H_2O-Al_2O_3-CO_2-(MgO)-SiO_2$ at $X_{CO_2} = 0.70$. Saturation limits: quartz (a), kyanite (b), corundum (c), andalusite (d), sillimanite (e).

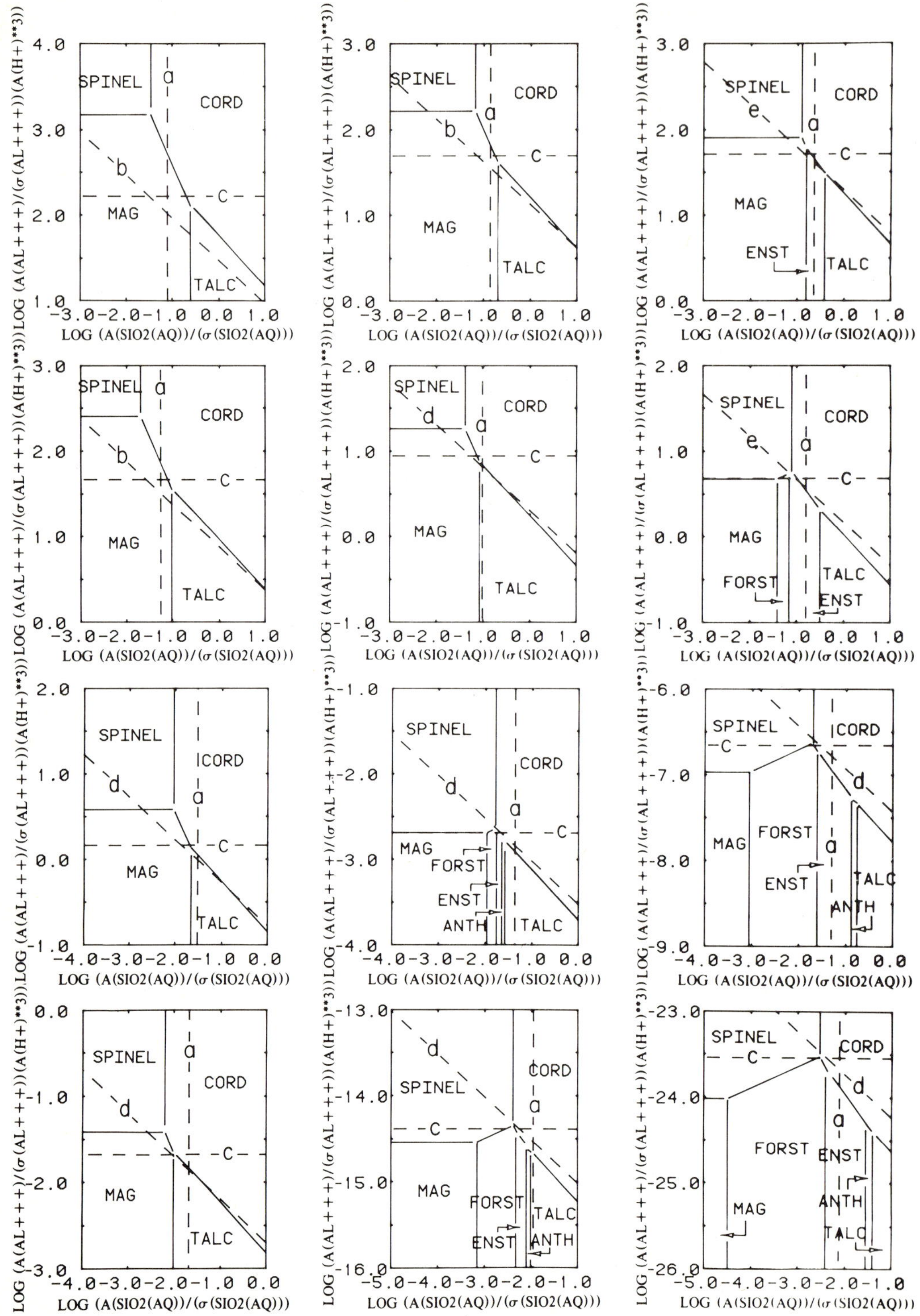

Phase relations in the system $HCl-H_2O-Al_2O_3-CO_2-(MgO)-SiO_2$ at $X_{CO_2} = 0.90$. Saturation limits: quartz (a), kyanite (b), corundum (c), andalusite (d), sillimanite (e).

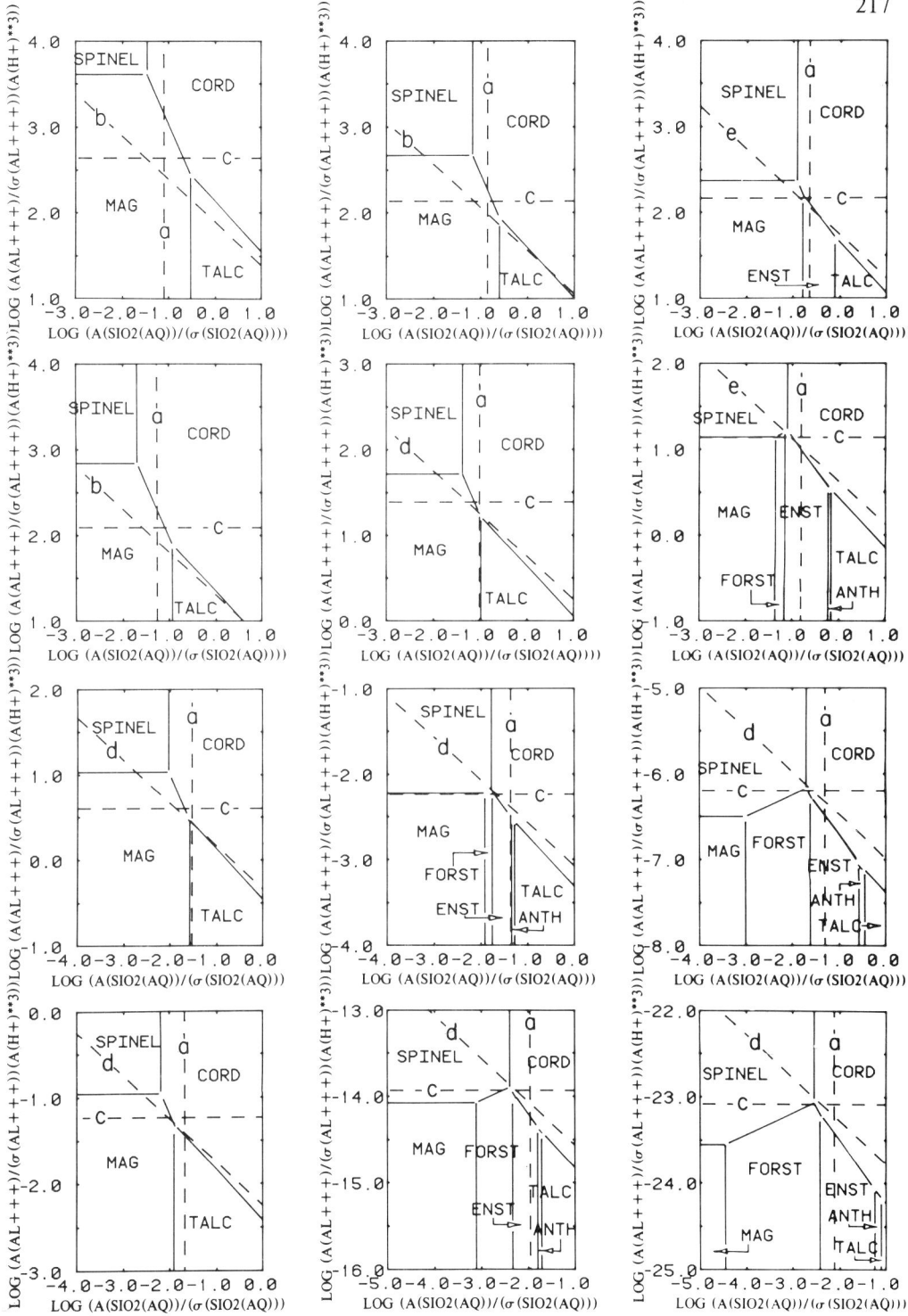

Phase relations in the system $HCl-H_2O-Al_2O_3-CO_2-(MgO)-SiO_2$ at $X_{CO_2} = 0.95$. Saturation limits: quartz (a), kyanite (b), corundum (c), andalusite (d), sillimanite (e).

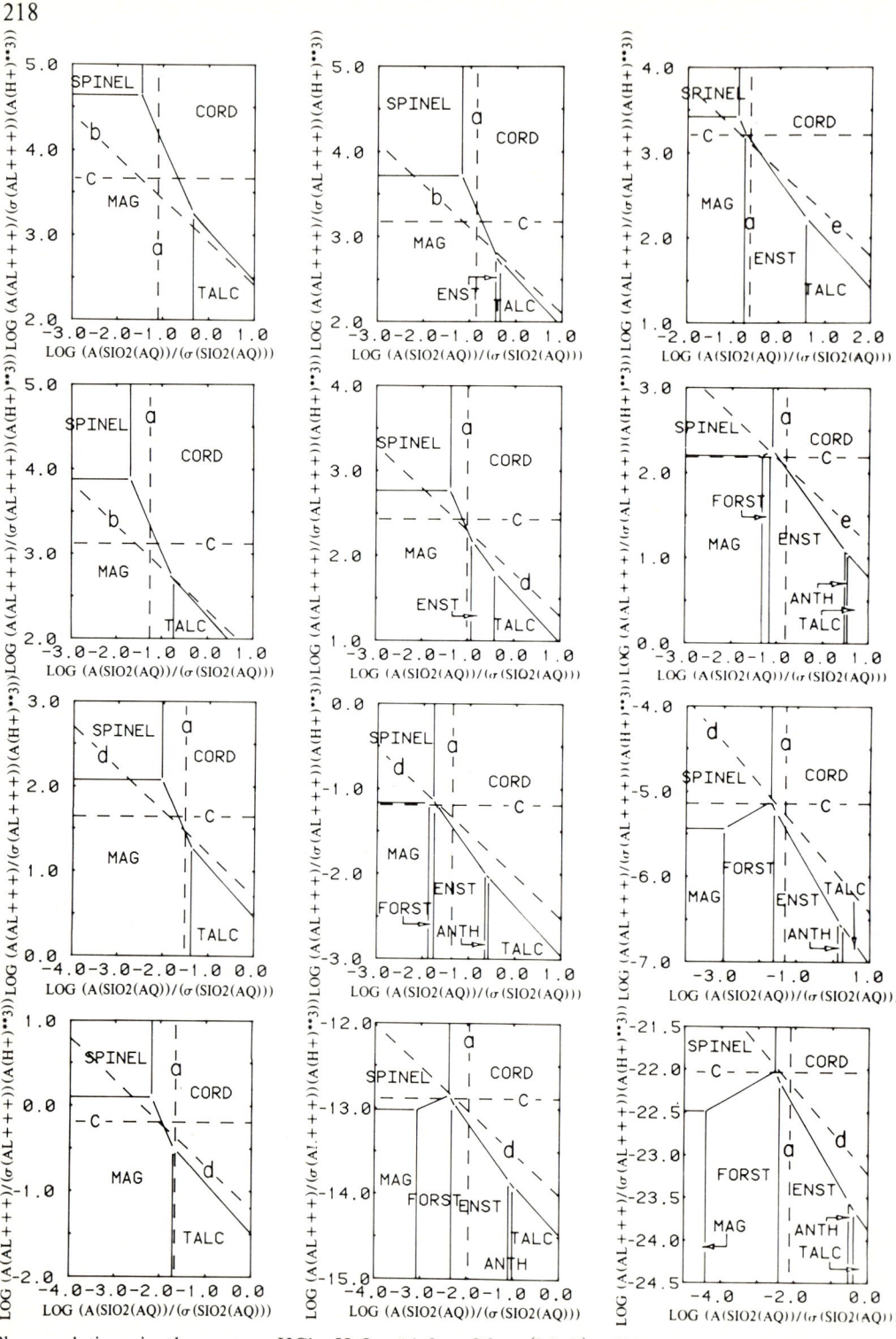

Phase relations in the system $HCl-H_2O-Al_2O_3-CO_2-(MgO)-SiO_2$ at $X_{CO_2} = 0.99$. Saturation limits: quartz (a), kyanite (b), corundum (c), andalusite (d), sillimanite (e).

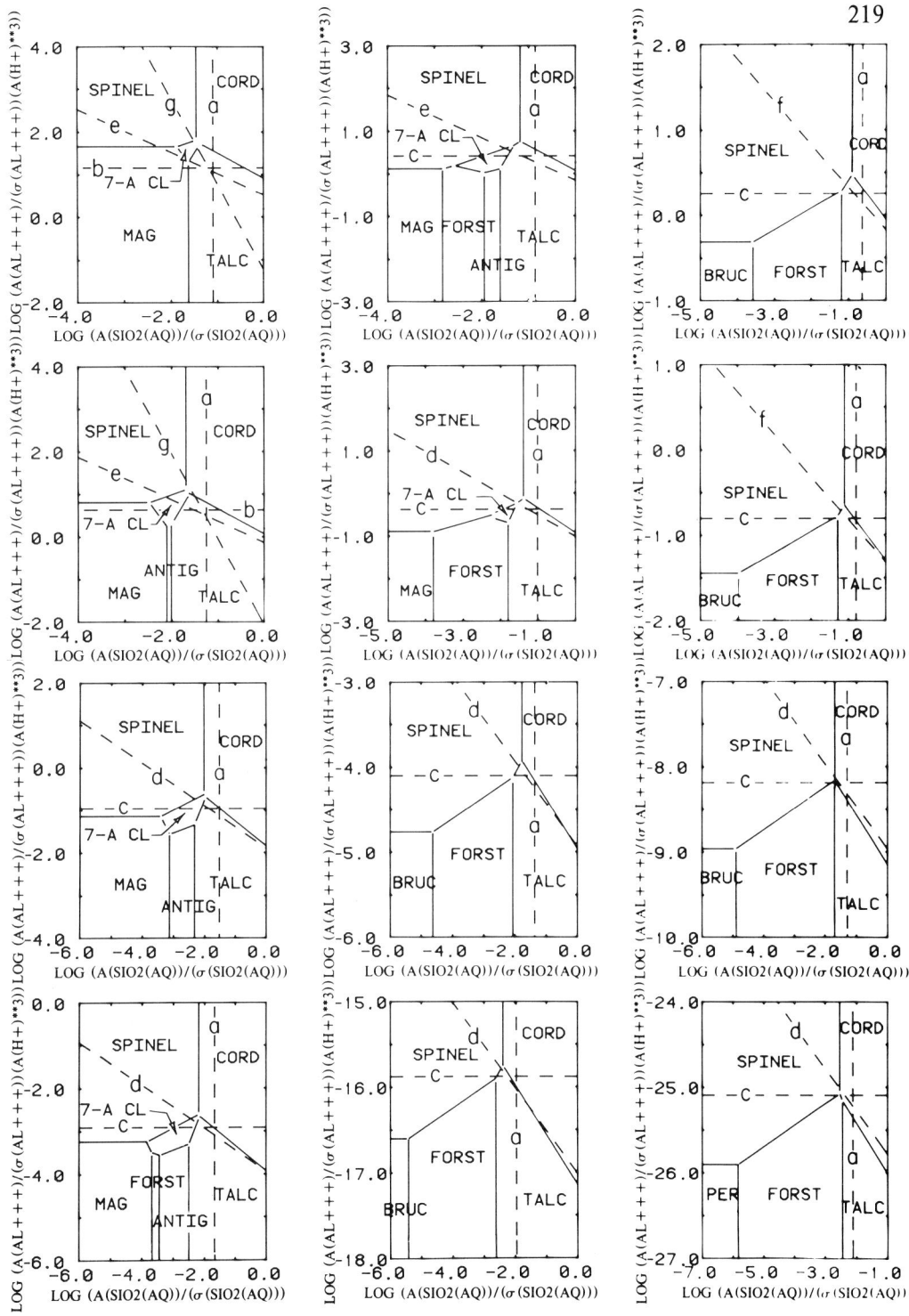

Phase relations in the system $HCl-H_2O-Al_2O_3-CO_2-(MgO)-SiO_2$ at $X_{CO_2} = 0.01$. Metastable 7-A clinochlore was considered instead of its stable counterpart, 14-A clinochlore. Saturation limits: quartz (a), diaspore (b), corundum (c), andalusite (d), kyanite (e), sillimanite (f), pyrophyllite (g).

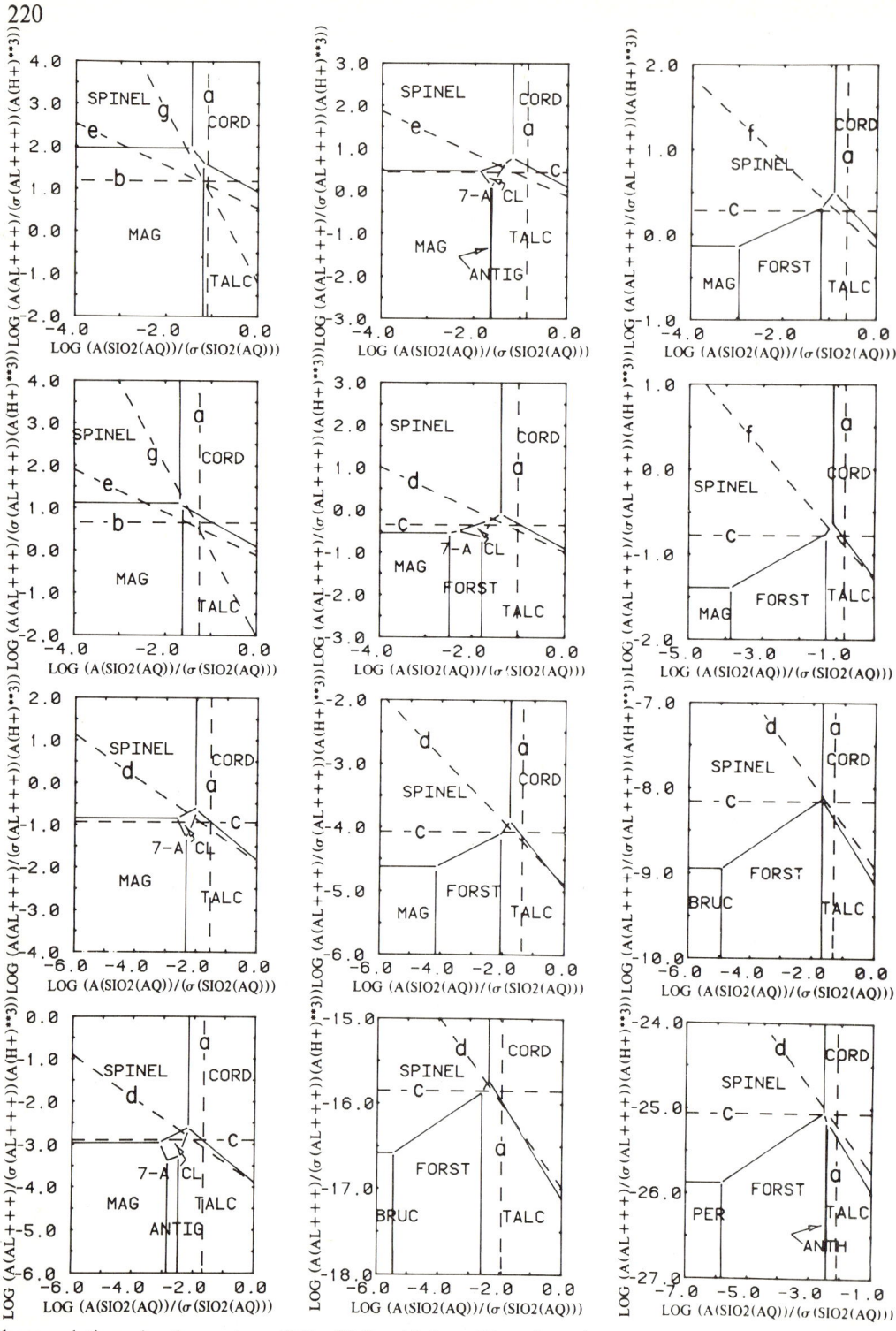

Phase relations in the system $HCl-H_2O-Al_2O_3-CO_2-(MgO)-SiO_2$ at $X_{CO_2} = 0.05$. Metastable 7-A clinochlore was considered instead of its stable counterpart, 14-A clinochlore. Saturation limits: quartz (a), diaspore (b), corundum (c), andalusite (d), kyanite (e), sillimanite (f), pyrophyllite (g).

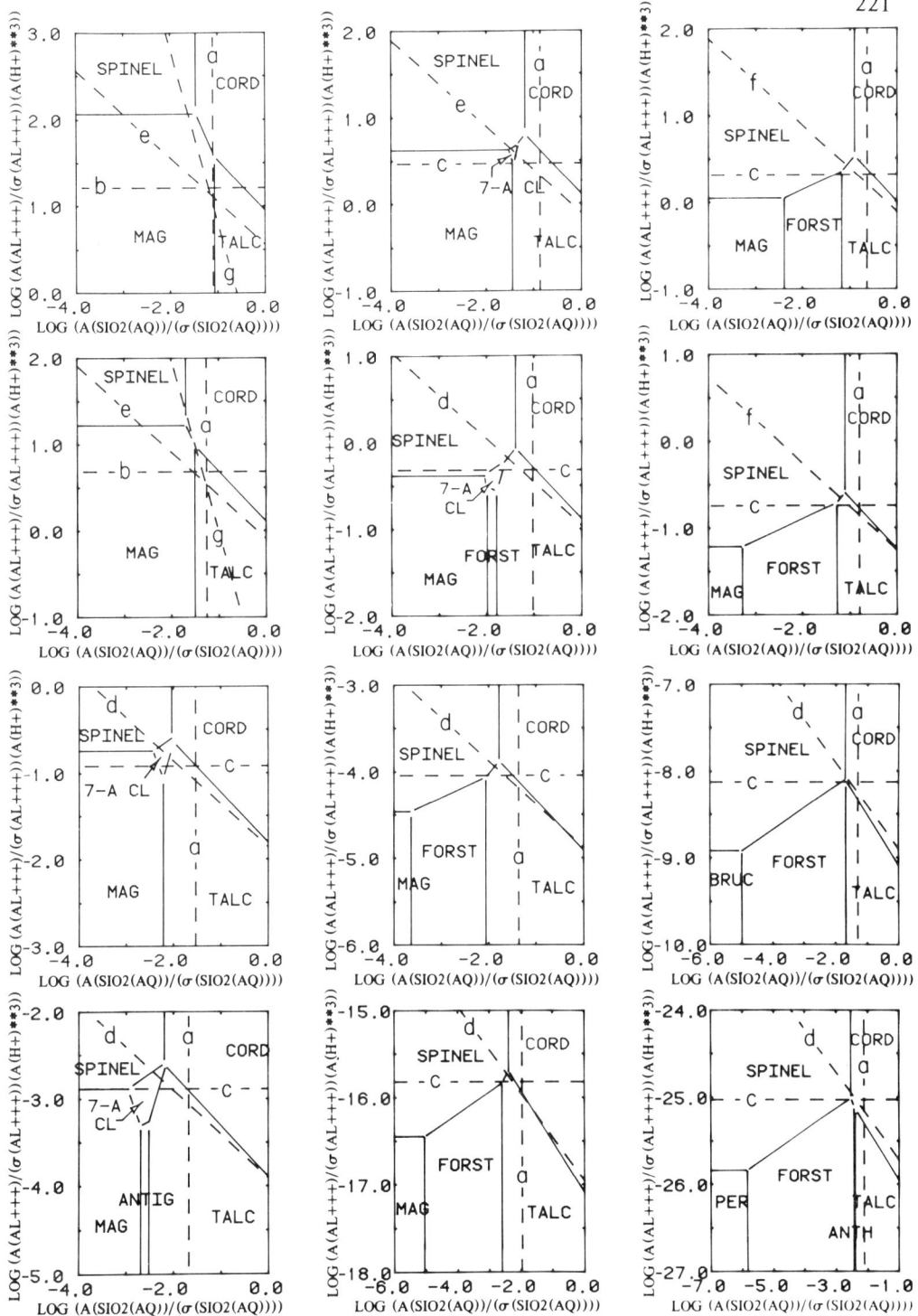

Phase relations in the system $HCl-H_2O-Al_2O_3-CO_2-(MgO)-SiO_2$ at $X_{CO_2} = 0.10$. Metastable 7-A clinochlore was considered instead of its stable counterpart, 14-A clinochlore. Saturation limits: quartz (a), diaspore (b), corundum (c), andalusite (d), kyanite (e), sillimanite (f), pyrophyllite (g).

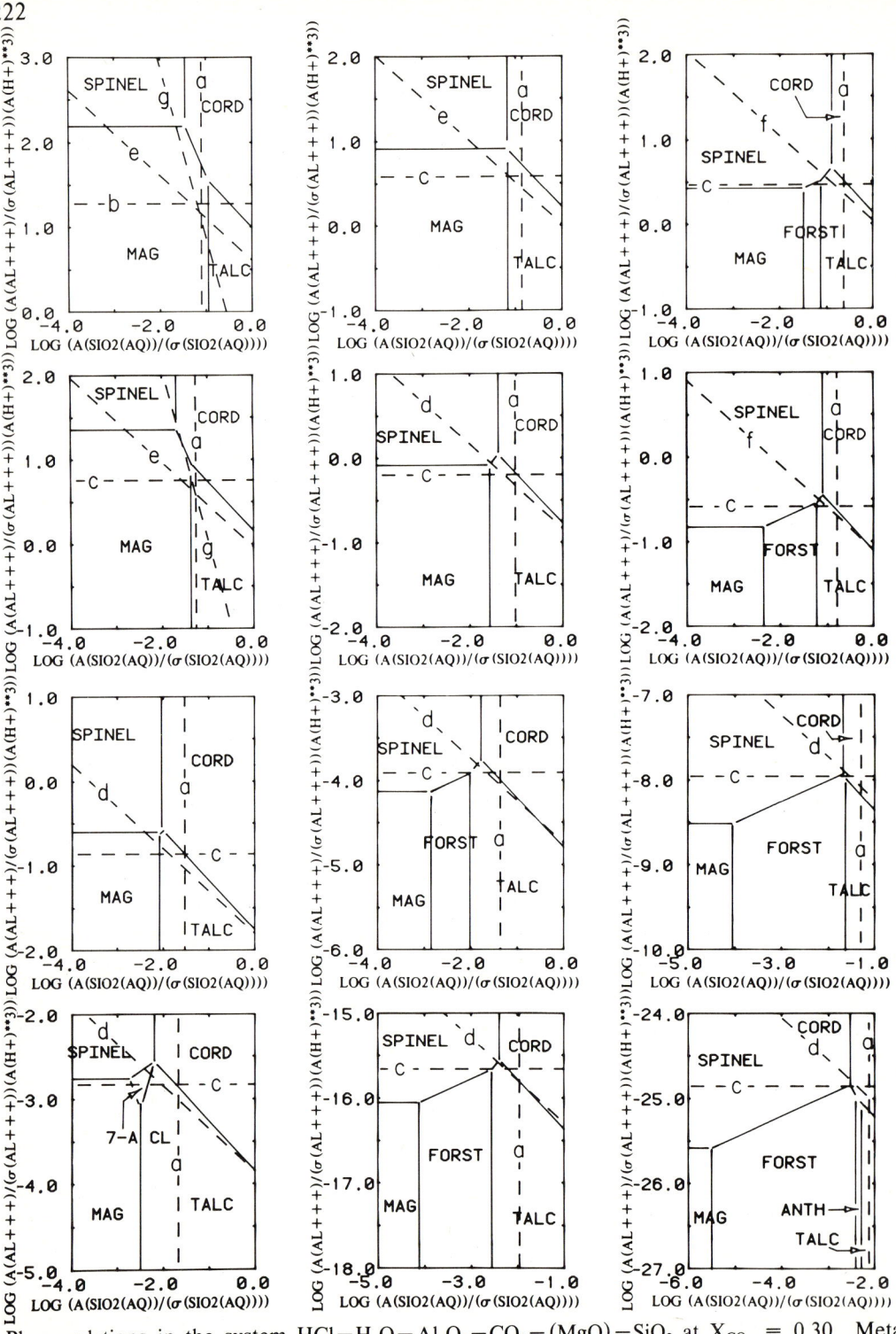

Phase relations in the system $HCl-H_2O-Al_2O_3-CO_2-(MgO)-SiO_2$ at $X_{CO_2} = 0.30$. Metastable 7-A clinochlore was considered instead of its stable counterpart, 14-A clinochlore. Saturation limits: quartz (a), diaspore (b), corundum (c), andalusite (d), kyanite (e), sillimanite (f), pyrophyllite (g).

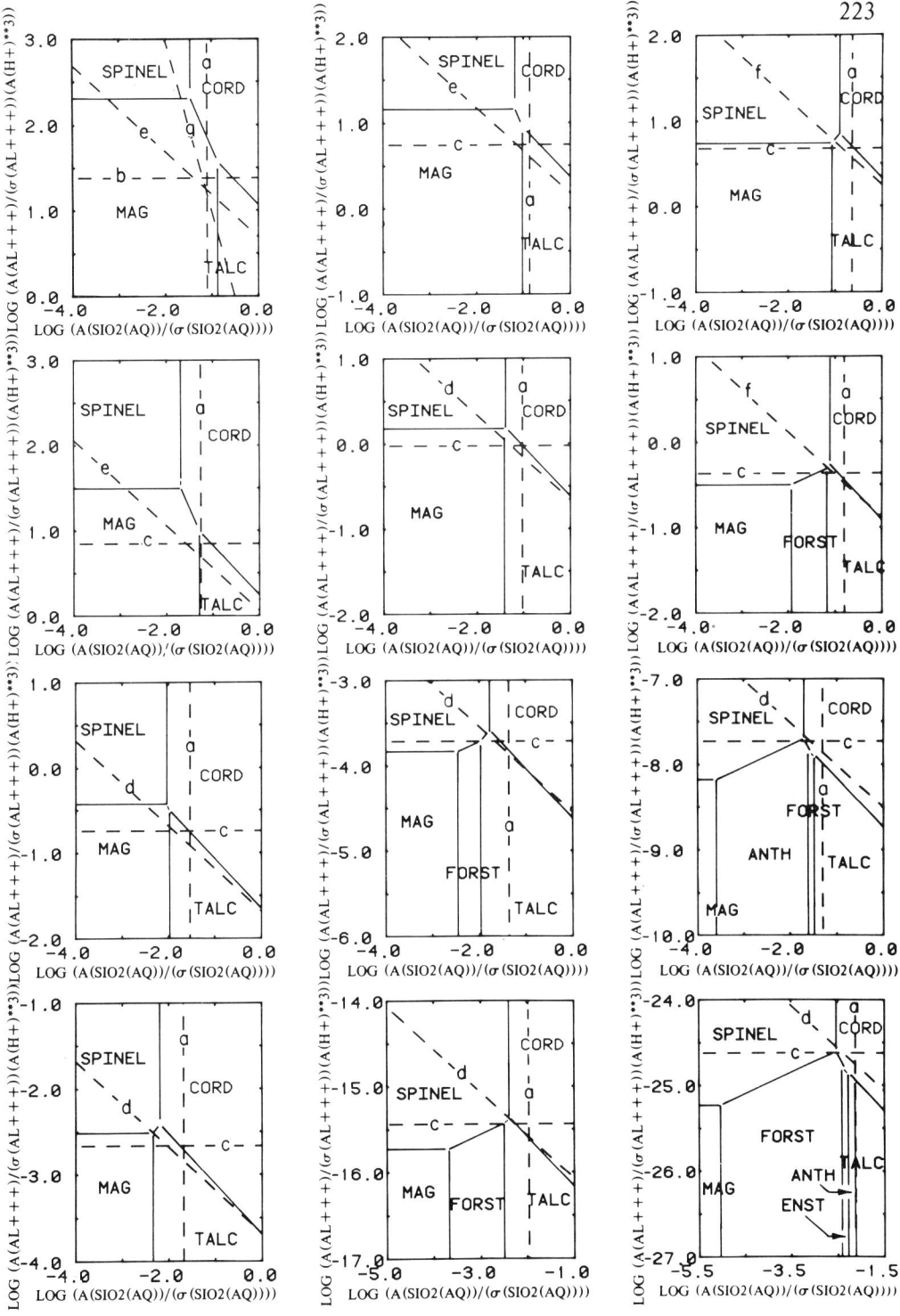

Phase relations in the system $HCl-H_2O-Al_2O_3-CO_2-(MgO)-SiO_2$ at $X_{CO_2} = 0.50$. Metastable 7-A clinochlore was considered instead of its stable counterpart, 14-A clinochlore. Saturation limits: quartz (a), diaspore (b), corundum (c), andalusite (d), kyanite (e), sillimanite (f), pyrophyllite (g).

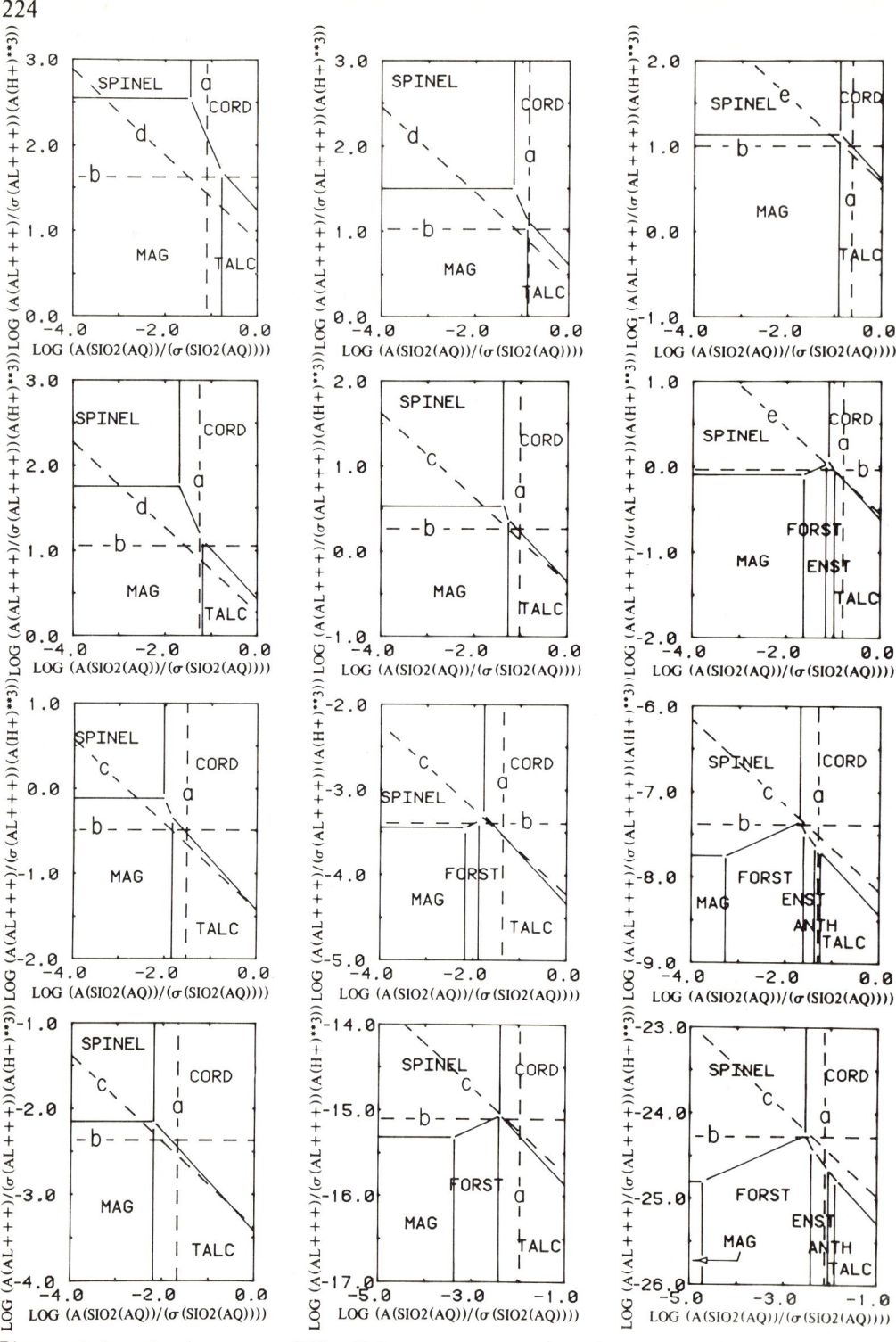

Phase relations in the system $HCl-H_2O-Al_2O_3-CO_2-(MgO)-SiO_2$ at $X_{CO_2} = 0.70$. Metastable 7-A clinochlore was considered instead of its stable counterpart, 14-A clinochlore. Saturation limits: quartz (a), corundum (b), andalusite (c), kyanite (d), sillimanite (e).

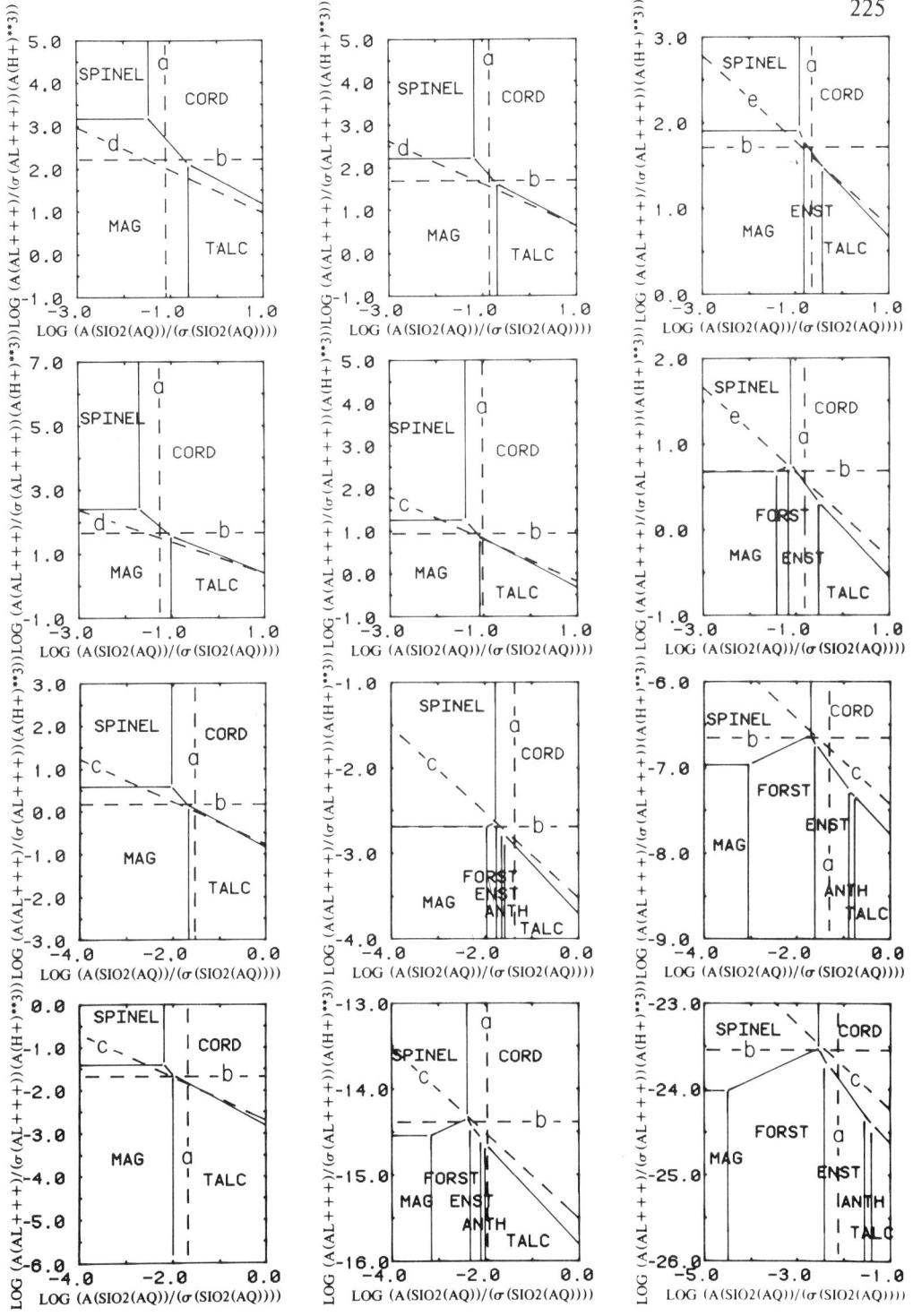

Phase relations in the system $HCl-H_2O-Al_2O_3-CO_2-(MgO)-SiO_2$ at $X_{CO_2} = 0.90$. Metastable 7-A clinochlore was considered instead of its stable counterpart, 14-A clinochlore. Saturation limits: quartz (a), corundum (b), andalusite (c), kyanite (d), sillimanite (e).

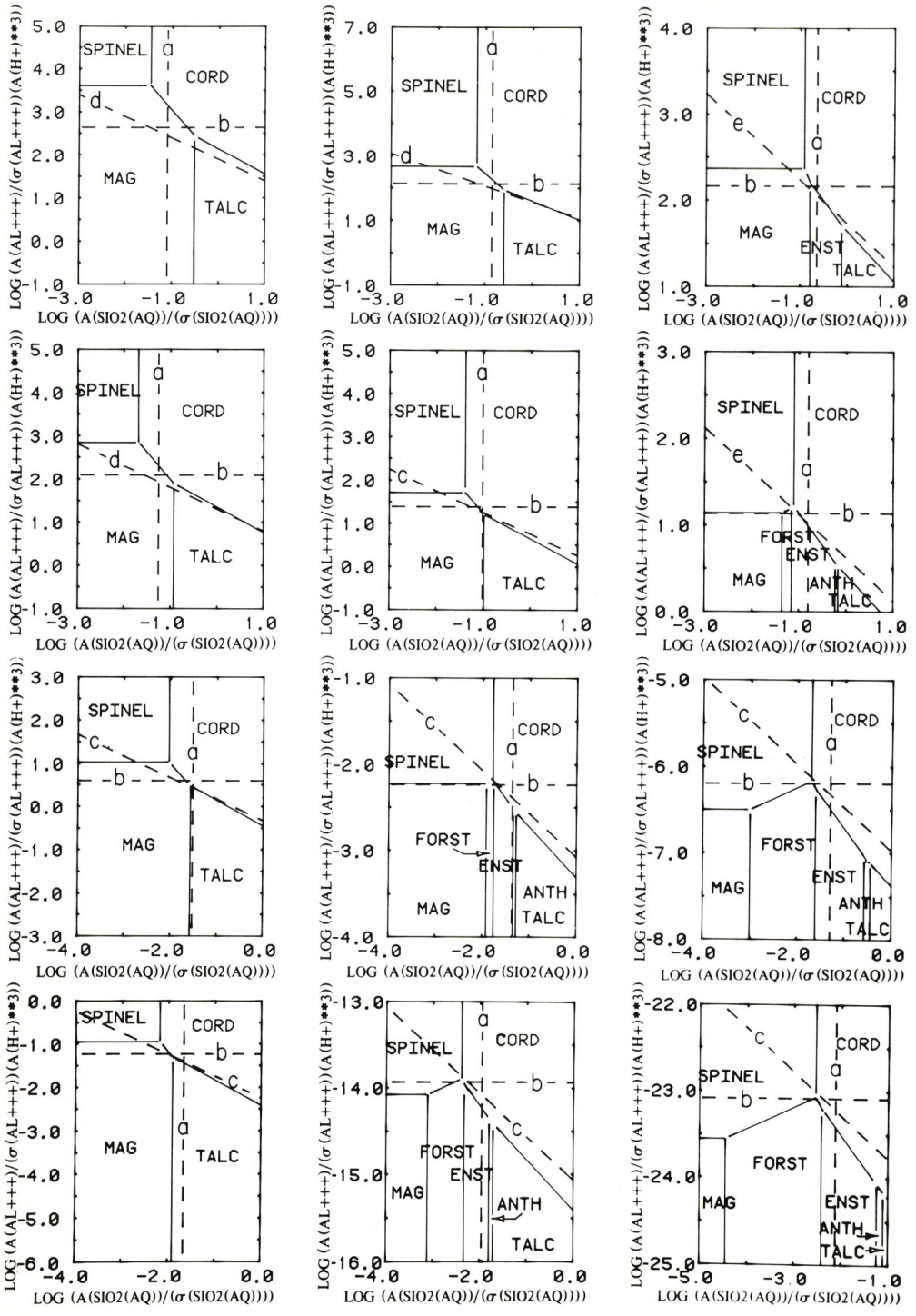

Phase relations in the system $HCl-H_2O-Al_2O_3-CO_2-(MgO)-SiO_2$ at $X_{CO_2} = 0.95$. Metastable 7-A clinochlore was considered instead of its stable counterpart, 14-A clinochlore. Saturation limits: quartz (a), corundum (b), andalusite (c), kyanite (d), sillimanite (e).

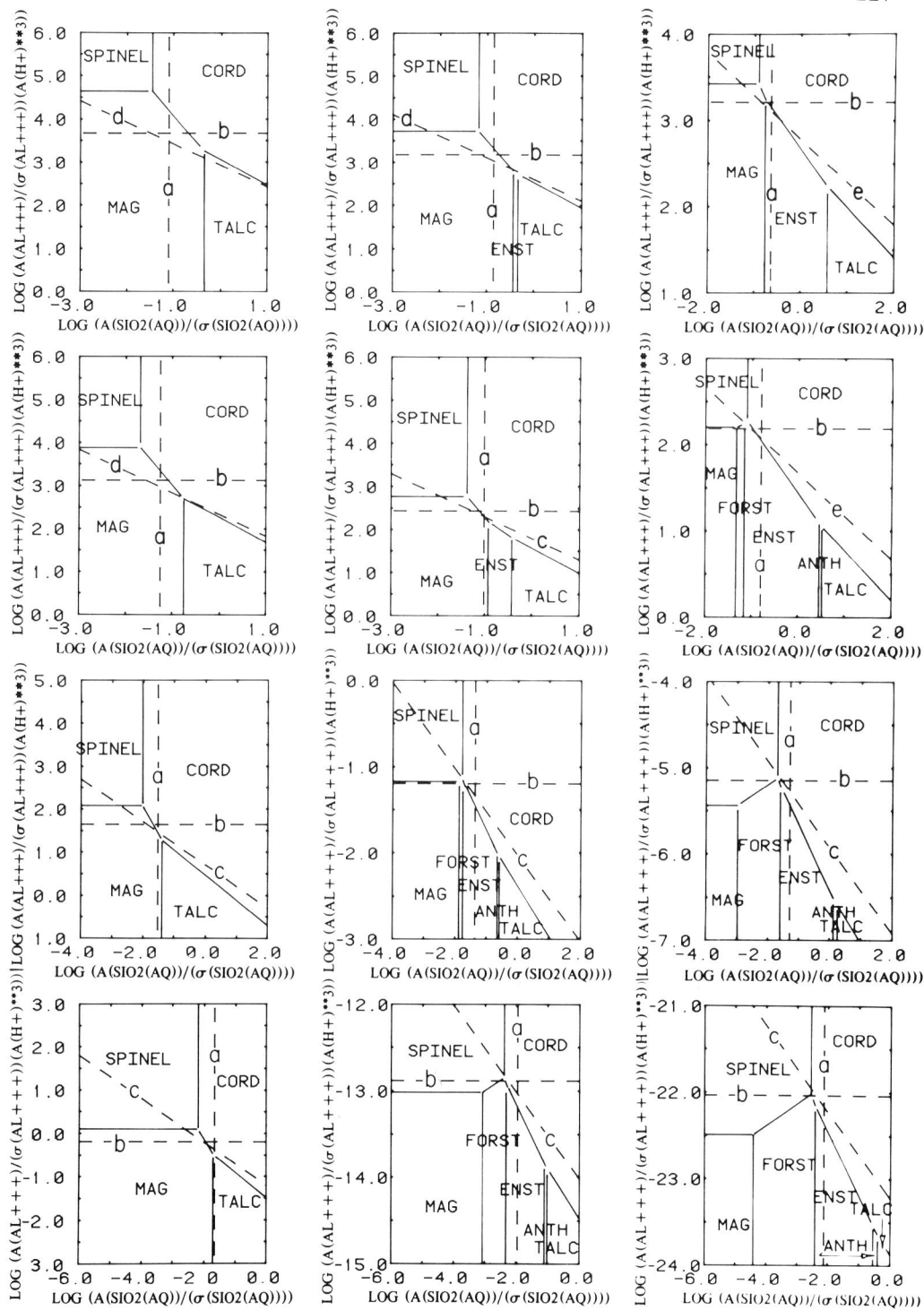

Phase relations in the system $HCl-H_2O-Al_2O_3-CO_2-(MgO)-SiO_2$ at $X_{CO_2} = 0.99$. Metastable 7-A clinochlore was considered instead of its stable counterpart, 14-A clinochlore. Saturation limits: quartz (a), corundum (b), andalusite (c), kyanite (d), sillimanite (e).

Phase relations in the system $HCl-H_2O-Al_2O_3-CO_2-MgO-(SiO_2)$ at $X_{CO_2} = 0.01$. Saturation limits: diaspore (a), corundum (b), magnesite (c), spinel (d), brucite (e), periclase (f).

229

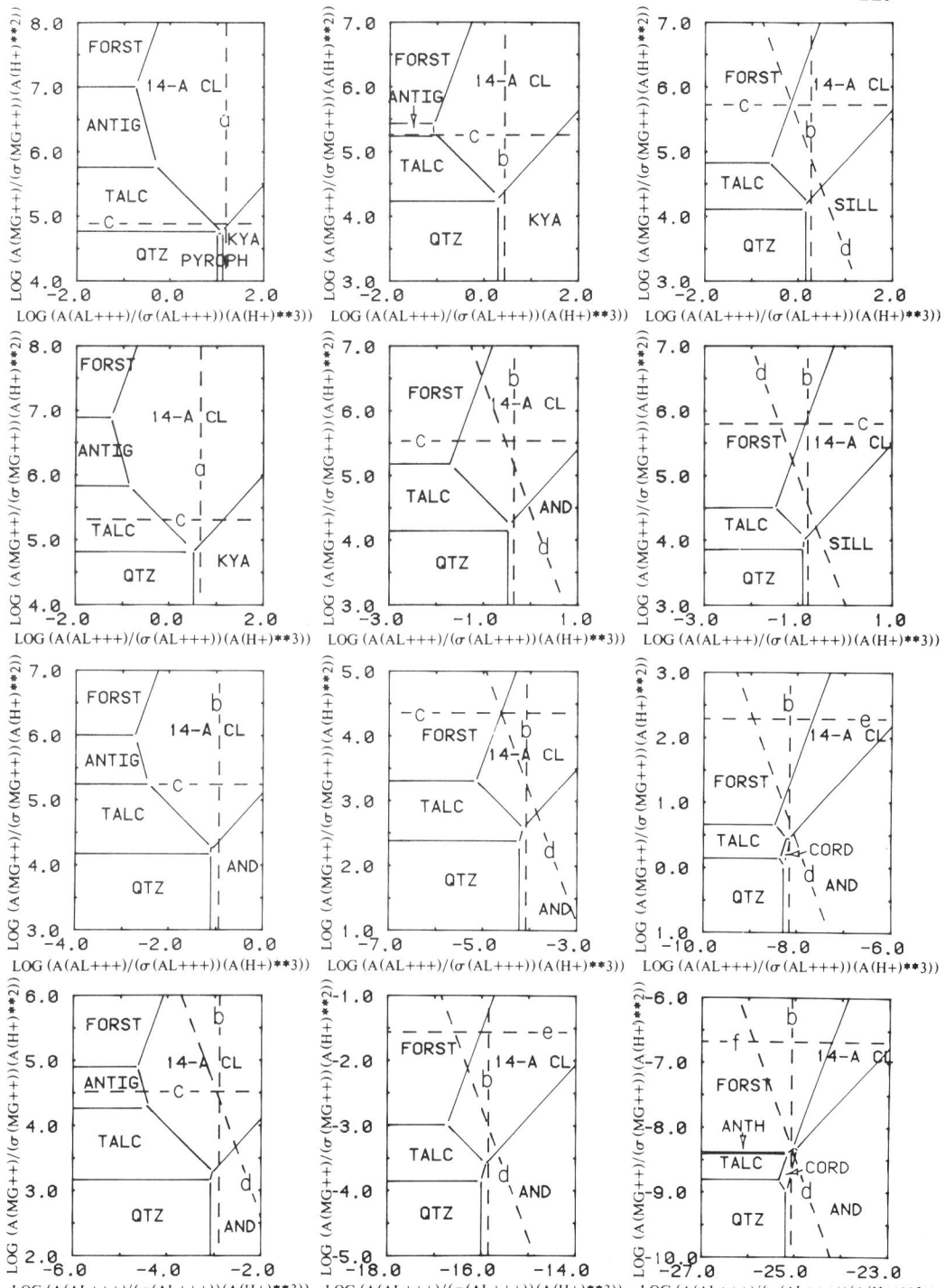

Phase relations in the system $HCl-H_2O-Al_2O_3-CO_2-MgO-(SiO_2)$ at $X_{CO_2} = 0.05$. Saturation limits: diaspore (a), corundum (b), magnesite (c), spinel (d), brucite (e), periclase (f).

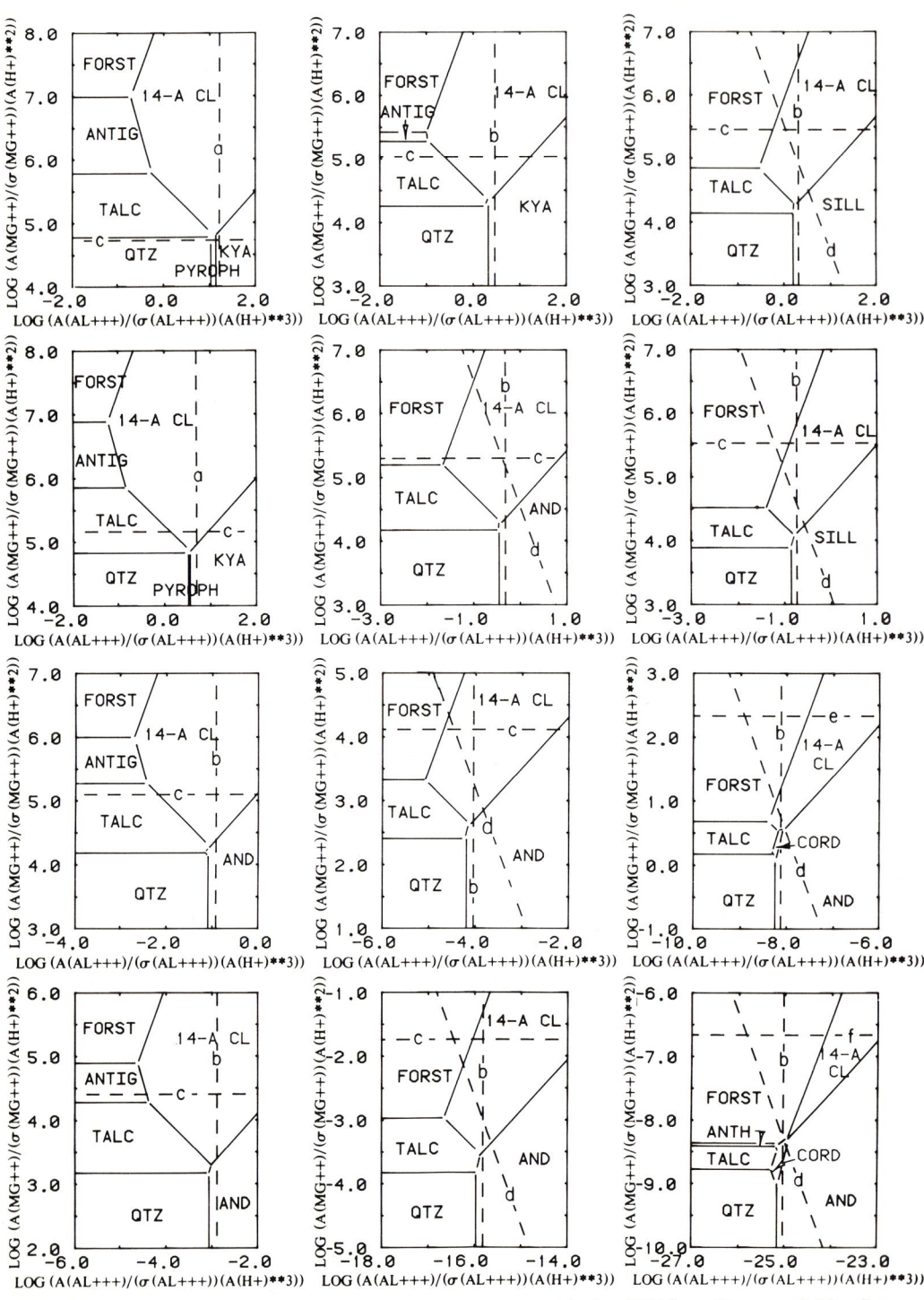

Phase relations in the system $HCl-H_2O-Al_2O_3-CO_2-MgO-(SiO_2)$ at $X_{CO_2} = 0.10$. Saturation limits: diaspore (a), corundum (b), magnesite (c), spinel (d), brucite (e), periclase (f).

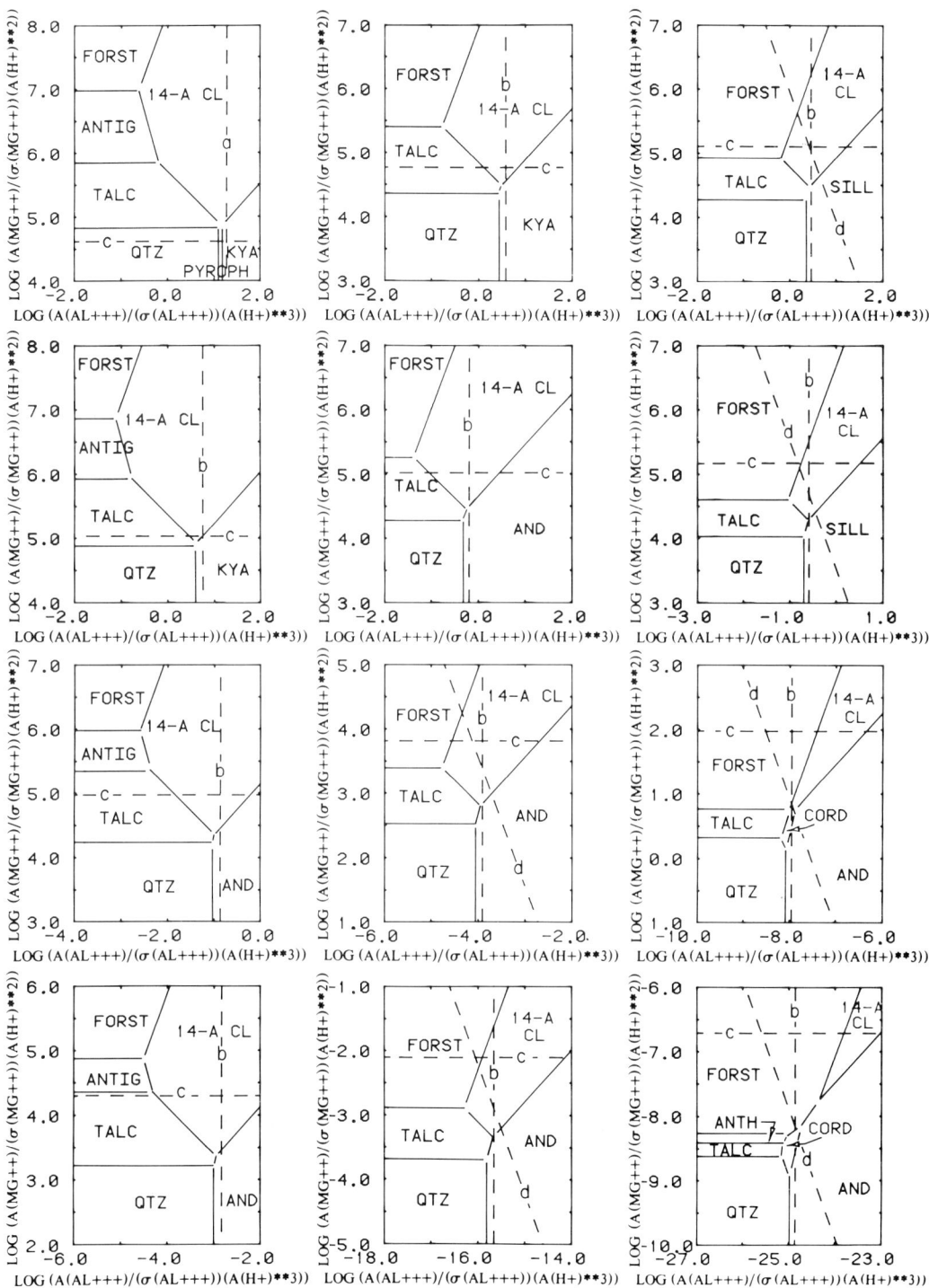

Phase relations in the system $HCl-H_2O-Al_2O_3-CO_2-MgO-(SiO_2)$ at $X_{CO_2} = 0.30$. Saturation limits: diaspore (a), corundum (b), magnesite (c), spinel (d).

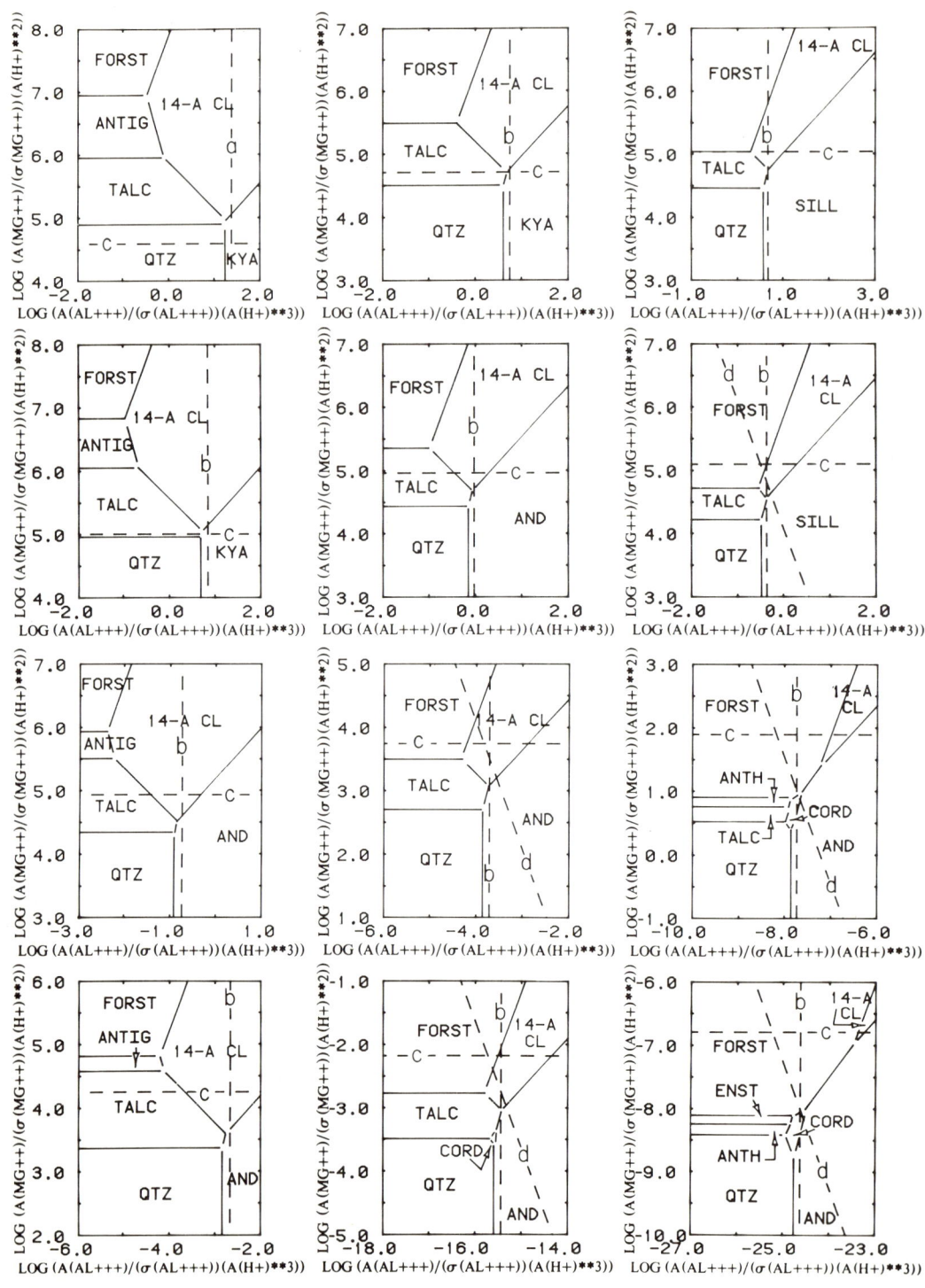

Phase relations in the system $HCl-H_2O-Al_2O_3-CO_2-MgO-(SiO_2)$ at $X_{CO_2} = 0.50$. Saturation limits: diaspore (a), corundum (b), magnesite (c), spinel (d).

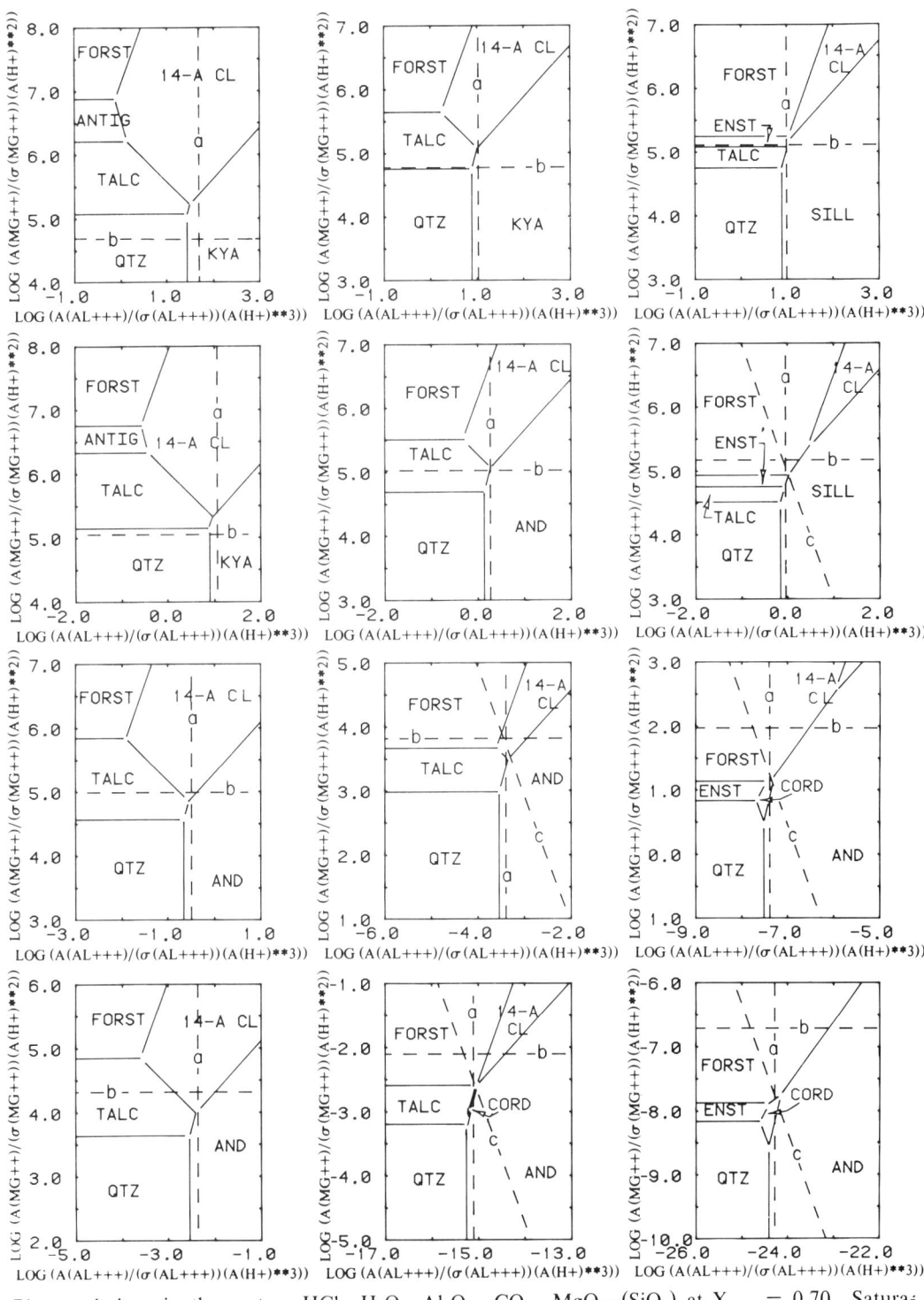

Phase relations in the system $HCl-H_2O-Al_2O_3-CO_2-MgO-(SiO_2)$ at $X_{CO_2} = 0.70$. Saturation limits: corundum (a), magnesite (b), spinel (c).

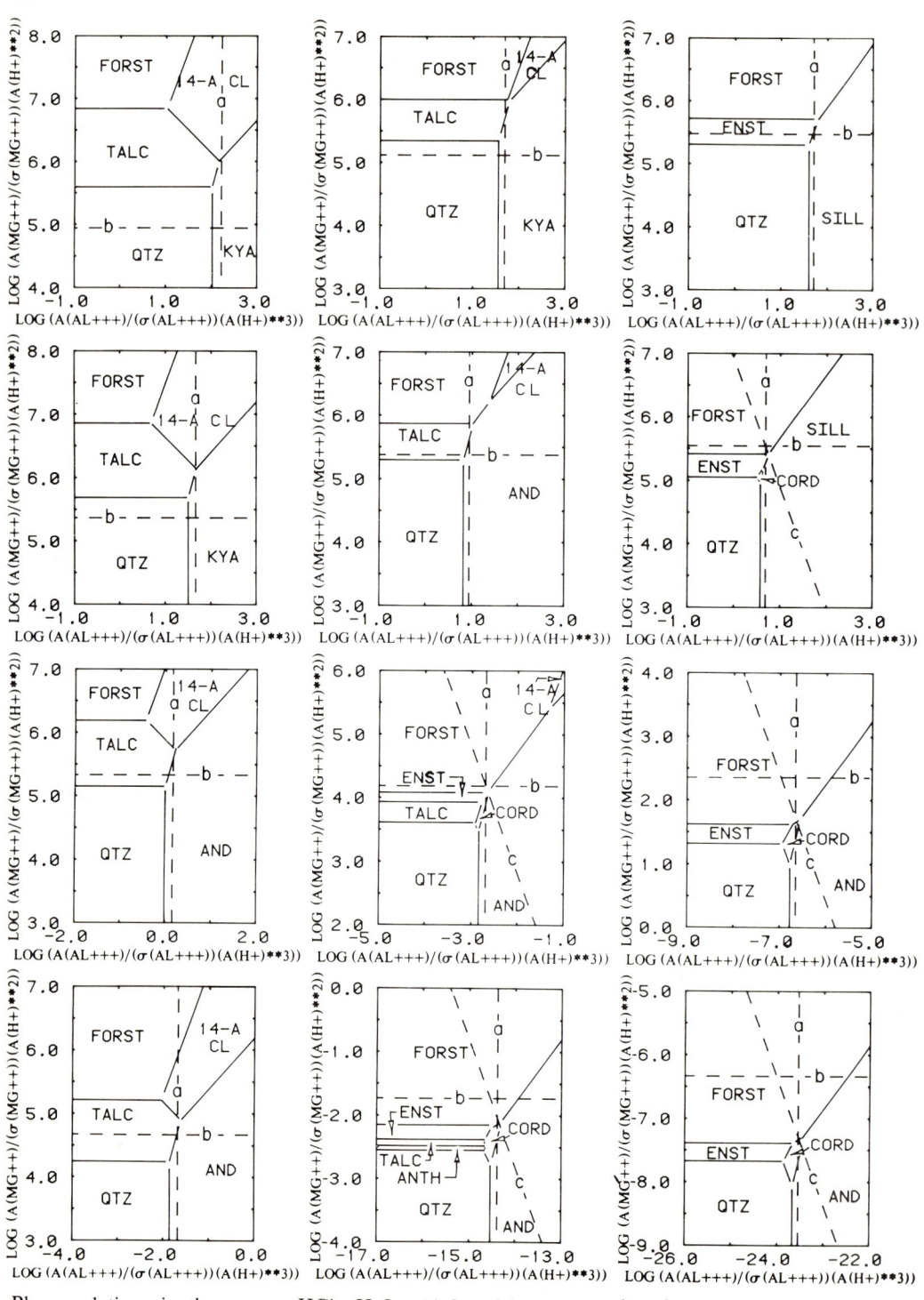

Phase relations in the system $HCl-H_2O-Al_2O_3-CO_2-MgO-(SiO_2)$ at $X_{CO_2} = 0.90$. Saturation limits: corundum (a), magnesite (b), spinel (c).

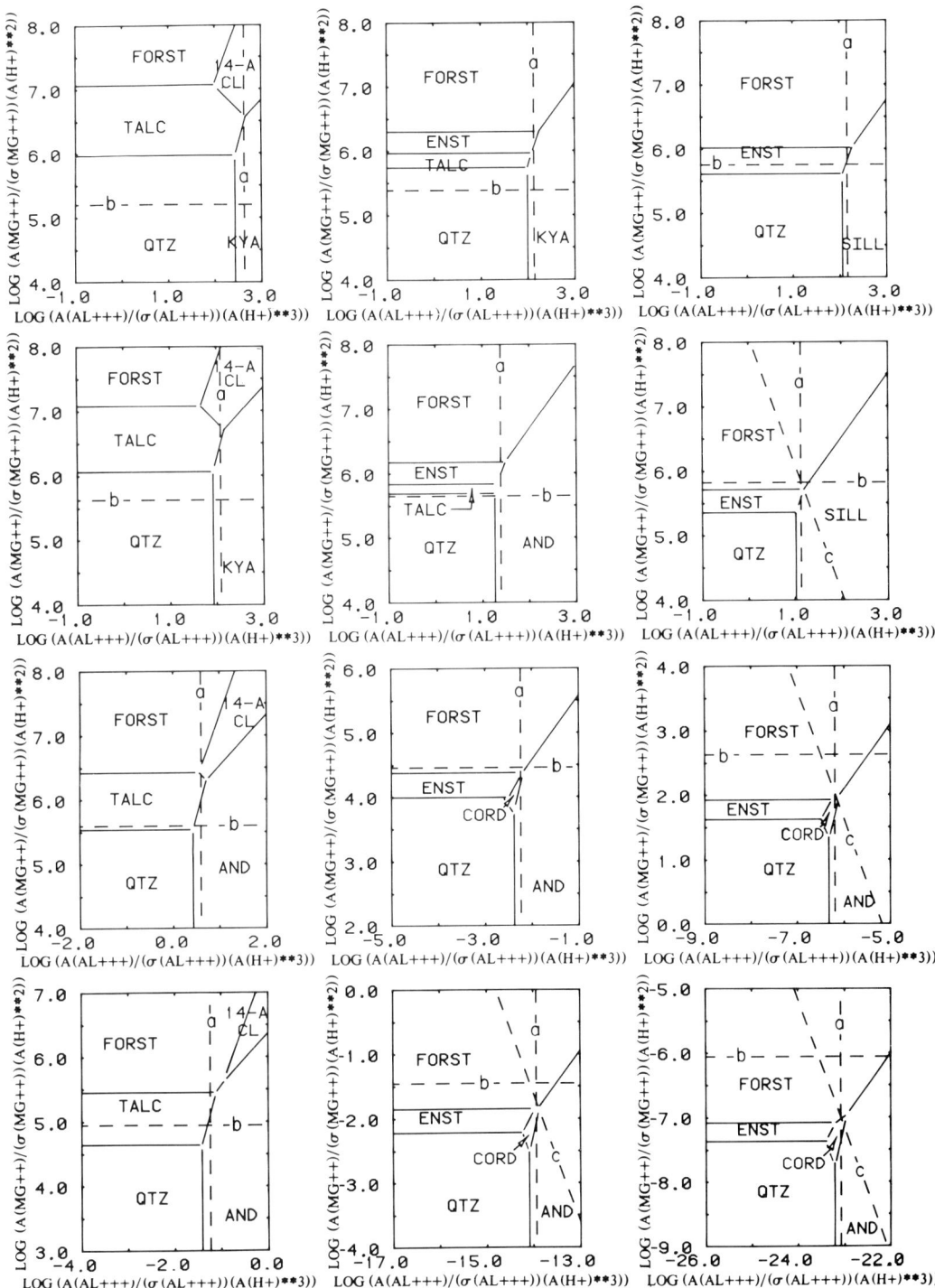

Phase relations in the system $HCl-H_2O-Al_2O_3-CO_2-MgO-(SiO_2)$ at $X_{CO_2} = 0.95$. Saturation limits: corundum (a), magnesite (b), spinel (c).

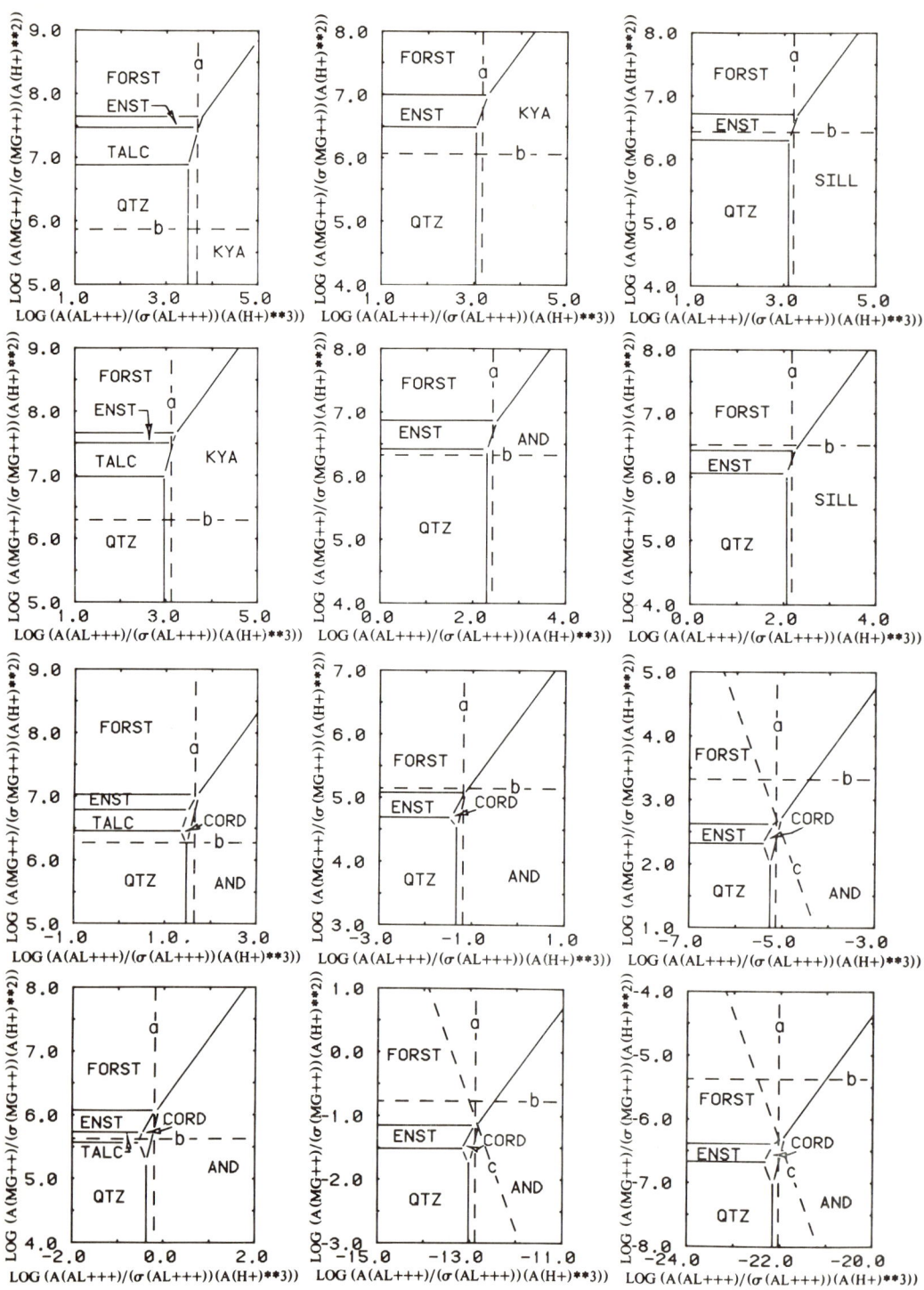

Phase relations in the system $HCl-H_2O-Al_2O_3-CO_2-MgO-(SiO_2)$ at $X_{CO_2} = 0.99$. Saturation limits: corundum (a), magnesite (b), spinel (c).

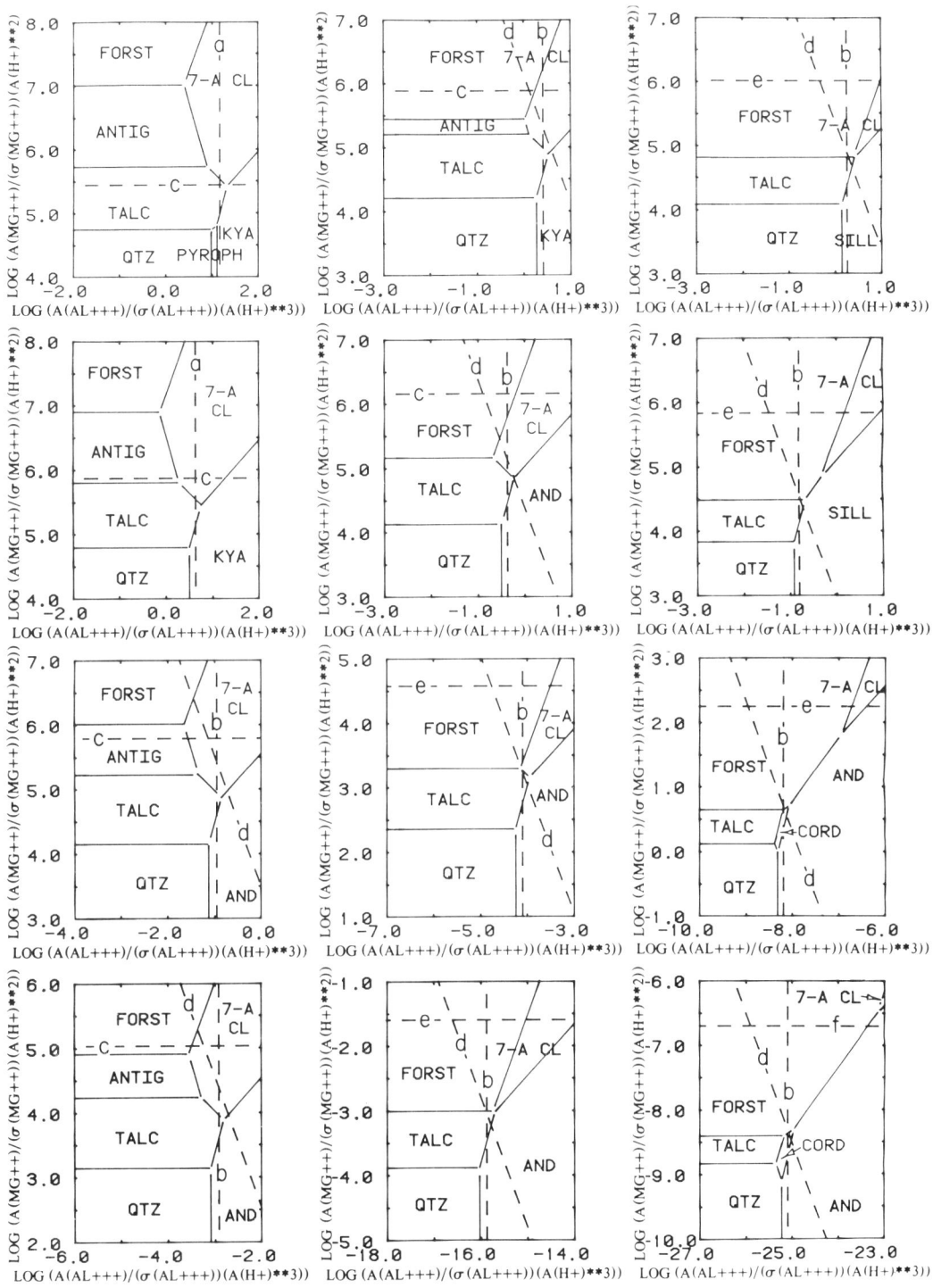

Phase relations in the system $HCl-H_2O-Al_2O_3-CO_2-MgO-(SiO_2)$ at $X_{CO_2} = 0.01$. Metastable 7-A clinochlore was considered instead of its stable counterpart, 14-A clinochlore. Saturation limits: diaspore (a), corundum (b), magnesite (c), spinel (d), brucite (e), periclase (f).

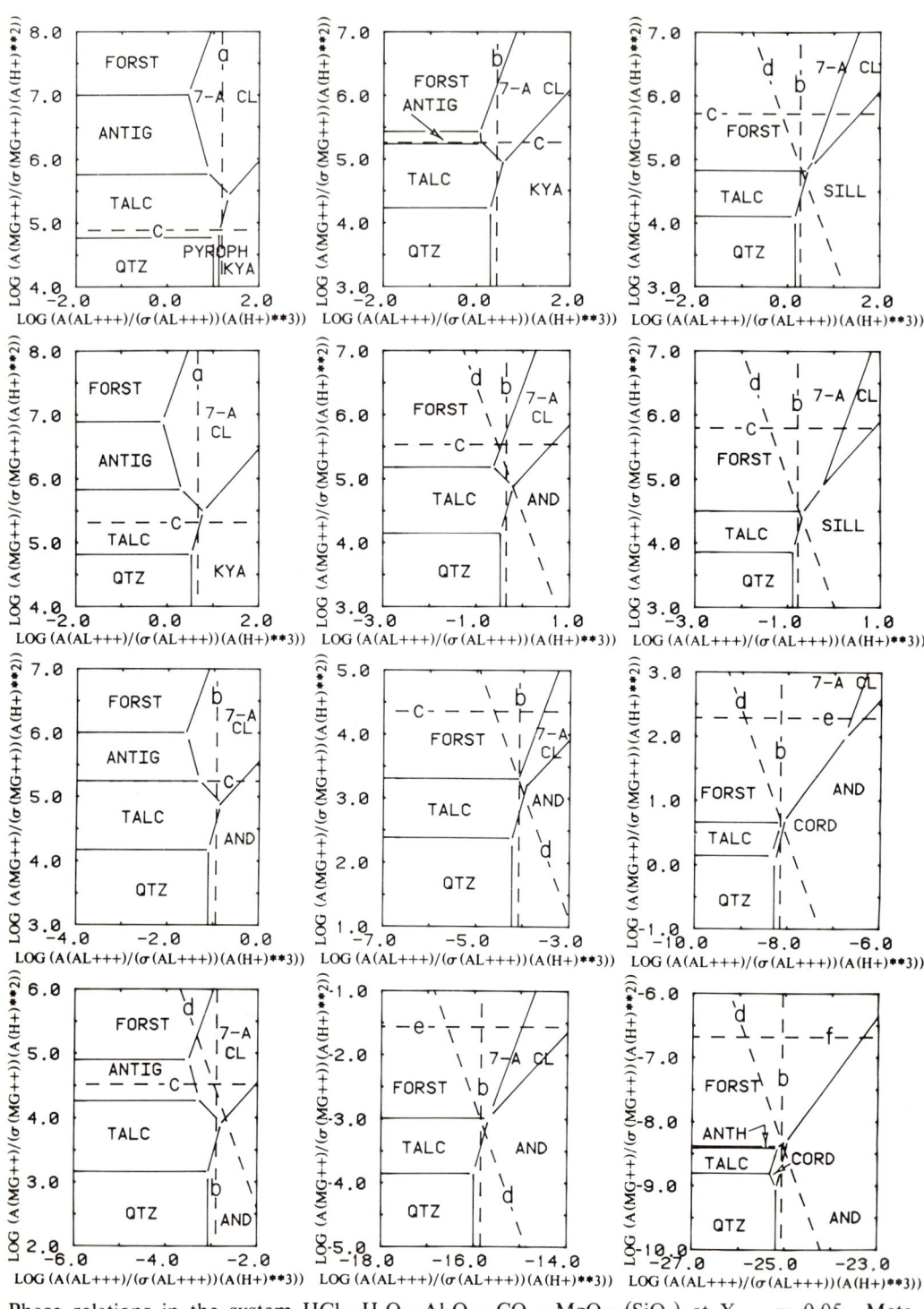

Phase relations in the system $HCl-H_2O-Al_2O_3-CO_2-MgO-(SiO_2)$ at $X_{CO_2} = 0.05$. Metastable 7-A clinochlore was considered instead of its stable counterpart, 14-A clinochlore. Saturation limits: diaspore (a), corundum (b), magnesite (c), spinel (d), brucite (e), periclase (f).

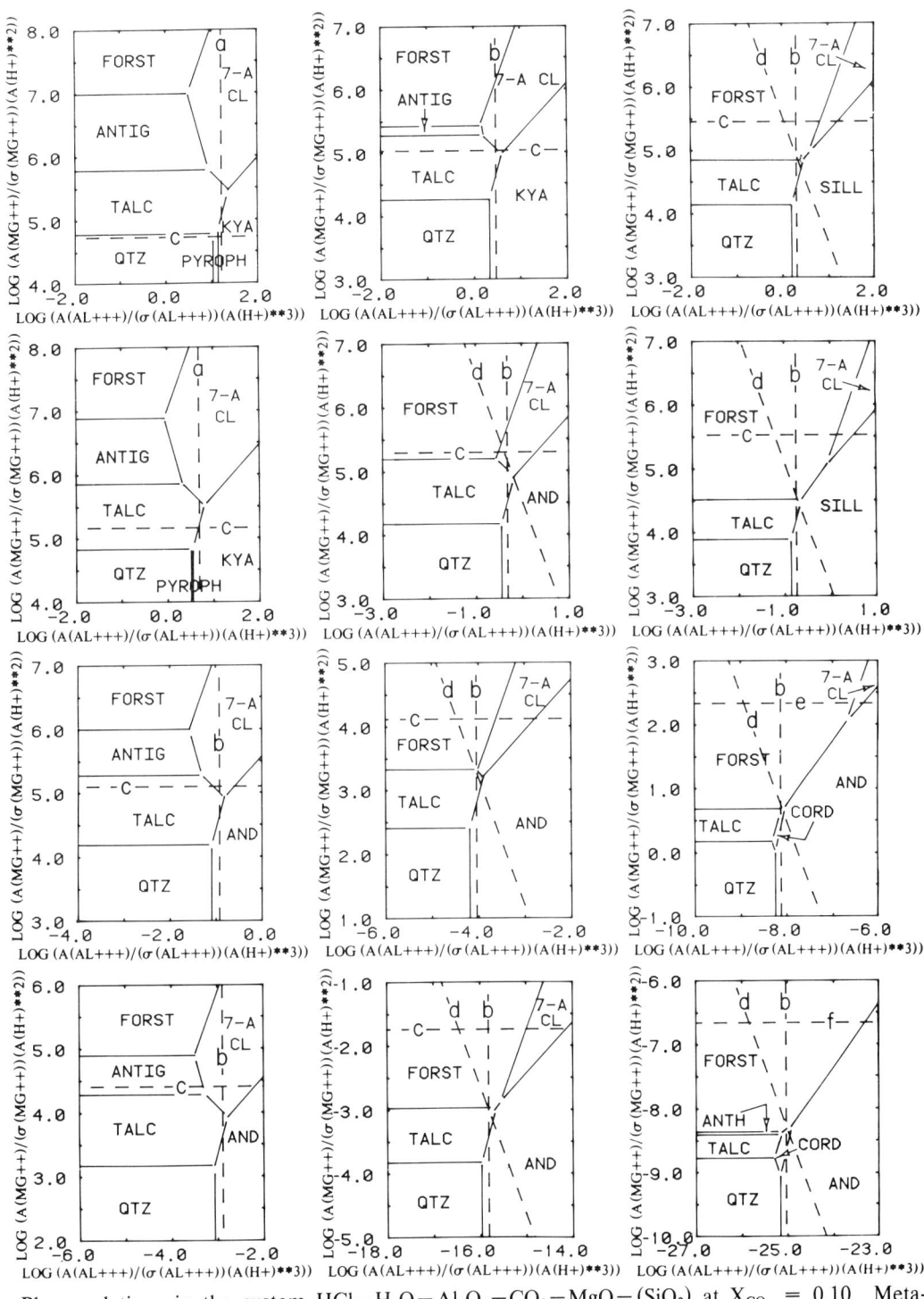

Phase relations in the system $HCl-H_2O-Al_2O_3-CO_2-MgO-(SiO_2)$ at $X_{CO_2} = 0.10$. Metastable 7-A clinochlore was considered instead of its stable counterpart, 14-A clinochlore. Saturation limits: diaspore (a), corundum (b), magnesite (c), spinel (d), brucite (e), periclase (f).

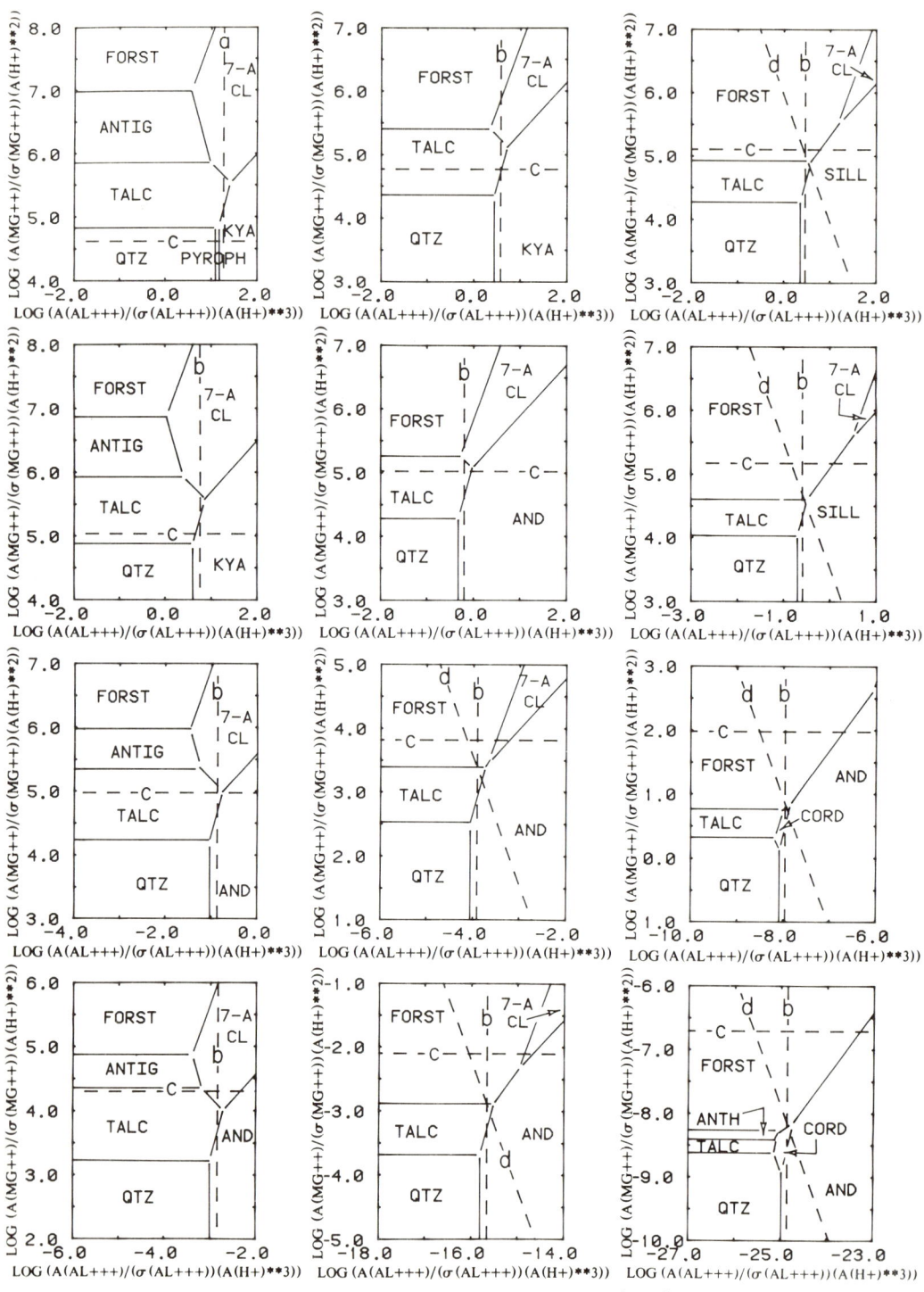

Phase relations in the system $HCl-H_2O-Al_2O_3-CO_2-MgO-(SiO_2)$ at $X_{CO_2} = 0.30$. Metastable 7-A clinochlore was considered instead of its stable counterpart, 14-A clinochlore. Saturation limits: diaspore (a), corundum (b), magnesite (c), spinel (d).

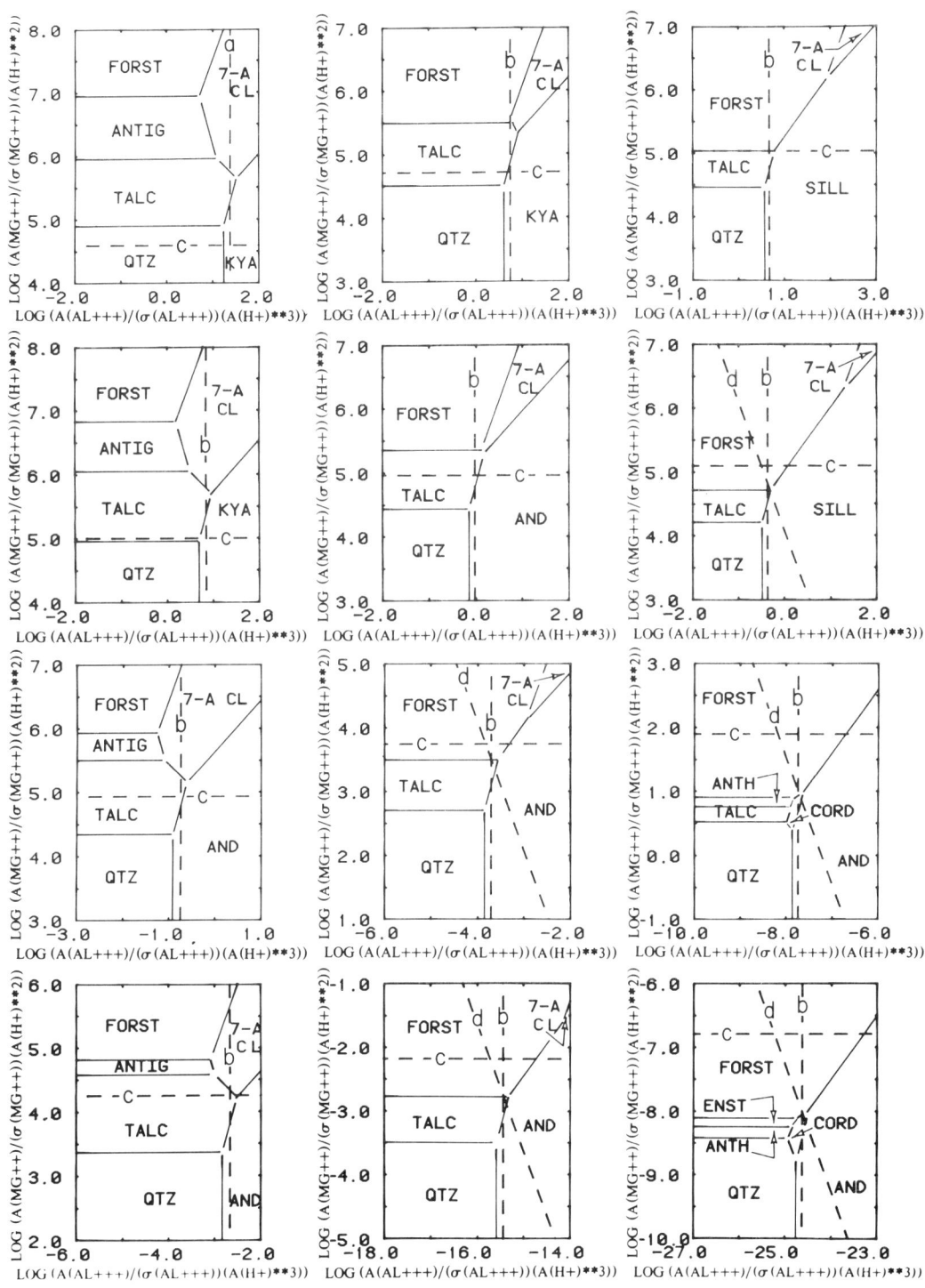

Phase relations in the system $HCl-H_2O-Al_2O_3-CO_2-MgO-(SiO_2)$ $X_{CO_2} = 0.50$. Metastable 7-A clinochlore was considered instead of its stable counterpart, 14-A clinochlore. Saturation limits: diaspore (a), corundum (b), magnesite (c), spinel (d).

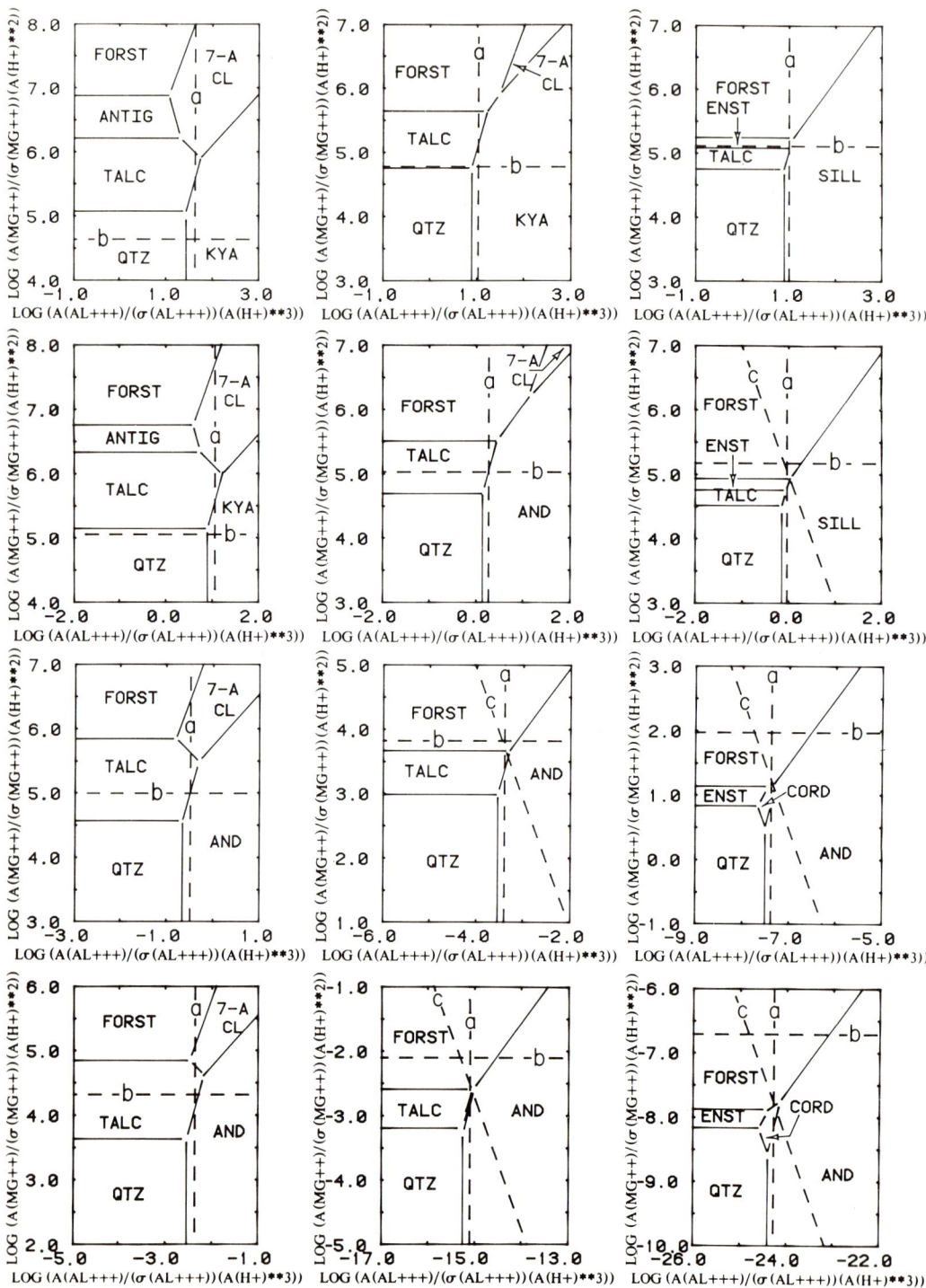

Phase relations in the system $HCl-H_2O-Al_2O_3-CO_2-MgO-(SiO_2)$ at $X_{CO_2} = 0.70$. Metastable 7-A clinochlore was considered instead of its stable counterpart, 14-A clinochlore. Saturation limits: corundum (a), magnesite (b), spinel (c).

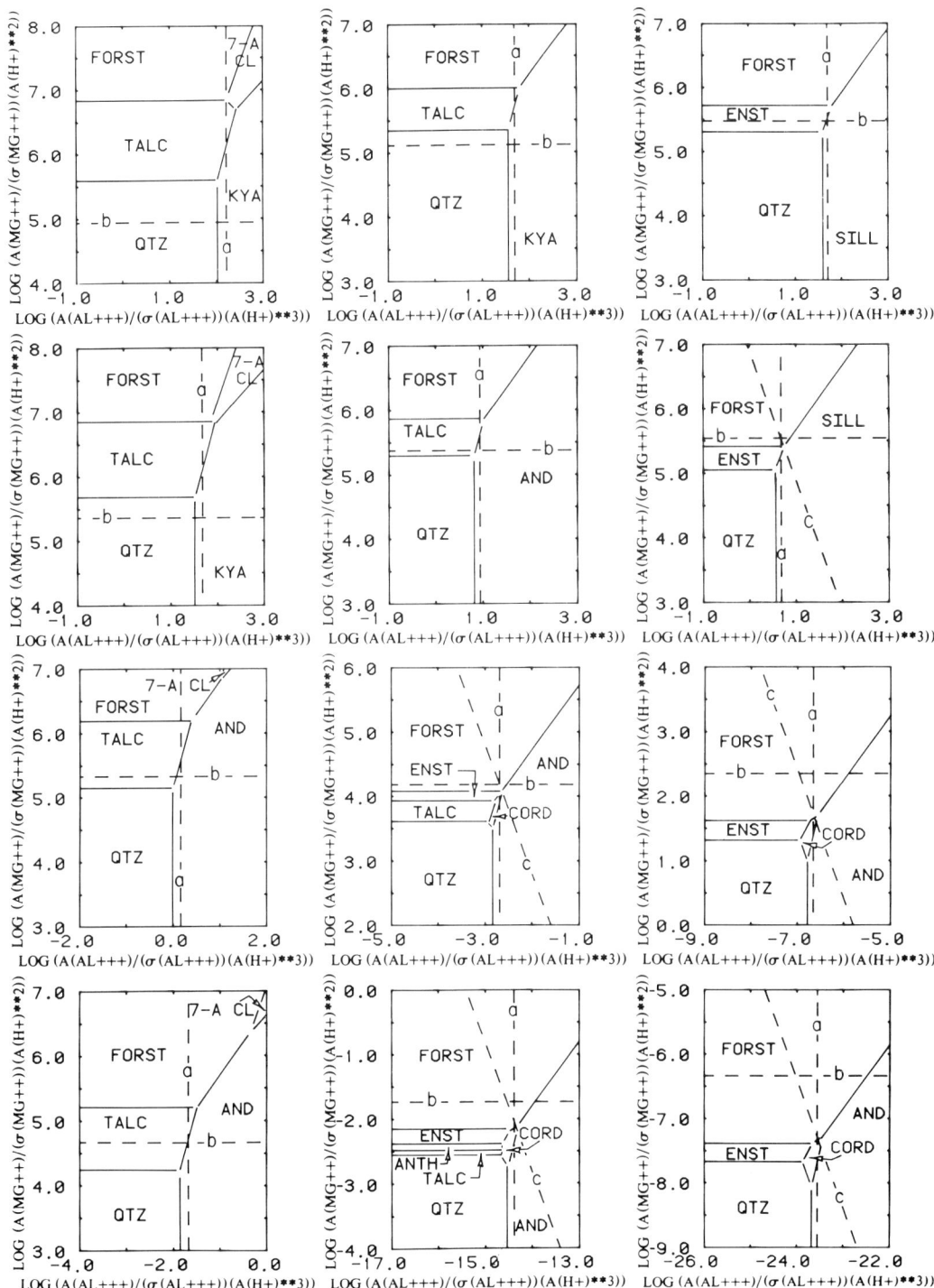

Phase relations in the system $HCl-H_2O-Al_2O_3-CO_2-MgO-(SiO_2)$ at $X_{CO_2} = 0.90$. Metastable 7-A clinochlore was considered instead of its stable counterpart, 14-A clinochlore. Saturation limits: corundum (a), magnesite (b), spinel (c).

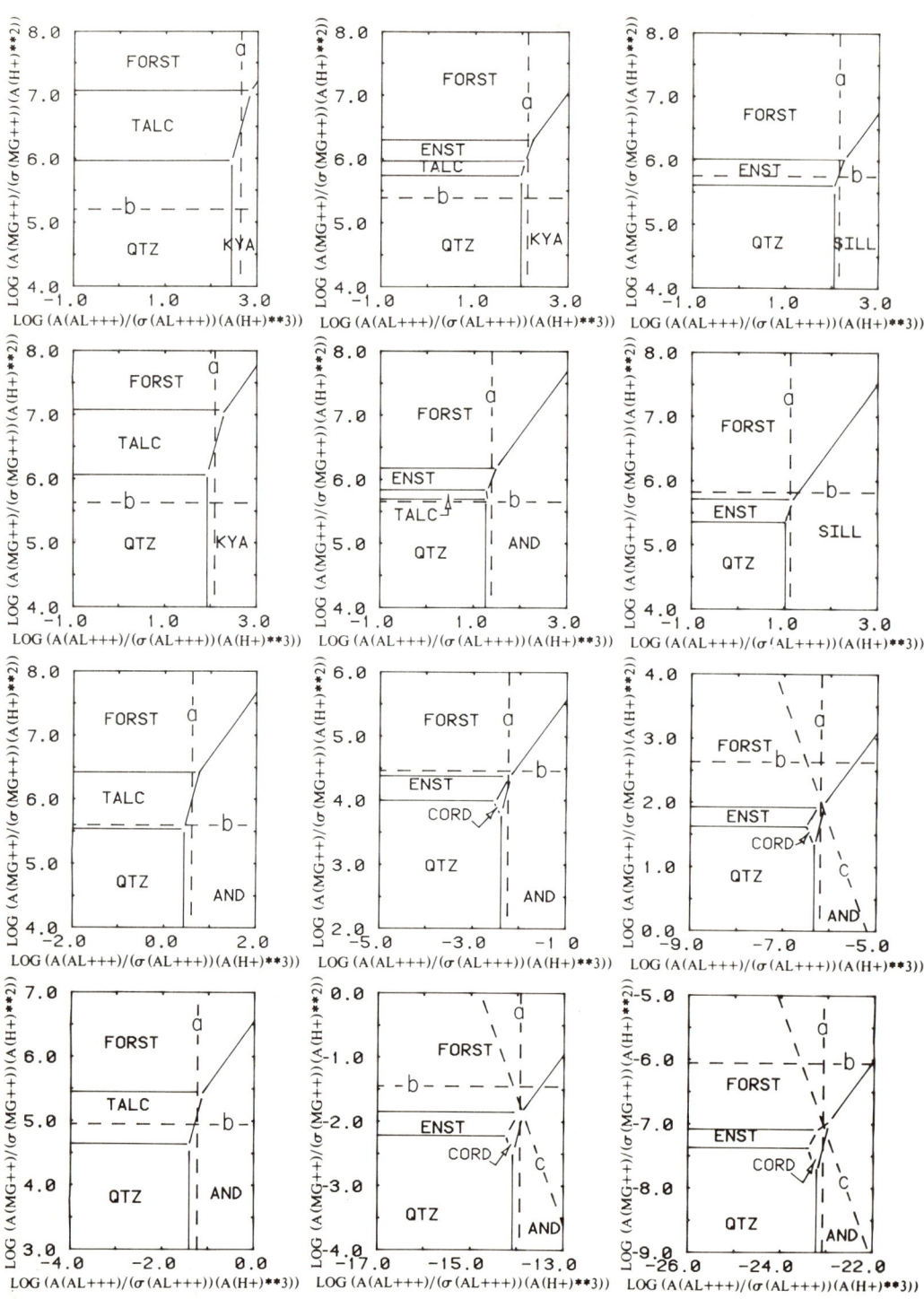

Phase relations in the system $HCl-H_2O-Al_2O_3-CO_2-MgO-(SiO_2)$ at $X_{CO_2} = 0.95$. Metastable 7-A clinochlore was considered instead of its stable counterpart, 14-A clinochlore. Saturation limits: corundum (a), magnesite (b), spinel (c).

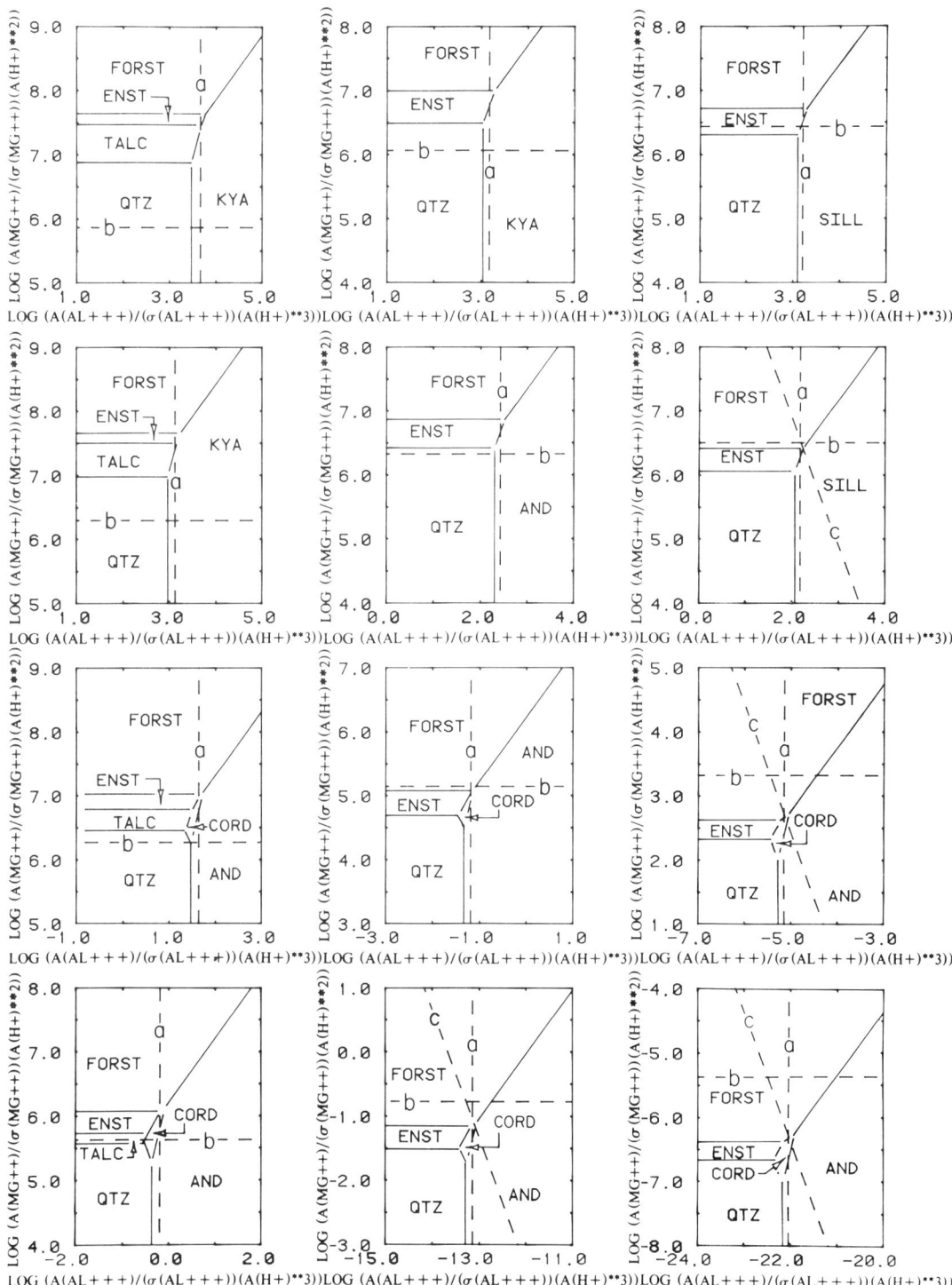

Phase relations in the system $HCl-H_2O-Al_2O_3-CO_2-MgO-(SiO_2)$ at $X_{CO_2} = 0.99$. Metastable 7-A clinochlore was considered instead of its stable counterpart, 14-A clinochlore. Saturation limits: corundum (a), magnesite (b), spinel (c).

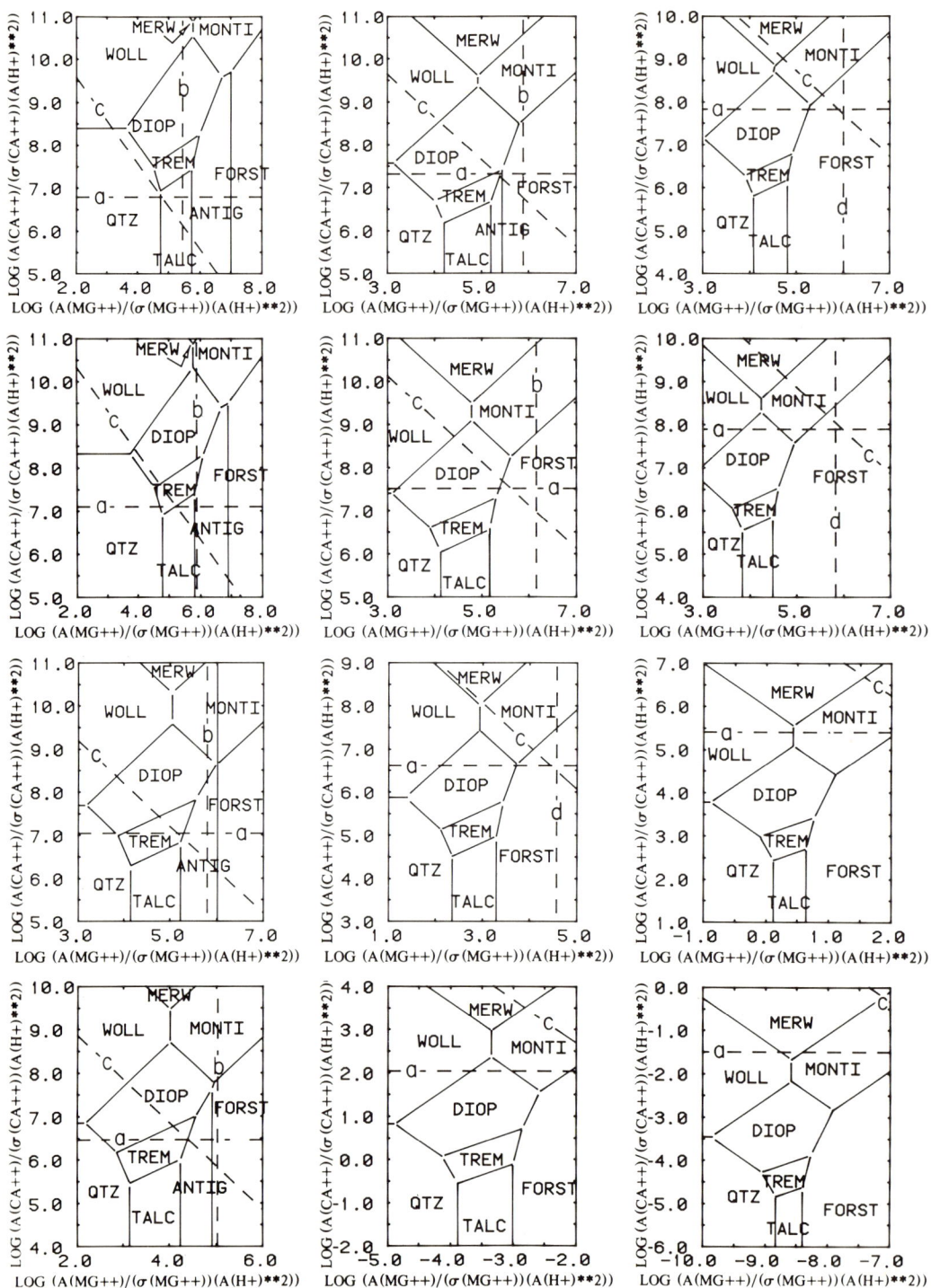

Phase relations in the system $HCl-H_2O-CaO-CO_2-MgO-(SiO_2)$ at $X_{CO_2} = 0.01$. Saturation limits: calcite (a), magnesite (b), dolomite (c), brucite (d).

247

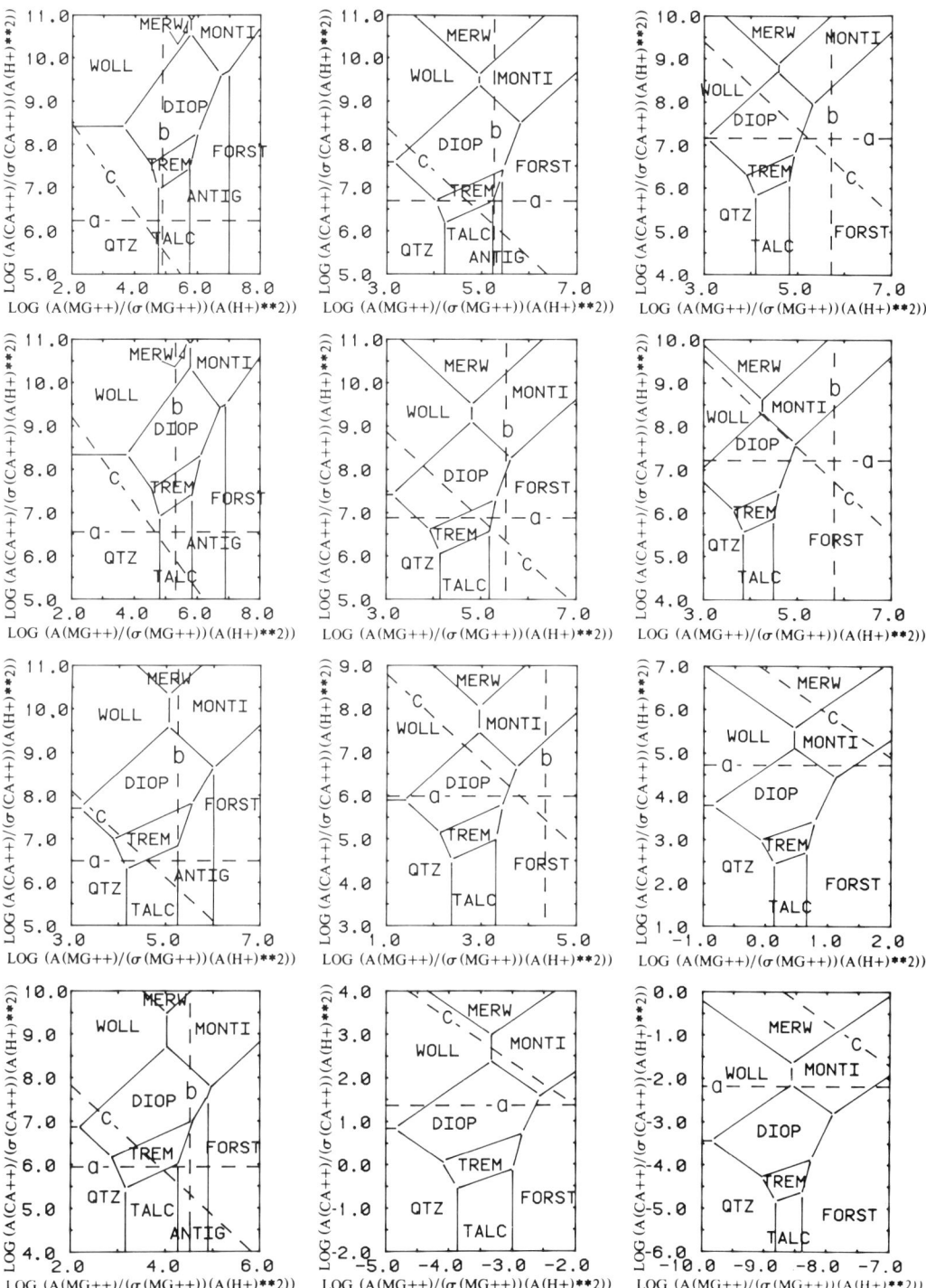

Phase relations in the system $HCl-H_2O-CaO-CO_2-MgO-(SiO_2)$ at $X_{CO_2} = 0.05$. Saturation limits: calcite (a), magnesite (b), dolomite (c).

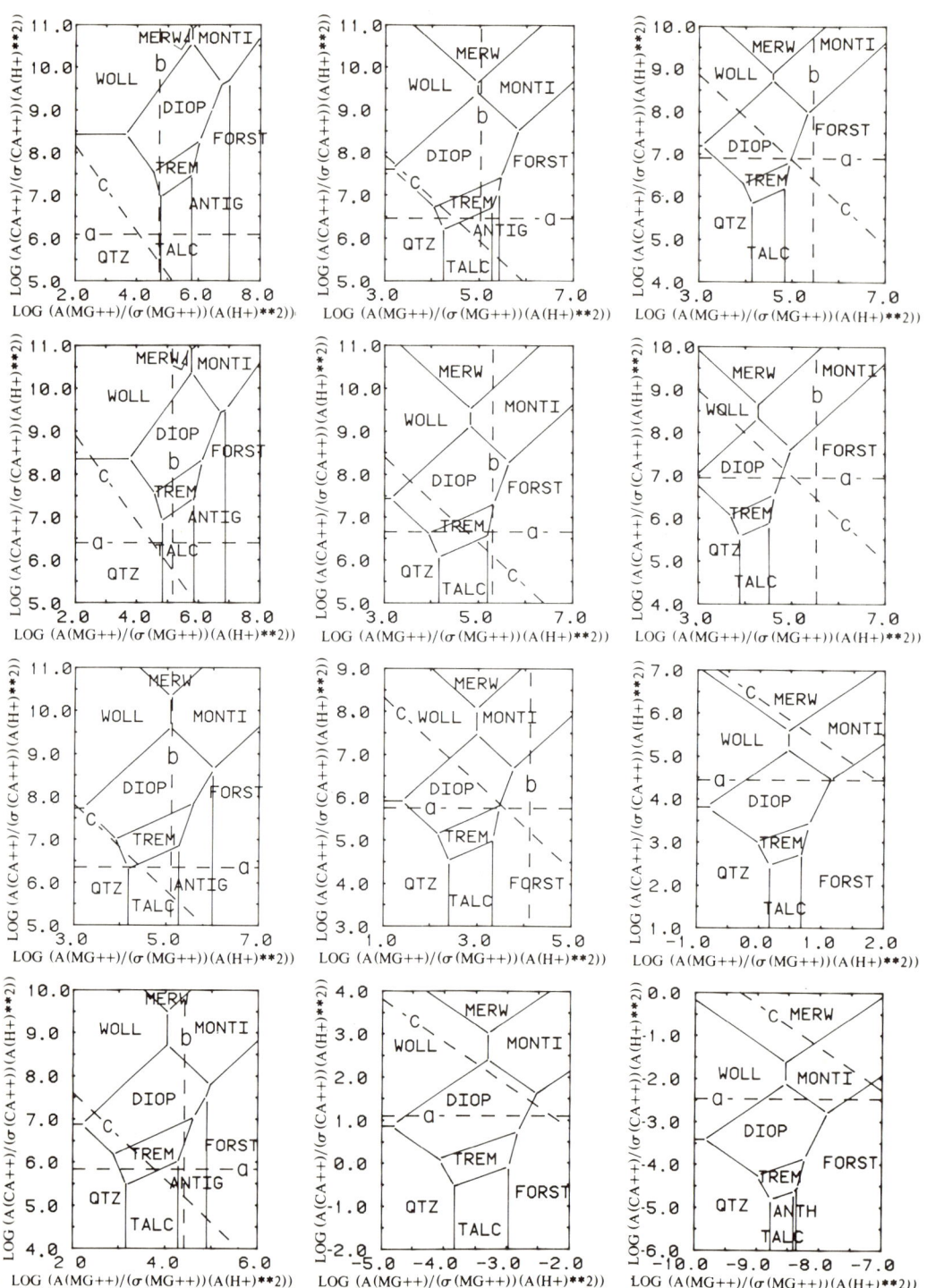

Phase relations in the system $HCl-H_2O-CaO-CO_2-MgO-(SiO_2)$ at $X_{CO_2} = 0.10$. Saturation limits: calcite (a), magnesite (b), dolomite (c).

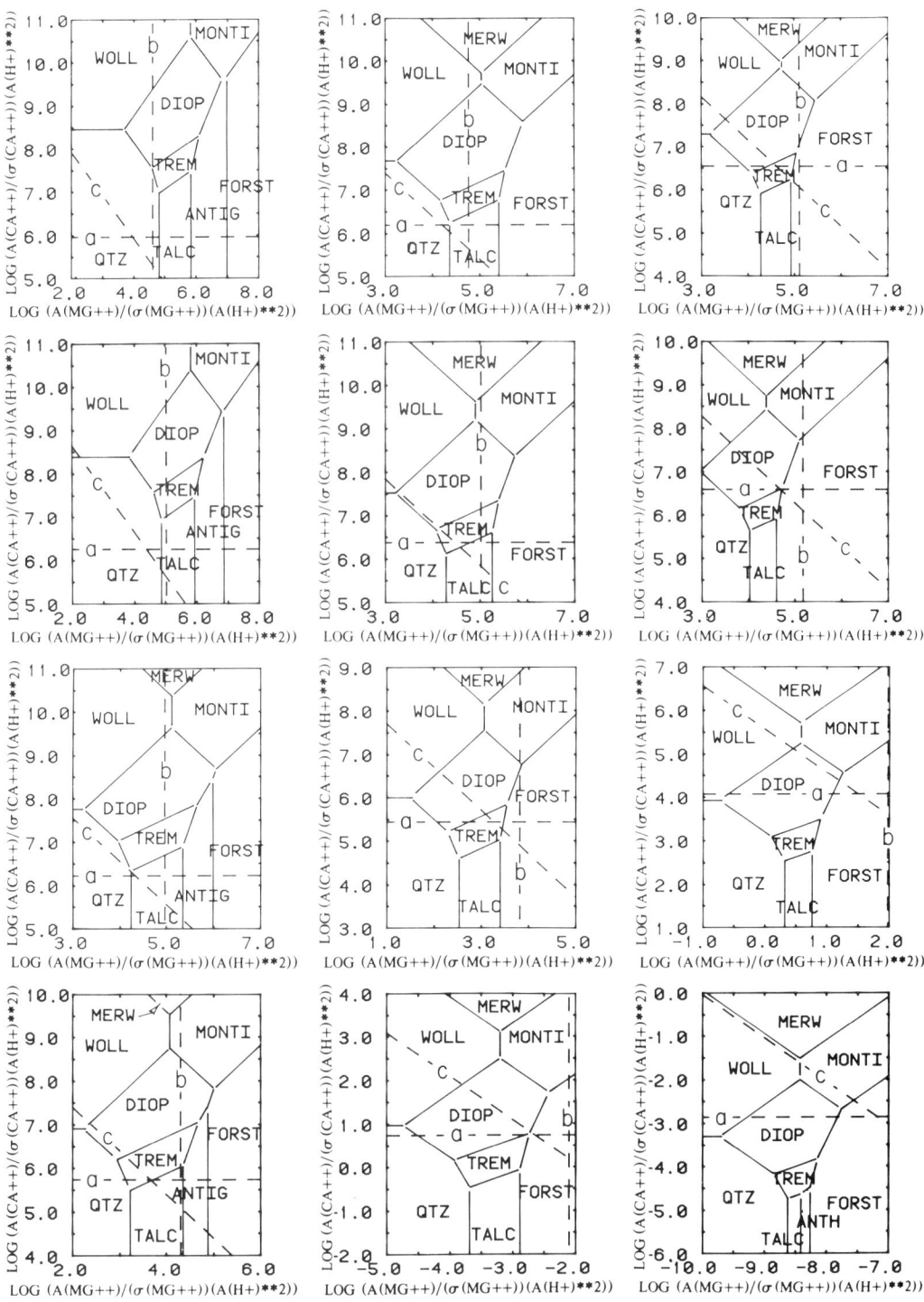

Phase relations in the system $HCl-H_2O-CaO-CO_2-MgO-(SiO_2)$ at $X_{CO_2} = 0.30$. Saturation limits: calcite (a), magnesite (b), dolomite (c).

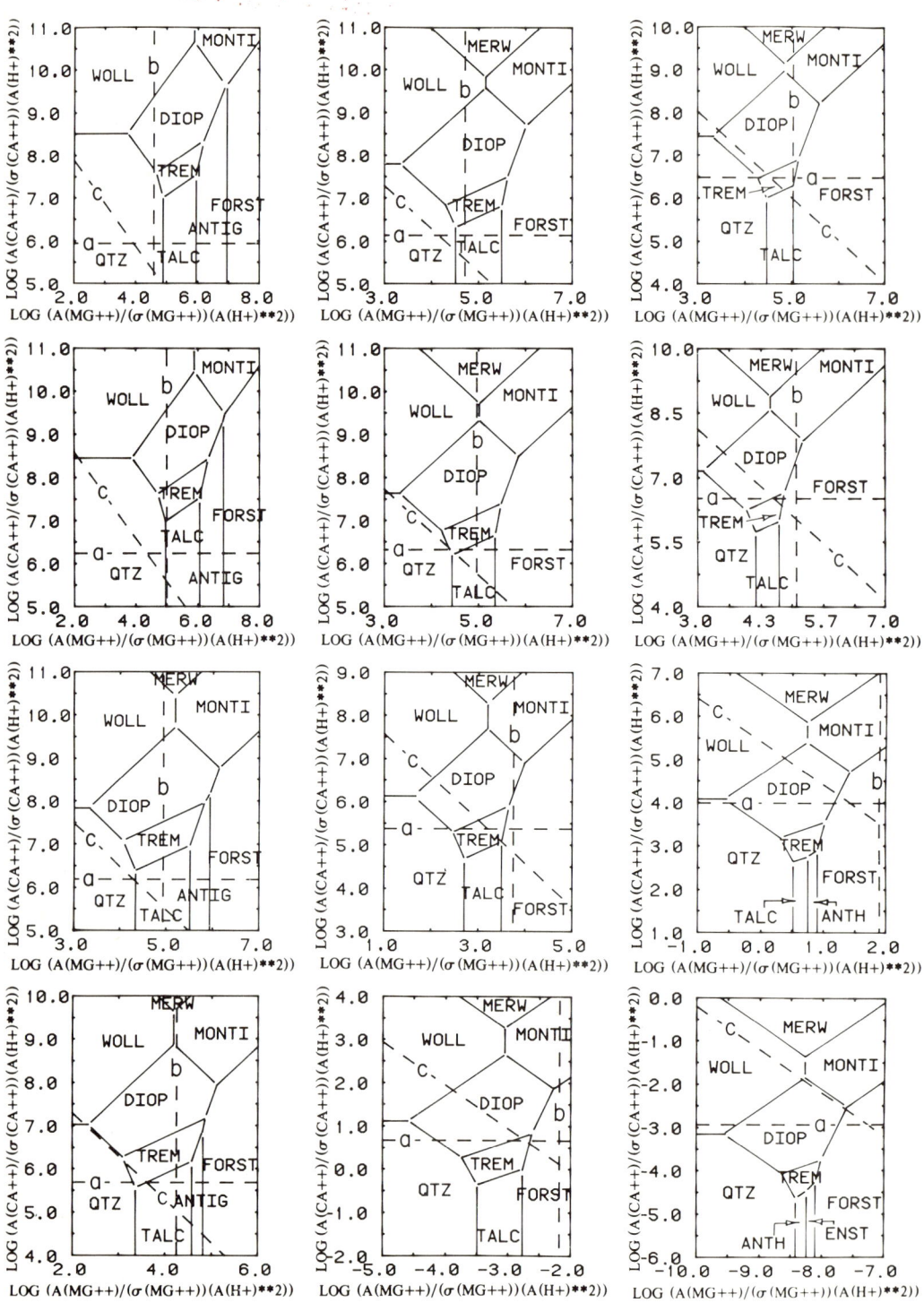

Phase relations in the system $HCl-H_2O-CaO-CO_2-MgO-(SiO_2)$ at $X_{CO_2} = 0.50$. Saturation limits: calcite (a), magnesite (b), dolomite (c).

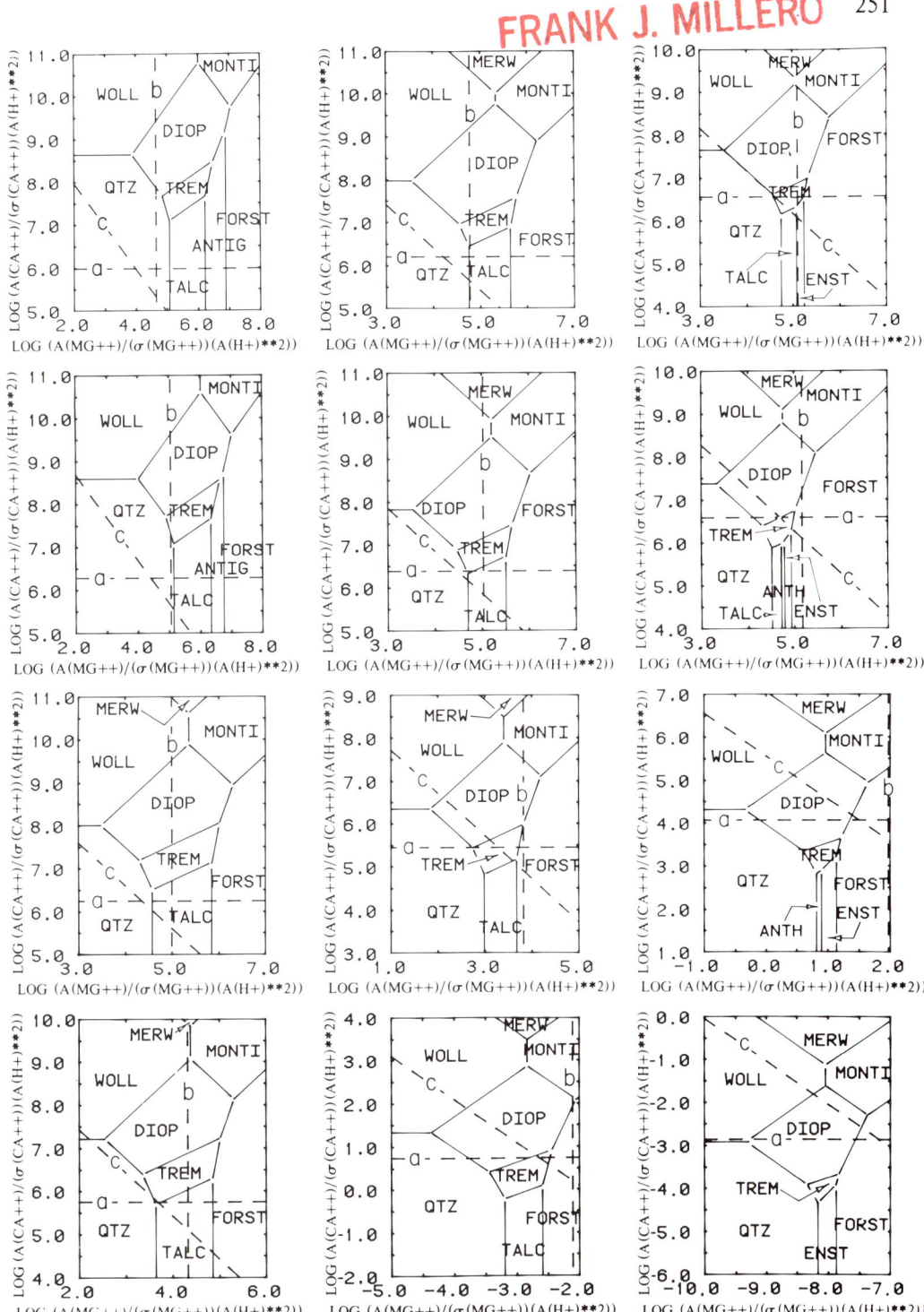

Phase relations in the system $HCl-H_2O-CaO-CO_2-MgO-(SiO_2)$ at $X_{CO_2} = 0.70$. Saturation limits: calcite (a), magnesite (b), dolomite (c).

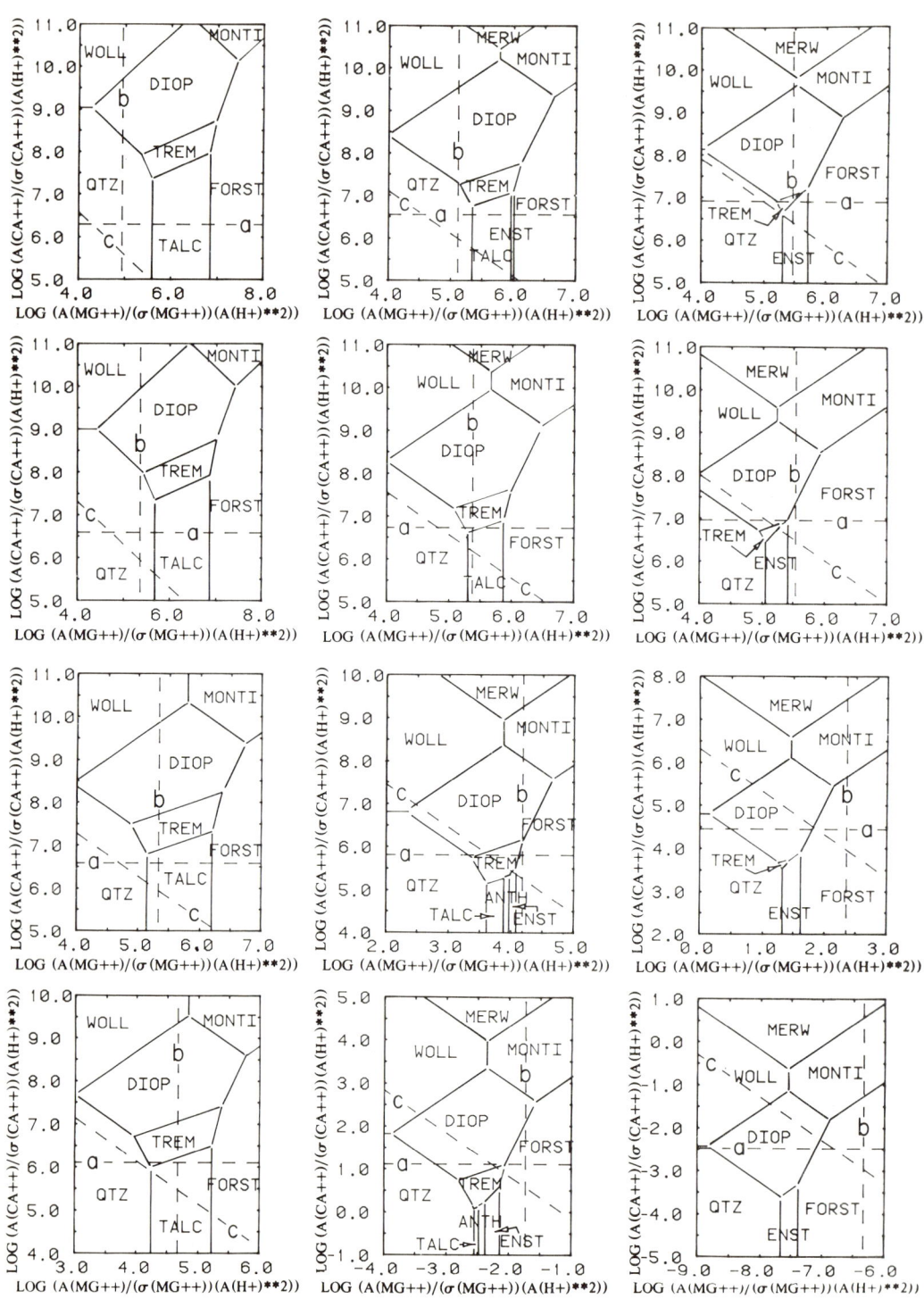

Phase relations in the system $HCl-H_2O-CaO-CO_2-MgO-(SiO_2)$ at $X_{CO_2} = 0.90$. Saturation limits: calcite (a), magnesite (b), dolomite (c).

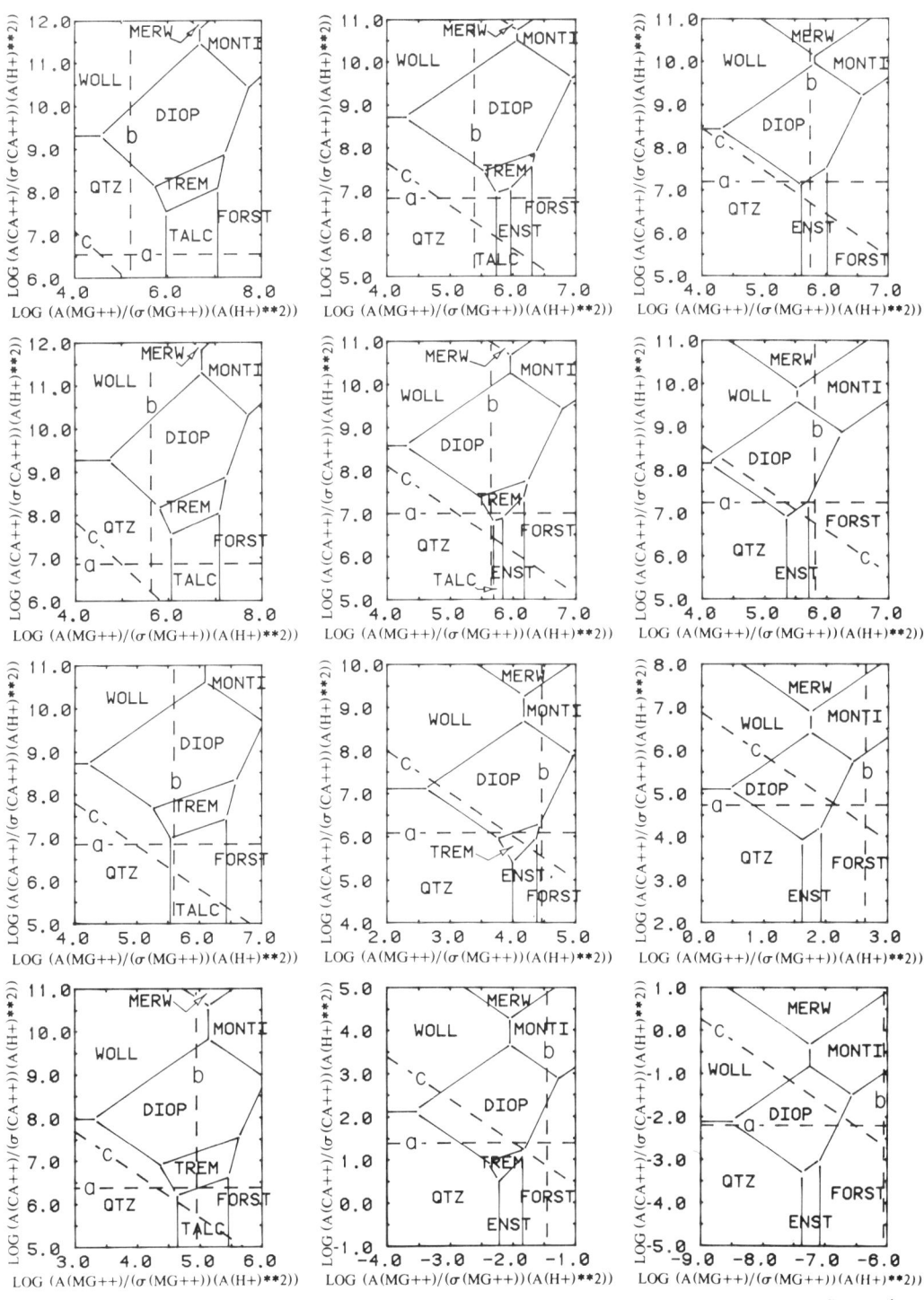

Phase relations in the system $HCl-H_2O-CaO-CO_2-MgO-(SiO_2)$ at $X_{CO_2} = 0.95$. Saturation limits: calcite (a), magnesite (b), dolomite (c).

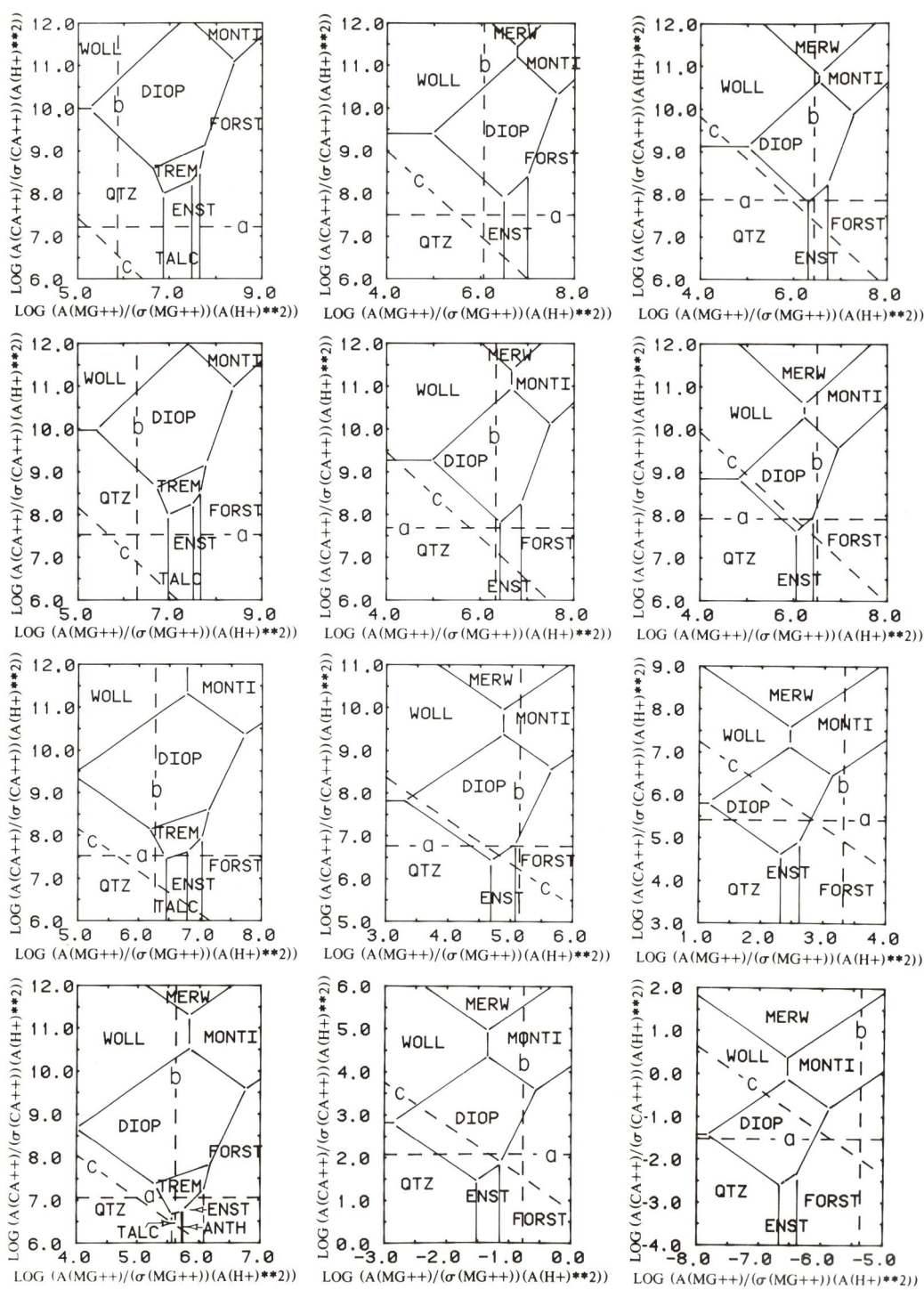

Phase relations in the system $HCl-H_2O-CaO-CO_2-MgO-(SiO_2)$ at $X_{CO_2} = 0.99$. Saturation limits: calcite (a), magnesite (b), dolomite (c).

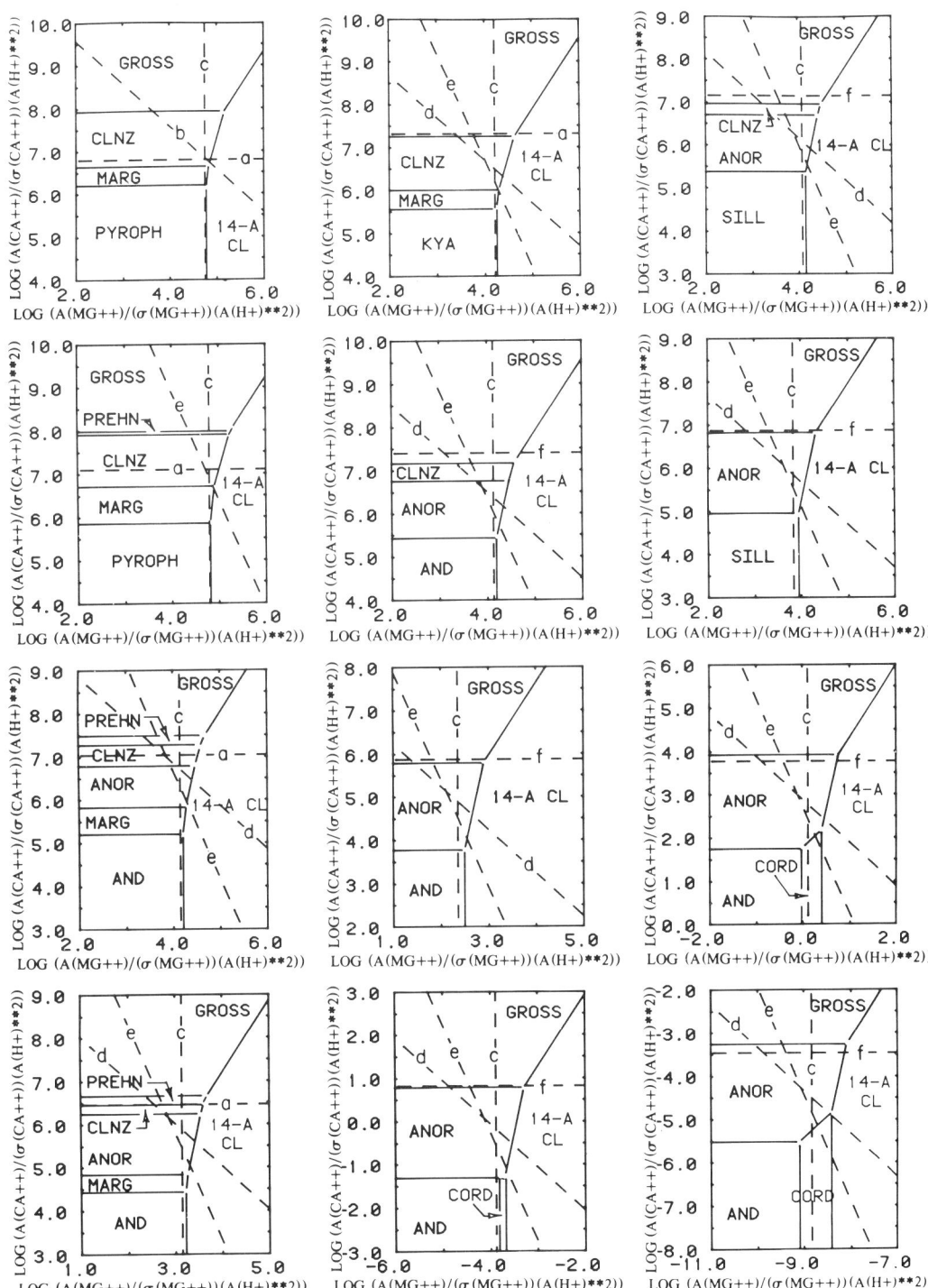

Phase relations in the system $HCl-H_2O-(Al_2O_3)-CaO-CO_2-MgO-SiO_2$, in equilibrium with quartz, at $X_{CO_2} = 0.01$. Saturation limits: calcite (a), dolomite (b), talc (c), diopside (d), tremolite (e), wollastonite (f).

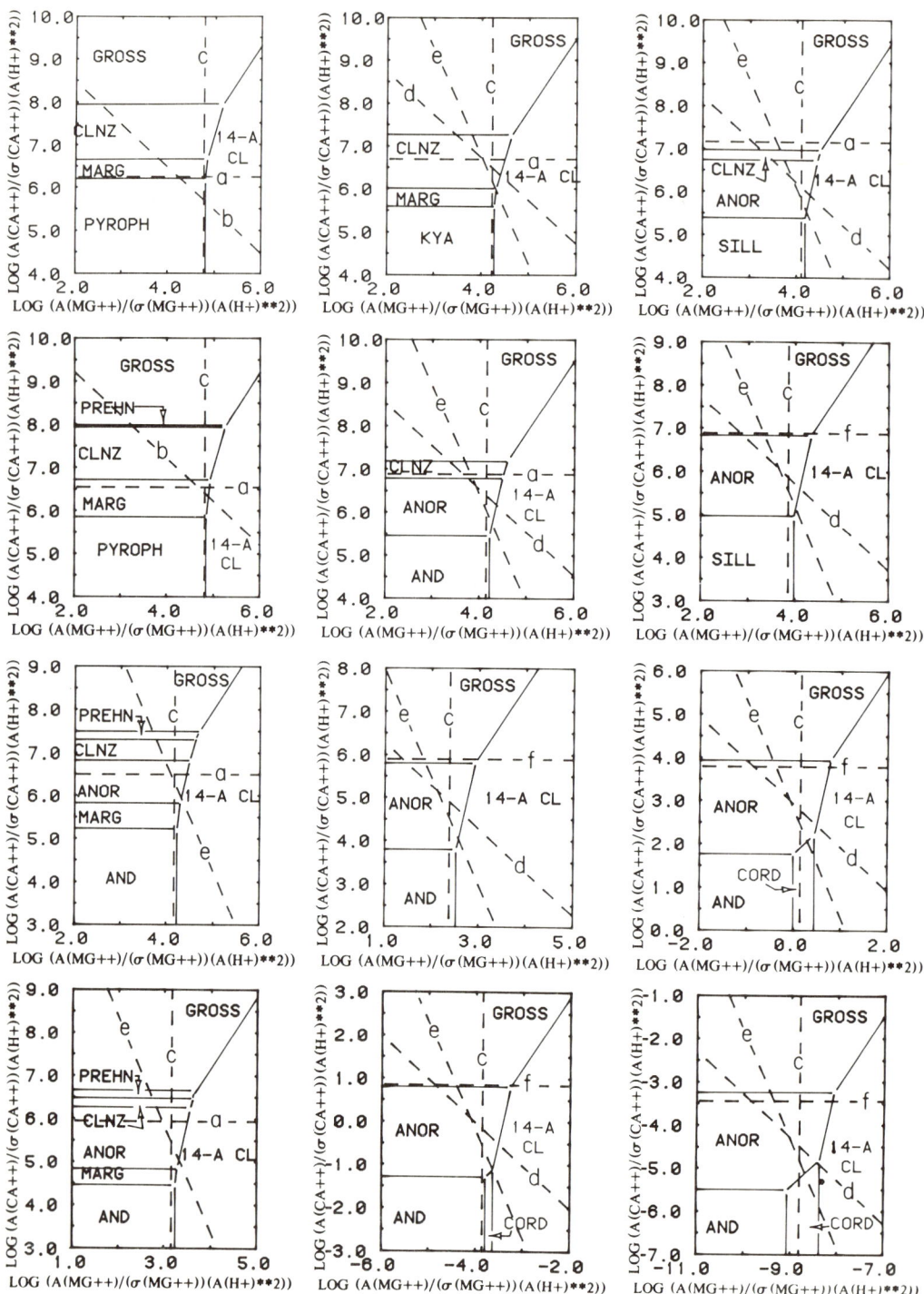

Phase relations in the system $HCl-H_2O-(Al_2O_3)-CaO-CO_2-MgO-SiO_2$, in equilibrium with quartz, at $X_{CO_2} = 0.05$. Saturation limits: calcite (a), dolomite (b), talc (c), diopside (d), tremolite (e), wollastonite (f).

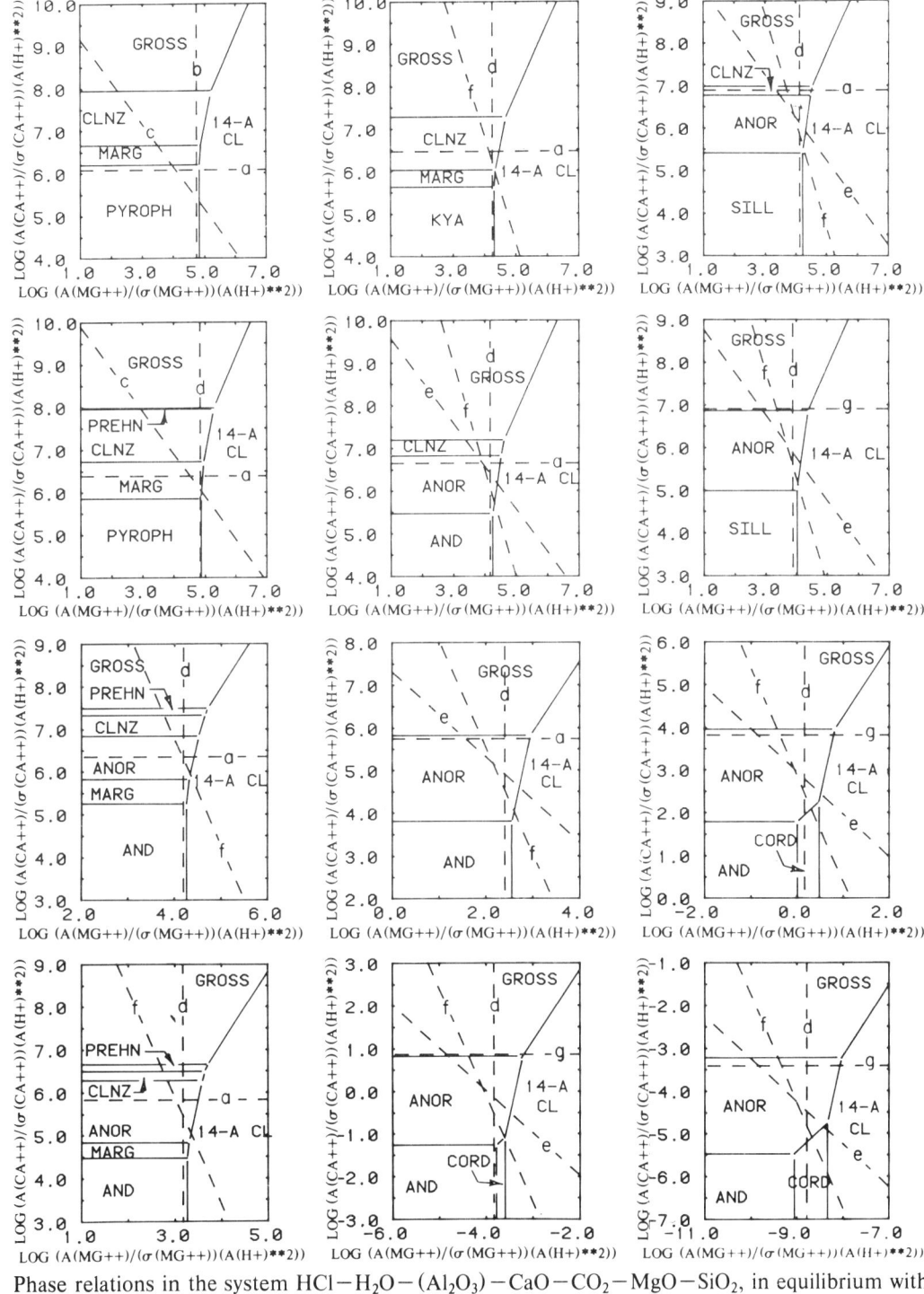

Phase relations in the system $HCl-H_2O-(Al_2O_3)-CaO-CO_2-MgO-SiO_2$, in equilibrium with quartz, at $X_{CO_2} = 0.10$. Saturation limits: calcite (a), magnesite (b), dolomite (c), talc (d), diopside (e), tremolite (f), wollastonite (g).

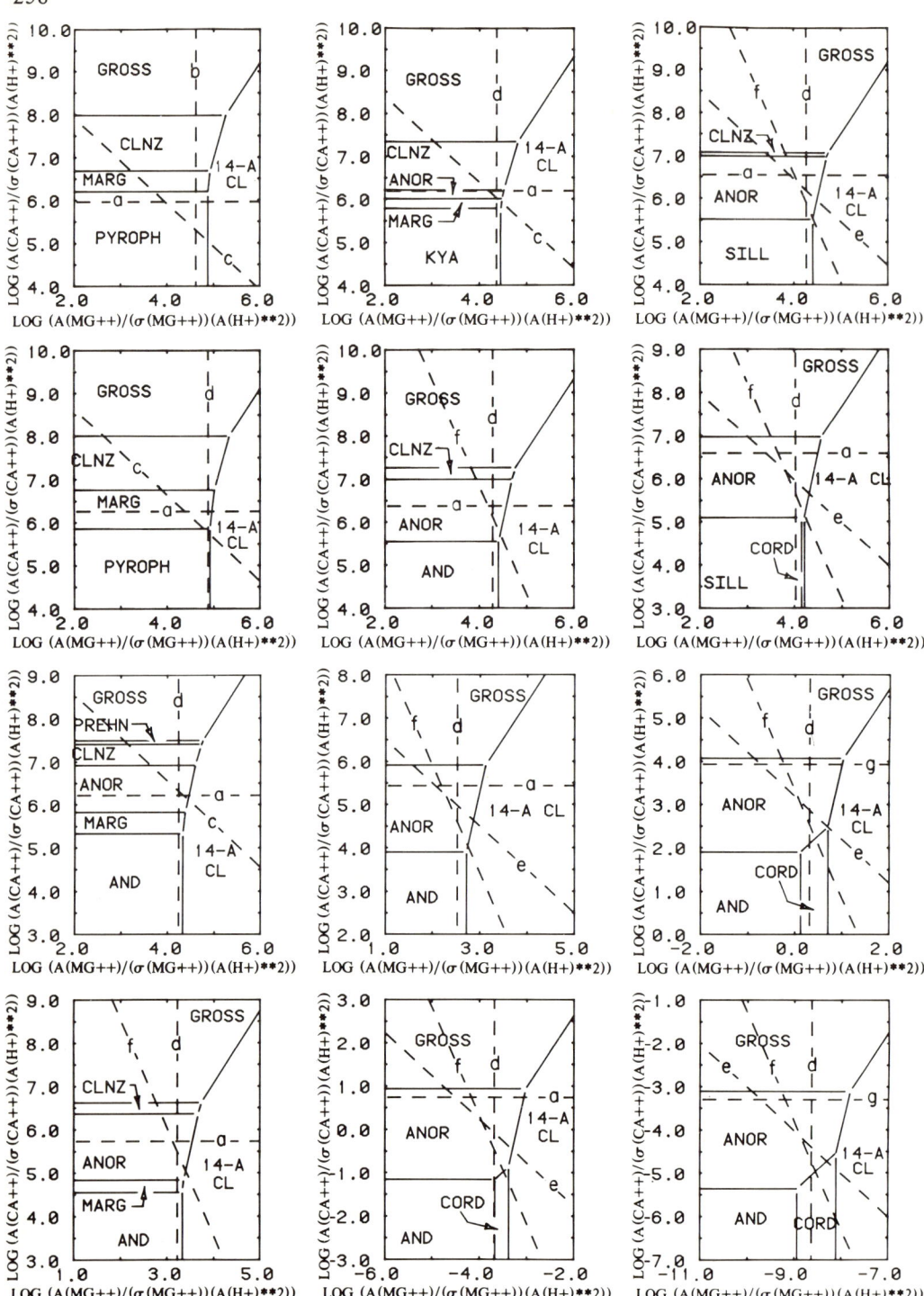

Phase relations in the system $HCl-H_2O-(Al_2O_3)-CaO-CO_2-MgO-SiO_2$, in equilibrium with quartz, at $X_{CO_2} = 0.30$. Saturation limits: calcite (a), magnesite (b), dolomite (c), talc (d), diopside (e), tremolite (f), wollastonite (g).

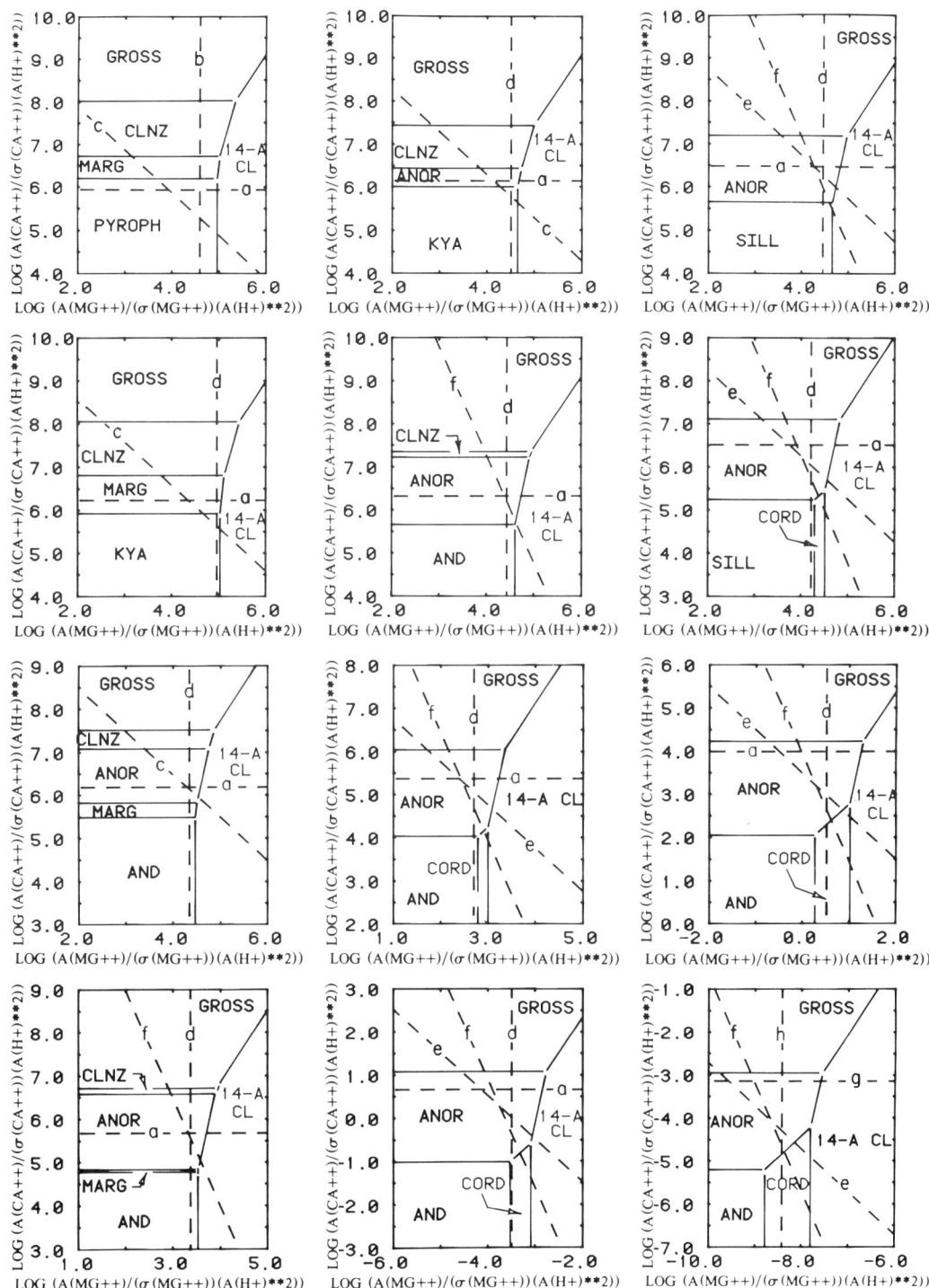

Phase relations in the system $HCl-H_2O-(Al_2O_3)-CaO-CO_2-MgO-SiO_2$, in equilibrium with quartz, at $X_{CO_2} = 0.50$. Saturation limits: calcite (a), magnesite (b), dolomite (c), talc (d), diopside (e), tremolite (f), wollastonite (g), anthophyllite (h).

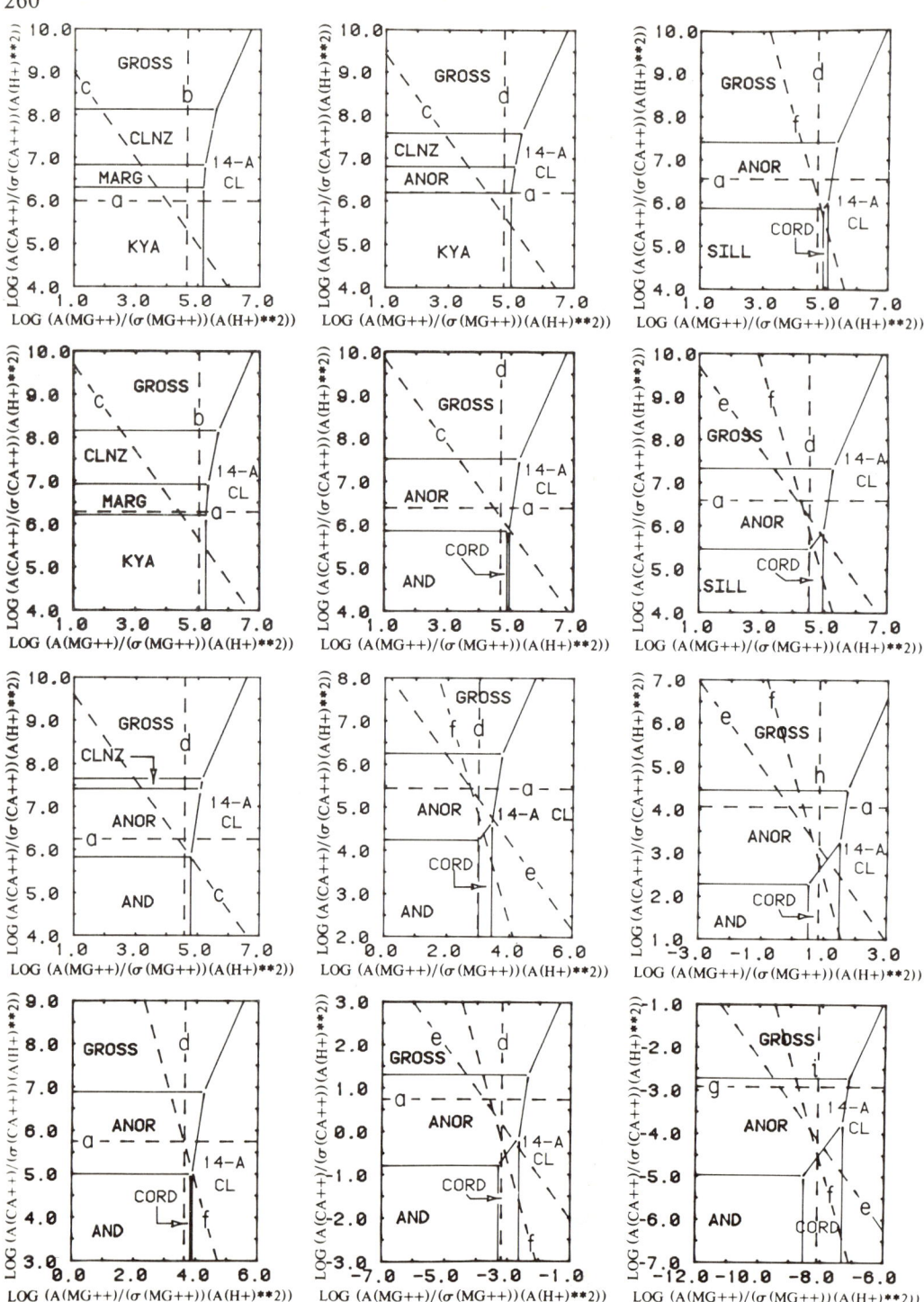

Phase relations in the system $HCl-H_2O-(Al_2O_3)-CaO-CO_2-MgO-SiO_2$, in equilibrium with quartz, at $X_{CO_2} = 0.70$. Saturation limits: calcite (a), magnesite (b), dolomite (c), talc (d), diopside (e), tremolite (f), wollastonite (g), anthophyllite (h), enstatite (i).

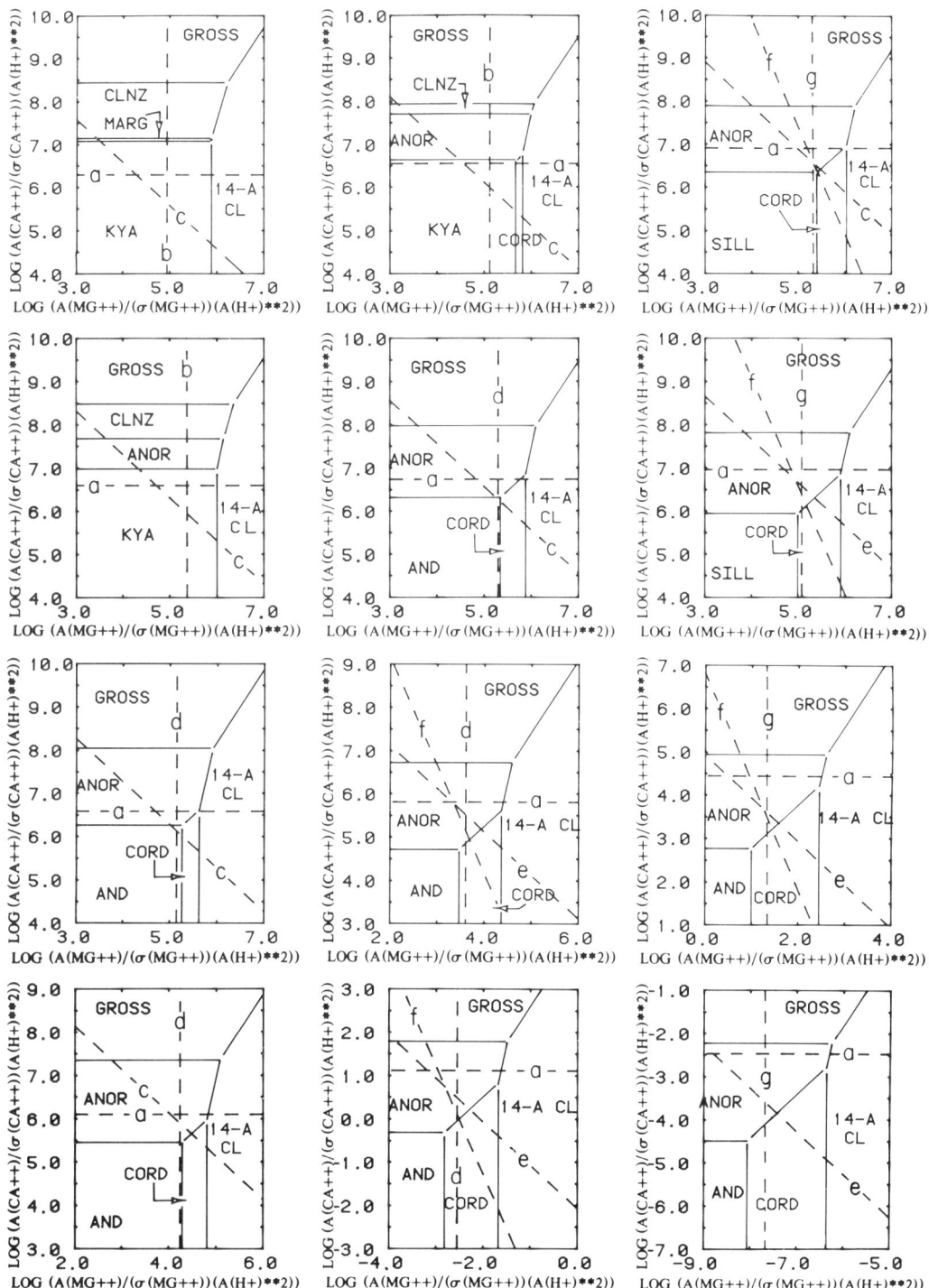

Phase relations in the system $HCl-H_2O-(Al_2O_3)-CaO-CO_2-MgO-SiO_2$, in equilibrium with quartz, at $X_{CO_2} = 0.90$. Saturation limits: calcite (a), magnesite (b), dolomite (c), talc (d), diopside (e), tremolite (f), enstatite (g).

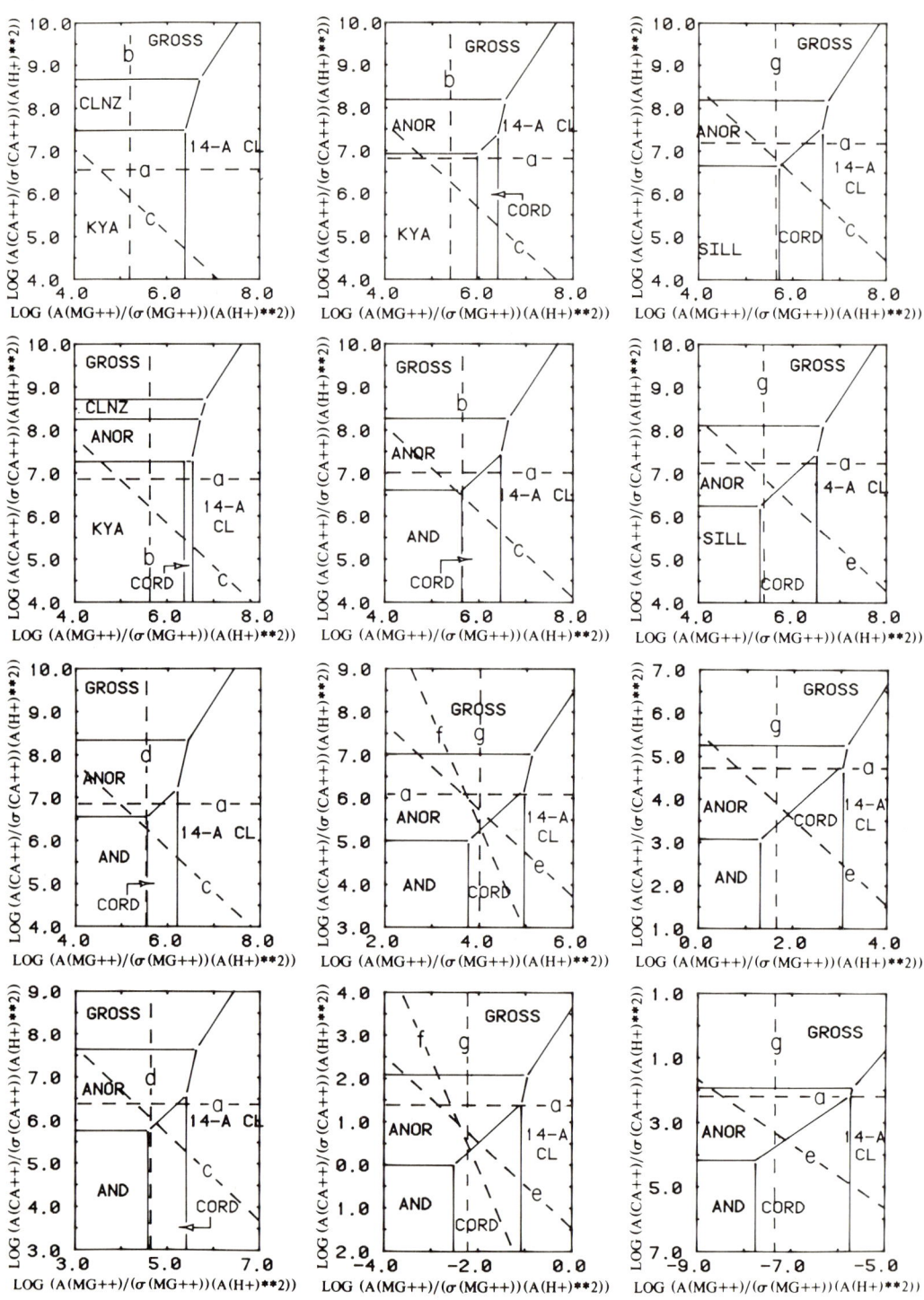

Phase relations in the system $HCl-H_2O-(Al_2O_3)-CaO-CO_2-MgO-SiO_2$, in equilibrium with quartz, at $X_{CO_2} = 0.95$. Saturation limits: calcite (a), magnesite (b), dolomite (c), talc (d), diopside (e), tremolite (f), enstatite (g).

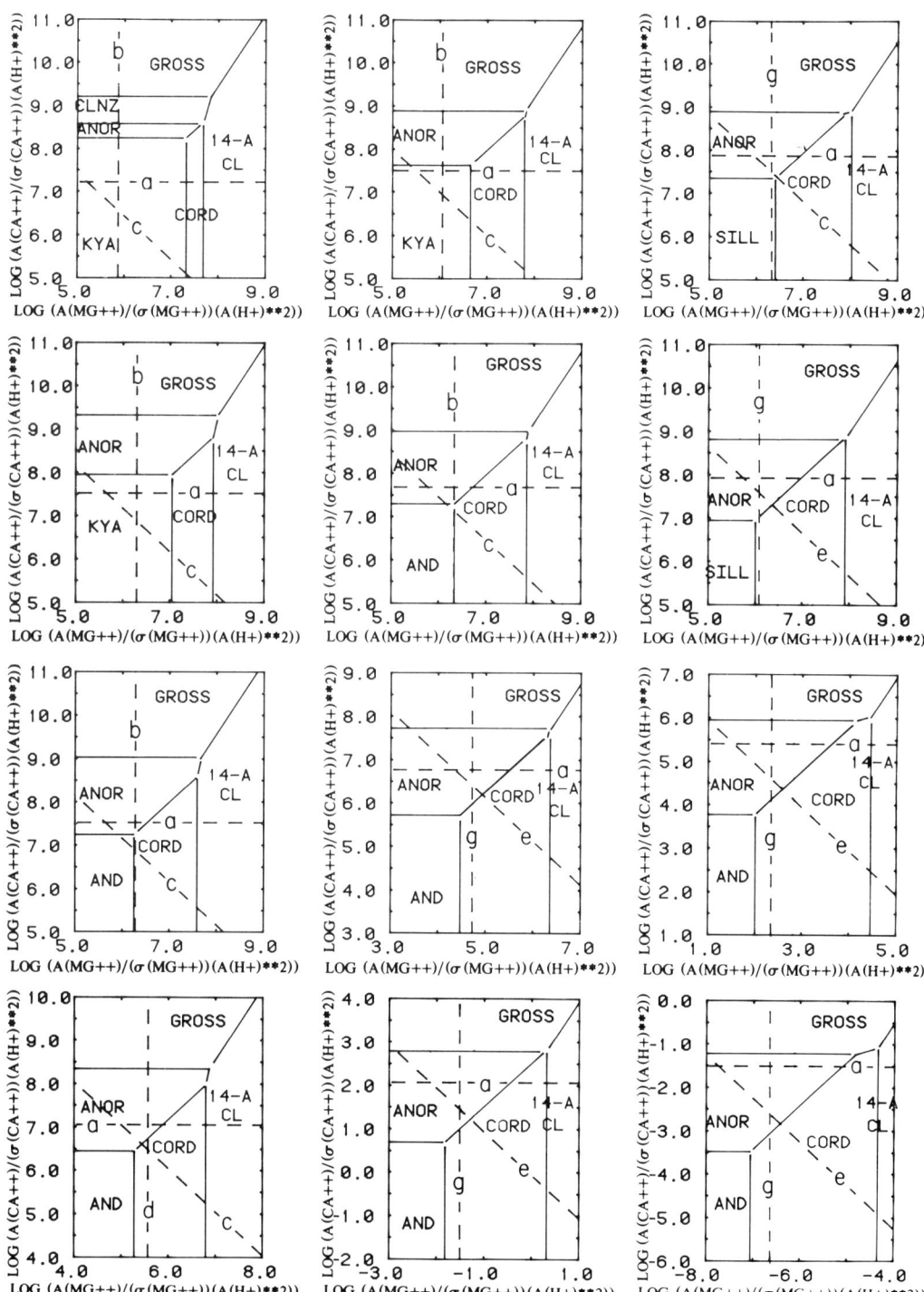

Phase relations in the system $HCl-H_2O-(Al_2O_3)-CaO-CO_2-MgO-SiO_2$, in equilibrium with quartz, at $X_{CO_2} = 0.99$. Saturation limits: calcite (a), magnesite (b), dolomite (c), talc (d), diopside (e), tremolite (f), enstatite (g).

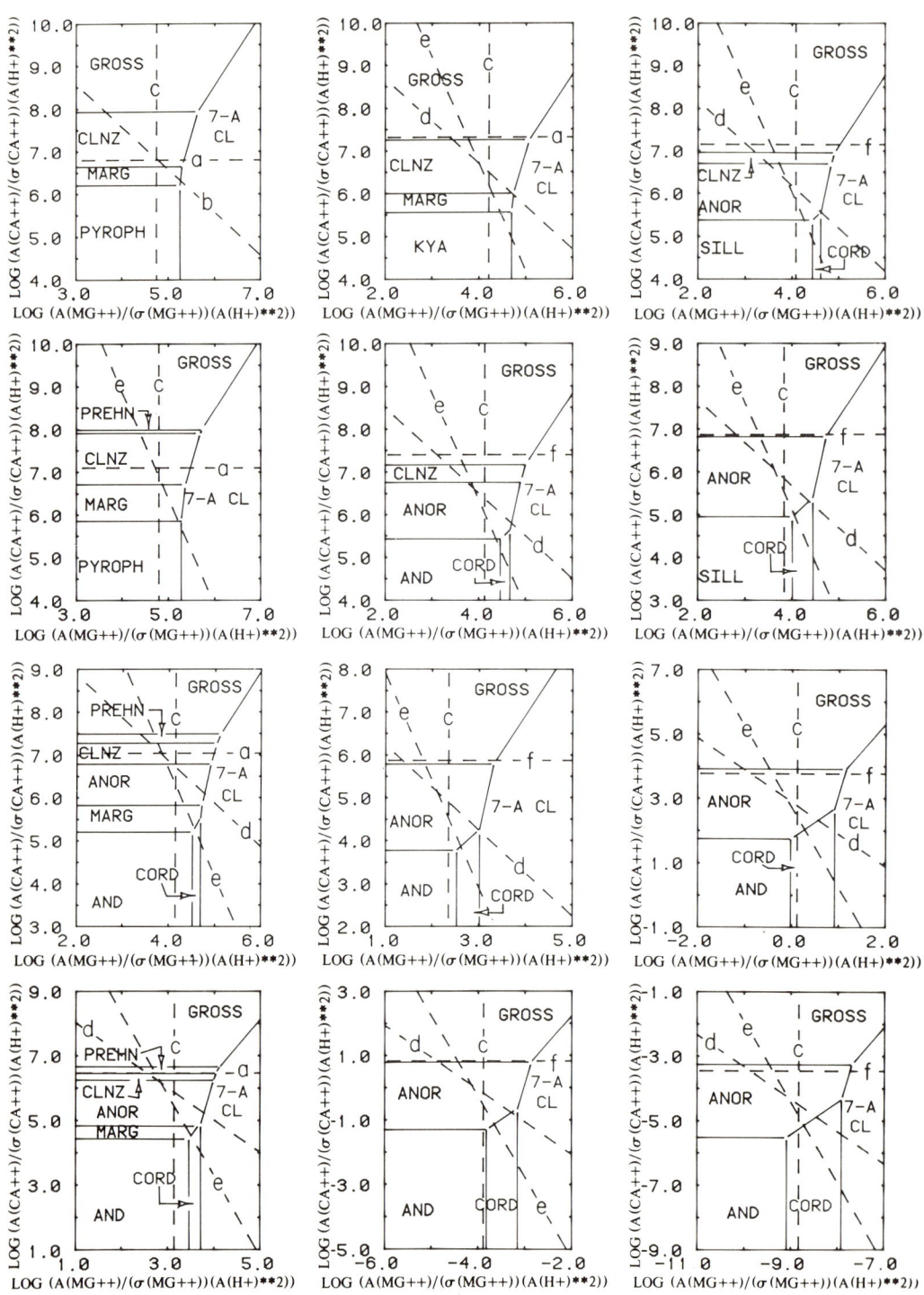

Phase relations in the system $HCl-H_2O-(Al_2O_3)-CaO-CO_2-MgO-SiO_2$, in equilibrium with quartz, at $X_{CO_2} = 0.01$. Metastable 7-A clinochlore was considered instead of its stable counterpart, 14-A clinochlore. Saturation limits: calcite (a), dolomite (b), talc (c), diopside (d), tremolite (e), wollastonite (f).

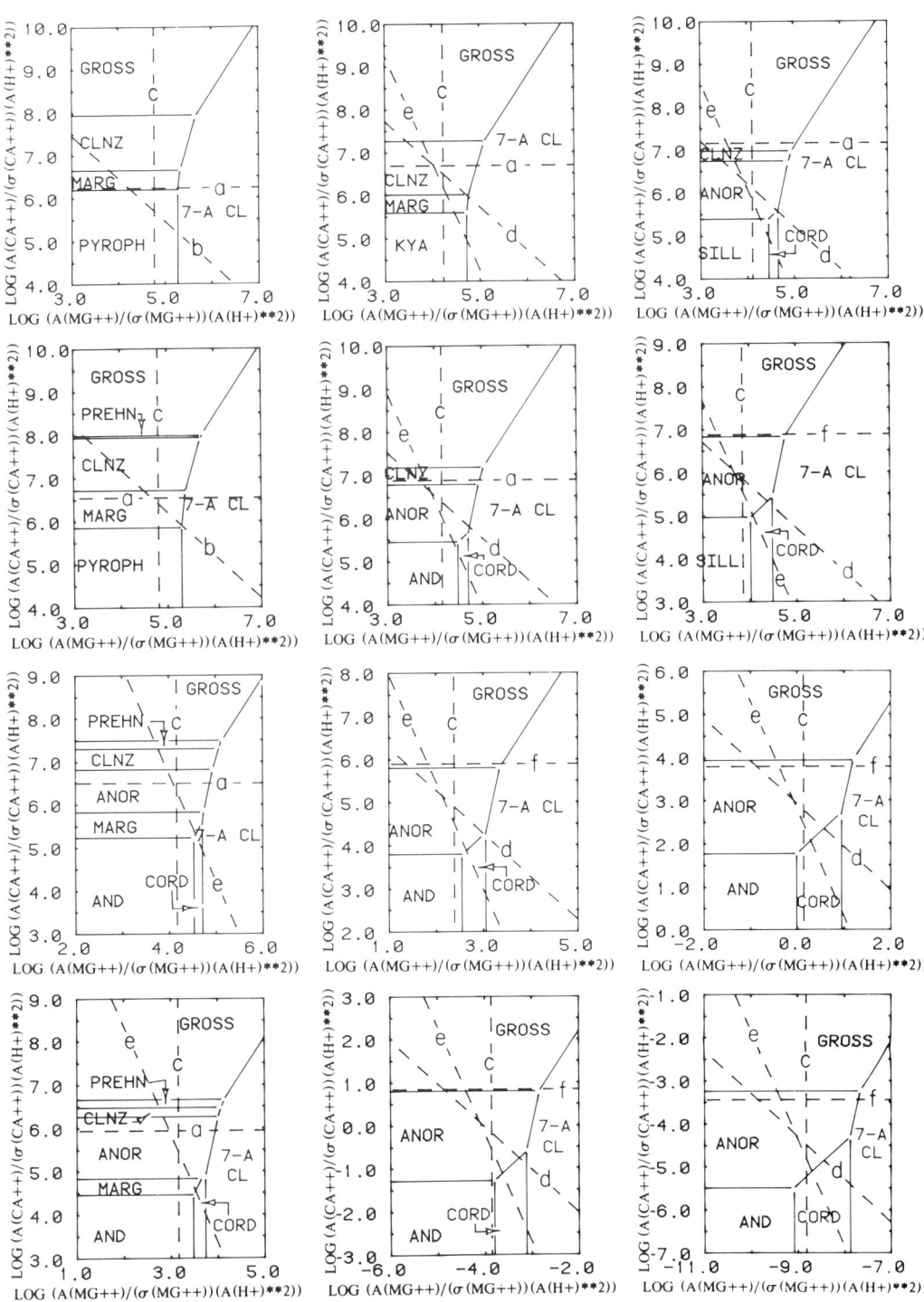

Phase relations in the system $HCl-H_2O-(Al_2O_3)-CaO-CO_2-MgO-SiO_2$, in equilibrium with quartz, at $X_{CO_2} = 0.05$. Metastable 7-A clinochlore was considered instead of its stable counterpart, 14-A clinochlore. Saturation limits: calcite (a), dolomite (b), talc (c), diopside (d), tremolite (e), wollastonite (f).

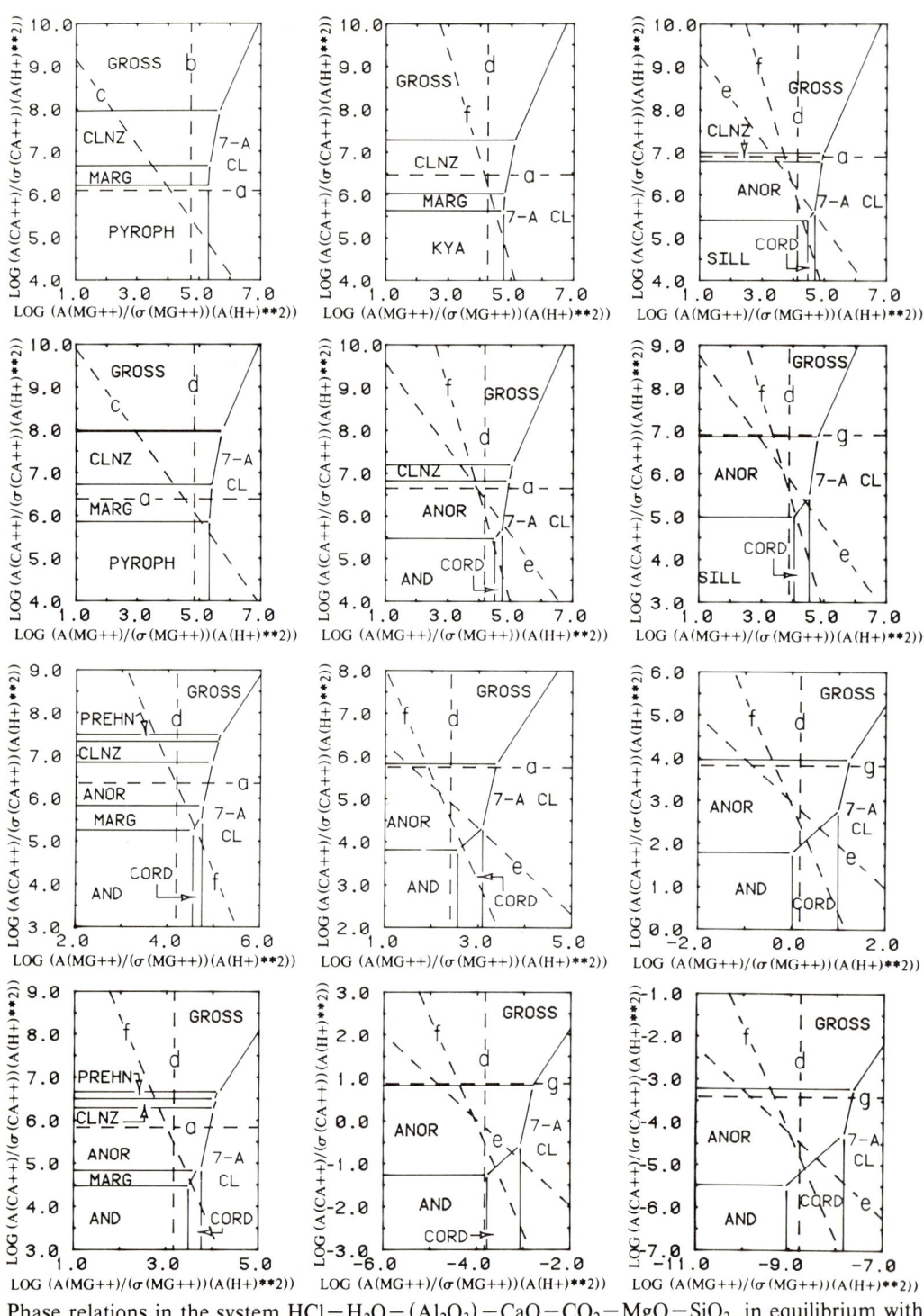

Phase relations in the system $HCl-H_2O-(Al_2O_3)-CaO-CO_2-MgO-SiO_2$, in equilibrium with quartz, at $X_{CO_2} = 0.10$. Metastable 7-A clinochlore was considered instead of its stable counterpart, 14-A clinochlore. Saturation limits: calcite (a), magnesite (b), dolomite (c), talc (d), diopside (e), tremolite (f), wollastonite (g).

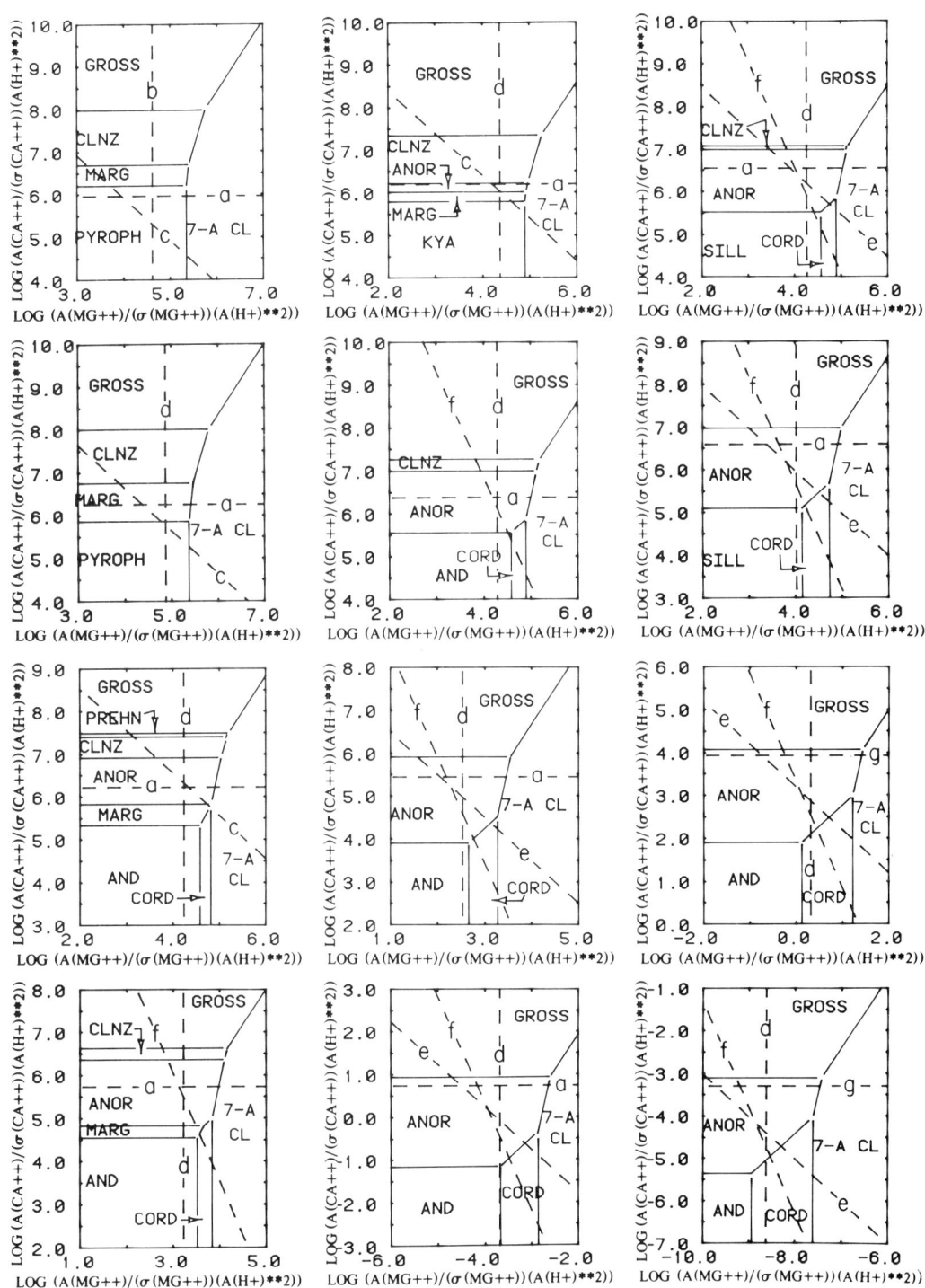

Phase relations in the system $HCl-H_2O-(Al_2O_3)-CaO-CO_2-MgO-SiO_2$, in equilibrium with quartz, at $X_{CO_2} = 0.30$. Metastable 7-A clinochlore was considered instead of its stable counterpart, 14-A clinochlore. Saturation limits: calcite (a), magnesite (b), dolomite (c), talc (d), diopside (e), tremolite (f), wollastonite (g).

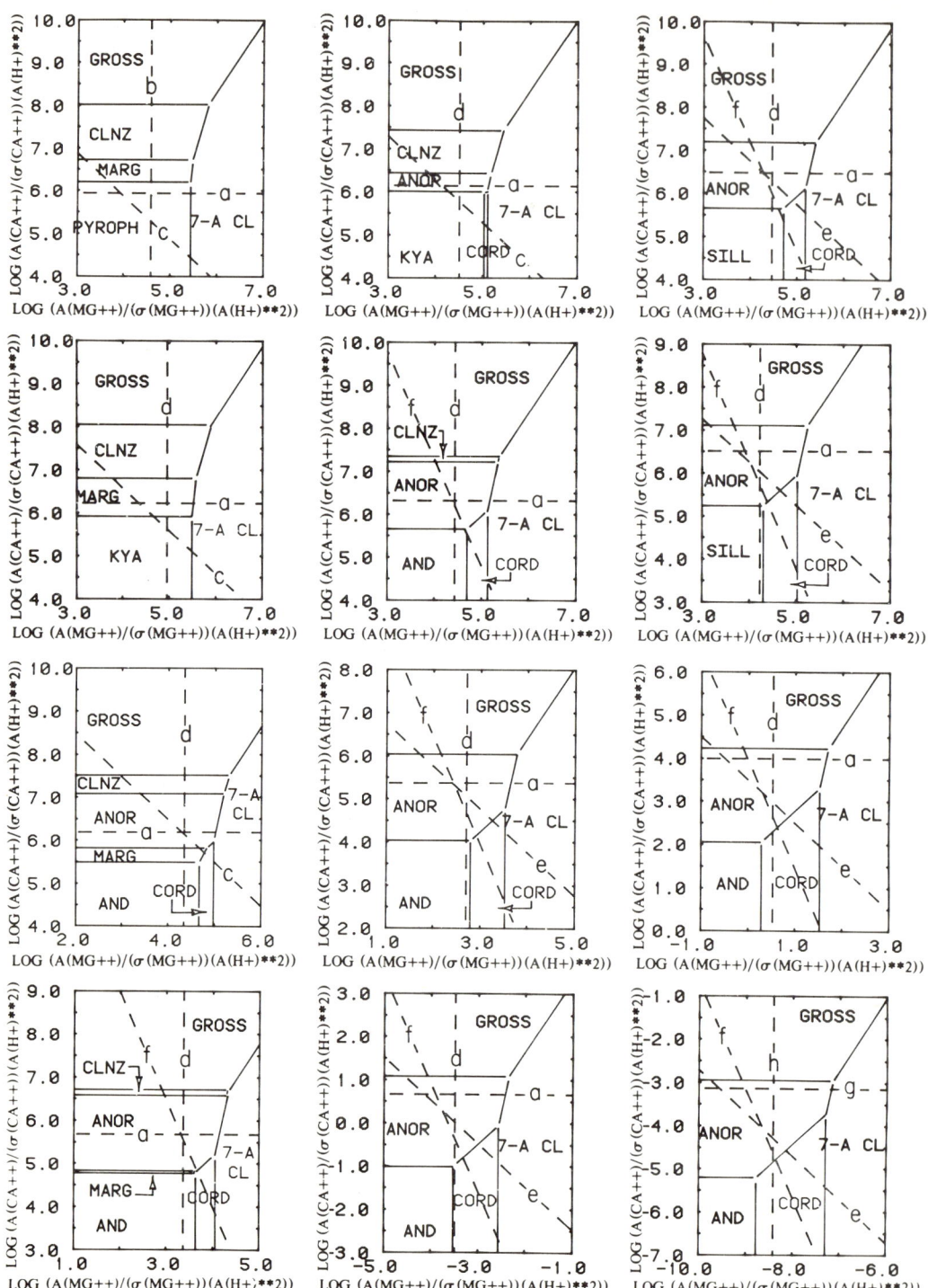

Phase relations in the system $HCl-H_2O-(Al_2O_3)-CaO-CO_2-MgO-SiO_2$, in equilibrium with quartz, at $X_{CO_2} = 0.50$. Metastable 7-A clinochlore was considered instead of its stable counterpart, 14-A clinochlore. Saturation limits: calcite (a), magnesite (b), dolomite (c), talc (d), diopside (e), tremolite (f), wollastonite (g), anthophyllite (h).

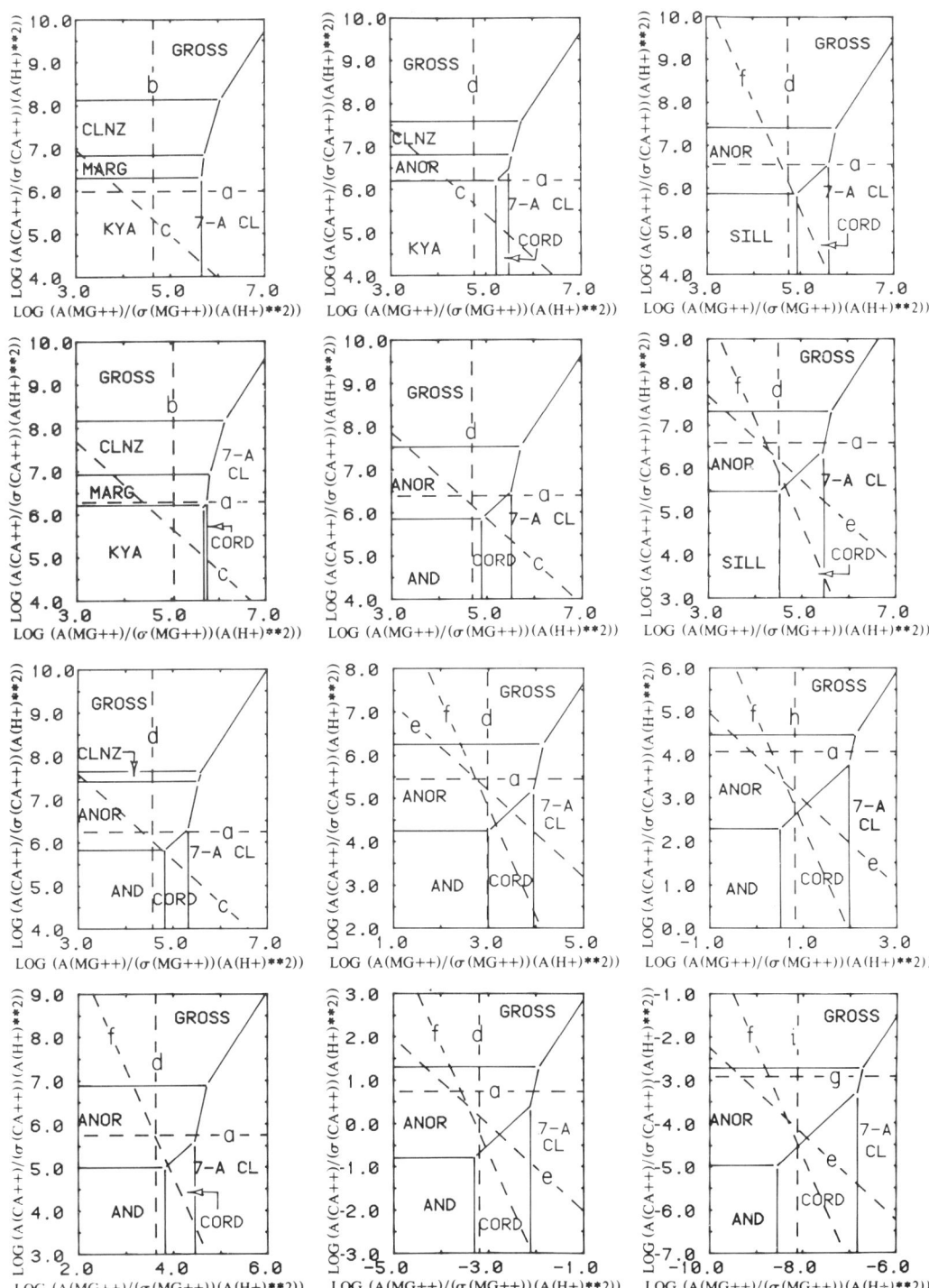

Phase relations in the system $HCl-H_2O-(Al_2O_3)-CaO-CO_2-MgO-SiO_2$, in equilibrium with quartz, at $X_{CO_2} = 0.70$. Metastable 7-A clinochlore was considered instead of its stable counterpart, 14-A clinochlore. Saturation limits: calcite (a), magnesite (b), dolomite (c), talc (d), diopside (e), tremolite (f), wollastonite (g), anthophyllite (h), enstatite (i).

Phase relations in the system $HCl-H_2O-(Al_2O_3)-CaO-CO_2-MgO-SiO_2$, in equilibrium with quartz, at $X_{CO_2} = 0.90$. Metastable 7-A clinochlore was considered instead of its stable counterpart, 14-A clinochlore. Saturation limits: calcite (a), magnesite (b), dolomite (c), talc (d), diopside (e), tremolite (f), enstatite (g).

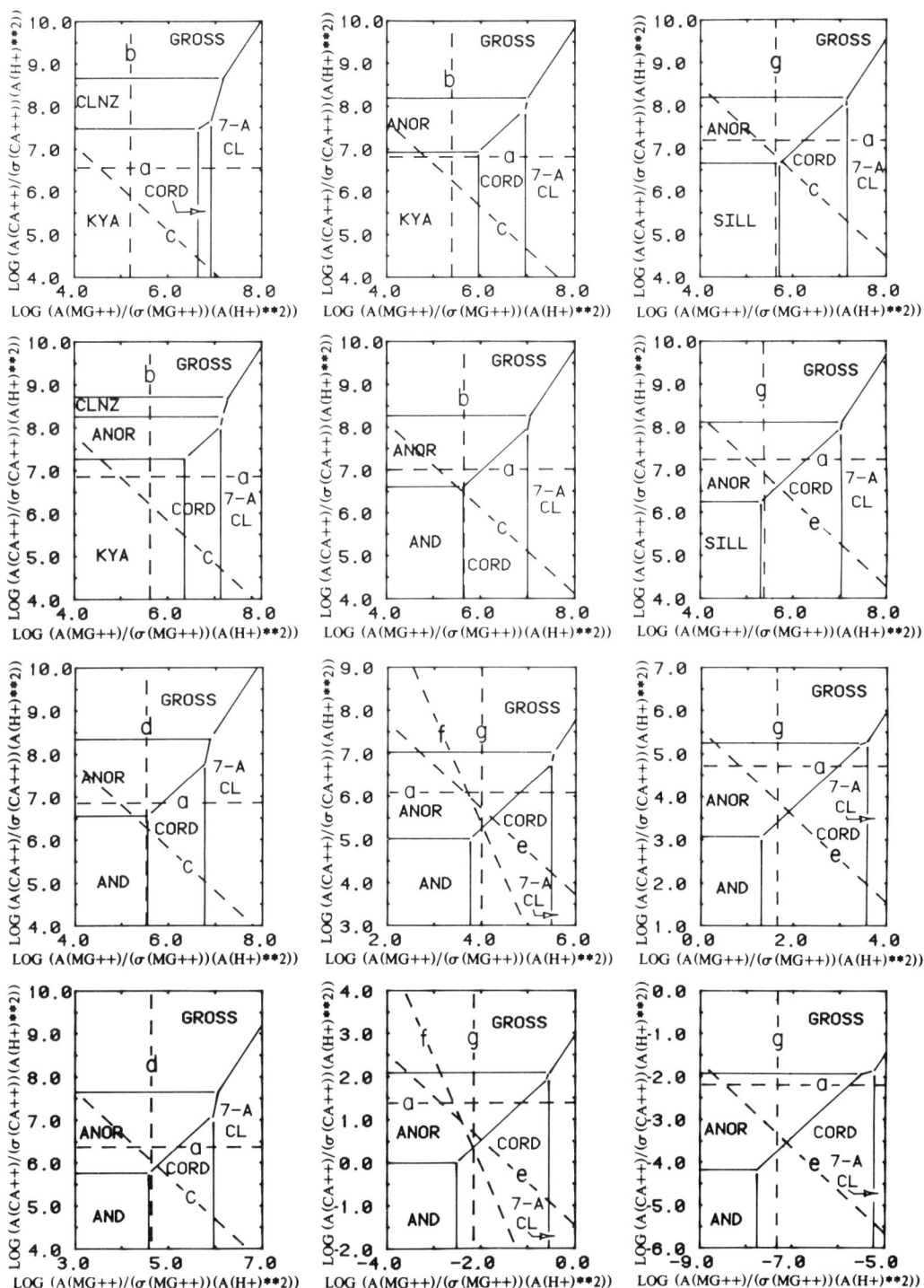

Phase relations in the system $HCl-H_2O-(Al_2O_3)-CaO-CO_2-MgO-SiO_2$, in equilibrium with quartz, at $X_{CO_2} = 0.95$. Metastable 7-A clinochlore was considered instead of its stable counterpart, 14-A clinochlore. Saturation limits: calcite (a), magnesite (b), dolomite (c), talc (d), diopside (e), tremolite (f), enstatite (g).

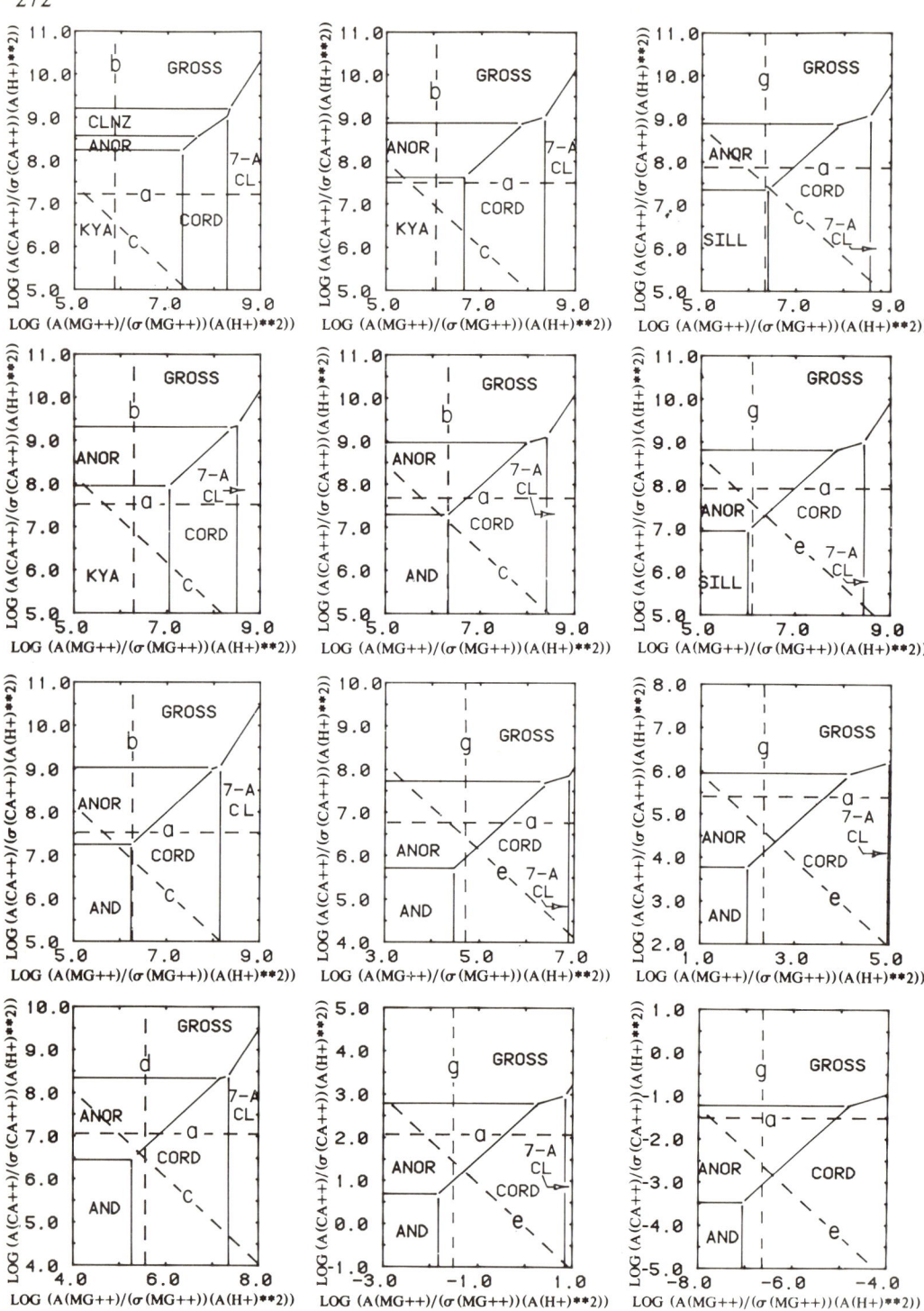

Phase relations in the system $HCl-H_2O-(Al_2O_3)-CaO-CO_2-MgO-SiO_2$, in equilibrium with quartz, at $X_{CO_2} = 0.99$. Metastable 7-A clinochlore was considered instead of its stable counterpart, 14-A clinochlore. Saturation limits: calcite (a), magnesite (b), dolomite (c), talc (d), diopside (e), tremolite (f), enstatite (g).

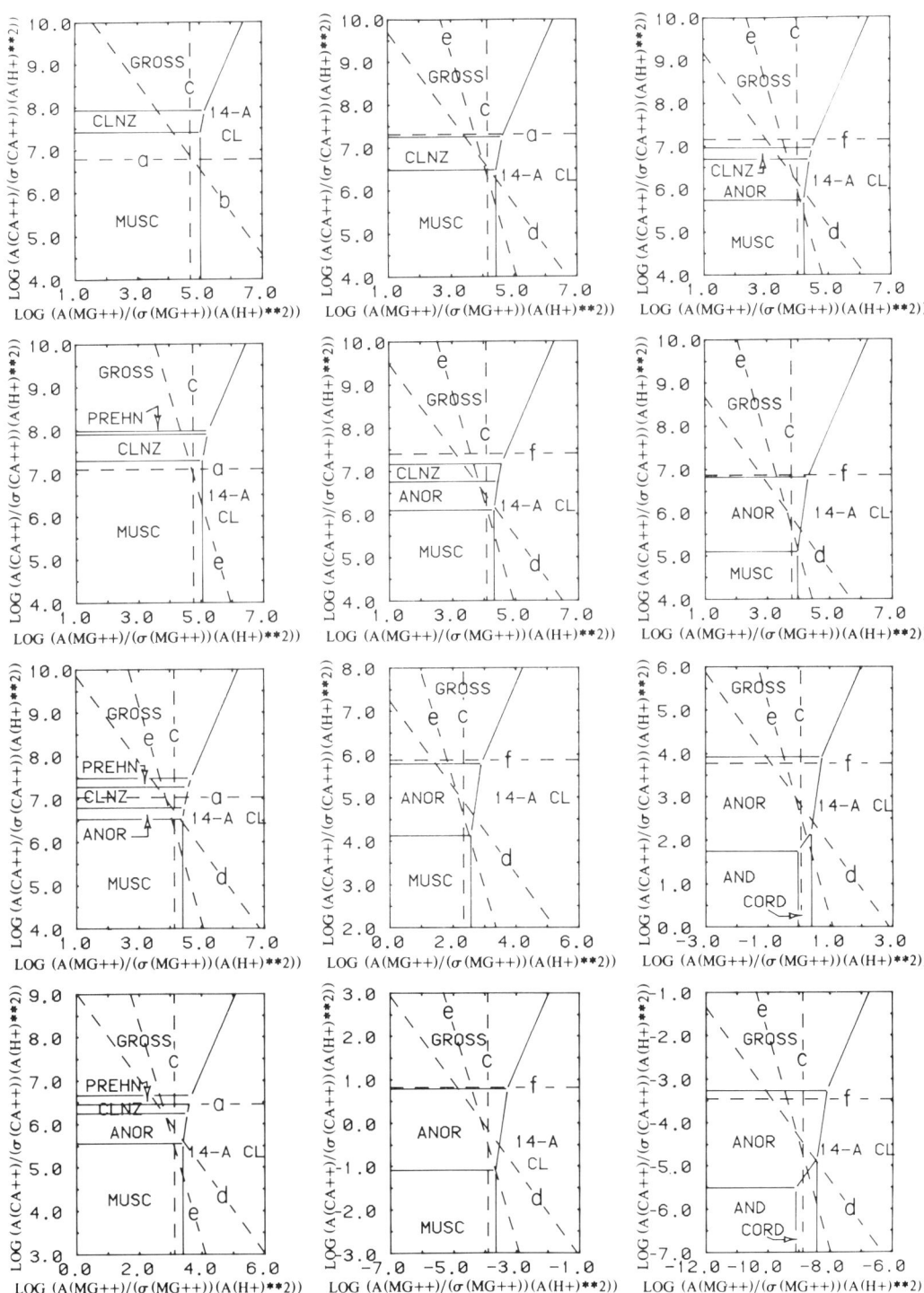

Phase relations in the system $HCl-H_2O-(Al_2O_3)-CaO-CO_2-K_2O-MgO-SiO_2$, in equilibrium with quartz and K-feldspar, at $X_{CO_2} = 0.01$. Saturation limits: calcite (a), dolomite (b), phlogopite (c), diopside (d), tremolite (e), wollastonite (f).

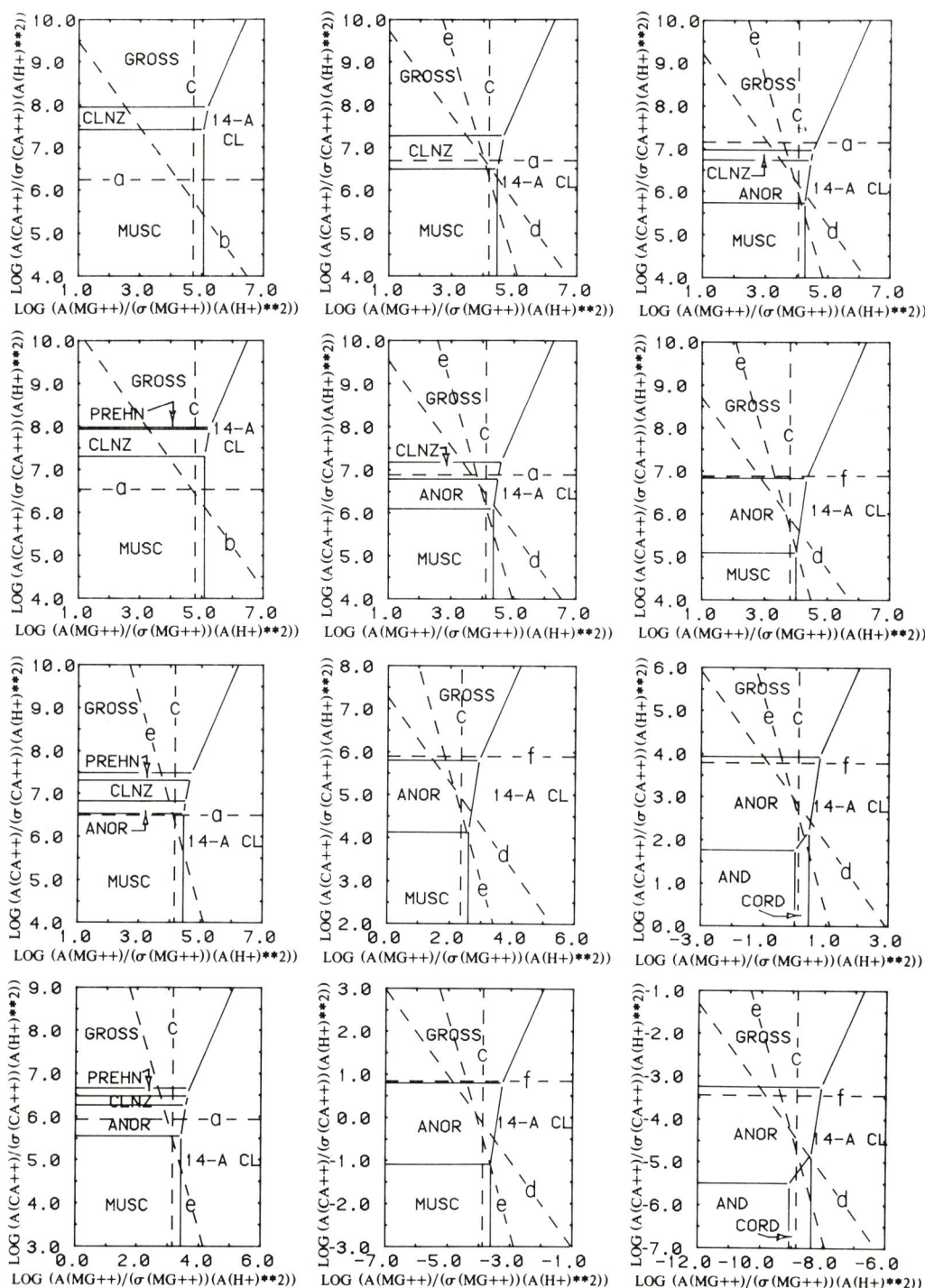

Phase relations in the system HCl−H$_2$O−(Al$_2$O$_3$)−CaO−CO$_2$−K$_2$O−MgO−SiO$_2$, in equilibrium with quartz and K-feldspar, at X$_{CO_2}$ = 0.05. Saturation limits: calcite (a), dolomite (b), phlogopite (c), diopside (d), tremolite (e), wollastonite (f).

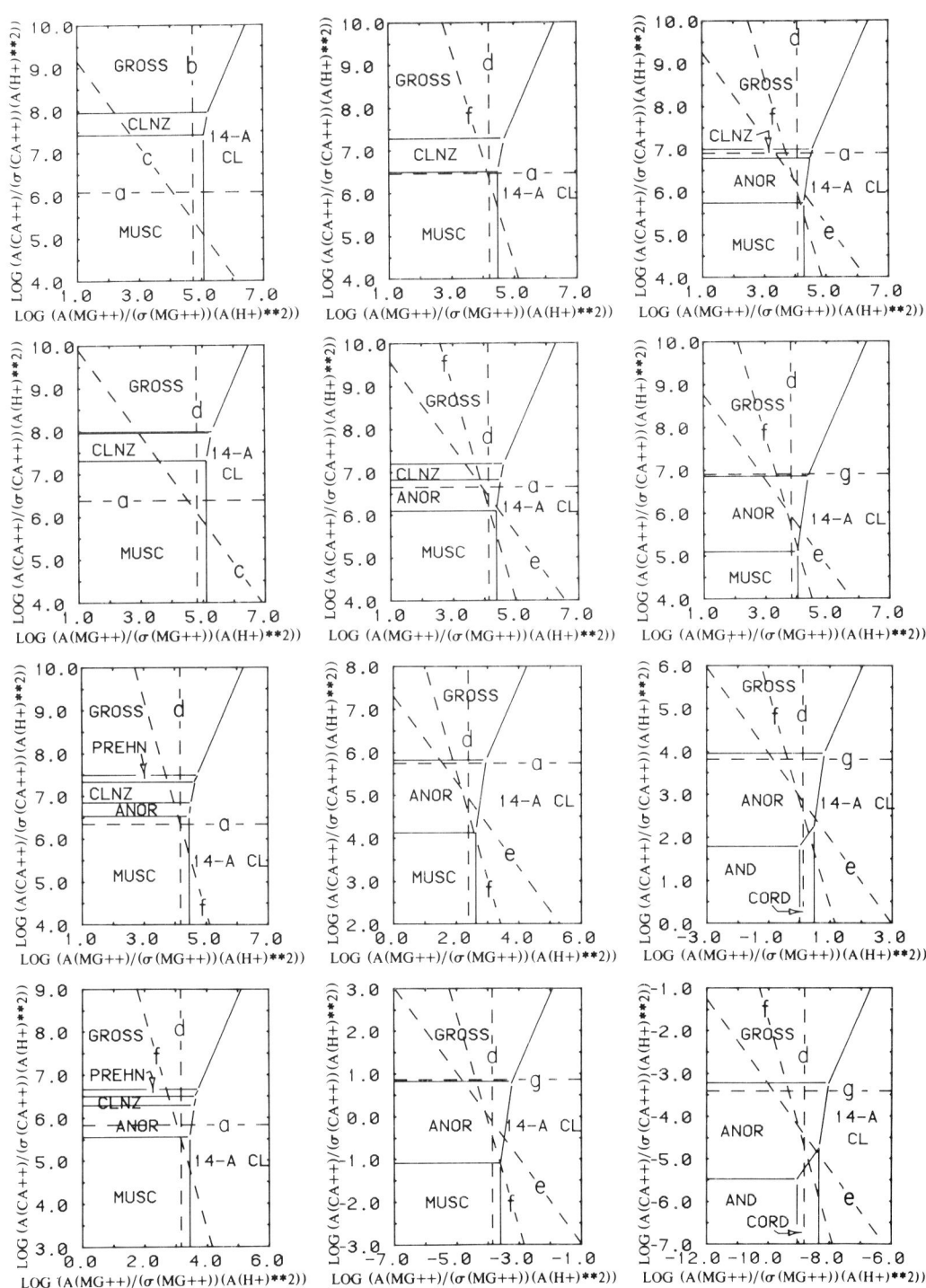

Phase relations in the system $HCl-H_2O-(Al_2O_3)-CaO-CO_2-K_2O-MgO-SiO_2$, in equilibrium with quartz and K-feldspar, at $X_{CO_2} = 0.10$. Saturation limits: calcite (a), magnesite (b), dolomite (c), phlogopite (d), diopside (e), tremolite (f), wollastonite (g).

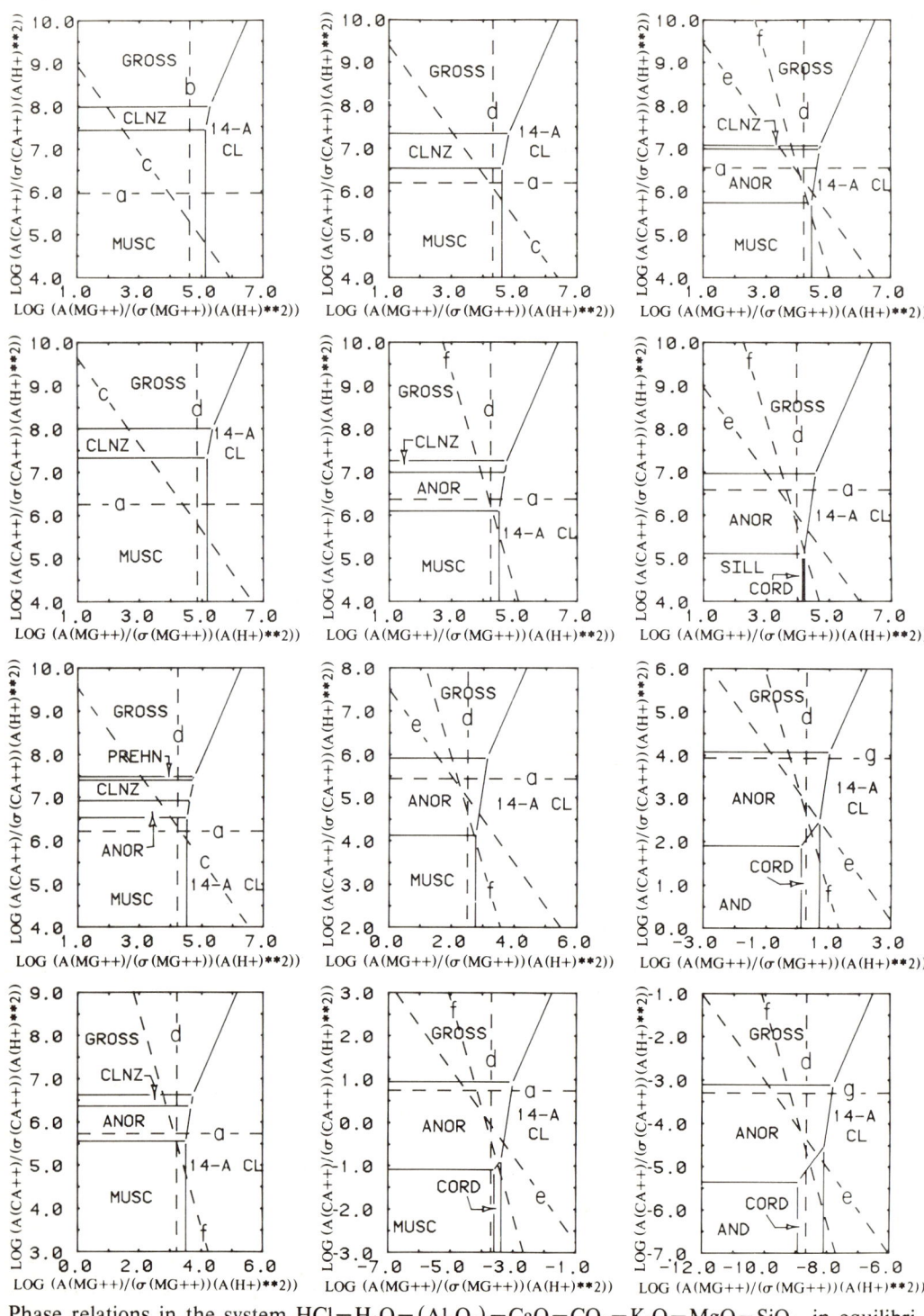

Phase relations in the system $HCl-H_2O-(Al_2O_3)-CaO-CO_2-K_2O-MgO-SiO_2$, in equilibrium with quartz and K-feldspar, at $X_{CO_2} = 0.30$. Saturation limits: calcite (a), magnesite (b), dolomite (c), phlogopite (d), diopside (e), tremolite (f), wollastonite (g).

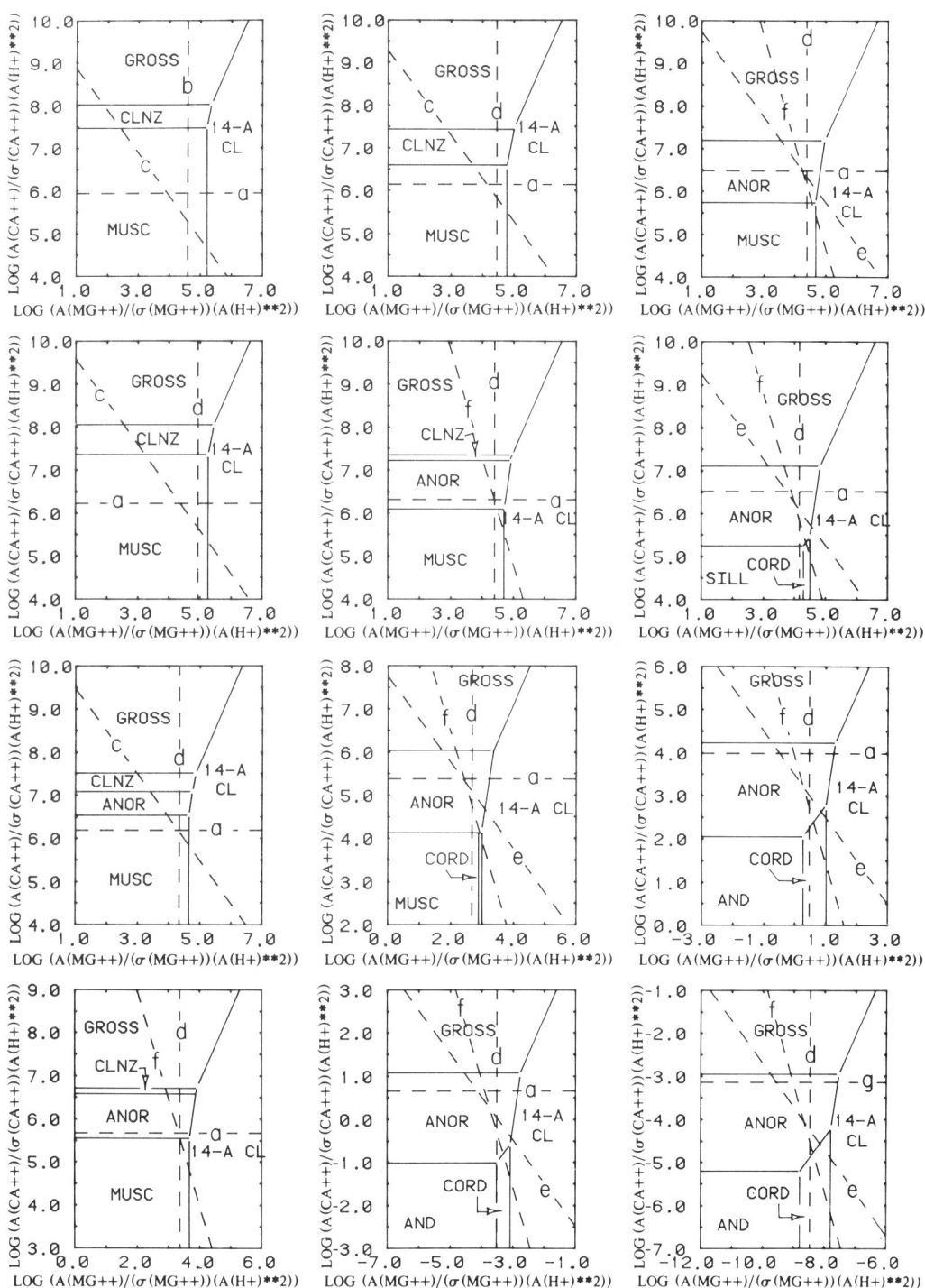

Phase relations in the system $HCl-H_2O-(Al_2O_3)-CaO-CO_2-K_2O-MgO-SiO_2$, in equilibrium with quartz and K-feldspar, at $X_{CO_2} = 0.50$. Saturation limits: calcite (a), magnesite (b), dolomite (c), phlogopite (d), diopside (e), tremolite (f), wollastonite (g).

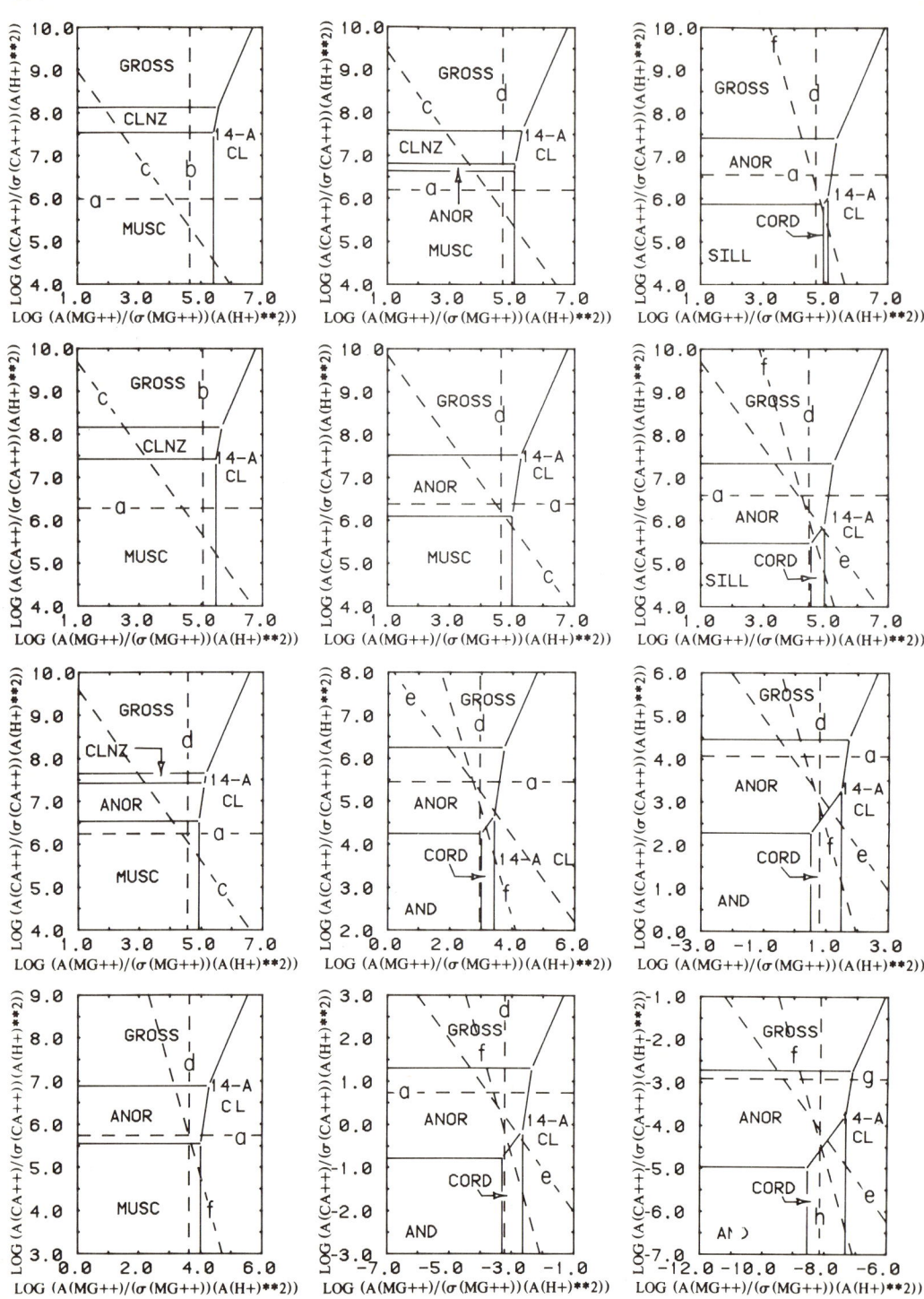

Phase relations in the system $HCl-H_2O-(Al_2O_3)-CaO-CO_2-K_2O-MgO-SiO_2$, in equilibrium with quartz and K-feldspar, at $X_{CO_2} = 0.70$. Saturation limits: calcite (a), magnesite (b), dolomite (c), phlogopite (d), diopside (e), tremolite (f), wollastonite (g), enstatite (h).

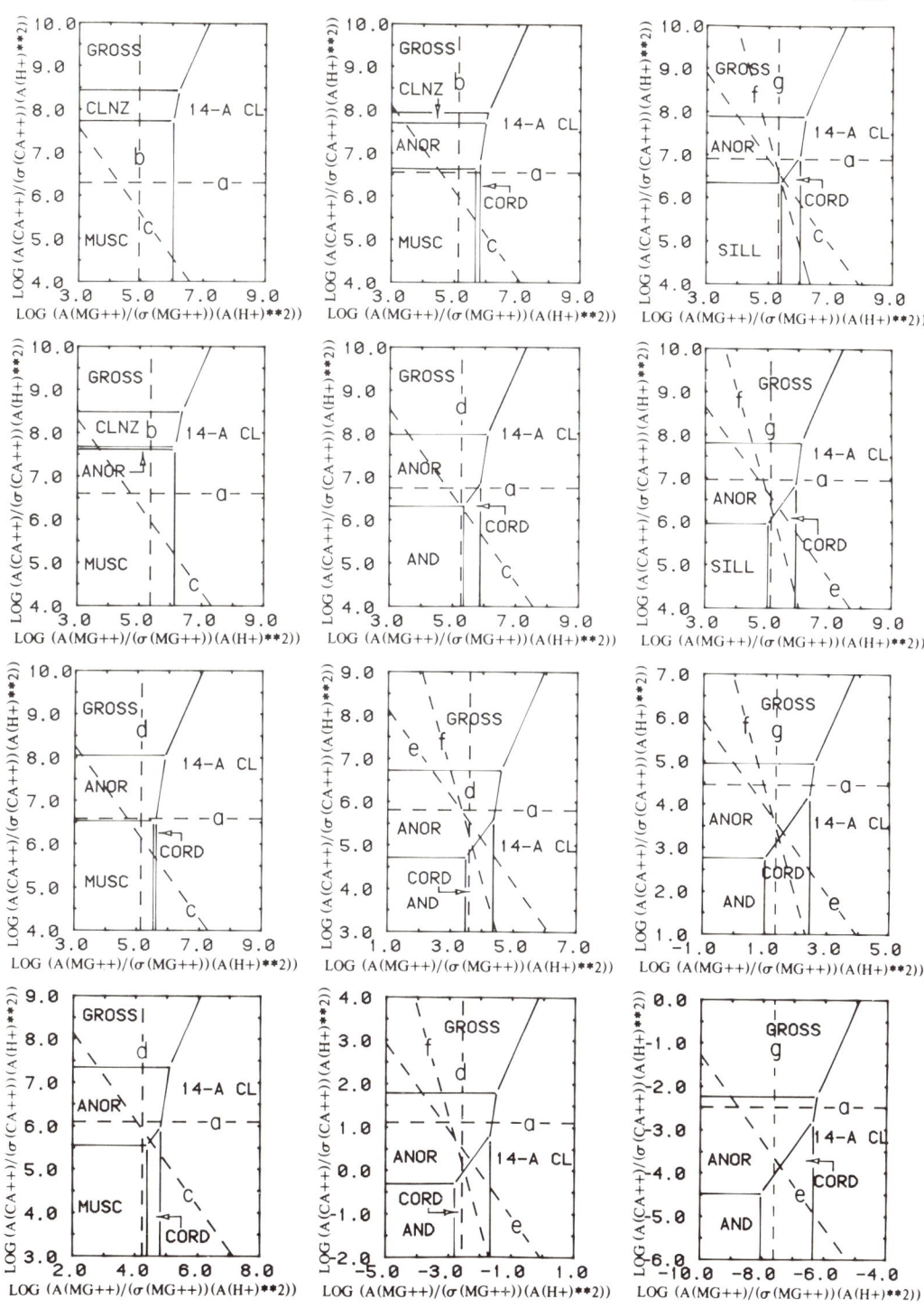

Phase relations in the system $HCl-H_2O-(Al_2O_3)-CaO-CO_2-K_2O-MgO-SiO_2$, in equilibrium with quartz and K-feldspar, at $X_{CO_2} = 0.90$. Saturation limits: calcite (a), magnesite (b), dolomite (c), phlogopite (d), diopside (e), tremolite (f), enstatite (g).

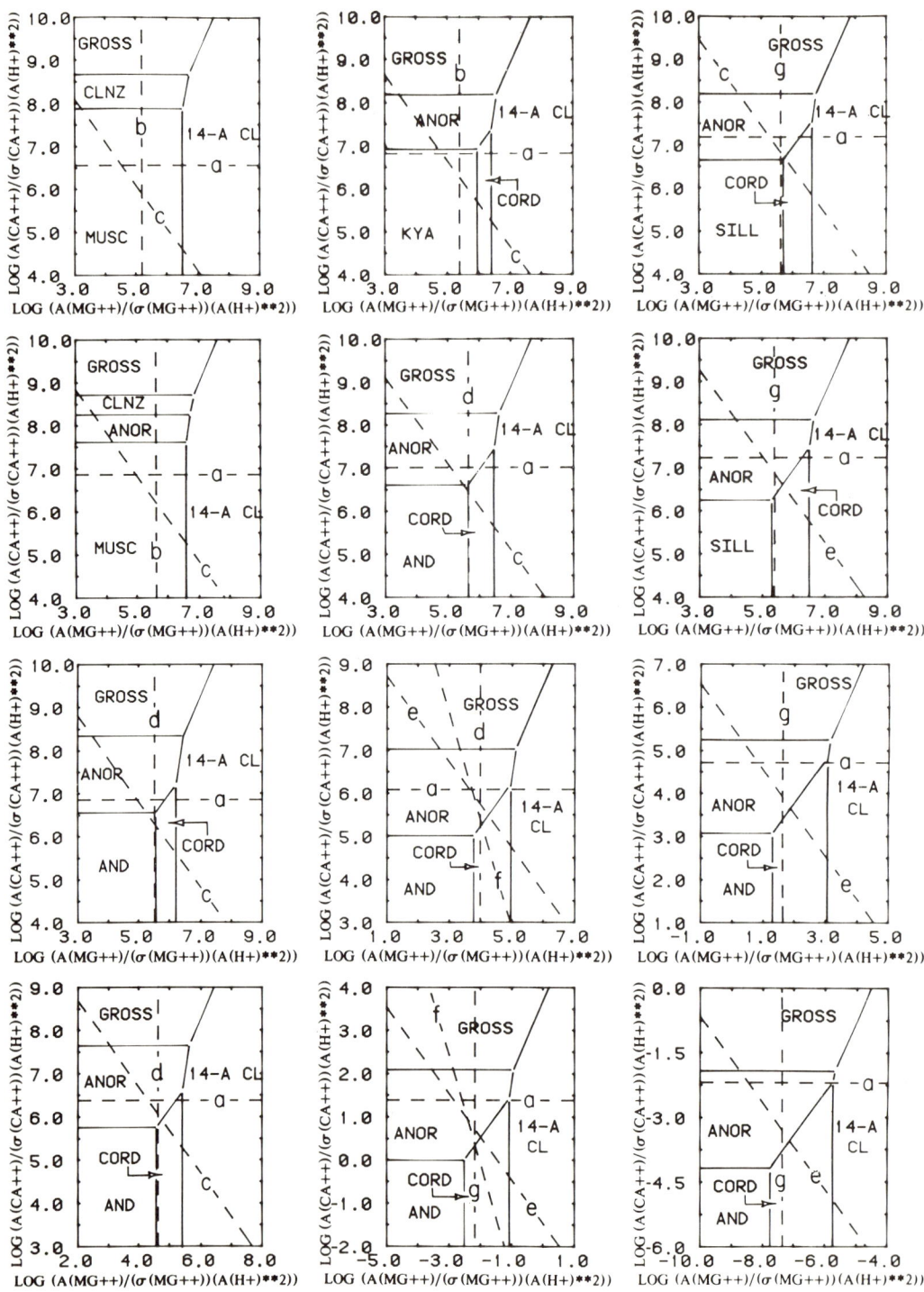

Phase relations in the system $HCl-H_2O-(Al_2O_3)-CaO-CO_2-K_2O-MgO-SiO_2$, in equilibrium with quartz and K-feldspar, at $X_{CO_2} = 0.95$. Saturation limits: calcite (a), magnesite (b), dolomite (c), phlogopite (d), diopside (e), tremolite (f), enstatite (g).

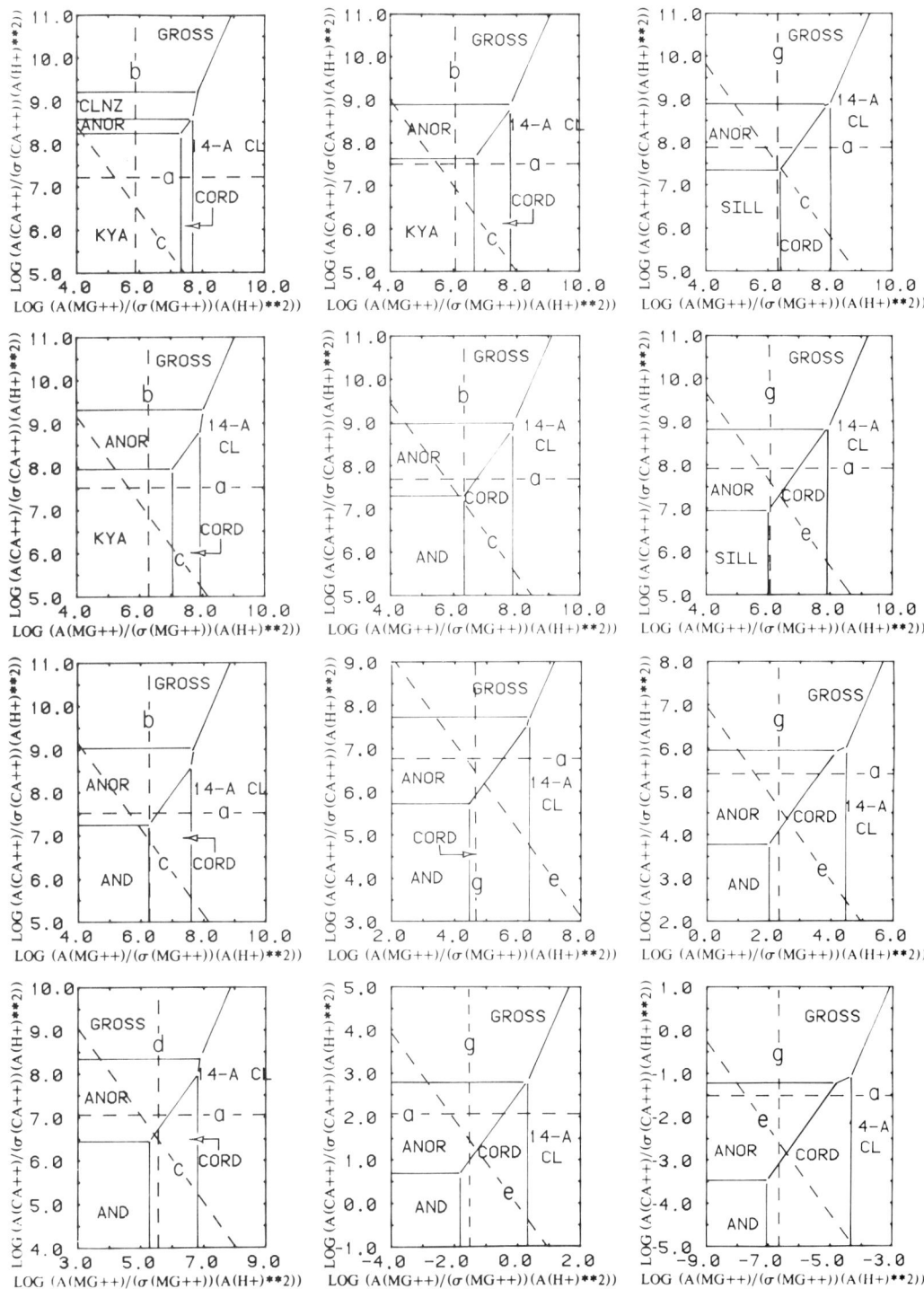

Phase relations in the system $HCl-H_2O-(Al_2O_3)-CaO-CO_2-K_2O-MgO-SiO_2$, in equilibrium with quartz and K-feldspar, at $X_{CO_2} = 0.99$. Saturation limits: calcite (a), magnesite (b), dolomite (c), phlogopite (d), diopside (e), tremolite (f), enstatite (g).

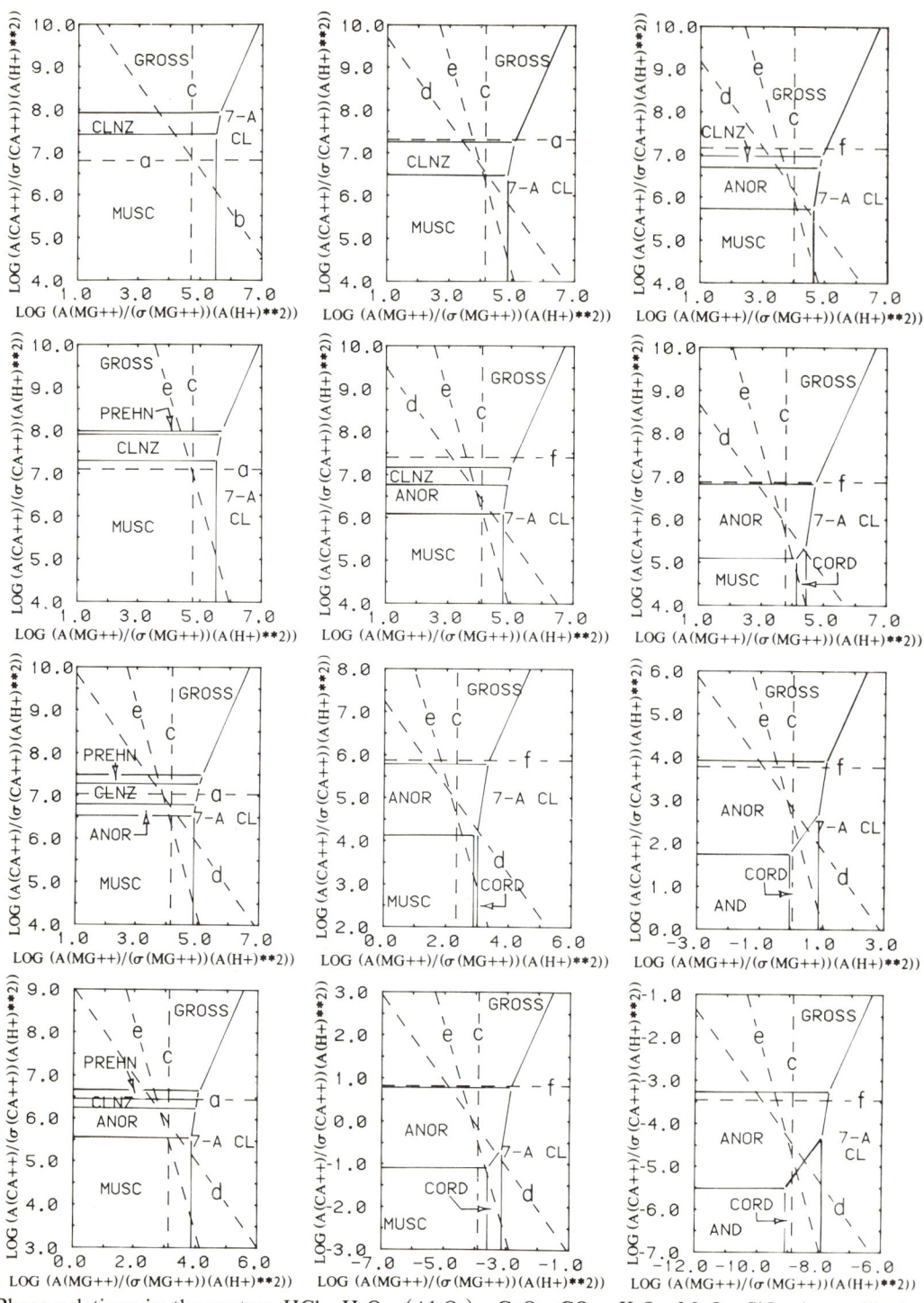

Phase relations in the system $HCl-H_2O-(Al_2O_3)-CaO-CO_2-K_2O-MgO-SiO_2$, in equilibrium with quartz and K-feldspar, at $X_{CO_2} = 0.01$. Metastable 7-A clinochlore was considered instead of its stable counterpart, 14-A clinochlore. Saturation limits: calcite (a), dolomite (b), phlogopite (c), diopside (d), tremolite (e), wollastonite (f).

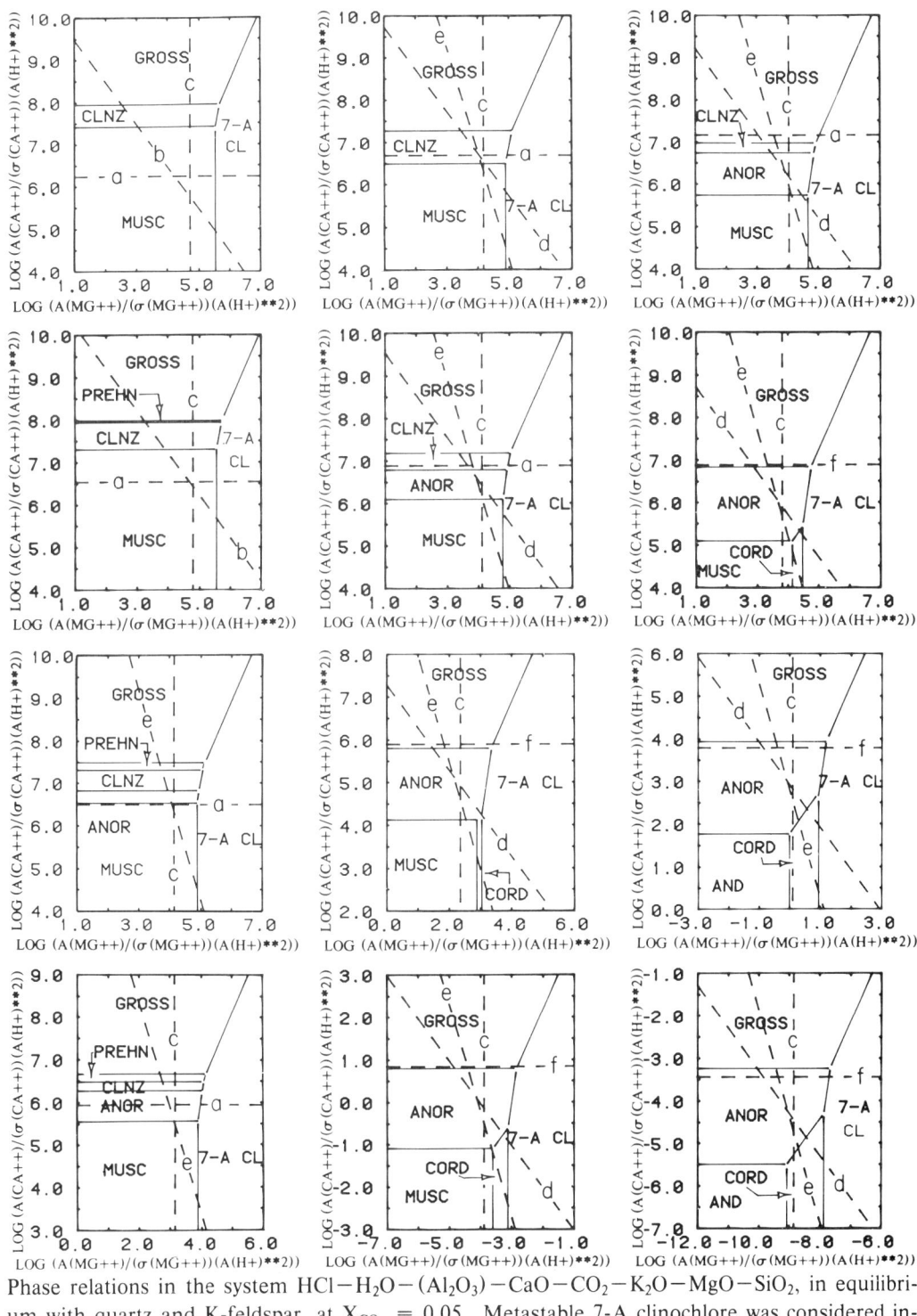

Phase relations in the system $HCl-H_2O-(Al_2O_3)-CaO-CO_2-K_2O-MgO-SiO_2$, in equilibrium with quartz and K-feldspar, at $X_{CO_2} = 0.05$. Metastable 7-A clinochlore was considered instead of its stable counterpart, 14-A clinochlore. Saturation limits: calcite (a), dolomite (b), phlogopite (c), diopside (d), tremolite (e), wollastonite (f).

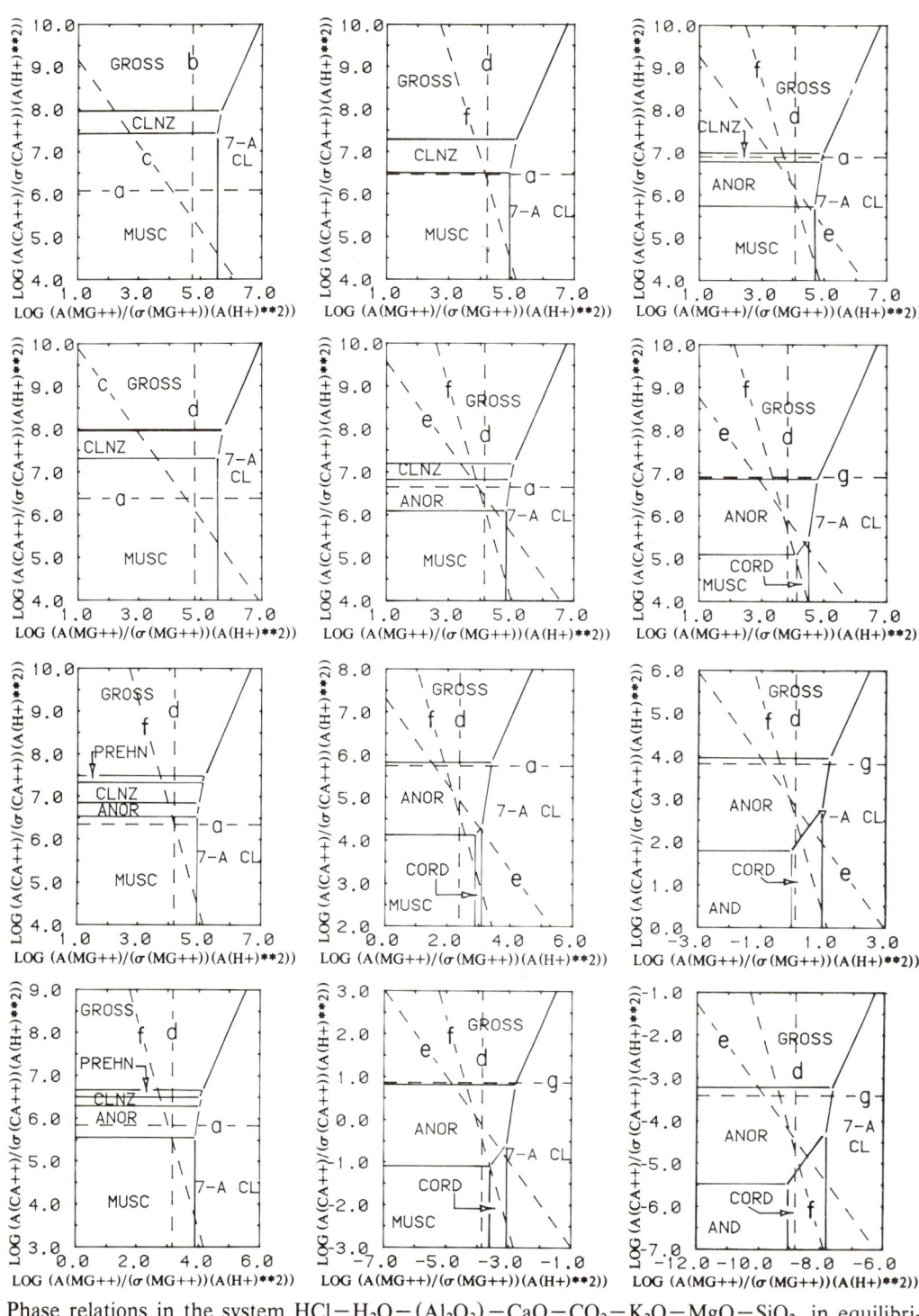

Phase relations in the system $HCl-H_2O-(Al_2O_3)-CaO-CO_2-K_2O-MgO-SiO_2$, in equilibrium with quartz and K-feldspar, at $X_{CO_2} = 0.10$. Metastable 7-A clinochlore was considered instead of its stable counterpart, 14-A clinochlore. Saturation limits: calcite (a), magnesite (b), dolomite (c), phlogopite (d), diopside (e), tremolite (f), wollastonite (g).

285

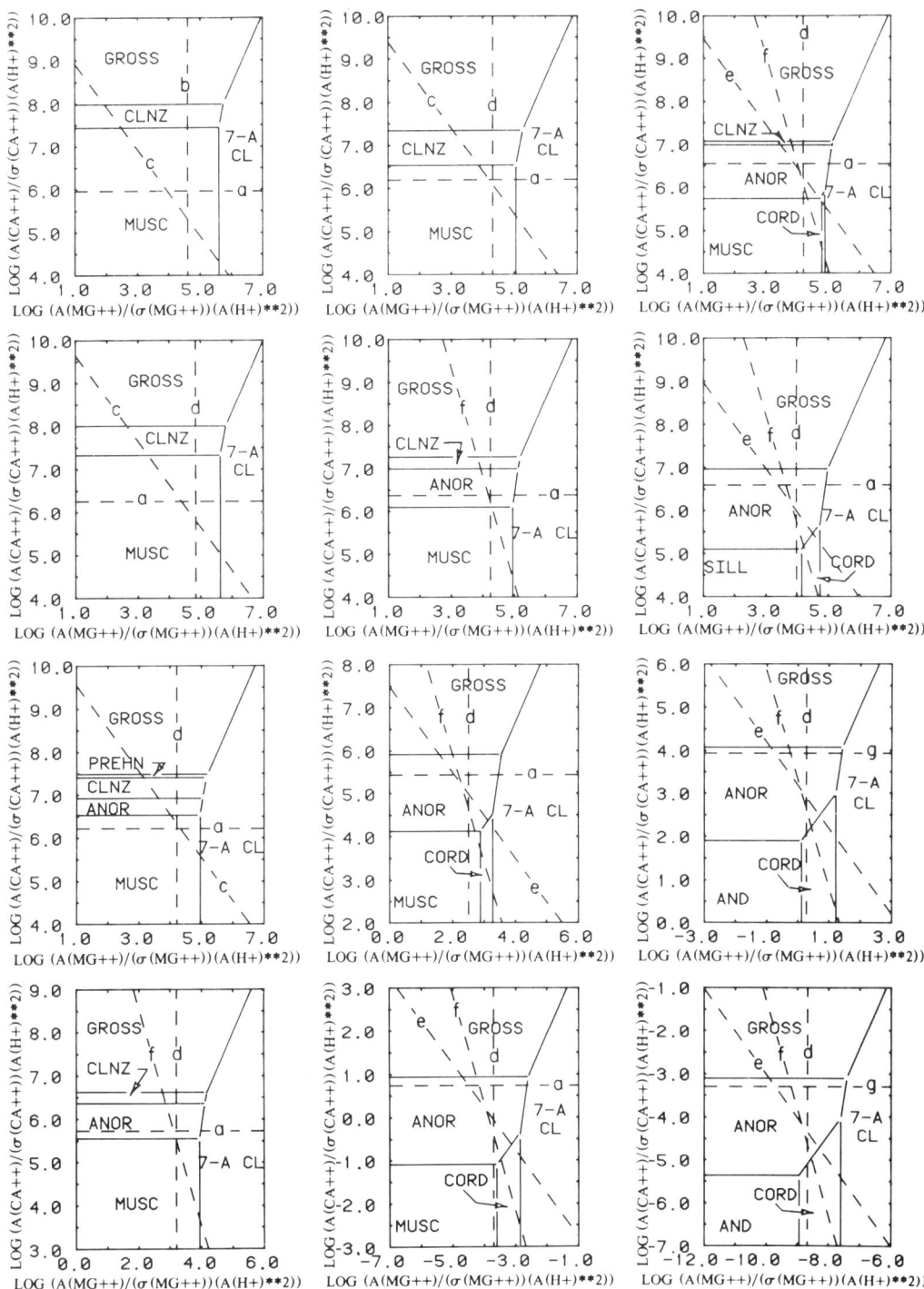

Phase relations in the system $HCl-H_2O-(Al_2O_3)-CaO-CO_2-K_2O-MgO-SiO_2$, in equilibrium with quartz and K-feldspar, at $X_{CO_2} = 0.30$. Metastable 7-A clinochlore was considered instead of its stable counterpart, 14-A clinochlore. Saturation limits: calcite (a), magnesite (b), dolomite (c), phlogopite (d), diopside (e), tremolite (f), wollastonite (g).

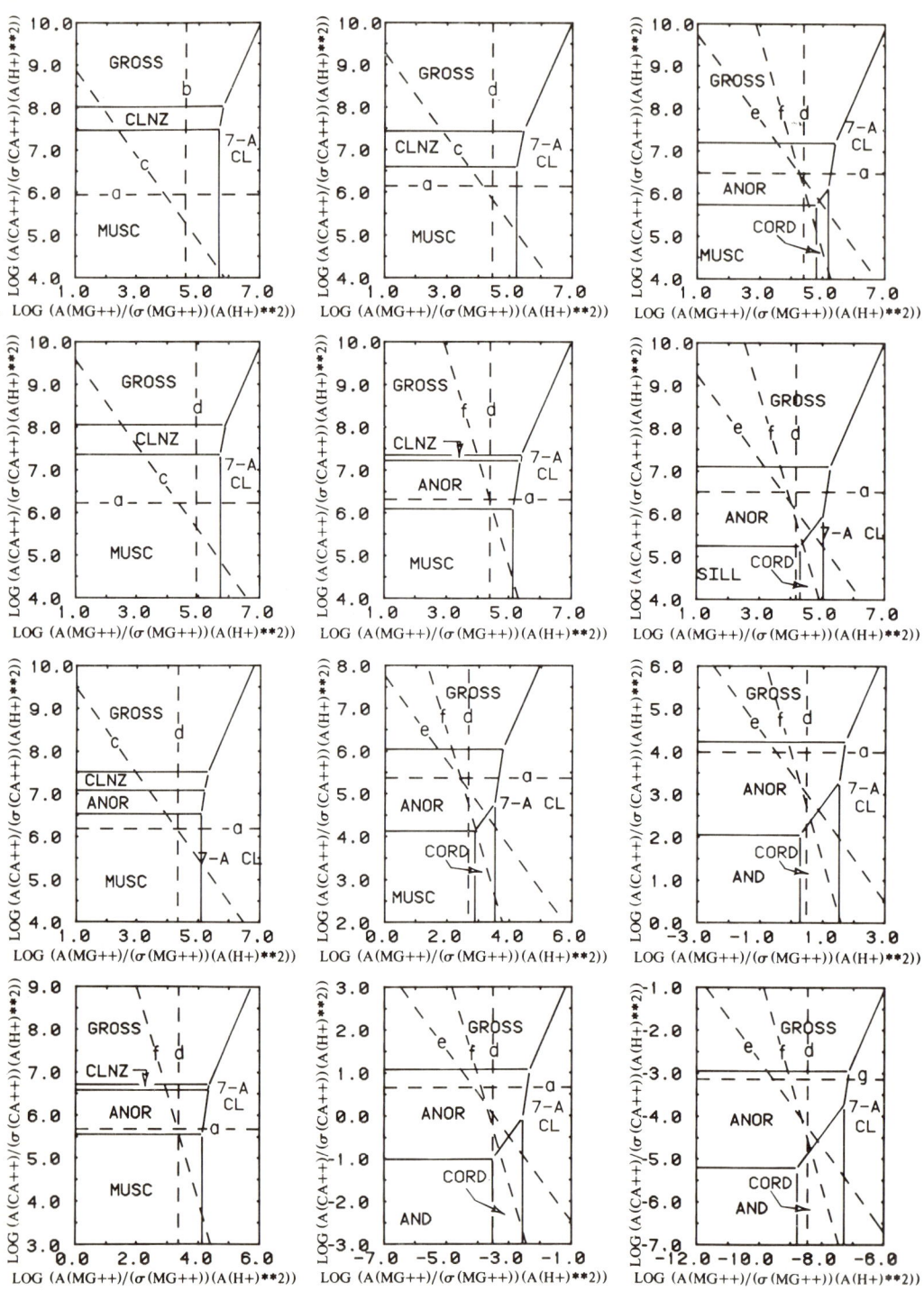

Phase relations in the system $HCl-H_2O-(Al_2O_3)-CaO-CO_2-K_2O-MgO-SiO_2$, in equilibrium with quartz and K-feldspar, at $X_{CO_2} = 0.50$. Metastable 7-A clinochlore was considered instead of its stable counterpart, 14-A clinochlore. Saturation limits: calcite (a), magnesite (b), dolomite (c), phlogopite (d), diopside (e), tremolite (f), wollastonite (g).

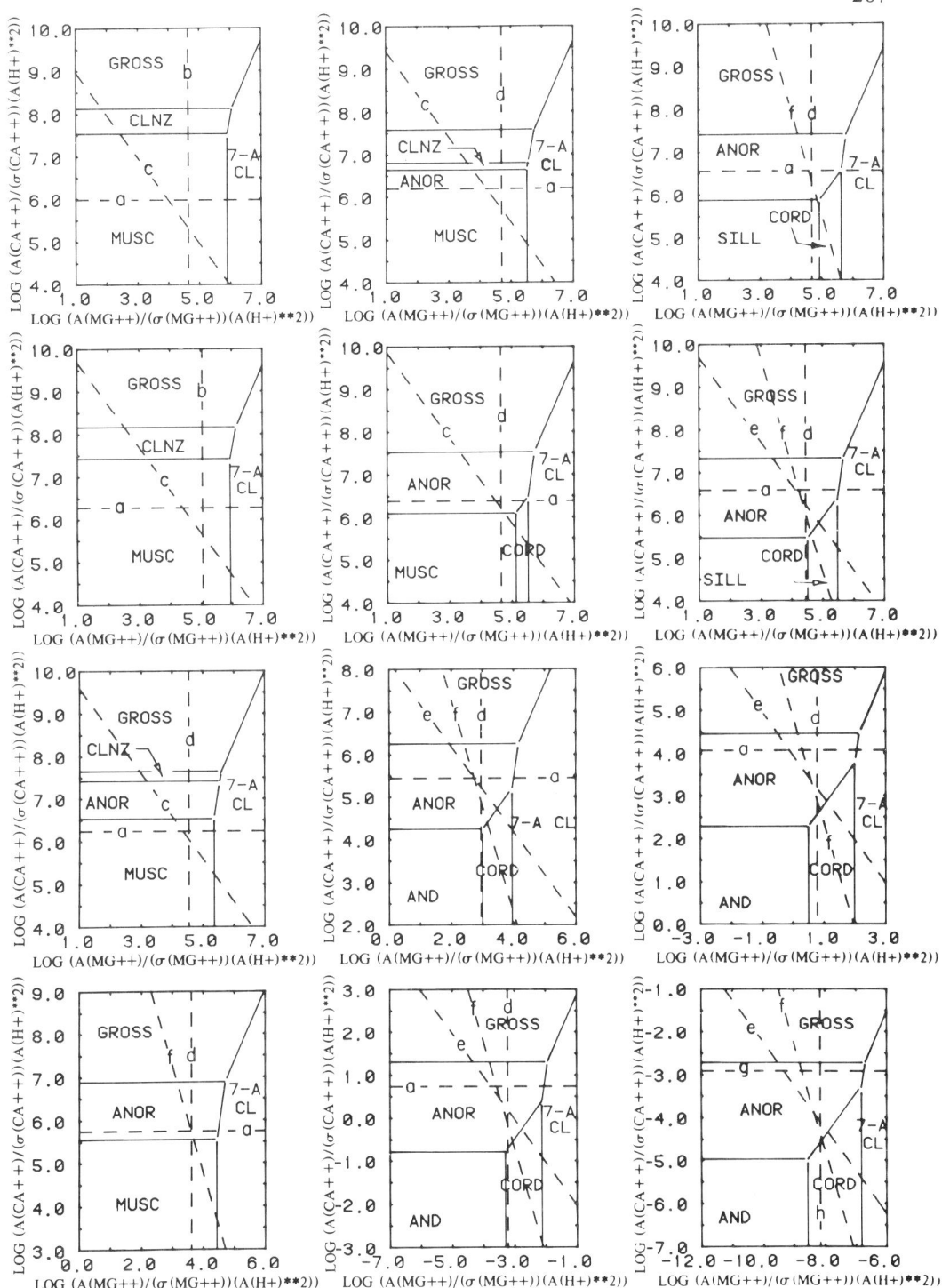

Phase relations in the system $HCl-H_2O-(Al_2O_3)-CaO-CO_2-K_2O-MgO-SiO_2$, in equilibrium with quartz and K-feldspar, at $X_{CO_2} = 0.70$. Metastable 7-A clinochlore was considered instead of its stable counterpart, 14-A clinochlore. Saturation limits: calcite (a), magnesite (b), dolomite (c), phlogopite (d), diopside (e), tremolite (f), wollastonite (g), enstatite (h).

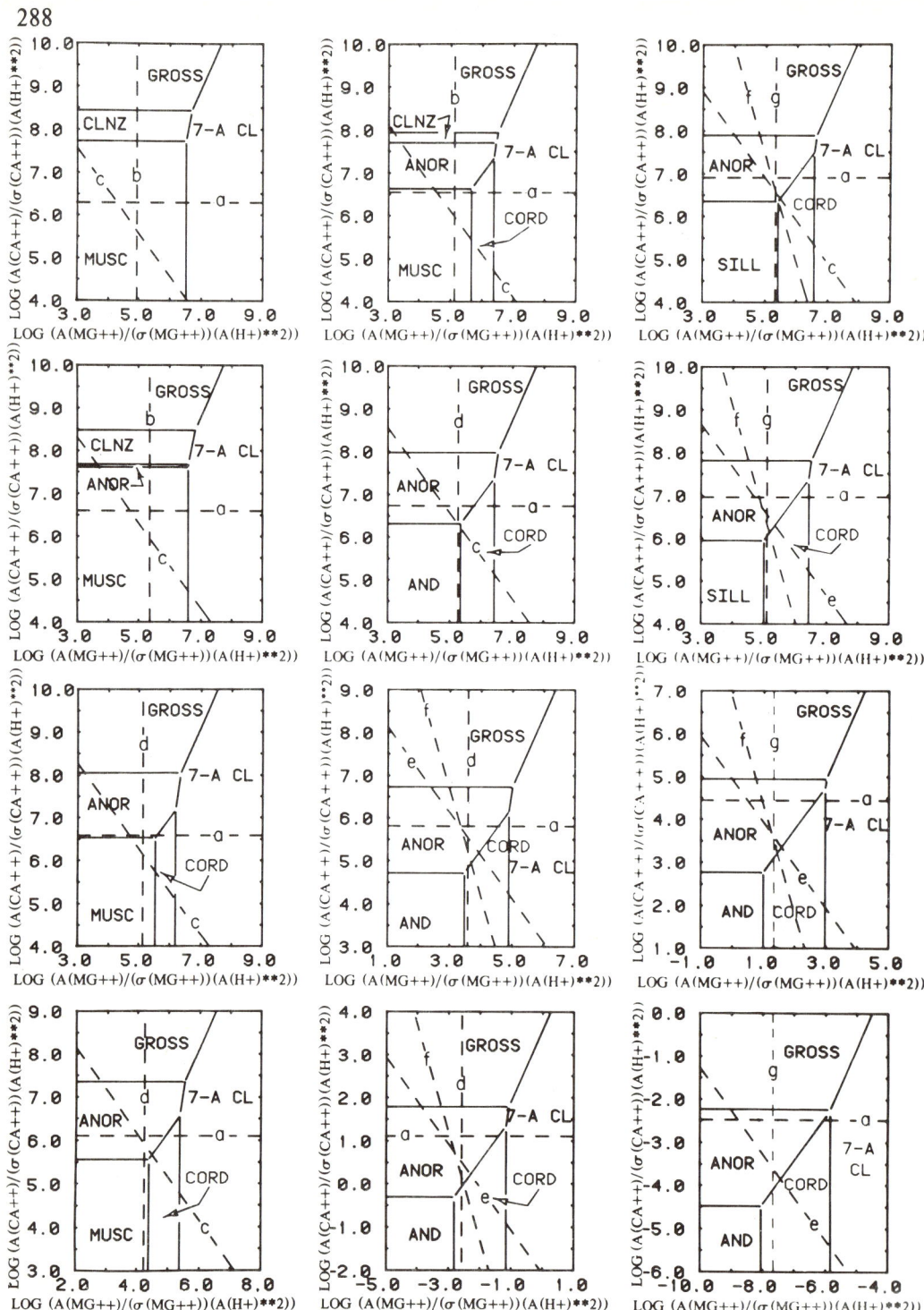

Phase relations in the system $HCl-H_2O-(Al_2O_3)-CaO-CO_2-K_2O-MgO-SiO_2$, in equilibrium with quartz and K-feldspar, at $X_{CO_2} = 0.90$. Metastable 7-A clinochlore was considered instead of its stable counterpart, 14-A clinochlore. Saturation limits: calcite (a), magnesite (b), dolomite (c), phlogopite (d), diopside (e), tremolite (f), enstatite (g).

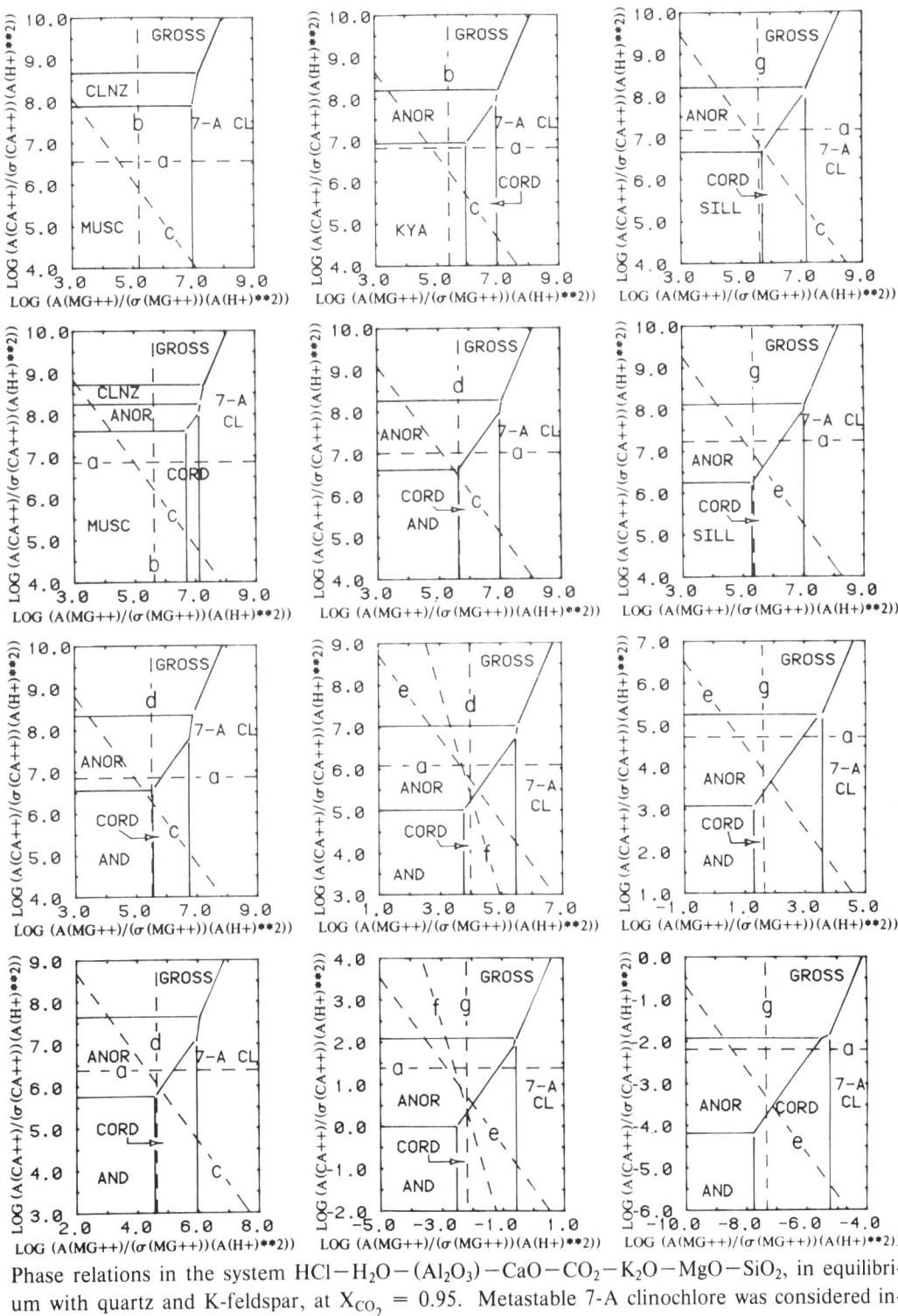

Phase relations in the system $HCl-H_2O-(Al_2O_3)-CaO-CO_2-K_2O-MgO-SiO_2$, in equilibrium with quartz and K-feldspar, at $X_{CO_2} = 0.95$. Metastable 7-A clinochlore was considered instead of its stable counterpart, 14-A clinochlore. Saturation limits: calcite (a), magnesite (b), dolomite (c), phlogopite (d), diopside (e), tremolite (f), enstatite (g).

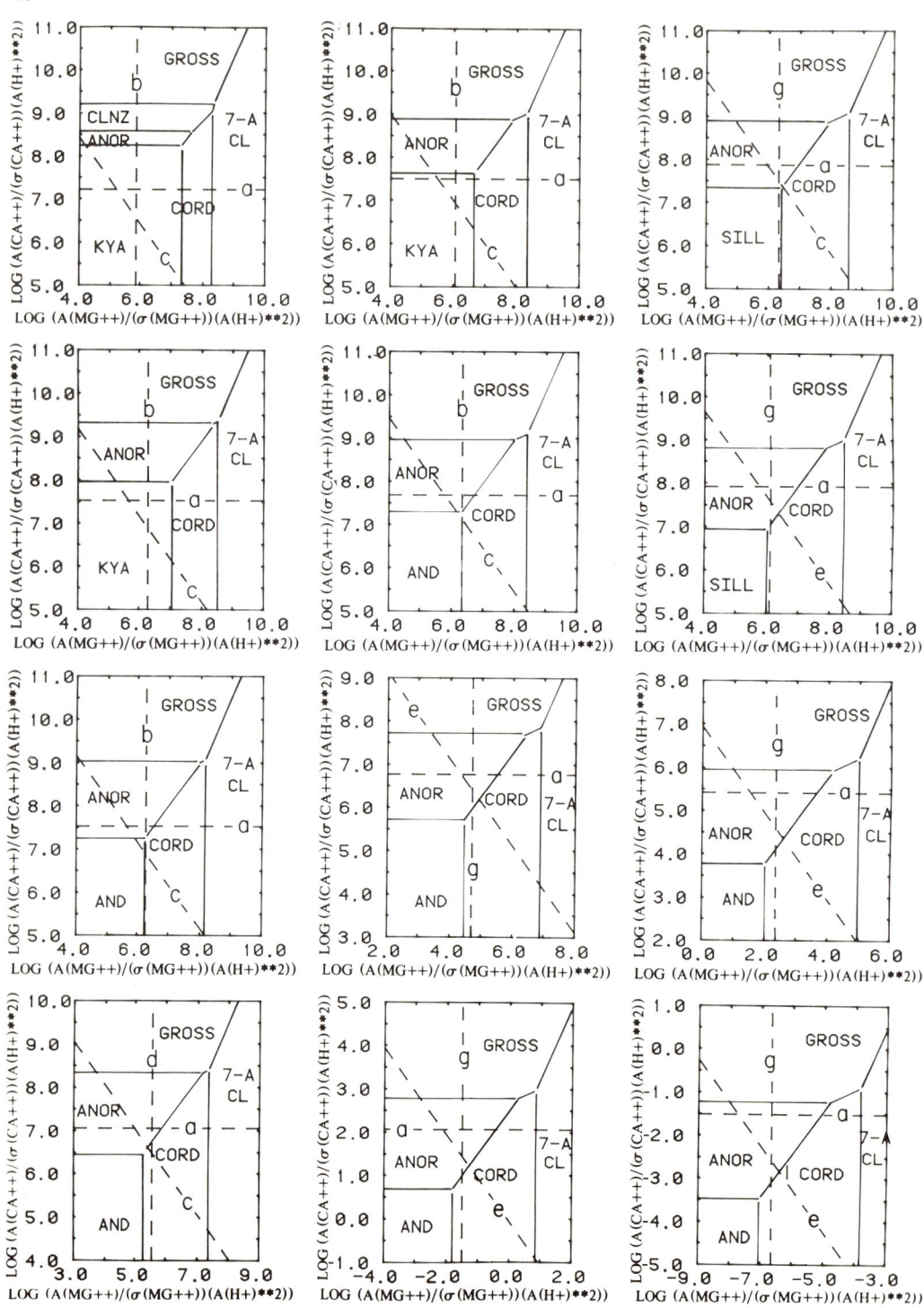

Phase relations in the system $HCl-H_2O-(Al_2O_3)-CaO-CO_2-K_2O-MgO-SiO_2$, in equilibrium with quartz and K-feldspar, at $X_{CO_2} = 0.99$. Metastable 7-A clinochlore was considered instead of its stable counterpart, 14-A clinochlore. Saturation limits: calcite (a), magnesite (b), dolomite (c), phlogopite (d), diopside (e), tremolite (f), enstatite (g).

FUGACITY DIAGRAMS

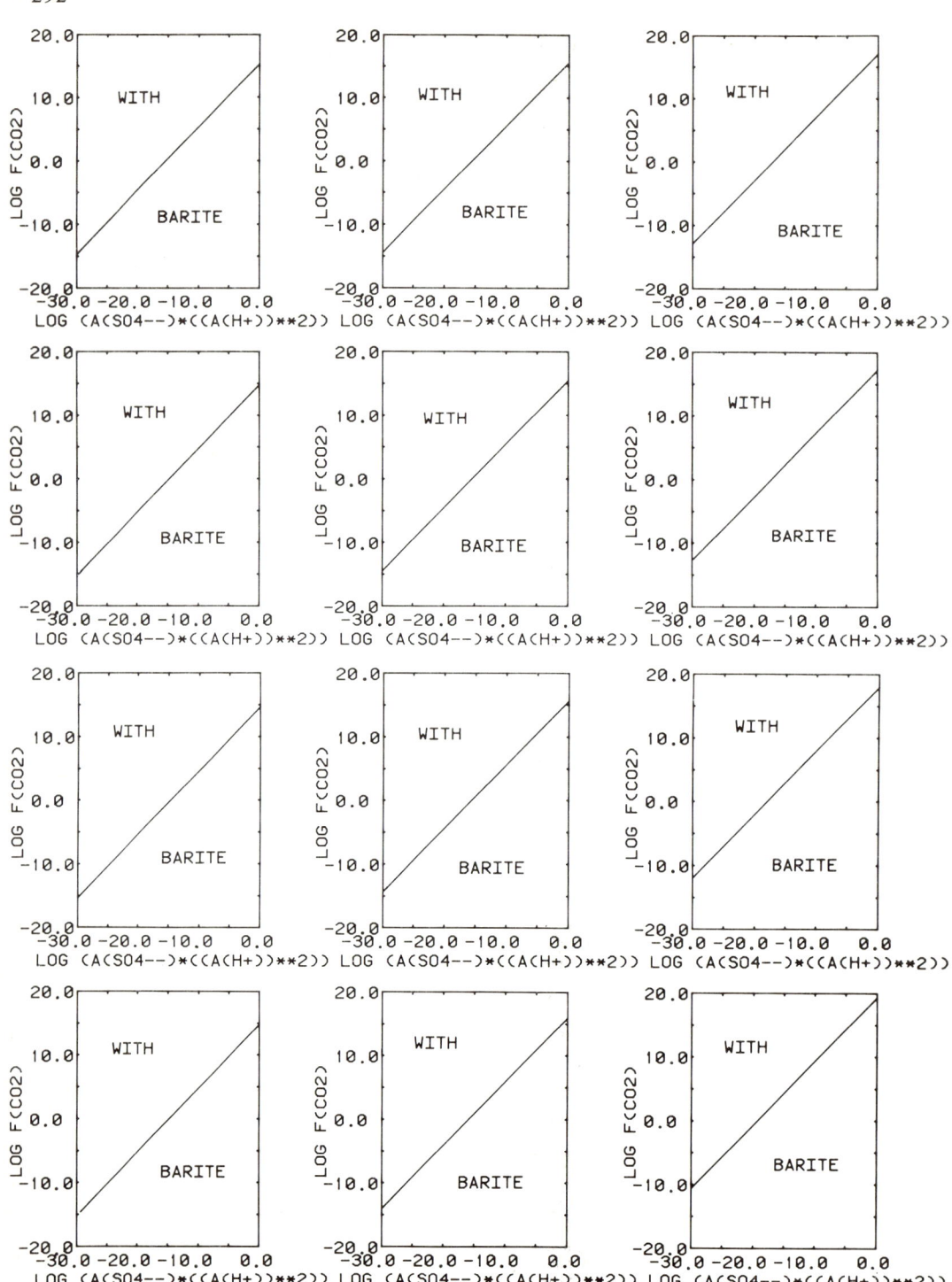

Phase relations in the system $HCl-H_2O-(BaO)-CO_2-H_2SO_4$.

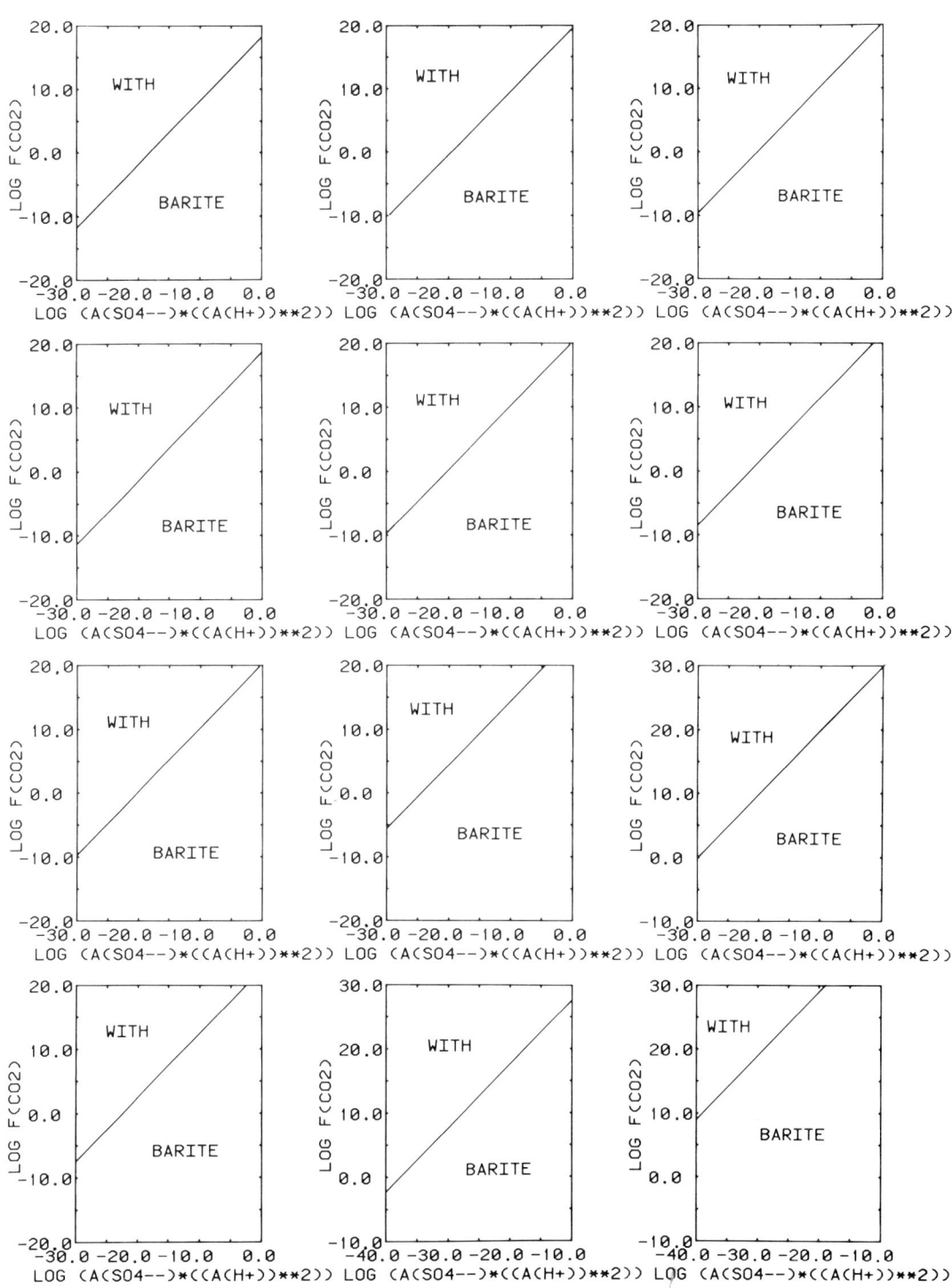

Phase relations in the system $HCl-H_2O-(BaO)-CO_2-H_2SO_4$.

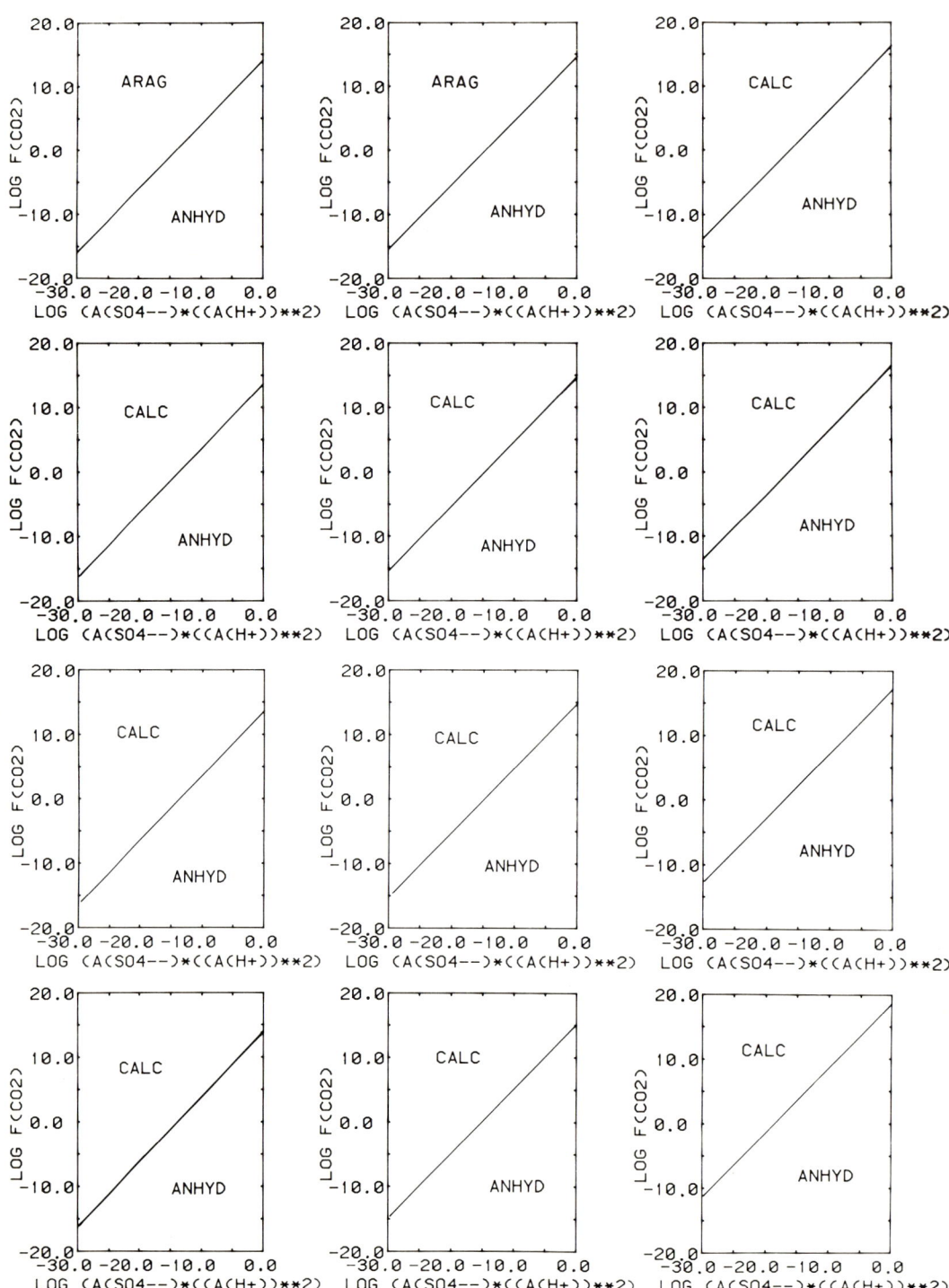

Phase relations in the system $HCl-H_2O-(CaO)-CO_2-H_2SO_4$.

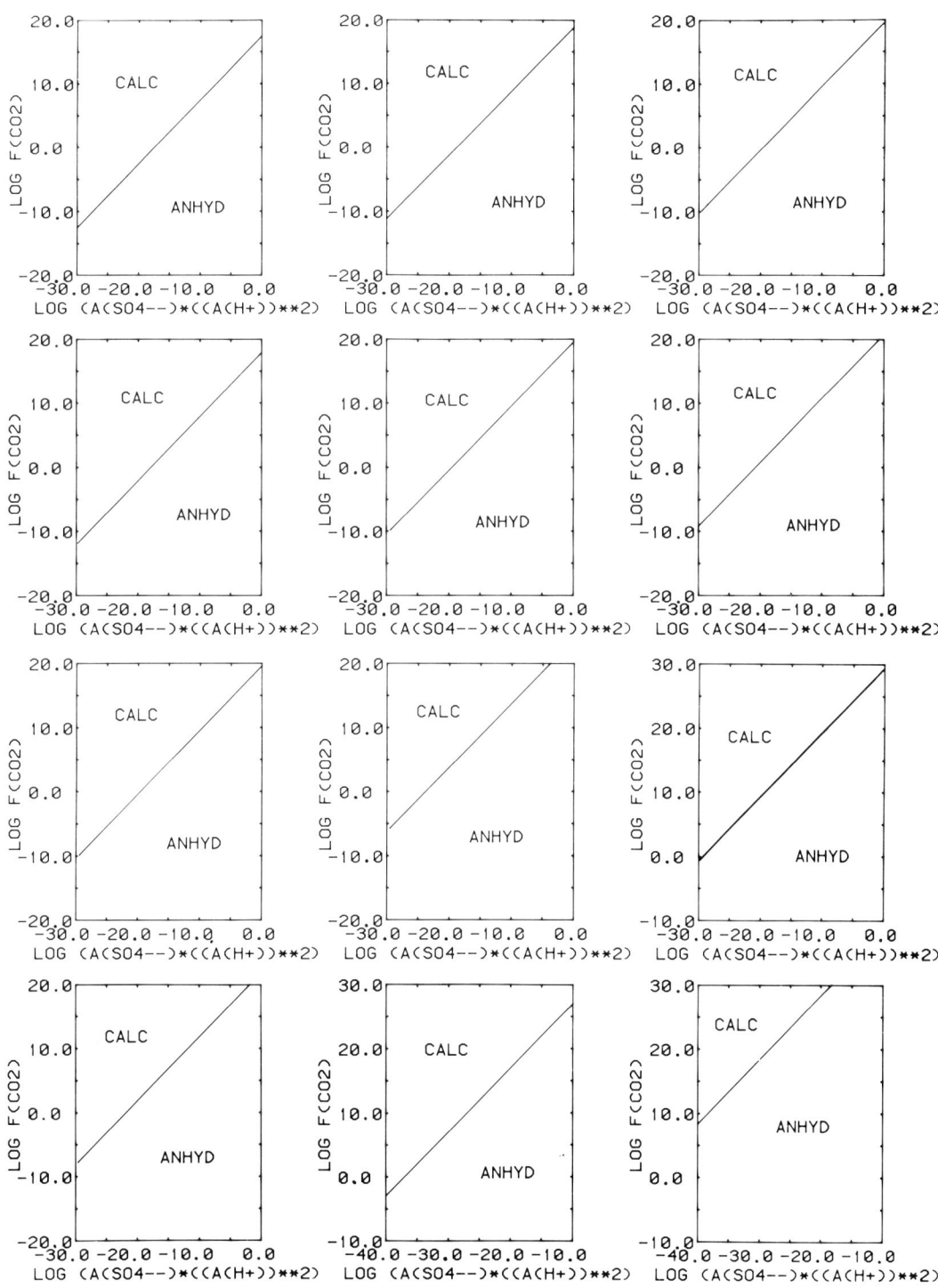

Phase relations in the system $HCl-H_2O-(CaO)-CO_2-H_2SO_4$.

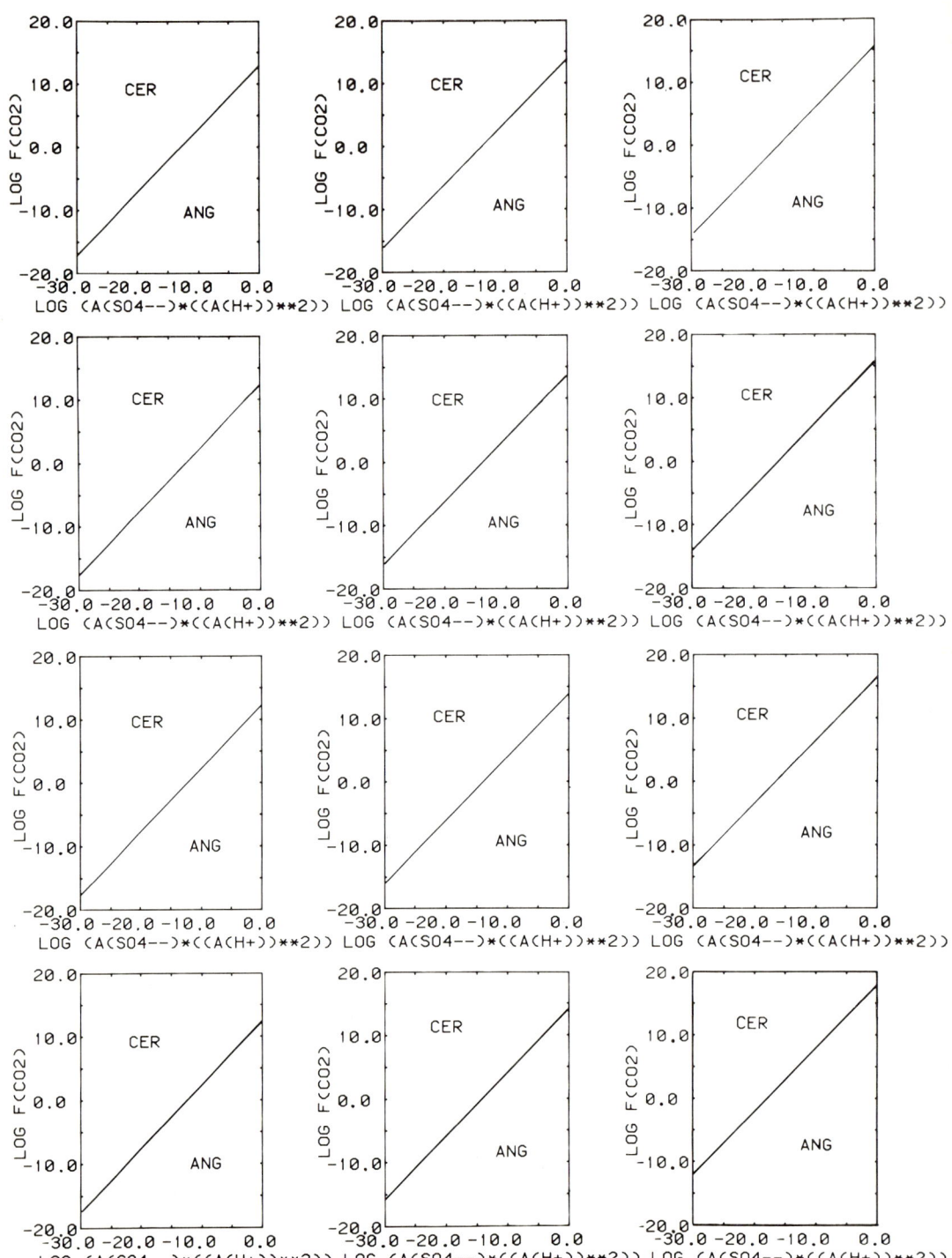

Phase relations in the system $HCl-H_2O-CO_2-H_2SO_4-(PbO)$.

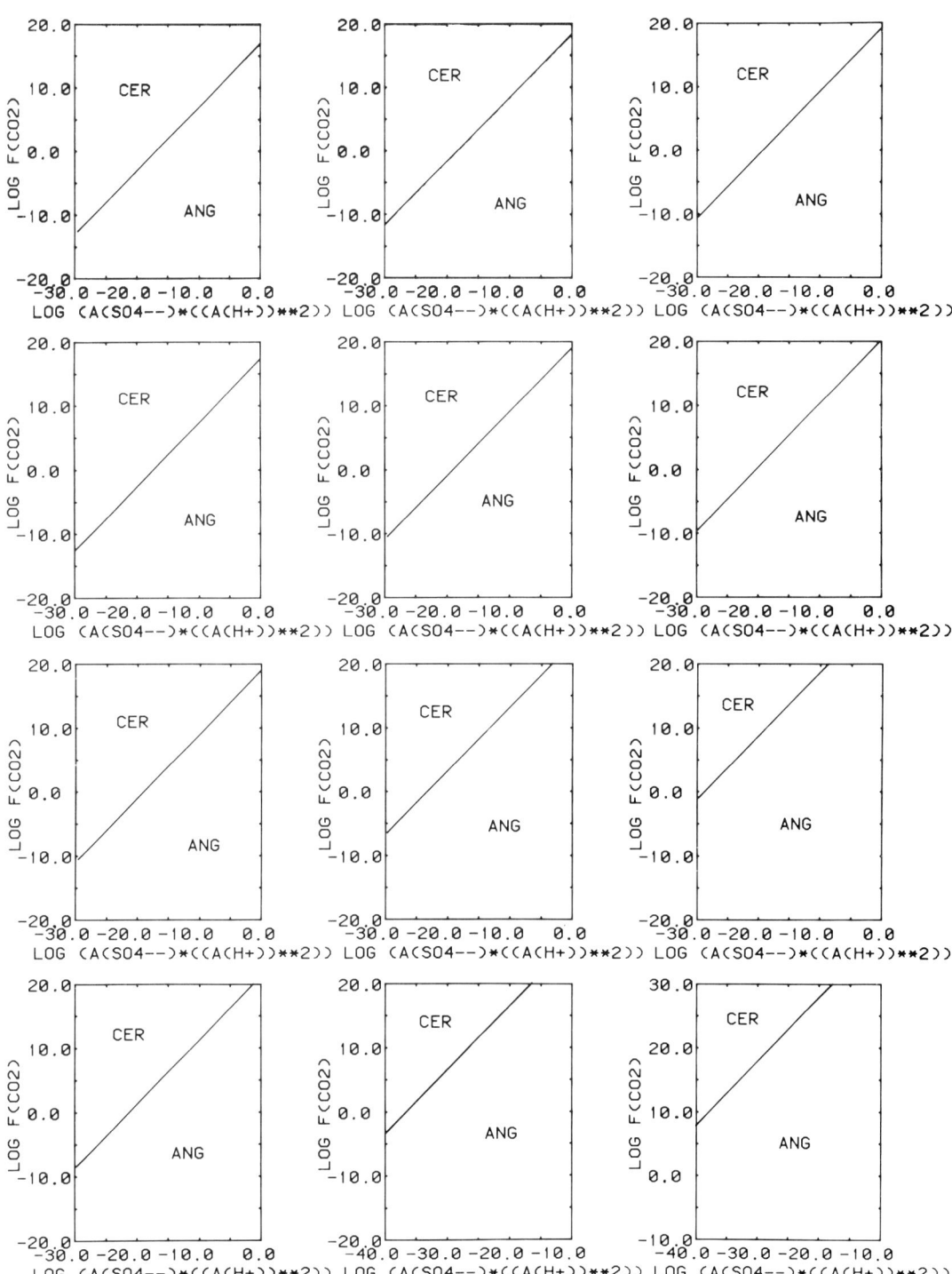

Phase relations in the system $HCl-H_2O-CO_2-H_2SO_4-(PbO)$.

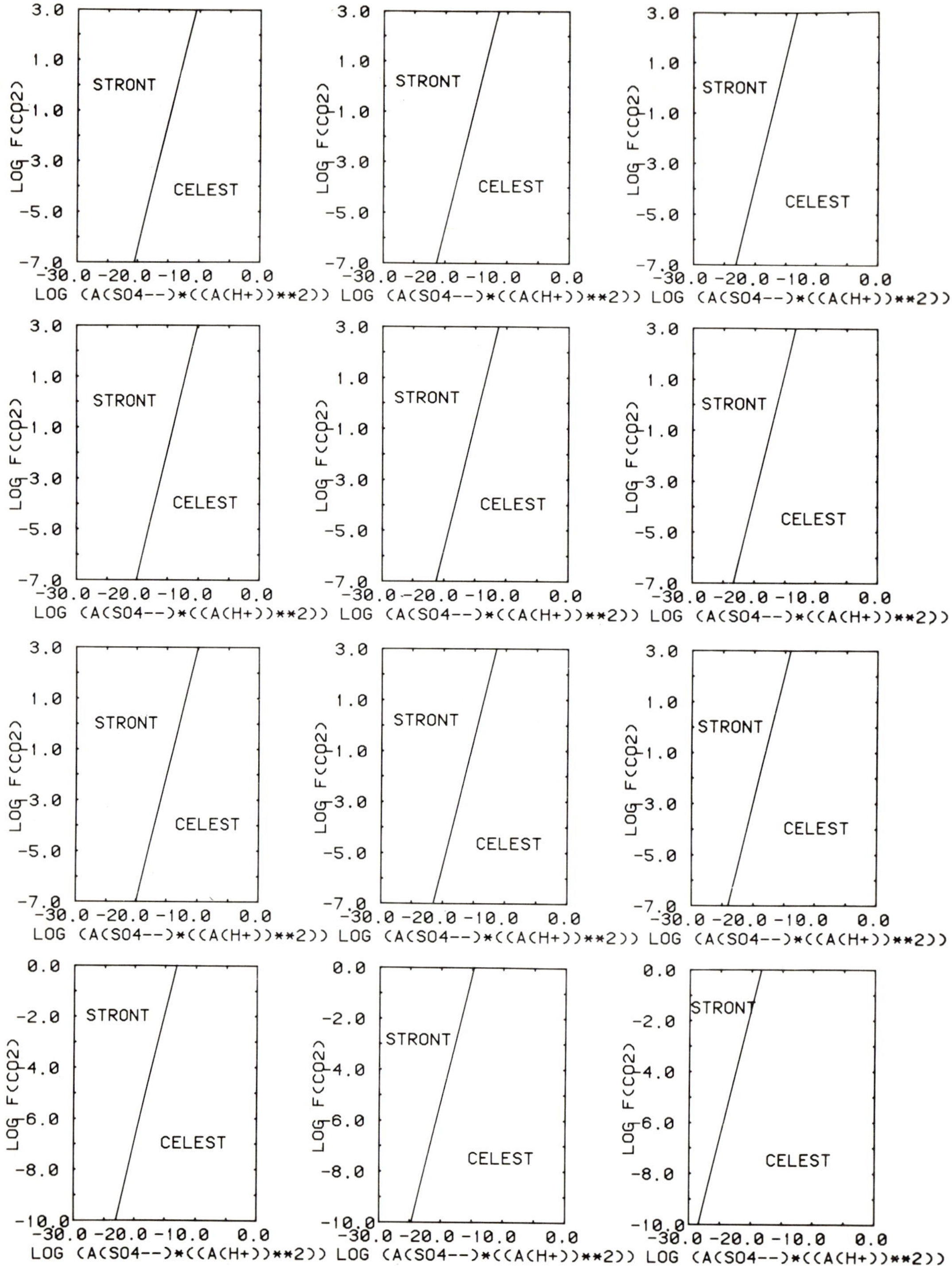

Phase relations in the system $HCl-H_2O-CO_2-H_2SO_4-(SrO)$.

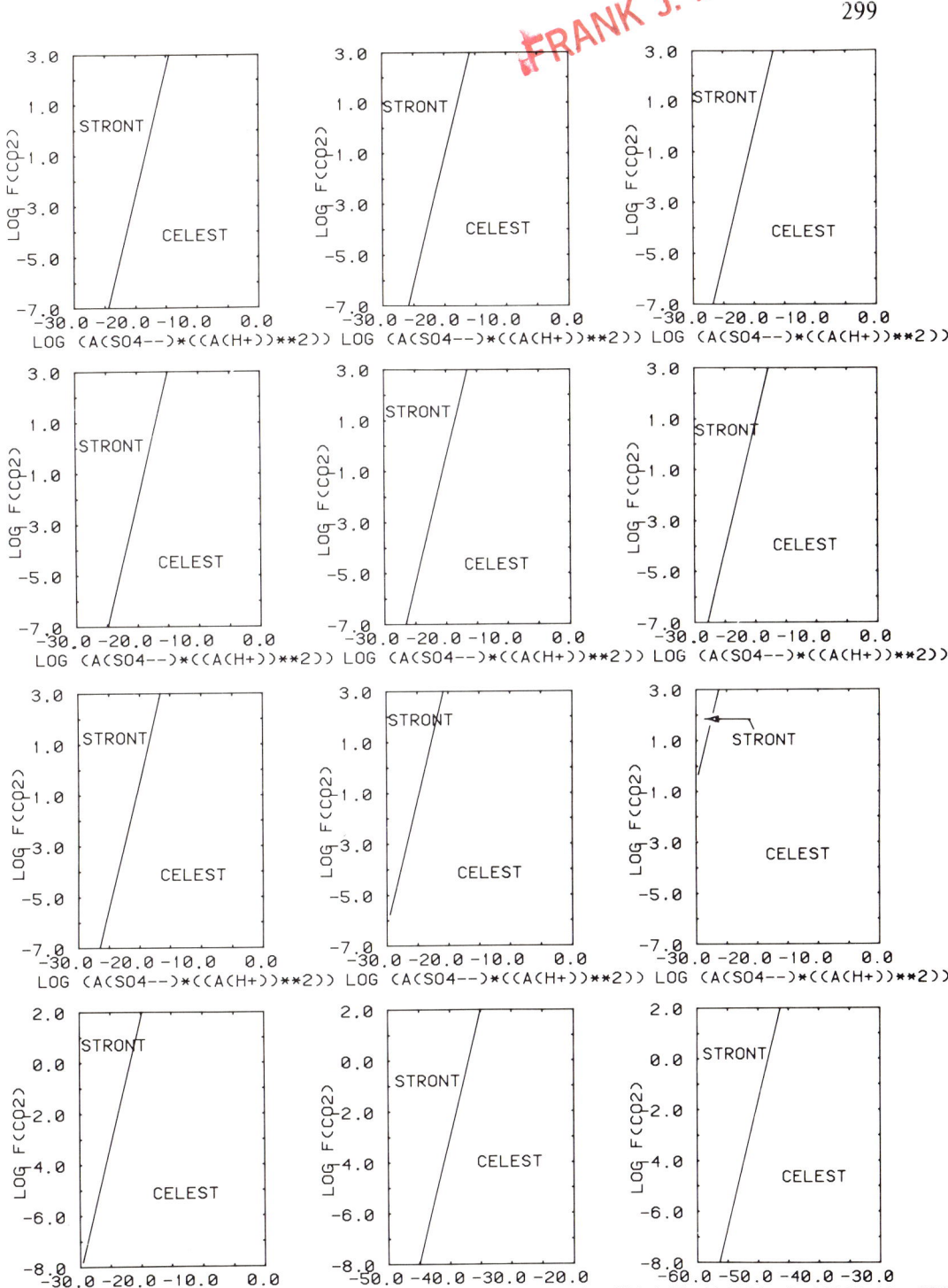

Phase relations in the system $HCl-H_2O-CO_2-H_2SO_4-(SrO)$.

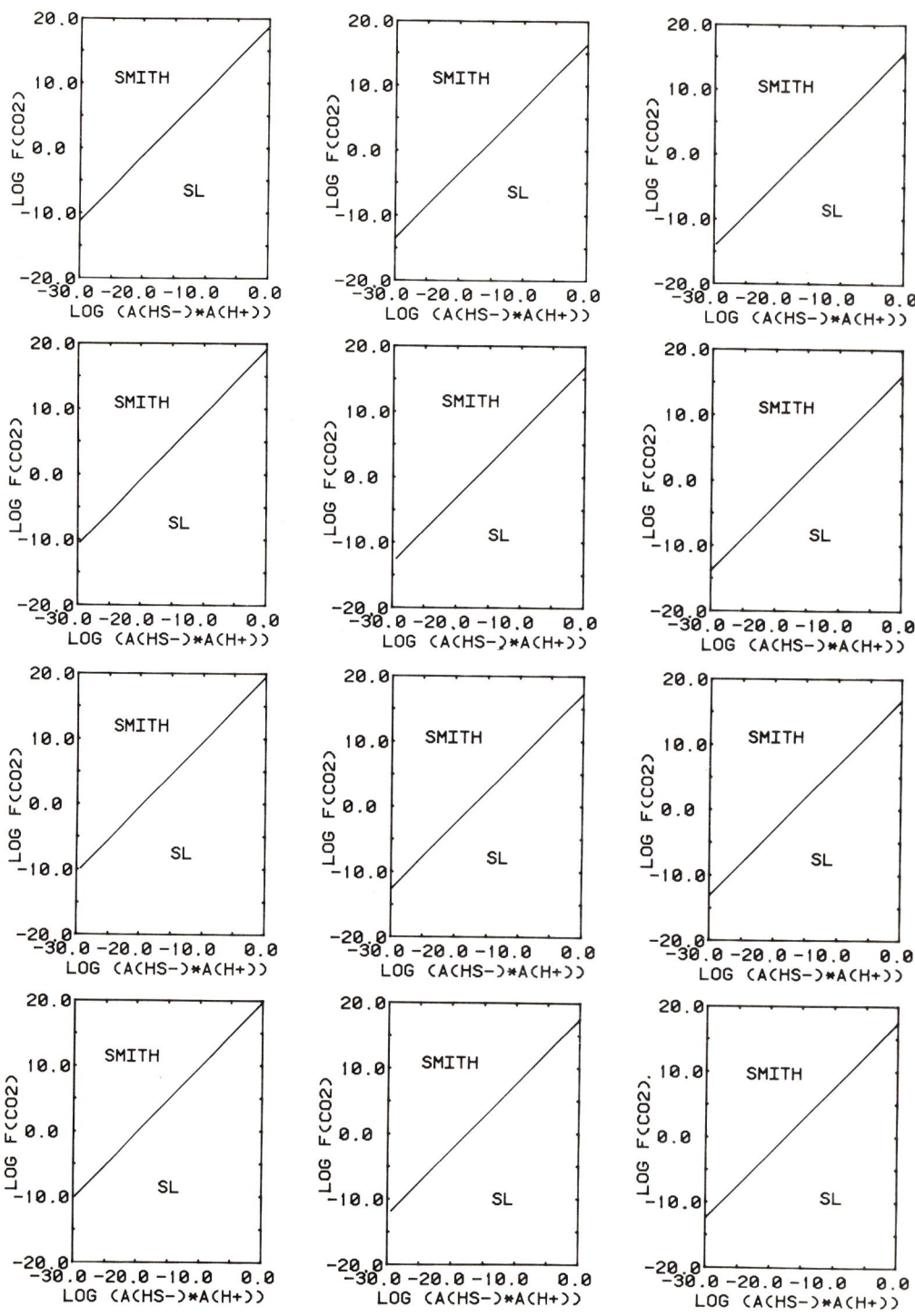

Phase relations in the system $HCl-H_2O-CO_2-H_2S-(ZnO)$.

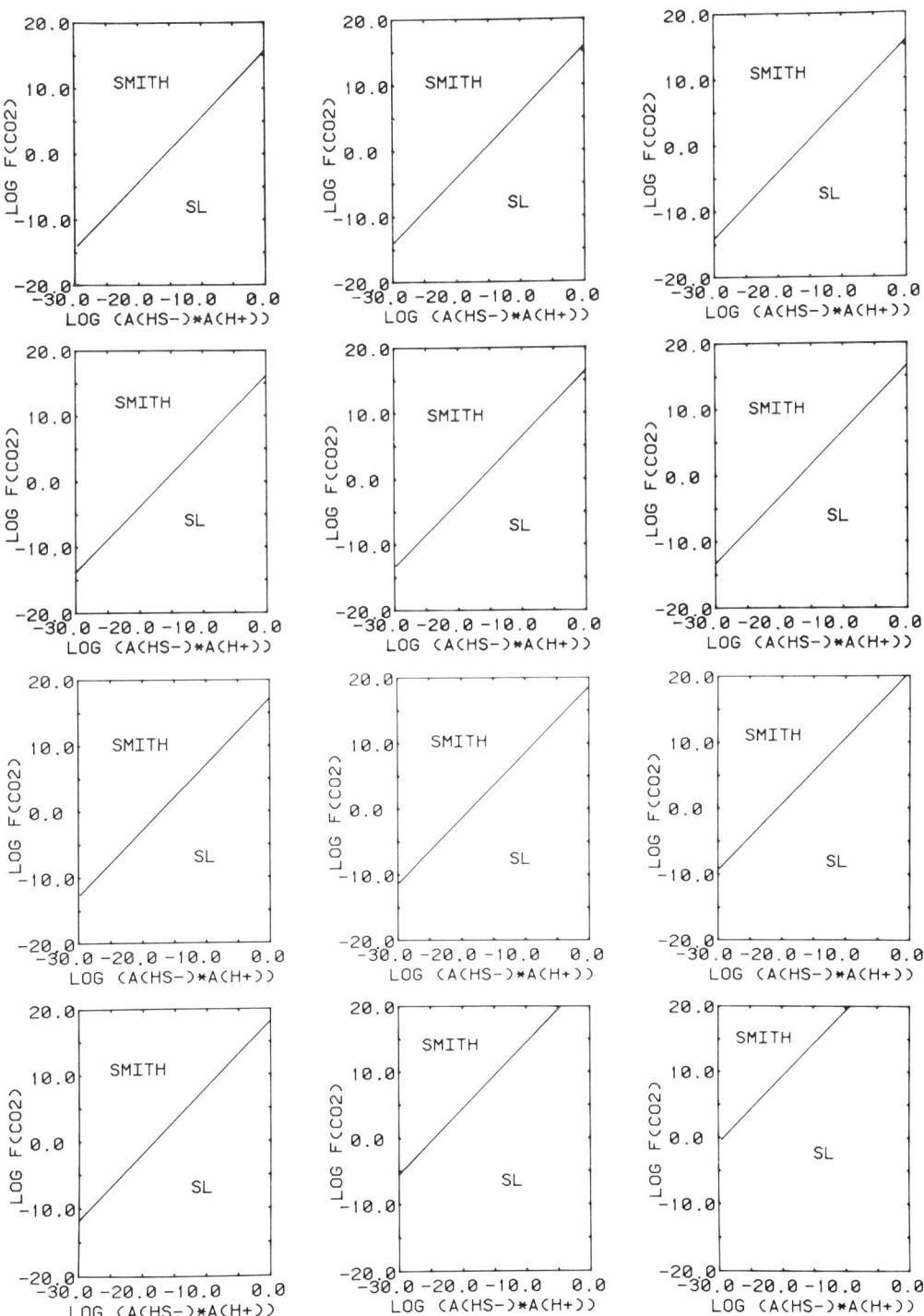

Phase relations in the system $HCl-H_2O-CO_2-H_2S-(ZnO)$.

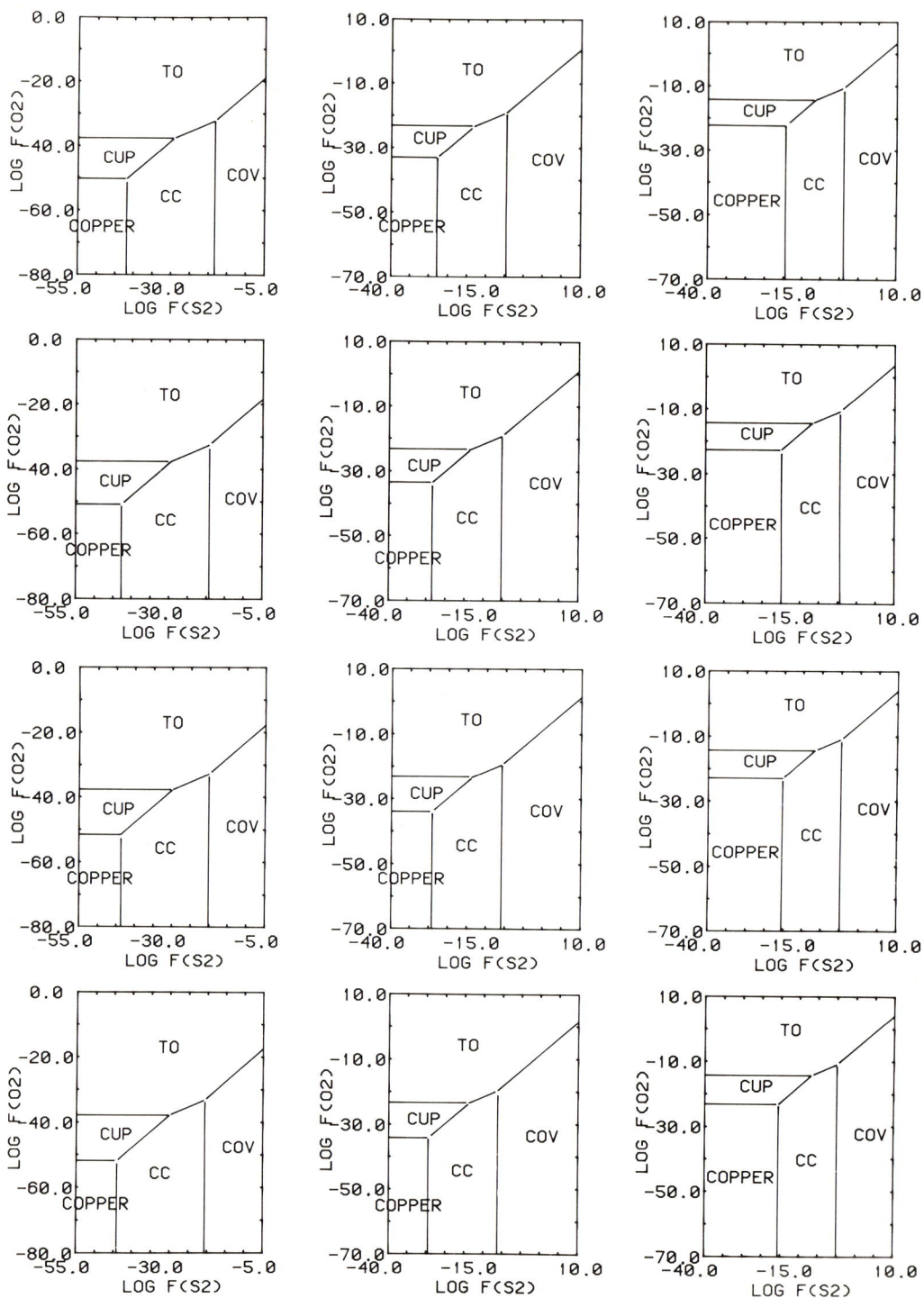

Phase relations in the system $HCl-H_2O-(CuO)-O_2-S_2$.

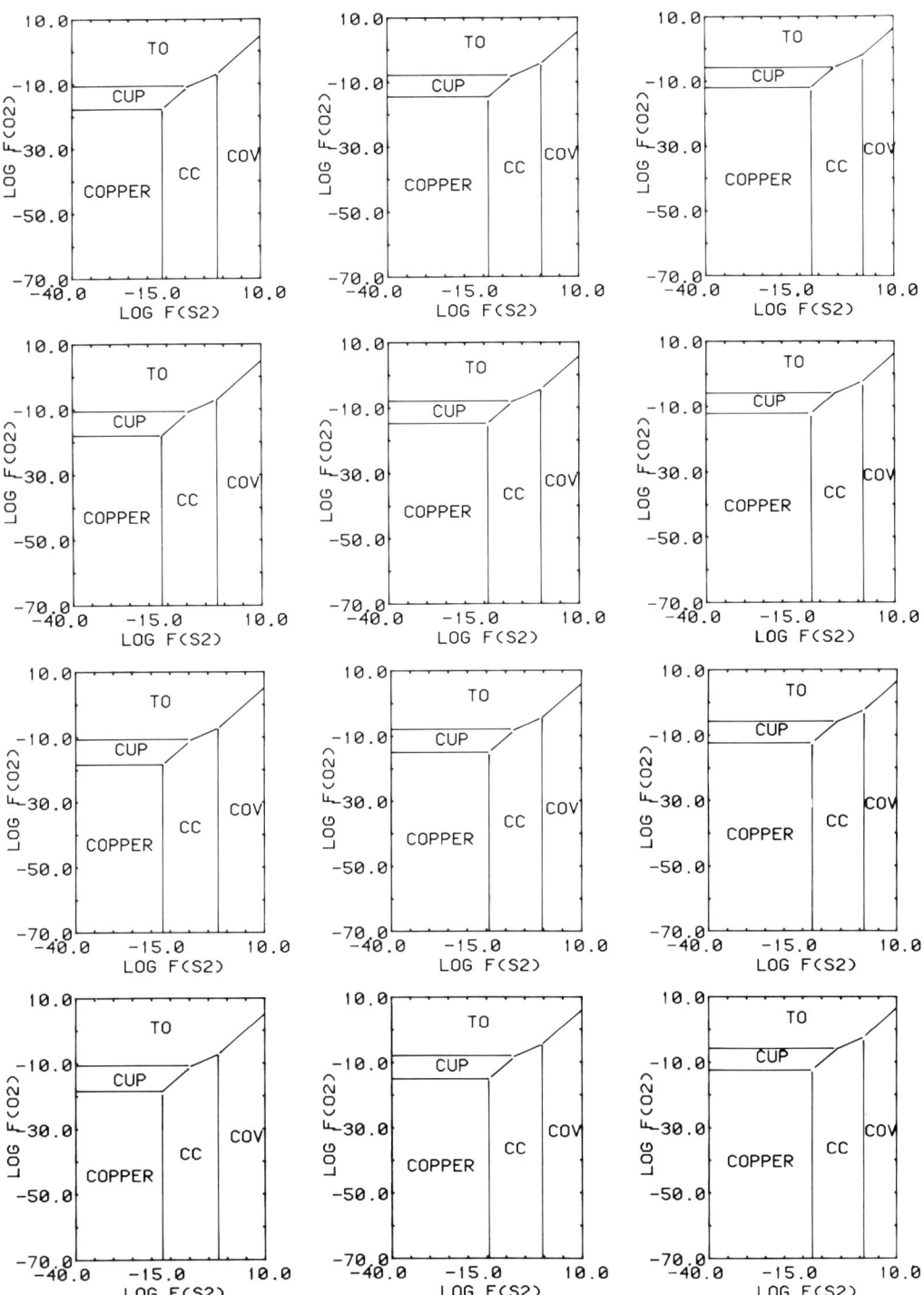

Phase relations in the system $HCl-H_2O-(CuO)-O_2-S_2$.

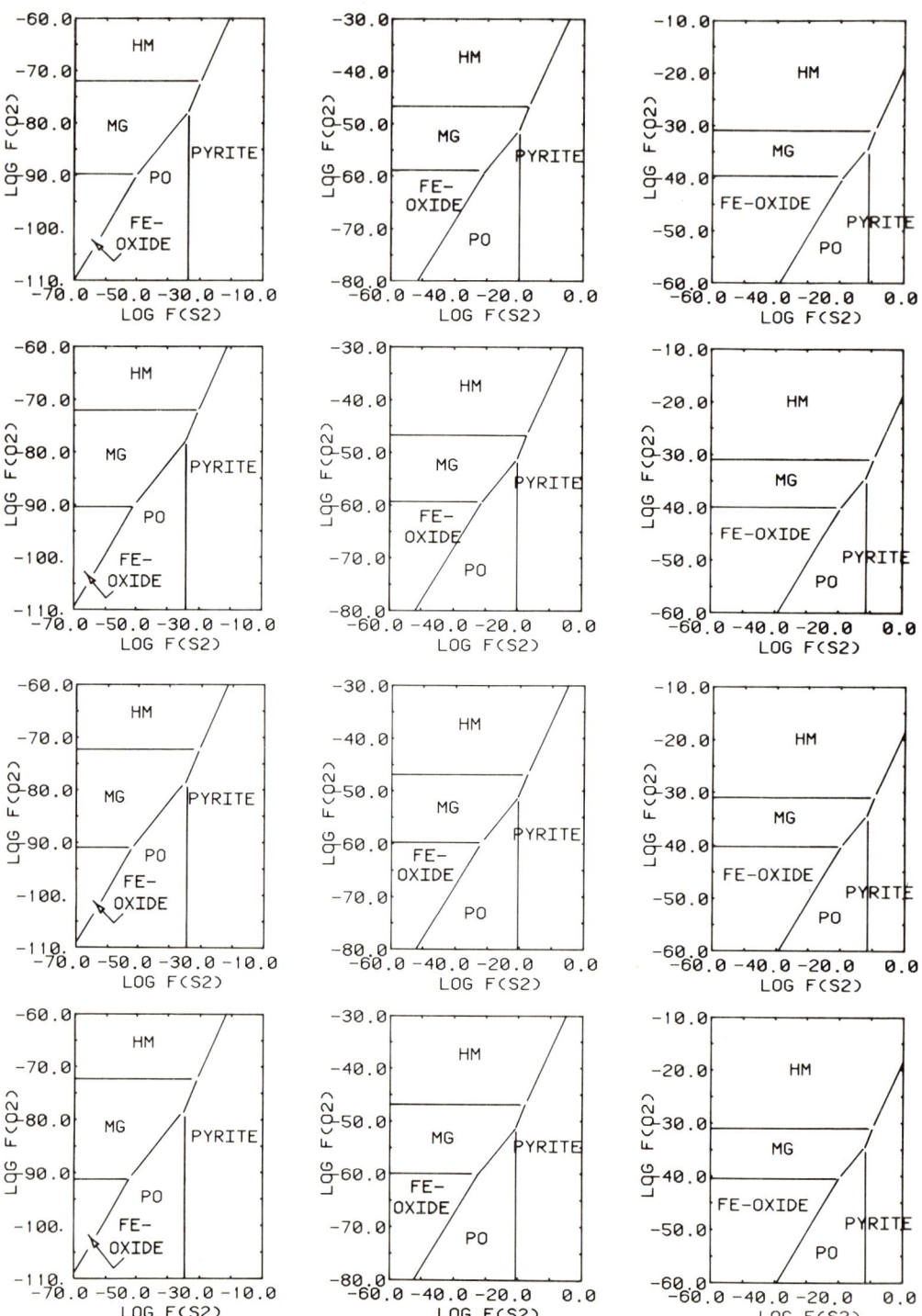

Phase relations in the system $HCl-H_2O-(FeO)-O_2-S_2$.

305

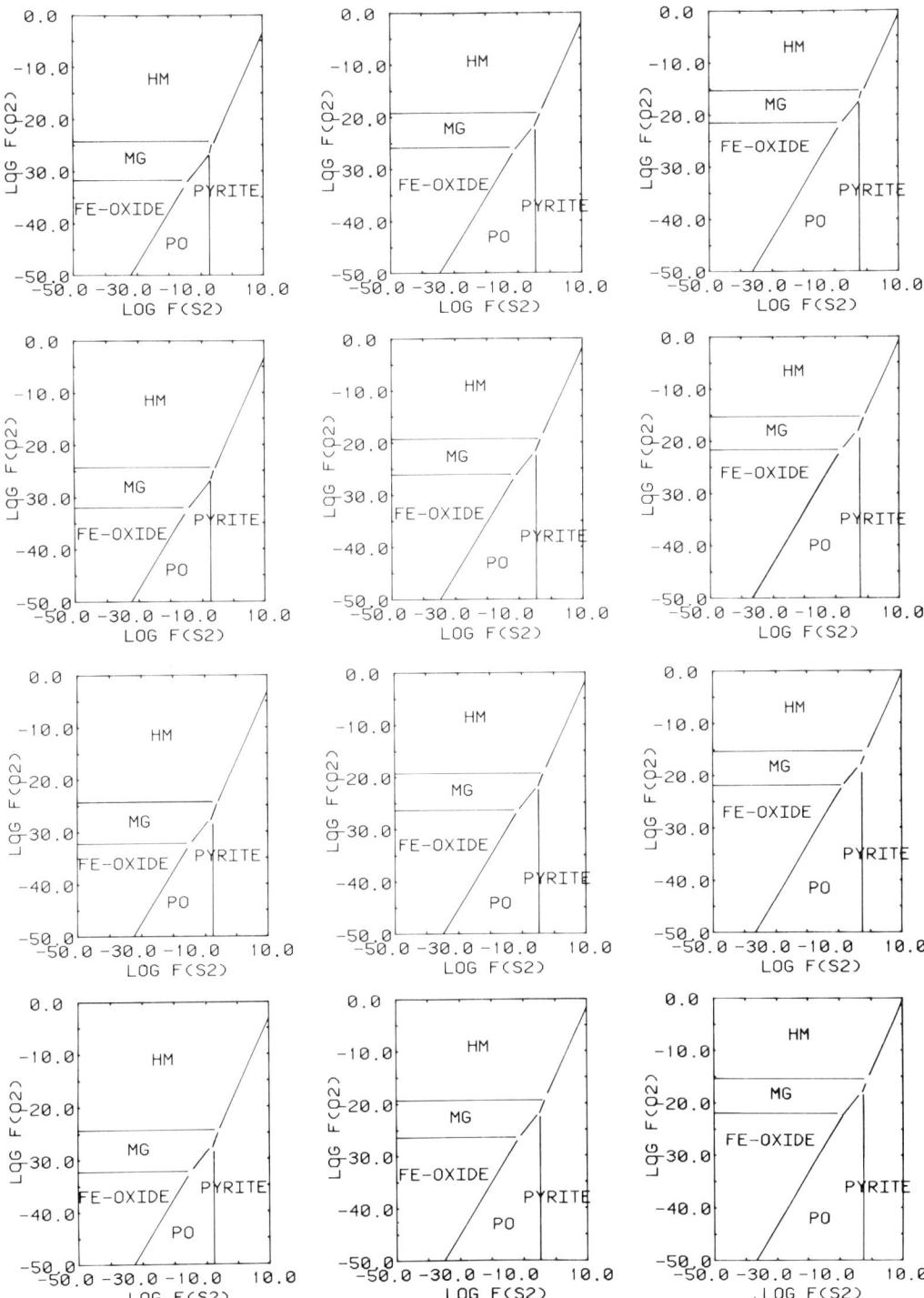

Phase relations in the system $HCl-H_2O-(FeO)-O_2-S_2$.

APPENDIX

The hydrolysis constants shown in the following pages correspond to those for the reactions shown in table 3. The reactions are designated below by the abbreviations given in table 2 for the minerals being hydrolyzed. The values of the equilibrium constants were generated using equations and data summarized by Helgeson and Kirkham (1974a), Walther and Helgeson (1977), Helgeson and others (1978), and Helgeson, Kirkham, and Flowers (1981), taking account of corrections given by Helgeson (1982b, 1984). P_{SAT} in the tables refers to pressures along the vapor-liquid equilibrium curve for H_2O.

Log K , P = P_{SAT} for T ⩽ 350°C ; P = 500 b for T > 350°C

NAME	INDEX	TEMPERATURE, °C							
		0	25	50	75	100	125	150	175
AND	1001	20.17	16.57	13.53	10.97	8.78	6.89	5.23	3.76
KYA	1002	19.83	16.30	13.32	10.80	8.65	6.80	5.18	3.74
SILL	1003	20.58	16.93	13.85	11.24	9.02	7.11	5.43	3.94
KAOL	1004	9.73	7.43	5.46	3.80	2.41	1.22	0.19	-0.72
CHRYS	1007	35.12	31.55	28.44	25.81	23.56	21.64	19.98	18.52
MUSC	1012	18.26	14.56	11.39	8.70	6.42	4.47	2.78	1.28
PARAG	1013	22.69	18.47	14.87	11.83	9.25	7.05	5.14	3.46
PHLOG	1014	43.20	38.22	33.90	30.22	27.08	24.39	22.06	20.00
ANNITE	1015	33.86	29.74	26.12	23.03	20.39	18.12	16.14	14.39
DEHYD-ANAL	1021	13.80	11.99	10.39	9.02	7.84	6.83	5.96	5.19
ANAL	1022	8.62	7.28	6.10	5.09	4.23	3.51	2.89	2.35
SEPIO	1023	33.74	31.00	28.45	26.24	24.38	22.81	21.49	20.36
LO-ALB	1025	3.92	3.10	2.29	1.57	0.95	0.41	-0.05	-0.45
MAX-MICRO	1027	0.45	0.08	-0.35	-0.76	-1.12	-1.44	-1.71	-1.95
K-SPAR	1028	0.45	0.08	-0.35	-0.75	-1.11	-1.42	-1.69	-1.93
HI-SAN	1029	1.83	1.28	0.70	0.16	-0.31	-0.73	-1.09	-1.41
ANOR	1030	31.75	27.06	23.05	19.63	16.70	14.18	11.97	10.00
NEPH	1031	16.23	14.13	12.35	10.84	9.55	8.46	7.51	6.67
WOLL	1035	14.85	13.62	12.53	11.57	10.75	10.04	9.41	8.86
DIOP	1039	23.15	20.97	19.02	17.33	15.87	14.61	13.51	12.54
CA-AL PYX	1040	42.06	36.46	31.73	27.72	24.30	21.34	18.76	16.46
AKER	1044	49.82	45.19	41.15	37.69	34.71	32.13	29.87	27.87
MERW	1045	75.13	68.25	62.31	57.23	52.87	49.08	45.78	42.85
MONTI	1046	32.91	29.75	27.02	24.69	22.69	20.96	19.45	18.11
GEHL	1047	64.29	56.65	50.19	44.72	40.04	36.01	32.48	29.35
FORST	1048	31.47	28.15	25.28	22.83	20.73	18.91	17.32	15.92
FAY	1049	21.50	19.05	16.92	15.10	13.53	12.17	10.98	9.93
HUNT	1050	13.36	10.75	8.32	6.11	4.10	2.24	0.49	-1.20
ART	1053	22.02	19.94	18.15	16.65	15.37	14.26	13.29	12.42
HED	1054	21.45	19.44	17.62	16.04	14.67	13.48	12.44	11.52
MALACH	1059	2.97	2.98	2.96	2.92	2.89	2.85	2.80	2.74
AZUR	1060	4.82	4.73	4.56	4.38	4.18	3.98	3.76	3.51
HYDRO-MAG	1062	36.11	31.56	27.53	24.02	20.94	18.20	15.70	13.37
SMITH	1064	1.01	0.56	0.12	-0.29	-0.67	-1.03	-1.37	-1.70
ANH-CORD	1065	63.61	53.83	45.41	38.23	32.06	26.74	22.07	17.92
HYD-CORD	1066	60.86	51.35	43.17	36.20	30.23	25.08	20.57	16.57
JADEITE	1067	10.04	8.72	7.55	6.54	5.68	4.94	4.31	3.76
CER	1068	-3.44	-3.19	-3.01	-2.87	-2.76	-2.67	-2.61	-2.57
STRONT	1069	-1.08	-1.08	-1.15	-1.24	-1.35	-1.47	-1.60	-1.75
DIS-DOL	1070	5.13	4.06	3.04	2.11	1.24	0.44	-0.33	-1.08

$\log K$, $P = P_{SAT}$ for $T \leqslant 350°C$; $P = 500$ b for $T > 350°C$

INDEX	TEMPERATURE, °C									
	200	225	250	300	350	400	450	500	550	600
1001	2.42	1.16	-0.09	-2.89	-8.07	-7.89	-18.22	-34.04	-45.90	(-52.58)
1002	2.43	1.19	-0.04	-2.81	-7.96	-7.77	-18.07	-33.87	-45.72	(-52.38)
1003	2.58	1.30	0.04	-2.79	-7.99	-7.83	-18.17	-34.00	-45.87	(-52.56)
1004	-1.55	-2.35	-3.16	-5.22	-9.79	-9.02	-18.99	-34.61	-46.21	(-52.56)
1007	17.22	16.02	14.86	12.38	8.01	8.10	-0.42	-13.41	-23.08	(-28.45)
1012	-0.10	-1.41	-2.74	-6.03	-13.17	-12.05	-27.45	-51.51	-69.41	(-79.26)
1013	1.93	0.48	-0.98	-4.49	-11.86	-10.81	-26.51	-50.98	-69.19	(-79.21)
1014	18.15	16.42	14.75	11.01	4.08	4.51	-9.44	-30.89	-46.86	(-55.73)
1015	12.81	11.33	9.88	6.55	5.96	0.71	-12.67	-33.34	-48.72	(-59.22)
1021	4.49	3.84	3.21	1.77	-1.06	-0.67	-6.57	-15.69	-22.48	(-26.23)
1022	1.87	1.42	0.97	-0.15	-2.70	-2.10	-7.75	-16.63	-23.19	(-26.73)
1023	19.37	18.47	17.58	15.43	10.48	11.71	0.66	-16.77	-29.52	(-36.21)
1025	-0.82	-1.17	-1.53	-2.52	-5.01	-4.21	-9.98	-19.15	-25.91	(-29.54)
1027	-2.18	-2.41	-2.67	-3.46	-5.75	-4.89	-10.37	-19.14	-25.61	(-29.08)
1028	-2.16	-2.39	-2.65	-3.46	-5.76	-4.92	-10.43	-19.21	-25.71	(-29.20)
1029	-1.70	-1.99	-2.30	-3.20	-5.56	-4.77	-10.32	-19.13	-25.65	(-29.16)
1030	8.22	6.55	4.93	1.32	-5.15	-4.94	-17.73	-37.26	-51.89	(-60.13)
1031	5.92	5.22	4.54	3.02	0.18	0.40	-5.35	-14.20	-20.79	(-24.45)
1035	8.36	7.91	7.48	6.60	5.23	5.17	2.66	-1.09	-3.90	(-5.49)
1039	11.66	10.85	10.08	8.43	5.63	5.66	0.27	-7.92	-14.03	(-17.45)
1040	14.38	12.45	10.58	6.56	-0.22	-0.44	-13.30	-32.75	-47.35	(-55.66)
1044	26.08	24.44	22.90	19.80	15.20	14.69	6.62	-5.33	-14.30	(-19.44)
1045	40.23	37.84	35.60	31.15	24.84	23.83	13.15	-2.50	-14.28	(-21.10)
1046	16.91	15.81	14.77	12.68	9.55	9.19	3.70	-4.42	-10.53	(-14.04)
1047	26.53	23.91	21.42	16.19	7.81	7.22	-8.17	-31.25	-48.60	(-58.53)
1048	14.65	13.49	12.40	10.16	6.76	6.39	0.36	-8.59	-15.32	(-19.19)
1049	8.98	8.10	7.26	5.48	2.54	2.45	-3.06	-11.40	-17.64	(-21.17)
1050	-2.88	-4.61	-6.47	-11.18	-20.78	-20.50	-40.64	-70.83	-94.00	(-107.50)
1053	11.61	10.83	10.04	8.17	4.47	4.58	-2.99	-14.65	-23.35	(-28.23)
1054	10.69	9.92	9.18	7.60	4.90	4.96	-0.28	-8.25	-14.20	(-17.53)
1059	2.65	2.54	2.38	1.80	0.10	0.51	-3.51	-9.87	-14.65	(-17.35)
1060	3.22	2.89	2.47	1.17	-2.10	-1.59	-8.98	-20.51	-29.25	(-34.29)
1062	11.14	8.92	6.61	0.97	-10.23	-10.01	-32.92	-68.06	-94.60	(-109.86)
1064	-2.04	-2.40	-2.80	-3.86	-6.17	-6.07	-10.95	-18.49	-24.21	(-27.53)
1065	14.15	10.63	7.19	-0.41	-14.03	-13.57	-40.48	-81.55	-112.32	(-129.66)
1066	12.93	9.54	6.22	-1.17	-14.60	-14.02	-40.74	-81.63	-112.23	(-129.40)
1067	3.25	2.78	2.31	1.13	-1.48	-0.95	-6.65	-15.61	-22.25	(-25.85)
1068	-2.55	-2.58	-2.66	-3.10	-4.65	-4.12	-8.00	-14.20	-18.86	(-21.47)
1069	-1.92	-2.13	-2.38	-3.15	-5.04	-4.78	-9.00	-15.59	-20.59	(-23.46)
1070	-1.83	-2.62	-3.47	-5.67	-10.27	-10.07	-19.69	-34.47	-45.70	(-52.24)

$\log K$, $P = P_{SAT}$ for $T \leq 350°C$; $P = 500$ b for $T > 350°C$

NAME	INDEX	TEMPERATURE, °C							
		0	25	50	75	100	125	150	175
ORD-DOL	1071	3.39	2.52	1.66	0.87	0.13	-0.57	-1.24	-1.91
ARAG	1072	2.24	1.88	1.51	1.17	0.85	0.54	0.24	-0.06
CALC	1073	2.07	1.71	1.35	1.01	0.69	0.38	0.08	-0.23
MAG	1074	3.12	2.44	1.80	1.23	0.71	0.23	-0.22	-0.65
DOL	1075	3.39	2.52	1.66	0.87	0.13	-0.57	-1.24	-1.90
SIDER	1076	0.25	-0.22	-0.68	-1.11	-1.51	-1.88	-2.25	-2.60
ANHYD	1078	-4.07	-4.27	-4.58	-4.95	-5.34	-5.77	-6.22	-6.70
FLUORITE	1079	-10.32	-10.04	-9.92	-9.90	-9.95	-10.06	-10.23	-10.45
BARITE	1080	-10.49	-9.96	-9.68	-9.54	-9.51	-9.56	-9.69	-9.89
ANG	1081	-8.09	-7.81	-7.71	-7.70	-7.75	-7.85	-8.00	-8.19
CELEST	1082	-6.51	-6.43	-6.53	-6.71	-6.95	-7.23	-7.56	-7.93
ALUN	1083	3.91	0.66	-2.26	-4.86	-7.20	-9.35	-11.38	-13.35
WITH	1084	-3.30	-3.00	-2.82	-2.70	-2.63	-2.60	-2.61	-2.65
RHODO	1085	0.11	-0.19	-0.50	-0.79	-1.06	-1.32	-1.57	-1.82
COV	1086	-26.58	-24.32	-22.48	-20.95	-19.66	-18.57	-17.64	-16.85
GALENA	1087	-16.20	-14.85	-13.76	-12.85	-12.08	-11.42	-10.85	-10.37
SL	1088	-11.95	-11.36	-10.93	-10.61	-10.36	-10.18	-10.05	-9.97
WURT	1089	-9.44	-9.06	-8.81	-8.64	-8.53	-8.46	-8.43	-8.45
M-CINN	1090	-41.89	-38.61	-35.89	-33.60	-31.64	-29.94	-28.47	-27.19
CINN	1091	-42.33	-38.98	-36.21	-33.87	-31.87	-30.13	-28.63	-27.32
ALA	1092	-0.07	-0.41	-0.77	-1.12	-1.45	-1.77	-2.08	-2.39
PYRITE	1093	-26.53	-24.70	-23.32	-22.26	-21.43	-20.80	-20.34	-20.02
GOLD	1100	-27.10	-24.38	-22.07	-20.08	-18.34	-16.81	-15.44	-14.22
SILVER	1101	-10.48	-9.30	-8.28	-7.39	-6.60	-5.89	-5.24	-4.66
COPPER	1102	-5.19	-4.55	-3.99	-3.48	-3.02	-2.59	-2.19	-1.81
GRAPHITE	1103	-5.68	-5.00	-4.44	-3.98	-3.60	-3.29	-3.02	-2.81
HALITE	1106	1.52	1.61	1.64	1.63	1.59	1.54	1.46	1.36
SYLVITE	1107	0.58	0.90	1.11	1.24	1.32	1.36	1.36	1.32
COR	1108	25.56	21.38	17.91	15.00	12.53	10.40	8.53	6.87
PER	1109	23.89	21.48	19.44	17.71	16.23	14.95	13.83	12.84
LIME	1110	35.70	32.60	29.98	27.74	25.82	24.15	22.69	21.40
TO	1111	6.61	6.18	5.82	5.52	5.28	5.08	4.91	4.76
CUP	1112	-2.35	-1.90	-1.50	-1.11	-0.73	-0.36	-0.01	0.33
FE-OXIDE	1113	15.19	13.50	12.07	10.85	9.81	8.91	8.12	7.41
GIBBS	1114	9.52	7.96	6.69	5.65	4.78	4.06	3.43	2.89
DIAS	1115	10.54	8.75	7.27	6.04	4.99	4.10	3.33	2.65
BOEH	1116	11.53	9.60	8.00	6.66	5.53	4.56	3.72	2.98
BRUC	1117	18.23	16.44	14.93	13.67	12.60	11.68	10.89	10.19
MANGAN	1119	19.85	17.92	16.28	14.90	13.73	12.72	11.84	11.07
SPINEL	1120	44.58	38.39	33.22	28.86	25.16	21.96	19.17	16.69

$\log K$, $P = P_{SAT}$ for $T \leqslant 350°C$; $P = 500$ b for $T > 350°C$

INDEX	TEMPERATURE, °C									
	200	225	250	300	350	400	450	500	550	600
1071	-2.58	-3.30	-4.09	-6.19	-10.70	-10.42	-19.97	-34.69	-45.88	(-52.37)
1072	-0.37	-0.70	-1.06	-2.03	-4.15	-4.01	-8.51	-15.45	-20.72	(-23.78)
1073	-0.53	-0.86	-1.23	-2.21	-4.33	-4.19	-8.70	-15.65	-20.93	(-24.00)
1074	-1.08	-1.53	-2.00	-3.21	-5.67	-5.60	-10.70	-18.52	-24.46	(-27.92)
1075	-2.58	-3.30	-4.09	-6.19	-10.70	-10.43	-19.98	-34.71	-45.90	(-52.39)
1076	-2.96	-3.34	-3.76	-4.86	-7.19	-7.07	-11.99	-19.56	-25.32	(-28.68)
1078	-7.21	-7.79	-8.44	-10.22	-14.02	-13.91	-21.92	-34.20	-43.58	(-49.11)
1079	-10.73	-11.09	-11.56	-13.05	-16.78	-16.30	-24.67	-37.69	-47.59	(-53.32)
1080	-10.16	-10.51	-10.96	-12.35	-15.72	-15.43	-22.84	-34.30	-43.04	(-48.16)
1081	-8.44	-8.76	-9.16	-10.46	-13.75	-13.29	-20.70	-32.25	-41.02	(-46.10)
1082	-8.35	-8.84	-9.41	-11.03	-14.65	-14.46	-22.22	-34.16	-43.28	(-48.63)
1083	-15.32	-17.36	-19.59	-25.46	-38.06	-37.33	-63.99	-105.15	-136.25	(-154.11)
1084	-2.72	-2.84	-3.02	-3.63	-5.33	-5.02	-8.94	-15.09	-19.75	(-22.43)
1085	-2.09	-2.37	-2.70	-3.63	-5.77	-5.53	-10.22	-17.53	-23.06	(-26.23)
1086	-16.18	-15.63	-15.19	-14.72	-15.35	-14.38	-17.17	-21.96	-25.55	(-27.49)
1087	-9.98	-9.66	-9.43	-9.35	-10.53	-9.61	-13.33	-19.52	-24.13	(-26.63)
1088	-9.95	-9.98	-10.09	-10.67	-12.61	-12.08	-16.83	-24.38	-30.08	(-33.31)
1089	-8.50	-8.61	-8.79	-9.48	-11.52	-11.08	-15.90	-23.51	-29.26	(-32.53)
1090	-26.08	-25.13	-24.33	-23.28	-23.69	-22.06	-25.32	-31.19	-35.51	(-37.71)
1091	-26.19	-25.22	-24.40	-23.31	-23.70	-22.05	-25.28	-31.14	-35.44	(-37.62)
1092	-2.71	-3.05	-3.44	-4.48	-6.79	-6.62	-11.56	-19.21	-25.01	(-28.35)
1093	-19.84	-19.79	-19.91	-20.81	-24.06	-23.29	-31.25	-43.87	-53.47	(-59.00)
1100	-13.12	-12.13	-11.21	-9.61	-8.21	-7.08	-5.96	-4.90	-3.98	(-3.17)
1101	-4.12	-3.63	-3.17	-2.34	-1.61	-0.98	-0.38	0.18	0.69	(1.15)
1102	-1.46	-1.13	-0.83	-0.26	0.19	0.71	0.98	1.12	1.30	(1.56)
1103	-2.63	-2.48	-2.37	-2.26	-2.41	-2.23	-2.85	-3.92	-4.75	(-5.24)
1106	1.24	1.09	0.90	0.32	-1.09	-0.96	-4.07	-8.91	-12.59	(-14.72)
1107	1.25	1.15	1.00	0.49	-0.81	-0.69	-3.59	-8.09	-11.53	(-13.54)
1108	5.37	3.96	2.58	-0.45	-5.77	-5.85	-16.14	-31.78	-43.52	(-50.20)
1109	11.95	11.13	10.38	8.90	6.93	6.44	3.30	-1.22	-4.64	(-6.67)
1110	20.24	19.19	18.22	16.42	14.32	13.50	10.61	6.66	3.64	(1.77)
1111	4.64	4.53	4.42	4.19	3.75	3.82	2.89	1.45	0.38	(-0.20)
1112	0.65	0.96	1.25	1.72	1.86	2.56	1.99	0.77	-0.04	(-0.29)
1113	6.79	6.21	5.66	4.55	2.93	2.68	-0.13	-4.25	-7.37	(9.18)
1114	2.39	1.92	1.46	0.34	-1.97	-1.77	-6.62	-14.16	-19.77	(-22.88)
1115	2.02	1.44	0.87	-0.45	-2.92	-2.82	-7.81	-15.48	-21.21	(-24.41)
1116	2.30	1.67	1.05	-0.34	-2.87	-2.82	-7.86	-15.57	-21.33	(-24.57)
1117	9.57	9.00	8.47	7.38	5.75	5.53	2.66	-1.59	-4.76	(-6.57)
1119	10.38	9.76	9.17	8.02	6.40	6.11	3.40	-0.58	-3.57	(-5.30)
1120	14.45	12.36	10.35	6.06	-1.05	-1.46	-14.76	-34.80	-49.87	(-58.49)

$\log K$, $P = P_{SAT}$ for $T \leq 350°C$; $P = 500$ b for $T > 350°C$

NAME	INDEX	TEMPERATURE, °C							
		0	25	50	75	100	125	150	175
K-OXIDE	1123	90.98	84.14	78.35	73.41	69.14	65.42	62.15	59.25
NA-OXIDE	1124	73.13	67.47	62.72	58.69	55.22	52.23	49.61	47.30
AMORPH SIL	1125	-2.99	-2.71	-2.51	-2.34	-2.20	-2.08	-1.98	-1.89
A-CRIST	1126	-3.89	-3.45	-3.13	-2.88	-2.68	-2.51	-2.36	-2.24
B-CRIST	1127	-3.37	-3.01	-2.74	-2.54	-2.38	-2.24	-2.12	-2.01
CHALCED	1128	-4.21	-3.73	-3.38	-3.10	-2.88	-2.69	-2.53	-2.38
WAIR	1130	22.26	18.56	15.29	12.47	10.05	7.96	6.15	4.54
LAUM	1132	17.01	14.15	11.61	9.43	7.58	6.00	4.65	3.46
ACAN	1501	-39.70	-36.05	-33.01	-30.45	-28.25	-26.35	-24.70	-23.26
CP	1502	-36.79	-34.10	-31.97	-30.23	-28.80	-27.62	-26.65	-25.86
BN	1503	-113.19	-104.00	-96.48	-90.15	-84.78	-80.17	-76.20	-72.80
CC	1504	-38.08	-34.75	-31.98	-29.63	-27.60	-25.88	-24.38	-23.07
QTZ	1505	-4.50	-4.00	-3.63	-3.33	-3.10	-2.89	-2.72	-2.56
MG	1506	31.78	27.99	24.75	22.01	19.65	17.61	15.81	14.21
KALS	1507	12.93	11.26	9.83	8.61	7.58	6.69	5.92	5.24
FERROSIL	1508	8.36	7.42	6.56	5.82	5.17	4.61	4.12	3.67
PYROPH	1509	2.61	1.06	-0.39	-1.67	-2.78	-3.73	-4.57	-5.31
TALC	1510	24.06	21.56	19.24	17.21	15.46	13.94	12.63	11.47
7-A CL	1512	81.89	71.94	63.46	56.31	50.24	45.04	40.53	36.55
14-A CL	1513	78.30	68.57	60.27	53.26	47.32	42.22	37.80	33.90
CLNZ	1515	51.00	43.91	37.86	32.72	28.33	24.55	21.26	18.33
LAWS	1516	26.47	22.70	19.48	16.77	14.47	12.50	10.80	9.29
TREM	1517	68.37	61.67	55.58	50.29	45.72	41.77	38.33	35.29
ANTH	1518	75.77	67.79	60.62	54.40	49.04	44.39	40.34	36.76
ZOIS	1519	51.05	43.96	37.90	32.75	28.36	24.58	21.28	18.35
GROSS	1529	58.87	52.13	46.33	41.38	37.14	33.49	30.30	27.47
ANDRA	1530	55.64	50.47	45.91	41.98	38.59	35.65	33.07	30.78
ALB	1531	3.92	3.10	2.29	1.57	0.95	0.41	-0.05	-0.45
PREHN	1532	38.12	33.28	29.09	25.52	22.47	19.85	17.56	15.54
ENST	1537	12.82	11.47	10.27	9.23	8.34	7.57	6.89	6.29
PARG	1538	117.20	103.24	91.23	81.01	72.26	64.73	58.17	52.36
ANTIG	1542	539.15	483.97	435.83	394.88	360.04	330.21	304.38	281.71
HM	1544	20.06	17.65	15.60	13.85	12.34	11.03	9.88	8.85
ORD-EP	1545	48.25	42.05	36.70	32.13	28.23	24.86	21.92	19.31
MARG	1551	50.19	42.17	35.42	29.73	24.89	20.74	17.12	13.90
MERCURY	1552	-22.14	-20.43	-18.96	-17.66	-16.49	-15.43	-14.47	-13.58
PO	1555	-3.67	-3.76	-3.92	-4.10	-4.30	-4.51	-4.75	-5.01
NESQ	1558	6.56	5.13	4.98	4.90	4.83	4.76	4.69	4.60
EPID	1559	48.25	42.05	36.70	32.13	28.23	24.86	21.91	19.29
HI-ALB	1560	5.41	4.42	3.46	2.61	1.88	1.24	0.70	0.22

$\log K$, $P = P_{SAT}$ for $T \leq 350°C$; $P = 500$ b for $T > 350°C$

INDEX	TEMPERATURE, °C									
	200	225	250	300	350	400	450	500	550	600
1123	56.66	54.34	52.24	48.57	45.32	42.84	40.08	37.25	34.91	(33.11)
1124	45.26	43.44	41.79	38.90	36.21	34.45	31.83	28.83	26.46	(24.82)
1125	-1.81	-1.74	-1.68	-1.60	-1.60	-1.46	-1.64	-1.96	-2.21	(-2.37)
1126	-2.12	-2.02	-1.93	-1.78	-1.70	-1.49	-1.58	-1.80	-1.95	(-1.99)
1127	-1.92	-1.83	-1.75	-1.63	-1.57	-1.37	-1.48	-1.71	-1.87	(-1.92)
1128	-2.26	-2.14	-2.04	-1.87	-1.78	-1.56	-1.64	-1.86	-1.99	(-2.03)
1130	3.08	1.72	0.37	-2.73	-8.87	-8.12	-20.89	-40.65	-55.35	(-63.46)
1132	2.38	1.37	0.35	-2.16	-7.78	-6.65	-18.97	-38.28	-52.55	(-60.26)
1501	-22.02	-20.96	-20.05	-18.74	-18.60	-17.12	-19.18	-23.17	-26.08	(-27.48)
1502	-25.25	-24.81	-24.56	-24.79	-27.50	-26.12	-33.78	-46.21	-55.60	(-60.89)
1503	-69.90	-67.48	-65.56	-63.47	-66.33	-61.86	-74.43	-96.10	-112.22	(-120.79)
1504	-21.92	-20.93	-20.08	-18.88	-18.92	-17.34	-19.77	-24.35	-27.69	(-29.31)
1505	-2.43	-2.30	-2.19	-2.01	-1.91	-1.68	-1.75	-1.96	-2.09	(-2.12)
1506	12.76	11.40	10.08	7.24	2.54	2.26	-6.58	-19.90	-29.94	(-35.71)
1507	4.62	4.04	3.47	2.13	-0.50	-0.23	-5.69	-14.14	-20.45	(-23.94)
1508	3.28	2.90	2.55	1.76	0.35	0.43	-2.36	-6.62	-9.80	(-11.58)
1509	-6.00	-6.67	-7.37	-9.26	-13.80	-12.70	-22.96	-39.13	-51.13	(-57.67)
1510	10.43	9.46	8.52	6.37	2.19	2.72	-5.97	-19.43	-29.42	(-34.90)
1515	15.68	13.19	10.76	5.31	-4.73	-4.30	-24.37	-55.14	-78.14	(-91.03)
1516	7.93	6.65	5.38	2.42	-3.50	-2.93	-15.23	-34.28	-48.46	(-56.28)
1517	32.56	30.03	27.59	22.24	12.52	13.20	-6.22	-36.01	-58.20	(-70.49)
1518	33.54	30.56	27.70	21.55	10.87	11.18	-9.62	-41.32	-64.97	(-78.18)
1519	15.70	13.21	10.78	5.32	-4.72	-4.29	-24.36	-55.13	-78.14	(-91.03)
1529	24.92	22.54	20.23	15.18	6.27	6.41	-11.00	-37.55	-57.42	(-68.59)
1530	28.71	26.78	24.92	20.90	13.99	14.00	0.67	-19.55	-34.70	(-43.24)
1531	-0.82	-1.17	-1.53	-2.52	-5.02	-4.24	-10.03	-19.21	-26.00	(-29.66)
1532	13.71	11.99	10.31	6.48	-0.81	-0.25	-15.11	-37.98	-55.05	(-64.54)
1537	5.75	5.25	4.78	3.79	2.17	2.12	-0.90	-5.46	-8.87	(-10.80)
1538	47.13	42.29	37.64	27.64	10.59	10.49	-22.23	-71.92	-109.09	(-130.02)
1542	261.41	242.72	224.76	185.78	116.53	118.62	-17.53	-225.72	-380.58	(-466.39)
1544	7.91	7.02	6.16	4.27	1.06	0.92	-5.20	-14.45	-21.43	(-25.43)
1545	16.94	14.72	12.54	7.67	-1.31	-0.90	-18.87	-46.43	-67.03	(-78.56)
1551	10.98	8.24	5.55	-0.53	-11.84	-11.35	-34.03	-68.83	-94.86	(-109.46)
1552	-12.77	-12.03	-11.35	-10.19	-9.49	-8.62	-9.12	-10.32	-11.09	(-11.28)
1555	-5.29	-5.61	-5.98	-7.03	-9.39	-9.23	-14.29	-22.12	-28.07	(-31.51)
1558	4.49	4.34	4.14	3.43	1.39	1.74	-3.00	-10.45	-16.05	(-19.20)
1559	16.62	14.69	12.51	7.63	-1.37	-0.97	-18.96	-46.53	-67.14	(-78.69)
1560	-0.21	-0.62	-1.04	-2.12	-4.70	-3.97	-9.80	-19.02	-25.84	(-29.52)

Log K , P = 1 kb

NAME	INDEX	TEMPERATURE, °C							
		0	25	50	75	100	125	150	175
AND	1001	21.29	17.67	14.60	12.01	9.81	7.93	6.31	4.91
KYA	1002	20.81	17.26	14.26	11.73	9.58	7.75	6.17	4.80
SILL	1003	21.67	18.00	14.89	12.26	10.03	8.13	6.49	5.06
KAOL	1004	10.69	8.49	6.53	4.87	3.47	2.29	1.29	0.45
CHRYS	1007	35.62	32.30	29.25	26.63	24.40	22.50	20.86	19.45
MUSC	1012	19.85	16.28	13.11	10.41	8.13	6.20	4.55	3.15
PARAG	1013	24.33	20.22	16.62	13.55	10.97	8.78	6.93	5.34
PHLOG	1014	44.34	39.64	35.36	31.69	28.56	25.90	23.61	21.64
ANNITE	1015	35.50	31.47	27.87	24.76	22.10	19.83	17.88	16.20
DEHYD-ANAL	1021	14.81	13.09	11.49	10.09	8.90	7.88	7.00	6.24
ANAL	1022	9.44	8.21	7.03	6.01	5.14	4.41	3.79	3.28
SEPIO	1023	34.27	32.26	29.90	27.76	25.92	24.37	23.07	21.98
LO-ALB	1025	4.72	4.13	3.37	2.65	2.02	1.49	1.03	0.65
MAX-MICRO	1027	1.22	1.09	0.72	0.32	-0.05	-0.37	-0.63	-0.86
K-SPAR	1028	1.23	1.09	0.72	0.32	-0.04	-0.35	-0.61	-0.83
HI-SAN	1029	2.61	2.30	1.77	1.24	0.76	0.35	-0.01	-0.30
ANOR	1030	33.37	28.72	24.68	21.23	18.28	15.76	13.58	11.68
NEPH	1031	16.98	14.88	13.07	11.54	10.25	9.15	8.21	7.41
WOLL	1035	15.17	14.02	12.94	11.99	11.17	10.45	9.83	9.28
DIOP	1039	23.52	21.56	19.67	18.01	16.56	15.31	14.23	13.28
CA-AL PYX	1040	43.38	37.73	32.95	28.91	25.48	22.55	20.01	17.79
AKER	1044	50.67	46.21	42.20	38.74	35.76	33.18	30.93	28.97
MERW	1045	76.16	69.43	63.51	58.43	54.06	50.29	47.01	44.13
MONTI	1046	33.42	30.35	27.63	25.30	23.30	21.57	20.07	18.76
GEHL	1047	66.07	58.34	51.81	46.29	41.59	37.56	34.08	31.03
FORST	1048	31.85	28.65	25.80	23.37	21.27	19.47	17.91	16.54
FAY	1049	22.20	19.76	17.63	15.80	14.22	12.86	11.68	10.65
HUNT	1050	15.23	12.40	9.85	7.60	5.60	3.80	2.15	0.63
ART	1053	22.38	20.28	18.47	16.96	15.68	14.60	13.66	12.84
HED	1054	22.01	20.15	18.38	16.81	15.44	14.26	13.23	12.33
MALACH	1059	3.50	3.50	3.44	3.38	3.33	3.29	3.26	3.22
AZUR	1060	5.90	5.76	5.51	5.28	5.07	4.86	4.66	4.46
HYDRO-MAG	1062	37.92	33.20	29.06	25.51	22.43	19.74	17.36	15.21
SMITH	1064	1.13	0.61	0.17	-0.23	-0.59	-0.91	-1.20	-1.48
ANH-CORD	1065	67.05	57.52	49.07	41.83	35.62	30.30	25.69	21.67
HYD-CORD	1066	64.12	54.87	46.67	39.65	33.65	28.50	24.05	20.18
JADEITE	1067	10.50	9.32	8.18	7.18	6.33	5.61	5.00	4.48
CER	1068	-2.84	-2.67	-2.54	-2.42	-2.32	-2.23	-2.14	-2.07
STRONT	1069	-0.45	-0.57	-0.68	-0.79	-0.90	-1.02	-1.13	-1.26
DIS-DOL	1070	6.12	4.92	3.84	2.88	2.01	1.23	0.51	-0.16

$\log K$, $P = 1$ kb

INDEX	TEMPERATURE, °C									
	200	225	250	300	350	400	450	500	550	600
1001	3.67	2.57	1.59	-0.18	-1.87	-3.81	-6.35	-9.88	-14.33	(-17.95)
1002	3.60	2.53	1.56	-0.16	-1.81	-3.71	-6.23	-9.74	-14.17	(-17.77)
1003	3.81	2.70	1.70	-0.09	-1.81	-3.75	-6.31	-9.85	-14.31	(-17.94)
1004	-0.28	-0.92	-1.48	-2.49	-3.54	-4.90	-6.96	-10.08	-14.17	(-17.46)
1007	18.23	17.16	16.20	14.55	13.04	11.39	9.28	6.40	2.79	(-0.11)
1012	1.94	0.88	-0.07	-1.76	-3.47	-5.67	-8.91	-13.76	-20.11	(-25.21)
1013	3.98	2.78	1.72	-0.16	-2.02	-4.34	-7.70	-12.67	-19.14	(-24.34)
1014	19.92	18.42	17.08	14.74	12.55	10.06	6.77	2.14	-3.73	(-8.44)
1015	14.74	13.46	12.32	10.32	8.41	6.18	3.13	-1.23	-6.82	(-11.31)
1021	5.59	5.02	4.52	3.64	2.82	1.86	0.54	-1.36	-3.80	(-5.75)
1022	2.84	2.47	2.15	1.60	1.06	0.35	-0.74	-2.42	-4.67	(-6.44)
1023	21.08	20.33	19.69	18.64	17.66	16.37	14.36	11.17	6.89	(3.58)
1025	0.34	0.07	-0.16	-0.55	-0.97	-1.58	-2.60	-4.26	-6.52	(-8.28)
1027	-1.04	-1.19	-1.32	-1.55	-1.84	-2.34	-3.27	-4.82	-6.96	(-8.64)
1028	-1.02	-1.17	-1.30	-1.55	-1.85	-2.37	-3.32	-4.90	-7.05	(-8.76)
1029	-0.55	-0.77	-0.95	-1.28	-1.65	-2.22	-3.21	-4.82	-6.99	(-8.71)
1030	10.01	8.54	7.21	4.86	2.64	0.17	-3.04	-7.43	-12.94	(-17.41)
1031	6.71	6.10	5.56	4.61	3.73	2.73	1.40	-0.49	-2.89	(-4.82)
1035	8.80	8.38	8.00	7.34	6.76	6.16	5.45	4.55	3.45	(2.57)
1039	12.46	11.73	11.08	9.94	8.91	7.81	6.42	4.57	2.26	(0.40)
1040	15.84	14.11	12.55	9.79	7.23	4.47	1.02	-3.55	-9.20	(-13.81)
1044	27.24	25.71	24.34	21.95	19.83	17.71	15.28	12.24	8.62	(5.68)
1045	41.60	39.34	37.33	33.82	30.72	27.67	24.27	20.11	15.23	(11.24)
1046	17.61	16.59	15.67	14.07	12.65	11.22	9.57	7.50	5.04	(3.03)
1047	28.35	25.97	23.83	20.05	16.61	13.00	8.66	3.03	-3.82	(-9.41)
1048	15.33	14.26	13.30	11.63	10.12	8.59	6.81	4.56	1.86	(-0.35)
1049	9.75	8.95	8.23	6.97	5.81	4.60	3.10	1.13	-1.29	(-3.25)
1050	-0.81	-2.18	-3.50	-6.15	-9.07	-12.75	-17.83	-24.94	-33.94	(-41.49)
1053	12.12	11.47	10.88	9.80	8.70	7.38	5.57	3.01	-0.27	(-2.94)
1054	11.54	10.84	10.22	9.13	8.14	7.07	5.73	3.92	1.68	(-1.23)
1059	3.18	3.14	3.08	2.93	2.68	2.20	1.39	0.10	-1.62	(-3.05)
1060	4.25	4.04	3.81	3.27	2.55	1.47	-0.21	-2.73	-6.01	(-8.76)
1062	13.25	11.43	9.72	6.44	2.97	-1.27	-7.05	-15.15	-25.38	(-33.87)
1064	-1.75	-2.00	-2.26	-2.78	-3.39	-4.21	-5.40	-7.10	-9.29	(-11.12)
1065	18.14	15.01	12.19	7.20	2.50	-2.74	-9.51	-18.77	-30.38	(-39.80)
1066	16.79	13.79	11.09	6.31	1.80	-3.26	-9.87	-18.98	-30.44	(-39.73)
1067	4.04	3.67	3.34	2.77	2.21	1.47	0.35	-1.37	-3.65	(-5.46)
1068	-2.01	-1.97	-1.94	-1.94	-2.07	-2.42	-3.11	-4.29	-5.90	(-7.23)
1069	-1.38	-1.51	-1.65	-1.98	-2.41	-3.04	-4.00	-5.44	-7.31	(-8.90)
1070	-0.79	-1.40	-2.00	-3.20	-4.56	-6.29	-8.71	-12.13	-16.47	(-20.12)

log K , P = 1 kb

NAME	INDEX	TEMPERATURE, °C							
		0	25	50	75	100	125	150	175
ORD-DOL	1071	4.37	3.37	2.46	1.64	0.90	0.23	-0.40	-0.98
ARAG	1072	2.76	2.32	1.92	1.56	1.24	0.94	0.66	0.40
CALC	1073	2.64	2.20	1.81	1.45	1.12	0.82	0.53	0.27
MAG	1074	3.54	2.81	2.15	1.57	1.06	0.60	0.18	-0.21
DOL	1075	4.37	3.37	2.46	1.64	0.90	0.23	-0.40	-0.98
SIDER	1076	0.84	0.26	-0.24	-0.69	-1.09	-1.45	-1.79	-2.11
ANHYD	1078	-3.20	-3.50	-3.85	-4.23	-4.63	-5.03	-5.44	-5.86
FLUORITE	1079	-9.53	-9.33	-9.24	-9.23	-9.27	-9.35	-9.47	-9.61
BARITE	1080	-9.59	-9.21	-8.98	-8.87	-8.84	-8.88	-8.97	-9.11
ANG	1081	-7.22	-7.05	-6.99	-6.99	-7.04	-7.13	-7.24	-7.37
CELEST	1082	-5.61	-5.67	-5.81	-6.00	-6.24	-6.51	-6.80	-7.11
ALUN	1083	6.26	2.67	-0.38	-3.03	-5.37	-7.45	-9.35	-11.10
WITH	1084	-2.66	-2.49	-2.35	-2.26	-2.20	-2.17	-2.17	-2.18
RHODO	1085	0.56	0.19	-0.15	-0.44	-0.71	-0.96	-1.18	-1.39
COV	1086	-26.25	-24.02	-22.20	-20.67	-19.38	-18.28	-17.32	-16.50
GALENA	1087	-15.70	-14.43	-13.37	-12.47	-11.70	-11.03	-10.44	-9.92
SL	1088	-11.46	-10.97	-10.57	-10.25	-9.99	-9.80	-9.64	-9.51
WURT	1089	-8.95	-8.67	-8.45	-8.28	-8.16	-8.08	-8.02	-7.99
M-CINN	1090	-41.33	-38.15	-35.47	-33.19	-31.22	-29.52	-28.02	-26.71
CINN	1091	-41.80	-38.55	-35.81	-33.48	-31.48	-29.73	-28.21	-26.86
ALA	1092	0.27	-0.13	-0.51	-0.85	-1.17	-1.47	-1.74	-2.01
PYRITE	1093	-25.99	-24.29	-22.93	-21.98	-21.01	-20.33	-19.81	-19.40
GOLD	1100	-27.34	-24.63	-22.31	-20.31	-18.56	-17.02	-15.65	-14.42
SILVER	1101	-10.47	-9.31	-8.29	-7.40	-6.61	-5.90	-5.26	-4.68
COPPER	1102	-5.11	-4.45	-3.90	-3.40	-2.94	-2.51	-2.12	-1.75
GRAPHITE	1103	-5.70	-5.05	-4.50	-4.04	-3.65	-3.33	-3.05	-2.82
HALITE	1106	1.72	1.75	1.76	1.75	1.72	1.67	1.62	1.55
SYLVITE	1107	0.79	1.05	1.25	1.38	1.47	1.52	1.54	1.53
COR	1108	26.60	22.29	18.75	15.81	13.32	11.21	9.39	7.80
PER	1109	24.08	21.67	19.63	17.90	16.42	15.14	14.03	13.06
LIME	1110	35.98	32.85	30.21	27.97	26.04	24.38	22.92	21.64
TO	1111	6.80	6.38	6.01	5.71	5.45	5.25	5.08	4.93
CUP	1112	-2.01	-1.52	-1.14	-0.78	-0.42	-0.06	0.28	0.61
FE-OXIDE	1113	15.53	13.79	12.33	11.11	10.06	9.16	8.37	7.92
GIBBS	1114	9.90	8.28	6.99	5.93	5.07	4.35	3.74	3.23
DIAS	1115	10.99	9.14	7.63	6.38	5.33	4.45	3.70	3.05
BOEH	1116	12.01	10.01	8.38	7.03	5.90	4.94	4.11	3.41
BRUC	1117	18.34	16.56	15.05	13.78	12.72	11.81	11.02	10.35
MANGAN	1119	20.05	18.11	16.46	15.08	13.90	12.90	12.03	11.28
SPINEL	1120	45.86	39.55	34.29	29.90	26.18	23.01	20.27	17.88

$\log K$, $P = 1$ kb

INDEX	TEMPERATURE, °C									
	200	225	250	300	350	400	450	500	550	600
1071	-1.54	-2.09	-2.62	-3.72	-4.98	-6.64	-9.00	-12.36	-16.65	(-20.25)
1072	0.14	-0.10	-0.35	-0.85	-1.44	-2.21	-3.32	-4.90	-6.92	(-8.61)
1073	0.01	-0.24	-0.49	-1.01	-1.60	-2.39	-3.50	-5.09	-7.12	(-8.82)
1074	-0.58	-0.93	-1.27	-1.94	-2.68	-3.62	-4.92	-6.74	-9.05	(-10.98)
1075	-1.54	-2.08	-2.62	-3.72	-4.98	-6.65	-9.00	-12.37	-16.67	(-20.27)
1076	-2.41	-2.70	-2.98	-3.58	-4.25	-5.12	-6.35	-8.10	-10.32	(-12.18)
1078	-6.28	-6.72	-7.16	-8.13	-9.27	-10.77	-12.86	-15.80	-19.51	(-22.65)
1079	-9.79	-10.00	-10.24	-10.83	-11.66	-12.92	-14.85	-17.73	-21.46	(-24.60)
1080	-9.29	-9.51	-9.76	-10.39	-11.25	-12.48	-14.30	-16.95	-20.34	(-23.22)
1081	-7.53	-7.71	-7.92	-8.44	-9.16	-10.26	-11.96	-14.50	-17.80	(-20.58)
1082	-7.44	-7.79	-8.16	-8.99	-10.01	-11.39	-13.37	-16.18	-19.75	(-22.77)
1083	-12.74	-14.31	-15.84	-18.92	-22.41	-26.96	-33.44	-42.71	-54.58	(-64.44)
1084	-2.21	-2.26	-2.33	-2.54	-2.86	-3.39	-4.25	-5.57	-7.30	(-8.75)
1085	-1.59	-1.79	-1.99	-2.41	-2.92	-3.64	-4.72	-6.31	-8.38	(-10.10)
1086	-15.79	-15.17	-14.64	-13.79	-13.23	-12.98	-13.13	-13.75	-14.80	(-15.66)
1087	-9.47	-9.08	-8.73	-8.19	-7.88	-7.85	-8.25	-9.18	-10.61	(-11.76)
1088	-9.42	-9.36	-9.33	-9.37	-9.59	-10.08	-10.99	-12.49	-14.51	(-16.19)
1089	-7.98	-7.99	-8.03	-8.18	-8.50	-9.07	-10.06	-11.61	-13.69	(-15.42)
1090	-25.16	-24.11	-23.16	-21.53	-20.22	-19.18	-18.41	-17.88	-17.48	(-16.92)
1091	-25.67	-24.62	-23.68	-22.12	-20.96	-20.24	-20.05	-20.50	-21.52	(-22.32)
1092	-2.26	-2.51	-2.75	-3.26	-3.87	-4.68	-5.86	-7.57	-9.76	(-11.60)
1093	-19.10	-18.89	-18.76	-18.75	-19.11	-19.99	-21.62	-24.22	-27.74	(-30.70)
1100	-13.31	-12.30	-11.39	-9.78	-8.40	-7.20	-6.14	-5.18	-4.32	(-3.55)
1101	-4.15	-3.65	-3.20	-2.38	-1.65	-1.01	-0.43	0.10	0.58	(1.03)
1102	-1.40	-1.07	-0.76	-0.19	0.32	0.78	1.19	1.55	1.87	(2.17)
1103	-2.62	-2.45	-2.31	-2.10	-1.97	-1.93	-1.98	-2.14	-2.41	(-2.64)
1106	1.47	1.39	1.29	1.06	0.74	0.26	-0.46	-1.53	-2.92	(-4.08)
1107	1.50	1.45	1.38	1.19	0.90	0.44	-0.25	-1.27	-2.59	(-3.71)
1108	6.41	5.16	4.04	2.05	0.16	-1.93	-4.60	-8.22	-12.72	(-16.39)
1109	12.20	11.44	10.75	9.55	8.49	7.47	6.36	5.06	3.57	(2.34)
1110	20.50	19.48	18.57	16.98	15.61	14.35	13.08	11.71	10.24	(9.00)
1111	4.82	4.72	4.63	4.49	4.36	4.22	4.02	3.74	3.37	(3.07)
1112	0.94	1.25	1.55	2.12	2.63	3.05	3.36	3.51	3.52	(3.59)
1113	7.08	6.54	6.06	5.20	4.44	3.67	2.80	1.70	0.42	(-0.62)
1114	2.79	2.41	2.07	1.46	0.85	0.10	-0.97	-2.54	-4.58	(-6.23)
1115	2.49	1.99	1.54	0.75	-0.01	-0.90	-2.08	-3.75	-5.88	(-7.60)
1116	2.79	2.24	1.75	0.88	0.05	-0.89	-2.13	-3.84	-6.00	(-7.75)
1117	9.75	9.23	8.76	7.95	7.23	6.50	5.66	4.59	3.31	(2.27)
1119	10.62	10.03	9.51	8.61	7.82	7.05	6.19	5.14	3.90	(2.90)
1120	15.77	13.90	12.22	9.23	6.47	3.50	-0.14	-4.94	-10.84	(-15.66)

log K , P = 1 kb

NAME	INDEX	TEMPERATURE, °C							
		0	25	50	75	100	125	150	175
K-OXIDE	1123	91.07	84.17	78.37	73.41	69.14	65.42	62.15	59.25
NA-OXIDE	1124	73.31	67.58	62.79	58.74	55.26	52.26	49.64	47.34
AMORPH SIL	1125	-2.85	-2.47	-2.23	-2.06	-1.92	-1.81	-1.71	-1.63
A-CRIST	1126	-3.81	-3.26	-2.91	-2.65	-2.45	-2.28	-2.14	-2.02
B-CRIST	1127	-3.26	-2.79	-2.50	-2.28	-2.12	-1.99	-1.87	-1.77
CHALCED	1128	-4.18	-3.59	-3.21	-2.92	-2.69	-2.50	-2.34	-2.20
WAIR	1130	24.02	20.57	17.34	14.50	12.06	9.96	8.16	6.60
LAUM	1132	18.50	15.91	13.42	11.23	9.37	7.78	6.44	5.31
ACAN	1501	-39.42	-35.86	-32.85	-30.28	-28.08	-26.16	-24.49	-23.02
CP	1502	-35.95	-33.38	-31.29	-29.57	-28.13	-26.92	-25.90	-25.04
BN	1503	-111.52	-102.46	-95.04	-88.75	-83.37	-78.73	-74.68	-71.15
CC	1504	-37.68	-34.35	-31.61	-29.27	-27.24	-25.51	-24.00	-22.66
QTZ	1505	-4.48	-3.87	-3.46	-3.15	-2.91	-2.70	-2.53	-2.38
MG	1506	32.98	29.02	25.72	22.93	20.55	18.51	16.74	15.18
KALS	1507	13.59	11.94	10.49	9.27	8.22	7.34	6.58	5.93
FERROSIL	1508	8.69	7.81	6.98	6.24	5.59	5.03	4.54	4.11
PYROPH	1509	3.60	2.37	1.01	-0.24	-1.34	-2.28	-3.08	-3.76
TALC	1510	24.61	22.59	20.40	18.41	16.68	15.19	13.90	12.79
7-A CL	1512	83.77	74.07	65.61	58.44	52.38	47.22	42.79	38.96
14-A CL	1513	80.10	70.62	62.34	55.33	49.39	44.34	40.01	36.26
CLNZ	1515	52.94	45.96	39.91	34.75	30.36	26.61	23.38	20.59
LAWS	1516	27.42	23.74	20.54	17.83	15.54	13.60	11.95	10.53
TREM	1517	69.75	63.96	58.12	52.91	48.38	44.47	41.09	38.16
ANTH	1518	77.01	70.00	63.09	56.97	51.66	47.07	43.09	39.63
ZOIS	1519	52.99	46.00	39.94	34.78	30.39	26.63	23.41	20.61
GROSS	1529	60.47	53.88	48.09	43.14	38.92	35.29	32.16	29.45
ANDRA	1530	57.04	52.04	47.53	43.60	40.21	37.28	34.74	32.52
ALB	1531	4.72	4.13	3.37	2.65	2.02	1.49	1.03	0.65
PREHN	1532	39.70	35.01	30.84	27.26	24.21	21.60	19.36	17.42
ENST	1537	12.98	11.75	10.59	9.57	8.68	7.91	7.25	6.67
PARG	1538	120.11	106.59	94.63	84.40	75.66	68.18	61.72	56.11
ANTIG	1542	546.89	496.05	448.89	408.27	373.63	344.06	318.71	296.84
HM	1544	20.87	18.36	16.26	14.48	12.96	11.65	10.51	9.52
ORD-EP	1545	50.10	44.01	38.66	34.08	30.18	26.83	23.94	21.44
MARG	1551	52.56	44.48	37.65	31.91	27.06	22.93	19.39	16.32
MERCURY	1552	-21.87	-20.20	-18.75	-17.46	-16.30	-15.25	-14.29	-13.41
PO	1555	-3.23	-3.41	-3.60	-3.78	-3.97	-4.16	-4.37	-4.58
NESQ	1558	6.86	5.40	5.22	5.13	5.06	5.01	4.95	4.90
EPID	1559	50.10	44.01	38.66	34.08	30.17	26.82	23.94	21.43
HI-ALB	1560	6.22	5.45	4.55	3.70	2.96	2.33	1.79	1.33

$\log K$, $P = 1$ kb

INDEX	TEMPERATURE, °C									
	200	225	250	300	350	400	450	500	550	600
1123	56.67	54.36	52.27	48.66	45.64	43.05	40.78	38.74	36.86	(35.24)
1124	45.31	43.50	41.89	39.11	36.82	34.85	33.10	31.46	29.91	(28.60)
1125	-1.56	-1.50	-1.44	-1.36	-1.31	-1.29	-1.30	-1.35	-1.45	(-1.53)
1126	-1.91	-1.81	-1.72	-1.57	-1.44	-1.33	-1.25	-1.20	-1.19	(-1.16)
1127	-1.69	-1.61	-1.53	-1.40	-1.29	-1.20	-1.14	-1.11	-1.10	(-1.09)
1128	-2.08	-1.97	-1.86	-1.69	-1.54	-1.41	-1.32	-1.27	-1.24	(-1.21)
1130	5.24	4.05	2.99	1.12	-0.67	-2.77	-5.66	-9.81	-15.14	(-19.41)
1132	4.33	3.49	2.76	1.48	0.20	-1.44	-3.91	-7.67	-12.64	(-16.57)
1501	-21.74	-20.61	-19.61	-17.95	-16.69	-15.85	-15.45	-15.54	-16.09	(-16.51)
1502	-24.32	-23.72	-23.23	-22.56	-22.34	-22.70	-23.88	-26.10	-29.27	(-31.95)
1503	-68.05	-65.35	-63.01	-59.25	-56.69	-55.49	-56.02	-58.68	-63.26	(-66.98)
1504	-21.47	-20.41	-19.47	-17.88	-16.68	-15.86	-15.51	-15.70	-16.37	(-16.88)
1505	-2.25	-2.13	-2.02	-1.82	-1.66	-1.53	-1.43	-1.37	-1.34	(-1.30)
1506	13.81	12.57	11.46	9.47	7.60	5.60	3.12	-0.12	-4.10	(-7.38)
1507	5.37	4.87	4.43	3.65	2.89	2.00	0.76	-1.02	-3.30	(-5.15)
1508	3.74	3.41	3.11	2.59	2.10	1.57	0.88	-0.06	-1.25	(-2.21)
1509	-4.35	-4.86	-5.31	-6.13	-7.01	-8.25	-10.24	-13.34	-17.48	(-20.78)
1510	11.83	10.99	10.25	8.96	7.75	6.35	4.43	1.65	-1.93	(-4.78)
1512	35.62	32.68	30.05	25.44	21.16	16.43	10.35	2.04	-8.38	(-16.80)
1513	33.00	30.12	27.55	23.04	18.84	14.19	8.18	-0.08	-10.45	(-18.82)
1515	18.14	15.97	14.03	10.60	7.34	3.64	-1.23	-8.02	-16.59	(-23.52)
1516	9.30	8.22	7.26	5.57	3.93	1.97	-0.78	-4.75	-9.87	(-13.98)
1517	35.60	33.35	31.36	27.89	24.68	21.13	16.50	10.05	1.90	(-4.63)
1518	36.60	33.93	31.55	27.41	23.61	19.51	14.30	7.21	-1.64	(-8.76)
1519	18.16	15.99	14.05	10.61	7.35	3.65	-1.23	-8.02	-16.59	(-23.52)
1529	27.07	24.96	23.07	19.75	16.65	13.23	8.85	2.86	-4.62	(-10.68)
1530	30.56	28.83	27.27	24.53	21.99	19.23	15.76	11.10	5.32	(0.63)
1531	0.34	0.07	-0.16	-0.56	-0.98	-1.60	-2.65	-4.33	-6.60	(-8.40)
1532	15.73	14.24	12.91	10.56	8.32	5.74	2.26	-2.66	-8.93	(-13.99)
1537	6.16	5.71	5.30	4.60	3.96	3.29	2.47	1.40	0.09	(-0.98)
1538	51.20	46.86	42.98	36.19	29.94	23.20	14.74	3.37	-10.76	(-22.19)
1542	277.84	261.20	246.44	220.82	197.33	171.55	138.30	92.53	34.94	(-11.24)
1544	8.63	7.84	7.11	5.82	4.59	3.25	1.57	-0.66	-3.40	(-5.65)
1545	19.25	17.31	15.57	12.49	9.56	6.25	1.88	-4.19	-11.86	(-18.06)
1551	13.63	11.26	9.13	5.34	1.72	-2.42	-7.91	-15.58	-25.28	(-33.14)
1552	-12.59	-11.83	-11.13	-9.85	-8.75	-8.14	-7.72	-7.46	-7.36	(-7.20)
1555	-4.80	-5.02	-5.26	-5.76	-6.38	-7.23	-8.46	-10.23	-12.50	(-14.41)
1558	4.85	4.79	4.71	4.51	4.16	3.57	2.58	1.03	-1.03	(-2.75)
1559	19.23	17.28	15.53	12.44	9.51	6.18	1.80	-4.29	-11.97	(-18.19)
1560	0.95	0.62	0.33	-0.15	-0.65	-1.33	-2.42	-4.14	-6.44	(-8.26)

Log K , P = 1.5 kb

NAME	INDEX	TEMPERATURE, °C							
		0	25	50	75	100	125	150	175
AND	1001	21.77	18.14	15.05	12.46	10.26	8.38	6.76	5.36
KYA	1002	21.21	17.66	14.66	12.12	9.98	8.15	6.58	5.21
SILL	1003	22.13	18.45	15.33	12.70	10.47	8.57	6.93	5.51
KAOL	1004	11.07	8.92	6.98	5.32	3.93	2.75	1.75	0.91
CHRYS	1007	35.66	32.52	29.50	26.90	24.68	22.79	21.16	19.76
MUSC	1012	20.47	16.97	13.83	11.13	8.85	6.92	5.29	3.89
PARAG	1013	24.96	20.92	17.34	14.28	11.70	9.52	7.67	6.09
PHLOG	1014	44.61	40.11	35.88	32.23	29.11	26.45	24.17	22.21
ANNITE	1015	36.05	32.14	28.56	25.46	22.80	20.53	18.58	16.90
DEHYD-ANAL	1021	15.20	13.55	11.96	10.56	9.36	8.33	7.45	6.69
ANAL	1022	9.75	8.59	7.43	6.41	5.54	4.80	4.18	3.66
SEPIO	1023	34.20	32.63	30.39	28.30	26.48	24.94	23.64	22.55
LO-ALB	1025	4.99	4.53	3.82	3.11	2.49	1.95	1.49	1.11
MAX-MICRO	1027	1.48	1.49	1.16	0.77	0.41	0.09	-0.18	-0.40
K-SPAR	1028	1.49	1.49	1.16	0.78	0.42	0.11	-0.16	-0.38
HI-SAN	1029	2.87	2.70	2.21	1.70	1.22	0.81	0.45	0.15
ANOR	1030	34.01	29.41	25.37	21.91	18.96	16.43	14.25	12.35
NEPH	1031	17.27	15.19	13.38	11.84	10.55	9.45	8.51	7.70
WOLL	1035	15.25	14.16	13.10	12.16	11.33	10.62	9.99	9.44
DIOP	1039	23.56	21.74	19.89	18.24	16.81	15.56	14.48	13.54
CA-AL PYX	1040	43.89	38.24	33.45	29.41	25.98	23.04	20.50	18.30
AKER	1044	50.89	46.56	42.58	39.13	36.15	33.57	31.33	29.37
MERW	1045	76.42	69.82	63.93	58.86	54.50	50.73	47.45	44.58
MONTI	1046	33.54	30.55	27.85	25.52	23.52	21.80	20.30	18.99
GEHL	1047	66.78	59.03	52.48	46.94	42.24	38.21	34.72	31.68
FORST	1048	31.89	28.79	25.97	23.54	21.45	19.66	18.10	16.74
FAY	1049	22.42	20.03	17.91	16.07	14.50	13.14	11.96	10.93
HUNT	1050	15.87	13.03	10.44	8.19	6.20	4.41	2.78	1.28
ART	1053	22.45	20.37	18.55	17.04	15.77	14.69	13.76	12.95
HED	1054	19.91	18.21	16.51	14.99	13.65	12.48	11.45	10.56
MALACH	1059	3.46	3.69	3.61	3.55	3.50	3.46	3.43	3.39
AZUR	1060	5.97	6.16	5.89	5.65	5.43	5.22	5.02	4.82
HYDRO-MAG	1062	38.50	33.80	29.61	26.06	22.99	20.32	17.95	15.83
SMITH	1064	1.11	0.60	0.16	-0.23	-0.57	-0.88	-1.17	-1.43
ANH-CORD	1065	68.42	59.05	50.62	43.37	37.16	31.81	27.19	23.18
HYD-CORD	1066	65.40	56.32	48.15	41.12	35.11	29.95	25.49	21.63
JADEITE	1067	10.62	9.53	8.42	7.44	6.60	5.88	5.27	4.76
CER	1068	-2.61	-2.45	-2.33	-2.22	-2.12	-2.03	-1.95	-1.87
STRONT	1069	-0.20	-0.36	-0.49	-0.60	-0.72	-0.83	-0.95	-1.07
DIS-DOL	1070	6.47	5.25	4.15	3.18	2.32	1.55	0.84	0.18

$\log K$, $P = 1.5$ kb

INDEX	TEMPERATURE, °C									
	200	225	250	300	350	400	450	500	550	600
1001	4.14	3.07	2.12	0.50	-0.92	-2.32	-3.90	-5.76	-7.73	(-9.06)
1002	4.03	2.98	2.06	0.49	-0.89	-2.26	-3.81	-5.64	-7.60	(-8.91)
1003	4.27	3.19	2.22	0.58	-0.85	-2.28	-3.87	-5.73	-7.72	(-9.05)
1004	0.19	-0.42	-0.94	-1.82	-2.59	-3.43	-4.52	-5.94	-7.53	(-8.49)
1007	18.54	17.49	16.57	15.03	13.74	12.52	11.20	9.68	8.09	(7.07)
1012	2.70	1.67	0.79	-0.68	-1.97	-3.36	-5.10	-7.34	-9.83	(-11.35)
1013	4.74	3.59	2.59	0.94	-0.49	-1.99	-3.82	-6.14	-8.70	(-10.27)
1014	20.52	19.05	17.77	15.62	13.80	12.03	10.05	7.71	5.22	(3.63)
1015	15.45	14.20	13.11	11.28	9.72	8.17	6.39	4.24	1.92	(0.46)
1021	6.04	5.48	4.99	4.19	3.52	2.85	2.09	1.17	0.18	(-0.42)
1022	3.22	2.86	2.56	2.08	1.69	1.28	0.75	0.03	-0.77	(-1.21)
1023	21.66	20.92	20.32	19.41	18.71	18.00	17.07	15.78	14.33	(13.63)
1025	0.80	0.54	0.33	0.02	-0.24	-0.54	-0.99	-1.63	-2.38	(-2.75)
1027	-0.58	-0.72	-0.83	-0.99	-1.13	-1.34	-1.71	-2.29	-2.98	(-3.31)
1028	-0.56	-0.71	-0.82	-0.99	-1.14	-1.37	-1.76	-2.36	-3.07	(-3.43)
1029	-0.10	-0.30	-0.47	-0.72	-0.94	-1.22	-1.65	-2.28	-3.01	(-3.39)
1030	10.70	9.25	7.97	5.79	3.90	2.06	0.05	-2.29	-4.76	(-6.41)
1031	7.01	6.41	5.89	5.03	4.30	3.59	2.80	1.86	0.85	(0.22)
1035	8.97	8.55	8.17	7.54	7.02	6.54	6.05	5.53	5.01	(4.65)
1039	12.72	12.01	11.38	10.32	9.42	8.59	7.70	6.70	5.67	(4.99)
1040	16.37	14.67	13.16	10.57	8.35	6.22	3.95	1.40	-1.26	(-3.09)
1044	27.65	26.13	24.78	22.49	20.57	18.82	17.09	15.25	13.42	(12.13)
1045	42.05	39.82	37.84	34.46	31.62	29.06	26.54	23.94	21.37	(19.51)
1046	17.84	16.83	15.93	14.41	13.12	11.94	10.77	9.52	8.27	(7.39)
1047	29.01	26.67	24.58	21.01	17.96	15.09	12.11	8.86	5.52	(3.17)
1048	15.54	14.49	13.55	11.96	10.61	9.36	8.11	6.77	5.41	(4.46)
1049	10.03	9.24	8.54	7.35	6.34	5.38	4.38	3.28	2.13	(1.35)
1050	-0.10	-1.40	-2.63	-4.97	-7.33	-9.98	-13.18	-17.08	-21.34	(-24.49)
1053	12.25	11.62	11.06	10.09	9.20	8.26	7.16	5.83	4.37	(3.39)
1054	9.78	9.09	8.48	7.45	6.57	5.74	4.85	3.86	2.83	(2.14)
1059	3.36	3.33	3.30	3.20	3.05	2.80	2.39	1.80	1.10	(0.62)
1060	4.63	4.43	4.23	3.81	3.30	2.60	1.64	0.36	1.10	(2.16)
1062	13.92	12.17	10.57	7.63	4.80	1.73	-1.90	-6.30	-11.07	(-14.50)
1064	-1.68	-1.91	-2.14	-2.57	-3.04	-3.60	-4.30	-5.20	-6.20	(-6.93)
1065	19.67	16.59	13.87	9.23	5.22	1.32	-2.96	-7.90	-13.11	(-16.61)
1066	18.26	15.31	12.71	8.28	4.47	0.73	-3.39	-8.18	-13.26	(-16.63)
1067	4.33	3.97	3.67	3.18	2.78	2.35	1.80	1.06	0.23	(-0.22)
1068	-1.81	-1.75	-1.70	-1.64	-1.66	-1.78	-2.08	-2.56	-3.15	(-3.54)
1069	-1.18	-1.30	-1.42	-1.68	-1.99	-2.40	-2.95	-3.68	-4.51	(-5.11)
1070	-0.43	-1.01	-1.56	-2.62	-3.70	-4.93	-6.44	-8.30	-10.34	(-11.84)

log K , P = 1.5 kb

NAME	INDEX	TEMPERATURE, °C							
		0	25	50	75	100	125	150	175
ORD-DOL	1071	4.73	3.71	2.77	1.95	1.21	0.54	-0.07	-0.65
ARAG	1072	2.96	2.50	2.09	1.73	1.40	1.10	0.83	0.57
CALC	1073	2.87	2.40	1.99	1.63	1.30	1.00	0.72	0.45
MAG	1074	3.67	2.95	2.28	1.70	1.19	0.73	0.32	-0.06
DOL	1075	4.73	3.71	2.77	1.95	1.21	0.54	-0.07	-0.64
SIDER	1076	1.06	0.46	-0.05	-0.50	-0.90	-1.27	-1.60	-1.91
ANHYD	1078	-2.89	-3.20	-3.56	-3.95	-4.34	-4.74	-5.15	-5.55
FLUORITE	1079	-9.27	-9.05	-8.97	-8.96	-8.99	-9.07	-9.18	-9.31
BARITE	1080	-9.30	-8.94	-8.72	-8.61	-8.58	-8.61	-8.70	-8.83
ANG	1081	-6.93	-6.75	-6.70	-6.71	-6.76	-6.84	-6.94	-7.07
CELEST	1082	-5.29	-5.38	-5.53	-5.73	-5.97	-6.23	-6.51	-6.82
ALUN	1083	7.16	3.46	0.37	-2.29	-4.61	-6.68	-8.56	-10.28
WITH	1084	-2.42	-2.29	-2.17	-2.09	-2.03	-2.00	-1.99	-2.00
RHODO	1085	0.71	0.34	0.00	-0.30	-0.57	-0.81	-1.03	-1.23
COV	1086	-26.27	-23.92	-22.10	-20.57	-19.27	-18.16	-17.21	-16.38
GALENA	1087	-15.54	-14.26	-13.21	-12.31	-11.54	-10.86	-10.27	-9.75
SL	1088	-11.31	-10.81	-10.42	-10.10	-9.85	-9.64	-9.47	-9.34
WURT	1089	-8.80	-8.51	-8.30	-8.13	-8.01	-7.92	-7.86	-7.82
M-CINN	1090	-41.15	-37.96	-35.29	-33.01	-31.05	-29.34	-27.84	-26.52
CINN	1091	-41.64	-38.37	-35.64	-33.32	-31.31	-29.57	-28.04	-26.69
ALA	1092	0.35	-0.03	-0.41	-0.75	-1.06	-1.35	-1.62	-1.87
PYRITE	1093	-25.83	-24.15	-22.79	-21.70	-20.84	-20.15	-19.61	-19.18
GOLD	1100	-27.49	-24.77	-22.44	-20.43	-18.68	-17.13	-15.75	-14.52
SILVER	1101	-10.51	-9.34	-8.32	-7.43	-6.64	-5.93	-5.29	-4.70
COPPER	1102	-5.21	-4.42	-3.87	-3.37	-2.91	-2.49	-2.10	-1.73
GRAPHITE	1103	-5.69	-5.06	-4.51	-4.05	-3.66	-3.33	-3.05	-2.81
HALITE	1106	1.77	1.80	1.80	1.79	1.77	1.73	1.68	1.62
SYLVITE ·	1107	0.84	1.10	1.30	1.44	1.53	1.58	1.60	1.60
COR	1108	27.05	22.68	19.11	16.16	13.67	11.56	9.74	8.16
PER	1109	24.11	21.73	19.68	17.95	16.47	15.20	14.09	13.13
LIME	1110	36.07	32.94	30.30	28.05	26.12	24.45	23.00	21.72
TO	1111	6.74	6.45	6.08	5.77	5.52	5.31	5.14	4.99
CUP	1112	-2.12	-1.39	-1.02	-0.66	-0.30	0.05	0.38	0.71
FE-OXIDE	1113	15.65	13.90	12.44	11.21	10.16	9.26	8.47	7.79
GIBBS	1114	10.07	8.42	7.11	6.05	5.19	4.47	3.86	3.36
DIAS	1115	11.18	9.30	7.78	6.52	5.48	4.60	3.85	3.21
BOEH	1116	12.22	10.19	8.55	7.19	6.05	5.09	4.27	3.57
BRUC	1117	18.34	16.58	15.07	13.80	12.74	11.83	11.05	10.38
MANGAN	1119	20.09	18.16	16.52	15.14	13.96	12.96	12.09	11.34
SPINEL	1120	46.38	40.01	34.73	30.32	26.61	23.43	20.69	18.32

$\log K$, $P = 1.5$ kb

INDEX	TEMPERATURE, °C									
	200	225	250	300	350	400	450	500	550	600
1071	-1.18	-1.69	-2.18	-3.13	-4.12	-5.28	-6.72	-8.52	-10.52	(-11.97)
1072	0.32	0.09	-0.13	-0.57	-1.02	-1.56	-2.24	-3.08	-4.02	(-4.70)
1073	0.20	-0.03	-0.26	-0.71	-1.18	-1.72	-2.41	-3.26	-4.21	(-4.90)
1074	-0.41	-0.74	-1.06	-1.65	-2.25	-2.92	-3.74	-4.73	-5.82	(-6.62)
1075	-1.18	-1.69	-2.18	-3.13	-4.12	-5.28	-6.73	-8.54	-10.53	(-12.00)
1076	-2.21	-2.48	-2.75	-3.26	-3.79	-4.41	-5.17	-6.12	-7.16	(-7.93)
1078	-5.96	-6.37	-6.78	-7.62	-8.54	-9.63	-10.97	-12.62	-14.44	(-15.82)
1079	-9.47	-9.65	-9.84	-10.30	-10.89	-11.70	-12.82	-14.30	-15.97	(-17.20)
1080	-9.00	-9.19	-9.41	-9.92	-10.57	-11.41	-12.53	-13.96	-15.57	(-16.78)
1081	-7.21	-7.37	-7.54	-7.94	-8.45	-9.15	-10.13	-11.43	-12.91	(-13.99)
1082	-7.13	-7.45	-7.79	-8.49	-9.30	-10.28	-11.53	-13.08	-14.80	(-16.10)
1083	-11.87	-13.35	-14.75	-17.43	-20.17	-23.34	-27.28	-32.20	-37.63	(-41.52)
1084	-2.03	-2.07	-2.12	-2.26	-2.47	-2.79	-3.27	-3.92	-4.67	(-5.21)
1085	-1.42	-1.60	-1.78	-2.12	-2.49	-2.96	-3.58	-4.39	-5.30	(-5.94)
1086	-15.66	-15.03	-14.47	-13.57	-12.90	-12.47	-12.28	-12.32	-12.52	(-12.60)
1087	-9.29	-8.88	-8.51	-7.91	-7.46	-7.21	-7.18	-7.40	-7.76	(-7.92)
1088	-9.24	-9.16	-9.11	-9.06	-9.13	-9.36	-9.79	-10.45	-11.24	(-11.78)
1089	-7.80	-7.79	-7.80	-7.88	-8.04	-8.35	-8.85	-9.57	-10.42	(-11.01)
1090	-25.35	-24.30	-23.36	-21.77	-20.51	-19.58	-18.98	-18.70	-18.65	(-18.44)
1091	-25.49	-24.42	-23.46	-21.83	-20.54	-19.58	-18.96	-18.66	-18.59	(-18.37)
1092	-2.11	-2.34	-2.56	-2.99	-3.45	-3.99	-4.70	-5.59	-6.58	(-7.29)
1093	-18.85	-18.60	-18.43	-18.28	-18.40	-18.82	-19.64	-20.86	-22.34	(-23.41)
1100	-13.41	-12.40	-11.48	-9.87	-8.48	-7.29	-6.23	-5.29	-4.45	(-3.70)
1101	-4.17	-3.67	-3.22	-2.40	-1.68	-1.04	-0.46	0.06	0.54	(0.98)
1102	-1.38	-1.06	-0.75	-0.18	0.34	0.81	1.23	1.62	1.97	(2.32)
1103	-2.61	-2.44	-2.29	-2.06	-1.91	-1.81	-1.80	-1.82	-1.90	(-1.95)
1106	1.55	1.48	1.40	1.22	0.99	0.69	0.27	-0.28	-0.90	(-1.36)
1107	1.58	1.54	1.49	1.35	1.14	0.84	0.43	-0.11	-0.73	(-1.19)
1108	6.78	5.57	4.49	2.64	1.03	-0.55	-2.27	-4.24	-6.32	(-7.76)
1109	12.27	11.52	10.84	9.68	8.70	7.81	6.96	6.10	5.25	(4.62)
1110	20.58	19.57	18.66	17.10	15.79	14.64	13.58	12.56	11.60	(10.84)
1111	4.87	4.77	4.69	4.56	4.46	4.36	4.26	4.13	3.99	(3.91)
1112	1.04	1.35	1.65	2.23	2.76	3.24	3.64	3.98	4.26	(4.59)
1113	7.19	6.65	6.18	5.36	4.66	4.03	3.39	2.72	2.04	(1.55)
1114	2.92	2.55	2.23	1.70	1.22	0.72	0.12	-0.65	-1.49	(-2.04)
1115	2.65	2.17	1.74	1.02	0.39	-0.23	-0.95	-1.80	-2.72	(-3.32)
1116	2.96	2.42	1.95	1.15	0.46	-0.22	-0.99	-1.88	-2.84	(-3.47)
1117	9.79	9.27	8.82	8.04	7.39	6.80	6.20	5.57	4.92	(4.48)
1119	10.69	10.11	9.60	8.73	8.01	7.37	6.74	6.08	5.44	(4.98)
1120	16.23	14.40	12.77	9.97	7.55	5.24	2.80	0.07	-2.75	(-4.73)

log K , P = 1.5 kb

NAME	INDEX	TEMPERATURE, °C							
		0	25	50	75	100	125	150	175
K-OXIDE	1123	91.05	84.15	78.34	73.39	69.12	65.40	62.13	59.23
NA-OXIDE	1124	73.33	67.59	62.79	58.74	55.26	52.26	49.64	47.35
AMORPH SIL	1125	-2.80	-2.36	-2.11	-1.94	-1.80	-1.69	-1.60	-1.52
A-CRIST	1126	-3.79	-3.18	-2.81	-2.55	-2.34	-2.18	-2.04	-1.92
B-CRIST	1127	-3.22	-2.70	-2.39	-2.17	-2.01	-1.87	-1.76	-1.67
CHALCED	1128	-4.19	-3.54	-3.14	-2.84	-2.61	-2.42	-2.26	-2.12
WAIR	1130	24.71	21.41	18.21	15.38	12.93	10.83	9.01	7.45
LAUM	1132	19.06	16.63	14.18	12.01	10.14	8.55	7.20	6.05
ACAN	1501	-39.42	-35.84	-32.81	-30.25	-28.03	-26.11	-24.43	-22.95
CP	1502	-35.77	-33.10	-31.02	-29.30	-27.86	-26.64	-25.61	-24.74
BN	1503	-111.57	-101.92	-94.51	-88.22	-82.83	-78.17	-74.11	-70.55
CC	1504	-37.79	-34.22	-31.48	-29.14	-27.11	-25.38	-23.86	-22.51
QTZ	1505	-4.49	-3.82	-3.39	-3.07	-2.82	-2.62	-2.45	-2.30
MG	1506	33.42	29.43	26.11	23.30	20.92	18.87	17.10	15.54
KALS	1507	13.84	12.21	10.77	9.54	8.50	7.61	6.86	6.21
FERROSIL	1508	8.78	7.96	7.14	6.41	5.76	5.20	4.71	4.29
PYROPH	1509	3.95	2.90	1.59	0.36	-0.72	-1.66	-2.46	-3.14
TALC	1510	24.62	22.90	20.79	18.84	17.13	15.64	14.36	13.26
7-A CL	1512	84.31	74.81	66.37	59.22	53.17	48.02	43.60	39.79
14-A CL	1513	80.59	71.33	63.07	56.08	50.16	45.12	40.80	37.07
CLNZ	1515	53.66	46.76	40.72	35.57	31.19	27.45	24.23	21.44
LAWS	1516	27.75	24.14	20.96	18.26	15.97	14.04	12.39	10.98
TREM	1517	69.87	64.66	58.98	53.84	49.35	45.46	42.09	39.17
ANTH	1518	77.05	70.66	63.91	57.87	52.60	48.03	44.06	40.62
ZOIS	1519	53.70	46.80	40.75	35.60	31.22	27.47	24.25	21.46
GROSS	1529	60.98	54.52	48.76	43.83	39.61	36.00	32.88	30.17
ANDRA	1530	57.42	52.59	48.11	44.20	40.82	37.90	35.36	33.15
ALB	1531	4.99	4.53	3.82	3.11	2.49	1.95	1.49	1.11
PREHN	1532	40.27	35.69	31.54	27.97	24.92	22.31	20.07	18.14
ENST	1537	12.98	11.83	10.69	9.68	8.80	8.04	7.37	6.80
PARG	1538	120.97	107.78	95.89	85.69	76.98	69.51	63.06	57.47
ANTIG	1542	547.33	499.46	452.93	412.64	378.20	348.75	323.48	301.73
HM	1544	21.17	18.64	16.52	14.73	13.21	11.90	10.76	9.76
ORD-EP	1545	50.74	44.76	39.44	34.87	30.97	27.62	24.74	22.25
MARG	1551	53.53	45.43	38.58	32.83	27.98	23.85	20.31	17.26
MERCURY	1552	-21.79	-20.11	-18.66	-17.38	-16.22	-15.18	-14.22	-13.34
PO	1555	-3.08	-3.28	-3.47	-3.65	-3.84	-4.02	-4.22	-4.43
NESQ	1558	6.96	5.49	5.30	5.21	5.15	5.09	5.04	5.00
EPID	1559	50.74	44.76	39.43	34.87	30.96	27.62	24.73	22.23
HI-ALB	1560	6.50	5.86	4.99	4.16	3.43	2.79	2.25	1.79

$\log K$, $P = 1.5$ kb

INDEX	TEMPERATURE, °C									
	200	225	250	300	350	400	450	500	550	600
1123	56.65	54.34	52.26	48.66	45.66	43.10	40.89	38.95	37.23	(35.75)
1124	45.31	43.51	41.90	39.14	36.88	34.97	33.32	31.87	30.59	(29.53)
1125	-1.45	-1.39	-1.34	-1.26	-1.20	-1.17	-1.17	-1.19	-1.23	(-1.27)
1126	-1.81	-1.72	-1.63	-1.48	-1.34	-1.23	-1.13	-1.05	-0.98	(-0.91)
1127	-1.58	-1.50	-1.43	-1.31	-1.19	-1.10	-1.02	-0.95	-0.90	(-0.83)
1128	-2.00	-1.89	-1.79	-1.61	-1.46	-1.33	-1.22	-1.13	-1.05	(-0.97)
1130	6.09	4.92	3.90	2.19	0.72	-0.74	-2.42	-4.46	-6.67	(-8.05)
1132	5.09	4.27	3.57	2.45	1.49	0.49	-0.79	-2.46	-4.33	(-5.39)
1501	-21.67	-20.52	-19.51	-17.79	-16.44	-15.42	-14.70	-14.26	-14.02	(-13.70)
1502	-24.00	-23.37	-22.83	-22.03	-21.56	-21.47	-21.82	-22.63	-23.73	(-24.47)
1503	-67.42	-64.66	-62.24	-58.23	-55.21	-53.18	-52.18	-52.22	-52.95	(-53.08)
1504	-21.32	-20.25	-19.29	-17.65	-16.34	-15.33	-14.62	-14.21	-13.99	(-13.67)
1505	-2.17	-2.05	-1.94	-1.75	-1.58	-1.44	-1.32	-1.23	-1.14	(-1.06)
1506	14.18	12.97	11.88	10.01	8.37	6.79	5.10	3.23	1.27	(-0.13)
1507	5.65	5.16	4.74	4.04	3.43	2.82	2.10	1.23	0.28	(-0.32)
1508	3.91	3.59	3.30	2.81	2.40	2.00	1.57	1.08	0.55	(0.21)
1509	-3.71	-4.20	-4.61	-5.28	-5.88	-6.58	-7.54	-8.87	-10.39	(-11.26)
1510	12.31	11.49	10.78	9.61	8.64	7.68	6.59	5.26	3.82	(2.96)
1512	36.49	33.61	31.08	26.80	23.16	19.65	15.82	11.41	6.76	(3.68)
1513	33.84	31.02	28.55	24.38	20.83	17.39	13.62	9.27	4.68	(1.65)
1515	19.02	16.91	15.05	11.90	9.17	6.49	3.50	-0.02	-3.77	(-6.23)
1516	9.77	8.73	7.82	6.30	5.00	3.67	2.11	0.18	-1.91	(-3.22)
1517	36.64	34.43	32.51	29.30	26.61	24.04	21.22	17.94	14.47	(12.28)
1518	37.62	35.01	32.71	28.85	25.60	22.53	19.24	15.50	11.59	(9.05)
1519	19.04	16.92	15.06	11.91	9.18	6.50	3.50	-0.02	-3.77	(-6.23)
1529	27.82	25.76	23.94	20.86	18.22	15.67	12.90	9.72	6.37	(4.14)
1530	31.21	29.50	28.00	25.43	23.23	21.13	18.88	16.33	13.68	(11.89)
1531	0.80	0.54	0.33	0.01	-0.25	-0.57	-1.03	-1.70	-2.47	(-2.87)
1532	16.47	15.01	13.74	11.59	9.75	7.92	5.84	3.35	0.69	(-1.03)
1537	6.29	5.85	5.46	4.79	4.23	3.71	3.17	2.57	1.95	(1.54)
1538	52.61	48.35	44.60	38.25	32.86	27.75	22.29	16.12	9.70	(5.40)
1542	282.93	266.62	252.39	228.64	208.71	189.73	169.11	145.31	120.25	(104.14)
1544	8.89	8.11	7.41	6.19	5.12	4.08	2.95	1.68	0.35	(-0.59)
1545	20.80	18.18	16.51	13.68	11.24	8.83	6.15	2.99	-0.36	(-2.56)
1551	14.60	12.28	10.24	6.78	3.76	0.77	-2.59	-6.56	-10.81	(-13.62)
1552	-12.52	-11.76	-11.05	-9.77	-8.64	-7.97	-7.43	-6.98	-6.60	(-6.17)
1555	-4.63	-4.84	-5.04	-5.47	-5.94	-6.52	-7.26	-8.20	-9.24	(-10.01)
1558	4.95	4.90	4.85	4.72	4.51	4.17	3.64	2.89	2.01	(1.38)
1559	20.06	18.16	16.48	13.64	11.18	8.76	6.06	2.90	-0.47	(-2.69)
1560	1.41	1.09	0.83	0.41	0.08	-0.29	-0.80	-1.51	-2.31	(-2.73)

Log K , P = 2 kb

NAME	INDEX	TEMPERATURE, °C							
		0	25	50	75	100	125	150	175
AND	1001	22.15	18.57	15.49	12.89	10.69	8.81	7.19	5.80
KYA	1002	21.52	18.03	15.04	12.50	10.36	8.53	6.96	5.61
SILL	1003	22.50	18.87	15.75	13.12	10.89	8.98	7.35	5.94
KAOL	1004	11.35	9.32	7.40	5.75	4.36	3.18	2.19	1.35
CHRYS	1007	35.52	32.64	29.68	27.10	24.89	23.00	21.38	19.99
MUSC	1012	20.90	17.59	14.49	11.81	9.53	7.60	5.97	4.59
PARAG	1013	25.41	21.55	18.01	14.96	12.39	10.21	8.36	6.80
PHLOG	1014	44.63	40.45	36.29	32.66	29.57	26.92	24.65	22.70
ANNITE	1015	36.38	32.70	29.18	26.09	23.44	21.16	19.21	17.54
DEHYD-ANAL	1021	15.51	13.96	12.40	11.00	9.80	8.76	7.88	7.12
ANAL	1022	9.97	8.92	7.79	6.78	5.90	5.16	4.54	4.02
SEPIO	1023	33.87	32.85	30.75	28.72	26.92	25.39	24.10	23.03
LO-ALB	1025	5.16	4.88	4.22	3.54	2.91	2.38	1.92	1.54
MAX-MICRO	1027	1.64	1.83	1.56	1.19	0.83	0.51	0.24	0.02
K-SPAR	1028	1.65	1.83	1.56	1.20	0.84	0.53	0.26	0.04
HI-SAN	1029	3.03	3.04	2.61	2.12	1.65	1.23	0.88	0.58
ANOR	1030	34.52	30.03	26.01	22.55	19.59	17.06	14.87	12.98
NEPH	1031	17.48	15.45	13.65	12.12	10.82	9.72	8.78	7.98
WOLL	1035	15.28	14.27	13.23	12.29	11.47	10.75	10.13	9.58
DIOP	1039	23.47	21.84	20.05	18.42	17.00	15.76	14.69	13.75
CA-AL PYX	1040	44.28	38.70	33.92	29.87	26.44	23.50	20.97	18.77
AKER	1044	50.95	46.82	42.89	39.45	36.47	33.90	31.66	29.70
MERW	1045	76.48	70.10	64.26	59.21	54.85	51.08	47.81	44.95
MONTI	1046	33.56	30.69	28.01	25.69	23.69	21.97	20.48	19.18
GEHL	1047	67.31	59.64	53.09	47.54	42.83	38.79	35.30	32.27
FORST	1048	31.82	28.88	26.08	23.66	21.58	19.79	18.24	16.89
FAY	1049	22.55	20.26	18.15	16.32	14.75	13.39	12.21	11.19
HUNT	1050	16.28	13.57	10.97	8.71	6.72	4.95	3.35	1.88
ART	1053	22.42	20.42	18.61	17.10	15.83	14.76	13.84	13.04
HED	1054	22.17	20.58	18.89	17.35	16.00	14.82	13.79	12.90
MALACH	1059	3.20	3.84	3.77	3.70	3.65	3.61	3.58	3.55
AZUR	1060	5.70	6.50	6.23	5.98	5.76	5.54	5.35	5.15
HYDRO-MAG	1062	38.79	34.29	30.10	26.53	23.47	20.81	18.47	16.38
SMITH	1064	1.03	0.57	0.14	-0.23	-0.57	-0.86	-1.14	-1.39
ANH-CORD	1065	69.46	60.44	52.06	44.80	38.58	33.22	28.60	24.59
HYD-CORD	1066	66.36	57.62	49.51	42.48	36.46	31.29	26.83	22.98
JADEITE	1067	10.64	9.68	8.62	7.66	6.83	6.12	5.52	5.02
CER	1068	-2.46	-2.25	-2.14	-2.03	-1.94	-1.84	-1.76	-1.68
STRONT	1069	-0.00	-0.17	-0.31	-0.43	-0.55	-0.66	-0.78	-0.89
DIS-DOL	1070	6.71	5.54	4.43	3.46	2.60	1.83	1.13	0.49

$\log K$, $P = 2$ kb

INDEX	TEMPERATURE, °C									
	200	225	250	300	350	400	450	500	550	600
1001	4.59	3.53	2.61	1.05	-0.25	-1.44	-2.65	-3.94	-5.14	(-5.68)
1002	4.43	3.41	2.51	1.00	-0.25	-1.41	-2.59	-3.85	-5.03	(-5.55)
1003	4.71	3.64	2.70	1.12	-0.19	-1.40	-2.63	-3.92	-5.13	(-5.68)
1004	0.64	0.05	-0.46	-1.27	-1.93	-2.56	-3.28	-4.13	-4.94	(-5.12)
1007	18.79	17.75	16.85	15.37	14.18	13.13	12.10	11.05	10.09	(9.71)
1012	3.41	2.41	1.56	0.18	-0.93	-1.98	-3.15	-4.51	-5.80	(-6.10)
1013	5.47	4.33	3.37	1.81	0.57	-0.59	-1.85	-3.27	-4.61	(-4.93)
1014	21.03	19.59	18.33	16.28	14.63	13.15	11.67	10.10	8.66	(8.15)
1015	16.11	14.87	13.80	12.05	10.64	9.37	8.06	6.65	5.35	(4.93)
1021	6.46	5.91	5.43	4.65	4.03	3.49	2.94	2.35	1.80	(1.65)
1022	3.59	3.23	2.93	2.48	2.15	1.86	1.53	1.14	0.78	(0.79)
1023	22.14	21.41	20.83	19.97	19.39	18.90	18.36	17.71	17.13	(17.30)
1025	1.23	0.98	0.77	0.49	0.30	0.12	-0.11	-0.41	-0.70	(-0.60)
1027	-0.16	-0.29	-0.40	-0.53	-0.61	-0.70	-0.86	-1.11	-1.35	(-1.24)
1028	-0.14	-0.28	-0.38	-0.53	-0.62	-0.73	-0.91	-1.18	-1.45	(-1.36)
1029	0.33	0.13	-0.03	-0.26	-0.42	-0.58	-0.80	-1.10	-1.39	(-1.31)
1030	11.34	9.91	8.65	6.54	4.79	3.21	1.64	-0.00	-1.52	(-2.21)
1031	7.30	6.70	6.19	5.36	4.70	4.11	3.53	2.90	2.34	(2.15)
1035	9.11	8.69	8.32	7.70	7.20	6.77	6.37	5.97	5.62	(5.45)
1039	12.94	12.24	11.62	10.59	9.76	9.03	8.34	7.63	6.99	(6.72)
1040	16.86	15.18	13.69	11.19	9.11	7.24	5.42	3.55	1.83	(0.95)
1044	27.99	26.48	25.15	22.91	21.07	19.47	17.99	16.57	15.29	(14.57)
1045	42.44	40.22	38.26	34.94	32.21	29.84	27.67	25.60	23.74	(22.62)
1046	18.04	17.03	16.15	14.65	13.42	12.35	11.36	10.39	9.53	(9.03)
1047	29.63	27.30	25.24	21.76	18.87	16.31	13.85	11.40	9.16	(7.93)
1048	15.71	14.67	13.75	12.19	10.91	9.79	8.73	7.70	6.77	(6.24)
1049	10.29	9.51	8.82	7.66	6.70	5.85	5.04	4.22	3.48	(3.10)
1050	0.54	-0.71	-1.89	-4.07	-6.17	-8.37	-10.85	-13.63	-16.39	(-18.01)
1053	12.35	11.74	11.20	10.28	9.48	8.71	7.88	6.96	6.07	(5.66)
1054	12.12	11.44	10.84	9.85	9.05	8.34	7.66	6.97	6.34	(6.07)
1059	3.52	3.50	3.47	3.41	3.31	3.15	2.89	2.54	2.16	(2.01)
1060	4.96	4.79	4.61	4.24	3.81	3.28	2.60	1.74	0.86	(0.38)
1062	14.51	12.82	11.28	8.51	5.97	3.40	0.59	-2.54	-5.59	(-7.28)
1064	-1.62	-1.83	-2.04	-2.43	-2.82	-3.26	-3.79	-4.40	-5.02	(-5.37)
1065	21.10	18.05	15.37	10.88	7.16	3.80	0.46	-3.02	-6.24	(-7.71)
1066	19.63	16.71	14.15	9.87	6.35	3.16	-0.03	-3.37	-6.46	(-7.80)
1067	4.60	4.25	3.96	3.51	3.18	2.88	2.54	2.14	1.76	(1.77)
1068	-1.61	-1.55	-1.49	-1.40	-1.36	-1.40	-1.54	-1.78	-2.05	(-2.11)
1069	-1.00	-1.11	-1.22	-1.45	-1.70	-2.01	-2.41	-2.90	-3.40	(-3.66)
1070	-0.10	-0.66	-1.18	-2.16	-3.12	-4.14	-5.29	-6.61	-7.92	(-8.68)

$\log K$, $P = 2$ kb

NAME	INDEX	TEMPERATURE, °C							
		0	25	50	75	100	125	150	175
ORD-DOL	1071	4.96	4.00	3.05	2.22	1.49	0.83	0.22	-0.34
ARAG	1072	3.10	2.65	2.24	1.88	1.55	1.25	0.98	0.73
CALC	1073	3.04	2.59	2.17	1.80	1.47	1.17	0.89	0.63
MAG	1074	3.74	3.06	2.39	1.82	1.31	0.86	0.45	0.08
DOL	1075	4.97	4.00	3.05	2.23	1.49	0.83	0.22	-0.33
SIDER	1076	1.24	0.65	0.13	-0.32	-0.73	-1.09	-1.42	-1.73
ANHYD	1078	-2.69	-2.94	-3.31	-3.70	-4.09	-4.49	-4.88	-5.28
FLUORITE	1079	-9.12	-8.79	-8.72	-8.71	-8.74	-8.81	-8.91	-9.03
BARITE	1080	-9.12	-8.72	-8.50	-8.39	-8.36	-8.39	-8.47	-8.59
ANG	1081	-6.76	-6.49	-6.44	-6.45	-6.50	-6.57	-6.67	-6.79
CELEST	1082	-5.09	-5.13	-5.28	-5.49	-5.72	-5.98	-6.26	-6.55
ALUN	1083	7.80	4.18	1.08	-1.58	-3.91	-5.97	-7.82	-9.50
WITH	1084	-2.25	-2.12	-2.01	-1.93	-1.87	-1.84	-1.83	-1.84
RHODO	1085	0.79	0.48	0.14	-0.17	-0.43	-0.66	-0.88	-1.08
COV	1086	-26.41	-23.84	-22.01	-20.48	-19.18	-18.06	-17.10	-16.27
GALENA	1087	-15.45	-14.11	-13.05	-12.16	-11.39	-10.71	-10.12	-9.59
SL	1088	-11.23	-10.68	-10.28	-9.96	-9.70	-9.49	-9.32	-9.18
WURT	1089	-8.72	-8.38	-8.16	-7.99	-7.87	-7.77	-7.71	-7.66
M-CINN	1090	-41.05	-37.79	-35.11	-32.84	-30.88	-29.17	-27.67	-26.34
CINN	1091	-41.55	-38.22	-35.49	-33.16	-31.16	-29.41	-27.88	-26.52
ALA	1092	0.36	0.05	-0.32	-0.66	-0.97	-1.25	-1.51	-1.75
PYRITE	1093	-25.77	-24.02	-22.65	-21.56	-20.69	-19.98	-19.42	-18.98
GOLD	1100	-27.67	-24.93	-22.59	-20.57	-18.80	-17.25	-15.87	-14.63
SILVER	1101	-10.59	-9.38	-8.36	-7.46	-6.67	-5.96	-5.32	-4.73
COPPER	1102	-5.40	-4.42	-3.86	-3.36	-2.90	-2.48	-2.09	-1.71
GRAPHITE	1103	-5.68	-5.06	-4.51	-4.05	-3.65	-3.32	-3.04	-2.80
HALITE	1106	1.78	1.83	1.84	1.84	1.82	1.78	1.74	1.69
SYLVITE	1107	0.87	1.15	1.35	1.49	1.58	1.64	1.67	1.68
COR	1108	27.43	23.04	19.47	16.50	14.00	11.89	10.07	8.50
PER	1109	24.09	21.76	19.71	17.98	16.51	15.24	14.13	13.17
LIME	1110	36.12	33.01	30.36	28.11	26.19	24.52	23.06	21.78
TO	1111	6.58	6.50	6.13	5.82	5.56	5.35	5.18	5.04
CUP	1112	-2.44	-1.31	-0.92	-0.56	-0.21	0.13	0.47	0.79
FE-OXIDE	1113	15.73	14.00	12.54	11.30	10.25	9.35	8.56	7.88
GIBBS	1114	10.20	8.55	7.23	6.17	5.30	4.58	3.98	3.48
DIAS	1115	11.34	9.45	7.92	6.67	5.62	4.74	3.99	3.35
BOEH	1116	12.39	10.36	8.71	7.34	6.21	5.25	4.43	3.73
BRUC	1117	18.29	16.58	15.07	13.81	12.74	11.84	11.06	10.39
MANGAN	1119	20.08	18.21	16.57	15.18	14.01	13.01	12.15	11.40
SPINEL	1120	46.77	40.43	35.14	30.72	26.99	23.82	21.09	18.72

log K , P = 2 kb

INDEX	TEMPERATURE, °C									
	200	225	250	300	350	400	450	500	550	600
1071	-0.86	-1.34	-1.80	-2.68	-3.54	-4.49	-5.58	-6.84	-8.10	(-8.81)
1072	0.49	0.27	0.06	-0.34	-0.74	-1.17	-1.68	-2.27	-2.87	(-3.20)
1073	0.39	0.16	-0.06	-0.47	-0.88	-1.33	-1.85	-2.45	-3.05	(-3.39)
1074	-0.26	-0.58	-0.88	-1.43	-1.96	-2.52	-3.15	-3.85	-4.56	(-4.96)
1075	-0.85	-1.34	-1.80	-2.68	-3.55	-4.49	-5.59	-6.85	-8.11	(-8.83)
1076	-2.01	-2.28	-2.53	-3.00	-3.47	-3.98	-4.56	-5.23	-5.90	(6.29)
1078	-5.67	-6.06	-6.45	-7.23	-8.05	-8.96	-10.02	-11.22	-12.44	(-13.21)
1079	-9.18	-9.33	-9.50	-9.90	-10.38	-10.99	-11.79	-12.79	-13.81	(-14.37)
1080	-8.74	-8.91	-9.11	-9.57	-10.11	-10.79	-11.63	-12.65	-13.69	(-14.32)
1081	-6.92	-7.06	-7.21	-7.55	-7.97	-8.50	-9.20	-10.07	-10.97	(-11.46)
1082	-6.85	-7.16	-7.47	-8.12	-8.82	-9.63	-10.59	-11.71	-12.84	(-13.55)
1083	-11.04	-12.47	-13.80	-16.28	-18.69	-21.27	-24.25	-27.66	-31.07	(-32.90)
1084	-1.86	-1.89	-1.93	-2.04	-2.20	-2.43	-2.76	-3.18	-3.62	(-3.85)
1085	-1.26	-1.43	-1.59	-1.89	-2.21	-2.57	-3.01	-3.54	-4.08	(-4.35)
1086	-15.54	-14.90	-14.33	-13.40	-12.69	-12.17	-11.85	-11.69	-11.62	(-11.42)
1087	-9.12	-8.70	-8.32	-7.68	-7.18	-6.83	-6.64	-6.60	-6.63	(-6.45)
1088	-9.07	-8.98	-8.91	-8.83	-8.83	-8.94	-9.18	-9.55	-9.96	(-10.10)
1089	-7.62	-7.61	-7.60	-7.64	-7.74	-7.93	-8.24	-8.68	-9.14	(-9.33)
1090	-25.16	-24.11	-23.16	-21.53	-20.22	-19.18	-18.41	-17.88	-17.48	(-16.92)
1091	-25.31	-24.23	-23.26	-21.60	-20.25	-19.19	-18.40	-17.85	-17.43	(-16.85)
1092	-1.98	-2.19	-2.39	-2.78	-3.17	-3.61	-4.12	-4.73	-5.34	(-5.66)
1093	-18.63	-18.35	-18.15	-17.92	-17.92	-18.15	-18.65	-19.39	-20.21	(-20.62)
1100	-13.52	-12.50	-11.58	-9.96	-8.57	-7.37	-6.32	-5.38	-4.55	(-3.80)
1101	-4.20	-3.70	-3.25	-2.43	-1.71	-1.07	-0.49	0.03	0.50	(0.94)
1102	-1.37	-1.05	-0.74	-0.17	0.35	0.82	1.25	1.64	2.01	(2.36)
1103	-2.59	-2.42	-2.26	-2.02	-1.85	-1.74	-1.68	-1.67	-1.68	(-1.67)
1106	1.63	1.56	1.49	1.34	1.16	0.93	0.63	0.26	-0.12	(-0.32)
1107	1.66	1.63	1.59	1.47	1.30	1.07	0.77	0.39	0.00	(-0.23)
1108	7.14	5.94	4.89	3.11	1.61	0.25	-1.11	-2.52	-3.84	(-4.50)
1109	12.32	11.57	10.90	9.76	8.82	7.99	7.24	6.53	5.89	(5.46)
1110	20.65	19.64	18.74	17.19	15.90	14.80	13.81	12.91	12.11	(11.52)
1111	4.92	4.82	4.74	4.61	4.52	4.45	4.37	4.30	4.23	(4.23)
1112	1.11	1.42	1.73	2.31	2.85	3.35	3.79	4.18	4.55	(4.96)
1113	7.28	6.75	6.28	5.48	4.81	4.23	3.68	3.15	2.66	(2.37)
1114	3.05	2.69	2.38	1.87	1.46	1.06	0.63	0.14	-0.33	(-0.50)
1115	2.80	2.33	1.92	1.23	0.66	0.14	-0.40	-0.97	-1.51	(-1.73)
1116	3.12	2.60	2.14	1.37	0.74	0.16	-0.43	-1.05	-1.62	(-1.88)
1117	9.81	9.30	8.85	8.10	7.48	6.94	6.44	5.95	5.50	(5.26)
1119	10.75	10.17	9.67	8.82	8.13	7.54	6.99	6.48	6.02	(5.74)
1120	16.66	14.85	13.24	10.53	8.27	6.23	4.25	2.23	0.37	(-0.63)

log K , P = 2 kb

NAME	INDEX	TEMPERATURE, °C							
		0	25	50	75	100	125	150	175
K-OXIDE	1123	90.99	84.10	78.30	73.36	69.09	65.37	62.10	59.21
NA-OXIDE	1124	73.31	67.57	62.78	58.72	55.25	52.25	49.64	47.34
AMORPH SIL	1125	-2.77	-2.27	-2.00	-1.82	-1.68	-1.57	-1.48	-1.41
A-CRIST	1126	-3.79	-3.12	-2.73	-2.46	-2.25	-2.09	-1.95	-1.83
B-CRIST	1127	-3.20	-2.62	-2.29	-2.07	-1.90	-1.77	-1.66	-1.57
CHALCED	1128	-4.22	-3.51	-3.08	-2.77	-2.54	-2.35	-2.19	-2.05
WAIR	1130	25.23	22.17	19.02	16.20	13.75	11.63	9.81	8.24
LAUM	1132	19.46	17.27	14.88	12.72	10.85	9.26	7.90	6.76
ACAN	1501	-39.53	-35.85	-32.81	-30.23	-28.01	-26.09	-24.40	-22.91
CP	1502	-35.79	-32.87	-30.78	-29.05	-27.60	-26.38	-25.34	-24.45
BN	1503	-112.24	-101.51	-94.05	-87.76	-82.36	-77.68	-73.60	-70.02
CC	1504	-38.12	-34.13	-31.37	-29.03	-27.00	-25.26	-23.74	-22.40
QTZ	1505	-4.52	-3.78	-3.33	-3.00	-2.75	-2.55	-2.38	-2.23
MG	1506	33.72	29.79	26.45	23.64	21.25	19.20	17.42	15.87
KALS	1507	14.01	12.44	11.02	9.79	8.75	7.87	7.11	6.46
FERROSIL	1508	8.82	8.08	7.28	6.55	5.91	5.35	4.86	4.44
PYROPH	1509	4.16	3.36	2.13	0.93	-0.15	-1.08	-1.88	-2.55
TALC	1510	24.43	23.09	21.08	19.17	17.48	16.01	14.74	13.64
7-A CL	1512	84.48	75.38	67.00	59.87	53.83	48.69	44.29	40.51
14-A CL	1513	80.72	71.86	63.66	56.69	50.79	45.76	41.46	37.76
CLNZ	1515	54.13	47.45	41.46	36.32	31.95	28.21	25.00	22.23
LAWS	1516	27.93	24.47	21.32	18.64	16.36	14.44	12.80	11.40
TREM	1517	69.55	65.11	59.62	54.56	50.12	46.25	42.90	40.01
ANTH	1518	76.62	71.05	64.50	58.55	53.32	48.78	44.85	41.44
ZOIS	1519	54.17	47.49	41.49	36.35	31.97	28.23	25.02	22.25
GROSS	1529	61.26	55.03	49.34	44.43	40.22	36.61	33.51	30.82
ANDRA	1530	57.54	52.99	48.59	44.70	41.33	38.41	35.88	33.68
ALB	1531	5.16	4.88	4.22	3.54	2.91	2.38	1.92	1.54
PREHN	1532	40.64	36.27	32.17	28.62	25.57	22.96	20.72	18.80
ENST	1537	12.92	11.88	10.76	9.76	8.89	8.13	7.47	6.90
PARG	1538	121.33	108.72	96.95	86.80	78.10	70.65	64.23	58.68
ANTIG	1542	544.98	501.37	455.69	415.79	381.55	352.26	327.15	305.58
HM	1544	21.37	18.88	16.75	14.96	13.43	12.12	10.98	9.99
ORD-EP	1545	51.13	45.39	40.12	35.57	31.67	28.33	25.46	22.98
MARG	1551	54.28	46.30	39.46	33.70	28.84	24.71	21.18	18.14
MERCURY	1552	-21.76	-20.03	-18.59	-17.31	-16.15	-15.11	-14.15	-13.27
PO	1555	-3.00	-3.17	-3.35	-3.53	-3.71	-3.89	-4.08	-4.28
NESQ	1558	7.00	5.57	5.38	5.29	5.22	5.17	5.12	5.08
EPID	1559	51.13	45.39	40.12	35.57	31.67	28.32	25.45	22.97
HI-ALB	1560	6.67	6.21	5.40	4.59	3.86	3.22	2.68	2.22

$\log K$, $P = 2$ kb

331

INDEX	TEMPERATURE, °C									
	200	225	250	300	350	400	450	500	550	600
1123	56.63	54.32	52.24	48.65	45.65	43.11	40.92	39.01	37.34	(35.91)
1124	45.32	43.51	41.90	39.16	36.91	35.02	33.42	32.03	30.83	(29.85)
1125	-1.34	-1.28	-1.24	-1.16	-1.11	-1.08	-1.07	-1.09	-1.11	(-1.14)
1126	-1.73	-1.63	-1.55	-1.40	-1.27	-1.15	-1.05	-0.96	-0.87	(-0.79)
1127	-1.48	-1.41	-1.34	-1.22	-1.11	-1.01	-0.93	-0.85	-0.78	(-0.70)
1128	-1.93	-1.82	-1.72	-1.54	-1.39	-1.26	-1.14	-1.04	-0.95	(-0.86)
1130	6.89	5.73	4.72	3.07	1.73	0.53	-0.71	-2.05	-3.29	(-3.69)
1132	5.79	4.98	4.30	3.23	2.41	1.65	0.81	-0.16	-1.07	(-1.16)
1501	-21.61	-20.46	-19.43	-17.68	-16.28	-15.19	-14.34	-13.71	-13.22	(-12.64)
1502	-23.69	-23.04	-22.48	-21.61	-21.02	-20.74	-20.77	-21.09	-21.52	(-21.60)
1503	-66.85	-64.05	-61.59	-57.45	-54.22	-51.83	-50.24	-49.37	-48.87	(-47.76)
1504	-21.19	-20.11	-19.14	-17.47	-16.11	-15.02	-14.17	-13.55	-13.05	(-12.45)
1505	-2.09	-1.98	-1.87	-1.68	-1.51	-1.37	-1.25	-1.14	-1.04	(-0.94)
1506	14.52	13.32	12.25	10.43	8.88	7.48	6.09	4.68	3.36	(2.60)
1507	5.91	5.44	5.02	4.35	3.81	3.31	2.79	2.22	1.69	(1.52)
1508	4.07	3.74	3.46	2.99	2.61	2.27	1.93	1.59	1.27	(1.14)
1509	-3.12	-3.59	-3.98	-4.60	-5.08	-5.55	-6.14	-6.87	-7.57	(-7.61)
1510	12.71	11.90	11.21	10.10	9.22	8.45	7.67	6.83	6.06	(5.87)
1512	37.24	34.40	31.92	27.80	24.44	21.43	18.45	15.36	12.52	(11.27)
1513	34.57	31.79	29.38	25.36	22.09	19.15	16.24	13.21	10.42	(9.22)
1515	19.84	17.76	15.94	12.91	10.42	8.16	5.88	3.47	1.23	(0.30)
1516	10.21	9.18	8.30	6.87	5.71	4.64	3.52	2.29	1.14	(0.79)
1517	37.50	35.34	33.45	30.37	27.89	25.72	23.59	21.38	19.37	(18.67)
1518	38.47	35.89	33.64	29.91	26.90	24.26	21.70	19.10	16.73	(15.76)
1519	19.85	17.77	15.95	12.92	10.43	8.17	5.88	3.47	1.23	(0.30)
1529	28.49	26.46	24.68	21.71	19.27	17.09	14.93	12.70	10.64	(9.73)
1530	31.76	30.08	28.60	26.11	24.06	22.23	20.44	18.61	16.93	(16.13)
1531	1.23	0.98	0.77	0.49	0.28	0.09	-0.15	-0.48	-0.79	(-0.71)
1532	17.14	15.71	14.46	12.41	10.73	9.21	7.65	5.99	4.45	(3.88)
1537	6.40	5.96	5.58	4.93	4.41	3.95	3.51	3.07	2.68	(2.49)
1538	53.86	49.65	45.97	39.83	34.81	30.37	26.05	21.65	17.63	(15.78)
1542	287.00	270.94	257.01	234.12	215.75	199.57	183.73	167.36	152.43	(146.60)
1544	9.12	8.35	7.66	6.48	5.48	4.56	3.64	2.70	1.81	(1.31)
1545	20.83	18.97	17.33	14.61	12.37	10.34	8.29	6.13	4.13	(3.30)
1551	15.51	13.23	11.24	7.91	5.16	2.65	0.09	-2.63	-5.17	(-6.24)
1552	-12.46	-11.70	-10.99	-9.70	-8.56	-7.87	-7.29	-6.78	-6.30	(-5.79)
1555	-4.47	-4.66	-4.85	-5.24	-5.64	-6.11	-6.66	-7.31	-7.96	(-8.33)
1558	5.05	5.01	4.97	4.86	4.72	4.48	4.12	3.65	3.13	(2.89)
1559	20.81	18.94	17.30	14.57	12.31	10.27	8.21	6.03	4.01	(3.17)
1560	1.84	1.53	1.27	0.89	0.62	0.37	0.08	-0.29	-0.62	(-0.57)

log K , P = 2.5 kb

NAME	INDEX	TEMPERATURE, °C							
		0	25	50	75	100	125	150	175
AND	1001	22.47	18.95	15.89	13.29	11.08	9.20	7.59	6.20
KYA	1002	21.77	18.35	15.37	12.84	10.70	8.87	7.31	5.97
SILL	1003	22.80	19.24	16.14	13.51	11.27	9.37	7.74	6.33
KAOL	1004	11.56	9.66	7.78	6.14	4.75	3.58	2.59	1.75
CHRYS	1007	35.23	32.66	29.76	27.21	25.01	23.14	21.53	20.15
MUSC	1012	21.20	18.10	15.07	12.41	10.83	8.22	6.59	5.22
PARAG	1013	25.71	22.07	18.59	15.57	13.00	10.83	8.99	7.44
PHLOG	1014	44.43	40.65	36.57	32.98	29.90	27.27	25.02	23.09
ANNITE	1015	36.52	33.15	29.70	26.63	23.99	21.72	19.77	18.11
DEHYD-ANAL	1021	15.73	14.31	12.79	11.41	10.20	9.16	8.27	7.51
ANAL	1022	10.11	9.20	8.11	7.11	6.24	5.50	4.88	4.36
SEPIO	1023	33.30	32.90	30.95	28.99	27.23	25.72	24.44	23.38
LO-ALB	1025	5.23	5.15	4.56	3.91	3.30	2.76	2.31	1.93
MAX-MICRO	1027	1.71	2.11	1.90	1.56	1.21	0.89	0.63	0.40
K-SPAR	1028	1.72	2.11	1.90	1.56	1.22	0.91	0.65	0.43
HI-SAN	1029	3.10	3.32	2.96	2.49	2.03	1.61	1.26	0.96
ANOR	1030	34.91	30.57	26.58	23.13	20.17	17.63	15.44	13.56
NEPH	1031	17.63	15.67	13.90	12.37	11.08	9.97	9.04	8.24
WOLL	1035	15.26	14.34	13.33	12.40	11.58	10.87	10.24	9.70
DIOP	1039	23.28	21.88	20.13	18.54	17.13	15.90	14.84	13.91
CA-AL PYX	1040	44.57	39.08	34.32	30.28	26.84	23.91	21.38	19.19
AKER	1044	50.87	46.98	43.10	39.68	36.72	34.15	31.91	29.97
MERW	1045	76.36	70.26	64.48	59.45	55.10	51.34	48.08	45.23
MONTI	1046	34.48	30.76	28.11	25.81	23.82	22.10	20.61	19.32
GEHL	1047	67.70	60.16	53.62	48.07	43.35	39.31	35.82	32.80
FORST	1048	31.65	28.89	26.13	23.73	21.66	19.88	18.34	17.00
FAY	1049	22.60	20.43	18.36	16.54	14.96	13.60	12.43	11.41
HUNT	1050	16.49	14.00	11.41	9.15	7.17	5.41	3.83	2.40
ART	1053	22.31	20.43	18.62	17.12	15.86	14.79	13.88	13.09
HED	1054	22.09	20.70	19.06	17.55	16.20	15.03	14.01	13.12
MALACH	1059	2.72	3.94	3.88	3.82	3.77	3.73	3.70	3.68
AZUR	1060	5.11	6.76	6.52	6.27	6.04	5.83	5.63	5.45
HYDRO-MAG	1062	38.85	34.65	30.46	26.91	23.85	21.21	18.89	16.84
SMITH	1064	0.89	0.52	0.11	-0.25	-0.57	-0.86	-1.12	-1.36
ANH-CORD	1065	70.25	61.63	53.33	46.09	39.86	34.49	29.87	25.87
HYD-CORD	1066	67.06	58.75	50.71	43.70	37.68	32.50	28.04	24.20
JADEITE	1067	10.59	9.79	8.78	7.85	7.03	6.33	5.74	5.25
CER	1068	-2.36	-2.07	-1.95	-1.85	-1.76	-1.67	-1.59	-1.51
STRONT	1069	1.47	0.01	-0.15	-0.28	-0.40	-0.51	-0.62	-0.73
DIS-DOL	1070	6.85	5.78	4.67	3.70	2.85	2.08	1.39	0.76

$$\log K, P = 2.5 \text{ kb}$$

INDEX	TEMPERATURE, °C									
	200	225	250	300	350	400	450	500	550	600
1001	5.00	3.96	3.05	1.53	0.30	-0.78	-1.82	-2.83	-3.66	(-3.83)
1002	4.80	3.80	2.91	1.45	0.27	-0.78	-1.78	-2.76	-3.57	(-3.72)
1003	5.11	4.06	3.13	1.60	0.35	-0.75	-1.79	-2.81	-3.65	(-3.29)
1004	1.06	0.47	-0.03	-0.80	-1.39	-1.92	-2.46	-3.04	-3.18	(-3.29)
1007	18.97	17.94	17.06	15.62	14.48	13.52	12.65	11.82	11.16	(11.08)
1012	4.06	3.08	2.25	0.94	-0.07	-0.96	-1.85	-2.79	-3.50	(-3.23)
1013	6.12	5.01	4.07	2.58	1.41	0.45	-0.53	-1.52	-2.28	(-2.02)
1014	21.44	20.02	18.80	16.81	15.25	13.92	12.69	11.50	10.57	(10.55)
1015	16.69	15.46	14.41	12.71	11.38	10.25	9.17	8.12	7.30	(7.35)
1021	6.86	6.30	5.82	5.06	4.48	3.99	3.53	3.10	2.77	(2.83)
1022	3.92	3.57	3.28	2.84	2.54	2.30	2.07	1.83	1.68	(1.90)
1023	22.50	22.02	21.22	20.41	19.51	19.51	19.16	18.80	18.63	(19.21)
1025	1.62	1.37	1.18	0.91	0.75	0.63	0.51	0.36	0.30	(0.62)
1027	0.23	0.09	-0.00	-0.12	-0.17	-0.20	-0.26	-0.36	-0.39	(-0.06)
1028	0.25	0.11	0.01	-0.12	-0.18	-0.23	-0.31	-0.43	-0.49	(-0.18)
1029	0.72	0.52	0.37	0.16	0.03	-0.08	-0.20	-0.35	-0.43	(0.14)
1030	11.92	10.50	9.26	7.19	5.52	4.08	2.72	1.41	0.35	(0.11)
1031	7.56	6.97	6.47	5.66	5.03	4.51	4.02	3.55	3.19	(3.21)
1035	9.22	8.81	8.44	7.83	7.35	6.94	6.57	6.24	5.98	(5.88)
1039	13.11	12.42	11.81	10.81	10.01	9.34	8.74	8.17	7.73	(7.64)
1040	17.29	15.63	14.17	11.72	9.72	7.99	6.38	4.85	3.58	(3.15)
1044	28.26	26.77	25.45	23.23	21.44	19.93	18.58	17.35	16.34	(15.88)
1045	42.73	40.53	38.59	35.31	32.64	30.38	28.38	26.56	25.05	(24.27)
1046	18.19	17.19	16.31	14.84	13.65	12.63	11.73	10.90	10.21	(9.90)
1047	30.17	27.86	25.82	22.40	19.61	17.20	15.00	12.94	11.22	(10.51)
1048	15.82	14.79	13.89	12.36	11.12	10.06	9.11	8.24	7.51	(7.18)
1049	10.52	9.47	9.05	7.91	6.99	6.20	5.48	4.80	4.25	(4.06)
1050	1.09	-0.12	-1.24	-3.31	-5.25	-7.22	-9.32	-11.55	-13.58	(-14.47)
1053	12.41	11.82	11.30	10.41	9.67	8.98	8.28	7.57	6.96	(6.82)
1054	12.35	11.67	11.09	10.11	9.34	8.69	8.09	7.54	7.10	(7.01)
1059	3.66	3.64	3.62	3.58	3.51	3.40	3.22	2.98	2.76	(2.76)
1060	5.27	5.10	4.94	4.60	4.23	3.79	3.24	2.59	1.98	(1.78)
1062	15.01	13.36	11.87	9.22	6.85	4.55	2.15	-0.35	-2.57	(-3.43)
1064	-1.57	-1.77	-1.96	-2.32	-2.67	-3.04	-3.47	-3.94	-4.37	(-4.54)
1065	22.40	19.37	16.73	12.33	8.77	5.68	2.79	0.09	-2.26	(-2.79)
1066	20.87	17.97	15.45	11.26	7.90	4.98	2.24	-0.40	-2.54	(-2.95)
1067	4.83	4.49	4.21	3.80	3.50	3.27	3.03	2.79	2.64	(2.86)
1068	-1.43	-1.36	-1.29	-1.18	-1.11	-1.10	-1.17	-1.30	-1.41	(-1.31)
1069	-0.84	-0.94	-1.04	-1.25	-1.47	-1.73	-2.04	-2.41	-2.75	(-2.85)
1070	0.18	-0.35	-0.85	-1.78	-2.66	-3.56	-4.54	-5.59	-6.54	(-6.95)

log K, P = 2.5 kb

NAME	INDEX	TEMPERATURE, °C							
		0	25	50	75	100	125	150	175
ORD-DOL	1071	5.11	4.23	3.29	2.46	1.73	1.07	0.48	-0.07
ARAG	1072	3.20	2.79	2.37	2.01	1.68	1.39	1.12	0.87
CALC	1073	3.16	2.74	2.32	1.95	1.62	1.32	1.04	0.79
MAG	1074	3.77	3.15	2.48	1.91	1.40	0.95	0.55	0.19
DOL	1075	5.11	4.23	3.29	2.47	1.73	1.07	0.48	-0.06
SIDER	1076	1.37	0.81	0.30	-0.16	-0.56	-0.92	-1.25	-1.55
ANHYD	1078	-2.59	-2.74	-3.10	-3.49	-3.88	-4.27	-4.66	-5.04
FLUORITE	1079	-9.07	-8.57	-8.50	-8.49	-8.53	-8.59	-8.69	-8.80
BARITE	1080	-9.06	-8.57	-8.33	-8.22	-8.18	-8.20	-8.28	-8.38
ANG	1081	-6.69	-6.27	-6.21	-6.22	-6.27	-6.34	-6.44	-6.54
CELEST	1082	-4.97	-4.94	-5.09	-5.29	-5.52	-5.78	-6.05	-6.33
ALUN	1083	8.24	4.78	1.70	-0.96	-3.28	-5.33	-7.16	-8.81
WITH	1084	-2.14	-1.99	-1.87	-1.79	-1.74	-1.71	-1.69	-1.70
RHODO	1085	0.83	0.59	0.25	-0.04	-0.30	-0.54	-0.75	-0.94
COV	1086	-26.67	-23.78	-21.94	-20.40	-19.10	-17.98	-17.01	-16.17
GALENA	1087	-15.43	-13.98	-12.92	-12.02	-11.25	-10.57	-9.97	-9.44
SL	1088	-11.21	-10.57	-10.16	-9.83	-9.58	-9.36	-9.18	-9.04
WURT	1089	-8.70	-8.27	-8.04	-7.87	-7.74	-7.64	-7.57	-7.51
M-CINN	1090	-41.01	-37.64	-34.96	-32.69	-30.72	-29.01	-27.51	-26.18
CINN	1091	-41.53	-38.04	-35.35	-33.02	-31.01	-29.27	-27.72	-26.37
ALA	1092	0.31	0.10	-0.25	-0.59	-0.89	-1.16	-1.41	-1.65
PYRITE	1093	-25.78	-23.94	-22.54	-21.44	-20.55	-19.84	-19.26	-18.79
GOLD	1100	-27.86	-25.10	-22.74	-20.71	-18.93	-17.37	-15.98	-14.74
SILVER	1101	-10.71	-9.44	-8.41	-7.50	-6.71	-6.00	-5.35	-4.77
COPPER	1102	-5.69	-4.43	-3.86	-3.36	-2.90	-2.48	-2.08	-1.72
GRAPHITE	1103	-5.65	-5.05	-4.50	-4.03	-3.64	-3.30	-3.02	-2.77
HALITE	1106	1.76	1.84	1.87	1.87	1.86	1.83	1.79	1.75
SYLVITE	1107	0.85	1.17	1.38	1.53	1.63	1.70	1.73	1.75
COR	1108	27.75	23.38	19.79	16.81	14.31	12.19	10.38	8.82
PER	1109	24.04	21.76	19.72	17.99	16.52	15.25	14.15	13.20
LIME	1110	36.14	33.05	30.41	28.16	26.23	24.56	23.11	21.83
TO	1111	6.31	6.52	6.15	5.85	5.59	5.38	5.21	5.07
CUP	1112	-2.95	-1.26	-0.86	-0.50	-0.15	0.19	0.52	0.85
FE-OXIDE	1113	15.77	14.07	12.61	11.38	10.33	9.42	8.64	7.96
GIBBS	1114	10.31	8.66	7.34	6.28	5.41	4.69	4.09	3.59
DIAS	1115	11.47	9.59	8.06	6.80	5.75	4.87	4.12	3.49
BOEH	1116	12.54	10.51	8.85	7.49	6.35	5.39	4.57	3.87
BRUC	1117	18.21	16.55	15.05	13.79	12.73	11.83	11.05	10.39
MANGAN	1119	20.02	18.23	16.59	15.21	14.04	13.04	12.18	11.44
SPINEL	1120	47.06	40.79	35.49	31.06	27.33	24.16	21.44	19.08

$\log K$, $P = 2.5$ kb

INDEX	TEMPERATURE, °C									
	200	225	250	300	350	400	450	500	550	600
1071	-0.57	-1.04	-1.47	-2.29	-3.08	-3.91	-4.82	-5.82	-6.72	(-7.08)
1072	0.64	0.42	0.23	-0.15	-0.51	-0.90	-1.32	-1.79	-2.21	(-2.38)
1073	0.55	0.33	0.13	-0.26	-0.64	-1.04	-1.47	-1.95	-2.38	(-2.56)
1074	-0.14	-0.45	-0.73	-1.25	-1.74	-2.23	-2.76	-3.33	-3.84	(-4.07)
1075	-0.57	-1.03	-1.47	-2.29	-3.08	-3.91	-4.83	-5.83	-6.74	(-7.10)
1076	-1.83	-2.09	-2.33	-2.78	-3.21	-3.66	-4.16	-4.69	-5.18	(-5.39)
1078	-5.42	-5.80	-6.17	-6.90	-7.66	-8.48	-8.39	-10.37	-11.30	(-11.78)
1079	-8.92	-9.06	-9.21	-9.55	-9.97	-10.48	-11.12	-11.88	-12.58	(-12.83)
1080	-8.52	-8.68	-8.86	-9.27	-9.76	-10.34	-11.05	-11.85	-12.62	(-12.98)
1081	-6.66	-6.79	-6.92	-7.22	-7.58	-8.02	-8.58	-9.24	-9.86	(-10.07)
1082	-6.61	-6.90	-7.20	-7.80	-8.45	-9.16	-9.98	-10.88	-11.73	(-12.16)
1083	-10.31	-11.69	-12.96	-15.30	-17.51	-19.79	-22.28	-24.96	-27.38	(-28.25)
1084	-1.71	-1.73	-1.76	-1.85	-1.98	-2.17	-2.42	-2.73	-3.02	(-3.10)
1085	-1.11	-1.27	-1.42	-1.70	-1.97	-2.27	-2.63	-3.03	-3.39	(-3.48)
1086	-15.43	-14.78	-14.21	-13.25	-12.51	-11.95	-11.56	-11.30	-11.10	(-10.77)
1087	-8.96	-8.54	-8.15	-7.48	-6.95	-6.55	-6.28	-6.12	-5.97	(-5.63)
1088	-8.91	-8.81	-8.73	-8.62	-8.58	-8.63	-8.77	-9.00	-9.22	(-9.18)
1089	-7.47	-7.44	-7.43	-7.43	-7.49	-7.62	-7.83	-8.13	-8.40	(-8.41)
1090	-24.99	-23.92	-22.97	-21.32	-19.97	-18.88	-18.03	-17.38	-16.81	(-16.08)
1091	-25.15	-24.06	-23.08	-21.39	-20.19	-18.90	-18.24	-17.34	-16.76	(-16.02)
1092	-1.86	-2.06	-2.25	-2.61	-2.96	-3.33	-3.75	-4.21	-4.64	(-4.77)
1093	-18.42	-18.13	-17.90	-17.62	-17.54	-17.66	-18.00	-18.50	-19.00	(-19.10)
1100	-13.62	-12.60	-11.68	-10.05	-8.66	-7.45	-6.40	-5.46	-4.63	(-3.88)
1101	-4.23	-3.74	-3.28	-2.46	-1.74	-1.10	-0.53	-0.01	0.47	(0.90)
1102	-1.37	-1.05	-0.74	-0.17	0.35	0.82	1.25	1.65	2.02	(2.38)
1103	-2.56	-2.38	-2.23	-1.98	-1.80	-1.68	-1.60	-1.56	-1.55	(-1.50)
1106	1.70	1.64	1.58	1.45	1.29	1.10	0.86	0.59	0.33	(0.24)
1107	1.73	1.71	1.68	1.58	1.43	1.24	0.99	0.70	0.42	(0.29)
1108	7.47	6.29	5.25	3.51	2.09	0.84	-0.34	-1.49	-2.44	(-2.74)
1109	12.35	11.61	10.95	9.82	8.90	8.10	7.40	6.76	6.22	(5.89)
1110	20.70	19.69	18.80	17.26	15.98	14.90	13.96	13.11	12.39	(11.88)
1111	4.95	4.85	4.77	4.65	4.57	4.50	4.44	4.39	4.36	(4.49)
1112	1.17	1.48	1.78	2.37	2.91	3.42	3.88	4.30	4.71	(5.16)
1113	7.36	6.83	6.37	5.58	4.93	4.37	3.87	3.40	3.01	(2.81)
1114	3.17	2.81	2.51	2.03	1.64	1.30	0.95	0.59	0.30	(0.31)
1115	2.95	2.48	2.07	1.41	0.88	0.41	-0.04	-0.48	-0.84	(-0.88)
1116	3.27	2.75	2.30	1.56	0.96	0.43	-0.06	-0.55	-0.94	(-1.02)
1117	9.81	9.31	8.86	8.12	7.53	7.02	6.56	6.14	5.80	(5.64)
1119	10.79	10.22	9.72	8.89	8.22	7.65	7.15	6.70	6.33	(6.15)
1120	17.03	15.24	13.66	11.01	8.84	6.94	5.19	3.51	2.12	(1.57)

log K , P = 2.5 kb

NAME	INDEX	TEMPERATURE, °C							
		0	25	50	75	100	125	150	175
K-OXIDE	1123	90.89	84.03	78.25	73.31	69.04	65.33	62.06	59.17
NA-OXIDE	1124	73.23	67.52	62.74	58.70	55.23	52.24	49.63	47.34
AMORPH SIL	1125	-2.74	-2.19	-1.90	-1.71	-1.57	-1.47	-1.38	-1.30
A-CRIST	1126	-3.80	-3.07	-2.65	-2.37	-2.16	-2.00	-1.86	-1.75
B-CRIST	1127	-3.20	-2.55	-2.20	-1.97	-1.81	-1.67	-1.57	-1.47
CHALCED	1128	-4.26	-3.48	-3.03	-2.71	-2.47	-2.28	-2.12	-1.98
WAIR	1130	25.61	22.82	19.75	16.94	14.49	12.36	10.54	8.97
LAUM	1132	19.73	17.82	15.50	13.36	11.50	9.90	8.54	7.39
ACAN	1501	-39.73	-35.91	-32.84	-30.25	-28.02	-26.08	-24.38	-22.88
CP	1502	-35.98	-32.69	-30.58	-28.84	-27.39	-26.15	-25.10	-24.20
BN	1503	-113.50	-101.23	-93.70	-87.39	-81.97	-77.28	-73.18	-69.56
CC	1504	-38.67	-34.09	-31.31	-28.95	-26.92	-25.18	-23.65	-22.29
QTZ	1505	-4.56	-3.75	-3.28	-2.94	-2.69	-2.48	-2.31	-2.16
MG	1506	33.91	30.08	26.75	23.93	21.53	19.47	17.70	16.16
KALS	1507	14.13	12.63	11.23	10.02	8.98	8.09	7.34	6.70
FERROSIL	1508	8.81	8.16	7.39	6.68	6.04	5.48	5.00	4.57
PYROPH	1509	4.28	3.74	2.60	1.43	0.37	-0.56	-1.35	-2.02
TALC	1510	24.05	23.15	21.25	19.39	17.73	16.27	15.02	13.94
7-A CL	1512	84.36	75.75	67.45	60.36	54.33	49.21	44.83	41.08
14-A CL	1513	80.56	72.19	64.08	57.14	51.26	46.25	41.97	38.30
CLNZ	1515	54.42	48.00	42.08	36.97	32.60	28.87	25.67	22.93
LAWS	1516	28.01	24.72	21.63	18.96	16.70	14.78	13.15	11.76
TREM	1517	68.84	65.28	60.00	55.05	50.66	46.83	43.51	40.64
ANTH	1518	75.79	71.16	64.83	58.98	53.82	49.32	45.42	42.04
ZOIS	1519	54.45	48.04	42.11	36.99	32.62	28.89	25.69	22.94
GROSS	1529	61.33	55.41	49.79	44.91	40.72	37.13	34.04	31.37
ANDRA	1530	57.43	53.25	48.93	45.08	41.72	38.82	36.30	34.11
ALB	1531	5.23	5.15	4.56	3.91	3.30	2.76	2.31	1.93
PREHN	1532	40.84	36.73	32.70	29.17	26.12	23.52	21.29	19.38
ENST	1537	12.79	11.88	10.79	9.80	8.94	8.19	7.54	6.97
PARG	1538	121.26	109.36	97.75	87.66	79.00	71.58	65.19	59.67
ANTIG	1542	540.17	501.61	456.93	417.50	383.54	354.46	329.56	308.20
HM	1544	21.49	19.07	16.95	15.15	13.62	12.31	11.17	10.18
ORD-EP	1545	51.32	45.88	40.69	36.16	32.27	28.94	26.08	23.62
MARG	1551	54.86	47.05	40.24	34.49	29.62	25.49	21.97	18.94
MERCURY	1552	-21.77	-19.97	-18.52	-17.24	-16.09	-15.05	-14.10	-13.22
PO	1555	-2.97	-3.08	-3.25	-3.42	-3.60	-3.78	-3.96	-4.15
NESQ	1558	6.99	5.63	5.43	5.34	5.28	5.23	5.19	5.16
EPID	1559	51.32	45.88	40.68	36.15	32.26	28.93	26.07	23.60
HI-ALB	1560	6.75	6.49	5.75	4.96	4.24	3.61	3.07	2.61

$\log K$, $P = 2.5$ kb

INDEX	TEMPERATURE, °C									
	200	225	250	300	350	400	450	500	550	600
1123	56.60	54.30	52.22	48.63	45.64	43.10	40.92	39.03	37.39	(35.97)
1124	45.31	43.51	41.91	39.17	36.92	35.05	33.47	32.11	30.95	(30.01)
1125	-1.24	-1.19	-1.14	-1.07	-1.02	-1.00	-0.99	-1.00	-1.02	(-1.04)
1126	-1.64	-1.55	-1.47	-1.32	-1.19	-1.08	-0.97	-0.88	-0.79	(-0.70)
1127	-1.39	-1.32	-1.25	-1.13	-1.03	-0.93	-0.85	-0.77	-0.70	(-0.61)
1128	-1.86	-1.75	-1.66	-1.48	-1.33	-1.20	-1.08	-0.98	-0.88	(-0.78)
1130	7.67	6.46	5.46	3.84	2.58	1.49	0.46	-0.54	-1.32	(-1.27)
1132	6.43	5.63	4.96	3.92	3.16	2.52	1.89	1.25	0.79	(1.14)
1501	-21.58	-20.41	-19.37	-17.60	-16.17	-15.03	-14.12	-13.40	-12.78	(-12.08)
1502	-23.43	-22.76	-22.18	-21.25	-20.59	-20.20	-20.07	-20.15	-20.27	(-20.03)
1503	-66.37	-63.54	-61.03	-56.80	-53.44	-50.85	-48.96	-47.65	-46.55	(-44.84)
1504	-21.08	-20.00	-19.02	-17.33	-15.94	-14.80	-13.88	-13.16	-12.52	(-11.78)
1505	-2.03	-1.91	-1.81	-1.62	-1.45	-1.31	-1.19	-1.07	-0.97	(-0.86)
1506	14.81	13.62	12.56	10.78	9.29	7.97	6.74	5.55	4.53	(4.08)
1507	6.15	5.68	5.28	4.62	4.11	3.67	3.25	2.83	2.50	(2.52)
1508	4.21	3.89	3.61	3.15	2.78	2.47	2.18	1.90	1.69	(1.65)
1509	-2.57	-3.04	-3.42	-3.99	-4.41	-4.77	-5.18	-5.62	-5.94	(-5.59)
1510	13.01	12.23	11.55	10.48	9.66	8.99	8.36	7.75	7.30	(7.43)
1512	37.84	35.04	32.61	28.60	25.40	22.65	20.09	17.64	15.66	(15.27)
1513	35.14	32.41	30.04	26.14	23.02	20.36	17.86	15.48	13.55	(13.21)
1515	10.59	9.58	8.72	7.34	6.26	5.33	4.43	3.53	2.48	(2.93)
1516	10.59	9.58	8.72	7.34	6.26	5.33	4.43	3.53	2.84	(2.93)
1517	38.17	36.04	34.19	31.20	28.86	26.90	25.10	23.41	22.11	(22.09)
1518	39.11	36.57	34.36	30.74	27.87	25.46	23.26	21.20	19.58	(19.34)
1519	20.57	18.51	16.73	13.78	11.42	9.38	7.44	5.57	4.05	(3.83)
1529	29.06	27.07	25.31	22.42	20.10	18.10	16.24	14.47	13.04	(12.74)
1530	32.21	30.55	29.09	26.66	24.70	23.00	21.44	19.96	18.75	(18.41)
1531	1.62	1.37	1.18	0.91	0.74	0.61	0.46	0.30	0.21	(0.51)
1532	17.74	16.32	15.10	13.10	11.52	10.15	8.85	7.59	6.58	(6.54)
1537	6.48	6.05	5.67	5.04	4.54	4.11	3.72	3.36	3.08	(2.99)
1538	54.89	50.74	47.12	41.12	36.32	32.32	28.47	24.93	22.07	(21.36)
1542	289.84	274.03	260.38	238.15	220.72	206.01	192.52	179.75	169.69	(168.66)
1544	9.32	8.55	7.87	6.72	5.76	4.90	4.09	3.30	2.63	(2.34)
1545	21.49	19.64	18.04	15.40	13.27	11.44	9.70	8.02	6.66	(6.47)
1551	16.34	14.09	12.13	8.90	6.29	4.02	1.86	-0.26	-1.89	(-2.25)
1552	-12.40	-11.64	-10.93	-9.64	-8.49	-7.80	-7.20	-6.66	-6.14	(-5.58)
1555	-4.33	-4.51	-4.69	-5.04	-5.41	-5.81	-6.26	-6.77	-7.24	(-7.42)
1558	5.12	5.09	5.06	4.98	4.87	4.69	4.42	4.07	3.74	(3.66)
1559	21.47	19.62	18.01	15.35	13.22	11.36	9.61	7.91	6.55	(6.34)
1560	2.24	1.93	1.68	1.31	1.07	0.88	0.69	0.49	0.38	(0.65)

Log K , P = 3 kb

NAME	INDEX	TEMPERATURE, °C							
		0	25	50	75	100	125	150	175
AND	1001	22.72	19.29	16.25	13.65	11.45	9.56	7.95	6.57
KYA	1002	21.95	18.63	15.67	13.15	11.01	9.19	7.63	6.29
SILL	1003	23.04	19.56	16.49	13.86	11.62	9.72	8.09	6.69
KAOL	1004	11.69	9.94	8.12	6.50	5.11	3.93	2.95	2.12
CHRYS	1007	34.80	32.58	29.74	27.22	25.05	23.19	21.60	20.24
MUSC	1012	21.36	18.52	15.57	12.94	10.68	8.77	7.15	5.78
PARAG	1013	25.88	22.49	19.10	16.11	13.56	11.39	9.56	8.01
PHLOG	1014	44.02	40.69	36.72	33.18	30.13	27.52	25.29	23.38
ANNITE	1015	36.47	33.47	30.12	27.09	24.45	22.19	20.25	18.60
DEHYD-ANAL	1021	15.88	14.61	13.14	11.77	10.57	9.53	8.63	7.87
ANAL	1022	10.18	9.42	8.39	7.41	6.54	5.80	5.18	4.66
SEPIO	1023	32.49	32.78	31.01	29.12	27.41	25.92	24.66	23.61
LO-ALB	1025	5.21	5.36	4.86	4.23	3.64	3.11	2.66	2.28
MAX-MICRO	1027	1.69	2.31	2.19	1.88	1.54	1.23	0.97	0.75
K-SPAR	1028	1.70	2.31	2.19	1.88	1.55	1.25	0.99	0.77
HI-SAN	1029	3.08	3.53	3.25	2.81	2.36	1.95	1.60	1.30
ANOR	1030	35.18	31.03	27.08	23.64	20.68	18.14	15.95	14.07
NEPH	1031	17.71	15.85	14.11	12.59	11.30	10.20	9.27	8.47
WOLL	1035	15.19	14.38	13.39	12.48	11.66	10.95	10.33	9.79
DIOP	1039	22.98	21.83	20.16	18.59	17.20	15.98	14.93	14.02
CA-AL PYX	1040	44.76	39.40	34.67	30.63	27.20	24.26	21.74	19.56
AKER	1044	50.64	47.05	43.23	39.83	36.88	34.32	32.10	30.16
MERW	1045	76.06	70.30	64.58	59.58	55.25	51.51	48.26	45.42
MONTI	1046	33.31	30.78	28.16	25.87	23.89	22.18	20.70	19.41
GEHL	1047	67.96	60.58	54.06	48.52	43.80	39.75	36.27	33.26
FORST	1048	31.39	28.84	26.11	23.73	21.68	19.91	18.38	17.05
FAY	1049	22.57	20.56	18.52	16.71	15.14	13.78	12.61	11.59
HUNT	1050	16.51	14.32	11.76	9.51	7.54	5.80	4.23	2.83
ART	1053	22.11	20.38	18.59	17.10	15.84	14.78	13.88	13.11
HED	1054	21.92	20.75	19.18	17.69	16.36	15.19	14.18	13.30
MALACH	1059	2.04	3.99	3.96	3.91	3.86	3.83	3.80	3.78
AZUR	1060	4.20	6.96	6.75	6.51	6.28	6.07	5.87	5.70
HYDRO-MAG	1062	38.68	34.87	30.72	27.18	24.14	21.51	19.21	17.19
SMITH	1064	0.71	0.45	0.06	-0.28	-0.59	-0.87	-1.12	-1.34
ANH-CORD	1065	70.77	62.65	54.45	47.24	41.00	35.63	31.01	27.01
HYD-CORD	1066	67.51	59.69	51.76	44.78	38.76	33.58	29.12	25.28
JADEITE	1067	10.46	9.83	8.90	8.00	7.20	6.51	5.93	5.44
CER	1068	-2.31	-1.91	-1.79	-1.69	-1.60	-1.51	-1.42	-1.34
STRONT	1069	0.25	0.12	-0.01	-0.14	-0.27	-0.38	-0.49	-0.59
DIS-DOL	1070	6.90	5.96	4.87	3.90	3.05	2.29	1.61	0.99

$\log K$, $P = 3$ kb

INDEX	TEMPERATURE, °C									
	200	225	250	300	350	400	450	500	550	600
1001	5.38	4.34	3.44	1.96	0.77	-0.25	-1.18	-2.04	-2.66	(-2.64)
1002	5.14	4.14	3.27	1.84	0.70	-0.27	-1.17	-2.00	-2.60	(-2.55)
1003	5.48	4.43	3.52	2.01	0.81	-0.22	-1.16	-2.03	-2.67	(-2.65)
1004	1.43	0.85	0.36	-0.38	-0.94	-1.40	-1.85	-2.27	-2.51	(-2.12)
1007	19.06	18.06	17.19	15.78	14.69	13.79	13.00	12.30	11.81	(11.88)
1012	4.64	3.67	2.86	1.60	0.66	-0.13	-0.87	-1.57	-1.97	(-1.40)
1013	6.71	5.62	4.69	3.25	2.17	1.29	0.47	-0.28	-0.72	(-0.16)
1014	21.74	20.35	19.15	17.21	15.72	14.50	13.40	12.43	11.78	(12.03)
1015	17.18	15.97	14.93	13.27	12.00	10.94	9.99	9.14	8.59	(8.89)
1021	7.22	6.66	6.19	5.43	4.87	4.41	4.01	3.65	3.44	(3.61)
1022	4.23	3.87	3.58	3.16	2.88	2.67	2.49	2.33	2.30	(2.63)
1023	22.75	22.05	21.50	20.72	20.24	19.93	19.69	19.49	19.54	(20.33)
1025	1.97	1.73	1.54	1.28	1.14	1.06	0.99	0.93	0.99	(1.43)
1027	0.57	0.44	0.35	0.24	0.22	0.21	0.20	0.19	0.27	(0.71)
1028	0.59	0.46	0.36	0.25	0.20	0.18	0.15	0.12	0.18	(0.59)
1029	1.06	0.87	0.72	0.52	0.41	0.34	0.27	0.20	0.24	(0.64)
1030	12.44	11.02	9.79	7.76	6.15	4.78	3.54	2.42	1.60	(1.60)
1031	7.79	7.21	6.72	5.92	5.32	4.83	4.39	4.01	3.76	(3.89)
1035	9.32	8.90	8.54	7.94	7.46	7.07	6.72	6.43	6.21	(6.16)
1039	13.23	12.54	11.94	10.96	10.20	9.57	9.01	8.53	8.19	(8.21)
1040	17.67	16.02	14.58	12.17	10.23	8.58	7.10	5.75	4.73	(4.53)
1044	28.46	26.98	25.67	23.48	21.72	20.26	18.98	17.87	17.01	(16.69)
1045	42.94	40.75	38.82	35.58	32.96	30.77	28.86	27.19	25.88	(25.28)
1046	18.29	17.30	16.43	14.98	13.81	12.83	11.98	11.23	10.65	(10.43)
1047	30.63	28.34	26.32	22.95	20.22	17.90	15.85	14.00	12.57	(12.14)
1048	15.89	14.87	13.97	12.48	11.27	10.25	9.36	8.57	7.96	(7.75)
1049	10.70	9.93	9.25	8.13	7.23	6.47	5.80	5.20	4.75	(4.66)
1050	1.55	0.38	-0.70	-2.68	-4.51	-6.32	-8.20	-10.12	-11.73	(-12.23)
1053	12.44	11.86	11.35	10.50	9.80	9.16	8.54	7.94	7.48	(7.49)
1054	12.53	11.86	11.28	10.32	9.57	8.95	8.41	7.93	7.59	(7.59)
1059	3.76	3.75	3.74	3.71	3.67	3.59	3.45	3.28	3.14	(3.23)
1060	5.52	5.37	5.21	4.91	4.58	4.19	3.72	3.19	2.73	(2.68)
1062	15.39	13.78	12.33	9.79	7.54	5.41	3.26	1.12	-0.64	(-1.06)
1064	-1.54	-1.73	-1.91	-2.23	-2.55	-2.88	-3.24	-3.63	-3.96	(-4.02)
1065	23.55	20.54	17.92	13.59	10.13	7.20	4.56	2.17	0.42	(0.40)
1066	21.96	19.09	16.58	12.47	9.21	6.45	3.96	1.71	0.09	(0.18)
1067	5.04	4.70	4.43	4.04	3.77	3.58	3.40	3.25	3.22	(3.55)
1068	-1.26	-1.19	-1.12	-0.99	-0.90	-0.86	-0.88	-0.94	-0.98	(-0.79)
1069	-0.70	-0.79	-0.89	-1.07	-1.27	-1.50	-1.77	-2.07	-2.32	(-2.34)
1070	0.43	-0.09	-0.57	-1.45	-2.28	-3.11	-3.98	-4.88	-5.63	(-5.85)

$\log K$, $P = 3$ kb

NAME	INDEX	TEMPERATURE, °C							
		0	25	50	75	100	125	150	175
ORD-DOL	1071	5.16	4.42	3.49	2.66	1.93	1.28	0.70	0.16
ARAG	1072	3.26	2.89	2.49	2.12	1.80	1.50	1.24	1.00
CALC	1073	3.25	2.87	2.46	2.09	1.75	1.45	1.18	0.93
MAG	1074	3.74	3.20	2.55	1.98	1.48	1.03	0.64	0.28
DOL	1075	5.16	4.42	3.49	2.67	1.94	1.28	0.70	0.17
SIDER	1076	1.46	0.96	0.45	-0.01	-0.41	-0.77	-1.10	-1.39
ANHYD	1078	-2.57	-2.59	-2.93	-3.31	-3.70	-4.09	-4.47	-4.85
FLUORITE	1079	-9.11	-8.39	-8.32	-8.31	-8.35	-8.41	-8.49	-8.59
BARITE	1080	-9.09	-8.47	-8.21	-8.09	-8.04	-8.06	-8.12	-8.22
ANG	1081	-6.73	-6.10	-6.02	-6.03	-6.08	-6.15	-6.24	-6.33
CELEST	1082	-4.95	-4.80	-4.93	-5.13	-5.36	-5.62	-5.88	-6.15
ALUN	1083	8.49	5.28	2.23	-0.42	-2.73	-4.77	-6.58	-8.20
WITH	1084	-2.08	-1.89	-1.76	-1.68	-1.62	-1.59	-1.57	-1.57
RHODO	1085	0.81	0.69	0.36	0.06	-0.20	-0.43	-0.63	-0.82
COV	1086	-27.04	-23.76	-21.89	-20.34	-19.03	-17.91	-16.94	-16.09
GALENA	1087	-15.47	-13.88	-12.80	-11.90	-11.12	-10.45	-9.84	-9.31
SL	1088	-11.25	-10.48	-10.06	-9.73	-9.47	-9.25	-9.06	-8.91
WURT	1089	-8.74	-8.18	-7.94	-7.76	-7.63	-7.53	-7.45	-7.39
M-CINN	1090	-41.04	-37.52	-34.82	-32.55	-30.58	-28.87	-27.37	-26.03
CINN	1091	-41.57	-37.98	-35.22	-32.89	-30.89	-29.14	-27.60	-26.23
ALA	1092	0.20	0.13	-0.20	-0.53	-0.82	-1.09	-1.33	-1.56
PYRITE	1093	-25.87	-23.88	-22.45	-21.33	-20.43	-19.70	-19.11	-18.64
GOLD	1100	-28.08	-25.29	-22.91	-20.86	-19.07	-17.50	-16.10	-14.85
SILVER	1101	-10.86	-9.52	-8.47	-7.57	-6.76	-6.05	-5.40	-4.81
COPPER	1102	-6.06	-4.45	-3.88	-3.37	-2.91	-2.49	-2.09	-1.73
GRAPHITE	1103	-5.61	-5.03	-4.47	-4.01	-3.61	-3.27	-2.98	-2.74
HALITE	1106	1.70	1.84	1.89	1.90	1.90	1.88	1.84	1.81
SYLVITE	1107	0.81	1.18	1.41	1.57	1.68	1.75	1.78	1.80
COR	1108	28.03	23.67	20.08	17.10	14.60	12.48	10.67	9.11
PER	1109	23.93	21.73	19.70	17.97	16.50	15.24	14.15	13.20
LIME	1110	36.11	33.07	30.43	28.18	26.25	24.59	23.13	21.86
TO	1111	5.95	6.52	6.16	5.86	5.61	5.40	5.23	5.09
CUP	1112	-3.65	-1.25	-0.84	-0.47	-0.12	0.22	0.56	0.88
FE-OXIDE	1113	15.78	14.13	12.68	11.44	10.39	9.48	8.70	8.02
GIBBS	1114	10.40	8.76	7.44	6.38	5.51	4.79	4.19	3.69
DIAS	1115	11.58	9.71	8.18	6.92	5.87	4.99	4.24	3.61
BOEH	1116	12.66	10.64	8.99	7.62	6.48	5.52	4.70	4.00
BRUC	1117	18.08	16.50	15.00	13.74	12.69	11.79	11.02	10.36
MANGAN	1119	19.92	18.23	16.60	15.23	14.06	13.06	12.20	11.46
SPINEL	1120	47.27	41.09	35.79	31.36	27.63	24.45	21.73	19.39

$\log K$, $P = 3$ kb

INDEX	TEMPERATURE, °C									
	200	225	250	300	350	400	450	500	550	600
1071	-0.32	-0.78	-1.20	-1.97	-2.71	-3.46	-4.27	-5.10	-5.81	(-5.98)
1072	0.77	0.57	0.37	0.02	-0.32	-0.67	-1.05	-1.44	-1.77	(-1.85)
1073	0.70	0.49	0.29	-0.09	-0.44	-0.80	-1.19	-1.59	-1.93	(-2.02)
1074	-0.04	-0.33	-0.61	-1.10	-1.56	-2.01	-2.49	-2.97	-3.38	(-3.50)
1075	-0.32	-0.77	-1.19	-1.97	-2.71	-3.46	-4.27	-5.12	-5.83	(-6.00)
1076	-1.67	-1.92	-2.15	-2.58	-2.99	-3.40	-3.84	-4.30	-4.69	(-4.80)
1078	-5.22	-5.58	-5.93	-6.64	-7.35	-8.10	-8.93	-9.79	-10.55	(-10.87)
1079	-8.71	-8.83	-8.97	-9.27	-9.63	-10.08	-10.63	-11.24	-11.77	(-11.85)
1080	-8.35	-8.49	-8.65	-9.03	-9.48	-10.00	-10.63	-11.31	-11.92	(-12.13)
1081	-6.44	-6.56	-6.68	-6.95	-7.26	-7.65	-8.13	-8.66	-9.13	(-9.19)
1082	-6.42	-6.70	-6.98	-7.55	-8.15	-8.81	-9.54	-10.32	-11.00	(-11.27)
1083	-9.67	-11.01	-12.24	-14.47	-16.55	-18.63	-20.83	-23.10	-24.98	(-25.32)
1084	-1.58	-1.60	-1.62	-1.69	-1.80	-1.96	-2.16	-2.41	-2.62	(-2.61)
1085	-0.98	-1.13	-1.27	-1.53	-1.78	-2.04	-2.34	-2.67	-2.93	(-2.92)
1086	-15.35	-14.69	-14.11	-13.13	-12.37	-11.78	-11.35	-11.04	-10.76	(-10.36)
1087	-8.83	-8.39	-8.00	-7.31	-6.76	-6.32	-6.00	-5.77	-5.54	(-5.11)
1088	-8.78	-8.67	-8.58	-8.44	-8.38	-8.38	-8.47	-8.62	-8.73	(-8.59)
1089	-7.34	-7.30	-7.28	-7.26	-7.29	-7.38	-7.54	-7.75	-7.91	(-7.82)
1090	-24.84	-23.77	-22.80	-21.14	-19.76	-18.64	-17.74	-17.01	-16.35	(-15.54)
1091	-25.01	-23.91	-22.92	-21.22	-19.81	-18.66	-17.74	-16.99	-16.30	(-15.48)
1092	-1.76	-1.96	-2.13	-2.46	-2.78	-3.11	-3.48	-3.86	-4.18	(-4.21)
1093	-18.25	-17.94	-17.69	-17.36	-17.23	-17.28	-17.51	-17.88	-18.20	(-18.12)
1100	-13.73	-12.71	-11.78	-10.14	-8.75	-7.54	-6.48	-5.54	-4.71	(-3.96)
1101	-4.27	-3.78	-3.32	-2.50	-1.78	-1.14	-0.56	-0.04	0.43	(0.86)
1102	-1.38	-1.05	-0.75	-0.18	0.34	0.81	1.25	1.65	2.02	(2.39)
1103	-2.53	-2.34	-2.19	-1.94	-1.75	-1.62	-1.53	-1.48	-1.44	(-1.38)
1106	1.76	1.71	1.66	1.54	1.41	1.24	1.04	0.82	0.62	(0.60)
1107	1.80	1.79	1.76	1.67	1.54	1.37	1.16	0.91	0.69	(0.62)
1108	7.76	6.59	5.57	3.87	2.49	1.31	0.23	-0.76	-1.52	(-1.62)
1109	12.36	11.62	10.97	9.85	8.95	8.17	7.50	6.91	6.42	(6.15)
1110	20.73	19.73	18.83	17.30	16.04	14.98	14.05	13.24	12.56	(12.09)
1111	4.97	4.87	4.79	4.68	4.60	4.54	4.49	4.45	4.44	(4.49)
1112	1.20	1.51	1.82	2.40	2.95	3.46	3.94	4.37	4.80	(5.27)
1113	7.43	6.90	6.44	5.66	5.02	4.48	4.00	3.58	3.24	(3.08)
1114	3.27	2.92	2.63	2.16	1.80	1.49	1.19	0.90	0.70	(0.80)
1115	3.07	2.61	2.21	1.56	1.05	0.62	0.22	-0.14	-0.40	(-0.35)
1116	3.41	2.89	2.45	1.72	1.14	0.65	0.21	-0.20	-0.50	(-0.48)
1117	9.79	9.29	8.85	8.13	7.55	7.06	6.63	6.25	5.96	(5.85)
1119	10.82	10.26	9.76	8.94	8.28	7.73	7.26	6.85	6.52	(6.39)
1120	17.35	15.58	14.02	11.41	9.30	7.50	5.87	4.39	3.25	(2.95)

log K , P = 3 kb

NAME	INDEX	TEMPERATURE, °C							
		0	25	50	75	100	125	150	175
K-OXIDE	1123	90.74	83.93	78.17	73.24	68.99	65.28	62.02	59.13
NA-OXIDE	1124	73.10	67.44	62.69	58.66	55.20	52.21	49.61	47.32
AMORPH SIL	1125	-2.74	-2.12	-1.81	-1.61	-1.48	-1.37	-1.28	-1.21
A-CRIST	1126	-3.82	-3.03	-2.59	-2.30	-2.09	-1.92	-1.79	-1.67
B-CRIST	1127	-3.21	-2.50	-2.13	-1.89	-1.72	-1.59	-1.48	-1.39
CHALCED	1128	-4.31	-3.47	-2.99	-2.66	-2.42	-2.22	-2.06	-1.93
WAIR	1130	25.85	23.37	20.38	17.60	15.15	13.03	11.19	9.62
LAUM	1132	19.86	18.26	16.04	13.93	12.07	10.47	9.11	7.96
ACAN	1501	-40.02	-36.01	-32.90	-30.29	-28.05	-26.10	-24.39	-22.88
CP	1502	-36.33	-32.55	-30.40	-28.66	-27.19	-25.96	-24.90	-23.98
BN	1503	-115.36	-101.09	-93.46	-87.10	-81.67	-76.96	-72.84	-69.20
CC	1504	-39.42	-34.08	-31.27	-28.91	-26.87	-25.12	-23.58	-22.22
QTZ	1505	-4.61	-3.74	-3.24	-2.89	-2.63	-2.42	-2.25	-2.10
MG	1506	33.99	30.31	26.99	24.16	21.76	19.71	17.93	16.39
KALS	1507	14.18	12.78	11.41	10.21	9.18	8.29	7.54	6.90
FERROSIL	1508	8.75	8.22	7.48	6.78	6.14	5.59	5.11	4.69
PYROPH	1509	4.28	4.03	3.00	1.87	0.82	-0.09	-0.88	-1.54
TALC	1510	23.49	23.09	21.30	19.51	17.87	16.44	15.20	14.14
7-A CL	1512	83.94	75.92	67.72	60.67	54.68	49.57	45.22	41.49
14-A CL	1513	80.09	72.32	64.31	57.43	51.57	46.58	42.33	38.69
CLNZ	1515	54.51	48.42	42.59	37.51	33.16	29.44	26.25	23.52
LAWS	1516	27.98	24.89	21.86	19.22	16.97	15.06	13.44	12.07
TREM	1517	67.73	65.16	60.14	55.30	50.97	47.19	43.91	41.08
ANTH	1518	74.54	70.98	64.90	59.17	54.08	49.63	45.77	42.43
ZOIS	1519	54.55	48.45	42.61	37.53	33.18	29.45	26.27	23.53
GROSS	1529	61.20	55.64	50.12	45.28	41.11	37.54	34.46	31.81
ANDRA	1530	57.08	53.35	49.13	45.33	42.00	39.11	36.61	34.44
ALB	1531	5.21	5.36	4.86	4.23	3.64	3.11	2.66	2.28
PREHN	1532	40.88	37.08	33.13	29.62	26.60	24.00	21.78	19.87
ENST	1537	12.62	11.85	10.79	9.82	8.97	8.22	7.58	7.02
PARG	1538	120.77	109.70	98.29	88.29	79.68	72.29	65.93	60.45
ANTIG	1542	532.89	500.18	456.65	417.78	384.17	355.36	330.68	309.55
HM	1544	21.54	19.22	17.11	15.31	13.78	12.46	11.33	10.34
ORD-EP	1545	51.31	46.23	41.13	36.64	32.77	29.45	26.60	24.15
MARG	1551	55.28	47.69	40.93	35.19	30.32	26.19	22.67	19.66
MERCURY	1552	-21.82	-19.92	-18.47	-17.19	-16.04	-15.00	-14.05	-13.17
PO	1555	-2.99	-3.01	-3.16	-3.33	-3.50	-3.67	-3.85	-4.03
NESQ	1558	6.94	5.66	5.47	5.38	5.32	5.27	5.24	5.21
EPID	1559	51.31	46.23	41.13	36.63	32.76	29.44	26.59	24.13
HI-ALB	1560	6.73	6.70	6.04	5.29	4.58	3.96	3.42	2.97

$\log K$, $P = 3$ kb

INDEX	TEMPERATURE, °C									
	200	225	250	300	350	400	450	500	550	600
1123	56.56	54.26	52.18	48.60	45.61	43.08	40.91	39.03	37.40	(36.00)
1124	45.30	43.51	41.90	39.17	36.93	35.07	33.50	32.16	31.02	(30.11)
1125	-1.15	-1.10	-1.06	-0.99	-0.94	-0.92	-0.91	-0.92	-0.94	(-0.96)
1126	-1.57	-1.48	-1.40	-1.25	-1.13	-1.01	-0.91	-0.82	-0.73	(-0.63)
1127	-1.31	-1.24	-1.17	-1.06	-0.96	-0.86	-0.78	-0.70	-0.62	(-0.54)
1128	-1.81	-1.70	-1.60	-1.43	-1.28	-1.15	-1.03	-0.92	-0.82	(-0.72)
1130	8.27	7.12	6.12	4.53	3.30	2.29	1.37	0.55	0.01	(0.30)
1132	7.00	6.20	5.54	4.52	3.80	3.23	2.71	2.25	2.03	(2.62)
1501	-21.57	-20.39	-19.34	-17.55	-16.10	-14.93	-13.98	-13.20	-12.51	(-11.74)
1502	-23.19	-22.51	-21.91	-20.94	-20.24	-19.78	-19.55	-19.49	-19.42	(-19.01)
1503	-65.97	-63.11	-60.57	-56.26	-52.81	-50.09	-48.02	-46.45	-45.02	(-42.99)
1504	-21.00	-19.91	-18.92	-17.21	-15.80	-14.63	-13.68	-12.89	-12.18	(-11.36)
1505	-1.97	-1.85	-1.75	-1.56	-1.40	-1.26	-1.13	-1.02	-0.91	(-0.80)
1506	15.05	13.87	12.83	11.07	9.62	8.36	7.20	6.14	5.29	(5.00)
1507	6.36	5.89	5.50	4.86	4.37	3.97	3.60	3.26	3.04	(3.17)
1508	4.32	4.00	3.73	3.27	2.92	2.62	2.36	2.13	1.96	(1.98)
1509	-2.09	-2.54	-2.91	-3.46	-3.83	-4.14	-4.44	-4.73	-4.83	(-4.28)
1510	13.23	12.46	11.80	10.76	9.99	9.37	8.83	8.36	8.09	(8.39)
1512	38.28	35.52	33.12	29.21	26.12	23.54	21.22	19.13	17.63	(17.69)
1513	35.56	32.86	30.53	26.73	23.73	21.22	18.98	16.96	15.51	(15.62)
1515	21.17	19.14	17.38	14.51	12.24	10.33	8.60	7.02	5.91	(6.07)
1516	10.90	9.91	9.07	7.74	6.72	5.87	5.09	4.38	3.93	(4.26)
1517	38.64	36.54	34.73	31.81	29.57	27.74	26.15	24.75	23.83	(24.19)
1518	39.54	37.04	34.87	31.33	28.57	26.31	24.32	22.58	21.37	(21.53)
1519	21.18	19.15	17.39	14.51	12.24	10.33	8.60	7.02	5.90	(6.06)
1529	29.52	27.54	25.82	22.99	20.75	18.88	17.20	15.68	14.59	(14.63)
1530	32.55	30.91	29.47	27.09	25.19	23.58	22.15	20.86	19.91	(19.83)
1531	1.97	1.73	1.54	1.28	1.13	1.03	0.95	0.87	0.90	(1.31)
1532	18.24	16.84	15.63	13.68	12.16	10.89	9.74	8.70	7.98	(8.22)
1537	6.53	6.10	5.73	5.12	4.63	4.22	3.87	3.55	3.33	(3.29)
1538	55.71	51.61	48.03	42.16	37.51	33.63	30.20	27.14	24.92	(24.83)
1542	291.43	275.86	262.47	240.81	224.06	210.31	198.22	187.50	180.14	(181.68)
1544	9.48	8.72	8.05	6.92	5.98	5.16	4.41	3.71	3.16	(2.98)
1545	22.04	20.21	18.63	16.05	14.01	12.29	10.73	9.32	8.31	(8.46)
1551	17.07	14.84	12.91	9.75	7.24	5.12	3.18	1.40	0.13	(0.29)
1552	-12.35	-11.60	-10.89	-9.59	-8.44	-7.74	-7.13	-6.57	-6.03	(-5.45)
1555	-4.21	-4.38	-4.55	-4.87	-5.21	-5.57	-5.97	-6.40	-6.75	(-6.83)
1558	5.18	5.16	5.13	5.08	4.98	4.83	4.62	4.34	4.11	(4.13)
1559	22.02	20.19	18.60	16.01	13.95	12.22	10.65	9.22	8.20	(8.34)
1560	2.59	2.29	2.04	1.69	1.47	1.31	1.18	1.06	1.07	(1.45)

Log K , P = 3.5 kb

NAME	INDEX	TEMPERATURE, °C							
		0	25	50	75	100	125	150	175
AND	1001	22.91	19.59	16.58	13.99	11.78	9.90	8.29	6.90
KYA	1002	22.07	18.86	15.94	13.43	11.29	9.47	7.92	6.58
SILL	1003	23.22	19.85	16.80	14.18	11.95	10.04	8.41	7.01
KAOL	1004	11.75	10.18	8.41	6.81	5.43	4.26	3.27	2.45
CHRYS	1007	34.20	32.40	29.64	27.16	25.01	23.17	21.59	20.24
MUSC	1012	21.39	18.84	15.99	13.41	11.17	9.26	7.65	6.29
PARAG	1013	25.91	22.82	19.53	16.59	14.05	11.89	10.07	8.53
PHLOG	1014	43.40	40.60	36.74	33.26	30.25	27.66	25.45	23.56
ANNITE	1015	36.23	33.67	30.44	27.46	24.84	22.59	20.66	19.01
DEHYD-ANAL	1021	15.94	14.85	13.44	12.10	10.90	9.86	8.97	8.20
ANAL	1022	10.17	9.59	8.62	7.67	6.81	6.08	5.46	4.93
SEPIO	1023	31.44	32.50	30.91	29.12	27.45	25.99	24.76	23.72
LO-ALB	1025	5.10	5.51	5.09	4.51	3.93	3.41	2.97	2.59
MAX-MICRO	1027	1.58	2.46	2.42	2.15	1.83	1.53	1.27	1.05
K-SPAR	1028	1.59	2.46	2.42	2.16	1.85	1.55	1.29	1.07
HI-SAN	1029	2.98	3.67	3.49	3.09	2.65	2.25	1.90	1.61
ANOR	1030	35.34	31.41	27.52	24.09	21.13	18.59	16.40	14.52
NEPH	1031	17.73	15.98	14.28	12.79	11.50	10.41	9.47	8.68
WOLL	1035	15.07	14.38	13.43	12.53	11.72	11.01	10.39	9.86
DIOP	1039	22.58	21.72	20.11	18.58	17.21	16.01	14.97	14.07
CA-AL PYX	1040	44.84	39.65	34.95	30.93	27.51	24.57	22.05	19.88
AKER	1044	50.27	47.01	43.26	39.90	36.97	34.42	32.21	30.28
MERW	1045	75.58	70.22	64.58	59.62	55.31	51.58	48.34	45.52
MONTI	1046	33.05	30.73	28.15	25.88	23.91	22.21	20.74	19.46
GEHL	1047	68.08	60.91	54.44	48.90	44.18	40.14	36.66	33.65
FORST	1048	31.02	28.73	26.04	23.68	21.65	19.89	18.37	17.05
FAY	1049	22.45	20.64	18.64	16.85	15.28	13.93	12.76	11.47
HUNT	1050	16.34	14.54	12.02	9.80	7.84	6.11	4.57	3.18
ART	1053	21.82	20.30	18.52	17.04	15.80	14.74	13.85	13.08
HED	1054	21.65	20.74	19.24	17.78	16.47	15.31	14.30	13.43
MALACH	1059	1.14	4.00	4.01	3.96	3.92	3.89	3.87	3.85
AZUR	1060	2.98	7.10	6.93	6.70	6.47	6.27	6.08	5.91
HYDRO-MAG	1062	38.27	34.98	30.88	27.35	24.33	21.72	19.44	17.44
SMITH	1064	0.47	0.36	0.00	-0.33	-0.63	-0.89	-1.13	-1.34
ANH-CORD	1065	71.04	63.49	55.42	48.25	42.02	36.65	32.02	28.02
HYD-CORD	1066	67.70	60.45	52.66	45.73	39.71	34.53	30.08	26.24
JADEITE	1067	10.26	9.83	8.97	8.11	7.33	6.65	6.08	5.60
CER	1068	-2.32	-1.76	-1.63	-1.53	-1.44	-1.36	-1.27	-1.19
STRONT	1069	0.32	0.24	0.10	-0.03	-0.15	-0.26	-0.37	-0.48
DIS-DOL	1070	6.86	6.10	5.03	4.07	3.22	2.47	1.79	1.18

$\log K$, $P = 3.5$ kb

INDEX	TEMPERATURE, °C									
	200	225	250	300	350	400	450	500	550	600
1001	5.71	4.69	3.79	2.33	1.17	0.20	-0.67	-1.44	-1.94	(-1.81)
1002	5.43	4.44	3.59	2.18	1.07	0.14	-0.69	-1.43	-1.90	(-1.74)
1003	5.81	4.77	3.86	2.38	1.20	0.22	-0.66	-1.44	-1.95	(-1.82)
1004	1.76	1.18	0.71	-0.02	-0.55	-0.98	-1.36	-1.70	-1.82	(-1.32)
1007	19.08	18.09	17.24	15.86	14.80	13.95	13.22	12.60	12.21	(12.38)
1012	5.15	4.19	3.40	2.16	1.27	0.54	-0.10	-0.66	-0.87	(-0.12)
1013	7.24	6.15	5.24	3.83	2.80	1.98	1.26	0.65	0.41	(1.14)
1014	21.95	20.57	19.39	17.50	16.06	14.90	13.91	13.06	12.59	(12.99)
1015	17.60	16.40	15.37	13.74	12.50	11.50	10.63	9.89	9.50	(9.55)
1021	7.55	6.99	6.52	5.77	5.21	4.77	4.40	4.09	3.94	(4.17)
1022	4.50	4.15	3.86	3.44	3.17	2.98	2.83	2.72	2.75	(3.14)
1023	22.88	22.20	21.65	20.90	20.46	20.20	20.02	19.93	20.11	(21.02)
1025	2.29	2.05	1.86	1.61	1.49	1.42	1.39	1.38	1.51	(2.01)
1027	0.88	0.75	0.66	0.56	0.54	0.56	0.58	0.62	0.77	(1.27)
1028	0.90	0.77	0.67	0.56	0.53	0.53	0.53	0.55	0.67	(1.15)
1029	1.37	1.18	1.03	0.84	0.74	0.69	0.65	0.63	0.74	(1.19)
1030	12.90	11.49	10.26	8.26	6.67	5.35	4.19	3.18	2.51	(2.64)
1031	8.00	7.43	6.94	6.15	5.56	5.09	4.70	4.36	4.18	(4.37)
1035	9.38	8.97	8.61	8.02	7.55	7.16	6.83	6.56	6.37	(6.34)
1039	13.29	12.61	12.02	11.06	10.32	9.71	9.20	8.77	8.50	(8.57)
1040	18.00	16.36	14.92	12.55	10.65	9.05	7.65	6.42	5.55	(5.49)
1044	28.59	27.12	25.82	23.65	21.92	20.49	19.27	18.23	17.45	(17.22)
1045	43.05	40.88	38.97	35.76	33.18	31.03	29.19	27.61	26.41	(25.93)
1046	18.34	17.36	16.50	15.07	13.92	12.97	12.15	11.45	10.93	(10.77)
1047	31.03	28.74	26.74	23.40	20.71	18.46	16.50	14.78	13.53	(13.26)
1048	15.90	14.89	14.01	12.54	11.35	10.37	9.52	8.79	8.25	(8.10)
1049	10.86	10.09	9.42	8.30	7.42	6.68	6.04	5.49	5.10	(5.07)
1050	1.93	0.78	-0.27	-2.17	-3.91	-5.61	-7.35	-9.07	-10.43	(-10.69)
1053	12.42	11.86	11.36	10.54	9.86	9.27	8.70	8.17	7.81	(7.89)
1054	12.67	12.00	11.43	10.49	9.75	9.15	8.64	8.21	7.93	(7.99)
1059	3.84	3.83	3.83	3.82	3.79	3.73	3.62	3.49	3.41	(3.54)
1060	5.75	5.59	5.45	5.16	4.86	4.51	4.09	3.63	3.27	(3.31)
1062	15.67	14.10	12.68	10.22	8.06	6.05	4.06	2.14	0.67	(0.52)
1064	-1.53	-1.71	-1.87	-2.17	-2.46	-2.76	-3.08	-3.42	-3.68	(-3.68)
1065	24.57	21.57	18.96	14.68	11.28	8.46	5.97	3.82	2.38	(2.64)
1066	22.92	20.06	17.57	13.51	10.31	7.65	5.32	3.30	1.99	(2.36)
1067	5.21	4.88	4.62	4.24	4.00	3.83	3.69	3.59	3.63	(4.03)
1068	-1.11	-1.03	-0.96	-0.82	-0.72	-0.66	-0.65	-0.67	-0.65	(-0.42)
1069	-0.57	-0.66	-0.75	-0.93	-1.11	-1.32	-1.56	-1.82	-2.01	(-1.98)
1070	0.63	0.13	-0.34	-1.19	-1.97	-2.75	-3.55	-4.35	-4.99	(-5.08)

$\log K$, $P = 3.5$ kb

NAME	INDEX	TEMPERATURE, °C							
		0	25	50	75	100	125	150	175
ORD-DOL	1071	5.11	4.55	3.65	2.83	2.10	1.46	0.88	0.36
ARAG	1072	3.27	2.98	2.58	2.22	1.90	1.60	1.34	1.10
CALC	1073	3.29	2.98	2.58	2.21	1.87	1.57	1.30	1.05
MAG	1074	3.66	3.23	2.59	2.03	1.53	1.09	0.70	0.36
DOL	1075	5.12	4.55	3.65	2.83	2.11	1.46	0.88	0.36
SIDER	1076	1.51	1.09	0.59	0.14	-0.27	-0.63	-0.95	-1.25
ANHYD	1078	-2.64	-2.48	-2.80	-3.18	-3.57	-3.95	-4.33	-4.69
FLUORITE	1079	-9.24	-8.24	-8.17	-8.16	-8.20	-8.26	-8.33	-8.43
BARITE	1080	-9.23	-8.44	-8.14	-8.00	-7.95	-7.96	-8.01	-8.10
ANG	1081	-6.86	-5.98	-5.87	-5.87	-5.92	-5.99	-6.07	-6.16
CELEST	1082	-5.01	-4.71	-4.82	-5.01	-5.24	-5.49	-5.74	-6.01
ALUN	1083	8.53	5.66	2.68	0.05	-2.25	-4.28	-6.07	-7.68
WITH	1084	-2.09	-1.83	-1.68	-1.58	-1.52	-1.49	-1.47	-1.47
RHODO	1085	0.74	0.76	0.45	0.16	-0.10	-0.33	-0.53	-0.71
COV	1086	-27.53	-23.75	-21.86	-20.30	-18.98	-17.86	-16.88	-16.02
GALENA	1087	-15.57	-13.80	-12.70	-11.79	-11.01	-10.33	-9.73	-9.19
SL	1088	-11.35	-10.42	-9.97	-9.64	-9.37	-9.14	-8.96	-8.80
WURT	1089	-8.84	-8.12	-7.85	-7.67	-7.53	-7.42	-7.34	-7.27
M-CINN	1090	-41.13	-37.42	-34.70	-32.42	-30.46	-28.74	-27.24	-25.90
CINN	1091	-41.68	-37.90	-35.12	-32.78	-30.77	-29.02	-27.48	-26.11
ALA	1092	0.04	0.14	-0.17	-0.48	-0.77	-1.03	-1.27	-1.49
PYRITE	1093	-26.03	-23.85	-22.38	-21.24	-20.33	-19.59	-18.99	-18.50
GOLD	1100	-28.33	-25.48	-23.08	-21.01	-19.21	-17.63	-16.23	-14.97
SILVER	1101	-11.05	-9.62	-8.55	-7.64	-6.83	-6.11	-5.46	-4.87
COPPER	1102	-6.53	-4.50	-3.91	-3.39	-2.93	-2.51	-2.11	-1.74
GRAPHITE	1103	-5.56	-4.99	-4.43	-3.96	-3.57	-3.23	-2.94	-2.69
HALITE	1106	1.61	1.82	1.90	1.93	1.93	1.92	1.89	1.86
SYLVITE	1107	0.74	1.18	1.44	1.61	1.72	1.79	1.84	1.86
COR	1108	28.25	23.94	20.36	17.37	14.86	12.74	10.93	9.38
PER	1109	23.79	21.68	19.65	17.93	16.47	15.21	14.12	13.18
LIME	1110	36.05	33.07	30.43	28.19	26.26	24.59	23.14	21.87
TO	1111	5.48	6.49	6.15	5.86	5.61	5.40	5.23	5.09
CUP	1112	-4.55	-1.28	-0.84	-0.47	-0.11	0.23	0.56	0.89
FE-OXIDE	1113	15.76	14.17	12.72	11.49	10.44	9.53	8.75	8.07
GIBBS	1114	10.46	8.85	7.54	6.47	5.60	4.88	4.28	3.78
DIAS	1115	11.66	9.81	8.29	7.03	5.97	5.09	4.35	3.72
BOEH	1116	12.76	10.76	9.11	7.74	6.60	5.63	4.82	4.12
BRUC	1117	17.91	16.42	14.93	13.68	12.63	11.74	10.97	10.32
MANGAN	1119	19.78	18.21	16.60	15.22	14.06	13.06	12.21	11.47
SPINEL	1120	47.37	41.32	36.04	31.61	27.88	24.71	21.99	19.65

$\log K$, $P = 3.5$ kb

INDEX	TEMPERATURE, °C									
	200	225	250	300	350	400	450	500	550	600
1071	-0.12	-0.56	-0.96	-1.71	-2.40	-3.10	-3.84	-4.58	-5.16	(-5.22)
1072	0.88	0.68	0.50	0.16	-0.17	-0.49	-0.84	-1.18	-1.46	(-1.48)
1073	0.83	0.62	0.42	0.07	-0.27	-0.61	-0.97	-1.33	-1.61	(-1.64)
1074	0.04	-0.24	-0.51	-0.98	-1.42	-1.85	-2.28	-2.71	-3.05	(-3.11)
1075	-0.12	-0.55	-0.96	-1.70	-2.40	-3.11	-3.85	-4.59	-5.18	(-5.24)
1076	-1.52	-1.76	-1.99	-2.41	-2.80	-3.19	-3.59	-4.00	-4.33	(-4.39)
1078	-5.05	-5.40	-5.75	-6.42	-7.10	-7.81	-8.58	-9.36	-10.02	(-10.25)
1079	-8.53	-8.65	-8.77	-9.04	-9.37	-9.77	-10.25	-10.78	-11.20	(-11.17)
1080	-8.21	-8.35	-8.50	-8.85	-9.26	-9.74	-10.31	-10.92	-11.43	(-11.56)
1081	-6.26	-6.37	-6.48	-6.72	-7.00	-7.35	-7.77	-8.24	-8.60	(-8.57)
1082	-6.27	-6.54	-6.80	-7.35	-7.91	-8.53	-9.20	-9.90	-10.49	(-10.67)
1083	-9.12	-10.43	-11.63	-13.78	-15.75	-17.70	-19.73	-21.74	-23.28	(-23.32)
1084	-1.47	-1.48	-1.50	-1.56	-1.65	-1.78	-1.96	-2.17	-2.33	(-2.28)
1085	-0.87	-1.01	-1.15	-1.39	-1.61	-1.85	-2.12	-2.39	-2.59	(-2.53)
1086	-15.27	-14.61	-14.02	-13.04	-12.26	-11.65	-11.19	-10.84	-10.51	(-10.07)
1087	-8.70	-8.26	-7.86	-7.17	-6.59	-6.14	-5.78	-5.51	-5.22	(-4.74)
1088	-8.66	-8.55	-8.45	-8.29	-8.21	-8.18	-8.24	-8.34	-8.38	(-8.18)
1089	-7.22	-7.18	-7.14	-7.11	-7.11	-7.18	-7.30	-7.46	-7.56	(-7.40)
1090	-24.71	-23.63	-22.66	-20.98	-19.59	-18.44	-17.51	-16.74	-16.02	(-15.15)
1091	-24.88	-23.78	-22.79	-21.06	-19.64	-18.47	-17.51	-16.72	-15.98	(-15.10)
1092	-1.68	-1.87	-2.04	-2.35	-2.64	-2.95	-3.27	-3.60	-3.85	(-3.83)
1093	-18.96	-17.62	-17.34	-16.95	-16.75	-16.72	-16.84	-17.06	-17.20	(-16.95)
1100	-13.84	-12.82	-11.88	-10.24	-8.84	-7.62	-6.56	-5.62	-4.79	(-4.04)
1101	-4.32	-3.83	-3.37	-2.54	-1.82	-1.18	-0.60	-0.08	0.39	(0.83)
1102	-1.40	-1.07	-0.76	-0.19	0.33	0.80	1.24	1.64	2.02	(2.38)
1103	-2.48	-2.30	-2.14	-1.88	-1.69	-1.56	-1.46	-1.40	-1.53	(-1.28)
1106	1.82	1.78	1.73	1.63	1.51	1.36	1.18	0.99	0.84	(0.85)
1107	1.86	1.85	1.83	1.75	1.64	1.48	1.29	1.08	0.89	(0.86)
1108	8.03	6.87	5.86	4.18	2.83	1.69	0.68	-0.22	-0.85	(-0.85)
1109	12.35	11.61	10.96	9.86	8.97	8.21	7.56	6.99	6.54	(6.30)
1110	20.74	19.75	18.85	17.33	16.08	15.02	14.11	13.33	12.67	(12.22)
1111	4.97	4.88	4.80	4.69	4.61	4.56	4.52	4.49	4.49	(4.55)
1112	1.21	1.53	1.83	2.42	2.97	3.49	3.97	4.41	4.86	(5.34)
1113	7.48	6.96	6.50	5.72	5.09	4.56	4.10	3.70	3.39	(3.26)
1114	3.37	3.02	2.73	2.27	1.92	1.63	1.36	1.12	0.98	(1.13)
1115	3.18	2.73	2.33	1.70	1.20	0.79	0.43	0.11	-0.09	(0.02)
1116	3.53	3.02	2.58	1.86	1.30	0.83	0.42	0.05	-0.19	(-0.11)
1117	9.75	9.26	8.83	8.11	7.54	7.07	6.66	6.31	6.05	(5.97)
1119	10.83	10.27	9.78	8.97	8.32	7.78	7.33	6.94	6.65	(6.54)
1120	17.62	15.86	14.32	11.74	9.68	7.93	6.39	5.03	4.04	(3.89)

$\log K$, $P = 3.5$ kb

NAME	INDEX	TEMPERATURE, °C							
		0	25	50	75	100	125	150	175
K-OXIDE	1123	90.55	83.80	78.07	73.16	68.92	65.21	61.96	59.08
NA-OXIDE	1124	72.93	67.33	62.61	58.60	55.16	52.18	49.58	47.30
AMORPH SIL	1125	-2.74	-2.06	-1.73	-1.53	-1.38	-1.28	-1.19	-1.12
A-CRIST	1126	-3.86	-3.00	-2.54	-2.24	-2.02	-1.85	-1.72	-1.60
B-CRIST	1127	-3.23	-2.45	-2.06	-1.81	-1.64	-1.51	-1.40	-1.31
CHALCED	1128	-4.38	-3.46	-2.96	-2.62	-2.37	-2.17	-2.01	-1.87
WAIR	1130	25.96	23.83	20.94	18.19	15.75	13.62	11.78	10.21
LAUM	1132	19.85	18.61	16.50	14.42	12.57	10.98	9.61	8.47
ACAN	1501	-40.41	-36.15	-33.00	-30.36	-28.10	-26.14	-24.42	-22.91
CP	1502	-36.85	-32.47	-30.27	-28.50	-27.03	-25.78	-24.71	-23.79
BN	1503	-117.80	-101.08	-93.31	-86.91	-81.44	-76.71	-72.57	-68.91
CC	1504	-40.38	-34.13	-31.27	-28.89	-26.84	-25.08	-23.54	-22.17
QTZ	1505	-4.68	-3.74	-3.21	-2.85	-2.59	-2.38	-2.20	-2.05
MG	1506	33.95	30.47	27.18	24.36	21.96	19.90	18.12	16.59
KALS	1507	14.17	12.89	11.56	10.38	9.35	8.47	7.72	7.09
FERROSIL	1508	8.64	8.24	7.54	6.85	6.23	5.68	5.19	4.78
PYROPH	1509	4.18	4.26	3.33	2.25	1.23	0.32	-0.46	-1.11
TALC	1510	22.75	22.90	21.25	19.51	17.92	16.51	15.29	14.25
7-A CL	1512	83.22	75.89	67.82	60.83	54.87	49.79	45.46	41.75
14-A CL	1513	79.33	72.25	64.37	57.55	51.73	46.77	42.54	38.93
CLNZ	1515	54.42	48.72	42.99	37.95	33.62	29.91	26.74	24.02
LAWS	1516	27.83	24.99	22.04	19.43	17.20	15.30	13.69	12.32
TREM	1517	66.22	64.78	60.03	55.33	51.08	47.35	44.11	41.31
ANTH	1518	72.89	70.52	64.72	59.13	54.12	49.73	45.92	42.62
ZOIS	1519	54.45	48.74	43.01	37.97	33.64	29.93	26.75	24.03
GROSS	1529	60.87	55.73	50.34	45.55	41.41	37.85	34.79	32.16
ANDRA	1530	56.51	53.31	49.22	45.47	42.17	39.30	36.81	34.66
ALB	1531	5.10	5.51	5.09	4.51	3.93	3.41	2.97	2.59
PREHN	1532	40.76	37.32	33.47	30.00	26.99	24.40	22.18	20.29
ENST	1537	12.38	11.77	10.75	9.80	8.96	8.22	7.59	7.03
PARG	1538	119.85	109.77	98.59	88.70	80.15	72.80	66.47	61.03
ANTIG	1542	523.14	497.11	454.88	416.67	383.49	354.98	330.57	309.67
HM	1544	21.50	19.33	17.23	15.44	13.90	12.59	11.45	10.47
ORD-EP	1545	51.09	46.45	41.46	37.01	33.17	29.86	27.02	24.58
MARG	1551	55.54	48.21	41.53	35.81	30.95	26.82	23.30	20.29
MERCURY	1552	-21.91	-19.88	-18.42	-17.14	-16.00	-14.96	-14.01	-13.13
PO	1555	-3.05	-2.96	-3.09	-3.25	-3.42	-3.58	-3.75	-3.93
NESQ	1558	6.85	5.67	5.49	5.41	5.35	5.30	5.27	5.24
EPID	1559	51.09	46.45	41.46	37.01	33.16	29.85	27.01	24.57
HI-ALB	1560	6.62	6.85	6.28	5.57	4.88	4.27	3.73	3.28

$\log K$, $P = 3.5$ kb

INDEX	TEMPERATURE, °C									
	200	225	250	300	350	400	450	500	550	600
1123	56.51	54.21	52.14	48.56	45.58	43.06	40.89	39.02	37.39	(36.00)
1124	45.28	43.49	41.89	39.17	36.94	35.08	33.52	32.19	31.06	(30.16)
1125	-1.06	-1.02	-0.97	-0.91	-0.87	-0.85	-0.84	-0.86	-0.88	(-0.89)
1126	-1.50	-1.41	-1.33	-1.19	-1.07	-0.96	-0.85	-0.76	-0.67	(-0.57)
1127	-1.23	-1.16	-1.10	-0.99	-0.89	-0.80	-0.72	-0.64	-0.56	(-0.47)
1128	-1.75	-1.65	-1.55	-1.38	-1.23	-1.10	-0.98	-0.88	-0.77	(-0.67)
1130	8.86	7.70	6.71	5.13	3.92	2.95	2.10	1.38	0.99	(1.41)
1132	7.50	6.70	6.04	5.04	4.34	3.81	3.36	3.00	2.92	(3.64)
1501	-21.58	-20.39	-19.33	-17.52	-16.05	-14.86	-13.88	-13.07	-12.33	(-11.53)
1502	-22.99	-22.29	-21.68	-20.68	-19.94	-19.44	-19.14	-19.00	-18.83	(-18.30)
1503	-65.66	-62.77	-60.20	-55.83	-52.31	-49.50	-48.31	-46.58	-43.93	(-41.71)
1504	-20.95	-19.85	-18.85	-17.13	-15.70	-14.50	-13.53	-12.70	-11.94	(-11.08)
1505	-1.92	-1.80	-1.70	-1.51	-1.35	-1.21	-1.08	-0.97	-0.86	(-0.75)
1506	15.25	14.07	13.04	11.30	9.87	8.56	7.55	6.57	5.82	(5.62)
1507	6.54	6.08	5.69	5.07	4.59	4.21	3.87	3.59	3.43	(3.62)
1508	4.41	4.10	3.83	3.38	3.03	2.75	2.50	2.29	2.16	(2.21)
1509	-1.66	-2.11	-2.47	-3.00	-3.35	-3.61	-3.85	-4.04	-4.02	(-3.36)
1510	13.35	12.59	11.95	10.94	10.21	9.63	9.16	8.77	8.61	(9.01)
1512	38.58	35.84	33.48	29.64	26.64	24.16	22.01	20.15	18.94	(19.27)
1513	35.83	33.16	30.87	27.13	24.23	21.84	19.75	17.95	16.80	(17.18)
1515	21.68	19.67	17.93	15.11	12.91	11.08	9.48	8.09	7.21	(7.60)
1516	11.17	10.19	9.36	8.06	7.08	6.28	5.59	4.99	4.69	(5.16)
1517	38.91	36.84	35.06	32.21	30.05	28.32	26.86	25.65	24.97	(25.55)
1518	39.77	37.30	35.17	31.71	29.04	26.88	25.04	23.50	22.54	(22.94)
1519	21.69	19.68	17.94	15.11	12.91	11.08	9.48	8.08	7.21	(7.59)
1529	29.89	27.92	26.22	23.44	21.27	19.47	17.90	16.55	15.67	(15.89)
1530	32.79	31.16	29.74	27.40	25.54	24.00	22.65	21.49	20.70	(20.76)
1531	2.29	2.05	1.86	1.61	1.47	1.40	1.34	1.32	1.42	(1.89)
1532	18.67	17.28	16.08	14.16	12.69	11.48	10.42	9.51	8.97	(9.38)
1537	6.55	6.13	5.76	5.16	4.69	4.30	3.96	3.68	3.49	(3.49)
1538	56.33	52.26	48.73	42.95	38.42	34.69	31.47	28.71	26.88	(27.14)
1542	291.79	276.47	263.31	242.15	225.96	212.89	201.74	192.32	186.64	(189.70)
1544	9.61	8.86	8.20	7.08	6.16	5.37	4.65	4.01	3.52	(3.42)
1545	22.49	20.68	19.12	16.58	14.60	12.95	11.51	10.26	9.48	(9.82)
1551	17.72	15.50	13.59	10.48	8.04	6.01	4.21	2.63	1.63	(2.04)
1552	-12.32	-11.58	-10.85	-9.55	-8.40	-7.69	-7.08	-6.51	-5.96	(-5.37)
1555	-4.10	-4.26	-4.42	-4.73	-5.05	-5.38	-5.74	-6.12	-6.41	(-6.42)
1558	5.22	5.21	5.19	5.14	5.07	4.94	4.76	4.53	4.35	(4.42)
1559	22.47	20.66	19.09	16.54	14.54	12.88	11.43	10.16	9.36	(9.69)
1560	2.91	2.61	2.36	2.02	1.81	1.68	1.57	1.51	1.59	(2.04)

Log K , P = 4 kb

NAME	INDEX	TEMPERATURE, °C							
		0	25	50	75	100	125	150	175
AND	1001	23.04	19.84	16.87	14.30	12.09	10.21	8.60	7.21
KYA	1002	22.13	19.04	16.17	13.68	11.55	9.73	8.18	6.85
SILL	1003	23.33	20.09	17.08	14.48	12.25	10.34	8.71	7.31
KAOL	1004	11.73	10.36	8.66	7.09	5.72	4.55	3.57	2.74
CHRYS	1007	33.46	32.11	29.44	27.01	24.89	23.07	21.51	20.18
MUSC	1012	21.28	19.07	16.35	13.81	11.60	9.70	8.09	6.74
PARAG	1013	25.79	23.05	19.90	17.01	14.49	12.35	10.53	9.00
PHLOG	1014	42.56	40.36	36.64	33.24	30.26	27.71	25.52	23.65
ANNITE	1015	35.79	33.75	30.66	27.74	25.16	22.92	20.99	19.35
DEHYD-ANAL	1021	15.94	15.04	13.70	12.39	11.20	10.17	9.27	8.51
ANAL	1022	10.08	9.71	8.82	7.90	7.06	6.33	5.70	5.18
SEPIO	1023	30.14	32.05	30.67	28.97	27.37	25.95	24.74	23.73
LO-ALB	1025	4.90	5.59	5.28	4.75	4.19	3.68	3.24	2.87
MAX-MICRO	1027	1.39	2.54	2.61	2.38	2.08	1.79	1.53	1.32
K-SPAR	1028	1.39	2.54	2.61	2.39	2.10	1.81	1.55	1.34
HI-SAN	1029	2.79	3.76	3.67	3.32	2.91	2.52	2.17	1.88
ANOR	1030	35.38	31.71	27.89	24.49	21.54	19.00	16.81	14.93
NEPH	1031	17.68	16.08	14.43	12.96	11.68	10.59	9.66	8.87
WOLL	1035	14.90	14.35	13.43	12.55	11.75	11.05	10.43	9.90
DIOP	1039	22.07	21.53	20.00	18.51	17.17	15.99	14.96	14.07
CA-AL PYX	1040	44.83	39.83	35.19	31.19	27.77	24.84	22.33	20.16
AKER	1044	49.76	46.88	43.22	39.90	36.99	34.45	32.25	30.33
MERW	1045	74.93	70.02	64.47	59.56	55.28	51.57	48.35	45.54
MONTI	1046	32.70	30.62	28.09	25.84	23.89	22.20	20.73	19.46
GEHL	1047	68.07	61.16	54.74	49.22	44.50	40.46	36.98	33.98
FORST	1048	30.55	28.55	25.91	23.58	21.56	19.82	18.31	17.01
FAY	1049	22.25	20.66	18.72	16.95	15.39	14.04	12.88	11.86
HUNT	1050	15.98	14.66	12.21	10.01	8.07	6.36	4.83	3.47
ART	1053	21.45	20.16	18.42	16.95	15.71	14.67	13.78	13.03
HED	1054	21.27	20.66	19.24	16.53	17.82	16.53	15.38	14.38
MALACH	1059	0.04	3.97	4.02	3.99	3.96	3.93	3.92	3.90
AZUR	1060	1.43	7.16	7.07	6.86	6.64	6.44	6.26	6.08
HYDRO-MAG	1062	37.64	34.96	30.93	27.44	24.44	21.85	19.58	17.61
SMITH	1064	0.17	0.25	-0.07	-0.38	-0.66	-0.92	-1.14	-1.35
ANH-CORD	1065	71.06	64.15	56.25	49.13	42.92	37.55	32.92	28.92
HYD-CORD	1066	67.64	61.05	53.42	46.54	40.55	35.38	30.92	27.08
JADEITE	1067	9.98	9.77	9.00	8.18	7.43	6.77	6.21	5.74
CER	1068	-2.38	-1.64	-1.49	-1.39	-1.30	-1.21	-1.13	-1.05
STRONT	1069	0.33	0.32	0.20	0.07	-0.05	-0.16	-0.27	-0.37
DIS-DOL	1070	6.73	6.19	5.15	4.20	3.36	2.61	1.95	1.35

$\log K$, $P = 4$ kb

INDEX	TEMPERATURE, °C									
	200	225	250	300	350	400	450	500	550	600
1001	6.02	5.00	4.11	2.66	1.52	0.57	-0.26	-0.97	-1.40	(-1.19)
1002	5.70	4.72	3.87	2.48	1.39	0.49	-0.31	-0.98	-1.38	(-1.15)
1003	6.11	5.07	4.17	2.70	1.54	0.58	-0.26	-0.97	-1.41	(-1.21)
1004	2.05	1.48	1.01	0.29	-0.22	-0.62	-0.97	-1.25	-1.30	(-0.73)
1007	19.04	18.06	17.22	15.87	14.84	14.02	13.33	12.76	12.45	(12.68)
1012	5.60	4.66	3.87	2.66	1.79	1.10	0.52	0.05	-0.04	(0.81)
1013	7.71	6.63	5.73	4.34	3.34	2.56	1.90	1.38	1.25	(2.10)
1014	22.05	20.69	19.53	17.68	16.29	15.18	14.25	13.49	13.13	(13.64)
1015	17.95	16.75	15.73	14.12	12.90	11.94	11.12	10.46	10.17	(10.71)
1021	7.85	7.29	6.82	6.07	5.52	5.08	4.73	4.45	4.35	(4.61)
1022	4.75	4.40	4.11	3.69	3.43	3.25	3.11	3.04	3.10	(3.53)
1023	22.90	22.23	21.70	20.98	20.57	20.33	20.20	20.18	20.45	(21.43)
1025	2.57	2.33	2.14	1.90	1.78	1.73	1.71	1.74	1.91	(2.45)
1027	1.15	1.02	0.93	0.84	0.83	0.85	0.90	0.96	1.16	(1.69)
1028	1.17	1.04	0.94	0.84	0.81	0.82	0.84	0.89	1.06	(1.57)
1029	1.64	1.45	1.30	1.11	1.02	0.98	0.96	0.98	1.12	(1.62)
1030	13.30	11.90	10.68	8.68	7.12	5.83	4.71	3.77	3.19	(3.41)
1031	8.19	7.62	7.13	6.36	5.78	5.32	4.94	4.64	4.50	(4.73)
1035	9.43	9.02	8.66	8.07	7.61	7.23	6.91	6.65	6.48	(6.47)
1039	13.30	12.63	12.05	11.11	10.39	9.80	9.32	8.92	8.69	(8.81)
1040	18.28	16.65	15.22	12.87	10.99	9.43	8.08	6.92	6.15	(6.17)
1044	28.66	27.19	25.90	23.76	22.05	20.65	19.46	18.46	17.75	(17.57)
1045	43.08	40.93	39.03	35.85	33.30	31.19	29.40	27.88	26.76	(26.34)
1046	18.35	17.38	16.53	15.11	13.98	13.05	12.25	11.59	11.11	(10.99)
1047	31.36	29.09	27.09	23.77	21.12	18.90	17.00	15.37	14.23	(14.06)
1048	15.87	14.87	14.00	12.55	11.39	10.43	9.61	8.92	8.43	(8.32)
1049	10.98	10.22	9.55	8.46	7.57	6.84	6.23	5.71	5.36	(5.36)
1050	2.23	1.11	0.09	-1.76	-3.44	-5.06	-6.70	-8.28	-9.47	(-9.58)
1053	12.38	11.82	11.33	10.53	9.88	9.32	8.79	8.31	8.00	(8.14)
1054	12.76	12.10	11.53	10.60	9.88	9.30	8.81	8.41	8.17	(8.27)
1059	3.90	3.89	3.89	3.89	3.88	3.83	3.75	3.64	3.59	(3.76)
1060	5.93	5.78	5.64	5.37	5.09	4.76	4.38	3.96	3.67	(3.76)
1062	15.87	14.32	12.92	10.53	8.44	6.51	4.64	2.87	1.58	(1.60)
1064	-1.53	-1.70	-1.85	-2.13	-2.40	-2.67	-2.97	-3.27	-3.49	(-3.44)
1065	25.47	22.47	19.87	15.61	12.26	9.50	7.11	5.10	3.86	(4.29)
1066	23.77	20.91	18.44	14.40	11.24	8.65	6.41	4.54	3.42	(3.97)
1067	5.35	5.03	4.78	4.41	4.18	4.03	3.92	3.86	3.94	(4.38)
1068	-0.96	-0.89	-0.81	-0.67	-0.56	-0.48	-0.46	-0.45	-0.40	(-0.14)
1069	-0.46	-0.55	-0.64	-0.80	-0.98	-1.17	-1.39	-1.62	-1.78	(-1.71)
1070	0.80	0.31	-0.15	-0.97	-1.73	-2.47	-3.22	-3.96	-4.51	(-4.53)

$\log K$, $P = 4$ kb

NAME	INDEX	TEMPERATURE, °C							
		0	25	50	75	100	125	150	175
ORD-DOL	1071	4.98	4.64	3.76	2.97	2.24	1.61	1.03	0.52
ARAG	1072	3.24	3.04	2.66	2.30	1.98	1.69	1.43	1.20
CALC	1073	3.29	3.07	2.68	2.31	1.98	1.68	1.41	1.16
MAG	1074	3.54	3.24	2.62	2.06	1.57	1.14	0.75	0.41
DOL	1075	4.98	4.64	3.77	2.96	2.25	1.61	1.04	0.52
SIDER	1076	1.51	1.20	0.72	0.27	-0.13	-0.49	-0.82	-1.11
ANHYD	1078	-2.79	-2.42	-2.71	-3.08	-3.46	-3.84	-4.21	-4.57
FLUORITE	1079	-9.46	-8.13	-8.04	-8.04	-8.07	-8.13	-8.20	-8.29
BARITE	1080	-9.47	-8.46	-8.12	-7.95	-7.89	-7.89	-7.93	-8.01
ANG	1081	-7.09	-5.90	-5.76	-5.75	-5.78	-5.85	-5.93	-6.01
CELEST	1082	-5.16	-4.66	-4.75	-4.93	-5.15	-5.39	-5.64	-5.90
ALUN	1083	8.39	5.95	3.06	0.46	-1.82	-3.84	-5.62	-7.21
WITH	1084	-2.15	-1.80	-1.62	-1.52	-1.45	-1.41	-1.38	-1.37
RHODO	1085	0.62	0.81	0.53	0.24	-0.01	-0.24	-0.43	-0.61
COV	1086	-28.14	-23.77	-21.84	-20.27	-18.95	-17.81	-16.83	-15.97
GALENA	1087	-15.74	-13.74	-12.61	-11.69	-10.91	-10.23	-9.62	-9.08
SL	1088	-11.51	-10.38	-9.90	-9.55	-9.28	-9.05	-8.86	-8.70
WURT	1089	-8.99	-8.08	-7.78	-7.58	-7.44	-7.33	-7.24	-7.17
M-CINN	1090	-41.29	-37.34	-34.60	-32.30	-30.34	-28.62	-27.12	-25.78
CINN	1091	-41.86	-37.83	-35.02	-32.68	-30.66	-28.91	-27.37	-25.99
ALA	1092	-0.19	0.12	-0.15	-0.45	-0.73	-0.98	-1.21	-1.43
PYRITE	1093	-26.26	-23.85	-22.32	-21.16	-20.23	-19.48	-18.87	-18.37
GOLD	1100	-28.59	-25.70	-23.26	-21.18	-19.37	-17.77	-16.36	-15.10
SILVER	1101	-11.27	-9.73	-8.64	-7.72	-6.90	-6.17	-5.52	-4.93
COPPER	1102	-7.10	-4.56	-3.95	-3.43	-2.96	-2.54	-2.14	-1.77
GRAPHITE	1103	-5.50	-4.95	-4.38	-3.91	-3.51	-3.17	-2.88	-2.64
HALITE	1106	1.49	1.79	1.90	1.95	1.96	1.96	1.94	1.91
SYLVITE	1107	0.64	1.17	1.45	1.64	1.76	1.84	1.88	1.91
COR	1108	28.43	24.17	20.60	17.62	15.11	12.99	11.18	9.62
PER	1109	23.61	21.60	19.58	17.87	16.41	15.16	14.08	13.14
LIME	1110	35.96	33.04	30.41	28.17	26.25	24.58	23.14	21.86
TO	1111	4.91	6.44	6.13	5.83	5.59	5.38	5.22	5.08
CUP	1112	-5.64	-1.36	-0.88	-0.50	-0.14	0.21	0.55	0.88
FE-OXIDE	1113	15.70	14.19	12.76	11.53	10.47	9.57	8.79	8.11
GIBBS	1114	10.50	8.92	7.62	6.55	5.68	4.96	4.36	3.86
DIAS	1115	11.72	9.90	8.39	7.13	6.08	5.19	4.45	3.82
BOEH	1116	12.84	10.87	9.23	7.86	6.71	5.75	4.93	4.24
BRUC	1117	17.70	16.31	14.84	13.59	12.55	11.66	10.91	10.25
MANGAN	1119	19.59	18.17	16.57	15.21	14.05	13.05	12.20	11.47
SPINEL	1120	47.39	41.50	36.24	31.82	28.09	24.92	22.21	19.87

$\log K$, $P = 4$ kb

INDEX	TEMPERATURE, °C									
	200	225	250	300	350	400	450	500	550	600
1071	0.05	-0.38	-0.77	-1.49	-2.15	-2.82	-3.51	-4.18	-4.68	(-4.66)
1072	0.98	0.79	0.61	0.27	-0.04	-0.35	-0.67	-0.98	-1.22	(-1.21)
1073	0.94	0.73	0.54	0.20	-0.13	-0.45	-0.79	-1.12	-1.37	(-1.37)
1074	0.10	-0.18	-0.43	-0.89	-1.31	-1.72	-2.13	-2.52	-2.82	(-2.84)
1075	0.05	-0.37	-0.77	-1.49	-2.15	-2.82	-3.51	-4.20	-4.70	(-4.69)
1076	-1.38	-1.62	-1.84	-2.25	-2.63	-3.00	-3.39	-3.77	-4.05	(-4.07)
1078	-4.92	-5.26	-5.60	-6.25	-6.91	-7.59	-8.31	-9.04	-9.63	(-9.80)
1079	-8.39	-8.49	-8.60	-8.86	-9.16	-9.52	-9.96	-10.44	-10.78	(-10.69)
1080	-8.11	-8.24	-8.38	-8.70	-9.09	-9.55	-10.07	-10.63	-11.08	(-11.15)
1081	-6.11	-6.21	-6.31	-6.53	-6.80	-7.12	-7.50	-7.91	-8.21	(-8.13)
1082	-6.16	-6.41	-6.67	-7.19	-7.73	-8.32	-8.95	-9.60	-10.12	(-10.24)
1083	-8.64	-9.93	-11.10	-13.20	-15.10	-16.96	-18.86	-20.71	-22.04	(-21.88)
1084	-1.37	-1.38	-1.40	-1.45	-1.53	-1.65	-1.81	-1.98	-2.11	(-2.03)
1085	-0.77	-0.91	-1.03	-1.26	-1.48	-1.70	-1.94	-2.18	-2.34	(-2.25)
1086	-15.22	-14.55	-13.96	-12.95	-12.16	-11.54	-11.06	-10.69	-10.33	(-9.86)
1087	-8.59	-8.15	-7.74	-7.04	-6.45	-5.98	-5.61	-5.31	-4.98	(-4.47)
1088	-8.56	-8.44	-8.33	-8.17	-8.06	-8.02	-8.05	-8.12	-8.12	(-7.87)
1089	-7.11	-7.07	-7.03	-6.98	-6.97	-7.02	-7.12	-7.24	-7.30	(-7.10)
1090	-24.58	-23.50	-22.52	-20.83	-19.43	-18.28	-17.32	-16.52	-15.77	(-14.87)
1091	-24.77	-23.66	-22.66	-20.93	-19.50	-18.31	-17.33	-16.51	-15.73	(-14.82)
1092	-1.62	-1.79	-1.96	-2.25	-2.53	-2.81	-3.11	-3.41	-3.62	(-3.55)
1093	-17.96	-17.62	-17.34	-16.95	-16.75	-16.72	-16.84	-17.06	-17.20	(-16.95)
1100	-13.96	-12.93	-11.99	-10.34	-8.93	-7.71	-6.64	-5.70	-4.86	(-4.11)
1101	-4.38	-3.88	-3.42	-2.59	-1.86	-1.22	-0.64	-0.12	0.35	(0.79)
1102	-1.42	-1.09	-0.78	-0.21	0.31	0.78	1.22	1.62	2.00	(2.37)
1103	-2.42	-2.24	-2.08	-1.82	-1.62	-1.49	-1.39	-1.32	-1.27	(-1.20)
1106	1.88	1.84	1.80	1.71	1.60	1.46	1.30	1.13	1.00	(1.04)
1107	1.92	1.91	1.89	1.82	1.72	1.58	1.40	1.20	1.04	(1.03)
1108	8.28	7.12	6.11	4.45	3.12	2.01	1.04	0.20	-0.36	(-0.28)
1109	12.31	11.58	10.94	9.85	8.97	8.22	7.59	7.04	6.61	(6.39)
1110	20.74	19.75	18.86	17.34	16.10	15.05	14.15	13.38	12.74	(12.31)
1111	4.96	4.87	4.80	4.68	4.61	4.56	4.53	4.51	4.51	(4.58)
1112	1.20	1.52	1.82	2.41	2.97	3.49	3.97	4.43	4.88	(5.37)
1113	7.52	7.00	6.54	5.77	5.15	4.63	4.18	3.79	3.50	(3.39)
1114	3.45	3.10	2.81	2.36	2.02	1.75	1.50	1.28	1.18	(1.37)
1115	3.29	2.83	2.44	1.81	1.33	0.93	0.59	0.30	0.13	(0.28)
1116	3.64	3.13	2.70	1.99	1.43	0.98	0.58	0.25	0.05	(0.16)
1117	9.69	9.20	8.78	8.07	7.51	7.05	6.66	6.32	6.08	(6.03)
1119	10.83	10.28	9.79	8.98	8.34	7.82	7.37	7.00	6.73	(6.64)
1120	17.85	16.10	14.56	12.02	9.98	8.27	6.78	5.51	4.62	(4.56)

log K, P = 4 kb

NAME	INDEX	TEMPERATURE, °C							
		0	25	50	75	100	125	150	175
K-OXIDE	1123	90.32	83.65	77.95	73.06	68.83	65.14	61.89	59.02
NA-OXIDE	1124	72.71	67.19	62.52	58.53	55.11	52.14	49.55	47.27
AMORPH SIL	1125	-2.76	-2.02	-1.66	-1.45	-1.30	-1.19	-1.11	-1.04
A-CRIST	1126	-3.91	-2.98	-2.49	-2.18	-1.96	-1.79	-1.66	-1.54
B-CRIST	1127	-3.26	-2.42	-2.00	-1.75	-1.57	-1.43	-1.33	-1.24
CHALCED	1128	-4.46	-3.47	-2.94	-2.59	-2.33	-2.13	-1.97	-1.83
WAIR	1130	25.92	24.18	21.41	18.71	16.28	14.15	12.31	10.73
LAUM	1132	19.72	18.87	16.88	14.85	13.02	11.42	10.06	8.91
ACAN	1501	-40.90	-36.33	-33.12	-30.45	-28.18	-26.20	-24.47	-22.95
CP	1502	-37.54	-32.42	-30.16	-28.36	-26.88	-25.63	-24.55	-23.62
BN	1503	-120.82	-101.19	-93.26	-86.78	-81.28	-76.53	-72.36	-68.68
CC	1504	-41.56	-34.21	-31.30	-28.89	-26.84	-25.07	-23.52	-22.15
QTZ	1505	-4.76	-3.75	-3.19	-2.82	-2.55	-2.33	-2.16	-2.01
MG	1506	33.79	30.58	27.32	24.51	22.10	20.04	18.28	16.74
KALS	1507	14.10	12.96	11.68	10.52	9.50	8.63	7.88	7.25
FERROSIL	1508	8.48	8.24	7.58	6.91	6.29	5.74	5.27	4.85
PYROPH	1509	3.98	4.40	3.61	2.58	1.59	0.69	-0.08	-0.73
TALC	1510	21.83	22.59	21.08	19.42	17.87	16.49	15.30	14.27
7-A CL	1512	82.20	75.68	67.75	60.83	54.91	49.87	45.57	41.88
14-A CL	1513	78.27	71.99	64.27	57.52	51.74	46.82	42.62	39.03
CLNZ	1515	54.14	48.88	43.29	38.31	34.00	30.31	27.15	24.44
LAWS	1516	27.58	25.02	22.15	19.59	17.37	15.48	13.88	12.53
TREM	1517	64.32	64.13	59.67	55.13	50.97	47.31	44.12	41.36
ANTH	1518	70.83	69.79	64.28	58.85	53.94	49.62	45.86	42.61
ZOIS	1519	54.17	48.91	43.31	38.32	34.02	30.32	27.16	24.45
GROSS	1529	60.35	55.70	50.44	45.71	41.61	38.07	35.03	32.41
ANDRA	1530	55.71	53.11	49.17	45.49	42.23	39.38	36.92	34.78
ALB	1531	4.90	5.59	5.28	4.75	4.19	3.68	3.24	2.87
PREHN	1532	40.48	37.44	33.71	30.29	27.30	24.73	22.52	20.63
ENST	1537	12.09	11.66	10.68	9.75	8.92	8.20	7.57	7.02
PARG	1538	118.51	109.55	98.65	88.89	80.42	73.12	66.83	61.42
ANTIG	1542	510.95	492.42	451.64	414.21	381.53	353.39	329.27	308.65
HM	1544	21.39	19.39	17.32	15.53	14.00	12.69	11.55	10.57
ORD-EP	1545	50.67	46.54	41.68	37.30	33.48	30.19	27.36	24.93
MARG	1551	55.64	48.64	42.05	36.36	31.51	27.38	23.86	20.86
MERCURY	1552	-22.04	-19.86	-18.39	-17.11	-15.96	-14.92	-13.97	-13.10
PO	1555	-3.17	-2.93	-3.03	-3.18	-3.34	-3.50	-3.67	-3.84
NESQ	1558	6.72	5.66	5.50	5.42	5.3	5.32	5.29	5.27
EPID	1559	50.68	46.54	41.68	37.29	33.47	30.18	27.35	24.91
HI-ALB	1560	6.42	6.93	6.47	5.81	5.14	4.54	4.01	3.56

$\log K$, $P = 4$ kb

INDEX	TEMPERATURE, °C									
	200	225	250	300	350	400	450	500	550	600
1123	56.46	54.16	52.09	48.52	45.54	43.02	40.86	39.00	37.38	(35.99)
1124	45.26	43.48	41.88	39.16	36.94	35.09	33.53	32.20	31.09	(30.20)
1125	-0.99	-0.94	-0.90	-0.84	-0.80	-0.78	-0.78	-0.79	-0.82	(-0.83)
1126	-1.44	-1.35	-1.28	-1.14	-1.01	-0.90	-0.80	-0.71	-0.62	(-0.52)
1127	-1.16	-1.10	-1.04	-0.93	-0.83	-0.74	-0.66	-0.58	-0.51	(-0.42)
1128	-1.71	-1.60	-1.51	-1.34	-1.19	-1.06	-0.94	-0.84	-0.73	(-0.63)
1130	9.38	8.22	7.23	5.65	4.46	3.50	2.70	2.04	1.74	(2.24)
1132	7.95	7.15	6.48	5.49	4.80	4.29	3.88	3.58	3.59	(4.38)
1501	-21.61	-20.42	-19.35	-17.52	-16.03	-14.82	-13.83	-12.99	-12.22	(-11.38)
1502	-22.81	-22.11	-21.48	-20.46	-19.70	-19.16	-18.82	-18.62	-18.37	(17.79)
1503	-65.41	-62.50	-59.91	-55.49	-51.92	-49.04	-46.77	-44.92	-43.14	(-40.80)
1504	-20.92	-19.81	-18.81	-17.07	-15.63	-14.42	-13.42	-12.57	-11.77	(-10.88)
1505	-1.88	-1.76	-1.65	-1.47	-1.31	-1.17	-1.04	-0.93	-0.82	(-0.71)
1506	15.41	14.24	13.21	11.48	10.07	8.88	7.81	6.88	6.20	(6.05)
1507	6.71	6.25	5.86	5.24	4.78	4.41	4.09	3.84	3.72	(3.95)
1508	4.49	4.18	3.90	3.46	3.12	2.84	2.61	2.42	2.30	(2.37)
1509	-1.28	-1.72	-2.08	-2.60	-2.93	-3.17	-3.37	-3.51	-3.40	(-2.67)
1510	13.39	12.65	12.02	11.04	10.33	9.79	9.36	9.03	8.94	(9.40)
1512	38.73	36.02	33.69	29.91	26.97	24.58	22.54	20.82	19.80	(20.30)
1513	35.96	33.32	31.05	27.38	24.55	22.24	20.26	18.62	17.65	(18.20)
1515	22.11	20.11	18.39	15.61	13.45	11.68	10.16	8.89	8.16	(8.69)
1516	11.38	10.41	9.60	8.31	7.36	6.60	5.96	5.43	5.23	(5.78)
1517	38.99	36.96	35.22	32.43	30.33	28.67	27.31	26.23	25.71	(26.43)
1518	39.80	37.37	35.28	31.89	29.29	27.21	25.48	24.07	23.29	(23.84)
1519	22.12	20.12	18.40	15.61	13.45	11.68	10.16	8.88	8.16	(8.68)
1529	30.16	28.21	26.53	23.79	21.66	19.91	18.42	17.18	16.43	(16.78)
1530	32.93	31.31	29.91	27.60	25.78	24.29	23.00	21.92	21.23	(21.38)
1531	2.57	2.33	2.14	1.90	1.77	1.70	1.67	1.68	1.82	(2.34)
1532	19.02	17.64	16.45	14.56	13.11	11.94	10.94	10.12	9.70	(10.20)
1537	6.54	6.13	5.77	5.18	4.71	4.34	4.02	3.76	3.59	(3.61)
1538	56.76	52.72	49.23	43.53	39.08	35.46	32.38	29.83	28.24	(28.73)
1542	291.01	275.92	262.99	242.29	226.57	214.03	203.55	195.03	190.45	(194.51)
1544	9.72	8.97	8.31	7.20	6.30	5.52	4.83	4.22	3.79	(3.72)
1545	22.85	21.06	19.51	17.00	15.06	13.47	12.10	10.96	10.31	(10.78)
1551	18.29	16.08	14.18	11.10	8.71	6.74	5.03	3.58	2.75	(3.32)
1552	-12.28	-11.53	-10.82	-9.52	-8.37	-7.66	-7.05	-6.47	-5.91	(-5.31)
1555	-4.00	-4.16	-4.31	-4.61	-4.91	-5.22	-5.56	-5.90	-6.15	(-6.13)
1558	5.25	5.24	5.22	5.19	5.13	5.02	4.85	4.65	4.51	(4.62)
1559	22.83	21.03	19.48	16.96	15.00	13.40	12.02	10.86	10.20	(10.66)
1560	3.19	2.89	2.65	2.31	2.11	1.98	1.90	1.87	1.99	(2.48)

Log K , P = 4.5 kb

NAME	INDEX	TEMPERATURE, °C							
		0	25	50	75	100	125	150	175
AND	1001	23.10	20.05	17.14	14.58	12.38	10.50	8.88	7.50
KYA	1002	22.12	19.19	16.38	13.91	11.79	9.97	8.42	7.09
SILL	1003	23.37	20.28	17.33	14.75	12.52	10.62	8.99	7.59
KAOL	1004	11.64	10.49	8.88	7.34	5.99	4.82	3.84	3.01
CHRYS	1007	32.57	31.73	29.15	26.77	24.69	22.90	21.36	20.04
MUSC	1012	21.03	19.21	16.63	14.16	11.97	10.09	8.49	7.14
PARAG	1013	25.54	23.19	20.19	17.36	14.88	12.75	10.94	9.42
PHLOG	1014	41.51	39.98	36.43	33.10	30.18	27.66	25.50	23.65
ANNITE	1015	35.17	33.72	30.80	27.95	25.40	23.18	21.26	19.63
DEHYD-ANAL	1021	15.85	15.18	13.93	12.65	11.48	10.45	9.55	8.79
ANAL	1022	9.92	9.78	8.98	8.09	7.27	6.55	5.93	5.41
SEPIO	1023	28.61	31.44	30.27	28.70	27.16	25.78	24.60	23.62
LO-ALB	1025	4.61	5.60	5.42	4.94	4.41	3.92	3.49	3.12
MAX-MICRO	1027	1.10	2.56	2.74	2.57	2.30	2.02	1.77	1.55
K-SPAR	1028	1.11	2.56	2.74	2.57	2.31	2.03	1.79	1.58
HI-SAN	1029	2.50	3.78	3.81	3.50	3.12	2.74	2.41	2.12
ANOR	1030	35.31	31.94	28.21	24.84	21.90	19.36	17.17	15.29
NEPH	1031	17.57	16.13	14.54	13.10	11.84	10.76	9.83	9.04
WOLL	1035	14.68	14.28	13.40	12.54	11.75	11.06	10.45	9.92
DIOP	1039	21.45	21.27	19.83	18.38	17.07	15.91	14.90	14.02
CA-AL PYX	1040	44.71	39.95	35.38	31.41	28.00	25.07	22.56	20.40
AKER	1044	49.11	46.66	43.08	39.81	36.93	34.42	32.23	30.32
MERW	1045	74.10	69.69	64.26	59.40	55.16	51.47	48.27	45.48
MONTI	1046	32.25	30.45	27.97	25.75	23.82	22.14	20.69	19.42
GEHL	1047	67.93	61.31	54.96	49.47	44.77	40.73	37.26	34.25
FORST	1048	29.99	28.30	25.72	23.42	21.43	19.70	18.21	16.92
FAY	1049	21.97	20.64	18.76	17.01	15.47	14.13	12.97	11.96
HUNT	1050	15.43	14.68	12.32	10.16	8.25	6.55	5.05	3.70
ART	1053	21.00	19.98	18.27	16.82	15.60	14.57	13.69	12.95
HED	1054	20.81	20.52	19.19	17.81	16.54	15.41	14.43	13.56
MALACH	1059	-1.27	3.90	4.00	4.00	3.97	3.95	3.94	3.93
AZUR	1060	-0.43	7.17	7.15	6.97	6.77	6.58	6.40	6.33
HYDRO-MAG	1062	36.78	34.83	30.89	27.45	24.48	21.90	19.67	17.71
SMITH	1064	-0.17	0.12	-0.15	-0.44	-0.71	-0.95	-1.17	-1.36
ANH-CORD	1065	70.83	64.65	56.93	49.89	43.71	38.35	33.72	29.72
HYD-CORD	1066	67.33	61.48	54.04	47.24	41.28	36.12	31.66	27.83
JADEITE	1067	9.62	9.66	8.99	8.23	7.50	6.86	6.31	5.85
CER	1068	-2.49	-1.53	-1.36	-1.25	-1.16	-1.07	-0.99	-0.91
STRONT	1069	0.31	0.39	0.28	0.16	0.04	-0.07	-0.18	-0.28
DIS-DOL	1070	6.50	6.23	5.23	4.30	3.47	2.74	2.08	1.48

$\log K$, $P = 4.5$ kb

NAME	INDEX	TEMPERATURE, °C							
		0	25	50	75	100	125	150	175
ORD-DOL	1071	4.76	4.68	3.85	3.06	2.36	1.73	1.16	0.65
ARAG	1072	3.17	3.08	2.72	2.38	2.06	1.77	1.51	1.28
CALC	1073	3.24	3.14	2.76	2.40	2.07	1.77	1.51	1.26
MAG	1074	3.36	3.22	2.62	2.07	1.59	1.17	0.79	0.45
DOL	1075	4.76	4.68	3.85	3.07	2.36	1.73	1.16	0.66
SIDER	1076	1.48	1.29	0.84	0.40	0.00	-0.36	-0.69	-0.98
ANHYD	1078	-3.03	-2.41	-2.66	-3.01	-3.38	-3.76	-4.12	-4.48
FLUORITE	1079	-9.77	-8.05	-7.95	-7.94	-7.97	-8.02	-8.10	-8.18
BARITE	1080	-9.81	-8.54	-8.14	-7.95	-7.86	-7.85	-7.89	-7.96
ANG	1081	-7.42	-5.85	-5.67	-5.64	-5.68	-5.74	-5.81	-5.89
CELEST	1082	-5.40	-4.66	-4.71	-4.88	-5.10	-5.33	-5.57	-5.82
ALUN	1083	8.05	6.14	3.37	0.82	-1.45	-3.45	-5.22	-6.80
WITH	1084	-2.26	-1.80	-1.59	-1.46	-1.39	-1.34	-1.31	-1.30
RHODO	1085	0.46	0.85	0.59	0.32	0.07	-0.15	-0.35	-0.52
COV	1086	-28.87	-23.82	-21.84	-20.26	-18.92	-17.78	-16.79	-15.93
GALENA	1087	-15.96	-13.69	-12.53	-11.60	-10.82	-10.13	-9.53	-8.98
SL	1088	-11.72	-10.36	-9.84	-9.48	-9.20	-8.97	-8.77	-8.60
WURT	1089	-9.21	-8.06	-7.72	-7.51	-7.36	-7.24	-7.16	-7.08
M-CINN	1090	-41.51	-37.28	-34.50	-32.20	-30.23	-28.51	-27.00	-25.66
CINN	1091	-42.10	-37.79	-34.94	-32.59	-30.57	-28.81	-27.26	-25.89
ALA	1092	-0.47	0.09	-0.14	-0.42	-0.69	-0.94	-1.17	-1.38
PYRITE	1093	-26.57	-23.87	-22.28	-21.09	-20.15	-19.39	-18.77	-18.25
GOLD	1100	-28.88	-25.92	-23.46	-21.35	-19.52	-17.92	-16.50	-15.22
SILVER	1101	-11.52	-9.86	-8.75	-7.81	-6.98	-6.25	-5.59	-4.99
COPPER	1102	-7.75	-4.64	-4.01	-3.48	-3.01	-2.58	-2.18	-1.80
GRAPHITE	1103	-5.42	-4.89	-4.32	-3.84	-3.44	-3.10	-2.81	-2.57
HALITE	1106	1.33	1.74	1.90	1.97	2.00	2.00	1.99	1.97
SYLVITE	1107	0.51	1.14	1.46	1.66	1.79	1.88	1.93	1.96
COR	1108	28.56	24.37	20.83	17.86	15.35	13.22	11.41	9.85
PER	1109	23.38	21.49	19.49	17.79	16.34	15.09	14.01	13.08
LIME	1110	35.83	32.99	30.37	28.14	26.22	24.56	23.11	21.84
TO	1111	4.24	6.37	6.08	5.79	5.55	5.35	5.19	5.05
CUP	1112	-6.93	-1.47	-0.96	-0.55	-0.19	0.17	0.51	0.84
FE-OXIDE	1113	15.61	14.19	12.77	11.55	10.50	9.60	8.82	8.14
GIBBS	1114	10.53	8.98	7.69	6.63	5.76	5.04	4.44	3.94
DIAS	1115	11.76	9.98	8.47	7.22	6.17	5.29	4.55	3.92
BOEH	1116	12.89	10.96	9.33	7.96	6.82	5.85	5.04	4.34
BRUC	1117	17.45	16.18	14.72	13.49	12.45	11.57	10.82	10.17
MANGAN	1119	19.36	18.11	16.53	15.18	14.02	13.03	12.19	11.45
SPINEL	1120	47.32	41.63	36.40	31.99	28.27	25.10	22.39	20.06

$\log K$, $P = 4.5$ kb

INDEX	TEMPERATURE, °C									
	200	225	250	300	350	400	450	500	550	600
1001	6.31	5.29	4.40	2.95	1.82	0.89	0.08	-0.59	-0.97	(-0.72)
1002	5.95	4.97	4.12	2.74	1.66	0.78	0.01	-0.62	-0.97	(-0.70)
1003	6.39	5.35	4.45	2.99	1.84	0.90	0.08	-0.60	-0.98	(-0.74)
1004	2.33	1.75	1.28	0.57	0.06	-0.32	-0.65	-0.90	-0.89	(-0.29)
1007	18.92	17.96	17.13	15.81	14.80	14.01	13.35	12.83	12.56	(12.83)
1012	6.01	5.07	4.28	3.08	2.23	1.57	1.02	0.61	0.60	(1.52)
1013	8.13	7.06	6.16	4.78	3.80	3.04	2.42	1.96	1.91	(2.83)
1014	22.08	20.73	19.59	17.78	16.41	15.34	14.46	13.77	13.48	(14.06)
1015	18.23	17.04	16.03	14.43	13.23	12.28	11.50	10.89	10.66	(11.26)
1021	8.13	7.57	7.09	6.34	5.79	5.36	5.02	4.76	4.67	(4.96)
1022	4.98	4.62	4.34	3.92	3.65	3.48	3.35	3.29	3.38	(3.83)
1023	22.81	22.16	21.65	20.95	20.56	20.36	20.26	20.29	20.61	(21.65)
1025	2.82	2.58	2.40	2.16	2.04	2.00	1.99	2.04	2.24	(2.80)
1027	1.39	1.26	1.17	1.08	1.07	1.10	1.16	1.24	1.46	(2.02)
1028	1.41	1.28	1.18	1.08	1.06	1.07	1.10	1.17	1.36	(1.90)
1029	1.88	1.69	1.54	1.36	1.27	1.23	1.22	1.26	1.43	(1.95)
1030	13.66	12.26	11.04	9.05	7.50	6.23	5.14	4.25	3.73	(4.00)
1031	8.37	7.80	7.31	6.54	5.96	5.51	5.15	4.87	4.76	(5.01)
1035	9.45	9.05	8.69	8.10	7.64	7.27	6.96	6.71	6.55	(6.55)
1039	13.26	12.60	12.03	11.11	10.40	9.84	9.37	9.00	8.81	(8.95)
1040	18.52	16.89	15.48	13.13	11.27	9.74	8.42	7.31	6.60	(6.68)
1044	28.66	27.20	25.93	23.80	22.11	20.73	19.56	18.60	17.93	(17.79)
1045	43.04	40.90	39.02	35.86	33.34	31.26	29.50	28.03	26.97	(26.60)
1046	18.32	17.36	16.52	15.11	14.00	13.08	12.31	11.66	11.21	(11.12)
1047	31.65	29.37	27.38	24.08	21.44	19.26	17.39	15.82	14.75	(14.64)
1048	15.79	14.80	13.94	12.51	11.37	10.43	9.64	8.98	8.52	(8.44)
1049	11.08	10.32	9.65	8.55	7.68	6.97	6.37	5.87	5.55	(5.58)
1050	2.48	1.38	0.37	-1.43	-3.06	-4.63	-6.19	-7.68	-8.76	(-8.76)
1053	12.31	11.75	11.28	10.50	9.86	9.32	8.81	8.37	8.10	(8.28)
1054	12.81	12.16	11.60	10.68	9.97	9.40	8.92	8.54	8.33	(8.45)
1059	3.93	3.93	3.94	3.95	3.94	3.91	3.84	3.75	3.73	(3.91)
1060	6.08	5.94	5.81	5.55	5.28	4.97	4.61	4.23	3.98	(4.10)
1062	15.99	14.46	13.10	10.74	8.71	6.84	5.04	3.38	2.22	(2.35)
1064	-1.54	-1.69	-1.84	-2.10	-2.35	-2.61	-2.89	-3.16	-3.35	(-3.28)
1065	26.26	23.27	20.67	16.43	13.10	10.37	8.04	6.13	5.01	(5.55)
1066	24.52	21.66	19.19	15.16	12.03	9.47	7.30	5.52	4.53	(5.19)
1067	5.47	5.16	4.91	4.55	4.33	4.19	4.10	4.06	4.18	(4.64)
1068	-0.83	-0.75	-0.67	-0.53	-0.41	-0.33	-0.29	-0.27	-0.20	(0.08)
1069	-0.37	-0.46	-0.54	-0.70	-0.86	-1.05	-1.25	-1.46	-1.60	(-1.51)
1070	0.95	0.46	0.01	-0.79	-1.52	-2.24	-2.96	-3.65	-4.15	(-4.12)

log K , P = 4.5 kb

INDEX	TEMPERATURE, °C									
	200	225	250	300	350	400	450	500	550	600
1071	0.19	-0.22	-0.61	-1.31	-1.95	-2.59	-3.25	-3.88	-4.32	(-4.26)
1072	1.07	0.88	0.70	0.37	0.07	-0.23	-0.53	-0.83	-1.04	(-1.01)
1073	1.04	0.84	0.65	0.31	-0.01	-0.32	-0.64	-0.95	-1.18	(-1.15)
1074	0.15	-0.13	-0.38	-0.82	-1.23	-1.62	-2.01	-2.38	-2.65	(-2.64)
1075	0.20	-0.22	-0.61	-1.31	-1.95	-2.59	-3.25	-3.89	-4.34	(-4.28)
1076	-1.24	-1.48	-1.71	-2.11	-2.48	-2.84	-3.21	-3.57	-3.83	(-3.83)
1078	-4.82	-5.16	-5.48	-6.12	-6.75	-7.42	-8.11	-8.80	-9.35	(-9.48)
1079	-8.27	-8.37	-8.47	-8.71	-8.99	-9.33	-9.74	-10.17	-10.47	(-10.33)
1080	-8.05	-8.16	-8.30	-8.60	-8.97	-9.40	-9.90	-10.42	-10.82	(-10.85)
1081	-5.98	-6.07	-6.17	-6.38	-6.63	-6.93	-7.29	-7.66	-7.92	(-7.79)
1082	-6.07	-6.32	-6.57	-7.07	-7.60	-8.16	-8.77	-9.38	-9.85	(-9.93)
1083	-8.21	-9.48	-10.64	-12.70	-14.56	-16.36	-18.18	-19.91	-21.10	(-20.82)
1084	-1.29	-1.30	-1.31	-1.35	-1.42	-1.53	-1.68	-1.84	-1.94	(-1.84)
1085	-0.67	-0.81	-0.93	-1.16	-1.36	-1.57	-1.80	-2.02	-2.15	(-2.03)
1086	-15.17	-14.50	-13.90	-12.89	-12.09	-11.46	-10.96	-10.57	-10.19	(-9.70)
1087	-8.49	-8.05	-7.64	-6.93	-6.33	-5.85	-5.46	-5.14	-4.79	(-4.26)
1088	-8.46	-8.33	-8.23	-8.05	-7.94	-7.89	-7.90	-7.94	-7.91	(-7.64)
1089	-7.02	-6.97	-6.92	-6.87	-6.85	-6.88	-6.96	-7.07	-7.10	(-6.87)
1090	-24.46	-23.38	-22.41	-20.71	-19.30	-18.14	-17.17	-16.35	-15.57	(-14.64)
1091	-24.66	-23.55	-22.55	-20.82	-19.37	-18.18	-17.18	-16.34	-15.53	(-14.60)
1092	-1.56	-1.73	-1.89	-2.17	-2.44	-2.71	-2.99	-3.26	-3.44	(-3.35)
1093	-17.83	-17.49	-17.20	-16.79	-16.57	-16.51	-16.60	-16.78	-16.87	(-16.57)
1100	-14.08	-13.04	-12.10	-10.44	-9.02	-7.80	-6.73	-5.78	-4.94	(-4.19)
1101	-4.45	-3.94	-3.48	-2.65	-1.92	-1.27	-0.69	-0.17	0.31	(0.74)
1102	-1.45	-1.13	-0.81	-0.24	0.28	0.78	1.20	1.60	1.98	(2.35)
1103	-2.35	-2.17	-2.01	-1.75	-1.56	-1.42	-1.32	-1.25	-1.19	(-1.11)
1106	1.94	1.90	1.87	1.78	1.68	1.55	1.40	1.24	1.14	(1.19)
1107	1.97	1.97	1.95	1.89	1.79	1.66	1.49	1.31	1.17	(1.17)
1108	8.51	7.36	6.35	4.69	3.38	2.28	1.33	0.53	0.03	(0.15)
1109	12.26	11.53	10.90	9.82	8.94	8.21	7.59	7.05	6.64	(6.44)
1110	20.73	19.73	18.85	17.34	16.10	15.06	14.16	13.40	12.78	(12.36)
1111	4.94	4.85	4.78	4.67	4.60	4.56	4.53	4.51	4.52	(4.59)
1112	1.17	1.49	1.80	2.39	2.95	3.47	3.96	4.42	4.88	(5.38)
1113	7.55	7.03	6.58	5.81	5.19	4.67	4.67	3.85	3.58	(3.48)
1114	3.53	3.18	2.89	2.44	2.11	1.84	1.60	1.40	1.32	(1.53)
1115	3.38	2.93	2.54	1.92	1.44	1.05	0.72	0.44	0.31	(0.47)
1116	3.75	3.24	2.80	2.10	1.55	1.10	0.72	0.40	0.22	(0.36)
1117	9.62	9.13	8.71	8.02	7.47	7.01	6.63	6.31	6.09	(6.05)
1119	10.82	10.27	9.78	8.98	8.35	7.83	7.40	7.04	6.78	(6.70)
1120	18.05	16.30	14.77	12.24	10.22	8.54	7.09	5.87	5.05	(5.04)

log K , P = 4.5 kb

NAME	INDEX	TEMPERATURE, °C							
		0	25	50	75	100	125	150	175
K-OXIDE	1123	90.05	83.47	77.82	72.95	68.74	65.06	61.82	58.95
NA-OXIDE	1124	72.44	67.03	62.40	58.45	55.04	52.09	49.51	47.24
AMORPH SIL	1125	-2.80	-1.98	-1.60	-1.38	-1.23	-1.12	-1.04	-0.97
A-CRIST	1126	-3.97	-2.97	-2.46	-2.14	-1.91	-1.74	-1.60	-1.49
B-CRIST	1127	-3.31	-2.40	-1.96	-1.69	-1.50	-1.37	-1.26	-1.18
CHALCED	1128	-4.55	-3.49	-2.93	-2.57	-2.30	-2.10	-1.93	-1.79
WAIR	1130	25.75	24.44	21.80	19.15	16.75	14.63	12.79	11.20
LAUM	1132	19.44	19.04	17.18	15.21	13.40	11.82	10.45	9.30
ACAN	1501	-41.48	-36.55	-33.27	-30.57	-28.27	-26.28	-24.54	-23.00
CP	1502	-38.38	-32.42	-30.08	-28.26	-26.76	-25.49	-24.41	-23.48
BN	1503	-124.42	-101.44	-93.29	-86.74	-81.19	-76.41	-72.22	-68.52
CC	1504	-42.94	-34.34	-31.36	-28.93	-26.86	-25.08	-23.52	-22.14
QTZ	1505	-4.86	-3.77	-3.19	-2.80	-2.52	-2.30	-2.12	-1.97
MG	1506	33.53	30.62	27.41	24.61	22.22	20.16	18.39	16.86
KALS	1507	13.98	12.99	11.77	10.63	9.63	8.76	8.02	7.39
FERROSIL	1508	8.28	8.20	7.59	6.94	6.33	5.79	5.32	4.91
PYROPH	1509	3.68	4.48	3.82	2.86	1.90	1.02	0.25	-0.40
TALC	1510	20.73	22.15	20.79	19.22	17.72	16.38	15.21	14.21
7-A CL	1512	80.89	75.27	67.51	60.68	54.82	49.82	45.54	41.89
14-A CL	1513	76.92	71.55	64.00	57.34	51.62	46.74	42.57	39.01
CLNZ	1515	53.68	48.92	43.48	38.57	34.30	30.63	27.48	24.78
LAWS	1516	27.21	24.97	22.21	19.69	17.50	15.63	14.04	12.69
TREM	1517	62.03	63.20	59.08	54.71	50.66	47.07	43.94	41.23
ANTH	1518	68.37	68.77	63.60	58.35	53.55	49.31	45.62	42.41
ZOIS	1519	53.70	48.94	43.50	38.59	34.32	30.64	27.49	24.79
GROSS	1529	59.62	55.52	50.42	45.77	41.71	38.20	35.18	32.58
ANDRA	1530	54.69	52.78	49.00	45.39	42.18	39.37	36.93	34.81
ALB	1531	4.61	5.60	5.42	4.94	4.41	3.92	3.49	3.12
PREHN	1532	40.04	37.46	33.86	30.50	27.55	24.99	22.79	20.91
ENST	1537	11.74	11.51	10.57	9.67	8.86	8.14	7.52	6.98
PARG	1538	116.75	109.05	98.47	88.87	80.49	73.25	67.02	61.65
ANTIG	1542	496.30	486.11	446.95	410.41	378.32	350.61	326.84	306.51
HM	1544	21.20	19.41	17.38	15.60	14.07	12.76	11.63	10.65
ORD-EP	1545	50.06	46.49	41.80	37.48	33.70	30.43	27.62	25.21
MARG	1551	55.58	48.95	42.48	36.84	32.01	27.89	24.37	21.37
MERCURY	1552	-22.22	-19.85	-18.37	-17.08	-15.94	-14.90	-13.95	-13.07
PO	1555	-3.33	-2.92	-2.98	-3.11	-3.27	-3.43	-3.59	-3.76
NESQ	1558	6.54	5.63	5.49	5.42	5.37	5.33	5.30	5.28
EPID	1559	50.06	46.49	41.80	37.47	33.69	30.42	27.61	25.19
HI-ALB	1560	6.14	6.95	6.61	6.01	5.37	4.78	4.26	3.81

$\log K$, P = 4.5 kb

INDEX	TEMPERATURE, °C									
	200	225	250	300	350	400	450	500	550	600
1123	56.39	54.10	52.04	48.47	45.50	42.98	40.83	38.96	37.35	(35.96)
1124	45.24	43.45	41.86	39.15	36.93	35.08	33.53	32.21	31.11	(30.22)
1125	-0.91	-0.87	-0.83	-0.78	-0.74	-0.73	-0.73	-0.74	-0.76	(-0.78)
1126	-1.39	-1.30	-1.22	-1.09	-0.96	-0.86	-0.76	-0.67	-0.58	(-0.48)
1127	-1.10	-1.03	-0.97	-0.87	-0.77	-0.69	-0.61	-0.53	-0.46	(-0.37)
1128	-1.67	-1.57	-1.47	-1.30	-1.15	-1.02	-0.91	-0.80	-0.70	(-0.59)
1130	9.85	8.68	7.69	6.11	4.92	3.97	3.19	2.57	2.32	(2.87)
1132	8.34	7.53	6.87	5.87	5.19	4.69	4.30	4.04	4.10	(4.94)
1501	-21.66	-20.46	-19.38	-17.54	-16.01	-14.81	-13.80	-12.94	-12.15	(-11.29)
1502	-22.66	-21.94	-21.31	-20.27	-19.48	-18.93	-18.56	-18.32	-18.02	(-17.39)
1503	-65.23	-62.30	-59.69	-55.23	-51.61	-48.69	-46.35	-44.43	-42.56	(-40.13)
1504	-20.90	-19.79	-18.78	-17.04	-15.58	-14.36	-13.34	-12.47	-11.66	(-10.74)
1505	-1.84	-1.72	-1.62	-1.43	-1.27	-1.13	-1.01	-0.89	-0.78	(-0.67)
1506	15.53	14.36	13.34	11.62	10.23	9.05	8.01	7.11	6.47	(6.36)
1507	6.85	6.39	6.01	5.39	4.94	4.58	4.27	4.04	3.95	(4.20)
1508	4.55	4.24	3.97	3.53	3.19	2.92	2.69	2.51	2.41	(2.49)
1509	-0.94	-1.38	-1.74	-2.25	-2.57	-2.80	-2.97	-3.07	-2.92	(-2.15)
1510	13.35	12.62	12.01	11.06	10.38	9.86	9.46	9.18	9.14	(9.65)
1512	38.76	36.08	33.77	30.04	27.16	24.83	22.86	21.25	20.36	(20.97)
1513	35.97	33.36	31.11	27.50	24.71	22.46	20.57	19.03	18.20	(18.86)
1515	22.47	20.48	18.77	16.01	13.88	12.15	10.69	9.49	8.87	(9.48)
1516	11.55	10.59	9.78	8.52	7.58	6.85	6.24	5.76	5.62	(6.23)
1517	38.90	36.90	35.19	32.47	30.43	28.83	27.54	26.55	26.15	(26.97)
1518	39.65	37.26	35.21	31.89	29.35	27.34	25.69	24.39	23.73	(24.38)
1519	22.47	20.49	18.77	16.01	13.88	12.15	10.68	9.49	8.87	(9.48)
1529	30.35	28.42	26.75	24.04	21.94	20.24	18.80	17.63	16.97	(17.40)
1530	32.97	31.38	29.99	27.71	25.92	24.46	23.22	22.20	21.58	(21.80)
1531	2.82	2.58	2.40	2.15	2.03	1.97	1.95	1.97	2.15	(2.68)
1532	19.30	17.93	16.75	14.87	13.45	12.31	11.35	10.58	10.23	(10.80)
1537	6.51	6.10	5.75	5.17	4.72	4.35	4.04	3.79	3.64	(3.69)
1538	57.01	53.02	49.55	43.91	39.54	35.99	33.02	30.60	29.19	(29.83)
1542	289.14	274.30	261.60	241.33	226.03	213.93	203.97	196.10	192.31	(197.06)
1544	9.79	9.05	8.39	7.29	6.40	5.64	4.96	4.38	3.97	(3.93)
1545	23.13	21.35	19.82	17.33	15.42	13.87	12.55	11.48	10.92	(11.47)
1551	18.80	16.60	14.71	11.64	9.27	7.34	5.69	4.33	3.61	(4.27)
1552	-12.26	-11.50	-10.79	-9.50	-8.35	-7.64	-7.02	-6.44	-5.87	(-5.27)
1555	-3.92	-4.07	-4.22	-4.51	-4.79	-5.09	-5.41	-5.73	-5.95	(-5.90)
1558	5.27	5.26	5.25	5.22	5.16	5.06	4.91	4.74	4.62	(4.75)
1559	23.11	21.32	19.78	17.29	15.36	13.80	12.46	11.38	10.81	(11.35)
1560	3.45	3.14	2.90	2.57	2.37	2.25	2.18	2.17	2.32	(2.83)

Log K , P = 5 kb

NAME	INDEX	TEMPERATURE, °C							
		0	25	50	75	100	125	150	175
AND	1001	23.10	20.22	17.37	14.83	12.65	10.77	9.15	7.77
KYA	1002	22.05	19.30	16.55	14.11	12.00	10.19	8.64	7.32
SILL	1003	23.36	20.44	17.55	14.99	12.78	10.88	9.25	7.85
KAOL	1004	11.47	10.58	9.05	7.56	6.22	5.07	4.09	3.26
CHRYS	1007	31.52	31.25	28.78	26.46	24.42	22.66	21.14	19.85
MUSC	1012	20.65	19.26	16.84	14.45	12.30	10.44	8.85	7.50
PARAG	1013	25.15	23.24	20.40	17.66	15.22	13.11	11.32	9.79
PHLOG	1014	40.25	39.46	36.09	32.86	30.00	27.52	25.39	23.57
ANNITE	1015	34.36	33.57	30.84	28.07	25.56	23.37	21.47	19.85
DEHYD-ANAL	1021	15.69	15.27	14.11	12.87	11.72	10.70	9.81	9.04
ANAL	1022	9.69	9.80	9.09	8.26	7.46	6.75	6.13	5.62
SEPIO	1023	26.83	30.66	29.73	28.28	26.82	25.50	24.36	23.40
LO-ALB	1025	4.23	5.56	5.50	5.09	4.60	4.12	3.70	3.34
MAX-MICRO	1027	0.73	2.52	2.82	2.71	2.47	2.21	1.97	1.76
K-SPAR	1028	0.74	2.52	2.82	2.72	2.48	2.22	1.99	1.78
HI-SAN	1029	2.14	3.74	3.89	3.65	3.30	2.94	2.61	2.32
ANOR	1030	35.13	32.09	28.46	25.13	22.21	19.68	17.49	15.61
NEPH	1031	17.40	16.14	14.63	13.22	11.98	10.91	9.98	9.20
WOLL	1035	14.41	14.18	13.35	12.50	11.73	11.05	10.44	9.91
DIOP	1039	20.73	20.93	19.59	18.20	16.92	15.78	14.79	13.93
CA-AL PYX	1040	44.50	40.01	35.51	31.57	28.18	25.27	22.76	20.60
AKER	1044	48.32	46.33	42.87	39.65	36.80	34.31	32.14	30.25
MERW	1045	73.09	69.25	63.94	59.15	54.95	51.29	48.12	45.35
MONTI	1046	31.71	30.22	27.80	25.61	23.70	22.04	20.60	19.35
GEHL	1047	67.64	61.38	55.12	49.67	44.98	40.95	37.48	34.48
FORST	1048	29.33	27.99	25.47	23.21	21.25	19.54	18.06	16.78
FAY	1049	21.60	20.57	18.76	17.04	15.52	14.91	13.03	12.02
HUNT	1050	14.70	14.61	12.35	10.25	8.37	6.70	5.21	3.88
ART	1053	20.46	19.75	18.09	16.66	15.45	14.44	13.57	12.83
HED	1054	20.24	20.32	19.08	17.75	16.51	15.40	14.43	13.57
MALACH	1059	-2.79	3.78	3.95	3.97	3.97	3.96	3.95	3.95
AZUR	1060	-2.61	7.11	7.19	7.05	6.87	6.68	6.51	6.35
HYDRO-MAG	1062	35.69	34.57	30.75	27.37	24.43	21.89	19.68	17.74
SMITH	1064	-0.56	-0.02	-0.25	-0.51	-0.77	-1.00	-1.20	-1.39
ANH-CORD	1065	70.34	64.98	57.48	50.52	44.38	39.04	34.42	30.42
HYD-CORD	1066	66.77	61.73	54.52	47.82	41.90	36.75	32.31	28.47
JADEITE	1067	9.18	9.49	8.94	8.24	7.55	6.93	6.39	5.94
CER	1068	-2.66	-1.44	-1.24	-1.12	-1.03	-0.94	-0.86	-0.78
STRONT	1069	0.24	0.43	0.34	0.23	0.11	0.00	-0.10	-0.20
DIS-DOL	1070	6.19	6.22	5.28	4.38	3.56	2.83	2.18	1.59

$\log K$, $P = 5$ kb

INDEX	TEMPERATURE, °C									
	200	225	250	300	350	400	450	500	550	600
1001	6.58	5.55	4.66	3.22	2.09	1.17	0.37	-0.27	-0.62	(-0.34)
1002	6.17	5.19	4.35	2.97	1.90	1.02	0.28	-0.33	-0.65	(-0.34)
1003	6.64	5.61	4.71	3.25	2.10	1.17	0.37	-0.29	-0.64	(-0.37)
1004	2.57	2.00	1.53	0.81	0.31	-0.07	-0.39	-0.61	-0.58	(0.06)
1007	18.74	17.79	16.98	15.69	14.71	13.93	13.30	12.81	12.58	(12.89)
1012	6.38	5.44	4.65	3.46	2.62	1.97	1.45	1.07	1.11	(2.07)
1013	8.52	7.45	6.55	5.18	4.20	3.46	2.86	2.44	2.44	(3.40)
1014	22.01	20.69	19.57	17.78	16.45	15.41	14.57	13.92	13.69	(14.32)
1015	18.46	17.28	16.27	14.67	13.49	12.56	11.80	11.22	11.03	(11.67)
1021	8.39	7.82	7.35	6.59	6.03	5.61	5.27	5.02	4.95	(5.25)
1022	5.18	4.83	4.54	4.12	3.85	3.68	3.56	3.51	3.62	(4.08)
1023	22.62	21.99	21.49	20.83	20.46	20.28	20.22	20.27	20.64	(21.72)
1025	3.05	2.81	2.62	2.38	2.27	2.23	2.23	2.29	2.50	(3.08)
1027	1.59	1.47	1.38	1.29	1.28	1.32	1.38	1.47	1.71	(2.28)
1028	1.61	1.48	1.39	1.29	1.27	1.29	1.32	1.40	1.61	(2.16)
1029	2.09	1.90	1.75	1.57	1.48	1.45	1.44	1.49	1.68	(2.21)
1030	13.98	12.58	11.36	9.37	7.83	6.56	5.49	4.63	4.15	(4.46)
1031	8.53	7.96	7.47	6.70	6.13	5.68	5.33	5.06	4.97	(5.23)
1035	9.45	9.05	8.70	8.12	7.66	7.29	6.99	6.75	6.60	(6.61)
1039	13.18	12.53	11.97	11.07	10.38	9.83	9.38	9.03	8.86	(9.02)
1040	18.73	17.11	15.69	13.36	11.51	9.99	8.69	7.62	6.95	(7.07)
1044	28.60	27.16	25.89	23.78	22.11	20.75	19.61	18.67	18.03	(17.91)
1045	42.92	40.80	38.93	35.80	33.30	31.25	29.52	28.09	27.06	(26.73)
1046	18.26	17.30	16.47	15.08	13.98	13.07	12.31	11.69	11.26	(11.19)
1047	31.88	29.61	27.62	24.33	21.70	19.54	17.70	16.16	15.14	(15.08)
1048	15.67	14.70	13.85	12.43	11.31	10.39	9.62	8.98	8.55	(8.50)
1049	11.15	10.39	9.72	8.63	7.77	7.06	6.47	5.99	5.68	(5.73)
1050	2.68	1.59	0.60	-1.16	-2.76	-4.28	-5.80	-7.22	-8.21	(-8.14)
1053	12.20	11.66	11.19	10.43	9.81	9.28	8.80	8.37	8.14	(8.35)
1054	12.83	12.19	11.63	10.72	10.02	9.46	9.00	8.63	8.44	(8.58)
1059	3.95	3.95	3.96	3.98	3.98	3.96	3.90	3.82	3.82	(4.02)
1060	6.20	6.08	5.94	5.69	5.43	5.13	4.79	4.44	4.21	(4.37)
1062	16.04	14.53	13.19	10.87	8.89	7.06	5.32	3.73	2.67	(2.88)
1064	-1.55	-1.70	-1.84	-2.09	-2.32	-2.57	-2.83	-3.08	-3.25	(-3.16)
1065	26.96	23.97	21.38	17.13	13.81	11.10	8.82	6.96	5.92	(6.54)
1066	25.17	22.31	19.84	15.82	12.70	10.16	8.03	6.31	5.40	(6.13)
1067	5.57	5.26	5.02	4.67	4.46	4.33	4.25	4.23	4.36	(4.85)
1068	-0.70	-0.62	-0.54	-0.40	-0.28	-0.19	-0.15	-0.12	-0.03	(0.27)
1069	-0.29	-0.38	-0.46	-0.61	-0.77	-0.94	-1.14	-1.34	-1.45	(-1.35)
1070	1.07	0.59	0.15	-0.64	-1.36	-2.06	-2.75	-3.41	-3.87	(-3.81)

$\log K$, $P = 5$ kb

NAME	INDEX	TEMPERATURE, °C							
		0	25	50	75	100	125	150	175
ORD-DOL	1071	4.45	4.67	3.89	3.14	2.44	1.82	1.27	0.77
ARAG	1072	3.06	3.10	2.77	2.43	2.12	1.84	1.58	1.35
CALC	1073	3.16	3.18	2.83	2.48	2.16	1.86	1.59	1.35
MAG	1074	3.14	3.17	2.60	2.07	1.60	1.18	0.81	0.48
DOL	1075	4.45	4.68	3.90	3.14	2.45	1.83	1.27	0.77
SIDER	1076	1.41	1.37	0.95	0.52	0.12	-0.23	-0.56	-0.85
ANHYD	1078	-3.36	-2.44	-2.65	-2.98	-3.34	-3.70	-4.06	-4.41
FLUORITE	1079	-10.18	-7.99	-7.88	-7.87	-7.89	-7.94	-8.01	-8.09
BARITE	1080	-10.26	-8.67	-8.20	-7.98	-7.87	-7.85	-7.87	-7.93
ANG	1081	-7.84	-5.85	-5.62	-5.57	-5.60	-5.65	-5.72	-5.80
CELEST	1082	-5.73	-4.71	-4.72	-4.87	-5.07	-5.30	-5.54	-5.78
ALUN	1083	7.52	6.24	3.60	1.11	-1.12	-3.10	-4.86	-6.43
WITH	1084	-2.44	-1.84	-1.57	-1.43	-1.34	-1.28	-1.25	-1.23
RHODO	1085	0.24	0.87	0.64	0.39	0.14	-0.07	-0.26	-0.44
COV	1086	-29.71	-23.89	-21.86	-20.25	-18.91	-17.79	-16.77	-15.90
GALENA	1087	-16.25	-13.67	-12.47	-11.53	-10.73	-10.04	-9.44	-8.89
SL	1088	-12.00	-10.35	-9.79	-9.41	-9.12	-8.89	-8.69	-8.52
WURT	1089	-9.49	-8.05	-7.67	-7.45	-7.29	-7.17	-7.07	-7.00
M-CINN	1090	-41.80	-37.24	-34.42	-32.10	-30.13	-28.41	-26.90	-25.56
CINN	1091	-42.40	-37.76	-34.88	-32.50	-30.48	-28.72	-27.17	-25.79
ALA	1092	-0.82	0.04	-0.15	-0.41	-0.67	-0.91	-1.13	-1.33
PYRITE	1093	-26.95	-23.92	-22.25	-21.03	-20.07	-19.30	-18.67	-18.15
GOLD	1100	-29.19	-26.16	-23.66	-21.53	-19.69	-18.07	-16.64	-15.36
SILVER	1101	-11.82	-10.00	-8.87	-7.91	-7.08	-6.34	-5.67	-5.07
COPPER	1102	-8.50	-4.74	-4.08	-3.54	-3.06	-2.63	-2.22	-1.85
GRAPHITE	1103	-5.34	-4.82	-4.24	-3.76	-3.36	-3.02	-2.73	-2.49
HALITE	1106	1.14	1.68	1.89	1.99	2.03	2.04	2.04	2.02
SYLVITE	1107	0.35	1.10	1.46	1.68	1.83	1.92	1.98	2.01
COR	1108	28.63	24.55	21.04	18.08	15.57	13.44	11.63	10.07
PER	1109	23.11	21.36	19.37	17.68	16.24	15.00	13.93	13.00
LIME	1110	35.67	32.91	30.31	28.09	26.17	24.52	23.08	21.81
TO	1111	3.46	6.28	6.01	5.74	5.50	5.31	5.15	5.02
CUP	1112	-8.42	-1.63	-1.06	-0.64	-0.26	0.10	0.45	0.79
FE-OXIDE	1113	15.48	14.17	12.78	11.56	10.51	9.61	8.84	8.16
GIBBS	1114	10.52	9.03	7.76	6.70	5.83	5.11	4.51	4.01
DIAS	1115	11.77	10.04	8.55	7.31	6.26	5.38	4.64	4.01
BOEH	1116	12.92	11.04	9.42	8.06	6.92	5.95	5.14	4.44
BRUC	1117	17.15	16.03	14.58	13.36	12.33	11.46	10.72	10.08
MANGAN	1119	19.08	18.03	16.48	15.13	13.98	13.00	12.15	11.43
SPINEL	1120	47.15	41.69	36.51	32.13	28.41	25.25	22.55	20.22

$\log K$, $P = 5$ kb

INDEX	TEMPERATURE, °C									
	200	225	250	300	350	400	450	500	550	600
1071	0.31	-0.10	-0.48	-1.16	-1.79	-2.41	-3.04	-3.67	-4.04	(-3.94)
1072	1.14	0.95	0.78	0.46	0.17	-0.13	-0.42	-0.70	-0.90	(-0.85)
1073	1.13	0.93	0.74	0.41	0.10	-0.21	-0.52	-0.82	-1.02	(-0.99)
1074	0.18	-0.09	-0.33	-0.77	-1.16	-1.54	-1.92	-2.27	-2.52	(-2.49)
1075	0.32	-0.10	-0.47	-1.16	-1.79	-2.41	-3.04	-3.65	-4.06	(-3.97)
1076	-1.12	-1.36	-1.58	-1.98	-2.34	-2.70	-3.06	-3.41	-3.65	(-3.63)
1078	-4.75	-5.08	-5.40	-6.02	-6.64	-7.28	-7.96	-8.62	-9.13	(-9.23)
1079	-8.18	-8.27	-8.37	-8.59	-8.86	-9.18	-9.57	-9.97	-10.23	(-10.07)
1080	-8.01	-8.12	-8.24	-8.53	-8.88	-9.30	-9.77	-10.27	-10.64	(-10.64)
1081	-5.88	-5.97	-6.06	-6.26	-6.49	-6.78	-7.12	-7.47	-7.70	(-7.54)
1082	-6.02	-6.26	-6.50	-6.99	-7.50	-8.04	-8.63	-9.22	-9.66	(-9.70)
1083	-7.83	-9.09	-10.24	-12.27	-14.10	-15.86	-17.62	-19.28	-20.37	(-20.00)
1084	-1.22	-1.23	-1.23	-1.27	-1.33	-1.44	-1.57	-1.72	-1.81	(-1.69)
1085	-0.59	-0.72	-0.84	-1.06	-1.26	-1.46	-1.67	-1.88	-2.00	(-1.86)
1086	-15.14	-14.46	-13.86	-12.84	-12.03	-11.39	-10.88	-10.48	-10.09	(-9.58)
1087	-8.40	-8.95	-7.54	-6.82	-6.23	-5.74	-5.34	-5.01	-4.64	(-4.09)
1088	-8.38	-8.25	-8.14	-7.96	-7.83	-7.77	-7.77	-7.80	-7.75	(-7.46)
1089	-6.93	-6.88	-6.83	-6.77	-6.74	-6.77	-6.84	-6.92	-6.93	(-6.69)
1090	-24.36	-23.27	-22.30	-20.60	-19.18	-18.01	-17.04	-16.20	-15.40	(-14.47)
1091	-24.56	-23.45	-22.45	-20.71	-19.26	-18.06	-17.06	-16.20	-15.38	(-14.43)
1092	-1.52	-1.68	-1.83	-2.08	-2.36	-2.62	-2.89	-3.14	-3.30	(-3.19)
1093	-17.72	-17.36	-17.07	-16.65	-16.41	-16.33	-16.40	-16.54	-16.60	(-16.27)
1100	-14.20	-13.16	-12.21	-10.54	-9.12	-7.89	-6.81	-5.86	-5.02	(-4.26)
1101	-4.52	-4.01	-3.55	-2.71	-1.97	-1.32	-0.74	-0.22	0.26	(0.70)
1102	-1.50	-1.17	-0.85	-0.27	0.25	0.73	1.17	1.57	1.96	(2.32)
1103	-2.27	-2.09	-1.93	-1.68	-1.49	-1.34	-1.24	-1.17	-1.11	(-1.03)
1106	1.99	1.96	1.93	1.85	1.75	1.63	1.49	1.34	1.25	(1.31)
1107	2.02	2.02	2.01	1.95	1.86	1.73	1.57	1.40	1.27	(1.28)
1108	8.73	7.57	6.57	4.91	3.60	2.52	1.59	0.81	0.34	(0.49)
1109	12.19	11.47	10.84	9.77	8.90	8.18	7.56	7.04	6.65	(6.45)
1110	20.69	19.70	18.82	17.32	16.08	15.05	14.16	13.41	12.79	(12.38)
1111	4.91	4.82	4.75	4.65	4.58	4.54	4.51	4.50	4.52	(4.59)
1112	1.12	1.44	1.75	2.35	2.91	3.44	3.94	4.40	4.86	(5.37)
1113	7.57	7.05	6.60	5.83	5.22	4.70	4.27	3.90	3.63	(3.54)
1114	3.60	3.25	2.97	2.52	2.19	1.92	1.69	1.50	1.44	(1.66)
1115	3.47	3.02	2.63	2.01	1.53	1.15	0.82	0.56	0.44	(0.62)
1116	3.85	3.34	2.90	2.20	1.65	1.21	0.83	0.53	0.37	(0.51)
1117	9.52	9.05	8.63	7.94	7.40	6.96	6.58	6.27	6.06	(6.04)
1119	10.80	10.25	9.77	8.97	8.34	7.83	7.40	7.05	6.81	(6.74)
1120	18.21	16.47	14.94	12.42	10.42	8.76	7.33	6.14	5.37	(5.40)

$\log K$, $P = 5$ kb

NAME	INDEX	TEMPERATURE, °C							
		0	25	50	75	100	125	150	175
K-OXIDE	1123	89.74	83.27	77.66	72.82	68.63	64.96	61.73	58.87
NA-OXIDE	1124	72.12	66.83	62.27	58.35	54.97	52.03	49.46	47.20
AMORPH SIL	1125	-2.85	-1.96	-1.55	-1.32	-1.16	-1.05	-0.97	-0.90
A-CRIST	1126	-4.05	-2.98	-2.44	-2.10	-1.87	-1.69	-1.55	-1.44
B-CRIST	1127	-3.38	-2.39	-1.92	-1.64	-1.45	-1.31	-1.21	-1.12
CHALCED	1128	-4.66	-3.53	-2.93	-2.55	-2.28	-2.07	-1.91	-1.77
WAIR	1130	25.44	24.61	22.12	19.53	17.15	15.04	13.21	11.62
LAUM	1132	19.03	19.11	17.40	15.50	13.72	12.15	10.80	9.64
ACAN	1501	-42.15	-36.81	-33.45	-30.71	-28.39	-26.39	-24.63	-23.08
CP	1502	-39.40	-32.47	-30.03	-28.17	-26.65	-25.38	-24.28	-23.34
BN	1503	-128.61	-101.81	-93.42	-86.77	-81.17	-76.34	-72.13	-68.41
CC	1504	-44.53	-34.50	-31.46	-28.99	-26.90	-25.11	-23.54	-22.15
QTZ	1505	-4.97	-3.81	-3.19	-2.79	-2.50	-2.28	-2.10	-1.94
MG	1506	33.16	30.60	27.45	24.68	22.29	20.24	18.47	16.94
KALS	1507	13.79	12.98	11.83	10.72	9.74	8.88	8.14	7.51
FERROSIL	1508	8.02	8.14	7.58	6.95	6.36	5.82	5.35	4.94
PYROPH	1509	3.27	4.47	3.97	3.09	2.16	1.30	0.54	-0.10
TALC	1510	19.45	21.58	20.40	18.92	17.48	16.18	15.04	14.06
7-A CL	1512	79.28	74.66	67.11	60.39	54.59	49.63	45.40	41.78
14-A CL	1513	75.27	70.91	63.56	57.01	51.36	46.53	42.40	38.87
CLNZ	1515	53.02	48.84	43.57	38.75	34.52	30.88	27.75	25.06
LAWS	1516	26.74	24.85	22.20	19.74	17.59	15.73	14.15	12.82
TREM	1517	59.34	62.01	58.25	54.07	50.15	46.64	43.58	40.92
ANTH	1518	65.49	67.48	62.67	57.62	52.95	48.81	45.18	42.04
ZOIS	1519	53.05	48.86	43.59	38.76	34.53	30.89	27.75	25.07
GROSS	1529	58.70	55.22	50.30	45.74	41.73	38.26	35.26	32.68
ANDRA	1530	53.44	52.28	48.70	45.19	42.03	39.25	36.84	34.75
ALB	1531	4.23	5.56	5.50	5.09	4.60	4.12	3.70	3.34
PREHN	1532	39.44	37.37	33.92	30.64	27.72	25.18	23.00	21.13
ENST	1537	11.34	11.32	10.43	9.55	8.76	8.06	7.45	6.92
PARG	1538	114.57	108.27	98.05	88.64	80.37	73.21	67.03	61.70
ANTIG	1542	479.19	478.17	440.80	405.27	373.85	346.66	323.30	303.31
HM	1544	20.93	19.39	17.40	15.63	14.11	12.81	11.68	10.70
ORD-EP	1545	49.24	46.32	41.80	37.57	33.83	30.59	27.80	25.40
MARG	1551	55.35	49.17	42.83	37.25	32.44	28.34	24.83	21.83
MERCURY	1552	-22.43	-19.86	-18.36	-17.07	-15.92	-14.88	-13.93	-13.05
PO	1555	-3.55	-2.93	-2.95	-3.06	-3.21	-3.36	-3.52	-3.68
NESQ	1558	6.31	5.58	5.46	5.40	5.36	5.32	5.30	5.28
EPID	1559	49.24	46.32	41.80	37.57	33.83	30.59	27.79	25.39
HI-ALB	1560	5.76	6.91	6.70	6.16	5.55	4.98	4.47	4.04

$\log K$, $P = 5$ kb

INDEX	TEMPERATURE, °C									
	200	225	250	300	350	400	450	500	550	600
1123	56.32	54.04	51.98	48.42	45.45	42.94	40.79	38.93	37.32	(35.93)
1124	45.21	43.43	41.84	39.14	36.92	35.08	33.53	32.22	31.11	(30.23)
1125	-0.85	-0.80	-0.77	-0.72	-0.68	-0.67	-0.67	-0.69	-0.71	(-0.73)
1126	-1.34	-1.25	-1.18	-1.04	-0.92	-0.82	-0.72	-0.63	-0.54	(-0.44)
1127	-1.04	-0.98	-0.92	-0.82	-0.72	-0.64	-0.56	-0.49	-0.41	(-0.33)
1128	-1.64	-1.54	-1.44	-1.27	-1.12	-1.00	-0.88	-0.77	-0.67	(-0.56)
1130	10.26	9.09	8.10	6.51	5.32	4.37	3.60	3.01	2.79	(3.37)
1132	8.68	7.87	7.20	6.20	5.52	5.02	4.64	4.40	4.50	(5.37)
1501	-21.72	-20.51	-19.43	-17.57	-16.05	-14.82	-13.79	-12.92	-12.11	(-11.24)
1502	-22.52	-21.80	-21.16	-20.11	-19.31	-18.73	-18.35	-18.07	-17.74	(-17.08)
1503	-65.10	-62.15	-59.52	-55.03	-51.38	-48.42	-46.04	-44.06	-42.12	(-39.62)
1504	-20.91	-19.79	-18.78	-17.02	-15.55	-14.32	-13.29	-12.41	-11.57	(-10.65)
1505	-1.81	-1.69	-1.59	-1.40	-1.24	-1.10	-0.98	-0.86	-0.75	(-0.64)
1506	15.61	14.45	13.43	11.72	10.34	9.17	8.15	7.27	6.66	(6.59)
1507	6.98	6.52	6.14	5.53	5.07	4.72	4.42	4.20	4.13	(4.39)
1508	4.59	4.28	4.01	3.58	3.24	2.97	2.75	2.58	2.49	(2.58)
1509	-0.63	-1.07	-1.43	-1.94	-2.26	-2.48	-2.64	-2.72	-2.53	(-1.73)
1510	13.23	12.52	11.92	11.00	10.34	9.86	9.48	9.23	9.23	(9.77)
1512	38.68	36.02	33.73	30.05	27.21	24.93	23.02	21.49	20.69	(21.39)
1513	35.86	33.27	31.06	27.48	24.75	22.55	20.71	19.25	18.51	(19.26)
1515	22.76	20.78	19.08	16.33	14.23	12.52	11.10	9.96	9.41	(10.08)
1516	11.69	10.73	9.93	8.67	7.75	7.04	6.45	6.01	5.91	(6.55)
1517	38.64	36.68	35.01	32.35	30.36	28.82	27.59	26.68	26.36	(27.26)
1518	39.32	36.98	34.96	31.72	29.24	27.30	25.71	24.49	23.92	(24.66)
1519	22.76	20.78	19.08	16.33	14.22	12.52	11.09	9.95	9.40	(10.07)
1529	30.46	28.55	26.89	24.21	22.14	20.46	19.06	17.94	17.35	(17.84)
1530	32.93	31.35	29.98	27.73	25.97	24.54	23.33	22.35	21.79	(22.05)
1531	3.05	2.81	2.62	2.38	2.26	2.20	2.18	2.22	2.41	(2.96)
1532	19.53	18.16	16.99	15.12	13.71	12.59	11.65	10.92	10.63	(11.24)
1537	6.46	6.06	5.71	5.14	4.69	4.34	4.04	3.80	3.67	(3.72)
1538	57.11	53.15	49.72	44.13	39.81	36.33	33.43	31.11	29.83	(30.58)
1542	286.23	271.66	259.20	239.37	224.46	212.76	203.22	195.86	192.68	(197.96)
1544	9.85	9.11	8.45	7.36	6.47	5.72	5.06	4.49	4.11	(4.09)
1545	23.34	21.57	20.05	17.59	15.69	14.16	12.88	11.86	11.37	(11.98)
1551	19.26	17.06	15.17	12.12	9.76	7.85	6.23	4.93	4.28	(5.01)
1552	-12.24	-11.48	-10.78	-9.48	-8.33	-7.62	-7.00	-6.42	-5.85	(-5.24)
1555	-3.84	-3.99	-4.13	-4.41	-4.69	-4.98	-5.29	-5.59	-5.79	(-5.72)
1558	5.27	5.26	5.26	5.23	5.19	5.09	4.95	4.79	4.69	(4.84)
1559	23.32	21.55	20.02	17.54	15.63	14.09	12.80	11.76	11.26	(11.85)
1560	3.67	3.37	3.13	2.79	2.60	2.48	2.42	2.42	2.58	(3.11)

INDEX

The systems considered in the present study are cross-indexed below in an alphabetical arrangement of components other than the ubiquitous components, HCl and H_2O. The latter components appear only at the top of each index page. The balancing component is shown in parentheses and the components represented by the descriptive variables appear in italics.

System HCl–H_2O–	Constraints	25°-300°C	400°-600°C
Ag_2O–(H_2S)–H_2SO_4		132	133
(Al_2O_3)–CaO–CO_2–K_2O–MgO–SiO_2	Quartz and K-feldspar saturation.		
	$X_{CO_2} = 0.01$		273
	$X_{CO_2} = 0.05$		274
	$X_{CO_2} = 0.10$		275
	$X_{CO_2} = 0.30$		276
	$X_{CO_2} = 0.50$		277
	$X_{CO_2} = 0.70$		278
	$X_{CO_2} = 0.90$		279
	$X_{CO_2} = 0.95$		280
	$X_{CO_2} = 0.99$		281
	Quartz and K-feldspar saturation. Metastable 7-A clinochlore was considered instead of its stable counterpart, 14-A clinochlore.		
	$X_{CO_2} = 0.01$		282
	$X_{CO_2} = 0.05$		283
	$X_{CO_2} = 0.10$		284
	$X_{CO_2} = 0.30$		285
	$X_{CO_2} = 0.50$		286
	$X_{CO_2} = 0.70$		287
	$X_{CO_2} = 0.90$		288
	$X_{CO_2} = 0.95$		289
	$X_{CO_2} = 0.99$		290
(Al_2O_3)–CaO–CO_2–MgO–SiO_2	Quartz saturation.		
	$X_{CO_2} = 0.01$		255
	$X_{CO_2} = 0.05$		256
	$X_{CO_2} = 0.10$		257
	$X_{CO_2} = 0.30$		258
	$X_{CO_2} = 0.50$		259
	$X_{CO_2} = 0.70$		260
	$X_{CO_2} = 0.90$		261
	$X_{CO_2} = 0.95$		262
	$X_{CO_2} = 0.99$		263
	Quartz saturation. Metastable 7-A clinochlore was considered instead of its stable counterpart, 14-A clinochlore.		

INDEX

System HCl–H$_2$O–	Constraints	25°-300°C	400°-600°C
(Al$_2$O$_3$)–CaO–CO$_2$–MgO–SiO$_2$ (cont'd)	$X_{CO_2} = 0.01$		264
	$X_{CO_2} = 0.05$		265
	$X_{CO_2} = 0.10$		266
	$X_{CO_2} = 0.30$		267
	$X_{CO_2} = 0.50$		268
	$X_{CO_2} = 0.70$		269
	$X_{CO_2} = 0.90$		270
	$X_{CO_2} = 0.95$		271
	$X_{CO_2} = 0.99$		272
(Al$_2$O$_3$)–CaO–CO$_2$–SiO$_2$	$X_{CO_2} = 0.01$		165
	$X_{CO_2} = 0.05$		166
	$X_{CO_2} = 0.10$		167
	$X_{CO_2} = 0.30$		168
	$X_{CO_2} = 0.50$		169
	$X_{CO_2} = 0.70$		170
	$X_{CO_2} = 0.90$		171
	$X_{CO_2} = 0.95$		172
	$X_{CO_2} = 0.99$		173
Al$_2$O$_3$–(CaO)–CO$_2$–SiO$_2$	$X_{CO_2} = 0.01$		174
	$X_{CO_2} = 0.05$		175
	$X_{CO_2} = 0.10$		176
	$X_{CO_2} = 0.30$		177
	$X_{CO_2} = 0.50$		178
	$X_{CO_2} = 0.70$		179
	$X_{CO_2} = 0.90$		180
	$X_{CO_2} = 0.95$		181
	$X_{CO_2} = 0.99$		182
Al$_2$O$_3$–CaO–CO$_2$–(SiO$_2$)	$X_{CO_2} = 0.01$		183
	$X_{CO_2} = 0.05$		184
	$X_{CO_2} = 0.10$		185
	$X_{CO_2} = 0.30$		186
	$X_{CO_2} = 0.50$		187
	$X_{CO_2} = 0.70$		188
	$X_{CO_2} = 0.90$		189
	$X_{CO_2} = 0.95$		190
	$X_{CO_2} = 0.99$		191

System HCl–H$_2$O–	Constraints	25°-300°C	400°-600°C
(Al$_2$O$_3$)–CaO–FeO–SiO$_2$	Amorphous silica saturation.	150	
	Quartz saturation.	54	55
Al$_2$O$_3$–CaO–FeO–(SiO$_2$)	Gibbsite, diaspore, or corundum saturation, depending on which mineral is stable at each pressure and temperature.	56	57
(Al$_2$O$_3$)–CaO–K$_2$O–SiO$_2$	Amorphous silica saturation.	151	
	Quartz saturation.	58	59
Al$_2$O$_3$–CaO–K$_2$O–(SiO$_2$)	Gibbsite, diaspore, or corundum saturation, depending on which mineral is stable at each pressure and temperature.	60	61
(Al$_2$O$_3$)–CaO–MgO–Na$_2$O–SiO$_2$	Albite and amorphous silica saturation.	163	
	Albite and amorphous silica saturation. Metastable 7-A clinochlore was considered instead of its stable counterpart, 14-A clinochlore.	164	
	Quartz and albite saturation.	122	123
	Quartz and albite saturation. Metastable 7-A clinochlore was considered instead of its stable counterpart, 14-A clinochlore.	124	125
(Al$_2$O$_3$)–CaO–MgO–SiO$_2$	Amorphous silica saturation.	152	
	Amorphous silica saturation. Metastable 7-A clinochlore was considered instead of its stable counterpart, 14-A clinochlore.	153	
	Quartz saturation.	62	63
	Quartz saturation. Metastable 7-A clinochlore was considered instead of its stable counterpart, 14-A clinochlore.	64	65
Al$_2$O$_3$–CaO–MgO–(SiO$_2$)	Gibbsite, diaspore, or corundum saturation, depending on which mineral is stable at each pressure and temperature.	66	67

System HCl–H$_2$O–	Constraints	25°-300°C	400°-600°C
Al$_2$O$_3$–CaO–MgO–(SiO$_2$) (cont'd)	Gibbsite, diaspore, or corundum saturation, depending on which mineral is stable at each pressure and temperature. Metastable 7-A clinochlore was considered instead of its stable counterpart 14-A clinochlore.	68	69
	Gibbsite, diaspore, or corundum saturation, depending on which mineral is stable at each pressure and temperature. Metastable chrysotile was considered instead of its stable counterpart antigorite.	70	71
	Gibbsite, diaspore, or corundum saturation, depending on which mineral is stable at each pressure and temperature. Metastable 7-A clinochlore and chrysotile were considered instead of their stable counterparts 14-A clinochlore and antigorite.	72	73
(Al$_2$O$_3$)–CaO–Na$_2$O–SiO$_2$	Amorphous silica saturation	154	
	Quartz saturation.	74	75
Al$_2$O$_3$–CaO–Na$_2$O–(SiO$_2$)	Gibbsite, diaspore, or corundum saturation, depending on which mineral is stable at each pressure and temperature.	76	77
(Al$_2$O$_3$)–CaO–SiO$_2$		2	3
Al$_2$O$_3$–(CaO)–SiO$_2$		4	5
Al$_2$O$_3$–CaO–(SiO$_2$)		6	7
(Al$_2$O$_3$)–CO$_2$–MgO–SiO$_2$	$X_{CO_2} = 0.01$		192
	$X_{CO_2} = 0.05$		193
	$X_{CO_2} = 0.10$		194
	$X_{CO_2} = 0.30$		195
	$X_{CO_2} = 0.50$		196
	$X_{CO_2} = 0.70$		197
	$X_{CO_2} = 0.90$		198
	$X_{CO_2} = 0.95$		199
	$X_{CO_2} = 0.99$		200

System HCl–H$_2$O–	Constraints	25°-300°C	400°-600°C
(Al$_2$O$_3$)–CO$_2$–MgO–SiO$_2$ (cont'd)	Metastable 7-A clinochlore was considered instead of its stable counterpart, 14-A clinochlore.		
	$X_{CO_2} = 0.01$		201
	$X_{CO_2} = 0.05$		202
	$X_{CO_2} = 0.10$		203
	$X_{CO_2} = 0.30$		204
	$X_{CO_2} = 0.50$		205
	$X_{CO_2} = 0.70$		206
	$X_{CO_2} = 0.90$		207
	$X_{CO_2} = 0.95$		208
	$X_{CO_2} = 0.99$		209
Al$_2$O$_3$–CO$_2$–(MgO)–SiO$_2$	$X_{CO_2} = 0.01$		210
	$X_{CO_2} = 0.05$		211
	$X_{CO_2} = 0.10$		212
	$X_{CO_2} = 0.30$		213
	$X_{CO_2} = 0.50$		214
	$X_{CO_2} = 0.70$		215
	$X_{CO_2} = 0.90$		216
	$X_{CO_2} = 0.95$		217
	$X_{CO_2} = 0.99$		218
	Metastable 7-A clinochlore was considered instead of its stable counterpart, 14-A clinochlore.		
	$X_{CO_2} = 0.01$		219
	$X_{CO_2} = 0.05$		220
	$X_{CO_2} = 0.10$		221
	$X_{CO_2} = 0.30$		222
	$X_{CO_2} = 0.50$		223
	$X_{CO_2} = 0.70$		224
	$X_{CO_2} = 0.90$		225
	$X_{CO_2} = 0.95$		226
	$X_{CO_2} = 0.99$		227
Al$_2$O$_3$–CO$_2$–MgO–(SiO$_2$)	$X_{CO_2} = 0.01$		228
	$X_{CO_2} = 0.05$		229
	$X_{CO_2} = 0.10$		230
	$X_{CO_2} = 0.30$		231
	$X_{CO_2} = 0.50$		232

INDEX

System HCl–H$_2$O–	Constraints	25°-300°C	400°-600°C
Al$_2$O$_3$–CO$_2$–MgO–(SiO$_2$) (cont'd)	$X_{CO_2} = 0.70$		233
	$X_{CO_2} = 0.90$		234
	$X_{CO_2} = 0.95$		235
	$X_{CO_2} = 0.99$		236
Al$_2$O$_3$–CO$_2$–MgO–(SiO$_2$)	Metastable 7-A clinochlore was considered instead of its stable counterpart, 14-A clinochlore.		
	$X_{CO_2} = 0.01$		237
	$X_{CO_2} = 0.05$		238
	$X_{CO_2} = 0.10$		239
	$X_{CO_2} = 0.30$		240
	$X_{CO_2} = 0.50$		241
	$X_{CO_2} = 0.70$		242
	$X_{CO_2} = 0.90$		243
	$X_{CO_2} = 0.95$		244
	$X_{CO_2} = 0.99$		245
(Al$_2$O$_3$)–FeO–K$_2$O–SiO$_2$	Amorphous silica saturation.	155	
	Quartz saturation.	78	79
Al$_2$O$_3$–FeO–K$_2$O–(SiO$_2$)	Gibbsite, diaspore, or corundum saturation, depending on which mineral is stable at each pressure and temperature.	80	81
(Al$_2$O$_3$)–FeO–MgO–SiO$_2$	Amorphous silica saturation.	156	
	Amorphous silica saturation. Metastable 7-A clinochlore was considered instead of its stable counterpart, 14-A clinochlore.	157	
	Quartz saturation.	82	83
	Quartz saturation. Metastable 7-A clinochlore was considered instead of its stable counterpart 14-A clinochlore.	84	85
Al$_2$O$_3$–FeO–MgO–(SiO$_2$)	Gibbsite, diaspore, or corundum saturation, depending on which mineral is stable at each pressure and temperature.	86	87

System HCl–H$_2$O–	Constraints	25°-300°C	400°-600°C
Al$_2$O$_3$–*FeO*–*MgO*–(SiO$_2$) (cont'd)	Gibbsite, diaspore, or corundum saturation, depending on which mineral is stable at each pressure and temperature. Metastable 7-A clinochlore was considered instead of its stable counterpart, 14-A clinochlore.	88	89
	Gibbsite, diaspore, or corundum saturation, depending on which mineral is stable at each pressure and temperature. Metastable chrysotile was considered instead of its stable counterpart, antigorite.	90	91
	Gibbsite, diaspore, or corundum saturation, depending on which mineral is stable at each pressure and temperature. Metastable 7-A clinochlore and chrysotile were considered instead of their stable counterparts, 14-A clinochlore and antigorite.	92	93
(Al$_2$O$_3$)–*K$_2$O*–*MgO*–SiO$_2$	Amorphous silica saturation	158	
	Amorphous silica saturation. Metastable 7-A clinochlore was considered instead of its stable counterpart, 14-A clinochlore.	159	
	Quartz saturation.	94	95
	Quartz saturation. Metastable 7-A clinochlore was considered instead of its stable counterpart, 14-A clinochlore.	96	97
Al$_2$O$_3$–*K$_2$O*–*MgO*–(SiO$_2$)	Gibbsite, diaspore, or corundum saturation, depending on which mineral is stable at each pressure and temperature.	98	99
	Gibbsite, diaspore, or corundum saturation, depending on which mineral is stable at each pressure and temperature. Metastable 7-A clinochlore was considered instead of its stable counterpart, 14-A clinochlore.	100	101

INDEX

System HCl–H$_2$O–	Constraints	25°-300°C	400°-600°C
Al$_2$O$_3$–K$_2$O–MgO–(SiO$_2$) (cont'd)	Gibbsite, diaspore, or corundum saturation, depending on which mineral is stable at each pressure and temperature. Metastable chrysotile was considered instead of its stable counterpart, antigorite.	102	103
	Gibbsite, diaspore, or corundum saturation, depending on which mineral is stable at each pressure and temperature. Metastable 7-A clinochlore and chrysotile were considered instead of their stable counterparts, 14-A clinochlore and antigorite.	104	105
(Al$_2$O$_3$)–K$_2$O–Na$_2$O–SiO$_2$	Amorphous silica saturation.	160	
	Quartz saturation.	106	107
Al$_2$O$_3$–K$_2$O–Na$_2$O–(SiO$_2$)	Gibbsite, diaspore, or corundum saturation, depending on which mineral is stable at each pressure and temperature.	108	109
(Al$_2$O$_3$)–K$_2$O–SiO$_2$		8	9
Al$_2$O$_3$–(K$_2$O)–SiO$_2$		10	11
Al$_2$O$_3$–K$_2$O–(SiO$_2$)		12	13
(Al$_2$O$_3$)–MgO–Na$_2$O–SiO$_2$	Amorphous silica saturation.	161	
	Amorphous silica saturation. Metastable 7-A clinochlore was considered instead of its stable counterpart, 14-A clinochlore.	162	
	Quartz saturation.	110	111
	Quartz saturation Metastable 7-A clinochlore was considered instead of its stable counterpart, 14-A clinochlore.	112	113
Al$_2$O$_3$–MgO–Na$_2$O–(SiO$_2$)	Gibbsite, diaspore, or corundum saturation, depending on which mineral is stable at each pressure and temperature.	114	115

System HCl–H$_2$O–	Constraints	25°–300°C	400°–600°C
Al$_2$O$_3$–MgO–Na$_2$O–(SiO$_2$) (cont'd)	Gibbsite, diaspore, or corundum saturation, depending on which mineral is stable at each pressure and temperature. Metastable 7-A clinochlore was considered instead of its stable counterpart, 14-A clinochlore.	116	117
	Gibbsite, diaspore, or corundum saturation, depending on which mineral is stable at each pressure and temperature. Metastable chrysotile was considered instead of its stable counterpart, antigorite.	118	119
	Gibbsite, diaspore, or corundum saturation, depending on which mineral is stable at each pressure and temperature. Metastable 7-A clinochlore and chrysotile were considered instead of their stable counterparts, 14-A clinochlore and antigorite.	120	121
(Al$_2$O$_3$)–MgO–SiO$_2$		14	15
	Metastable 7-A clinochlore was considered instead of its stable counterpart, 14-A clinochlore.	16	17
Al$_2$O$_3$–(MgO)–SiO$_2$		18	19
	Metastable 7-A clinochlore was considered instead of its stable counterpart, 14-A clinochlore.	20	21
	Metastable chrysotile was considered instead of its stable counterpart, antigorite.	22	23
	Metastable 7-A clinochlore and chrysotile were considered instead of their stable counterparts, 14-A clinochlore and antigorite.	24	25
Al$_2$O$_3$–MgO–(SiO$_2$)		26	27
	Metastable 7-A clinochlore was considered instead of its stable counterpart, 14-A clinochlore.	28	29

INDEX

System HCl–H$_2$O–	Constraints	25°-300°C	400°-600°C
Al$_2$O$_3$–MgO–(SiO$_2$) (cont'd)	Metastable chrysotile was considered instead of its stable counterpart, antigorite.	30	31
	Metastable 7-A clinochlore and chrysotile were considered instead of their stable counterparts, 14-A clinochlore and antigorite.	32	33
(Al$_2$O$_3$)–Na$_2$O–SiO$_2$		34	35
Al$_2$O$_3$–(Na$_2$O)–SiO$_2$		36	37
Al$_2$O$_3$–Na$_2$O–(SiO$_2$)		38	39
(BaO)–CO$_2$–H$_2$SO$_4$		292	293
CaO–(Al$_2$O$_3$)–CO$_2$–K$_2$O–MgO–SiO$_2$	Quartz and K-feldspar saturation. $X_{CO_2} = 0.01$		273
	$X_{CO_2} = 0.05$		274
	$X_{CO_2} = 0.10$		275
	$X_{CO_2} = 0.30$		276
	$X_{CO_2} = 0.50$		277
	$X_{CO_2} = 0.70$		278
	$X_{CO_2} = 0.90$		279
	$X_{CO_2} = 0.95$		280
	$X_{CO_2} = 0.99$		281
	Quartz and K-feldspar saturation. Metastable 7-A clinochlore was considered instead of its stable counterpart, 14-A clinochlore. $X_{CO_2} = 0.01$		282
	$X_{CO_2} = 0.05$		283
	$X_{CO_2} = 0.10$		284
	$X_{CO_2} = 0.30$		285
	$X_{CO_2} = 0.50$		286
	$X_{CO_2} = 0.70$		287
	$X_{CO_2} = 0.90$		288
	$X_{CO_2} = 0.95$		289
	$X_{CO_2} = 0.99$		290
CaO–(Al$_2$O$_3$)–CO$_2$–MgO–SiO$_2$	Quartz saturation. $X_{CO_2} = 0.01$		255
	$X_{CO_2} = 0.05$		256
	$X_{CO_2} = 0.10$		257

System HCl–H$_2$O–	Constraints	25° -300°C	400° -600°C
CaO–(Al$_2$O$_3$)–CO$_2$–MgO–SiO$_2$ (cont'd)	$X_{CO_2} = 0.30$		258
	$X_{CO_2} = 0.50$		259
	$X_{CO_2} = 0.70$		260
	$X_{CO_2} = 0.90$		261
	$X_{CO_2} = 0.95$		262
	$X_{CO_2} = 0.99$		263
	Quartz saturation. Metastable 7-A clinochlore was considered instead of its stable counterpart, 14-A clinochlore.		
	$X_{CO_2} = 0.01$		264
	$X_{CO_2} = 0.05$		265
	$X_{CO_2} = 0.10$		266
	$X_{CO_2} = 0.30$		267
	$X_{CO_2} = 0.50$		268
	$X_{CO_2} = 0.70$		269
	$X_{CO_2} = 0.90$		270
	$X_{CO_2} = 0.95$		271
	$X_{CO_2} = 0.99$		272
CaO–(Al$_2$O$_3$)–CO$_2$–SiO$_2$	$X_{CO_2} = 0.01$		165
	$X_{CO_2} = 0.05$		166
	$X_{CO_2} = 0.10$		167
	$X_{CO_2} = 0.30$		168
	$X_{CO_2} = 0.50$		169
	$X_{CO_2} = 0.70$		170
	$X_{CO_2} = 0.90$		171
	$X_{CO_2} = 0.95$		172
	$X_{CO_2} = 0.99$		173
(CaO)–Al$_2$O$_3$–CO$_2$–SiO$_2$	$X_{CO_2} = 0.01$		174
	$X_{CO_2} = 0.05$		175
	$X_{CO_2} = 0.10$		176
	$X_{CO_2} = 0.30$		177
	$X_{CO_2} = 0.50$		178
	$X_{CO_2} = 0.70$		179
	$X_{CO_2} = 0.90$		180
	$X_{CO_2} = 0.95$		181
	$X_{CO_2} = 0.99$		182

INDEX

System HCl–H$_2$O–	Constraints	25°–300°C	400°–600°C
$CaO-Al_2O_3-CO_2-(SiO_2)$	$X_{CO_2} = 0.01$		183
	$X_{CO_2} = 0.05$		184
	$X_{CO_2} = 0.10$		185
	$X_{CO_2} = 0.30$		186
	$X_{CO_2} = 0.50$		187
	$X_{CO_2} = 0.70$		188
	$X_{CO_2} = 0.90$		189
	$X_{CO_2} = 0.95$		190
	$X_{CO_2} = 0.99$		191
$CaO-(Al_2O_3)-FeO-SiO_2$	Amorphous silica saturation.	150	
	Quartz saturation.	54	55
$CaO-Al_2O_3-FeO-(SiO_2)$	Gibbsite, diaspore, or corundum saturation, depending on which mineral is stable at each pressure and temperature.	56	57
$CaO-(Al_2O_3)-K_2O-SiO_2$	Amorphous silica saturation.	151	
	Quartz saturation.	58	59
$CaO-Al_2O_3-K_2O-(SiO_2)$	Gibbsite, diaspore, or corundum saturation, depending on which mineral is stable at each pressure and temperature.	60	61
$CaO-(Al_2O_3)-MgO-Na_2O-SiO_2$	Albite and amorphous silica saturation.	163	
	Albite and amorphous silica saturation. Metastable 7-A clinochlore was considered instead of its stable counterpart, 14-A clinochlore.	164	
	Quartz and albite saturation.	122	123
	Quartz and albite saturation. Metastable 7-A clinochlore was considered instead of its stable counterpart, 14-A clinochlore.	124	125
$CaO-(Al_2O_3)-MgO-SiO_2$	Amorphous silica saturation.	152	
	Amorphous silica saturation. Metastable 7-A clinochlore was considered instead of its stable counterpart, 14-A clinochlore.	153	
	Quartz saturation.	62	63

System HCl–H$_2$O–	Constraints	25°-300°C	400°-600°C
CaO–(Al$_2$O$_3$)–MgO–SiO$_2$ (cont'd)	Quartz saturation. Metastable 7-A clinochlore was considered instead of its stable counterpart, 14-A	64	65
CaO Al$_2$O$_3$–MgO–(SiO$_2$)	Gibbsite, diaspore, or corundum saturation, depending on which mineral is stable at each pressure and temperature.	66	67
	Gibbsite, diaspore, or corundum saturation, depending on which mineral is stable at each pressure and temperature. Metastable 7-A clinochlore was considered instead of its stable counterpart, 14-A clinochlore.	68	69
	Gibbsite, diaspore, or corundum saturation, depending on which mineral is stable at each pressure and temperature. Metastable chrysotile was considered instead of its stable counterpart, antigorite.	70	71
	Gibbsite, diaspore, or corundum saturation, depending on which mineral is stable at each pressure and temperature. Metastable 7-A clinochlore and chrysotile were considered instead of their stable counterparts, 14-A clinochlore and antigorite.	72	73
CaO–(Al$_2$O$_3$)–Na$_2$O–SiO$_2$	Amorphous silica saturation.	154	
	Quartz saturation.	74	75
CaO–Al$_2$O$_3$–Na$_2$O–(SiO$_2$)	Gibbsite, diaspore, or corundum saturation, depending on which mineral is stable at each pressure and temperature.	76	77
CaO–(Al$_2$O$_3$)–SiO$_2$		2	3
(CaO)–Al$_2$O$_3$–SiO$_2$		4	5
CaO–Al$_2$O$_3$–(SiO$_2$)		6	7
CaO–(CO$_2$)–FeO		126	127
(CaO)–CO$_2$–H$_2$SO$_4$		294	295

INDEX

System HCl–H$_2$O–	Constraints	25° -300°C	400° -600°C
CaO–(CO$_2$)–MgO		128	129
CaO–CO$_2$–MgO–(SiO$_2$)	$X_{CO_2} = 0.01$		246
	$X_{CO_2} = 0.05$		247
	$X_{CO_2} = 0.10$		248
	$X_{CO_2} = 0.30$		249
	$X_{CO_2} = 0.50$		250
	$X_{CO_2} = 0.70$		251
	$X_{CO_2} = 0.90$		252
	$X_{CO_2} = 0.95$		253
	$X_{CO_2} = 0.99$		254
CaO–(MgO)–SiO$_2$		40	41
	Metastable chrysotile was considered instead of its stable counterpart, antigorite.	42	43
(CaO)–MgO–SiO$_2$		44	45
CaO–MgO–(SiO$_2$)		46	47
	Metastable chrysotile was considered instead of its stable counterpart, antigorite.	48	49
CO$_2$ and other components--see listing under other components.			
Cu$_2$O–FeO–H$_2$S–(H$_2$SO$_4$)	$\log(a_{H^+}a_{HS^-}) = -10.0$	144	145
	$\log(a_{H^+}a_{HS^-}) = -11.0$	146	147
	$\log(a_{H^+}a_{HS^-}) = -12.0$	148	149
CuO–H$_2$S–(H$_2$SO$_4$)		134	135
CuO–(H$_2$S)–H$_2$SO$_4$		136	137
(CuO)–O$_2$–S$_2$		302	303
FeO–(Al$_2$O$_3$)–CaO–SiO$_2$	Amorphous silica saturation.	150	
	Quartz saturation.	54	55
FeO–Al$_2$O$_3$–CaO–(SiO$_2$)	Gibbsite, diaspore, or corundum saturation, depending on which mineral is stable at each pressure and temperature.	56	57
FeO–(Al$_2$O$_3$)–K$_2$O–SiO$_2$	Amorphous silica saturation.	155	

System HCl–H₂O–	Constraints	25°–300°C	400°–600°C
$FeO-(Al_2O_3)-K_2O-SiO_2$ (cont'd)	Quartz saturation.	78	79
$FeO-Al_2O_3-K_2O-(SiO_2)$	Gibbsite, diaspore, or corundum saturation, depending on which mineral is stable at each pressure and temperature.	80	81
$FeO-(Al_2O_3)-MgO-SiO_2$	Amorphous silica saturation	156	
	Amorphous silica saturation. Metastable 7-A clinochlore was considered instead of its stable counterpart, 14-A clinochlore.	157	
	Quartz saturation.	82	83
	Quartz saturation. Metastable 7-A clinochlore was considered instead of its stable counterpart, 14-A clinochlore.	84	85
$FeO-Al_2O_3-MgO-(SiO_2)$	Gibbsite, diaspore, or corundum saturation, depending on which mineral is stable at each pressure and temperature.	86	87
	Gibbsite, diaspore, or corundum saturation, depending on which mineral is stable at each pressure and temperature. Metastable 7-A clinochlore was considered instead of its stable counterpart, 14-A clinochlore.	88	89
	Gibbsite, diaspore, or corundum saturation, depending on which mineral is stable at each pressure and temperature. Metastable chrysotile was considered instead of its stable counterpart, antigorite.	90	91
	Gibbsite, diaspore, or corundum saturation, depending on which mineral is stable at each pressure and temperature. Metastable 7-A clinochlore and chrysotile were considered instead of their stable counterparts, 14-A clinochlore and antigorite.	92	93

INDEX

System HCl–H$_2$O–	Constraints	25° -300°C	400° -600°C
$FeO-CaO-(CO_2)$		126	127
$FeO-(CO_2)-MgO$		130	131
$FeO-Cu_2O-H_2S-(H_2SO_4)$	$\log(a_{H^+}a_{HS^-}) = -10.0$	144	145
	$\log(a_{H^+}a_{HS^-}) = -11.0$	146	147
	$\log(a_{H^+}a_{HS^-}) = -12.0$	148	149
$FeO-H_2S-(H_2SO_4)$		138	139
$FeO-MgO-(SiO_2)$		50	51
	Metastable chrysotile was considered instead of its stable counterpart, antigorite.	52	53
$(FeO)-O_2-S_2$		304	305
$HgS-(H_2S)-H_2SO_4$		140	141
$(H_2S)-Ag_2O-H_2SO_4$		132	133
$H_2S-CO_2-(ZnO)$		300	301
$H_2S-Cu_2O-FeO-(H_2SO_4)$	$\log(a_{H^+}a_{HS^-}) = -10.0$	144	145
	$\log(a_{H^+}a_{HS^-}) = -11.0$	146	147
	$\log(a_{H^+}a_{HS^-}) = -12.0$	148	149
$H_2S-CuO-(H_2SO_4)$		134	135
$(H_2S)-CuO-H_2SO_4$		136	137
$H_2S-FeO-(H_2SO_4)$		138	139
$(H_2S)-HgS-H_2SO_4$		140	141
$(H_2S)-PbS-ZnS$		142	143
$H_2SO_4-Ag_2O-(H_2S)$		132	133
$H_2SO_4-(BaO)-CO_2$		292	293
$H_2SO_4-(CaO)-CO_2$		294	295
$H_2SO_4-CO_2-(PbO)$		296	297
$H_2SO_4-CO_2-(SrO)$		298	299
$(H_2SO_4)-Cu_2O-FeO-H_2S$	$\log(a_{H^+}a_{HS^-}) = -10.0$	144	145
	$\log(a_{H^+}a_{HS^-}) = -11.0$	146	147
	$\log(a_{H^+}a_{HS^-}) = -12.0$	148	149
$(H_2SO_4)-CuO-H_2S$		134	135
$H_2SO_4-CuO-(H_2S)$		136	137
$(H_2SO_4)-FeO-H_2S$		138	139

System HCl–H$_2$O–	Constraints	25°–300°C	400°–600°C
H$_2$SO$_4$–HgS–(H$_2$S)		140	141
K$_2$O–(Al$_2$O$_3$)–CaO–CO$_2$–MgO–SiO$_2$	Quartz and K-feldspar saturation. $X_{CO_2} = 0.01$ $X_{CO_2} = 0.05$ $X_{CO_2} = 0.10$ $X_{CO_2} = 0.30$ $X_{CO_2} = 0.50$ $X_{CO_2} = 0.70$ $X_{CO_2} = 0.90$ $X_{CO_2} = 0.95$ $X_{CO_2} = 0.99$		273 274 275 276 277 278 279 280 281
	Quartz and K-feldspar saturation. Metastable 7-A clinochlore was considered instead of its stable counterpart, 14-A clinochlore. $X_{CO_2} = 0.01$ $X_{CO_2} = 0.05$ $X_{CO_2} = 0.10$ $X_{CO_2} = 0.30$ $X_{CO_2} = 0.50$ $X_{CO_2} = 0.70$ $X_{CO_2} = 0.90$ $X_{CO_2} = 0.95$ $X_{CO_2} = 0.99$		282 283 284 285 286 287 288 289 290
K$_2$O–(Al$_2$O$_3$)–CaO–SiO$_2$	Amorphous silica saturation.	151	
	Quartz saturation.	58	59
K$_2$O–Al$_2$O$_3$–CaO–(SiO$_2$)	Gibbsite, diaspore, or corundum saturation, depending on which mineral is stable at each pressure and temperature.	60	61
K$_2$O–(Al$_2$O$_3$)–FeO–SiO$_2$	Amorphous silica saturation.	155	
	Quartz saturation.	78	79
K$_2$O–Al$_2$O$_3$–FeO–(SiO$_2$)	Gibbsite, diaspore, or corundum saturation, depending on which mineral is stable at each pressure and temperature.	80	81
K$_2$O–(Al$_2$O$_3$)–MgO–SiO$_2$	Amorphous silica saturation.	158	

System HCl–H$_2$O–	Constraints	25° -300°C	400° -600°C
K$_2$O–(Al$_2$O$_3$)–MgO–SiO$_2$ (cont'd)	Amorphous silica saturation. Metastable 7-A clinochlore was considered instead of its stable counterpart, 14-A clinochlore.	159	
	Quartz saturation.	94	95
	Quartz saturation. Metastable 7-A clinochlore was considered instead of its stable counterpart, 14-A clinochlore.	96	97
K$_2$O–Al$_2$O$_3$–MgO–(SiO$_2$)	Gibbsite, diaspore, or corundum saturation, depending on which mineral is stable at each pressure and temperature.	98	99
	Gibbsite, diaspore, or corundum saturation, depending on which mineral is stable at each pressure and temperature. Metastable 7-A clinochlore was considered instead of its stable counterpart, 14-A clinochlore.	100	101
	Gibbsite, diaspore, or corundum saturation, depending on which mineral is stable at each pressure and temperature. Metastable chrysotile was considered instead of its stable counterpart, antigorite.	102	103
	Gibbsite, diaspore, or corundum saturation, depending on which mineral is stable at each pressure and temperature. Metastable 7-A clinochlore and chrysotile were considered instead of their stable counterparts, 14-A clinochlore and antigorite.	104	105
K$_2$O–(Al$_2$O$_3$)–Na$_2$O–SiO$_2$	Amorphous silica saturation.	160	
	Quartz saturation.	106	107
K$_2$O–Al$_2$O$_3$–Na$_2$O–(SiO$_2$)	Gibbsite, diaspore, or corundum saturation, depending on which mineral is stable at each pressure and temperature.	108	109
K$_2$O–(Al$_2$O$_3$)–SiO$_2$		8	9

System HCl–H$_2$O–	Constraints	25° -300°C	400° -600°C
(K$_2$O)–Al$_2$O$_3$–SiO$_2$		10	11
K$_2$O–Al$_2$O$_3$–(SiO$_2$)		12	13
MgO–(Al$_2$O$_3$)–CaO–CO$_2$–K$_2$O–SiO$_2$	Quartz and K-feldspar saturation. $X_{CO_2} = 0.01$		273
	$X_{CO_2} = 0.05$		274
	$X_{CO_2} = 0.10$		275
	$X_{CO_2} = 0.30$		276
	$X_{CO_2} = 0.50$		277
	$X_{CO_2} = 0.70$		278
	$X_{CO_2} = 0.90$		279
	$X_{CO_2} = 0.95$		280
	$X_{CO_2} = 0.99$		281
	Quartz and K-feldspar saturation. Metastable 7-A clinochlore was considered instead of its stable counterpart, 14-A clinochlore. $X_{CO_2} = 0.01$		282
	$X_{CO_2} = 0.05$		283
	$X_{CO_2} = 0.10$		284
	$X_{CO_2} = 0.30$		285
	$X_{CO_2} = 0.50$		286
	$X_{CO_2} = 0.70$		287
	$X_{CO_2} = 0.90$		288
	$X_{CO_2} = 0.95$		289
	$X_{CO_2} = 0.99$		290
MgO–(Al$_2$O$_3$)–CaO–CO$_2$–SiO$_2$	Quartz saturation. $X_{CO_2} = 0.01$		255
	$X_{CO_2} = 0.05$		256
	$X_{CO_2} = 0.10$		257
	$X_{CO_2} = 0.30$		258
	$X_{CO_2} = 0.50$		259
	$X_{CO_2} = 0.70$		260
	$X_{CO_2} = 0.90$		261
	$X_{CO_2} = 0.95$		262
	$X_{CO_2} = 0.99$		263

INDEX

System HCl–H$_2$O–	Constraints	25°–300°C	400°–600°C
$MgO-(Al_2O_3)-CaO-CO_2-SiO_2$ (cont'd)	Quartz saturation. Metastable 7-A clinochlore was considered instead of its stable counterpart, 14-A clinochlore.		
	$X_{CO_2} = 0.01$		264
	$X_{CO_2} = 0.05$		265
	$X_{CO_2} = 0.10$		266
	$X_{CO_2} = 0.30$		267
	$X_{CO_2} = 0.50$		268
	$X_{CO_2} = 0.70$		269
	$X_{CO_2} = 0.90$		270
	$X_{CO_2} = 0.95$		271
	$X_{CO_2} = 0.99$		272
$MgO-(Al_2O_3)-CaO-Na_2O-SiO_2$	Albite and amorphous silica saturation.	163	
	Albite and amorphous silica saturation. Metastable 7-A clinochlore was considered instead of its stable counterpart, 14-A clinochlore.	164	
	Quartz and albite saturation.	122	123
	Quartz and albite saturation. Metastable 7-A clinochlore was considered instead of its stable counterpart, 14-A clinochlore.	124	125
$MgO-(Al_2O_3)-CaO-SiO_2$	Amorphous silica saturation.	152	
	Amorphous silica saturation. Metastable 7-A clinochlore was considered instead of its stable counterpart, 14-A clinochlore.	153	
	Quartz saturation.	62	63
	Quartz saturation. Metastable 7-A clinochlore was considered instead of its stable counterpart, 14-A clinochlore.	64	65
$MgO-Al_2O_3-CaO-(SiO_2)$	Gibbsite, diaspore, or corundum saturation, depending on which mineral is stable at each pressure and temperature.	66	67

System HCl–H$_2$O–	Constraints	25° -300°C	400° -600°C
MgO–Al_2O_3–CaO–(SiO_2) (cont'd)	Gibbsite, diaspore, or corundum saturation, depending on which mineral is stable at each pressure and temperature. Metastable 7-A clinochlore was considered instead of its stable counterpart, 14-A clinochlore.	68	69
	Gibbsite, diaspore, or corundum saturation, depending on which mineral is stable at each pressure and temperature. Metastable chrysotile was considered instead of its stable counterpart, antigorite.	70	71
	Gibbsite, diaspore, or corundum saturation, depending on which mineral is stable at each pressure and temperature. Metastable 7-A clinochlore and chrysotile were considered instead of their stable counterparts, 14-A clinochlore and antigorite.	72	73
MgO–(Al_2O_3)–CO_2–SiO_2	$X_{CO_2} = 0.01$		192
	$X_{CO_2} = 0.05$		193
	$X_{CO_2} = 0.10$		194
	$X_{CO_2} = 0.30$		195
	$X_{CO_2} = 0.50$		196
	$X_{CO_2} = 0.70$		197
	$X_{CO_2} = 0.90$		198
	$X_{CO_2} = 0.95$		199
	$X_{CO_2} = 0.99$		200
	Metastable 7-A clinochlore was considered instead of its stable counterpart, 14-A clinochlore.		
	$X_{CO_2} = 0.01$		201
	$X_{CO_2} = 0.05$		202
	$X_{CO_2} = 0.10$		203
	$X_{CO_2} = 0.30$		204
	$X_{CO_2} = 0.50$		205
	$X_{CO_2} = 0.70$		206
	$X_{CO_2} = 0.90$		207

INDEX

System HCl–H$_2$O–	Constraints	25° –300°C	400° –600°C
MgO–(Al_2O_3)–CO_2–SiO_2 (cont'd)	$X_{CO_2} = 0.95$		208
	$X_{CO_2} = 0.99$		209
(MgO)–Al_2O_3–CO_2–SiO_2	$X_{CO_2} = 0.01$		210
	$X_{CO_2} = 0.05$		211
	$X_{CO_2} = 0.10$		212
	$X_{CO_2} = 0.30$		213
	$X_{CO_2} = 0.50$		214
	$X_{CO_2} = 0.70$		215
	$X_{CO_2} = 0.90$		216
	$X_{CO_2} = 0.95$		217
	$X_{CO_2} = 0.99$		218
	Metastable 7-A clinochlore was considered instead of its stable counterpart, 14-A clinochlore.		
	$X_{CO_2} = 0.01$		219
	$X_{CO_2} = 0.05$		220
	$X_{CO_2} = 0.10$		221
	$X_{CO_2} = 0.30$		222
	$X_{CO_2} = 0.50$		223
	$X_{CO_2} = 0.70$		224
	$X_{CO_2} = 0.90$		225
	$X_{CO_2} = 0.95$		226
	$X_{CO_2} = 0.99$		227
MgO–Al_2O_3–CO_2–(SiO_2)	$X_{CO_2} = 0.01$		228
	$X_{CO_2} = 0.05$		229
	$X_{CO_2} = 0.10$		230
	$X_{CO_2} = 0.30$		231
	$X_{CO_2} = 0.50$		232
	$X_{CO_2} = 0.70$		233
	$X_{CO_2} = 0.90$		234
	$X_{CO_2} = 0.95$		235
	$X_{CO_2} = 0.99$		236
	Metastable 7-A clinochlore was considered instead of its stable counterpart, 14-A clinochlore.		
	$X_{CO_2} = 0.01$		237
	$X_{CO_2} = 0.05$		238

System HCl–H$_2$O–	Constraints	25° -300°C	400° -600°C
MgO–Al$_2$O$_3$–CO$_2$–(SiO$_2$) (cont'd)	$X_{CO_2} = 0.10$		239
	$X_{CO_2} = 0.30$		240
	$X_{CO_2} = 0.50$		241
	$X_{CO_2} = 0.70$		242
	$X_{CO_2} = 0.90$		243
	$X_{CO_2} = 0.95$		244
	$X_{CO_2} = 0.99$		245
MgO–(Al$_2$O$_3$)–FeO–SiO$_2$	Amorphous silica saturation.	156	
	Amorphous silica saturation. Metastable 7-A clinochlore was considered instead of its stable counterpart, 14-A clinochlore.	157	
	Quartz saturation.	82	83
	Quartz saturation. Metastable 7-A clinochlore was considered instead of its stable counterpart, 14-A clinochlore.	84	85
MgO–Al$_2$O$_3$–FeO–(SiO$_2$)	Gibbsite, diaspore, or corundum saturation, depending on which mineral is stable at each pressure and temperature.	86	87
	Gibbsite, diaspore, or corundum saturation, depending on which mineral is stable at each pressure and temperature. Metastable 7-A clinochlore was considered instead of its stable counterpart, 14-A clinochlore.	88	89
	Gibbsite, diaspore, or corundum saturation, depending on which mineral is stable at each pressure and temperature. Metastable chrysotile was considered instead of its stable counterpart, antigorite.	90	91
	Gibbsite, diaspore, or corundum saturation, depending on which mineral is stable at each pressure and temperature. Metastable 7-A clinochlore and chrysotile were considered instead of their stable counterparts, 14-A clinochlore and antigorite.	92	93

INDEX

System HCl–H$_2$O–	Constraints	25°–300°C	400°–600°C
MgO–(Al$_2$O$_3$)–K$_2$O–SiO$_2$	Amorphous silica saturation.	158	
	Amorphous silica saturation. Metastable 7-A clinochlore was considered instead of its stable counterpart, 14-A clinochlore.	159	
	Quartz saturation.	94	95
	Quartz saturation. Metastable 7-A clinochlore was considered instead of its stable counterpart, 14-A clinochlore.	96	97
MgO–Al$_2$O$_3$–K$_2$O–(SiO$_2$)	Gibbsite, diaspore, or corundum saturation, depending on which mineral is stable at each pressure and temperature.	98	99
	Gibbsite, diaspore, or corundum saturation, depending on which mineral is stable at each pressure and temperature. Metastable 7-A clinochlore was considered instead of its stable counterpart, 14-A clinochlore.	100	101
	Gibbsite, diaspore, or corundum saturation, depending on which mineral is stable at each pressure and temperature. Metastable chrysotile was considered instead of its stable counterpart, antigorite.	102	103
	Gibbsite, diaspore, or corundum saturation, depending on which mineral is stable at each pressure and temperature. Metastable 7-A clinochlore and chrysotile were considered instead of their stable counterparts, 14-A clinochlore and antigorite.	104	105
MgO–(Al$_2$O$_3$)–Na$_2$O–SiO$_2$	Amorphous silica saturation.	161	
	Amorphous silica saturation. Metastable 7-A clinochlore was considered instead of its stable counterpart, 14-A clinochlore.	162	
	Quartz saturation.	110	111

System HCl–H$_2$O–	Constraints	25° -300°C	400° -600°C
MgO–(Al$_2$O$_3$)–Na$_2$O–SiO$_2$ (cont'd)	Quartz saturation. Metastable 7-A clinochlore was considered instead of its stable counterpart, 14-A clinochlore.	112	113
MgO–Al$_2$O$_3$–Na$_2$O–(SiO$_2$)	Gibbsite, diaspore, or corundum saturation, depending on which mineral is stable at each pressure and temperature.	114	115
	Gibbsite, diaspore, or corundum saturation, depending on which mineral is stable at each pressure and temperature. Metastable 7-A clinochlore was considered instead of its stable counterpart, 14-A clinochlore.	116	117
	Gibbsite, diaspore, or corundum saturation, depending on which mineral is stable at each pressure and temperature. Metastable chrysotile was considered instead of its stable counterpart, antigorite.	118	119
	Gibbsite, diaspore, or corundum saturation, depending on which mineral is stable at each pressure and temperature. Metastable 7-A clinochlore and chrysotile were considered instead of their stable counterparts, 14-A clinochlore and antigorite.	120	121
MgO–(Al$_2$O$_3$)–SiO$_2$		14	15
	Metastable 7-A clinochlore was considered instead of its stable counterpart, 14-A clinochlore.	16	17
(MgO)–Al$_2$O$_3$–SiO$_2$		18	19
	Metastable 7-A clinochlore was considered instead of its stable counterpart, 14-A clinochlore.	20	21
	Metastable chrysotile was considered instead of its stable counterpart, antigorite.	22	23

INDEX

System HCl–H$_2$O–	Constraints	25° -300°C	400° -600°C
(MgO)–Al$_2$O$_3$–SiO$_2$ (cont'd)	Metastable 7-A clinochlore and chrysotile were considered instead of their stable counterparts, 14-A clinochlore and antigorite.	24	25
MgO–Al$_2$O$_3$–(SiO$_2$)		26	27
	Metastable 7-A clinochlore was considered instead of its stable counterpart, 14-A clinochlore.	28	29
	Metastable chrysotile was considered instead of its stable counterpart, antigorite.	30	31
	Metastable 7-A clinochlore and chrysotile were considered instead of their stable counterparts, 14-A clinochlore and antigorite.	32	33
MgO–CaO–(CO$_2$)		128	129
MgO–CaO–CO$_2$–(SiO$_2$)	$X_{CO_2} = 0.01$		246
	$X_{CO_2} = 0.05$		247
	$X_{CO_2} = 0.10$		248
	$X_{CO_2} = 0.30$		249
	$X_{CO_2} = 0.50$		250
	$X_{CO_2} = 0.70$		251
	$X_{CO_2} = 0.90$		252
	$X_{CO_2} = 0.95$		253
	$X_{CO_2} = 0.99$		254
(MgO)–CaO–SiO$_2$		40	41
	Metastable chrysotile was considered instead of its stable counterpart, antigorite.	42	43
MgO–(CaO)–SiO$_2$		44	45
MgO–CaO–(SiO$_2$)		46	47
	Metastable chrysotile was considered instead of its stable counterpart, antigorite.	48	49
MgO–(CO$_2$)–FeO		130	131
MgO–FeO–(SiO$_2$)		50	51

System HCl–H$_2$O–	Constraints	25°–300°C	400°–600°C
MgO–FeO–(SiO$_2$) (cont'd)	Metastable chrysotile was considered instead of its stable counterpart, antigorite.	52	53
Na$_2$O–(Al$_2$O$_3$)–*CaO–MgO–*SiO$_2$	Albite and amorphous silica saturation.	163	
	Albite and amorphous silica saturation. Metastable 7-A clinochlore was considered instead of its stable counterpart, 14-A clinochlore.	164	
	Quartz and albite saturation.	122	123
	Quartz and albite saturation. Metastable 7-A clinochlore was considered of its stable counterpart, 14-A clinochlore.	124	125
Na$_2$O–(Al$_2$O$_3$)–*CaO–*SiO$_2$	Amorphous silica saturation.	154	
	Quartz saturation.	74	75
Na$_2$O–Al$_2$O$_3$–*CaO–*(SiO$_2$)	Gibbsite, diaspore, or corundum saturation, depending on which mineral is stable at each pressure and temperature.	76	77
Na$_2$O–(Al$_2$O$_3$)–*K$_2$O–*SiO$_2$	Amorphous silica saturation.	160	
	Quartz saturation.	106	107
Na$_2$O–Al$_2$O$_3$–*K$_2$O–*(SiO$_2$)	Gibbsite, diaspore, or corundum saturation, depending on which mineral is stable at each pressure and temperature.	108	109
Na$_2$O–(Al$_2$O$_3$)–*MgO–*SiO$_2$	Amorphous silica saturation.	161	
	Amorphous silica saturation. Metastable 7-A clinochlore was considered instead of its stable counterpart, 14-A clinochlore.	162	
	Quartz saturation.	110	111
	Quartz saturation. Metastable 7-A clinochlore was considered instead of its stable counterpart, 14-A clinochlore.	112	113

INDEX

System HCl–H$_2$O–	Constraints	25°–300°C	400°–600°C
Na$_2$O–Al$_2$O$_3$–MgO–(SiO$_2$)	Gibbsite, diaspore, or corundum saturation, depending on which mineral is stable at each pressure and temperature.	114	115
	Gibbsite, diaspore, or corundum saturation, depending on which mineral is stable at each pressure and temperature. Metastable 7-A clinochlore was considered instead of its stable counterpart, 14-A clinochlore.	116	117
	Gibbsite, diaspore, or corundum saturation, depending on which mineral is stable at each pressure and temperature. Metastable chrysotile was considered instead of its stable counterpart, antigorite.	118	119
	Gibbsite, diaspore, or corundum saturation, depending on which mineral is stable at each pressure and temperature. Metastable 7-A clinochlore and chrysotile were considered instead of their stable counterparts, 14-A clinochlore and antigorite.	120	121
Na$_2$O–(Al$_2$O$_3$)–SiO$_2$		34	35
(Na$_2$O)–Al$_2$O$_3$–SiO$_2$		36	37
Na$_2$O–Al$_2$O$_3$–(SiO$_2$)		38	39
O$_2$–(CuO)–S$_2$		302	303
O$_2$–(FeO)–S$_2$		304	305
(PbO)–CO$_2$–H$_2$SO$_4$		296	297
PbS–(H$_2$S)–ZnS		142	143
S$_2$–(CuO)–O$_2$		302	303
S$_2$–(FeO)–O$_2$		304	305
SiO$_2$ and other components--see listing under other components.			
(SrO)–CO$_2$–H$_2$SO$_4$		298	299
(ZnO)–CO$_2$–H$_2$S		300	301
ZnS–(H$_2$S)–PbS		142	143

Advances in Physical Geochemistry

Editor-in-Chief: S. K. Saxena

Volume 2

Editor: S. K. Saxena

With contributions by numerous experts

1982. 113 figures. X, 353 pages.
ISBN 3-540-90644-4

The second volume of the series **Advances in Physical Geochemistry** critically reviews recent data and presents new information on the crystal chemistry, thermodynamics, and kinetics of the Earth's crust and mantle. Among the topics treated are:

- intracrystalline reactions, including demonstrations of future applications to thermodynamics of solid solutions and the cooling history of rocks

- melts, incorporating silicate density structure relationships and phase equilibrium data on feldspars coexisting with fluids

- thermodynamic methods of calculating phase equilibria

- Gibbs free energy of formation of substances in laterites and bauxites
  the crystal chemistry of pyroxenes and perovskite

Summarizing contemporary topics and new crystal-chemical and thermodynamic data, this volume will be extremely useful geochemists, geophysicists, and other material scientists.

Springer-Verlag
Berlin
Heidelberg
New York
Tokyo

Advances in Physical Geochemistry

Editor-in-Chief: **S. K. Saxena**

Volume 3

Kinetics and Equilibrium in Mineral Reactions

Editor: **S. K. Saxena**

With contributions by numerous experts

1983. 99 figures. X, 273 pages
ISBN 3-540-90865-X

Contents: Compositional Zoning of Crystals: A Record of Growth and Reaction History. – Exsolution and Fe^{2+}-Mg Order-Disorder in Pyroxenes. – Geospeedometry: An Extension of Geothermometry. – Mg-Fe Fractionation in Metamorphic Environments. – Geobarometry in Granulites. – The Cordierite-Garnet-Sillimanite-Quartz Equilibrium: Experiments and Applications. – Experimental Investigation of Exchange Equilibria in the System Cordierite-Garnet-Biotite. – Thermodynamics of Complex Phases. – Index.

Volume **3** of **Advances in Physical Geochemistry** continues this annual series of reviews of contemporary research on thermodynamics and reaction kinetics in geological systems.
The authors, all experts in their specialized fields, consider that equilibrium data are important in measuring pressure and temperature of the paleonenvironment of rock formation, but they also emphasize that the kinetics of mineral reactions is of utmost significance in revealing the thermal history of rocks.
Each chapter in this volume not only reviews the state-of-the-art and the authors' own original work, but also considers the directions which future research will take.

Springer-Verlag
Berlin
Heidelberg
New York
Tokyo

FRANK J. MILLERO